CIVIL ENGINEERING AND URBAN PLANNING III

PROCEEDINGS OF THE 3RD INTERNATIONAL CONFERENCE ON CIVIL ENGINEERING AND URBAN PLANNING (CEUP 2014), WUHAN, CHINA, 20–22 JUNE 2014

Civil Engineering and Urban Planning III

Editors

Kouros Mohammadian
Department of Civil and Materials Engineering, University of Illinois at Chicago, Chicago, USA

Konstadinos G. Goulias
Professor of Transportation and Director of GeoTrans Laboratory, Department of Geography, Division of Mathematical, Physical, and Life Sciences. University of California Santa Barbara, Santa Barbara, USA

Elif Cicek
Department of Civil Engineering, Ataturk University, Erzurum, Turkey

Jieh-Jiuh Wang
Department of Architecture, Ming Chuan University, Taoyuan, Taiwan

Chrysanthos Maraveas
Department of Structural Engineering, C. Maraveas Partnership-Consulting Engineers, Athens, Greece

CRC Press is an imprint of the
Taylor & Francis Group, an **informa** business

A BALKEMA BOOK

CRC Press/Balkema is an imprint of the Taylor & Francis Group, an informa business

© 2014 Taylor & Francis Group, London, UK

Typeset by MPS Limited, Chennai, India
Printer details: Printed and bound in Great Britain by CPI Group (UK) Ltd, Croydon, CR0 4YY.

All rights reserved. No part of this publication or the information contained herein may be reproduced, stored in a retrieval system, or transmitted in any form or by any means, electronic, mechanical, by photocopying, recording or otherwise, without written prior permission from the publishers.

Although all care is taken to ensure integrity and the quality of this publication and the information herein, no responsibility is assumed by the publishers nor the author for any damage to the property or persons as a result of operation or use of this publication and/or the information contained herein.

Published by: CRC Press/Balkema
P.O. Box 11320, 2301 EH Leiden, The Netherlands
e-mail: Pub.NL@taylorandfrancis.com
www.crcpress.com – www.taylorandfrancis.com

ISBN: 978-1-138-00125-1 (Hardback)
ISBN: 978-1-315-74300-4 (eBook PDF)

Table of contents

A critical comparison between CPM and PERT with Monte Carlo simulation in project
management and scheduling 1
Y. Liu & P.W. Zhang

A discrete-continuous agent model for fire evacuation modeling
from multistory buildings 5
E. Kirik & A. Malyshev

A ground passenger transport network optimization method on the basis of urban rail transit 9
Z. Guo, H.Y. Wu & L. Tong

A study on structural strengthening technology 15
G.H. Qiao & Y.B. Cheng

A systematic solution to the smart city – distinguished from the intelligent city 21
Y.H. Mao, H.Y. Li & B. Yang

A total bridge model study of Jiangdong Bridge in Hangzhou 27
D.C. Qi & P.F. Li

An automatic FE model generation system used for ISSS 33
J. Duan, X.M. Chen, H. Qi & Y.G. Li

Analysis of genetic mechanism and stability of Shaixiping landslide 37
M.H. Su, L. Ma & W. Li

Analysis of green design in outdoor game space for children 43
X. Li

Analysis of influencing factors on urbanization: A case study of Hebei Province, China 47
G.B. Lai & B.H. Zhao

Analysis of static characteristics of a typical gravity dam 53
K. He, J.K. Chen, Y.L. Li, Z.Y. Wu & Y.Y. Zhen

Analysis of the campus design based on the color idea of landscape space 59
Y. Wang & L. Yuan

Analysis of the consolidation effect of ultra-soft ground via vacuum preloading 63
B. Xu, T. Noda & S. Yamada

Buckling model of reinforcing bars with imperfections based on concrete lateral expansion 69
P.C. Mo & B. Kan

Characteristics and patterns of land development in hilly and urban fringe
areas: A case study of Fangshan District in Beijing 73
J. Xu & Y. Yao

Comparative analysis of CO_2 emissions of steel bar and synthetic fiber
used in shield tunnel segment 79
Q.S. Li, Y. Bai, A. Ridout & Z.W. Jiang

Comparative study on exploitations of Konpira Grand Theatre in Japan and
Sichuan Qijian Ancient Stages in China 83
L. Yuan & M. Yamasaki

Comparative study on street space comparative study of the Western Sichuan and Huizhou folk houses X.K. Li, Y.Z. Li, Y.H. Tang & Y. Cao	87
Comparison and suggestions on roof structure schemes of single-storey pure basement H.W. Kuang, H.Y. Ma & W. Pan	93
Concept and practice of reconstruction of European old buildings Y.T. Wang	97
Consolidation grouting Karst foundation treatment strategy for existing buildings Y.J. Xiang & Y. He	101
Constructing the evaluation system of urban planning implement under the guideline of new urbanization—taking Tongchuan as an example H. Yang & P. Zhang	105
Construction situation and prospect of high-speed rail earthquake monitoring and early warning system in China J. Yang, J. Huang & Z.G. Chen	109
Danger prediction system for dangerous goods transportation T.J. Zhang, Y. Cao & X.W. Mu	113
Deformation characteristics of a typical high rockfill dam built on a thick overburden foundation Z.J. Zhou, J.K. Chen, W.Q. Wang, Y.L. Li & Z.Y. Wu	117
Deformation monitoring and regression analysis of surrounding rock during the construction of Zhegu Mountain Tunnel H.S. Zhang & S.G. Chen	123
Details & decoration—an interpretation of Carlo Scarpa's architecture work C. Huang	129
Detecting the interface defects of steel-bonded reinforcement concrete structure by infrared thermography techniques Y.W. Yang, J.Y. Pan & X. Zhang	133
Detection, appraisal and strengthening of the support structure of a boiler fan in part II of a power plant M.X. Tan	137
Developing an Artificial Filter Bank (AFB) for measuring the dynamic response of civil structures G. Heo & J. Jeon	141
Development and prospects of structural vibration control L. Yue	147
Discussion on construction technology of asphalt pavement on expressway X.-P. Ou, K. Ding, W.-L. Xu, F. Gao & L. Jiang	151
Effects of elevation difference on GPS monitoring accuracy W. Qiu, T. Huang & X.M. Wang	155
Estimation of fault rupture length of Wenchuan earthquake using strong-motion envelopes D.Y. Yin, Q.F. Liu & M.S. Gong	159
Evaluation and analysis of bridges with encased filler beams V. Kvocak, D. Dubecky, R. Kocurova, P. Beke & M. Al Ali	165
Evaluation of cultural heritage resources of Yanhecheng in Beijing R.L. Miao, M. Li, Y.L. Xu & H.N. Zhang	169
Evaluation on cultural heritage resources of Gubeikou fort in Beijing Y.L. Xu, R.L. Miao, M. Zhang & M. Li	173

Experimental study on the deformation characteristics and strength parameters of
rock-soil of landslide of levee's bank slope
R.Y. Ma, H. Liang & S.Q. Dai 177

Fabric-formed concrete: A novel method for forming concrete structures
R.P. Schmitz 183

Feasibility of discharging tailing in the open pit of Fengshan copper mine
using transforming open-pit into underground mining
L. Xia & S.Y. Xia 191

Flow around the 3D square cylinder and interference effects on double square cylinders
L.J. Meng, X.W. Ma, W.Y. Ma & L.Q. Hou 195

Identification of highway dangerous driving behavior based on Hidden Markov Model
J. Wang, J. Li & X.W. Hu 199

Impact of the climate conditions on waterproofing materials in long-term use in
industrially polluted regions
V. Bartošová, D. Katunský, M. Labovský & M. Lopušniak 203

Implementation of PMS at a local level-case study based on StreetSaver®
G. Wang, D. Frith & D. Morian 207

Inheritance and creation of regional architecture features in urbanization process
Y. Han & Q. Wang 213

Inhibitory effect from house price rising expectation on adverse selection:
Evidences from resale house trading experiments
H. Zhang, X. Sun & Y. Zhang 219

Instability warning model of open-pit mine slope based on BP neural network
Z.H. Xie, S.S. Liang & T.T. Luan 225

Investigating urban transportation planning using IC card trip data
S.Z. Zhao, Y.F. Gao, Q.F. Tian & F. Zong 229

ISM-based identification of factors influencing pedestrian violations at
signalized intersections
M.M. Zheng & Z.Y. Zuo 235

Mechanical behavior of water diversion tunnel lining before and after consolidation grouting
Y.S. Li, S.G. Chen & Z.L. Zhou 239

Numerical analysis of Hefei metro shield tunnel construction crossing pile
foundation of building
K.W. Ding & D.W. Man 245

Numerical analysis of soil deformation in shielding tunneling considering the effect of
shield gap and grouting patterns
Z.L. Zhou & S.G. Chen 249

Numerical analysis of soil nailing's instability in deep foundation pit
H.G. Dong & D.K. Chen 253

Numerical simulation of the characteristic of air motion in the spillway tunnel with aerators
S.B. Yue, M.J. Diao & C. Qiu 257

One-dimensional consolidation settlement numerical analysis of a large-scale land
reclamation project
J. Zhao, Y.H. Jin & J. Yi 261

Optimization of clapboard with wind box type construction ventilation in super-long
railway tunnel
Z.M. Cao, Q.X. Yang & C. Guo 267

Participation in urban planning
Y.Y. Liu, Z.C. Zhang, D.S. Gao & M.Y. Wang 275

Performance investigation of steel slag porous asphalt treated mixture *C.H. Li, X.D. Xiang, Q. Guo & L.X. Jiao*	279
Probe into the landscape planning of Chaoshou Village, Chengbei Town, Giange County, Guangyuan City *C.X. Ma, A.M. Liu & L. Yuan*	283
Rational performance tests for permanent deformation evaluation of asphalt mixtures *Q. Li & F. Ni*	287
Reliability analysis of asphalt pavement considering two failure modes *H.L. Liu & X. Xu*	291
Reliability analysis of geotechnical engineering problems based on an RBF metamodeling technique *Q. Wang, J. Lin, J. Ji & H. Fang*	297
Research of Chinese healthy city construction practice *F. Lv, Y. Zhang & X.Y. Song*	301
Research on the function hierarchies of MA rail transit network system based on the time goal *M. Yang, X.C. Guo, X.J. Ling & X.Y. Bi*	307
Research on the spatial morphology and holistic structural preservation of East Asian city moat area—sample: Haohe river historic area of Nantong city *L.L. Sun & H. Gu*	313
Research on urban ecological water system construction: A study case of Wulijie, China *S. Ye*	319
Research on vibration characteristics of building construction while tunnel crossing underneath *Z. Li & Z.Q. Li*	325
Review and assessment of urban Car Free Day program in China *Y. Zhang*	329
Review of prestressed concrete technology in flexural members *Y.X. Yin, Y.B. Cheng, Z.L. Guo & G.H. Qiao*	335
Roads in urban areas: Limits to regulations and design criteria *A. Annunziata & F. Annunziata*	339
Route choice behavior model under time pressure *F. Gao*	349
Self-vibration characteristic test and finite element analysis of spring vibration isolated turbine-generator foundation *T.J. Qu, H.Z. Yang & G. Zhao*	355
Shaking table test study on bottom-business multi-storey structure *Y.D. Liang, X. Guo, F. He, Y. Zhou & W.S. Yang*	361
Simulation of wireless sensor for monitoring corrosion of reinforcement steel in concrete *G.W. Xu, J. Wu, Z. Wang & Z. Wang*	365
Slope's automatic monitoring and alarm system based on TDR technology *C.Y. Lin*	371
Speed variations of left-turning motor vehicles from minor road approach while merging with mainline motor vehicles at non-signalized at-grade intersection *G.Q. Zhang, Q.Y. Zhang & Y.L. Qi*	377
Stability impact analysis about the layout of tie beam for thin-walled pier *X.M. Dong*	383
Studies concerning the guidance in curves of bogies with elastic driven wheelsets *I. Sebeşan & M.A. Spiroiu*	387

Study on bicycle flow characteristics at intersections X.H. Deng & J. Xu	393
Study on characteristics of typical vernacular dwellings in Mount Emei K.Z. Chen, Y.Z. Li & Y.T. Jiang	401
Study on space structure change of land used for carbon source and carbon sink in urban fringe areas M.L. Qin, H. Li, J. Ya, J. Zhao & W.C. Chen	407
Study on the mechanical performance of bridges affected by the creep of concrete in bridge widening Q.K. Zhang, G.Y. Zhang & M.Q. Li	413
Study on the traffic prediction of Dezhou-yu'e province boundaries highway M.L. Li & H.B. Liu	417
Study on traffic safety risk assessment of expressway in cold areas J. Wang, K. Zhou & X.W. Hu	421
Study on urban spatial development and traffic flow characteristics in Tianjin Y. Wang & K.M. Chen	425
Survey and analysis of energy consumption of shopping mall buildings in Chengdu Z. Liu & Y. Xiang	429
Sustainable high-rise buildings L.X. Li, Z.D. Wang & R. Qian	433
Temperature deformations of the ultralong frame structure of MIXC-Qingdao N.N. Han, S. Ke, Y.J. Ge & Q. Li	439
Terminal departure passenger traffic forecast based on association rule S.W. Cheng, H.B. Zhang, J. Xu & Y.P. Zhang	443
Test and analysis of the vacuum degree transport within plastic drainage board T.Y. Liu, H.Z. Kang & A.M. Liu	447
The analysis of anti-seismic safety behavior of split columns with a core of reinforced concrete X.J. Chu & Y. Liang	451
The analysis of urban-rural transit trip characteristics using a Structural Equation Model P.F. Li, X.H. Chen & Y.F. Tu	457
The application of creeper and overhang plants on exterior facades of Guangfu dwellings S.Y. Zhang	463
The comparison of multi-ribbed frame structure filled with phosphogypsum with conventional frame structure F. Liu & F.J. Wang	467
The correlation of P-wave velocity and strength of solidified dredged marine soil C.-M. Chan, K.-H. Pun & L.-S. Hoo	473
The discussion of multi-channel emergency management pattern in the north bank of Wenzhou Ou River Y.H. Yao & Y.F. Li	479
The Florence charter and the conservation of Chinese classical gardens Y. Zhong & J.J. Cheng	483
The impact of the rocking wall layout on the structural seismic performance S.B. Yang, Y.B. Zhao, Z.T. Wei & J.H. Jia	487
The loads variations on the locomotives axles I. Sebeşan & M.M. Călin	493
The performance of ground resource heat pump unit in a long time operation in summer hot and winter warm district Y.N. Hu, F. Ruan & S.S. Hu	499

The planning bases on landscape-oriented proposal stems *L.M. Bai & J.M. Hou*	503
The research and application of reinforced truss slab in steel framework *W.X. Xia & H.Y. Wan*	507
The research of construction engineering life *W.S. Zhao*	513
The research on structural design to resist progressive collapse *Z.L. Guo, Y.B. Cheng, Y.X. Yin & G.H. Qiao*	517
The study of closed surface-water-source heat pump system in hot summer warm winter zone *W.B. Lian, Y. Jiang & Y.N. Hu*	523
Urban study on physical environment and social migrants networking focused on Russian-speaking Town in Seoul, Korea *E. Shafray & K. Seiyong*	531
Waste water treatment through public-private partnerships: The experience of the regional government of Aragon (Spain) *S. Carpintero & O.H. Petersen*	535
Welding process from the civil engineering point of view *M. Al Ali, M. Tomko & I. Demjan*	541
Author index	545

A critical comparison between CPM and PERT with Monte Carlo simulation in project management and scheduling

Yang Liu
School of Urban Construction and Management, Yunnan University, Yunnan, China

Puwei Zhang
School of Civil Engineering, Kunming University of Science and Technology, Yunnan, China

ABSTRACT: Project management and scheduling control are important issues in construction and management science. It has been proved that CPM and PERT are most suitable methods to solve the problem of project management and scheduling in practice. A very important problem has been neglected for a long time. The problem is how big the differences between CPM and PERT. The papers made a classic case of textbooks as an example and calculate parameters of all works. After 5000 times' simulation, the results showed that scheduling with PERT was not as optimistic as CPM, the scheduling risk is so big that we must find out the largest risk points. Through the sensitivity analysis, the work which has large influence could be found and should be treat as key control points. It is proved that PERT with Monte Carlo simulation was better to solve the problem of project management and scheduling. And, it is meaningful to compare CPM and PERT in project management and scheduling.

1 INTRODUCTION

Schedule, quality and cost are three key goals in construction project management. As everyone knows, quality and cost are largely determined by schedule of a project. So, how to define the schedule of a project has become a very important and popular issue in construction project management. Generally speaking, there have been some important methods to define the schedule of a project such as CPM, PERT, GERT, VERT, SCERT, ID and some improvements in themselves. For example, someone combined fuzzy with CPM in order to get better results on schedule control [1]. Some one used fuzzy to make sure the probability instead of traditional triangular distribution [2]. As the development and application of computer technology, large-scale calculation became a reality gradually. Simulation and Monte Carlo by computer became more and more popular on this issue [3]. Monte Carlo simulation was a very good way to solve uncertain problem especially the problem such as construction schedule management [4]. The theoretical basis of Monte Carlo was law of large numbers and the central-limit theorem, the practice method of it was repeated sampling and high-speed computation. GERT was a good method to solve the problem such as quite uncertain. For example, when you don't know the relationships and time parameters of all works, GERT will be a better choice for you [5~6]. VERT (Venture Evaluation and Review Techniques) was used on problem of defining the risk and venture. SCERT (Synergistic Contingency Evaluation and Response Techniques) was a method to solve the problem of evaluation on uncertain [7].

As everyone knows, construction procedures were defined in construction management, so, the relationships between procedures are defined. For example, if someone wants to produce reinforced concrete, the order of 'template support', 'assembling reinforcement', 'concreting' and 'maintenance reinforced concrete' were couldn't be changed, everybody must do it according to the order or it would be cause lots of problems. Another problem was the time parameters of all procedures. According to CPM, duration of every work was certain parameter, for example, duration of template support was 3 days etc. In practice, duration was impossible to know accurately before we do the work, so, 3 days was an estimation number. In reality, some factors such as construction efficiency and construction conditions would influence duration time a lot. The results of doing template support may be 3.12 days or 2.87 days. MCCAULLEY, JW introduce the two methods in 1969 [13]. Some researchers discuss how to define the parameters with CPM and PERT [8~9]. Lots of researcher used the two methods in many subjects [10]. But, few researchers pay close attention to how much difference between PERT and CPM, especially on the same project.

2 METHOD

2.1 Get the AOA network

According to practice and relationships between the procedures of project, draw the network by AOA rule correctly.

2.2 Calculate the network by CPM

Firstly, calculate parameters of all procedures. There are 6 time parameters need to be calculate in the method. They are ES (Earliest Start), EF (Earliest Finish), LS (Latest Start), LF (Latest Finish), TF (Total Float), FF (Free Float). According to the calculation rule of CPM. Secondly, define the critical path. According to CPM, critical path was combined by the procedures whose durations were the maximum. Thirdly, find out the key procedures and promote specific measures to ensure key procedures complete on schedule.

2.3 Calculate the network by PERT

Firstly, define the distribution and parameters of all procedures by Delphi Method. In practice, during time was impossible to know accurately before we do the work, so, 3 days was an estimation number. In reality, some factors such as construction efficiency and construction conditions would influence during time. The result of doing support template was 3.12 days or 2.87 days maybe. According to lots of construction practice, during time of works in construction fit 3 distributions as triangular distribution, normal distribution and beta PERT- distribution. Their functions and parameters were shown in Table 1. And then, define the parameters with Delphi Method. Secondly, find all path of the network. As the duration of all procedures was not defined, so, it was probably that some un-critical path would become into critical path. So, it is necessary to list all path of the work and calculate the duration respectively. Thirdly, simulate all paths by Monte Carlo Method. After finishing the Delphi questionnaire survey and define the parameters of selective distribution, we need to do Monte Carlo simulation by repeated sampling. According to Law of Large Numbers, the number of sampling times should be at least 1000~5000 and even more. Fourthly, analyze the simulation results.

2.4 Compare the results and make decision

After the simulation by Monte Carlo, compare the results with CPM, and check the differences between two results. The first is checking the total duration of construction and judge whether it meet the requirement. The second is to find the work which had biggest change after the comparison. The works was key control points during the construction procedure. The third is to check whether the critical path change or not and define the new one if there was a changing.

3 CASE STUDY

3.1 Get the AOA network

The network program was shown in Fig. 1. The name and duration of works were shown above and below of the arrow respectively. For example, the name of work 1–2 was A, and the duration was 6 days.

3.2 Calculate the network by PERT

According to CPM method, all parameters of works had been calculated and shown in Fig. 1, the parameters icon was shown at the top right corner in Fig. 2.

According to the results, total duration of the project was 36 days. There were 2 critical paths in the network, one is A-B-E-H-J, and another is A-C-F-H-J. Procedure B and C are most important procedures.

3.3 Calculate the network by PERT

Define the duration of works by beta PERT distribution with Delphi Method, though 3 rounds questionnaire, the results was shown in Table 1. Do Monte Carlo simulation with Excel and @Risk soft wares. The distribution of all parameters were PERT, the iterations of simulation was 5000, the simulation was 1.

4 RESULTS

There are 4 paths in PERT network; they are A-B-E-H-J, A-C-F-H-J, A-D-I-J and A-C-F-GI-J. The total duration of network is equal to the maximum of the 4 paths'. The simulation results was shown in Fig. 3.

According to the results, in the condition of PERT, total duration has changed from 36 days with CPM into

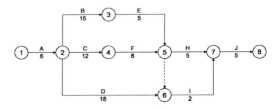

Figure 1. The AOA network and duration of all procedures.

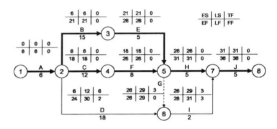

Figure 2. The CPM network program and its parameters calculation results.

a distribution of Lognormal. The mean of distribution was 36.4440; the probability of total duration between 34.54 and 38.35 was 90%. So, the manager should consider the probability of total duration under 36days. The result was shown in Fig. 4. The probability was 35.3%. It means the manager has a big risk to finish the work within 36 days.

The simulation results of 4 paths was shown in Fig. 5, total duration of path A-B-E-H-J and A-D-I-J were in accordance with Weibull distribution; total duration of path A-C-F-H-J was in accordance with Lognormal distribution and total duration of path

Table 1. PERT parameters of works estimated by Delphi Method

Parameters	Minimum	Most likely	Maximum
A	4.9	6.33	8.3
B	12.5	14.97	17.3
C	9.8	11.84	13.65
D	16.5	18.22	20
E	3.85	4.81	6
F	7.2	8.27	10
G	3.85	4.81	6
H	1.2	2.1	3
I	3.85	4.81	6
J	4.9	6.33	8.3

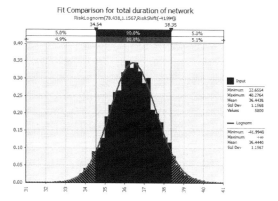

Figure 3. The simulation results of total duration of network.

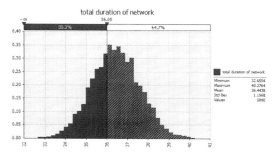

Figure 4. The simulation result of probability of total duration within 36 days.

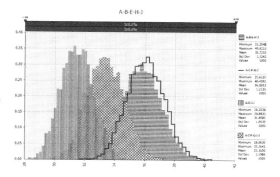

Figure 5. The comparison of total duration between 4 paths.

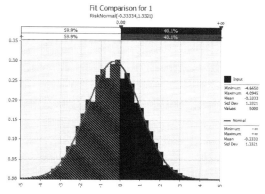

Figure 6. The simulation results of total duration of path A-B-E-H-J minus Path A-C-F-H-J.

A-C-F-G-I-J was in accordance with Beta General distribution.

According to the results, Path A-C-F-H-J was most likely be the critical path in the network, but the total duration of path A-B-E-H-J was longer than Path A-C-F-H-J in some area. It means that path A-B-E-H-J maybe was the critical path the same. The problem is what the probability for that probability is. We use total duration of path A-B-E-H-J minus total duration of Path A-C-F-H-J and simulate the results with 5000 times. The results was shown in Fig. 6.

According to the results, total duration of path A-B-E-H-J minus Path A-C-F-H-J was in accordance with Normal distribution. The probability of total duration of path A-B-E-H-J longer than Path A-C-F-H-J was 40.1%. It means the manager should prepare that path A-B-E-H-J was also become critical path the same. So, the next problem was which procedure should the manager to care about especially? There were 4 different procedures in two paths; they were B, C, E, and F. We made a sensitivity analysis with procedure B, C, E, and F, and find out which procedure would influence the result most. The result was shown in Fig. 7.

According to the results, work B has greatest influence in the network; the next order was C, F and E. The order is the same as their durations' order. So, it

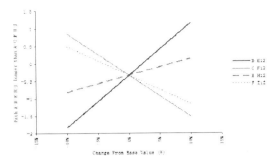

Figure 7. The sensitivity analysis results of procedure B, C, E, and F.

was very necessary to control the works in order to control the total duration.

5 CONCLUSION

It is proved that Monte Carlo simulation was a very good method to solve the problem on uncertain. In construction management practice, more and more people realized that the problem of schedule management and risk control need uncertain tools to do it. With Monte Carlo simulation, there are 3 points need to be pay attention to. The first is to make clear the problem you want to solve. The second is to make sure and define the duration parameters with scientific and reasonable method such as Delphi or something else. In practice, expert engineers, scholars, project managers and professors are good information channel to obtain the useful information. The last is to do simulation as much as possible, the times be better 1000 times at least.

ACKNOWLEDGMENT

We express our sincere thanks to Science and Engineering Major Project Foundation of Yunnan University (NO.2013CG032), who gave us invaluable help and fund for this study.

REFERENCES

S. Chanas, P. Zieliński, Critical path analysis in the network with fuzzy activity times, Fuzzy Sets and Systems. 122 (2001), p. 195–204.

S. Chanas, J. Kamburoswk. The use of fuzzy variables in PERT, Fuzzy Sets and Systems 5 (1981) p. 1–9.

K. Kurihara, N. Nishiuchi, Efficient Monte Carlo simulation method of GERT-type network for project management, Computers & Industrial Engineering 42 (2002), p. 521–531.

Kenzo Kurihara, Nobuyuki Nishiuchi. Efficient Monte Carlo simulation method of GERT-type network for project management. Computers & Industrial Engineering. Volume 42, Issues 2–4, 11 April 2002, P. 521–531.

Whitehouse, Gary E. GERT, A Useful Technique for Analyzing Reliability Problems. Technimetrics. 12 (1970), No.1, P. 33.

Lin, KP; Wu, MJ; Hung, KC; Kuo, YY. Developing a T(omega) (the weakest t-norm) fuzzy GERT for evaluating uncertain process reliability in semiconductor manufacturing. Applied soft computing. 2011, Vol. 11. No. 8. P. 5165–5180.

Herman, DJ. Scert. Scert – Computer evaluation tool – Computer evaluation tool. Datamation. 1967. Vol. 13. No. 2. P. 26–28.

McCaulley, JW. An introduction to pert and CPM. Communications of the ACM. 1963. Vol. 7. No. 6. P. 364–365.

Zhiwei Zhu; Heady, R.B. A simplified method of evaluating PERT/CPM network parameters Engineering Management, IEEE Transactions on. 1994. Vol. 41. No. 4. P. 426–430.

Ming Lu; S.M. AbouRizk. Simplified CPM/PERT Simulation Model. Journal of Construction Engineering and Management. 2000. Vol. 126. No. 3. P. 219–226.

Weber, A. Project-management with CPM, PERT and precedence diagramming, 3rd edition – Moder, J, Phillips, C, Davis, E. Data management. 1984. Vol. 22. No. 5. P. 8.

A discrete-continuous agent model for fire evacuation modeling from multistory buildings

E. Kirik & A. Malyshev

Institute of Computational Modeling SB RAS, Krasnoyarsk, Akademgorodok, Russia

ABSTRACT: In the paper a discrete-continuous model of pedestrian movement is presented. The model is implemented in SigmaEva evacuation module that is a part of "SigmaFS" software for fire safety engineering applications. Some case study of fire evacuation simulation is presented. This work has got partial financial support by the Integration project of SB RAS, number 49/2012.

Keywords: evacuation, multistory buildings, fire safety, simulation.

1 INTRODUCTION

Modeling of pedestrian dynamics is actual problem at present days. Simulations are used in many fields from organization of mass events to fire safety of buildings, ships, aircrafts. The main task of applying such simulations is to estimate evacuation time in different scenarios to provide safe conditions for visitors, passengers in emergency situations.

Different approaches from the social force model based on differential equations to stochastic CA models are developed; see Schadschneider et al. 2009 and references therein. They reproduce many collective properties including lane formation, oscillations of the direction at bottlenecks, the so-called "faster-is-slower" effect.

The most popular approach from practical applications is individual one mining that each person is considered as individual agent, and a model gives coordinates of each person. Each person may be assigned with individual properties: free movement speed, evacuation starting time, projection size, evacuation way. This allows to solve variety of evacuation tasks, including fire evacuation tasks.

At present time there are two main approaches to simulate individual people movement: continuous and discrete. There was developed a discrete-continuous model SIgMA.DC (Kirik et al. 2014) which was motivated by advantages of the continuity of a modeling space and the intuitive clarity of update rules in discrete models. In this model agents move in a continuous space (in this sense model is continuous), but the number of directions for the particle to move is limited and predetermined by a user (in this sense model is discrete). So in the model a size of the modelling space is considered as it is (it is very important for such narrow places as doors). At the same time mathematics of the model is simpler comparing with continuous models.

The article is organized as follows. In the next section, model is presented. It is followed by the case study section when some fire evacuation example is considered under different initial conditions and conclusion.

2 DESCRIPTION OF THE MODEL

2.1 Space and initial conditions

A continuous modeling space $\Omega \in R^2$ and an infrastructure (obstacles) are known. (Here and below under "obstacle" we mean only walls, furniture. Agents are never called "obstacle".) There is a unified coordinate system, and all data are given in this system. Agents may move in a free space. To orient in the space agents use the static floor field S, Schadschneider & Seyfried (2009). The nearest exit is assumed as a target point.

Shape of each agent is disk with diameter d_i, $i = \overline{1, N}$, N – number of agents, $\vec{x}_i(0) = (x_i^1(0), x_i^2(0))$, $i = \overline{1, N}$ – the initial positions of agents mining coordinates of disks centers (it is supposed that they are coordinates of body's mass center projection). Each agent is assigned with the free movement speed v_i^0, $i = \overline{1, N}$, the square of projection. We assume that the free movement speed is random normal distributed value with some mathematical expectation and variation, Kholshevnikov & Samoshin (2009). It is supposed that while moving people do not exceed maximal speed (the free movement speed), and persons control speed according to a local density.

Each time step t each agent i may move in one of predetermined directions $\vec{e}_i(t) \in \{\vec{e}^\alpha(t), \alpha = \overline{1, q}\}$, q – number of directions, model parameter (here a set of directions uniformly distributed around the circle $\{\vec{e}^\alpha(t), \alpha = \overline{1, q}\} = \{(\cos \frac{2\pi}{q}\alpha, \sin \frac{2\pi}{q}\alpha), \alpha = \overline{1, q}\}$

is considered). Agents who cross target line leave the modeling space.

2.2 Preliminary calculations

To model directed movement the "map" which stores the information on the shortest distance to the nearest exit is used. This distance is measured in meters, [m]. Such map is saved in the static floor field S imported from a floor field Cellular Automata approach which provides pedestrians with information about ways to exits, Schadschneider & Seyfried (2009). This field increases radially from the exit and it is zero in the exit(s) line(s), Kirik et al. 2011. It is not changeable with time and is independent on the presence of the particles. To calculate the field S the modeling space $\Omega \in R^2$ is covered by a discrete orthogonal grid with cells 10–40 cm in size, and, the Dijkstra's algorithm with 16-nodes pattern is used, for instance. A distance to the exit from arbitrary point is given by bidirectional interpolation among nearest nodes.

2.3 Movement equation

A movement equation for each agent is derived from a finite-difference expression of velocity $v(t)\vec{e}(t) \approx (\vec{x}(t) - \vec{x}(t-\Delta t))/\Delta t$. This expression allows to present the new position of the agent as a function of the previous position and current velocity of the agent. Thus for each time t the coordinates of i-th agent are given by the formula:

$$\vec{x}_i(t) = \vec{x}_i(t-\Delta t) + v_i(t)\vec{e}_i(t)\Delta t, \quad i=\overline{1,N}, \quad (1)$$

where $\vec{x}_i(t-\Delta t)$ denotes the particle's position at time $(t-\Delta t)$; $v_i(t)$, [m/s], is the particle's speed; $\vec{e}_i(t)$ is the unit direction vector. The time shift $\Delta t = 0.25$, [s] is assumed to be fixed, Figure 1.

Unknown values in (1) for each time step for each particle are the speed $v_i(t)$ and the direction $\vec{e}_i(t)$. A value of the speed is obtained from experimental data (fundamental diagram), for example, Kholshevnikov & Samoshin (2009), in accordance with a local density. The direction $\vec{e}_i(t)$ of the next step is proposed to be stochastic with the probabilities distribution calculated, this idea adopted from the discrete CA approach.

2.4 Choosing movement direction

This discrete-continuous model was inspired by a previously presented stochastic CA FF model, Kirik et al. 2011. All predetermined directions for each agent each time step are assigned with some probabilities to move, and the direction is chosen according to the probability distribution obtained.

Personal probabilities to move in each direction each time step have contributions: a) the main driven force (given by the destination point), b) an interaction with other pedestrians, c) an interaction with

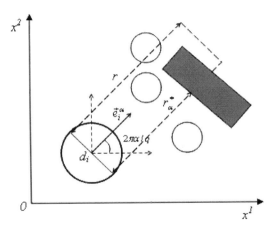

Figure 1. Scheme of visibility area of particle in direction $\vec{e}_i^\alpha(t)$.

an infrastructure (non movable obstacles). The highest probability (mainly with value > 0.9) is given to the direction that has most preferable conditions for movement considering other agents and obstacles and strategy of the people movement (the shortest path and/or the shortest time).

2.5 Choosing movement direction

Let i-th agent has got coordinate $\vec{x}_i(t-\Delta t)$. The probability to move from this position to the direction $\vec{e}^\alpha(t) = (\cos\frac{2\pi}{q}\alpha, \sin\frac{2\pi}{q}\alpha)$, $\alpha=\overline{1,q}$ during next time step is the following:

$$p_\alpha^i(t) = \frac{\hat{p}_\alpha^i(t)}{Norm} = \frac{\exp\left(-k_W^i(1-\frac{r_\alpha^*}{r})1(\Delta S_\alpha)\right)\exp\left(-k_P^i F(r_\alpha^*)\right)\exp\left(k_S^i \Delta S_\alpha\right)}{Norm} W\left(r_\alpha^* - \frac{d_i}{2}\right), \quad (2)$$

where $Norm = \sum_{\alpha=1}^q \hat{p}_\alpha^i(t)$.

The visibility radius r ($r \geq \max\{d_i/2\}$), [m], is model parameter representing the maximum distance at which people and obstacles influence on the probability in the given direction. Obstacles can reduce visibility radius r to value r_α^* (see Figure 1). The people's density $F(r_\alpha^*) \in [0,1]$ is estimated in the visibility area, see Kirik et al. 2014. Function $1(\cdot)$ is Heaviside unit step function. There are model parameters: $k_S^i > 0$ — field S-sensitive parameter; $k_W^i > k_S^i$ — wall-sensitive parameter; $k_P^i > 0$ — density-sensitive parameter. Information on parameters one can find in Kirik et al. 2011, Kirik et al. 2014.

$\Delta S_\alpha = S(t-\Delta t) - S_\alpha$, where $S(t-\Delta t)$ — static floor field in the coordinate $\vec{x}_i(t-\Delta t)$, S_α — static floor field in the coordinate $\vec{x} = \vec{x}_i(t-\Delta t) + 0.1\vec{e}_i^\alpha(t)$. With ΔS_α moving to the target point is controlled.

Function $W(r_\alpha^* - \frac{d_i}{2}) = \begin{cases} 1, & r_\alpha^* - \frac{d_i}{2} > w; \\ 0, & r_\alpha^* - \frac{d_i}{2} \leq w \end{cases}$ controls approaching to obstacles[1], model parameter $0 \leq w \leq 0.1$, [m], – coefficient of inadherence to obstacles.

If $Norm = 0$ than particle does not leave present position[2]. If $Norm \neq 0$ than required direction $\vec{e}_i(t)$ is considered as discrete random value with distribution that is given by transition probabilities obtained. Exact direction $\vec{e}_i(t) = \vec{e}_i^{\hat{\alpha}}(t) = \left(\cos\frac{2\pi}{q}\hat{\alpha}, \sin\frac{2\pi}{q}\hat{\alpha}\right)$ is determined in accordance with standard procedure for discrete random values.

As in cellular automata models a parallel update is used here. Decision rules to choose direction and final conflict resolution procedure are presented in (Kirik et al. 2011, Kirik et al. 2014).

2.6 Speed calculation

Person's speed is density dependent (Kholshevnikov & Samoshin (2009), Schadschneider et al. 2009). We assume that only conditions in front of the person influence on speed. It is motivated by a front line effect (that is very well pronounced while flow moves in open boundary conditions) in a dense people mass when front line people move with free movement velocity while middle part is waiting a free space available for movement. As a result it leads to a diffusion of the flow. Otherwise simulation will be slower then real process. Thus only density $F(r_\alpha^*)$ in direction chosen $\vec{e}_i(t) = \vec{e}_i^{\hat{\alpha}}(t)$ is required to determine speed. According [3, 4] current speed of the agent is

$$v_i(t) = v_i^{\hat{\alpha}}(t) = \begin{cases} v_i^0(1 - a_1 \ln\frac{F(r_\alpha^*)}{F^0}), & F(r_\alpha^*) > F^0; \\ v_i^0, & F(r_\alpha^*) \leq F^0, \end{cases} \quad (3)$$

where F^0 – limit people density until which free people movement is possible (density does not influence on speed of people movement); $a_1 = 0.295$ is for horizontal way; $a_2 = 0.4$, for down stairs; $a_3 = 0.305$, for upstairs.

Numerical procedures that is used to estimate local density is presented in Kirik et al. 2014. Area where density is determined is reduced by direction chosen and visibility area, see Figure 1.

3 SIMULATION

The model presented was realized in the computer program module SigmaEva©. There was performed validation of the module under open and periodic

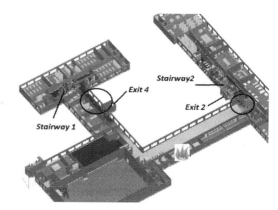

Figure 2. Fire position, exits and stairways in 3-store school building.

boundary conditions with fundamental diagram (the specific flow versus the density), Kirik et al. 2014.

The validation shown good dynamical properties: maintaining the speed according to local "directed" density, initial density and free movement speed maintains approximately till 0.5 [pers/m], flow diffusion realizes if it is possible, a model full flow rises with the bottleneck width increasing. A comparison with experimental data (Kholshevnikov & Samoshin (2009), Schadschneider et al. 2009, Seyfried et al. 2009) says model results are within an existing conception of the speed-density dependence.

The module SigmaEva is a part of software SigmaFS©. This software consist of 3D-building editor, CFD (computational fluid dynamics) simulation module for modeling of spreading of dangerous fire factors, simulation module for evacuation modeling, and 3D-visualization module.

SigmaFS is a tool for simulations that may be used in many fields from organization of mass events to fire safety of buildings, ships, aircrafts. One may vary a combustible material and its mass, places of fire, systems of smoke removal, pressurization systems, doors conditions, number of people, their initial positions, individual properties (free movement speed, evacuation starting time, projection size, evacuation way), furniture positions.

The main task of applying such simulations is to estimate evacuation time in different scenarios in order to provide safe conditions for visitors, passengers in emergency situations.

Let us consider and compare two evacuation scenarios from a school under fire conditions. According to both scenarios people evacuate from 3-store building using nearest stairways (Stw) number 1 and 2 and exits 4 and 2 correspondingly. A fire started in a marked room, Figure 2. While simulating it was assumed that doors in corridors were open.

The difference in scenarios is in initial conditions concerning a delay of the starting evacuation: 30 and 120 seconds later than the fire started.

---

[1] Note, function $W(\cdot)$ "works" with nonmovable obstacles only.

[2] Actually this situation is impossible. Only function $W(\cdot)$ may give (mathematical) zero to probability. If $Norm = 0$ then particle is surrounded by obstacles from all directions.

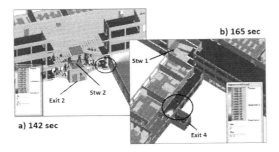

Figure 3. Evacuation from a school building under fire conditions, an evacuation stated 30 seconds later then the fire started.

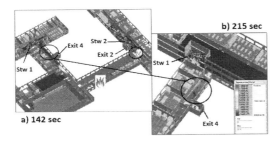

Figure 4. Evacuation from school building under fire conditions, evacuation stated 120 seconds later then fire started.

Fire conditions in the pictures are presented as a 2D-slice for height 1.7 m over a floor that is extracted from 3D-simulation of smoke. Blue color is for safe areas for people health, red color is for dangerous areas (boundary value is from the Russian fire safety legislation).

Joint simulation of the evacuation and the fire spread helps to show and estimate dangerous consequences of some initial conditions. Here the evacuation start delay is considered. In Figures 3 and 4 the most important areas are marked.

For the case of 120 second delay one can see that people moving to exit 2 are in risk area on 142 second. People moving to exit 4 are in a dangerous area on 215 second. Over 100 people evacuated in unsafe conditions if they start with 120 seconds delay.

If people start 30 seconds after fire started they evacuate in comfortable conditions, Figure 3.

4 CONCLUSION

These scenarios were simulated in educational purposes and are a part of special software – fire safety simulator. The idea is to show influence of different conditions on evacuation result.

Such comparative simulation with some parameters may be used for many purposes – evacuation plans design, an investigating of different conditions of building operating, a building safety design, etc.

REFERENCES

Kirik, E., Yurgel'yan, T., Krouglov, D. 2011. On realizing the shortest time strategy in a CA FF pedestrian dynamics model. *Cybernetics and Systems* 42(01): 1–15.

Kirik E., Malyshev A., Popel E. 2014. Fundamental diagram as a model input – direct movement equation of pedestrian dynamics. In U. Weidmann, U. Kirsch, M. Schreckenberg (Eds.): Pedestrian and Evacuation Dynamics, Zurich, 5–7 June 2010; Proc. of the Int. conference. Springer: 691–703.

Kholshevnikov, V.V., Samoshin, D.A. 2009. *Evacuation and human behavior in fire.* Moscow: Academy of State Fire Service, EMERCOM of Russia. (Rus.)

Schadschneider, A., Klingsch, W., Kluepfel, H., Kretz, T., Rogsch, C., Seyfried, A. 2009. Evacuation Dynamics: Empirical Results, Modeling and Applications. In, *Encyclopedia of Complexity and System Science.* Springer.

Schadschneider, A., Seyfried, A. 2009. Validation of CA models of pedestrian dynamics with fundamental diagrams. *Cybernetics and systems* 40(5): 367–389.

Seyfried, A., Rupprecht, T., Passon, O., Steffen, B., Klingsch, W., Boltes, M. 2009. New insights into pedestrian flow through bottlenecks. *Transportation Science* 43: 395–406.

A ground passenger transport network optimization method on the basis of urban rail transit

Zhong Guo & Hong Yang Wu
China Urban Sustainable Transport Research Centre (CUSTReC), China Academy of Transport Sciences, MOT, P.R. China

Lu Tong
Beijing Jiao tong University, Beijing, China

ABSTRACT: As major cities in China entered the era of urban rail transit, the integration of multi-modal public transport has become the inevitable way to improve public transport attractiveness. This study carried out technical and economic characteristics of two public transport modes, which were used for functional positioning. A bus network optimization method from point to line then to surface was proposed, and different types of feeder station layout form and scale were determined. Existing bus line network adjustment method is based on spatial relationship and passenger travel cost. In the new feeder bus line layout optimization process, with minimizing the total travel time of passengers as the planning objective, a feeder efficiency model was built. Last but not least, the example of subway line 1 in a city was used as a case study; by analyzing the status of the bus operation along subway line 1, suggestions were given on bus network optimization.

1 INTRODUCTION

As a solution to the problems of urban traffic, especially the traffic congestion in major cities and megacities, urban rail transit is widely implemented. It plays a significant role in promoting the urban social and economic development. The operation of urban rail transit certainly has a strong impact on the ground transit. Therefore, adjusting the layout of conventional bus line networks for the rail has now become the objective of urban development. There are two main research methods of feeder bus system. Martins CL (Martins C L, Pato M V. 1989) proposed return analysis method; Kuah, et al. (G.K. Kuah, J. Perl. 1989) built network optimization analysis model. These studies pay more attention to theoretical research and lack of practicality. This paper proposes a ground passenger transport network optimization method from the aspects of point, line and surface.

2 DEFINITION OF URBAN RAIL TRANSIT AND GROUND PASSENGER TRANSPORT

2.1 Comparison of the technical and economic characteristics

The technical characteristics of urban rail transit and bus system are different, and they were contrastive analyzed in terms of transport capacity, level of service, the right-of-way form, and operation cost. Table 1 shows the construction and operation costs. Urban rail transit is designed to cover the main passenger corridors and hubs to maximize its passenger capacity. BRT has similar design objective. However, bus system is designed to provide basic service. Accessibility and station distance indicate that urban rail transit is to meet the needs of long-distance travel, with its mass capacity, and fast, punctual and comfortable services. On the other hand, ground passenger transport is designed to provide service to middle- and short-distance travel needs.

2.2 Functional orientation analysis

Comparison between technical characteristics of urban rail transit and ground passenger transport indicated that urban rail transit and ground passenger transport together constitute the urban transport system to meet diversified travel needs. The relationship between them is both competition and cooperation. BRT can be built as the supplementary service in the area where is not covered by rail transit network. Bus system, on the other hand, can compete with urban rail transit with its advantages of short station distance and low fare. It also connects urban rail transit to ground passenger transport system, ensuring the coverage of public transport system. The functional orientation of urban rail transit and ground passenger transport is shown in Fig. 1 (Wang Wei, Yang Xin Miao, Chen Xue-wu. 2002).

Table 1. Comparison between technical characteristics of urban rail transit and ground passenger transport.

Item		Urban rail transit	Ground passenger transport	
			BRT	Bus
Capacity (thousand people/one-way peak hour)		30–60	15–25	6–10
Service level	Average running speed (km/h)	30–60	20–25	15–20
	Punctuality rate	good	good	discrepancy
	Comfort level	good	good	discrepancy
	Safety level	good	good	preferably
Station distance (km)		1–1.5	0.8–1	0.3–0.5
Right-of-way form		dedicated rights-of-way	dedicated rights-of-way	mixed or dedicated rights-of-way
Accessibility		acceptable	acceptable	good
Construction cost (million yuan/km)		300–500	20–70	low
Operating cost		High	low	low

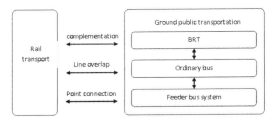

Figure 1. Line relationship between rail transportation and ground passenger transport.

3 GROUND PASSENGER TRANSPORT NETWORK OPTIMIZATION METHOD

3.1 Principles and procedures

Ground passenger transport network optimization includes two situations. One is to adjust the lines that have conflict with rail transit lines. The other is to increase the feeder lines in order to expand the coverage and solve the "last kilometer trip" problem according to the construction of the bus transfer points. Specific optimization principles are as follows.

A. Connected bus line form. The layout in central area should be "S"-shaped in order to increase the connection level. The layout near edges should be radial, to disperse the passenger flow of subway. The direction of the lines must be consistent with the travel

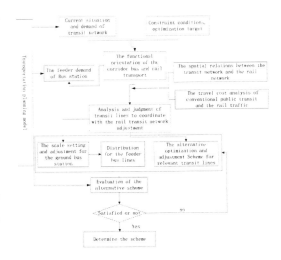

Figure 2. The ground transit network optimization procedures on the basis of rail transportation.

flows, and the feeder bus station distance should be carefully determined in order to reduce the total travel time for passengers.

B. Make the feeder lines evenly distributed in rail transit coverage. As the rail – feeder bus travel patterns already have 1–2 times of transfers; bus blank area should be eliminated by the form of direct transport.

C. Land use, residential and employment distribution should also be considered in optimizing feeder bus network, and different settings (in terms of length, speed, station distance) can be used.

Optimization procedures are shown in Fig. 2.

3.2 Bus network optimization method

3.2.1 Set of bus stations

Under the situation of public transport integration, the positioning and setting of ground bus stations connected to rail transit has a great influence on the connecting efficiency, and it is directly related to the urban public transport integration service level. Different feeder site scales have different passenger flows, and the sizes and forms of ground passenger transport connecting points are different as well. Corresponding to different scales of feeder sites, bus connection points are classified as hub station, large bay station, bay station or crossed station.

(1) Determining the size of bus station

The passenger volume of rail transit varies from site to site, and the connected bus lines should be set accordingly. There are three types of bus stations:

– As part of the integrated hub station, 90–100 m^2/standard unit (vehicle) are required for buses in order to provide nighttime parking. Three-dimensional channel is used to separate passenger and traffic, make it convenient for passengers to transfer and allow traffic entering and leaving smoothly, which makes it easy to manage.

- Large bus station, as part of a large feeder station, is parking station.
- Bus station, corresponding to the general interchange point, should be placed as close as possible to the rail transit entrance/exit, with distance of 50–80 m. Bus station can be set as bay station (on main road) or crossed station (on branch road).

(2) Determining access capacity of ground bus station

Bus station should be set on the basis of the safety of waiting passengers, the convenience for passenger to transfer and go across the street, making buses docked easily, and also minimizing the impact of road sections and intersections capacity. The number of stations is affected by the number of vehicles. However, analysis of the number of vehicles, which is the demand analysis of the feeder transfer transport capacity, is mainly to identify the time of reaching the main transport passenger flow peak, set the vehicle number of the most efficient rail transit of passenger as (units/minute), analyze the number of feeder buses in rail sites, and get the length of feeder site, by the following formula:

$$N_v = \frac{\left[(T_k/I+1)P_m\alpha/(P_bT_k)\right]}{(1-\beta)}, \quad (1)$$

where T_k is the duration of the peak passenger flow (minutes), I is the average departure interval of rail traffic in the peak hours, P_m is the average number of up and down passengers in rail site in peak hours, P_b is the average passenger capacity of a ground bus, α is the percentage of transfer passenger flow to total track sites up and down passenger traffic, and β is the bus load factor.

Then, we get the effective number of berths for feeder station buses:

$$N = N_v + 1. \quad (2)$$

Take the integer part of N, according to Eq. (2), calculate the number of berths for continuous arrangement set number, which is shown in Fig. 3, there is no space superposition between berth #1 and berth #2. However, when more of the bus lines pass, in order to save the trunk stations sites, berth often overlap set. As shown in Fig. 3 berth #3 is set between berth #1 and berth #2; when there is no vehicle parked between berth #1 and berth #2, bus docked in berth #3 can be pitted.

3.2.2 *Existing bus line network adjustment method*
(1) Line adjustment method based on spatial relationship

Using cluster analysis to consider the operational status of the rail lines, the index of line area layout, the current line length, the spatial relationship between passenger flow, bus routes and rail line (cross line length, cross line site, parallel length and parallel site) are chosen to get different categories set according to the clustering index, which is the existing line reservations, adjustment and cancellation of the line set.

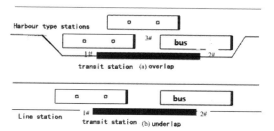

Figure 3. Station berth arrangement diagram.

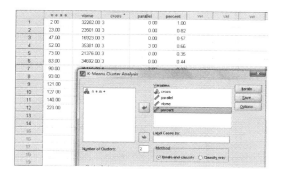

Figure 4. Analysis of clustering in spatial relation.

(2) Line network adjustment method based on travel cost (Wu Hong Yang. 2013)

Let ρ_{21} be the travel cost ratio of track – feeder one-time transfer bus to ground direct bus, ρ_{32} be the travel cost ratio of ground one-time transfer bus to ground direct bus, ρ_{41} be the travel cost ratio of rail transportation to ground direct bus, the following formula can be obtained:

$$\rho_{21} = \frac{F_{T2}}{F_{T1}}, \quad \rho_{32} = \frac{F_{T3}}{F_{T1}}, \quad \rho_{41} = \frac{F_{T4}}{F_{T1}}. \quad (3)$$

Taking ρ_{21}, ρ_{32}, ρ_{41} as the optimized transit network to reasonably adjust the quantization index, combined with qualitative analysis in Table 2, the ground bus adjustment implement method is given out.

3.2.3 *New feeder bus line layout area model*
(1) Basic assumptions

In the precondition of not influencing the problem analysis, in order to facilitate the establishment of model, the following hypothesis were made.

In the plane of entire network layout, urban terrain without severe fluctuation, there is no mountain, lake and other natural barrier to make a traffic mode difficult to reach.

Assume c is a non-linear coefficient, considering the coverage of feeder lines, let c = 1.4.

The average speeds of bus and rail transit are respectively 20 and 40 km/h.

Due to the same comparison conditions, the transfer time, which is the time to walk to the bus stop and time to wait for the train, was not considered.

Table 2. Optimization of public transportation network adjustment program implementation.

Number	Adjustment mode	Applicable conditions
1	Reasonable line fully	Quantify: $\rho_{21} > 1$, $\rho_{32} > 1$, $\rho_{41} > 1$. Qualitative: Bus rapid transit or regular bus reserved service range is not affected by the impact of rail transit. For instance, for the peripheral area or suburban with appropriate length and not yet covered by rail lines, or where bus rapid transit or ordinary bus lines have small overlap with rail transit line, competition between them is very weak in terms of passenger flow.
2	The line merged or extraction	Quantify: $\rho_{21} \leq 1$, $\rho_{32} \leq 1$, $\rho_{41} \leq 1$. Qualitative: Bus rapid transit or ordinary bus lines is coincide with rail transit lines, bridging the competition between the passenger flow of bus rapid transit or ordinary bus and rail transit, to extract or merger of such bus lines.
3	Line is truncated and extend (or shorten)	Quantify: $\rho_{21} \leq 1$, $\rho_{32} \leq 1$, $\rho_{41} \leq 1$. Qualitative: Most of the bus rapid transit or ordinary bus lines are overlapped with rail transit lines, and repetition coefficient is relatively large; part of the line can be cut, in order to encourage passengers to take trail transit; transfer the truncated bus line into the feeder line, extend or shorten the line according to the standards to expand the coverage of services.

Fare difference was not considered; only bus and rail transit were included in the analysis.

(2) Constraint conditions

Starting from the simplest single case, the model establishment was explored.

Assume that coordinates $A(X_A, X_A)$, $B(X_B, X_B)$, $C(X_C, X_C)$, $D(X_D, X_D)$ are known in Fig. 5, suppose point $P(X_P, X_P)$ is out of the attractive area of the rail transit station, passengers at point P intend to take feeder bus and then transfer to rail transit to the destination. To simplify the model, start point and end point are set O and D. Let t_A, t_B, t_C be the minimum travel time of rail station A, B, C, which can be transferred from feeder bus stations P, in order to minimize the total travel time, Equation (6) should satisfy the following constraints:

$$t_C \leq t_B, t_C \leq t_A \quad (4)$$

On the contrary, if passengers at the point P intend to go to point O by taking feeder bus and then transfer to rail transit. The travel time is minimum if passengers get on the bus at point B and the feeder bus is connected

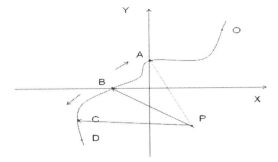

Figure 5. Schematic diagram of feeder line.

to point B. Equation (5) should satisfy the following constraints:

$$t_B \leq t_C, t_B \leq t_A \quad (5)$$

(3) Establishment of models

From the analysis above, city rail transit feeder bus line layout area model can be established. Assume there is a start point P_i, passengers at point P_i intend to go to point δ_k by getting on the bus at station j, and the travel time for all the train from station j is the shortest, set the collection of all rail station site as A, then:

$$\min t_j = t_{p_i R_j} + t_{R_j \delta_k}, j \in A \quad (6)$$

$$t_{p_i R_j} = \frac{l_{p_i R_j}}{V_F} \quad (7)$$

$$t_{R_j \delta_k} = \frac{l_{R_j \delta_k}}{V_R} \quad (8)$$

$$l_{p_i R_j} = cl_{ij} \quad (9)$$

$$ST \; t_j \leq t_r, (j \neq r, r \in A) \quad (10)$$

$$i, j, k > 0$$

In the equations, $t_{p_i R_j}$ is the feeder bus travel time, $t_{R_j \delta_k}$ is the rail transit travel time, $l_{p_i R_j}$ is the feeder bus travel distance, $l_{R_j \delta_k}$ is rail transit travel distance, l_{ij} is the shortest way from point i to point j, V_F is the feeder bus operating speed, V_R is the trail transit operating speed, and δ_k is the travel destination, $k = 1, 2$, which are the start and end points of the orbit OD.

4 CASE STUDIES (GUO ZHONG. 2013)

Through cluster analysis of bus lines based on line space relation, six categories were obtained in Table 3. According to the passenger flow, overlapped and parallel length of bus line and rail transit line, six categories are corresponded with three sets of bus line

Table 3. Line 1 adjustment options based on the spatial relationships.

Set category	Set adjustment	Corresponding relationship	Adjust the amount	Adjust the implementation
Five	Cancel line	Passenger flow is not large, length of cross line and parallel is longer than 12 km, and site points are more than 12.	4	A decrease in initial frequency or merging line
Two, Six	Adjust line	Passenger flow is not big, of the line is not big, length of cross line and parallel is longer than 6 km, and site points are more than 4.	17	Direction and operation adjustment
One, Three, Four	Keep line	Length of cross line and parallel is shorter than 6 km.	31	Part of the line take "small" adjustment programs, according to the transit hub and site, also direction and operation ajustments.

adjustments; adjustment of 17 lines is needed, 4 lines are canceled, and 31 lines remain the same.

Note: Cancelation of the lines is based on comprehensive consideration of various factors; lines to cancel or merge were selected from the withdrawal line collection class.

According to the rail transit passenger distribution requirements, the heaviest passenger flow sections in the city are near the interchange stations. However, for inception point, in order to make section flow distributed uniformly, it was advised to set corresponding feeder bus lines at Xiaowangsiying station and Shijicheng station.

Chenggongnan station, the station to the south of Line 1, is to distribute passengers quickly and uniform cross-section traffic. Set the railway station area mentioned above as the center of radiation, and select the surrounding important site as the feeder bus station, the sites includes a typical residential area, bazaars, stadiums, commercial buildings, schools etc.

5 CONCLUSIONS

According to the demotic and international development of urban public transport, rail transit and ground transportation are the two most important components in the metropolis public transport system, must coordinate the development in order to achieve comprehensive urban transportation sustainability. In fact, the integrated development between rail transit and ground transportation of big cities in China still has many problems. In this paper, from the angle of rail transit and ground bus network integration, bus line adjustment method was discussed, the form and size of feeder line region emplacement model was established and different modes of connection site layout were determined. The methodology is applicable for early operation of rail transit system. However, because the adjustment of bus lines involves a wide range of complex researches, it is the initial attempt of the early formation of the rail network of the public transportation network adjustment. Due to limited time and capacity, there are many areas that need to further improvement, including in-depth research of travel costs of the various travel modes, the conditions of mass rail transit, new feeder bus lines laid utility model and algorithm, etc.

REFERENCES

G.K. Kuah, J. Perl. Optimaiztion of feeder bus routes and busstop spacing [J]. Journal of Transportation Engineering, 1989, 114(3):751–767.

Guo Zhong. Kun Ming Transit Metropolis Development and Construction Planning (2011–2020) [R]. China Academy of Transportation Sciences, 2013.

Martins C L, Pato M V Seatch Strategies for the Feeder Bus Network Design Problem European. Journal of Operation Research, 1989(106):425–430.

Wu Hong Yang, rail transit network layout and ground public transportation network adjust the layout of the research report [R]. China Academy of Transportation Sciences, 2013.

Wang Wei, Yang Xin Miao, Chen Xue-wu urban public transport system planning methods and management techniques [M] Science Press, 2002.

A study on structural strengthening technology

Guanghua Qiao & Yuanbing Cheng
North China University of Water Resources and Electric Power, Zhengzhou, China

ABSTRACT: Most buildings in service in China have reached, and some even exceeded, their designing period. Many of these buildings are unable to meet architectural requirements or even have safety problems. If all of these buildings are removed or rebuild, it will cost a lot of money. Therefore, study on reinforcing techniques of structure is of great significance. This paper mainly introduces the reinforcement methods and the scope of applications of reinforced concrete structure. Analysis was performed on problems of reinforcement technologies that need to be solved in China.

Keywords: strengthening; structure; reinforced concrete; safety

1 INVESTIGATION AND ANALYSIS OF THE IMPORTANCE OF STRENGTHENING

With the rapid development of China's economic and large-scale urbanization, which is bound to cause the urban resources available compression, city construction development tends to be smooth. Due to the building's service life, natural disasters, the underestimation in original design and many other factors, buildings have appeared different degrees of damage. Some buildings, because of the particularity of history, the human activities, or the influence of the surrounding environment and the transition of the use function, have become poor constructions. All the demolition and reconstruction, on one hand, will cause a huge waste of resources and investment increase, while on the other hand, after the demolition of the old buildings, will produce building garbage, which not only seriously pollute the surrounding environment, and requires a lot of manpower, material resources, financial resources to deal with. In order to maintain the continuity of the culture and construction of normal use or improve its functions, will gradually highlight the importance of engineering structure reinforcement and reconstruction. In addition, with the development of economic construction, many existing buildings require increased height, improved service load, increased space, added layers, etc. So the identification, maintenance, reinforcement and reconstruction have been gradually put on the agenda.

2 THE GENERAL PROVISIONS OF BUILDING STRUCTURE REINFORCEMENT

Before strengthening, the reliability evaluation and strengthening design must be carried out by professionals. The scope of reinforcement can be as small as a component, or extended to the entire building. Before the strengthening, the safety of the building shall be determined jointly by the client and the designer, and strengthening design should work closely with the construction, to ensure that the new and old components work together (Haijun Yuan and Hong Jiang 2003). Economic effect should be considered in the strengthening design and unnecessary removal or replacement should be avoided. During strengthening, necessary temporary security measures should be taken. The service period of structure after reinforcement should be determined by the designer and the client, and in general should be determined as 30 years. Structures use bonding or polymer in strengthening should be checked periodically, generally every 10 years. Without the identification and approval from the designer, the function of the building after reinforcement cannot be changed.

In the reinforcement calculation, linear elastic analysis method is adopted generally, the bearing capacity and service ability need to be checked according to the provisions, and the strain hysteresis characteristics and degree of coordinate work of new and old structures should be considered. The load on structure should be determined according to field investigation and detection, the calculating sketch should be consistent with the actual, the effect of combination, combination coefficient and partial coefficient are determined according to the load code, and the additional internal force of eccentric load, structural distortion and temperature effect should be considered. During designing, measured values are adapted to the original part and nominal values are adapted to the new part. The grade of concrete and steel should adopt the testing results of standard values, and the quality and performance of reinforcement materials should comply with the relevant provisions. When using adhesive or polymer (modified asphalt, polymer mortar etc.) in strengthening, structure calculation should be performed to the original structure, and appropriate safety guarantee is needed for the original structures to ensure the structure safety in case of accident lead to failure of the reinforced part. When the structure weight increase or load transfer path changes, related components and foundations must be checked. Strengthening design of structures in seismic zones should comply with the build seismic design code and seismic strengthening code.

3 THE PRINCIPLES OF BUILDING STRUCTURE REINFORCEMENT

3.1 Identification before consolidation

For buildings in service, due to the influence of environment, the structure materials will gradually age. Due to abnormal use, the structure may be gradually damaged, and the reliability of the structure will gradually reduce. If no maintenance was carried out, the service life of the building is relatively short; on the contrary, if repair and reinforcement are carried out normally, the service life will be prolonged. Under other abnormal conditions such as design errors, poor quality of construction, improper use or architectural function changed, the actual state of the structure can not reach the expectations, and reinforcement is a good way to solve these problems.

(Shangmu Zhuo et al. 1997) In order to find out both the reason and extent of structure function deterioration, and to make a scientific, reasonable and economical repair, reinforcement and strengthening plan, test and analysis of the reliability of the building are needed.

3.2 The principle of overall structural system effect

When making the reinforcement scheme, in addition to consider the reliability appraisal conclusion and reinforcement content, the overall effect after reinforcement should also be considered. For housing, for example, a layer of the pillar or wall reinforcement sometimes will change the dynamic characteristics of the whole structure, resulting in a weak layer, and bringing bad influence on the seismic (Chunhua Huang and Xinli Hu 2008). Therefore, in the formulation of a reinforcement scheme, analysis of the force of the whole structure should be comprehensive and detailed; the "no overall plan for a fundamental transformation" concept cannot be used.

3.3 Principle of reinforcement scheme optimization

In the reinforcement, to make every reinforcement material useful, unnecessary waste of resources should be avoided. In addition, in the numerous reinforcement schemes, the best one should be chosen according to the reason for reinforcement.

3.4 Combination of the seismic code

China is a country prone to earthquakes, grade 6 and above earthquake zones almost distribute throughout the country (Chunhua Huang and Xinli Hu 2008). Most buildings constructed before 1976 had no seismic. In the seismic code before 1989, only buildings in grade 7 were seismic. In order to grant these buildings with appropriate safety, when reinforcing their bearing capacity and durability, current seismic code should be taken into account.

3.5 Principle of material selection and determination of their strength value

3.5.1 In the reinforcement design, the strength value of materials should be determined according to the following provisions

(Chunhua Huang and Xinli Hu 2008) If the material types and properties are consistent with the original design, use the original design (or code) values; otherwise, use the current values.

3.5.2 The selection of reinforcing materials

Reinforcement materials should be high strength, light weight, and be able to work together with the original structure. Steels should be level I or II, cement should be ordinary Portland cement with grade not less than 42.5. The concrete strength for reinforcement should be higher than that of the original, and the concrete should not be mixed with fly ash, or blast furnace slag. The bond strength of bonding material and chemical grouting material should be higher than the tensile strength and shear strength of structure concrete. Bonding material and chemical grouting material generally should be finished or semi-finished products. Before preparation by the constructer, experiment and test are necessary.

3.6 The load calculation principle

For the reinforcement structure, load value should be chosen based on field survey. Normally, load value should be chosen according to the old edition "load code for the design of building structure" when the original structure was designed according to the code. In the reinforcement calculation, load value should be chosen according to the new edition "load code for the design of building structures".

If there are no relevant provisions in the current load code for the dead load, the value can be decided based on sampling. (Chunhua Huang and Xinli Hu 2008) Sampling number should not be less than 5, and the average value should be multiplied by 1.1 as the standard values of dead load. Values of process load and crane load should be based on the data provided by the plant.

3.7 The overall structural system effect principle

(Chunhua Huang and Xinli Hu 2008) In the bearing capacity calculation, structure calculation diagram should be based on the actual stress and size of the structure. The cross-sectional area of members should use the actual cross-sectional area, and the damage of structure, defects, rust and other adverse effects should be taken into account. In checking, designers should consider the actual stress degree in strengthening and the force of reinforcing part of hysteresis characteristics, as well as the degree of reinforcement part and the original structure to work. The design strength of the strengthening part should be appropriately reduced, and additional internal force caused by the actual load eccentricity, structural deformation, local damage, and temperature effect should also considered (RTL Alan and SC Edwards 1992). When the mass of the structure increase after reinforcement, the structures and foundations should be checked.

3.8 Other principles

For the damages caused by high temperature, corrosion, freeze-thaw, vibration, uneven subsidence of foundation and so on, corresponding countermeasures in reinforcement design should be put forward firstly. Structure reinforcement should also consider the economy, not to damage the original structure, and retain the useful elements, avoiding unnecessary removal or replacement of members.

4 REINFORCEMENT METHODS OF BUILDING STRUCTURE

4.1 Direct reinforcement method

Direct reinforcement method is to improve the bearing capacity of structural elements or nodes directly. (RTL Alan and SC Edwards 1992) Such methods include enlarged cross section method, concrete replacement method, outsourcing external steel method, sticking steel plate method, viscose fiber composite materials, wire and wire mesh-polymer mortar strengthening method, etc. Following are their brief introduction.

4.1.1 Method of enlarging cross section

In order to improve the bearing capacity and meet the normal use, enlarge cross section method is to increase the cross section area of member. This method can be widely used in concrete structure beam, slab, column and other general component (Guoxiong Li 2006; Zhongnan Song 2002). In the strengthening, if the original concrete strength is not less than C10, the new concrete should be one level higher than the original concrete, which means not less than C20. When consolidating the plate, the thickness of the new concrete should not be less than 40 mm; when reinforcing beam or column, the thickness should not be less than 60 mm in artificial casting; if using shot concrete technology,

Figure 1. Enlarging cross section method for column.

Figure 2. Section enlargement method for consolidate beam.

the thickness should not be less than 50 mm. Hot rolled steel bar should be adopted. Diameter of steel bars in plate reinforcing should not be less than 8 mm, diameter of steel bars in beam reinforcing should not be less than 12 mm, diameter of steel bars in column reinforcing should not be less than 14 mm, and diameter of stirrup should not less than 8 mm. U-shaped stirrup diameter should be the same as the original stirrup, and the diameter of distributed stirrup should not be less than 6mm. The interface processing and paste quality should comply with the rules. These reinforcements are as illustrated in Figs. 1 and 2.

4.1.2 Reinforcement method of outsourcing steel

This method is to strengthen the members by wrapping with steel (dry or wet). It is suitable for where the concrete cross section is limited (Guoxiong Li 2006; Zhongnan Song 2002), and greatly improve of bearing capacity is needed. When using chemical grouting on extension-story reinforcement, surface temperature of steel should not be higher than 60°; when members are in corrosive environment, reliable protections should be made. When glued steel method in reinforced concrete beam is used, corner paste angle must be placed;

Figure 3. Reinforced beam with outsourcing steel.

Figure 5. Pre-stressed reinforcement method in reinforcement of beam.

Figure 4. Reinforced column with outsourcing steel.

if the beam has compressive zone with flange or floor, the internal corner should be placed by steel strip. These reinforcements are as illustrated in Figs. 3 and 4.

4.1.3 Reinforcement method of replacement of concrete

This method is suitable for members with poor compression ability or serious defects in compression area of the section. Unloading is needed in this method. Effective brace is needed when strengthening beam component. In the strengthening of column and wall, bearing capacity check and supervisory control in the whole loading history are needed, and no tensile stress should appear in the new section; otherwise effective braces should be taken. The replacement part should be located in the compression zone, and defective concrete should be eliminated according to the direction of stress (Guoxiong Li 2006). For avoiding uneven stress or load eccentricity, elimination should be done along the entire width of one or both sides of the section. The strength grade of new concrete should be higher than that of the old, and not lower than C25; in order to avoid bolt pin effect, the strength grade of new concrete should not be too high. The requirements of minimum depth of replaced concrete are the same as the thickness of the new concrete in enlarging cross section method. Replacement length should be determined according to the test and calculation results of concrete strength and defect; in the partly replacement, both ends shall be extended not less than 100 mm respectively.

4.2 Indirect reinforcement method

Based on the rationality or integrity of the structure, this method strengthens the structure by adding new members to change the overall layout and the way of force transmission. It can reduce the internal force, increase the structural stiffness and ductility (Guoxiong Li 2006). The method mainly includes pre-stressed reinforcement method, new-built shear wall and lateral support method, adding damper, adding protection and Rachel connecting method, etc.

4.2.1 Reinforcement method of adding pre-stress

This method is, by adding pre-stress steel rod (horizontal tie, under rod and combined rod) or poles, to reinforce the structure. It is suitable for the concrete structures when increasing load-bearing capacity, stiffness and crack resistances are required, and this method occupies small space. This method should not be used in concrete structure under temperature of 60°C, otherwise protective treatment should be done. (Guoxiong Li 2006; Zhongnan Song 2002) It also does not apply to the concrete structure with large shrinkage and creep of concrete. These reinforcements are illustrated in Figs. 5 and 6.

4.2.2 Reinforcement method of change structure of force transmission way

(1) Add fulcrums

The method can reduce the span and deformation and improve the bearing capacity of the structure. According to the mechanical performance of the support structure, the method can be divided into rigid support method and elastic support method.

Figure 6. Pre-stressed reinforcement method in plant.

(2) Reinforcement method of support beam and eliminated column

This method is used for the condition where the upper structure is unable to be removed or replaced. According to different construction skills, this method can be divided into supporting beam method with brace, supporting beam method with no brace, no supporting jack beam and pulled-apart column and double beam plate bracket supporting jack beam and pulled-apart column, which is suitable to plant functional change, increasing the space of the old factory (Zhongnan Song 2002).

5 CONCLUSIONS

Reinforcement technologies are developed and varied, each has its advantages and disadvantages. For example, the direct reinforcement method of increasing section, the concrete displacement method and the external sticking steel method are suitable for bending and compression members; the steel plate pasting method is suitable for bending; large eccentric compression and tension component, fiber composite material pasting method and the indirect reinforcement method of the pre-stressed method are suitable for bending, tension, compression and shear components. Enlarging cross section method and replacement of concrete method have no restrict to environmental conditions, and the technology is simple and mature; durable performance by this method is good, but the curing time is long, and the reinforcement occupies space. Method of pre-stress, steel bond, plate paste and fiber composite materials paste have short construction period and occupy little space, but they are only valid in the environment of temperature lower than 60°C. The external bond section steel method and paste steel plate method also need regular maintenance, and the construction of pre-stressed strengthening method is of great difficulty. (Abdel Baky et al. 2006; Anyong Li 2000; Xiaofang Qin 2007; Gang Wu et al. 2002; Zhongnan Song 2002) Fiber composite material has excellent physical and mechanical performance, with tensile strength a dozen times high than steel. It is light weight, high strength and corrosion resistance, especially suitable for irregular or uneven surface component. But fiber composites are expensive because it is mostly made from non-renewable resources, and most of the fiber composite materials used are universal. It is urgent to create new variety of high-tech fibers in short term. Therefore, choosing the reinforcement method should be based on local conditions and comparison of multiple solutions, conforming to advantage and reliability of technology, and reasonable economy principle. When structure strengthening and member strengthening coexist, priority should be given to the structure, followed by the members. In the strengthening of different types of members, priority should be given to the important members and key members. When a single method has poor performance, give priority to synthesis.

REFERENCES

Abdel Baky, USAMA A Ebead & Kenneth W. Flexural and interfacial behavior of FRP strengthened reinforced concrete beams [J]. *Journal of Composites for Construction.* ASCE; 2006(06):629–639.

Anyong Li. The present and future of reinforced concrete structure technology [J]. *Journal of Suzhou Urban Construction and Environmental Protection Institute,* 2000.

Chunhua Huang & Xinli Hu. Current situation and discussion of structure reinforcement [N]. *Journal of Guangdong building materials.* 2008.

Gang Wu, An Lin & Lv Zhitao. Experimental study on the flexural strengthening of reinforced concrete beams strengthened with CFRP [D]. *Engineering Structures*; 2002.

Guoxiong Li. Status and prospect analysis of concrete structure reinforcement technology [J]. *Science and Technology of Overseas Building Materials,* 2006, 27(3):65–68.

Haijun Yuan & Hong Jiang. Detection and identification of structure reinforcement handbook. *China Architecture & Building Press*; 2003.

RTL Alan & SC Edwards. Concrete building repair [M]. *Water Conservancy Electric Power Press.* 1992.

Shangmu Zhuo, Zhicang Ji & Changzhi Zhuo. Analysis and reinforcement of reinforced concrete structure of the accident [M]. *China Architecture & Building Press*; 1997.

Xiaofang Qin & Yanmin Zhao. Reinforcement technology of building structure and development trend [J]. *ShanXi Architecture,* 2007.

Zhongnan Song. Present status and development countermeasure of strengthening technique for concrete in our country [J]. *Concrete*; 2002, 10(156):10–11.

[Author]: Guanghua Qiao
[Address]: North Ring of Jinshui District, Zhengzhou City, Henan Provice, No. 36 North China University of Water Resources and Electric Power
[Telephone]: 15038287062

A systematic solution to the smart city – distinguished from the intelligent city

Yihua Mao, Hongyu Li & Bing Yang
College of Civil Engineering and Architecture, Zhejiang University, Hangzhou, Zhejiang, China

ABSTRACT: Smart Cities, not only being intelligent but also Economy-Ecology-Society coordinated, are the future of cities. Basing on the smart city theory, this paper proposes a systematic method, namely, the "12345" model, which requires the cities to make up a tailored proposal for cities and citizens, and build "ONE urban management system, TWO supporting and guarantee measures, THREE information infrastructure platforms, FOUR urban management modes, FIVE application service systems". Then the cities will get the ability of self-adjustment and self-improvement. The systematic construction method of the smart City not only solves the problems encountered by cities at current stage, but also promotes the all-round development of cities.

1 INTRODUCTION

With the rapid development of economy, ecological resources and social problems in modern cities have become increasingly prominent. Due to the absence of systematic planning and construction, urban construction has been a "progress in groping stage". Cities, developing in traditional model, have been much criticized, and require new development method.

According to the research by Abdoulleav et al. (2011), the present problems encountered in the development of urban construction are largely due to the lack of the ability of self-adjustment and sustainable development. Therefore, in November 2008, IBM proposed the theory of "Smarter Planet", which was based on the sustainable development theory, and led to a global craze in the research and construction of smart cities (Hollands, 2008). Deakin and Wright proposed the framework of smart city from technical aspects (Deakin & Waer 2011, Wright & Steventon 2004), and Wu and his assistants defined the smart city by digital city (Wu et al. 2010). However, researchers rarely took the layers of society, ecology and management into account, and some of their theories might be partial.

On macro level, the smart city is the integration of digital city, knowledge city, eco-city and innovation city. The smart city should be developed in an Economy-Ecology–Social coordinated way, and have the ability of self-adjustment and self-improvement (Xu et al. 2012). Specifically, smart cities are required to get the vigor of self-growth. By self-adjustment, cities can promote sustainable economic development, improve the ecological environment, provide digital and intelligent services, contribute to the better quality of the material and spiritual life of citizens, and enhance city happiness.

2 THE CONNOTATION OF SYSTEMATIC METHOD FOR SMART CITY CONSTRUCTION

2.1 Smart city contains more than digital city

In recent years, many cities have carried out "smart city" construction, but most of them just equipped themselves with information technology. Beijing, the capital of China, defined "smart city" as: Integrate and share the information that was closely related with the citizens by using information technology, to digitize the elements of the urban life, which are networked, intelligent, interactive and collaborative, and ameliorate the quality of citizens' lives.

The city, only equipped by information technology, is called digital city or intelligent city, and far from smart city. Smart cities should apply advanced urban management philosophy and information technology to the construction, operation, and development of urban system and optimize the allocation of resources, to achieve the self-improvement and coordinated development of urban functions.

2.2 The theoretical model of smart city

In the theoretical system of smart cities, the demand of citizens is the basis, the information technology is the means, and the Urban Think Tank, which consists of policy set in the city, experts in various areas and representatives of citizens, is the operation center. A platform for information, sharing, innovation, and management, supported by information technology, e.g. Internet of things (IOT) and Cloud Computing, will be built. Through the platform, the Urban Think Tank perceives the indicators of economy, ecology and

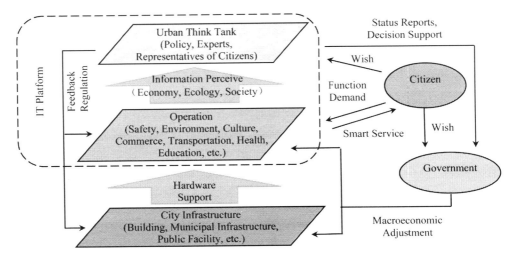

Figure 1. The Theoretical Model of Smart City.

society, and the demand of citizens, monitors the status of urban operation and development, and co-ordinates the city operation for various status to promote the healthy operation and sustainable development of the city (as illustrated in Fig. 1). A smart city system provides smart service for the citizens, and offers real-time status reports to the government to provide a reference for the government's intelligent decisions.

2.3 The construction method for smart city

The connotation of smart cities construction should be "People-oriented, Development-oriented and Wisdom-oriented". Specifically, smart city construction is people-oriented, its goal is promoting coordinated development of cities, and its features and advantages is wisdom.

1) People-oriented construction requests the construction, development, and renewal of the city to be loyal to the wish of citizens, fulfilling the demand of service, and improving the quality of life.
2) Development-oriented construction emphasizes that the construction of a smart city should be Economy-Ecology–Society coordinated.
3) Wisdom-oriented construction indicates that the concept of smart city should be absorbed into urban design, construction, and operation, in order to endow the city with the ability of self-growth.

3 SMART CITY CONSTRUCTION IN CHINA

3.1 The mode of smart city construction in China

A national craze for smart city construction can be found in China at present. By 2012, more than 230 cities have put forward the target of smart city construction in their reports on government's work and

Table 1. The Framework of Smart Hangzhou.

Layers	Function
Layer of basic support	The basis of smart city, provides industrial support and scientific and technological support.
Layer of information base	Provides communication support.
Layer of public service	Provides public service support for the intelligent applications which are used by government, enterprise, and citizens.
Layer of intelligent application	Responds to the application demands of government, enterprise, and citizens, and builds intelligent application system to improve management and service.

"12th Five-Year Plan", and the investment is expected to exceed 2 trillion (RMB). As pilot cities of the smart city construction, Beijing, Hangzhou, and Shanghai have accumulated much experience.

According to the *Smart Beijing Platform for Action (2012)*, eight action plans were proposed for Beijing, namely Intelligent Operation Plan for City, Digital Life Plan for Citizens, Network Operation Plan for Enterprise, Integration Services Plan for Government, Enhancing Plan for Information Infrastructure, Building Plan for Public Wisdom Platform, Docking Plan between Applications and Industries, and Innovation Plan for environment development.

According to the *Overall Planning for Smart Hangzhou (2012–2015)*, the framework of smart city construction of Hangzhou can be divided into four layers: the layer of basic support, the layer of information base, the layer of public service, and the layer of intelligent application (as illustrated in Table 1). Generally, by constructing information infrastructure and public service platform, Hangzhou integrates interactive form, ameliorates the service for citizens, the management of urban operation and the development of intelligence industry, and improves the administrative efficiency.

According to the *Promotion Plan for Smart City of Shanghai (2011–2013)*, by building the system of information perception and intelligent application, Shanghai focuses on promoting the urban construction and management, the safety of urban operation, intelligent transportation, social programs and public services, e-government, and the development and utilization of information resources, and improving the urban operation, management, economic development, public services and life quality of citizens.

3.2 The problems in smart city construction

The smart city construction in China, which just focuses on IT aspects, improves the efficiency of urban operation in a way, but is deviating from the original intention of "economy-environment-society" coordinated development. Problems as follows exist.

1) Lacking the ability of macro analysis of urban operation, the "smart city" at present is just the information gathering tools for the government. It can only provide relatively primitive data, and is useless for improving the efficiency of policy.
2) Lacking the ability of adjusting, the current "smart city" only serves as a platform of resource allocation, and is useless to the urban problems.
3) A smart city should not only be intelligent, but also be a new urban form with the ability of self-growth. The current framework of "smart city" is far away from that.

Smart city is the next goal for the urban development, and it will be the theme of city construction. In March 2014, the State Council issued *The State Plan for New-type Urbanization (2014–2020)*, which puts forward the development direction of smart city construction, namely, broadband information network, digital planning and management, intelligent infrastructure, convenient public service, modernization of industry development, and detailed social governance. The document covers all the connotations of smart city, and will guide the smart city construction in China in the following decades.

4 THE "12345" MODEL FOR THE CONSTRUCTION OF SMART CITY

As elaborated above, smart city construction, based on the application of information technology, could ameliorate the existing problems, optimize the resources allocation, urban management and urban decision, and incarnates the "People-oriented, Development-oriented, and Wisdom-oriented" construction ultimately.

The construction of smart city can be summarized as a "12345" model: By building "ONE urban management system, TWO supporting and guarantee measures, THREE information infrastructure platforms, FOUR urban management modes, and FIVE application service systems", the city may achieve the goal of developing intelligently (Fig. 2).

4.1 ONE urban management system

To gain the ability of self-growth, smart cities, distinguished from traditional cities, chiefly requires an urban management system, which supports urban design, management, operation, and self-adjustment.

The pre-analysis of the city status is necessary to the smart city construction. Based on the achievements of investigation and analysis, the Urban Think Tank conducts feasibility studies on the construction of smart city, and develops strategic planning. By combing the characteristics of the city with a suitable management system for the city, the strategic planning includes smart city construction mechanisms, guidelines, standards, evaluation, etc.

4.2 TWO supporting and guarantee measures

The definite positioning, scientific and reasonable scheme, environment-friendly urban hardware and intelligent management system, and the conception of wisdom community management and social service are essential in the construction of smart city, and powerful guarantee and supervision for it are indispensable in the meantime. Thus, two types of supporting guarantee measures should be applied for smart city construction: the measures of guarantee and supervision and the measures of assessment and adjustment.

4.2.1 Measures of guarantee and supervision

The measures are deemed to be one of the supports for smart city construction. During the process of planning, construction, operation and development, a smart city must equip powerful policy guarantee and supervision system.

While planning, under the targeted policy, the scheme, appraised by both citizens and experts (As shown in Fig. 3), should be people-oriented and environment-harmonious. And it is required to ameliorate the status of the area and life of citizens, and prevent the culture and environment from damage.

While attracting investment, with friendly measures, the government helps attract advanced experience and capital in the world vigorously. On the other hand, the government should promulgate a series of policies to select suitable construction contractors to undertake the smart city projects.

In the construction stage, under the strict supervision, urban construction should be strictly enforced in accordance with the planning scheme. In addition, some norms should be given for construction sites about material-transport, noise, and discharge of the waste that may affect the environment.

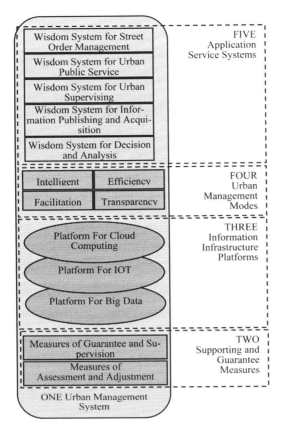

Figure 2. The "12345" Model of Smart City Construction.

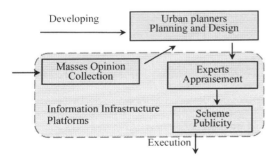

Figure 3. The Process of Developing A Smart City Construction Scheme.

In the operation and management stage, related regulation and supervision systems are needed to maintain the operation of the smart city. For instance, the regulations of facility protection, wisdom service system maintenance, and intelligent transportation are essential.

4.2.2 *Measures of assessment and adjustment*
The measures of assessment and adjustment are deemed to be the "Compass" of the smart city construction. In the process of urban development, disequilibrium development, which deviates from the development goal for urban economy, society, and environment, may exist. In allusion to the problems above, macroscopic measures of assessment and adjustment are required to guarantee that the construction and development of the city do not deviate from the "channel", by supervising and analyzing the indicators of the city regularly, and modulating the urban operation accordingly.

4.3 *THREE information infrastructure platforms*

Many articles about the "smart city" construction, focus on the building of intelligent information platform, which is only one part of smart city construction. Three information infrastructure platforms, namely platform for cloud computing, platform for IOT, and platform for data sharing, are the base technology for the construction of smart city. With the platforms, a smart city can implement the functions of resource allocation, information detection and release, self-adjusting, etc.

4.3.1 *Platform for cloud computing*
By building Infrastructure-as-a-Service (IaaS), Platform-as-a-Service (PaaS), and Software-as-a-Service (SaaS) of Cloud Computing, the platform supports the service, analysis, decision-making for urban construction, urban operation, and urban management.

4.3.2 *Platform for IOT*
By building the awareness and network system of IOT, and combining with Geographic Information System (GIS), Radio Frequency Identification (RFID), GPS, sensor technology, and remote sensing technique, the platform perceives various data of the city, and supports intelligent recognition, intelligent allocation, and data acquisition of the smart city.

4.3.3 *Platform for big data*
By building the dig data sharing system, the platform acquires the data and issues information perceived from IOT and resulted from Cloud Computing. It provides data and network for the functions of the city, and supports data sorting, information release and resource sharing of the smart city.

4.4 *FOUR urban management modes*

Simultaneously, in order to improve the operation efficiency and quality of life in the city, the urban management mode of a smart city should be intelligent, efficient, convenient, and transparent.

Intelligent management means that the urban management implements intellisense, intelligent analysis and adjustment, and processes the problems of the city in operation timely and optimally.

Efficient management means that with the integration of urban resources, the efficiency of society will be improved. The coordination and collaboration of urban operation will be enhanced largely. Existing work and assessment mode will be optimized.

Convenient management means to optimize the functions of government services, build a convenient mode of communication between the Government and citizens, simplify the work processes, provide long-term services and emergency response mechanism, and ultimately enhance the service function of the city, through the three information platforms.

Transparent management means to establish an open, fair and impartial decision-making model for the city. Through the data sharing platform, the administrative organizations release information, collect public opinions, notice the development plan, announce urban indicators, and ultimately make urban decision-making process "people-oriented".

4.5 *FIVE application service systems*

To realize the four urban management modes, five application service systems are required accordingly, namely Wisdom System for Street Order Management, Wisdom System for Urban Public Service, Wisdom System for Urban Supervising, Wisdom System for Information Publishing and Acquisition, and Wisdom System for Decision and Analysis.

The Wisdom System for Street Order Management is able to perceive and modulate the street order. For instance, the system can regulate the city tide-lane intelligently and automatically, with the support of information infrastructure platforms and the Urban Think Tank.

The Wisdom System for Urban Public Service is well connected with all aspects of the city, and can provide citizens with intelligent social services. It will intelligently response to the basic social service demand of business, health, culture and education, and improve the resource utilization.

The Wisdom System for Urban Supervising supervises the urban indicators of economy, ecology, and society, provides guidance and adjustment timely, and promotes the coordinated development. Specifically, the wisdom supervision and management can be embodied in the wisdom allocation, wisdom development guidance, and wisdom city security, etc. In routine urban operation, the system can help to achieve on-demand deployment of resources, intelligent traffic guidance, and efficient response to urban emergency, e.g., crime, fire, and emergency medical treatment. With urban development, when some resources limit the urban development, the urban intelligent management system will start the city improving mechanism. For example, when perceived air pollution or reduction of green area, the system can make additional scheme to balance the urban ecological environment or fine-tune the existing area sectorization to promote efficient land use.

The Wisdom System for Information Publishing and Acquisition is the application services that correspond to transparent management mode. Basing on the information infrastructure platforms, it improves public involvement in city administration by using network interactive, mobile applications as carriers.

The Wisdom System for Decision and Analysis, by analyzing urban development based on the indicators and measuring the phase of urban development, offers the Urban Think Tank important reference sources for adjustment schemes.

5 CONCLUSION

Smart city is the direction of urban development. On the basis of the connotation of smart city, this paper proposed the "12345" model of smart city construction. Firstly, the model defines the urban development expectation, and coordinates economic, ecological and social interests. Secondly, it establishes scientific and reasonable planning scheme, which is appraised by both citizens and experts, and builds "ONE urban management system, TWO supporting and guarantee measures, THREE information infrastructure platforms, FOUR urban management modes, FIVE application service systems". Thirdly, smart cities provide residents with intelligence services, manage and respond to various problems in urban construction and operation, support the government decisions, optimize the urban operation, and promote the urban sustainable development.

REFERENCES

Abdoullaev, A. 2011. A Smart World: A Development Model for Intelligent Cities. *The 11th IEEE International Conference on Computer and Information Technology.*

Deakin, M. & Waer, H. 2011. From intelligent to smart cities. *Intelligent Buildings International*, 3(3): 140–152.

Hollands, R.G. 2008. Will the real smart city please stand up? Intelligent, progressive or entrepreneurial? *City*, 12(3): 303–320.

Wright, S. & Steventon, A. 2004. Intelligent spaces—the vision, the opportunities and the barriers. *BT technology journal*, 22(3): 15–26.

Wu, X.B. & Yang, Z.G. 2010. The Concept of Smart City and Future City Development. *Urban Studies*, 2010 (011): 56–60.

Xu Q.R. et al. 2012. The Vision, Architecture and Research Models of Smart City. *Journal of Industrial Engineering/Endineering Management*, 26(4): 1–7.

A total bridge model study of Jiangdong Bridge in Hangzhou

Dongchun Qi & Pengfei Li
College of Civil Engineering and Architecture, China Three Gorges University, Yichang, China

ABSTRACT: Jiangdong Bridge is a self-anchored suspension bridge with spatial cables. In order to investigate the structural system transformation, static and dynamic characteristics by model test, a total bridge test model was designed according to geometric scaled ration of 1:16 and force scaled ration of 1:4. In this paper, the model design, test system, research method and main test results are introduced. The results show that the model test really reflects the mechanical properties of the real bridge. The test results are in good agreement with the calculated results.

1 INTRODUCTION

Jiangdong Bridge is a self-anchored suspension bridge with a span arrangement of (83 + 260 + 83) m. The main cable of the middle span is spatial cable plane. The distance between the two main cables in the midspan is from 3.0 m at the top point of the tower to 42.5 m at the bottom. The two main cables of side span are in parallel arrangement. There isn't any hanger in the side span. There are 26 pairs of hangers in the mid-span. There is a tilt angle in the transverse direction. The slope is approximate 2.9:1. The stiffening girder was designed as a separated twin-box girder connected by the box section beam. The width of the stiffening girder is 47 m. The type of this bridge was a new structural system. This type was used for the first time at home and abroad. It is necessary to study the system transformation process of the structure by the total bridge model tests. The static and dynamic mechanical properties of the structure after the completion of bridge were studied. The purpose was to check the correctness of the design theory, and understand the characteristics of this structure by the model test, in order to ensure the safety and reliability of the structure and provide guidance for the practical construction process. This paper introduces the design and production, test system, system transformation tests, and static and dynamic load tests after the completion of the bridge.

2 THE DESIGN OF THE TEST MODEL

Jiangdong Bridge test model was based on similarity theory to simulate the real bridge. Geometric similarity, similar stiffness and similar boundary conditions were used to design the test model (Liu Ziming. 1999). Considering the requirements of structural parameters, model materials, fabrication technology and laboratory conditions, the geometric scaled ratio of the model was 1:16, and the force scale ratio was 1:4. Full length

Figure 1. Images of the Jiangdong Bridge model for test.

of the model was 26.625 m. In the model design, the geometric relation between the main cable and the stiffening girder should strictly satisfy the similarity ratio requirement (Hu Jianhua et al. 2007). The test model was composed of double main cables, 26 pairs of hangers, steel box, horizontal connection box, anchoring beams at the end, pylons, saddles, temporary piers and test system. Fig. 1 shows the general view of the test model.

2.1 Structure design of the stiffening girder model

According to the scale ration planned, if the shape of the stiffening girder was designed similar to the real stiffening girder, the cross-sectional thickness of the plate would be too small. On the one hand, there is not a suitable material can be used; On the other hand, overly thin plate is difficult to meet the requirements of local stability (Fang Zhenzheng et al. 2011). Therefore, the content of the main consideration was as follows. First, the vertical force characteristic of the stiffening girder was the emphasis of the test. So the test must meet the similarity of vertical bending stiffness strictly. Second, it was tried to meet the similarity of the width. It is conducive to study the mechanical characteristics of separate box girder with the large span, the effective width and the stress distribution of stiffening girder, and mechanical characteristics of the transverse connection box in the middle of the stiffening girder. Finally, torsional performance of the structure was an

Table 1. Model parameters of the stiffening girder cross-section.

Number	Sectional area (cm^2)			Vertical bending moment of inertia (cm^4)			Torsional moment of inertia (cm^4)		
	Theoretical value	Actual value	Error (%)	Theoretical value	Actual value	Error (%)	Theoretical value	Actual value	Error (%)
C	1.12E+03	4.25E+03	277.8	9.32E+06	9.56E+06	2.6	2.11E+07	2.98E+07	41.0
D	1.05E+03	4.97E+03	371.7	8.66E+06	9.14E+06	5.5	2.00E+07	2.92E+07	45.7
E/F/G	9.74E+02	4.74E+03	387.1	8.36E+06	8.30E+06	−0.7	1.86E+07	2.59E+07	39.7

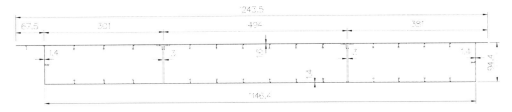

Figure 2. The standard section sizes of stiffening girder.

important part to study. Therefore, on the basis that the first two points have been met, it was tried to ensure that torsional stiffness was similar. In order to simulate the different characteristics of every section of the actual bridge girders accurately, the sections model were divided into five types, including types C, D, E, F, and G. The D was a variable cross-section beam; the cross-sectional size of the E, F, and G were identical, except for their lengths. The model parameters of stiffening girder cross-section are shown in Table 1.

Stiffening girder bottom plate, inside and outside walls of the model were made up of a steel plate and then stamped with a cantilever roof plate after the internal construction members were completed of box girder. A transverse bulkhead was set at 187.5 m interval. They were disconnected when the transverse bulkhead met vertical webs. Stiffener on the floor through the transverse bulkhead was disconnected when the stiffener on the roof met the transverse bulkhead. Transverse bulkhead, floor and sidewalls were connected by welding, and connected with the top plate by riveting. Steel arm was set in the place with permanent hanger and temporary hanger. The steel arm was connected with the hanger. In order to simulate the real mechanical properties of steel boxes for the real bridge, the vertical and horizontal separator plates in box girder, truss webs and stiffeners were simulated according to the similar ration. The standard section sizes of stiffening girder are shown in Fig. 2.

2.2 *The structure of the model bridge tower design*

The towers of the real bridge were reinforced concrete structures. The steel towers were used for the test model. The model towers were designed according to the similarity of vertical bending stiffness, lateral bending stiffness and axial compressive stiffness. The model towers will be up to 2 m in accordance with the geometric scaled ration. It is not convenient for test operation. In accordance with the design principle of anti-push rigidity, the height of the tower column was reduced to 0.9675 m. Considering that the saddle pushing of the tower top and main cable linear space was limited, the bridge tower roof was wide, the main span side roof asymmetric form was used, the side spans were wide, and main span was narrow. The vertical stiffener was set at the top of the tower to meet the requirements of saddle force transmission. The design of the beam for bridge tower met the similarity requirement of the vertical bending stiffness. Four seats were set on each tower beam. The beam support plate provided two transverse stiffeners to strengthen the local bearing capacity of the structure.

2.3 *The design of the main cable and cable model*

For the main cables and hangers, the most important was to meet the similarity of the axial stiffness. There were two main cables for the model. Each main cable consisted of 37 high-strength spring steel wires, and the diameter of the steel wire is 1.6 mm. The possible inelastic deformation was eliminated through tensioning. The wires were arranged in parallel, the materials were prepared according to the needed length, and the main cable with anchor head of the model were produced in the factory. The hanger consisted of two high-strength spring steel wires, and the diameter of the steel wire is 1.6 mm. The design ensures that the model area was similar to the hanger area of the real bridge. The top of the hanger used the wire rope clamps. The bottom and the anchor plate in the stiffening girder arm were connected with nuts through the threaded bolt.

(a) Cable force sensor (b) Main cable anchorage force sensor

(c) The force sensor in side piers (d) The force sensor of beam

Figure 3. Force sensors for the model.

3 INSTALLATION AND DEBUGGING OF THE MODEL

The stiffening girder segments from factory were shipped to the test site. Pre-assembly was completed on full framing. The stiffening girder segment elevation was adjusted through adjusting the height of the full framing. Appropriate camber was set according to a similar proportion. For the main cable tension and adjustment, at the top of the tower the saddle was put on deflection firstly, and the elevation of main cable was adjusted according to the theory of main cable to make it reach the theoretical design value. Then, the main cable and saddle were fixed on the top of the tower, and the position of the anchor head nut was adjusted, so that the side span line reached the design requirements. For the hanger tension and adjustment, the hanger force can be increased or decreased through tightening and loosening the nut.

4 TEST SYSTEMS AND ARRANGEMENT

The test systems of the model include force test system, displacement test system and stress test system.

4.1 The test of force

The strain force sensors were used for the test of the hanger force, main cable anchoring force and the counterforce used. The ranges of the sensor were different according to the magnitude of the force. A cable force sensor was fixed with the middle of every hanger. The top of the sensor was connected with a hanger through the suspension loops, while the bottom was connected with the anchor point through the tensioning screws, with a total of 26 cable force sensors. The force sensors were fixed with the end of the main cable to test the anchoring force of the main cable. Sensor was put at the supports of the side pier, the temporary pier and the beam to test the reaction force. The arrangements of all kinds of force sensors are shown in Fig. 3.

4.2 The test of stress

The stress of the stiffening girder and the bridge tower were tested by resistance strain gauge and FBG sensor systems. Ten sections of stiffening girders were selected to arrange resistance strain gauges. There were six ordinary sections, with 24 measurement points on each section; there were four encryption sections, with 40 measurement points on each section. The resistance strain gauges were arranged on the bridge tower and transverse connection boxes, with a total of 406 resistance strain measuring points.

4.3 The displacement test of stiffening girder

The vertical displacement of stiffening girder was tested with dial indicators. Seven sections were chosen, with four dial indicators were arranged on each section. Dial gauges were arranged at the supports of Bridge tower, beam and pier to monitor the stiffening girder vertical displacement. Dial gauges were set at the ends of the stiffening girder to monitor their horizontal displacement. In the middle of the mid-span, dial gauges were set to monitor the lateral displacement of stiffening girder.

4.4 The displacement test of the main cable

The vertical deflection and longitudinal displacement of the main cable in side span were measured by the method of projection. A level gauge was fixed on the ground, with calibration on the level gauge. A plumb was hanged on the main cable; it drove the plumb line to move when the main cable had horizontal displacement. The value of the movement can be read from the horizontal ruler. The steel rule was hanged on the main cable. The vertical displacement can be got from the intersection point value of the vertical ruler and the horizontal ruler when vertical deformation of the main cable was occurring. The displacement of the main cable for middle span was measured by total station.

5 THE CONTENT OF EXPERIMENTAL RESEARCH

With the test model, system transformation, static and dynamic characteristics of the structure after the completion of the bridge were studied (Shen Ruili et al. 2011).

5.1 The research of system transformation model

5.1.1 The experimental research methods of the system transformation model

Jiangdong Bridge is a self-anchored suspension bridge with the main cable shape of vertical plane. There is a transformation process from the vertical plane to the spatial plane. This process must be implemented through system transformation. From the empty ropes to the real bridge, the main cable changes for about 20°

in the transverse direction of bridge. The angle change is big, and the horizontal angle control and main cable reverse control are difficult in hanger installation and tensioning process. The method of temporary hanger was used for Jiangdong Bridge. Before the system transformation, firstly five pairs of temporary hangers are tensioned in the mid-span. The main cable is pulled into space polyline. There are little differences with the real bridge. The transverse angle of the hanger is reduced greatly. Then, 26 pairs of permanent hangers are installed symmetrically from the bridge tower to the mid-span. The system transformation tests were carried out by three tensioning methods. The third method adjusted the hanger installation order, and solves the problem that the hanger installation angle is too big comparing with the real bridge. The system transformation process in detail is shown in the literature.

5.1.2 *The main research conclusions of the system transformation model*

(1) Model test and calculation analysis showed that five pairs of temporary hangers were tensioned to make the main cable shape become a space curve, then cable clamp and tension hanger were installed in turn (cable clamp was fixed after the installation of two hangers). This method is the effective and reliable measure to solve the main cable twist that the cable clamp's angle changes before and after fastening.

(2) In the transformation process of the system, if the tension is in accordance with the hanger control force, the magnitude of the temporary cable force is the key to ensure that a permanent cable reach reasonable force for the completion of bridge.

(3) Theoretical calculations and model tests show that adjusting the hanger tensioning sequence near the middle of the span in the process of system transformation can effectively reduce the change of the rope clip's angle before and after the hanger is tensioned. The construction process proves that the optimization scheme of the hanger tensioning sequence is reasonable and effective.

5.2 *The static characteristic research*

5.2.1 *The methods of static characteristic research for the structure*

The deformation, and the distribution of internal force and counteracting force reflect the ability of the structure to bear external effect caused by external load. This ability is the static performance expression of the structure. After the completion of system transformation in the total bridge model, weight was added to simulate the load and study the static performance of the structure. The test included the influence line test of the structure, and the most unfavourable loading test of the typical cross-section. (Shen Ruili et al. 2008).

Studying the influence line or the influence range of structure is an important means to study the mechanical properties of the structure. In the test, according to the definition of influence line, concentrated load along the span was applied on the specified cross-section of the side span and the mid-span. Comprehensive test was conducted when load was applied once. The test results were arranged sequentially in a row to get the influence line of the force. A total of seven influence line tests of the section deflection and 26 pairs of the cable force influence line tests were carried out. Because the width-span ratio of the model was big, the methods of symmetrical loading and eccentric loading were used.

In order to research the static characteristics of the structure further, the middle of the side span, bridge tower, the middle of the mid-span, the 1/4 span sections in the model test were chosen, and the maximum and minimum bending moments of the lane load test were obtained. Symmetrical loading and eccentric loading were tested. The load range of uniformly distributed load and the loading position of concentrated load were determined according to the bending moment influence line. Fourteen groups of the most unfavourable moment load tests were completed.

5.2.2 *The main research conclusions of the structure static characteristic model*

(1) The nonlinearity of the structure is not obvious because of the live load after the completion of the bridge. The deformation, the internal force (Sling force, anchoring force) and the reaction of the structure has a linear relation with the loading. The superposition principle can be applied. The method of the influence line was used for the design. Design precision can be guaranteed consistent with the conclusion of the literature (Zhang Zhe et al. 2005).

(2) There is an obvious shear-lag effect in the wide thin-walled box girder, the stress distribution and deformation of the box girder section are influenced by the shear-lag, and the distribution of the cable force is slightly influenced by the shear-lag of the stiffening girder.

(3) The curve shapes of all hangers force influence lines are the same. The maximum appears in the middle of the span basically. The load applied to the mid-span makes the hanger force increase, and the minimum value of cable force in the mid-span is the dead-load force. The influence of the concentration force applied to a location of the hanger is limited for the hanger. This shows that the stress amplitude of the hanger is not very high, and it does not make the hanger force increase a lot because of the large concentration.

5.3 *The dynamic characteristic research*

5.3.1 *The research methods of dynamic characteristic model for the structure*

The main purpose of the dynamic load test is to determine the dynamic characteristics of the bridge parameters, including natural frequency, vibration mode, and damping coefficient. Evaluations of the dynamic response of the bridge were performed, and the test results were important indicators for determining bridge operation state and Load-bearing characteristics. The incentives of the test used initial displacement excitation predominantly, adding the free vibration attenuation of pulse excitation method. The particular method was to use wire rope to hang a 32 kg weight vertically on the incentive points, to make the total bridge out of shape. Then ropes were cut off suddenly to stimulate vibration mode of the girder in vertical, torsional, and mutual coupling conditions. The excitation spectrum of initial displacement and initial pulse was a continuous spectrum with a certain bandwidth. The excitation spectrum within the bandwidth of the test covered the important modal frequency of the total bridge model.

5.3.2 *The main research conclusions of the structure dynamic characteristic model*

(1) The choice of the test point and the incentives are reasonable. They can inspire a variety of vertical bending, torsion and coupling vibration model, and achieve sparse to dense modal in the measure phase. They ensure the identification precision of modal parameters when the modal is intensive. The location and number of test points are reasonable; they ensure the accurate identification of the overall vibration mode of the model.

(2) Through the modal analysis of the experimental data, a total of 18 vibration modal are identified in the bridge model, including 5 vertical bending, 4 reversing, and 9 coupling mode. The natural frequency, the average damping ratio and the vibration parameter of each mode provide the reference for the dynamic characteristics analysis of the structure and the revise of the actual bridge computing model.

6 CONCLUSION

The test scale of the total bridge model is large. The design of the model, production of the model and the design of the measurement system are difficult. The test results are in good agreement with the calculated results. The test fills the gap of the self-anchored suspension bridge with spatial cables in China.

REFERENCES

Fang Zhenzheng, Chen Yongjian, Zhang Chao. 2011. Experimental Study on Self-anchored Suspension Bridge with Three Towers Model. *Fuzhou University Journal (Natural Science)*, 39(4): 568–574.

Hu Jianhua, Shen Ruili, Zhang Guiming. 2007. Experimental Study on Pingsheng Bridge Model. *China Civil Engineering Journal*, 40(5): 17–25.

Liu Ziming. 1999. Experimental Study on the Bridge Structure Model. *Bridge Construction*, (4): 1–12.

Shen Ruili, Qi Dongchun, Tang Maolin. 2011. Experimental study on static characteristics of a full-bridge model Hangzhou Jiangdong Bridge. *China Civil Engineering Journal*, 44(1): 74–80.

Shen Ruili, Wang Zhicheng. 2008. Mechanical properties deflection theory of self-anchored suspension. *Highway Traffic Science and Technology*, 25(4): 94–98.

Zhang Zhe, Zhang Hongjin, Qiu Wenliang. 2005. Experimental Study on Concrete Self-anchored Suspension Bridge Model. *Dalian University of Technology Journal*, 45(4): 575–579.

An automatic FE model generation system used for ISSS

J. Duan, X.M. Chen, H. Qi & Y.G. Li
China State Construction Technical Center, Beijing, PR China

ABSTRACT: In this paper, an automatic FE model generation system is presented for the ISSS, abbreviated from "integrated simulation system for structures", which has previously been developed by the authors. According to the FE model transformation, there are mainly two issues that need to be considered seriously and managed carefully. The first one is the general meshing of the geometric structures, i.e. the discretization of the geometric structures into FE models with the building particularities ignored. The other one is the special handling according to building structures, i.e. the simplification of the structural model to increase the efficiency of the subsequent FE analysis.

1 INTRODUCTION

With the rapid development of high-rise buildings and long-span structures in the recent years, high performance simulation (HPS) is becoming more and more important for the structure design and building construction. The traditional design software solutions, such as PKPM, ETABS, MIDAS, YJK, etc., cannot meet the advanced requirements of HPS. On the other hand, although the large-scale general finite element analysis (FEA) software solutions, such as ABAQUS, ANSYS, ADINA, etc., have powerful computational abilities, they cannot be applied directly to the architectural structure analysis and design, due to the fact that their preprocessors are inconvenient for building modeling and their postprocessors cannot present computational results according to the building structure specifications, for example the civil codes and the custom of engineering.

To satisfy the above engineering requirements, the authors and their co-workers have developed an integrated simulation system for structures (ISSS for short) (Duan et al. 2014). This system is an integration of traditional design software and general FEA software, together with abundant secondary software developments. The interface of ISSS is shown in Fig. 1, in which it can be seen that (1) The building structures would be modeled by the popular design software, imported into the simulation system and transformed into FE models automatically, including FE meshing and boundary conditions setting; (2) The general FE software would be invoked as a computing kernel to solve the FE models; (3) Using the calculated results of FEA, the structural damage would be assessed according to the theory of mechanics and building structure specifications.

In this study, an automatic FE model generation system was developed and discussed. Obviously, it is one of the key components of the ISSS. Generally,

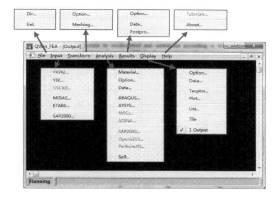

Figure 1. Interface of ISSS.

there are mainly two issues that need to be considered seriously and managed carefully for the FE model transformation. The first one is the general meshing of the geometric structures, i.e. the discretization of the geometric structures into FE models with the building particularities ignored. The other issue is the special handling according to building structures, i.e. the simplification of the structural model to increase the efficiency of the subsequent FE analysis.

2 GENERAL MESHING (PAVING METHOD)

Due to the facts that the geometry of building structures is theoretically arbitrary and the quadrilateral mesh is the most suitable for FEA, a quad-free-meshing method, i.e. the paving method was selected and improved to discretize building structures.

2.1 Paving method

Paving method is developed by Blacker and Stephenson (1991) and Blacker, Stephenson and

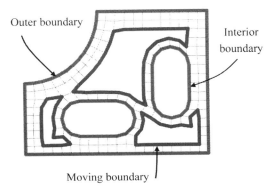

Figure 2. Illustration of the paving method.

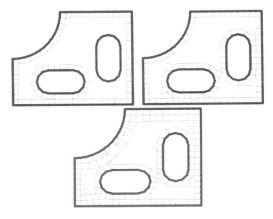

Figure 3. An example of the paving method.

Canann (1991). Since then, it has been applied to many FEA software solutions, such as ANSYS, MSC, FLUENT, etc. The paving method is based on iteratively layering or paving rows of elements from the outer boundary to the interior region, and simultaneously from the interior boundary(s) to the outer region, as shown in Figs. 2 and 3. As rows begin to overlap or coincide at the interior of the geometry, they are carefully connected together to form a valid quadrilateral mesh. Obviously, there are two major features of the paving method. First, it is suitable for any plane geometric boundary, and second, the mesh quality near the boundary(s) is primarily guaranteed in the meshing progress, which is especially important for FEA.

2.2 Developable surface meshing

In building structures, developable surface geometries, such as cylindrical surface, conical surface, etc., are not rare. The paving method described in the above section cannot be used to mesh the curved surface directly. This issue can be solved heuristically by mapping the curved surface into a plane one, meshing it and mapping back into curved one (Duan et al. 2014). Figure 4 gives an illustration of conical surface meshing.

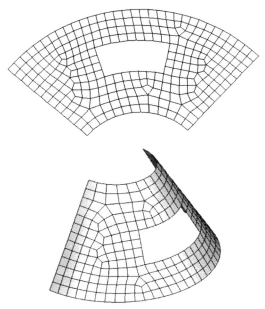

Figure 4. An example of conical surface meshing.

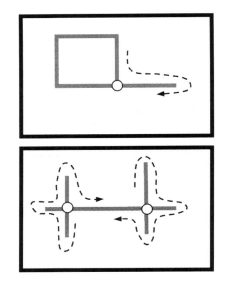

Figure 5. Illustration of constraints meshing.

2.3 Constraints meshing

Randomly distributed line constraints are very common in raft foundations and flat slabs. The prime paving method presented by Blacker and Stephenson (1991) is not suitable for this issue. Based on that method, Park et al. (2007) has developed a new technique for the constraints meshing. As illustrated in Figure 5, the lines can form various shapes composed of both a closed boundary and an open boundary by intersecting with the boundary(s) and each other. By linking the connected nodes at both boundaries and

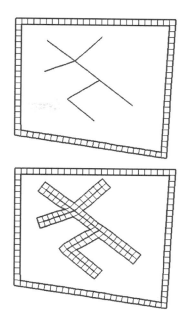

Figure 6. An example of constraints meshing.

lines, a loop or loops can be obtained. Then, the direct paving method can be used to generate quadrilaterals on the loops, as Fig. 6 shows. It is worth pointing out that the actual meshing method used in this study was derived from the technique of Blacker and Stephenson (1991) independently, while it is quite similar to the work of Park et al. (2007).

3 SPECIAL HANDLING ACCORDING TO BUILDING STRUCTURES

When a structural model is transformed to FE model, some simplification has to be accepted to increase the efficiency of the subsequent FE analysis. For example, some components, such as beams, columns and braces, will be simulated by one-dimensional elements and other components, such as walls and slabs, will be simplified to two-dimensional elements. Those simplifications will inevitably produce some inharmonious problems, which have to be managed carefully, or the calculated results will be unreliable.

3.1 Connection between beams and walls

In building structures, the shear-walls will be discretized into shell elements, while the frames will be discretized into beam elements, as shown in Fig. 7. Obviously, for the structural model, the connection region is an area, while for the FE model the connection region is merely a point. To minish the difference between the FE model and the structural model, some constraints should be added near the connection region. Those constraints to be accepted should be decided by the actual stiffness of the connection. If the

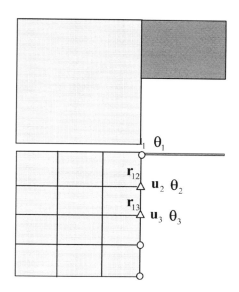

Figure 7. Illustration of beam-wall connection.

connection stiffness is assumed rigid, the constraints of Equations (1) and (2) should be enforced in the FE model.

$$\mathbf{u}_2 = \mathbf{u}_1 - \mathbf{S}(\mathbf{r}_{12})\boldsymbol{\theta}_1, \quad (1\text{-}a)$$

$$\boldsymbol{\theta}_2 = \boldsymbol{\theta}_1, \quad (1\text{-}b)$$

$$\mathbf{S}(\mathbf{r}) = \begin{bmatrix} 0 & -r_z & r_y \\ r_z & 0 & -r_x \\ -r_y & r_x & 0 \end{bmatrix} \quad (2)$$

3.2 Off-center of components

The off-center of components is a very common phenomenon in building structures. This problem results from the eccentric components, such as columns, beams, walls, and the structural transfer, as shown in Fig. 8. The central axes of those components are not on line, leading to the result that the adjacent one-dimensional elements have no shared nodes. To simulate the connection exactly, the constraints of Equation (3) should be enforced in the FE model.

$$\mathbf{u}_2 = \mathbf{u}_1 - \mathbf{S}(\mathbf{r}_{12})\boldsymbol{\theta}_1, \quad (3\text{-}a)$$

$$\boldsymbol{\theta}_2 = \boldsymbol{\theta}_1, \quad (3\text{-}b)$$

3.3 Rigid slabs

Rigid slab, as shown in Figure 9, is among the most popular hypothesizes of building structural analysis. This hypothesis would simplify the structural computation remarkably and make the concept of structure-floor very clear, which is important for the data

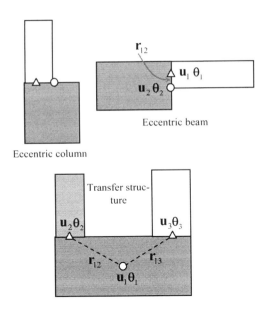

Figure 8. Illustration of components' off-center.

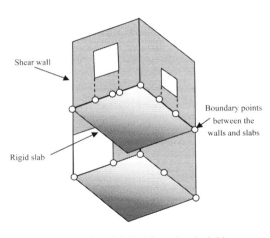

Figure 9. Illustration of rigid slabs and node rigid zone.

post-processing according to the civil codes and the manner of engineering. The constraints condition of rigid slabs is very similar to the components off-center, except that rigid slabs only take into account the in-plane degree of freedoms.

3.4 *Different meshing sizes according to different components*

To increase the efficiency of computation and avoid some side effects of uniform FE meshes, the simulation system accepts different meshing sizes according to different components. That means the mesh sizes of walls, slabs and frames can be set identical or different, decided by the actual engineering, as shown in Fig. 10.

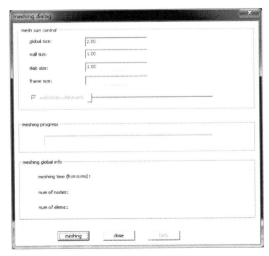

Figure 10. Mesh size control.

4 CONCLUSION

In this paper, an automatic FE model generation system was presented for the ISSS, abbreviated from "integrated simulation system for structures", which has previously been developed by the authors (Duan et al. 2014). Two key issues of FE model transformation were discussed. The first one is the general meshing of the geometric structures, i.e. the discretization of the geometric structures into FE models with the building particularities ignored. The other issue discussed is the special handling according to building structures, i.e. the simplification of the structural model to increase the efficiency of the subsequent FE analysis.

REFERENCES

Duan J., Chen X.M. & Li Y.G., 2014. Automatic Mesh Generation for Building Structures. *Journal of Building Structures*. (in Chinese, submitted).

Duan J., Chen X.M., Qi H. & Li Y.G. 2014. An Integrated Simulation System for Building Structures. *Proceedings of the 4th International Conference on Civil Engineering, Architecture and Building Materials*, 24–25 May, 2014, Haikou, China. (Accepted).

Blacker T.D. & Stephenson M.B. 1991. Paving: A New Approach to Automated Quadrilateral Mesh Generation, *International Journal for Numerical Methods in Engineering*, 32:811–847.

Blacker T.D., Stephenson M.B. & Canann S.A. 1991. Analysis automation with paving: a new quadrilateral meshing technique, *Proceedings of Design Productivity Institute and International Conference*, University of Missouri-Rolla, Honolulu.

Park C., Noh J.S., Jang I.S. & Kang J.M. 2007. A new automated scheme of quadrilateral mesh generation for randomly distributed line constraints. *Computer-Aided Design*, 39:258–267.

Analysis of genetic mechanism and stability of Shaixiping landslide

Minghui Su
Wuhan University of Technology Huaxia College, Wuhan, China

Lei Ma
The 606 Brigade of Geological Exploration and Resources Development, Chengdu, China

Wei Li
Chongqing Geological and Mineral Resource Exploration and Development Bureau 607 Geological Team, Chongqing, China

ABSTRACT: On the basis of field geological survey, this study expounded the geological environmental conditions and characteristics of Shaixiping landslide, and analyzed its genetic mechanism systematically. Based on the results of field test and experiment, the landslide stability with 3 kinds of working conditions was calculated. At last, feasible treatment measures were put forward for Shaixiping landslide.

Keywords: landslide; genetic mechanism; stability; prevention engineering; geological environment

1 INTRODUCTION

Shaixiping landslide was located in Cangxi county, Sichuan province (Fig. 1). Since 2005, the landslide began to deform and suffered several times of heavy rain, then landslide broke out, which was 125 m in length and 150 m in width, covering an area of $1.9 \times 10^4 \, m^2$ with a volume of $7.5 \times 10^4 \, m^3$. The landslide had seriously affected people's lives, property and infrastructure. Based on field survey and experiment, the characteristics, genetic mechanism and stability of the landslide were analyzed in this study, and relevant engineering treatments were put forward.

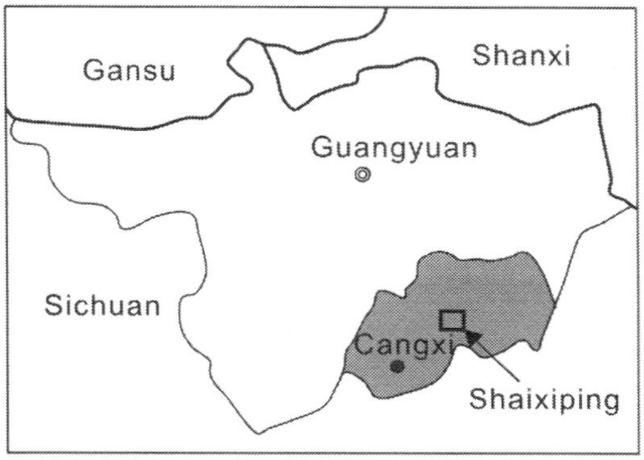

Figure 1. Location map of Shaixiping landslide.

2 GEOLOGICAL ENVIRONMENTAL CONDITIONS

2.1 Topography

The study area was a low mountain deep hilly area located in northern part of Sichuan Basin. Landslide overall sloped (Fig. 2a), trailing edge sloped steeply (Fig. 2c), visible bedrock exposed, the area had an elevation of 437–440 m, and the leading edge was dry land scarps (Fig. 2b), of which height was 421 m. Both

Figure 2. Images of Shaixiping landslide characteristics (a-panorama; b-trailing edge; c-leading edge; d-cracks; e-building deformation).

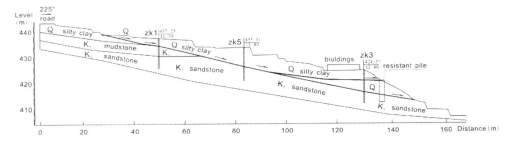

Figure 3. Engineering geological profile of cross-section 1-1' of Shaixiping landslide.

sides of the landslide were gullies, and the surface was in the shape of a ladder due to the transformation of farmland.

2.2 Lithology and structure

Exposed strata in the study area were the Quaternary (Q) and the Lower Cretaceous (K_1) strata. Exploration results showed that the Q eluvial soil distributed widely in the range of the landslide and was 1.5–8.4 m in thickness, and easy to form slope collapse. K_1 sandstone and mudstone were buried deeply in the front and low-lying, and shallowly in the lateral, central and slope ridge (Fig. 3). The landslide area was located in Central Yangtze platform, with undeveloped fault. But the structure was changed by the "5.12" Earthquake.

2.3 Hydrogeology

Groundwater was mainly pore water and bedrock fissure water, and there charge source was atmospheric precipitation. Pore water occurred in the residual loose soil, with instable water level; in the landslide center the water depth was 16–21 m. Rock fissure water was exposed in the form of spring, and spring flow was generally less than 0.2 l m/s with depth of more than 35 m.

3 CHARACTERISTICS OF LANDSLIDE

3.1 Spatial development

The landslide was long tongue in planar (Fig. 3 and Fig. 5a), the length of main sliding direction was 125 m, and it was 150 min width, 19012 m² in range. The overall slope was 20°, average thickness of landslide 4.5 m, while leading edge was thicker (6.8 m). Slip direction was 225°, and sliding volume was 75020 m³. It was classified as a shallow landslide.

3.2 Slip deformation

Results of field observation showed that the mode of slope failure was ground crack and partial sliding (Fig. 2d, 2e). On the surface of the slope, cracks distribute mainly in the middle, the right and the leading edge, there were eight cracks in the middle extending far and deforming obviously. Crack lengthened 4.20–40.6 m, widened 5–12 cm, and was 5–20 cm in depth, with extending directions roughly divided into three groups. In addition, the ground surface fell down, and the maximum depth was 35 cm.

4 GENETIC MECHANISM

4.1 Mechanism analysis

The mechanism of Shaixiping landslide was affected by several factors, including internal factors and external factors. The former were mainly controlled by topography and material aspects, where the terrain was steep and prone to form adverse discontinuities. The material in the sliding body was high porosity; the weak zone came to be under infiltration of rainwater, and then induced instability and deformation of the landslide.

The external factors were dominated by rainfall, earthquake and human activities. The structure of the sliding body destroyed by "5.12" earthquake was conductive to the infiltration of surface water (Li and Zhou, 2008), increasing the sliding force of slope soil. An artificial slope in leading edge, with height of 2.5–3.5 m, built by local residents in the front of the shear outlet section, exacerbated the deformation of the landslide. Frequent rainfall supplied surface water, which infiltrated along structure surface, to provide potential landslide "lubricant" (Chen and Xu, 2009; Peng et al., 2011). Under the long-term gravity and weathering, the slope body deformed and loosened. When storm happened, the strength of slip loose soil was reduced by the infiltration of a large amount of rain (Rahardjo et al., 2001), and then the slip surface formed. When some external force suddenly increased, the slope suffered damage and landslide happened. According to the mechanical properties of Shaixiping landslide, it was a traction loess landslide.

4.2 Development trend

Results of field investigation suggested that Shaixiping landslide began to deform in April, 2005;

Figure 4. Geological images of zk5 (a) and zk2 (b) boreholes.

after the "5.12" earthquake in 2008, the area suffered six successive rainfalls in 2009. Under the soak of the rain, the shear capacity of slide body and bedrock weak surface decreased, slope material separated from parent and generated the slope sliding, and strongly tensile fracture formed in the middle and leading edge.

Through comprehensive analysis, Shaixiping landslide was in the creeping deformation stage. Further deformation under various unfavorable loads and continuous heavy rain would decrease the landslide stability, and cause local deformation and instability of the whole landslide.

5 STABILITY CALCULATION

5.1 Experimental analysis

Drilling investigation suggested that (Fig. 4) the material in the slip zone was soft plastic silty clay with high water content. Meanwhile, the interior of undisturbed and remolded soil in sliding zone was analyzed, and the corresponding parameters were determined (Table 1).

5.2 Method and results

According to the analysis of the deformation and failure modes of the landslide mentioned above, when the slide zone generated along the rock and soil interface, the surface structure was the broken line.

Table 1. Stability calculation results for Shaixiping landslide.

Silty clay	Native state gravity			Water saturated state gravity		
	/kN·m^{-3}	C/kPa	Φ/°	/kN·m^{-3}	C/kPa	Φ/°
Slide body	19.6	44.8	14.8	20.5	15.0	12.0
Slide zone	19.2	41.4	14.0	19.5	10.6	7.2

Table 2. Stability calculation results for Shaixiping landslide.

Calculation cross-section	Working conditions	Safety coefficient	Sliding force
1-1'	I	1.77	0
	III	1.04	277
	IV	1.44	0
3-3'	I	2.77	0
	III	1.63	0
	IV	2.15	0
4-4'	I	2.81	0
	III	1.66	0
	IV	2.20	0

The rigid body limit equilibrium analysis has been commonly used in slope stability analysis (Zhang et al., 2007). In this paper, the transmission coefficient of sliding surface was calculated by broken line, with the formula as follows.

$$K_f = \frac{\sum_{i=1}^{n-1}((W_i((1-r_U)\cos\alpha_i - A\sin\alpha_i) - R_{Di})\tan\phi_i + c_i L_i)\prod_{j=i}^{n-1}\psi_j) + R_n}{\sum_{i=1}^{n-1}((W_i(\sin\alpha_i + A\cos\alpha_i) + T_{Di})\prod_{j=i}^{n-1}\psi_j) + T_n} \quad (1)$$

Based on the characteristics of the landslide body deformation and without the stable groundwater, three kinds of working conditions (Table 2) were selected, including I (weight), III (weight and rain) and IV (weight and earthquake). Conditions I and IV were chose to use natural gravity, internal friction angle and cohesion, while III was in water saturated state. The safety coefficient (K) for residual thrust calculation was 1.10.

As the calculation results showed, the safety coefficients of cross-section 3-3' and 4-4' under all working conditions were greater than 1.63 (Fig. 5a), indicating that both sides of the landslide body were stable on the whole. Meanwhile, the safety coefficient of cross-section 1-1' was 1.04, which presented a less stable state. The main stress of the slide body was situated in the middle; there was still a possibility of sudden slide. On the whole, the limit equilibrium state of slow creeping deformation was consistent suggested that the biggest stress was from Northeast, thus the direction of the landslide was southwest.

6 TREATMENT

In accordance with characteristics of the geological environment, sliding force, influence of human engineering, and fully based on the stability calculation results, feasible solution was selected (Yin, 2010). The stability calculation results showed that both sides of the slide body were in relatively stable states, while the middle was in a less stable state (S 1-1′), indicating that the main stress released in center position, and the two wings were in passive slump deformation. Therefore, feasible engineering control measure was to set resistant slid pile between zk3 and zk5 in leading edge.zk3 was an manually dug deep pit located in the northern 20 m away, on the forward scarp (Fig. 3). Considering the above factors, setting the pile in front of zk3 was more feasible (Fig. 5a). At the same time, monitoring inspections on the slope should be installed. Dynamic monitoring should be carried out in the whole process of treatment, in order to ensure that the purpose of disaster prevention and reduction was realized.

7 CONCLUSIONS

The genetic mechanism of Shaixiping landslide was a comprehensive result of factors including locating in steep slope, high porosity in landslide formation, prone to form weak zone, frequent rainfalls, and impact of earthquake and human engineering activities. Under the working condition III, stability coefficient was 1.04, which indicates a less stable state. Once storms invaded, Shaixiping landslide would happen. Resistant slide pile was designed in front of zk3, and monitoring inspections would be installed on the slope.

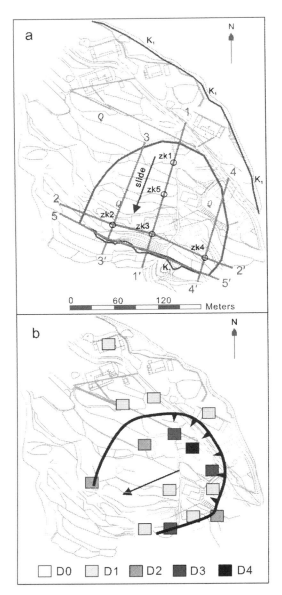

Figure 5. The project layout plan (a) and landslide damage map (b) (D0-no damage; D1-non-structural damage, slight non-structural damage; D2-moderate non-structural damage, slight structural damage; D3-moderate structural damage, heavy non-structural damage with large and extensive cracks in most walls; D4-very heavy structural damage, buildings have suffered partial collapses with serious failure of (walls).

with the field observation and the analysis of the causes.

In order to reflect the landslide effects of damage on structures, (Lazzari 2009) divided damages into five degrees (D1 to D5). Shaxiping landslide was still in a relatively stable period, thus did not cause D5 damages. The damage level distribution shown in Fig. 5b

REFERENCES

Chen, S., Xu, G. L. 2009.Research on engineering geology characteristics of rail in sliding zone of Huangtupo landslide in Three Gorges Reservoir area. *Rock and Soil Mechanics*, 30(10), 3048–3052.

Code for investigation of geotechnical engineering. GB50021-2001. Beijing: Chinese Standard Press, 2001.

Lazzari, M. 2009. The Bosco Piccolo snow-melt triggered-landslide (southern Italy): a natural laboratory to apply integrated techniques to mapping, monitoring and damage assessment. Proceedings of the Conference on <Landslide processes: from geomorphologic mapping to dynamic modelling>, a tribute to prof. Theo Van Asch, 6–7 February 2009. Strasbourg, France, 163–168.

Li, X. J., & Zhou, Z. H. 2008. Preliminary Analysis of Strong Motion Recordings from the Magnitude 8.0 Wenchuan, China, Earthquake of 12 May 2008. Seismological Research Letters, 79: 844–854.

Peng, Y. L, Guo, D. L., Hu, X. W., Gu, C. Z., Zhou, J. 2011. Characteristics and stability analysis of a landslide in Kuancheng country, Hebei province. Journal of Mountain Science, 29(5): 591–597.

Rahardjo, H., Li, X. W., Toll, D. G. 2001. The effect of antecedent rainfall on slope stability. Geotechnica land Geological Engineering, 19 (3): 371–399.

Specification of geological investigation for landslide stabilization. DZ/T0218-2006. Beijing: Chinese Standard Press, 2006.

Yin, Y. P. 2010. Mechanism of apparent dip slide of inclined bedding rockslide-a case study of Jiweishan rockslide in Wulong, Chongqing. Chinese Journal of Rock Mechanics and Engineering, 29(2): 217–226.

Zhang, X. Z., He, J. D. 2007. The application and compare of the rigid limit balance method and finite element approach in the slope stability analysis. Jilin Water Research, 8: 32–35.

Analysis of green design in outdoor game space for children

Xin Li
Tianjin University of Technology, Tianjin, China

ABSTRACT: With the outdoor game facilities for children and the design of outdoor environment as the research topic, this paper points out that only the design with meticulous planning and with nature can be accepted, which help to achieve the ultimate aim of improving the human settlements in urban communities and the quality of outdoor recreation for children, and to motivate the vitality and frequency in outdoor social contacts among people, and to benefit the healthy development of the ecological environment in urban communities.

Outdoor activities in communities are often spontaneous and interactive. People can improve their various social relations through these activities and improve the certain area they live. Outdoor activities are relaxing, natural and healthy, making people feel less lonely in their lives (Jan Gehl and translated by He Renke. 2002. Communication and Space (M)). Good environmental condition of the outdoor space is necessary for the external environment and the inevitable outcome of sophisticated human civilization.

1 DIVERSIFICATION OF OUTDOOR CONTACT SPACE AND VARIETY OF THE DESIGN CONTENT

There are a variety of outdoor activity contact spaces. Among all these, outdoor game space for children is the most general space for social activities. The material planning design of these open space is important, the pattern of manifestation affects directly the using effect of a serious of social contact activities. So, as the designer, he/she should aim at improving the quality of the behavior's activity, and leading to promoting the vision and using effect of the regional environment of the used space as a final result.

So how to tell whether the design of the outdoor game space environment for children is successful? Children's activities in outdoor game space is a process of self-reinforcement, when a child is having fun in the space he chose, other children would come and join him by watching and being enlightened. Then more children would join in this to make the group larger. What happened in this space is a stimulated and persistent social contact process, that is to say, the relaxing and long-term outdoor staying means the environment design of the public outdoor space is successful, vigorous and attractive. From the above, we can see that the designer should do perfect in interior detail designing in the whole independent public space, for keeping and even lengthening the time people stay there. In general speaking, hollowness and desolation should be eliminated, every detail should be done perfectly in the space, such as guiding signs, richness of the ecological communities around, waiting area for parents to communicate with each other, cultural background wall, sculptures, area and manifestation of water, highlights in the entrance, attractive appearance and function of trash bins, surface pavement and soil protection and enclosure for generating the regional microclimate. All these designs need to be taken into account, besides the unusual and interesting looking, environmentally friendly and durable game facilities.

2 ENVIRONMENT DESIGN OF OUTDOOR CONTACT SPACE SHOULD BE COMBINED WITH NATURE

2.1 Design derived from nature is what children need

Delicate design with nature, getting back to ecology and nature is what contemporary designers' dream of.

Nowadays, with the establishment of metropolis and rapid industrialization, the science stays together with destroy, for instance, developing of air pollution and bad microclimate, extinction of insects and some small animals, and the emerging of some other severe problems. Therefore, financial condition is an inevitable factor of determining the design scale and long-term maintenance if everything needs to be changed. Protect and change the nature with nature itself, do development and innovation in nature should be supported largely on finance, which is a sustainable design strategy. So, designing and enlarging outdoor game space for children in present situation should be focused on taking natural evolution into account. Better natural environment would benefit people a lot finally, so it would cost some in protecting and changing the nature. (Farr and translated by Huang Jing. 2013. Sustainable Urbanism-Urban Design with Nature (M)).

In urban area, with progress over time and environmental deterioration, the demand for more large open space remains, which is based on protection and maintenance of the natural conditions and is important for improving the urban climate and living conditions. The designer should apply the view of ecology to the planning and construction theory of outdoor game space for children. For example, at present, more designers are inclined to use plants as a theme decoration, which is a good method for not being replaced by new design concepts with time passes by. The soil and rock with good quality, good drainage system, frondent trees, lush grass and beautiful flowers need to be kept and utilized. According to the value of the ecological environment, select the superior and eliminate the inferior, and then rearrange them, design and preserve with long-term work. All these items are important parts and are extremely good for the future development of children, which are also determined by the special demand of keeping the independence and safety of the game space. (Low and translated by Wei Zesong. 2013. Rethinking Urban Parks (M)).

2.2 "Regionalization" is a necessity in designing with nature

Besides the preserving and utilizing of natural resources in the design above, outdoor game space for children, as the social contact space, should be regional, thus it is easier to be accepted and like by the public.

Division of the border, selection of the landform, feature of the ecological plant and microclimate of the whole outdoor game space for children is important external conditions for regionalization. And in this region, the design is relatively different and special which is mainly restricted by the ecology. First, we should choose the region with good microclimate, moderate temperature and comfortable breezing; the land covering should be soft and better be organic matters mostly; second, there would be no erosion, walking on the land should be comfortable and quiet (MacHargand translated by Huang Jingwei. 2006. Design with Nature [M]) like cork dust, silver sand and lawn. When having fun in the space, children would feel relaxing physically and psychologically in the natural, and with all green in from of them. The development of these ecological environment helps with the improvement and adjustment of the regional environment. For instance, tall trees combined with styled walls to form an enclosure or semi-enclosure to make the space more nature integrated and independent. The more plants there are, the more birds and insects children like would live there, which makes the space a fictitious land of peace. Meanwhile, dust, voice and highlight are avoided and lowed. Children would have fun in a safer environment. If buildings around are tall and the number is large, it is crowded there, the selection of the location of game space for children should be carefully, more sunshine and less noise should be taken into account. Background noise should be below 40 DB to make sure that children's outdoor activities are in a quiet environment. (Nicolas T Dines and translated by Liu Yujie. 2002. Landscape Architect's Portable Handbook (M)).

3 DESIGN OF CHILDREN'S FACILITIES CANNOT GOWITHOUT "GREEN"

3.1 Material selection is the safety factor in design

The outdoor game facilities have direct contact with children when being used, so materials shall not be used unless it is confirmed that their supplies have relevant national certifications in design, and materials that are poisonous, hard or polluted shall never be used, such as wood containing arsenic or aluminum or scrap tires with volatile chemicals including remaining gas, and so on. It is thus clear that green design is the guarantee for the health of children, and un-polluted material including sands, water, stones, concrete, woods, stainless steel, hemp ropes, etc. shall be used in the design as the best choices for design of children's facilities. For example, water and sand are both natural materials with good plasticity and without pollution, the combination of which is perfect and valuable for the children who have the same good plasticity.

3.2 Modeling and function design of facilities require creativity and cohesiveness

See the following two figures: these are drafts of two game facilities for children to climb and speel, which cover large areas. Figure 1, it can hold several children to play in it at the same time. It is modeled by large three-dimensional grid mountains composed by crossing lines, like a huge irregular spider's web, and the whole structure is fixed and supported by four strong metal columns. Firm and strong weaving hemp ropes are used as the main material, the connections of which are inter-fixed by metal ball components, so as to increase the bearing capacity and safety of the whole facility. Each grid space is controlled within 400 to 500 mm, enough for a child of 4 to 6 years old to climb up and down. The program will greatly develop the muscle strength of the children's hands and arms, the sticking capacity of the feet and the balance of bodies. The just and brave Spiderman's cartoons are set in the surroundings to accompany the children. Activities in such situations can cultivate the bravery and will of the children when playing games.

Figure 2 is a climbing facility modeled by a strong and heavy cartoon, with concrete and stones as the main material. The interior of the facility is modeled as a cave and the exterior has a slide, an inlet and an outlet. Children can go in and out through the cave, play hide-and-seek, and climb and slide on the

Figure 1. Draft 1 for design of a children's outdoor game facility.

Figure 2. Draft 2 for design of a children's outdoor game facility.

mountain, thus greatly improving the strain capacity, organization, association and cohesiveness capacities of children.

The above two design models seem to be simple, but they completely get away from the old straight game instruments made of plastic and polymeric sheets, and fully achieve the function of "one thing for different playing". For the main patterns to organize the design, one is to expand a huge three-dimensional "mountains" by the crossing lines, and the other is to be demonstrated with a cartoon image as the basic form. Their materials are natural with fun, creativity and challenge. Both the schemes choose "climbing mountains" as the theme of the game, satisfying the children's psychology to climb and explore in group. On the design basis of improving the self-physical ability and early social association of children, the design concept is realized to seek the harmonious unity of ecology and natural.

4 CONCLUSION

In a word, the design is not enclosed for a complete and enjoyable children's outdoor activityspace in urban community, but it is breathable, and is created and expanded by taking advantage of the existing natural space resources. It is the guarantee and key for the healthy growing and early education of children, and a sustainable and long-term career requiring investment and maintenance in the beautifying and building campaign of the city.

REFERENCES

- Farr and translated by Huang Jing. 2013. Sustainable Urbanism-Urban Design with Nature (M). China Architecture & Building Press.
- Low and translated by Wei Zesong. 2013. Rethinking Urban Parks (M). China Architecture & Building Press.
- MacHarg and translated by Huang Jingwei. 2006. Design with Nature (M). Tianjin University Press.
- Nicolas T Dines and translated by Liu Yujie. 2002. Landscape Architect's Portable Handbook (M). China Architecture & Building Press.
- Jan Gehl and translated by HeRenke. 2002. Communication and Space (M), China Architecture & Building Press.

Note: Fun project/No.: Culture art No. d12005.

Analysis of influencing factors on urbanization: A case study of Hebei Province, China

Guangbao Lai
Hebei vocational and technical college of building materials, Qinhuangdao, China
Agricultural University of Hebei, Baoding, China

Banghong Zhao
Agricultural University of Hebei, Baoding, China

ABSTRACT: Urbanization, which can be impacted by many factors in all aspects, is a complex and integrated project. The urbanization of Hebei Province, which is determined by the government's macro-economic policy, migration and immigration, education level, urban and rural differences in natural population growth, industrial structure, the associated system, etc., is at the period of accelerating development and has been becoming one of the key strategies for Hebei's economic and social development.

Keywords: urbanization, impact factors, Hebei province

1 INTRODUCTION

By the end of 2012, Central Economic Work Conference pointed out that "Urbanization is the historical task of China's modernization, but also the greatest potential for expanding domestic demand. We should focus on improving the quality of urbanization, capitalizing on the trend while avoiding disadvantages, and actively guiding the healthy development of urbanization." In the 18th National Congress of the Communist Party of China, it was proposed to put forward the industrialization, information technology development, urbanization and agricultural modernization as the carrier of building a well-off society; and to further focus on the quality of urbanization. Since the reform and opening-up, Hebei has made rapid development in urbanization. The province has made great progress in enlarging the number and the sizes of the newly built towns, as well as in improving the towns' quality and economic efficiency. However, compared with the national average level, especially with that in the eastern coastal provinces, the development level of urbanization in Hebei is still backward, and is incompatible with the economic and social development. Therefore, analysis of the impact factors that affect the development of urbanization is of great significance to promote a healthy and rapid development of urbanization in Hebei Province.

2 THE GOVERNMENT'S MACRO-ECONOMIC POLICY

Urbanization strategy, regional development strategy, industrial layout planning and the government's efforts to promote the development of urbanization in Hebei Province are the main factors affecting the process of urbanization. The government's macro-economic policy is immediately related to urbanization.

(1) The change by the government in administrative settings and rural/urban management system is the main driving force of urbanization. Several large-scale changes in urban and rural populations are closely related to the external push by the government. For example, in 1978, Tangshan and Shijiazhuang were set to be provincial municipalities; Langfang, Hengshui, Botou and other places were set to be cities. Then, the urbanization rate increased from 11.3% in 1976 to 13.69% in 1982. In 1983, the government implemented the "city governing county" system and the "county to city" policy. Since the government lowered the standards to establish a town and relaxed the restrictions on household management in 1984, the urban population is increasing rapidly.[1]

(2) The adjustment of government's regional development strategy directly affects the spatial layout of urban areas. In the past three decades, Hebei has adjusted its regional development strategy for a number of times, such as the "Ring of Beijing and Tianjin" development strategy in 1986; the "developing rural area with city, along with coastal and railway line" strategy in 1988; "one line (coastal), two areas (Shi Jia Zhuang and Lang Fang development zone) with multiple points (the high-tech development zones, high-tech industrial parks, tourism development

---
[1]Reference to Xu Gaofei "Hebei Administrative Divisions Change Ceremony" (1949–2005), local press, 2006.

Table 1. Population, urbanization rate and per capita GDP in Hebei Province and main cities in Hebei Province.

(province/cities)	Population (ten-thousand)	Urbanization rate (%)	Population from outland (ten-thousand)	Population to outland (ten-thousand)	Net inflow population (ten-thousand)	Ratio of net inflow population (%)	Per capita GDP (ten-thousand RMB per person)
Hebei Province	7185.42	43.94	829.73	797.18	32.55	0.45	2.86
Shijiazhuang	1016.38	50.62	154.25	108.05	46.20	4.55	3.36
Tangshan	757.73	50.82	95.64	67.51	28.13	3.71	5.93
Qinhuangdao	298.76	47.51	49.53	34.70	14.83	4.96	3.13
Handan	917.47	43.44	71.82	89.56	−17.74	−1.93	2.62
Xingtai	710.41	40.02	51.56	59.43	−7.87	−1.11	1.72
Baoding	1119.44	38.77	111.73	116.63	−4.90	−0.44	1.84
Zhangjiakou	434.55	45.16	94.10	120.05	−25.95	−5.97	2.25
Chengde	347.32	38.67	44.04	64.73	−20.69	−5.96	2.57
Cangzhou	713.41	40.83	59.21	60.08	−0.87	−0.12	3.12
Langfang	435.88	48.53	68.46	44.41	24.05	5.52	3.14
Hengshui	434.08	38.19	29.39	32.03	−2.64	−0.61	1.81

Note: The net inflow population is the population migrating from outland (living longer than six months) minus the population moving out for more than six months. The net outflow of population is the population moving out for more than six months minus the population migrating from outland (living longer than six months). Data are from the 6th census in Hebei Province.

zones and free trade zones)" development strategy in 1992; "two rings (around Beijing and Tianjin, and Bohai Sea)" development strategy in 1993; "two rings (around Beijing and Tianjin, the Bohai Sea) opening and driving strategy" in 1995; "one line (Including Shijiazhuang, Baoding, Langfang, Tangshan, Qinhuangdao) two areas (Including North area composed of Zhangjiakou, Chengde and South area composed of Handan, Xingtai, Hengshui, Cangzhou)" regional economic development strategy in 2004; emphasis on "Coastal consciousness" and the development of heavy chemical industry base at Caofeidian and coastal heavy chemical industrial belt in recent years; the programmatic objectives on "strong development of economy in the coastal province" and the goal on creating a "coastal economic uplift". These strategic decisions and adjustments ultimately determine the spatial layout of urban development trend.[2]

3 MIGRATIONS AND IMMIGRATIONS

There are two reasons for the increasing urbanization rate: the urban population's natural growth and the increasing rate of immigration. The population movements mainly consist of local rural population movements towards local towns, as well as outland to the local town's population flows.

The floating population registered by the census refers to the migration population and the immigration population for more than six months. The sixth census data of Hebei Province (Table 1) shows that in 2010, 797.18 million people moved out of the province for more than 6 months, and 829.73 million people from outland lived in the province for more than 6 months. A net inflow of 32.55 million people moved into the province, accounting for 0.45% of the province's overall population. The cities with positive net inflows of population are Shijiazhuang, Tangshan, Langfang and Qinhuangdao. Their net inflows of population are 462,000, 281,400, 240,600 and 140,830, accounting for 4.55%, 3.71%, 5.52% and 4.96% of the total population, respectively. The top three cities with highest net outflows of population are Zhangjiakou (259,500), Chengde (206,900) and Handan (177,400), with ratios of 5.97%, 5.96% and 1.93% of total population. It can be seen that the cities with positive net inflows of population are exactly the top four cities with highest urbanization rates. They are also the top four cities in terms of per capita GDP. This shows that population mobility, urbanization rate and per capita GDP (economic development level) have a strong positive correlation.

According to the 6th census data, the top three motivations for the immigration with household registration within Hebei Province are: job and business relocation (21.18%), education and training (18.39%) and family based relocation (16.54%). In total 56.11% of overall inflow population immigrated because of these three causes. The top three motivations for the immigration with household registration outside of Hebei Province are: job and business relocation (51.82%), family based relocation (13.63%) and education and training (9.43%). In total 74.88% of overall inflow population immigrated because of these three causes. Therefore, job and business relocation, education and training and family based relocation are the three main motivations leading to immigration. Since most of the family based immigrations are due to job and business relocation of family members, the main factors for immigration are economic factor and cultural and education factor. Therefore,

---

[2]Reference to Zhou Liqun, Xie Siquan: Bohai Rim Regional Economy Development Report (2008)- regional coordination with economic and social development. Beijing, social sciences academic press, 2008, page 288.

Table 2. Educated population in Hebei Province and in cities of Hebei in 2010.

Region	Urbanization (%)	Never attended school (%)	Elementary school (%)	Junior middle school (%)	Senior middle school (%)	Tech. College (%)	Under graduate (%)	Graduate (%)	Univ. & Above (%)
Hebei Prov.	43.94	3.26	26.79	48.23	13.80	5.03	2.74	0.15	7.93
Shijiazhuang	50.62	2.91	21.05	43.96	18.40	8.28	5.03	0.37	13.68
Tangshan	50.82	2.68	26.05	46.04	15.84	5.86	3.40	0.14	9.40
Qinhuangdao	47.51	2.18	25.40	44.13	15.95	6.86	5.15	0.34	12.35
Handan	43.44	4.19	27.68	50.36	12.44	3.61	1.64	0.08	5.33
Xingtai	40.02	3.58	25.48	53.45	12.52	3.54	1.37	0.07	4.97
Baoding	38.77	2.82	28.17	50.16	12.11	4.10	2.46	0.18	6.74
Zhangjiakou	45.16	4.64	31.73	41.91	14.46	4.98	2.22	0.06	7.26
Chengde	38.67	4.38	31.69	43.26	13.09	5.14	2.37	0.08	7.58
Cangzhou	40.83	3.55	29.44	50.33	11.21	3.99	1.44	0.05	5.48
Langfang	48.53	2.49	26.19	48.45	12.71	6.02	3.97	0.17	10.16
Hengshui	38.19	2.49	26.68	53.96	12.35	3.07	1.42	0.04	4.53

Note: Data are from the 6th census in Hebei Province.

in order to attract more immigration, Hebei should provide more job opportunities by developing economy to attract migrant workers. Hebei should also improve the investment environment to attract more investors. On the other hand, Hebei should improve schooling conditions, enhance the quality of education, and develop bigger and stronger educational and training institutions.

4 EDUCATIONAL ATTAINMENT OF POPULATION

According to the 6th census result (Table 2), 5.2425 million people that consist of 7.93% of the province's population are with college equivalent education level. It also shows that 9.1317 million people, consisting of 13.80% of total population, are with senior middle school equivalent education level. The percentage of population with junior middle school level is 48.23%, including 3.1903 million people. There are 1.77197 million people, consisting of 26.79% of total population, with elementary education level. (The numbers in the statistic result include graduates, dropouts and students in all kinds of schools).

Fig. 1 shows that the urbanization rate is strongly correlated to the percentage of populations with education levels of high school, college and above. The results in Table 3 are based on correlation analysis of the data in Table 2. The correlations coefficients shown in Table 3 are significant at alpha level 0.05. Elementary education level and junior high school education level have a negative correlation with the urbanization rate. High school education level and college and above education level show a positive correlation with the urbanization rate. College and higher education level has stronger correlation with urbanization rate than that of high school education level. These facts indicate that in total population, the higher percentage of population with elementary and junior high school education levels is, the lower the urbanization rate is. On the other hand, the higher percentage of population with high school and college and above education levels is, the higher the urbanization rate will be.

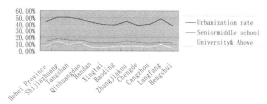

Figure 1. The urbanization rate and educated population situation of Hebei Province in 2010.
Note: Data are from the 6th census in Hebei Province.

Table 3. The correlation analysis results between urbanization and education.

Educated level	Urbanization rate
Elementary school	−0.5384
Junior middle school	−0.5359
Senior middle school	0.7781
University & Above Education	0.8075

Overall, compared to the people who only received compulsory education (elementary and junior high school education), the people who received higher education has higher social status, earn higher salary, and are more open-minded. Therefore, they are more willing and more capable of staying and living in cities and towns; and hence, the relative majority of them stay eventually. By contrast, people who only received compulsory education are less willing and less capable of staying. The majority of them eventually return to the rural areas.

Table 4. The comparison of urbanization level, industrial structure and employment structure of Hebei Province.

Year	Urbanization rate	GDP structure			Employment structure		
		Primary industry	Secondary industry	Tertiary industry	Primary industry	Secondary industry	Tertiary industry
2000	26.33	16.35	49.86	33.79	49.56	26.20	24.24
2001	...	16.56	48.88	34.56	49.17	26.39	24.24
2002	...	15.90	48.38	35.72	48.40	27.05	24.55
2003	33.51	15.37	49.38	35.25	48.19	27.17	24.64
2004	35.83	15.73	50.74	33.53	45.86	28.23	25.91
2005	37.69	13.98	52.66	33.36	43.84	29.24	26.92
2006	38.77	12.75	53.28	33.97	42.24	29.99	27.77
2007	40.25	13.26	52.93	33.81	40.42	30.96	28.62
2008	41.90	12.71	54.34	32.95	39.76	31.41	28.83
2009	43.73	12.81	51.98	35.21	39.00	31.73	29.27
2010	44.50	12.57	52.50	34.93	37.88	32.36	29.76

From: Hebei Statistical Yearbook.

Table 5. The relationship between the labor force of the inter-industry employment structure and urbanization process in different countries.

Country type	Primary industry			Secondary industry			Tertiary industry			Urbanization level		
	1960	1965	1980	1960	1965	1980	1960	1965	1980	1960	1965	1980
Low-income countries	77	77	71	10	9	15	14	14	15	13	17	17
Middle-income countries	61	57	44	15	16	22	24	27	34	33	36	45
Market economy industrial countries	18	14	6	38	39	38	44	48	56	68	71	78

From: World Development Report.

5 NATURAL POPULATION GROWTH DIFFERENCES BETWEEN URBAN AND RURAL AREAS

The natural population growth of an area is determined jointly by fertility rate and mortality rate, and is calculated as: natural population growth rate = (number of births during the year − number of deaths during the year)/annual average population × 1000‰ = birth rate − mortality rate. Comparing the natural population growth rates in urban and in rural areas, if the natural population growth rate in urban areas is larger than that in rural areas, the urbanization is accelerated. Otherwise, the urbanization is decelerated.

In 2010, the urban population birth rate was 9.87‰; and the mortality rate was 4.53‰; and the natural growth rate was 5.33‰. In rural areas, the birth rate was 14.11‰; and the mortality rate was 7.61‰ and the natural growth rate was 6.50‰. The rural population natural growth rate was 1.17‰ higher than that in urban areas, thereby constraining the urbanization process in Hebei Providence.[3]

---

[3]Data are from the 6th census in Hebei Province.

6 INDUSTRIAL AND EMAPLOYMENT STRUCTURE

Urbanization requires a strong industry support. It is the only way to provide adequate financial security for urbanization development, to provide employment security, thereby to enhance the competitiveness of urban areas. Agriculture is the prerequisite and foundation for the development of urbanization. Industry is an important driving force of urbanization. The degree of development of the tertiary industry is a sign of urbanization.

It can be seen in Table 4 that: 1) the greater the proportion of non-agricultural industries is, the higher the urbanization rate is; 2) the higher employment rate of non-agricultural industries is, the higher the urbanization rate is. Urbanization rate changes in the same direction as the non-agricultural industries change and in the same direction of the employment structure change of the second and tertiary industries. As non-agricultural industries, especially the tertiary industry, develop better, the urbanization rate gets higher (Table 5).

Hebei's industrial structure and urbanization development are in general positively correlated. But it has a large gap to the average level of the world. The urbanization rate and urban employment structure in Hebei in 2010 is roughly equivalent to that in the world's middle-income countries in 1980 (see Table 5).

7 RELATED SYSTEMS

The main factors that affect urbanization progress are: employment system, household registration system, land system and social security system.

The impact of household registration system on urbanization: Existing household registration system has negative effects on urbanization progress. First, the divided household registration management in urban and rural areas impedes the progress of converting peasants to citizens. Second, social welfare system behind the household registration system results in a semi-urbanization phenomenon of "working in urban, household registered in rural; labor in urban, families in rural; earning in urban, accumulating in rural; living in urban, rooted in rural".

The impact of employment system on urbanization: Legacy and outdated employment opinions and employment system still exist and are effective. In some areas and regions, there are still substantial barriers to employment between urban and rural areas. The rural-urban employment information and the employment channel are not yet clear. There are substantial difference in employability between peasants and urban citizens.

The impact of land system on urbanization: Efficient and flexible land transfer system has not yet been established. Peasants do not get necessary asset benefit from the land. Landless peasants do not finish capital accumulation towards the urban areas.

The impact of social security system on urbanization: The social security system includes unemployment insurance, medical care, pension system and minimum living standard security system. Currently, the social security system is prevailingly imbalanced between urban and rural areas, or has too large gap and urban-rural split issues. Improvement and enhancement of social security system will improve the quality of urbanization and will promote acceleration of the healthy development of urbanization.

REFERENCES

Cao Guangzhong, Wang Chunjie, Qi Yuanjing. The Space differences of Urbanization–Level Affecting Factors in China's Eastern Coastal Provinces and the of among [J]. Geographical Research, No. 11, 2008;1399–1406.

Jiang Yihua. Influencing Factors Analysis of County Population Urbanization, [J]. Statistics and Decision, No. 6, 2012:109–111.

Li Suqin, New Round of Constraints Analysis in China's Urbanization Process [J]. Modern Economy Explore, No. 4, 2011;79–83.

Luo Zhigang. Urbanization rate and discuss related research [J]. Urban Planning, No. 6, 2007:60–66.

Yang Gang, Xie Hongzhong. Agricultural land Circulation in Urbanization Process [J]. Urban Problems, No. 4, 2010;59–62.

Analysis of static characteristics of a typical gravity dam

K. He, J.K. Chen, Y.L. Li & Z.Y. Wu
State Key Laboratory of Hydraulic and Mountain River Engineering, School of Water Resources and Hydropower, Sichuan University, Chengdu, China

Y.Y. Zhen
Guang An Chengping Port Service Co., Guangan, China

ABSTRACT: Operational risks during the construction period and first impoundment time of gravity dams are the dam engineering industry's primary concern. In this paper, taking the typical section of a gravity dam as an example, a 3D model of a dam and foundation was built. Through the finite element method, the static characteristics of the dam, dam foundation and weak intercalations during completion, and the working conditions of a normal storage level were analyzed. Also, the characteristics of the dam deformation and stress distribution, the change processes, and the corresponding structural weaknesses under two kinds of working conditions were summarized. All this can provide a reference to ensure the safe operation of engineering, and reduce the risks associated with it.

1 INTRODUCTION

The construction period and the first impoundment time are sensitive stages during gravity dam operations (Yusof, et al., 2004). The main reason is that the dam and foundation system readjust the structure form under a water load (Zhang, et al., 2005). The process of unpredictability often leads to the risk of dam safety operation (Bowles, et al., 2003). Statistics show that about 60% of dam crashes occur during the construction period and the first impoundmengt time (Xie, et al., 2009). Therefore, how to accurately understand the dam in the completion stage, the static characteristics of the water stage, and what appropriate engineering measures need to be taken are the key problems in the trade (Mark, et al., 1977), and based on the finite element method, the three-dimensional numerical simulation technology provides an efficient way to accomplish this.

This is based on a typical gravity dam project. Through a simulation process, calculation of the dam and dam foundation's static state in the completion period and initial storage period will be performed, an analysis done of the dam and the dam foundation's deformation and stress distribution at these stages, an examination carried out of the weak spots of the dam's operations, and a reference to reduce the risks involved with the dam's operation will be provided.

2 OVERVIEW OF THE PROJECT

A typical project is located on the Yalong River, in Liangshan Yi Autonomous Prefecture of Sichuan

Figure 1. A general view of the dam.

Province, China. The total capacity is $7.6 \times 10^9 \, m^3$ water and electricity produced is 2.4×10^3 MW. As shown in Figure 1, the water retaining structure is a roller-compacted concrete gravity dam. This structure features a crest elevation of 1334 m, normal pool level of 1330 m, crest length of 516 m, minimum foundation elevation of 1166 m maximum height of 168 m, and dam bottom width of 153.2 m. Engineering construction began in 2007, and it first operated in March 2012, with full completion in March 2013.

The project is located in the mountain valley area, "V" type of valley is asymmetrical, geological structure is relatively complex. Outcropping strata is hard, and has a good integrity, Mainly to P2.β15, In addition to F8 faults on the left bank and weak interlayer of the river bed, there is no large-scale faults and other fracture zone, but tectonic fault zone relatively developed, mostly in 0.3m width, generally less than 0.6 m.

Table 1. Material parameters in different zones.

Zone	ρ (kg/m³)	E (GPa)	μ	C (MPa)	Φ (°)
C I	2500	35.5	0.167	2.19	62.85
C IV	2500	35.1	0.167	2.19	62.85
C V	2500	39.4	0.167	2.19	62.85
R I	2520	36.4	0.2	2.65	57.51
R II	2520	33.5	0.2	2.65	57.51
R III	2520	29.5	0.2	2.65	57.51
R IV	2520	37.5	0.2	2.65	57.51
R V	2520	34.4	0.2	2.65	57.51
rock II	2850	18	0.22	1.7	52.43
fxh01	2100	0.3	0.35	0.04	24.23
fxh05	2100	0.01	0.35	0.07	28.81

3 INFINITE ELEMENT METHOD

The core idea of the finite element method is area discretion (Zhu, et al., 2006). That is, instead of a continuous space with a group of finite numbers of discrete units, and units are connected by nodes, and can send each nodal force and displacement. In the calculation process, according to the relationship between displacement and strain and the virtual displacement principle, establish the strain matrix and stiffness equation of any point, formula is shown below:

$$\{\varepsilon\} = [B]\{\delta_e\} \quad (1)$$

$$\{V_e\} + \{P_{ep}\} = [K_e]\{\delta_e\} \quad (2)$$

$$[K_e] = \int_{V_e} [B]^T [D][B] dv \quad (3)$$

Accumulation for the finite element equations is done of all units in the area of according to certain principles and general finite element equations is got, namely:

$$[K]\{\Delta\} = \{P\} \quad (4)$$

where [K], {Δ}, {P} is whole structural stiffness matrix, the whole displacement matrix, integrated equivalent node load matrix, respectively.

On the basis of the general finite element equations, assuming the approximate solution of the cell model, the appropriate method is adopted to establish the relationship between the quantity of the unit interior point and unit node. Using suitable numerical calculation method for general finite element equations of modified boundary conditions, then function values can be obtained of each node.

4 SOLVE OF STRESS AND STRAIN

4.1 Generation of three-dimension

According to Table 1, the real concrete parameters, here we generate a three-dimension model about the

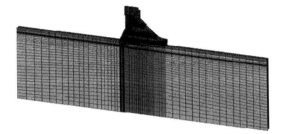

Figure 2. 3-D finite element model of the dam.

13# overflow dam contained weak intercalated layers fxh01 and fxh05. As Figure 2 shows, the model gridding is eight nodes and hexahedron isoparametric element entity unit, contains 152219 notes and 128315 units. And the coordinate system origin lies at the dam heel of upstream. The height, crest width and base width of the model are 168 m, 21.5 m and 153.2 m, respectively, extends about twice height (336 m) of the model to upstream and downstream, and extends down about twice height (336 m) of the model at the dam base.

4.2 Assumptions and boundary conditions

The stress state is complex during dam operation, allowing for numerical computation is limited, it is believed that the initial crustal stress is produced only by the rock mass self-weight, but not consider the impact of geotechnic stress (Zhao, et al., 1984). Suppose that the dam body is imperviousness, while the dam rock base is permeable, the seepage body force along the river in dam body simplied as linear surface force on the downstream dam surface (Zheng, et al., 2002). And the vertical seepage body force simplied as uplift pressure on the dam base surface. For the silt pressute which has a complex impact, is superimposed with the water pressure of upstream, the computational formula is as shown below, where silt elevation is 1212.9 m, buoyant unit weight is 6.0 kN/m³, Internal friction angle is 12° (Huang, et al., 2009).

$$p_n = \gamma_n h_n tg^2\left(45° - \frac{\varphi_n}{2}\right) \quad (5)$$

In the model, the two vertical face perpendicular to the axis of dam, the vertical face of upstream and downstream of the base and the base are all applied the normal constraints, while the upstream and downstream surface and the crest are free.

5 ANALYSIS OF STRESS AND DEFORMATION UNDER THE COMPLETION CONDITIONS

5.1 Analysis of deformation characters

Figure 3 shows the distribution of horizontal and vertical deformation under the completion conditions,

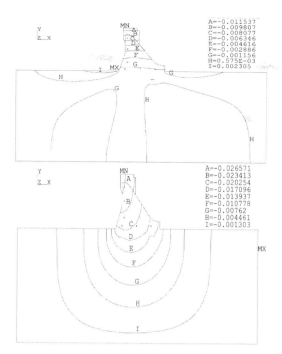
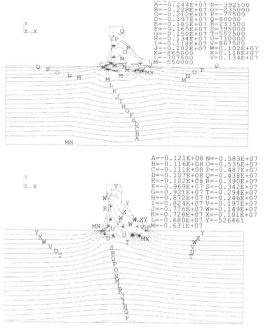

Figure 3. Deformation contour map under the completion conditions (Up is horizontal deformation, downstream as the positive direction; Down is vertical deformation, upward as the positive direction; unit: m).

Figure 4. Major and minor principal stress Isogram at the stage of the completion conditions (Up is δ_1, Down is δ_3, tensile stress is positive; units: Pa).

deducting the deformation caused by the dam base self-weight. In the horizontal direction, as the barycenter derotates upstream direction, the dam inclines upstream, then the dam heel rock base is squeezed. The horizontal deformations in all height derotates upstream direction, and the magnitude increases parallely with height, the maximum value is 12.3 mm at the crest of upstream. With the impact of dam self-weight, the dam base of upstream and downstream has horizontal deformation to upstream and downstream direction. The weak intercalated layers has dislocations, either.

In the vertical direction, the dam body and rock base shows up as bulk settling, and the settling increases parallely with height from upstream to downstream. The maximum settlement is 28.1 mm also at the crest of upstream. The rock base shows that the middle part has the maximum settlement, and decreases parallely along the upstream and downstream direction, and has significant relationship with dam load.

5.2 Stress characterization

Major and minor principal stress distribution under the completion conditions are shown in Figure 4. At the beginning of the completion, in addition to the corridor around the dam, spillway weir at the top position, pier, weak interlayer, and other parts of the river bed with magnitude smaller tensile stress (The maximum value is 1.34 MPa occur in weak interlayer fxh05), the dam and rock is basically under state of compression and the stress level is generally low, the maximum compressive stress is 12.1 MPa appear at the dam heel. Overall, the stress values of dam and bedrock decreases with increasing elevation, as the center of gravity tend to upstream, stress level of the upstream face is greater than the downstream face at the same elevation of the dam.

6 ANALYSIS OF STRESS AND DEFORMATION UNDER THE NORMAL STORAGE CONDITIONS

6.1 Analysis of deformation characters

Figure 5 shows the distribution of horizontal and vertical deformation under the normal storage conditions, it's not hard to see that under the combined action of hydrostatic pressure of upstream and downstream, silt pressute and uplift pressure at the dam base, the horizontal and vertical deformation of dam change significantly. In the horizontal direction, as all the deformation toward to the dam upstream during the complication period changes to the dam downstream, at the same time the magnitude increases parallely with height from upstream to downstream, the maximum value is 28.2 mm at the crest of downstream. The horizontal deformation in dam rock base also has obvious

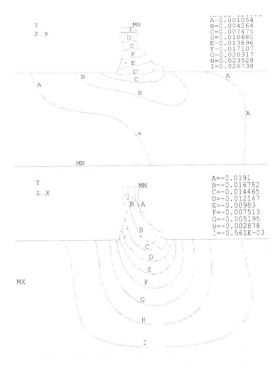

Figure 5. Deformation contour map under the normal storage conditions (Up is horizontal deformation, downstream as the positive direction; Down is vertical deformation, upward as the positive direction; unit: m).

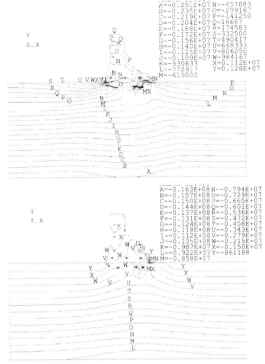

Figure 6. Major and minor principal stress Isogram at the stage of the normal storage conditions (Up is δ_1, Down is δ_3, tensile stress is positive; units: Pa).

changes, changing from complication displacement filed to overall settlement status, as a whole presents that as a whole presents that with the farther away from the dam, the deformation tendency becomes gradually smaller, the main reason is the dam downstream displacement at the same time led to the downstream displacement of dam base.

In the vertical direction, the dam body shows up as bulk settling downward, but the value reduces, the settlement at downstream bigger than that of upstream at the same height, the maximum value is 20.1 mm at the crest of downstream. The vertical direction of dam heel at the dam base reduces a little, but that at the damsite increase, the difference is large. Rock base shows up as bulk settling, the value is smaller than the completion conditions.

6.2 Stress characteristic

Major and minor principal stress distribution under the normal storage conditions are shown in Figure 6. Under external loads, the dam body and dam foundation stress distribution characteristics are changed, the change of dam body stress distribution is especially significant, the tensile stress of dam body appear area decreased significantly, only appearing in the local area of the pier, the tensile stress value decreases. The maximum principal stress of dam heel level greatly reduced and appeared tensile stress, the maximum principal tensile stress is 0.64 MPa, the stress levels of dam site rises greater, resulting in stress concentration at the dam site. The maximum principal stress is 16.3 MPa. Stress distribution correspond to the dam, stress of rock in dam heel significantly reduced. The rock stress increased significantly at the dam site, in addition to the smaller magnitude of the tensile stress occurs near the upper reaches of the river bed, the rest did not appear tensile stress.

7 CONCLUSION

This work has analysis the static performance of a typical gravity dam section, conclusions are as follows:

1. Under the impact of self-weight, the dam total deformation present to upstream and settle to downstream. While at the water load in front of the dam, the horizontal deformation during the normal storage conditions is squeezed to downstream, the bulk settling increases a little;
2. After the dam began filling, the dam stress condition improved, tensile stress area decrease obviously, values reduced. The values of dam heel and dam site changes greatly, dam heel significantly reduced from the maximum compressive stress

levels and generated the maximum tensile stress 0.64 MPa, the level of stress at the site of the dam increased significantly to 16.3 MPa;
3. Dam heel and dam site have produced a greater level of stress, especially at the dam heel pressure after the first pull of the stress state and the uneven settlement of weak structure bring the risk of dam safety operation. That should be focused on in the design and construction period.

REFERENCES

Bowles, D.S., Anderson, L.R. 2003. Risk-informed Dam Safety Decision-making. *ANCOLD*.

Huang, Y.Y., Shen, Z.Z., Tian, B. 2009. Study on the Effect of the Couple of Seepage Field and Stress Field on the Displacement of Concrete Dam. *Water Power* 35(8): 18–21.

Mark, R.K. 1977. Stuart-Alexander D E, Disasters as a necessary part of benefit-cost analysis. *Science* (17): 1160–1162.

Xie, J.B., Sun, D.Y. 2009. Statics of dam failures in china and analysis on failure causations. *Water Resources and Hydropower Engineering* 40(12):124–128.

Yusof, G. 2004. Failure modes approach to safety evalution of dams. *13th World Conference on Earthquake Engineering*.

Zhang, X.H. 2005. About the stability of concrete gravity dam on rock foundation against sliding. *Journal of Hydroelectric Engineering* 24(1):127–128.

Zhu, Y.F., Hao, Z., Yang, Z.T. 2006. Application of ANSYS to numerical simulation of dam. *Rock and Soil Mechanics* 27(6):965–972.

Zhao, D.S. 1984. Analysis of gravity dam water load. *Journal of Hydraulic Engineering* (7):53–61.

Zheng, D.J., Cao, Q.L., Wang, J. 2002. Analysis of utmost uplift pressure below concrete dam. *Journalof Yangtze River Scientific Research Institute* 19(3):13–15.

Analysis of the campus design based on the color idea of landscape space

Ying Wang & Li Yuan
Urban Planning and Design, Southwest University of Science and Technology Mianyang, China

ABSTRACT: Since the 21th century, our educational system has been improving and developing constantly, while each university is constructing and evolving. The campus landscape color, the most intuitive visual element, has great impact on the psychology, physiology and the creation of environment and atmosphere. It includes the color of campus landscape, architecture, floor covering, and night lighting which affect people's daily life (Wei Zhou, 2011). From the perspective of landscape color, this paper has a positive effect on university campus' environment space color design and planning.

Keywords: Campus space color of landscape; space environment; color psychology

1 INTRODUCTION OF THE LANDSCAPE COLOR OF CAMPUS

With the continuous enhancement of people's aesthetic consciousness and psychological needs, many universities have higher requirement of the color design of space landscape. The campus should be a cultural space which is full of natural beauty and an ecological environment with humane spirit. The campus environment' space color influences teachers and students directly, and indicates the overall image of the university on a large scale. In addition, it embodies the life styles of teachers and students and affects their daily life at the same time. The buildings' color during the day and light irradiation at night play a positive role in teachers and students' campus life.

As an important part of campus culture, landscape design of a campus has positive effects of, not only shaping a good image of the university, but also forming the perfect personality of the students and promoting physical and mental health and harmonious development. Campus environment is closely related to the level of landscape color design. The campus landscape color is the most intuitive visual element that should not be ignored for the psychology, physiology and the creation of environment and atmosphere (Li Ma, 2012). This paper, taking the campus of Southwest University of Science and Technology as an example, probes into the basic principle and environment design of the university campus based on the idea of landscape color space.

2 CURRENT PROBLEMS OF THE CAMPUS LANDSCAPE COLOR

Landscape color holds a special place in the campus culture. Color and landscape represent the image of the campus which should express the campus environment simply and directly. As long as the element of campus landscape resorting to vision, they must be influenced by the color. When we enter a school, we begin with the overall harmonious and orderly campus space environment color, and then the landscape, and they all play a guidance role on students' consciousness and feeling. So the good campus landscape color plays an important role to improve the comprehensive strength for the universities nowadays (Wei Zhou, 2011).

For now, in terms of color, many universities and colleges have various problems to be solved. These problems are analyzed below:

2.1 The deficiency of campus landscape color hierarchy

If a picture has the sense of hierarchy, it should focus on the close view, distant view and the view in between. At present, many universities and colleges are engaged in their own unique style of architecture which leads to ignore the relationship between the styles and the surrounding landscape colors. Therefore most of them seem dull and lack of the sense of hierarchy. It dose not matter how abundant in the color of the buildings or the variety of styles of buildings, the landscape whether matches the surrounding environment or not does the matter. The style and color of the buildings in old campus of Southwest University of Science and Technology seem monotonous, especially the design of landscape color which does not form close shot, medium shot and distant view. So the visual sense is deficient. Sadly, the plants on both sides of North Gate Plaza only play a role as separation and street trees, because it is short of various colors to beautify the campus space.

Figure 1. Comparison of the old campus and the new campus.

Figure 2. The ground surface pavement of the old and new campuses.

2.2 The color of old and new architecture is not uniform

Taking the old campus of Southwest University of Science and Technology as an example, from the North Gate to the axle wire front of the main building of Xiyi, there are street trees and teaching buildings which are lack of a relation on colors. There is a building which is called QINGHUA building built in 1952, and the exterior wall is made of stone and decorated with the lush creepers. So the building has an antique flavor. Other teaching buildings are all modern, and their exterior walls are made of light colored tiles. It is too harsh visually and it lacks of harmonious beauty. Although the buildings of the new campus add a lot of colors to the campus, they do not match the color of the old campus. So people will think there are two universities if there is not a scene road between the two campuses. We must pay more attention to the selection of wall materials to prevent the color pollution of faded walls (Zhuo Yu, 2010) (Fig. 1).

2.3 The deficiency of colors of ground surface pavement

Along with the enrollment expansion of colleges, new problems of campus planning arise. We can enrich the space environment by paving ground surface to improve visual effects. The colors of the pavement material can be different from the traditional graystone or other colors. By doing this, the students' activities are satisfied and a lively and colorful campus landscape environment is created. The ground surface pavements of the sculpture in the square facing the North-gate Plaza and the Wutong Avenue in front of the library are monotonous which make color deficient and it dose not achieve a certain visual effect. In order to change the situation, we could probably just make pavement the same color. For example, red bricks make the east-west pavement; blue bricks make the north-south pavement; skid resistant bricks make the transition between the new and old campus. Moreover, taking the color of the pavement as direction indicator is probably a good choice (Xiaoqian Yang, 2012) (Fig. 2).

2.4 The deficiency of campus lighting color at night

The campus night scene is a manifestation of cultural environment. With the campus cultural life becoming rich and colorful, people are willing to go out at night. But the number of campus lighting does not get an increase. At the same time that would lead to safety issues and teachers and students will feel unsafe. The road lamp is dim at night, the quarters of both teaching and living are lack of lighting colors, and it makes our campus lost vitality visually. If the light system can be increased and we use various color lighting technology to establish the campus' own image, the colorful lighting meets teachers and students' activity system in this way. And it not only creates more performance's space and supplies more performance's theme and scenes for the campus, which enriches teachers and students' daily life, but also creates a beautiful night scene at night. As a result, it is urgent to install more colorful lighting system at night (Wei Zhou, 2011).

3 THE BASIC METHOD OF CAMPUS LANDSCAPE COLOR DESIGN

In order to create a place which combines with leisure, entertainment, learning and a comfortable, pleasant, natural environment for vigorous, and talented students, much importance should be attached to the color design of campus landscape. The surface of the space is an important element of landscape's substance that means surface and space rely on each other and they are indivisible. Surface is existent which can be touched; space is nonexistent which only can be felt. Surface and space are intertwined. They convey connotation of the times and the culture. In this way, the color design of campus landscape can achieve overall shaping well (Bin Zhang, 2006).

3.1 The overall planning of campus landscape space color

The campus colors should be unified and changed at the same time. There should not be more than three colors for the teaching buildings. The colors should be in low purity and low brightness that makes buildings majestic and steady-going. Big buildings should be dim color, and small building should be in bright color. They maintain a balance both on volume and visual color. Along with decoration on roads, the colors of buildings set the basic tone for the campus (Xiuzhe Liu, 2009).

3.2 The relationship between the elements of function and the coordination of colors

After setting the basic stone, different universities and colleges should set their own the main color, auxiliary color and the ornament color according to the different departments of different universities and colleges' characteristics. The new campus of Southwest University of Science and Technology is used as an example.

The administrative office buildings and teaching buildings: the two districts represent the image of a school. Because they seem both solemn and quiet, combing three kinds of colors with low in purity, low in brightness is a good choice on the visual color, to avoid that people feel too excited or nervous by the too hash colors. Taking the orange color in low purity and low bright can highlight the mature and stable sense of a school.

Living and accommodation quarter: because the quarter is supposed to be a place where people feel relaxed and comfortable. By contrast, the color of the quarter should be warm color. The warm color will ease tensions of teachers and students, and it makes a balance on the visual color.

Historic and memorial buildings: the quarter should be like the teaching building in color which gives people a warm and knowledgeable atmosphere. Do not underestimate the color design of the nodes sculpture on the square, because the various nodes combine the different buildings together (Xiuzhe Liu, 2009).

3.3 The landscape color of campus plants

3.3.1 The choice of plant species

Each plant has its own growth characteristic, and there is a close relationship among geography, environment and geology. So it is a good choice to select the plants adapted to the local conditions. It is also appropriate to introduce foreign characteristic plants adapted to the local conditions to create a unique and charismatic culture. The plants not only have charm on colors, but also send out fragrant scent. It is the essence of landscape plants to make a colorful, aromatic environment. There are nearly one thousand kinds of plant species in Southwest University of Science and Technology: plum flower in January, apricot in February, peach blossom in March, cherry blossom in April, durian flower in May, lotus flower in June, impatiens in July, osmanthus in August, chrysanthemum in September, lotus in October, daffodils in November and wintersweet in December. Each month has its own characteristic which makes people enjoy themselves very well.

Plants have similarities with buildings. There are main and basic plants. The basic plants are widely distributed in campus, adapting to the climate environment of campus and matching to the culture of campus. Main plants are different from other plants. They are precious, unique, scented and can be used for landscape. In order to create a unique landscape,

Figure 3. The landscape color of the new and old campuses.

it is important for designers to well-design (Xiaoqian Yang, 2012).

3.3.2 Pay attention to choose colors of plants

Plants' color stone is green which alleviates visual fatigue. But plants can not perform their own charm if all plants are in green. It is better to have colorful plants to enhance each other's beauty. Colored plants refer to that all or part of the leaves, stem shows in non-green. The buttonwood trees, on the both sides of Wutong Avenue of Southwest University of Science and Technology's old campus, are beautiful and have four colors a year. The various plants in the Science and Technology plaza show a unique charm. There are pink peaches and green shrubs which enhance each other's beauty along with green lawns, all of which make people feel comfortable. For beautifying the environment, there are supposed to be more green plants to alleviate visual fatigue in teaching buildings, more colored plants in Scenery Avenue and living quarter (Fig. 3).

4 CONCLUSIONS

The design of landscape space color is one of key elements in campus environment. Multiple interactions among teachers, students and the society, as well as the changing of ideas to landscape color bring new cognition and reconstruction to the campus landscape color. The material and spiritual function of color are becoming more important, and the ultimate goal for the construction of campus culture is creating a very atmosphere. Because our universities and colleges symbolize the highest modern culture level, its epoch should be very strong. The design color of landscape space is a new area of study and should be attached more importance. In order to protect and develop the regional culture and create the excellent campus landscape, we should attach importance to the color design (Xiuzhe Liu, 2009).

REFERENCES

Bin Zhang. 2006. Brief probe into the basic principle of scene in university campus. Journal of Chu Zhou college 8(1):126–128.
Li Ma. 2012. Brief probe into the color psychology of Architecture color planning in campus. Art Panorama (4):179.
Wei Zhou. 2011. Brief probe into the campus color design 10(1):69–70.
Xiaoqian Yang. 2012. The study of the ground surface pavement 5(10):82–83.
Xiuzhe Liu. 2009. Probe into the scene planning and design of the campus color. Architecture of Fu Zhou (2):104–105.
Zhuo Yu. 2010. The new study of the old campus's landscape. Chinese horticultural Abstracts 26(10):102–105.

Analysis of the consolidation effect of ultra-soft ground via vacuum preloading

B. Xu
Tianjin Port Engineering Institution Ltd. of CCCC, Tianjin, China Key Lab. of Geotechnical Engineering of Tianjin, Tianjin, China
Key Lab. of Geotechnical Engineering, Ministry of Communication, Tianjin, China

T. Noda & S. Yamada
Nagoya University, Nagoya, Japan

ABSTRACT: A soil-water coupled finite element analysis incorporating the macro-element method is carried out to investigate the consolidation effect of ultra-soft soils using the vacuum preloading method. SYS Cam-clay model is employed as the constitutive model. Taking the one-dimensional compression tests and in-situ vane shear tests into consideration, the soil parameters and initial values can be determined correspondingly. The calculation results show that there is a good accordance between the observational and numerical settlements, and the degree of consolidation after 90 days' application of the vacuum pressure is around 61% by predicting the long-term settlement. By reproducing the unconfined compression tests numerically, the undrained shear strength can be determined using the state variables. The variation of undrained shear strength at a depth of 1 meter at different stages of the vacuum preloading shows that the strength of ultra-soft soils increases gradually, which is also very close to the in-situ vane strength.

1 BACKGROUND

Land reclamation has been the primary response to solving land shortages in coastal regions. The dredged soils, mainly consisting of ultra-soft soils with very high water content, low strength, and small permeability, are often used for the reclamation. Due to its low bearing capacity and small coefficient of consolidation, it is usually impossible to use additional preloading to consolidate ultra-soft soils. However, it has been proved that the vacuum preloading method is effective for the consolidation of ultra-soft soils.

There are many literatures related to the analysis of the consolidation effect of vacuum preloading (Wu and Zhao, 1999, Li and Zang, 2005, Yan, et al., 2004, Yang and Zhang, 2009, Xie, et al., 2000). Most of the analyses put emphasis on the comparison of observational datum such as the vane strength, and water content before and after the application of vacuum pressure. Observational settlement and pore water pressure are also used to determine the degree of consolidation. Meanwhile, numerical and theoretical analyses have also been extensively carried out to predict the consolidation results of vacuum preloading (Qiu, et al., 2013, Indraratna, et al., 2005, Chai, et al., 2005). However, the focus in the numerical analyses is put on the curve fit of the observational settlement and prediction of long-term settlement. Few numerical researches are involved with the analysis of the consolidation effect of vacuum preloading in the aspect of strength variation.

In this paper, a soil-water coupled finite element analysis is carried out numerically to investigate the settlement and strength variation during the vacuum preloading. Chapter 1 provides the research background. In Chapter 2, the calculation conditions, including the mesh partition, the boundary conditions and the determination of soil parameters and initial values, are presented. A brief introduction of the macro-element method that is incorporated in the finite element analysis is also given. In Chapter 3, the calculation results, including the settlement at the ground surface, the distribution of specific volume and the variation of undrained strength, are demonstrated in detail. In addition, how to determine the undrained strength of the ground is also introduced. The conclusions are given in Chapter 4.

2 CALCULATING CONDITIONS

2.1 Mesh partitions and boundary conditions

For simplicity, a one-dimensional condition is employed for the ultra-soft ground, as shown in Figure 1. The depth of the ground is 4 meters and the length of the plastic vertical drain is 3.5 meters. There are 16 partitions in the vertical direction.

For the displacement boundary, the horizontal and vertical directions are fixed at the bottom and only the horizontal direction is fixed for the nodes on the side surfaces.

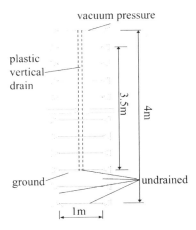

Figure 1. Sketch of the mesh partition and boundary conditions.

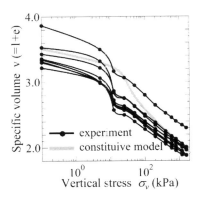

Figure 2. Curve fit of experimental and numerical results for one-dimensional compression test.

For the hydraulic condition, the undrained boundary is applied at the bottom and two side surfaces and the vacuum pressure is applied at the ground surface. The broken line represents the plastic vertical drain and it should be noted that there are no specific elements for the plastic vertical drains. The elements with the plastic vertical drain are called macro-elements and details will be shown in Section 2.2. At the bottom of the plastic vertical drain, the undrained condition is assigned.

2.2 Brief introduction of the macro-element method

The difficulty in numerical calculation of the vacuum consolidation is that if specific meshes are designated to the plastic vertical drains, because of the large difference between the sizes of the ground and the vertical drain, the mesh will become too fine to accomplish the calculation in a reasonable period. Based on the soil-water continuity equation proposed by Akai & Tamura (1978), Seikiguchi, et al., (1986) introduced the concept of the macro-element method for two-dimensional consolidation problems. A pore water pressure is added to the element with the plastic vertical drain and by changing the pressure, the pore water can flow into and out of the element freely. Therefore, the volume of the element is variable as if there were plastic vertical drains. Details can be seen in Yamada, et al., (2013). According to the field experiment, the well spacing d_e, the well diameter d_w and the discharge capacity k_w are 0.6 m, 6.75 cm and 4.0 cm/s, respectively, and the drain spacing ratio n is approximately 9.

2.3 Determination of soil parameters and initial values

The SYS Cam-clay model is used as the constitutive model of the ultra-soft soils. By introducing the concept of structure, consolidation, and anisotropy based on the Cam-clay model, it is able to describe the different extents of disturbance of the soils (Asaoka, et al., 2002). Through the curve fit of the one-dimensional compression experiment, the soil parameters and initial values, including the specific volume v, stress ratio η, degree of structure $1/R^*$, degree of overconsolidation $1/R$ and degree of anisotropy ς, can be determined accordingly (Xu, et al., in press). However, the shear strength of the ultra-soft soils is excluded in the determination process, and it is difficult to say how precise the result is. Here the shear strength of the ultra-soft soils is also taken into consideration with the results of the oedometer tests. Figure 2 illustrates the experimental and numerical results of the one-dimensional compression test. The numerical results are obtained from the constitutive calculation by giving the corresponding deformation conditions. It can be seen that there is a sudden decrease around 10 kPa vertical stress.

In addition, based on the field-survey datum, the distribution of specific volume and the vane strength along the depth is known. According to these two known conditions, the unconfined compression test is carried out numerically. By assigning other initial values besides the initial specific volume, the undrained strength of the ultra-soft soil is obtained to compare with the initial vane strength. If these two strengths are close enough, the initial values can be regarded as being accurate. The unconfined results are first given in Figure 3 and the details about the calculation of the unconfined compression test will be given in Section 3.3. According to the unconfined results, the undrained strengths of the ultra-soft soil are approximately 2, 3, 4, and 6 kPa at the depths of 1, 2, 3, and 4 m, respectively, which is very close to the vane strength (≤ 5 kPa). In addition, the undrained strength gradually increases as the depth becomes larger.

Based on the results of Figures 2 and 3, the soil parameters can be determined, as shown in Table 1. The distribution of initial values is demonstrated in Figure 4. The degrees of structure $1/R^*$ and overconsolidation $1/R$ are assumed to be uniform along the depth and are 15 and 2.2, respectively. While the

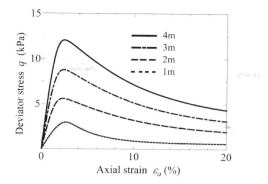

Figure 3. Unconfined compression result of ultra-soft soils at different depths.

Table 1. Soil parameters of the ultra-soft soil.

Elasto-plastic parameters	
Compression index $\tilde{\lambda}$	0.205
Swelling index $\tilde{\kappa}$	0.045
Critical state constant M	1.4
Intercept of NCL* N	2.4
Poisson's ratio ν	0.4
Evolutional parameters	
Degradation index of structure a	1.4
Degradation index of structure b	1.0
Degradation index of structure c	1.0
Degradation index of overconsolidation m	0.9
Soil particle density ρ (g/cm^3)	2.76
Permeability coefficient k (cm/s)	5.5×10^{-8}
Variation rate of permeability c_k	0.37

*Specific volume on NCL of fully remolded soil when $p' = 98.1$ kPa, $q = 0$.

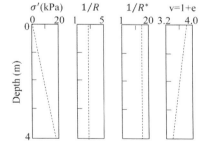

Figure 4. Distribution of initial values in the ground before the vacuum preloading.

specific volume shows larger near the upper part, it is smaller near the bottom. Considering the short reclamation history of the ultra-soft soils, the initial stress ratio η and degree of anisotropy ς are assumed to be zero, namely the coefficient of earth pressure at rest K_0 equals 1 and the soil is in an isotropic state.

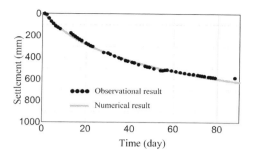

Figure 5. Observational and numerical settlements at the ground surface.

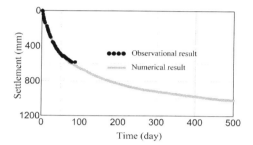

Figure 6. Prediction of long-term settlement at the ground surface.

3 CALCULATION RESULTS

3.1 Settlement

As pointed out by Xu, et al., (in press), when carrying out the consolidation analysis of the ultra-soft soil, there are significant variations of the specific volume, which would result in the decrease of the permeability coefficient. A linear relationship between the specific volume of the ultra-soft soil and the permeability coefficient should be used as shown in Equation (1).

$$v - v_i = c_k \left(\ln k - \ln k_i \right) \quad (1)$$

where c_k is the variation rate of permeability as shown in Table 1; v and v_i are the current and initial specific volumes, respectively; k and k_i are the current and initial permeability coefficients, respectively.

The observational and numerical settlements at the surface of the ground are shown in Figure 5. As can be seen, there is a good accordance between these two settlements, which proves that the determined soil parameters and initial values are reliable. After 90 days' application of vacuum preloading, the settlement at the ground surface is around 620 mm. If the consolidation time is long enough, the long-term settlement can be predicted correspondingly, as shown in Figure 6. As can be seen, after 500 days of the application of vacuum pressure, the settlement gradually becomes convergent and the final settlement is approximately 1017 mm. Therefore, the degree of

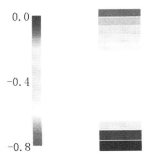

Figure 7. Distribution of specific volume change in the ground when 90d vacuum preloading.

consolidation after 90 days' application of vacuum preloading is about 61%.

3.2 Distribution of specific volume change

The distribution of the specific volume change is shown in Figure 7. In the color legend, the blue end (0.0) represents that there is no specific volume change and the red end (−0.8) indicates that there is specific volume decrease, namely volumetric compression. As the figure shows, the largest variation of specific volume occurred at the top layer, and the volume change gradually decreases from the top to the bottom. Beyond the length of the plastic vertical drain, there is no volumetric change.

3.3 Consolidation effect due to vacuum preloading

In order to analyze the consolidation effects of the vacuum preloading, vane shear tests are often carried out to obtain the vane strength of the ground at different consolidation stages. For numerical results, the undrained shear strength obtained from the unconfined compression tests is one of the ways to evaluate the consolidation effect.

In order to fulfill the unconfined compression test numerically, first analyze the procedure of soil sampling. Figure 8 presents the usual sampling procedure. Where $q = \sigma'_1 - \sigma'_3$, $p' = (\sigma'_1 + 2\sigma'_3)/3$ and $\sigma'_3 = K_0\sigma'_1$; σ'_1 and σ'_3 are vertical and horizontal effective stresses. As can be seen, the soil initially is in the K_0 state. Due to sampling, the stress state moves from the K_0 state to the isotropic state along the undrained path (with no volumetric change). Then without any consolidation, the undrained triaxial compression test is carried out numerically. During the above procedure, there is no consolidation (volumetric change) and therefore the obtained peak value equals twice the undrained shear strength of the soil. The undrained shear strength in Figure 3 is obtained by carrying out the above procedures.

During the calculation, the state variables such as the degrees of structure $1/R^*$ and over-consolidation $1/R$, the specific volume v, the stress ratio η, and degree of anisotropy ς, can be picked up at different stages of the vacuum preloading, as listed in Table 2.

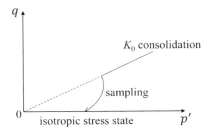

Figure 8. Sampling procedure.

Table 2. State variables at the depth of 1 m at different stages.

Days	0	20	40	60
$1/R$	2.2	1.52	1.21	1.07
$1/R^*$	15	9.3	5.7	3.4
v	3.47	3.2	3.06	2.97
η	0.0	0.553	0.545	0.543
ς	0.0	0.0161	0.0172	0.0208

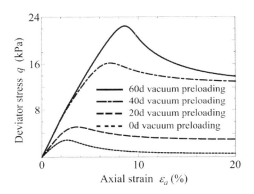

Figure 9. Unconfined compression tests of soil at the depth of 1 m at different vacuum preloading stages.

Taking the state variables as the departure, following the unconfined compression test numerically, the undrained shear strength can be determined. Figure 9 illustrates the unconfined compression test at different stages of the vacuum preloading. Where $q = \sigma'_1$ and ε_a is the vertical strain. It can be seen that as the duration of vacuum preloading becomes longer, the undrained shear strength also increases correspondingly and the consolidation effect can be confirmed. The undrained shear strengths at different stages are 2, 3, 8, and 11 kPa, respectively, which are also very close to the in-situ vane strengths of 2, 4, 9, and 13 kPa (Liang, et al., 2013).

4 CONCLUSIONS

Based on the soil-water coupled finite deformation analysis and the macro-element method, a simple one-dimensional vacuum preloading analysis was carried out to investigate the consolidation effect at different stages. The conclusions are as follows:

(1) The SYS Cam-clay model can be used to describe the mechanical behavior of ultra-soft soils. The soil parameters and initial values can be determined considering the compression characteristics of ultra-soft soils from one-dimensional compression tests and shear characteristics from unconfined compression tests.
(2) Due to the considerable variations of specific volumes of the ultra-soft soils, it was necessary to take the decrease of permeability coefficient accompanied with the reduction of specific volume into consideration.
(3) The observational and numerical settlements were very comparable. The long-term settlement can be predicted and the degree of consolidation after 90 days' application of vacuum pressure was around 61%.
(4) According to the distribution of specific volume changes, the largest variation occurs at the first layer and the variation of specific volume gradually decreases towards the bottom. There is no volumetric change for the elements without the installation of the plastic vertical drains.
(5) The unconfined compression test was reproduced numerically based on the SYS Cam-clay constitutive model and the undrained shear strengths obtained at a depth of one meter at different stages of the vacuum preloading were calculated according to the state variables. The undrained shear strength increased gradually as the duration of the vacuum preloading became longer.

REFERENCES

Akai, K. & Tamura, T. 1978. Study of two dimensional consolidation accompanied by an elastoplastic constitutive equation. *Proceeding of JSCE*: 98–104. (In Japanese)

Asaoka, A., Noda, T., Yamada, E., Kaneda, K. & Nakano, M. 2002. An elasto-plastic description of two distinct volume change mechanisms of soils. *Soils and Foundations* 42(5): 47–57.

Chai, J.C., Carter, J. & Hayashi, S. 2005. Ground deformation induced by vacuum consolidation. *Journal of Geotechnical and Geoenvironmental Engineering* 131(12): 1552–1561.

Indraratna, B., Sathananthan, I., Rujikiatkamjorn, C. & Balasubramaniam, A. 2005. Analytical and numerical modeling of soft soil stabilized by prefabricated vertical drains incorporating vacuum preloading. *International Journal of Geomechanics* 5(2): 114–124.

Li, J. & Zang, S.H. 2005. Effect analysis of soft ground consolidation using vacuum preloading method in Nansha. *Port and Waterway Engineering* 9: 101–105. (In Chinese)

Liang, A.H., Liu, J.J., Li, M.Y. & Li, W. 2013. Vacuum preloading technology for ultra-soft soils with footed-in filter tubes and its field tests, *Chinese Journey of Geotechnical Engineering* 35(2): 712–716. (In Chinese)

Qiu, C.L., Yan, S.W., Sun, L.Q. & Ji, Y.C. 2013. Effect of varying void on consolidation of dredger fill under vacuum preloading. *Rock and Soil Mechanics* 34(3): 631–638. (In Chinese)

Sekiguchi, H., Shibata, T., Fujimoto, A. & Yamaguchi, H. 1986. A macro-element approach to analyzing the plane strain behavior of soft foundation with vertical drains. *Proceedings of the 31st Symposium on JGS*: 111–116. (In Japanese)

Wu, Y.D. & Zhao, W.B. 1999. Study on strengthening of soft foundations of express highway by vacuum-surcharge preloading. *Journal of Hohai University* 27(6): 77–81. (In Chinese)

Xie, X.Y., Zhu, X.R., Pan, Q.Y., & Zeng, G.X. 2000. Effect of Zhoushan airport runway improved by surcharge precompression combined with vertical drains. *China Civil Engineering Journal* 33(3): 60–65. (In Chinese)

Xu, B., Liang, A.H., Fu, J.B., Hou, J.F. & Li, J.S. Numerical reproducing of self-sealing vacuum preloading method. *Hydro-science and Engineering*, under review. (In Chinese)

Xu, B., Liang, A.H. & Li, M.Y. Effect of variable permeability coefficient on consolidation of ultra-soft ground under vacuum preloading. *4th International Conference on Civil Engineering Architecture and Building Materials*, accepted. (In Chinese)

Yamada, S., Noda, T., Tashiro, M. & Son, N. 2013. Simulation of well resistance phenomena using macro elements with water pressure within drains as unknowns. *Proceedings of the 48th JGS*: 815–816. (In Japanese)

Yan, S.W., Zhu, P. & Liu, R. 2004. Case study on consolidation of soft foundation by means of vacuum preloading. *Journal of Hydraulic Engineering* 3: 87–93. (In Chinese)

Yang, F.L. & Zhang, Z.X. 2009. Effect analysis of surface hydraulic soil's reinforcement by vacuum preloading method in Nansha. *Port and Waterway Engineering* 3: 132–135. (In Chinese)

Buckling model of reinforcing bars with imperfections based on concrete lateral expansion

Pengcheng Mo & Bo Kan
Department of Civil Engineering, Chongqing University, Chongqing, China

ABSTRACT: Unsupported bar length (L/d ratio), yield strength fy, and initial imperfections (e/d ratio) are believed to be the buckling factors in existing models for concrete column failure. Only the bar length and yield strength have been considered in existing buckling models, while the initial imperfections have been ignored, as it is so complex to quantify. The e/d ratio is used as a measure to take the initial imperfection of a reinforcing bar into consideration. Tests show that when the e/d ratio increases, the extremum of the skeleton curve gets lower. Through demonstration, the e/d ratio can be inserted into buckling models by quantifying the lateral expansion of the concrete columns. When a longitudinal bar has imperfections, its stiffness depends not only on axial stiffness but also on bending stiffness. Numerical models show that geometry is not the most significant difference between bars with initial imperfections and those without. The three numerical examples show that the most remarkable difference is not a geometrical factor.

1 INTRODUCTION

Analysis of the seismic performance of structures is becoming more crucial because of the increasing frequency of earthquakes. As a column is the most important element to transport the gravity load in a frame structure, its seismic performance is directly related to the seismic performance of the structure. Earthquake disaster investigations show that the decline of the load capacity of concrete columns is mostly because of plastic hinges. So, due to the cyclic effects of seismic forces, the plastic hinge longitudinal reinforcements fail, and once the failure happens, the load capacity of the concrete column decreases significantly. Therefore, it is of great importance to figure out the capacity of a column under seismic loading and establish an accurate buckling model.

2 BACKGROUND

Finite element programs were used by Dhakal (2002) in the study of buckling factors, and the results show that the average compressive behavior depends on only one parameter: $L/d \sqrt{fy}$. With the increase of this parameter, buckling will happen at some point in time. In an experiment consisting of 162 specimens, Bae (2005) got the reinforcement skeleton curves under different L/d ratios, different e/d ratios and different fy. Then Bae discovered that initial imperfections can occur during the construction stage or by lateral volumetric increases of the concrete core in a reinforcing cage at large inelastic deformation levels. As human factors play a significant role in the construction quality, it is difficult to form quantitative research, so studies on the onset point of buckling are less common. By studying the effects of the L/d ratio, Rodriguez (1999) proposed a method to define the onset period of buckling.

Monti and Nuti (1992) proposed a reinforcing model for the buckling specimens that considered the effect of buckling to predict the hysteretic behavior, and the parameters in this model were determined by experimental results. In 1997, based on the famous Mennegotto-Pinto model, Gomes (1997) proposed a model called the GA buckling model which consisted of the Baushinger effect, isotropic strain hardening, decreases of the curvature, and reductions of the yield stress after a reverse. After the amendment by Kunnath (2009), three parameters were applied in the GA model. These three parameters are β, γ, and r. Then Zhang (2013) suggested the recommended values of these three parameters. Based on the Mennegotto-Pinto model, a numerical analysis buckling model called the DM model was proposed by Dhakal (2002). However, all the models mentioned above hadn't considered the buckling factors sufficiently.

In terms of software simulation, OpenSees is an open source simulation software. The official website of OpenSees provides information on software architecture and the access route of source code. It is a software framework for simulating the seismic response of structural and geotechnical systems. So, the constitutive model of material is the foundation of numerical analysis. The accuracy of the material constitutive model determines the similarity between the mathematical simulation and real life conditions. In the analysis of reinforced concrete fiber models, OpenSees has many reinforcement models: *1.* Steel01 model, which uses a bilinear model and considers the isotropic strain hardening; *2.* Steel02 model, based on the Giuffré-Menegotto-Pinto model (1994), and also considers the isotropic strain hardening; *3.* Hysteretic material model, bilinear hysteretic model; *4.* Reinforcing Steel model, based on the Chang-Mander (1994) model, considers the buckling and the fatigue damages (Brown, 2000) of reinforcement; *5.* Dodd-Restrepo design, considers the coordinate conversion between engineering coordinates and natural coordinates recommended by Dodd. This design can establish a uniform tension and compression skeleton curve, but it has not considered the buckling of reinforcement.

3 THE SIGNIFICANCE OF BUCKLING

Today, research on the factors of buckling has become more established, and the conclusions are widely agreed upon. The buckling factors are the unsupported bar length (L/d ratio), yield strength fy, initial imperfection (e/d ratio), and buckling models. Even though the buckling of reinforcements has been taken into account, many existing reinforcement models consider the nonlinear analysis. Once used, nonlinear analysis of reinforced concrete columns can overestimate the load capacity of the column. And it is reasonable to believe that is because of the ill-considered buckling factors.

4 COMPARISON OF EXISTING BUCKLING MODELS

The reinforcement models proposed by Monti and Nuti (1992) is based on a single bar's data, while the parameters in this model are determined by experimental results. The GA model suggested by Gomes (1997) is based on a theoretical derivation of the simplified model and only considers the affection of L/d ratio and fy. Moreover, Dhakal (2002) proposed a numerical analysis buckling model based on the Menegotto-Pinto model. Among the three models mentioned above, GA is the most simple and efficient model. Even so, none of the models have elaborated the reason for choosing buckling factors.

5 A BUCKLING MODEL CONSIDERING THE LATERAL EXPANSION OF CONCRETE

As the seismic design of buildings in China are based on the *Code for seismic design of building* (GB50011-2010), the stirrups in the plastic hinge zone hardly reach the yield stress, so there will be only one buckling model for longitudinal bars. As for the exclusion of the e/d ratio, there may be two reasons: *1.* The calculation method is simplified; *2.* In real-life situations, steel is believed to be straight and therefore, the e/d ratio is rarely considered. However, in practice, longitudinal bars in reinforced concrete columns do not work alone. Bae (2005) has proved that the e/d ratio can influence the skeleton curve of longitudinal bars, and also pointed out that one of the e/d ratio factors is the lateral expansion of concrete. Braga (2006) has recommended a calculation method of the lateral load on the longitudinal bars produced by the lateral expansion of concrete. However, this process is limited to the elastic stage.

As buckling and fatigue damage are both related to stress amplitude and the cycling number, it is necessary to confirm the value of the fatigue damage before establishing the buckling model. Then the experimental results can be used to subtract the fatigue damage to achieve the pure buckling results. Finally, a buckling model can be produced with accurate buckling results.

The buckling model that considers the lateral expansion of concrete should start with the following aspects:

① Find the buckling factors according to experimental results.
② Obtain the values of fatigue damage parameters through experimental data.
③ Take the e/d ratio into account using the OpenSees reinforcement buckling models.
④ Program a simulation, and if the e/d ratio makes a significant difference to the column's hysteresis curve, then find a simple process to use the e/d ratio to measure the lateral expansion of concrete.

Based on the research ideas mentioned above, one method to consider the e/d ratio is as follows:

Using the GA buckling model as an example, Figure 1 shows the calculation diagram where 'e' represents the initial imperfection of the longitudinal reinforcement and 'L' is the length of the unsupported longitudinal bar. According to the Pythagorean Theorem shown in Figure 2, the real unsupported length of the longitudinal bar is:

$$L^* = 2 * \sqrt{(e^2 + (\frac{L}{2})^2)} \quad (1)$$

If L^* is used to redefine the length of the unsupported longitudinal bar, then the calculation diagram will be the same as Figure 3.

This calculation process starts from the length of the unsupported longitudinal bar, and using parameter 'e' to calculate the actual length between the two stirrups, the e/d ratio has been considered in the GA buckling model. The advantage of this process is that it is very simple and straightforward, and has little change from the original model. However, the disadvantage is that when one longitudinal bar has imperfections, its stiffness depends on both axial stiffness and bending stiffness, but in the new calculation method mentioned above, its stiffness only depends on axial stiffness. Therefore, to figure out the pros

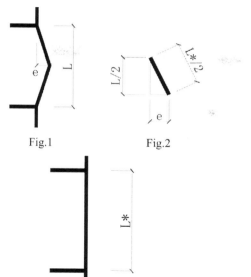

Fig.1 Fig.2

Fig. 3.

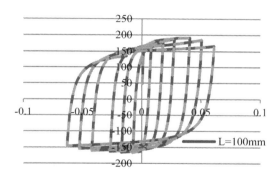

Fig. 4.

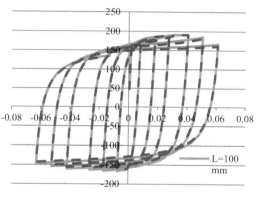

Fig. 5.

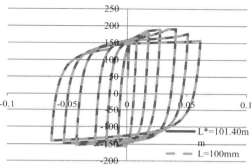

Fig. 6.

and cons of this calculation method, there are three numerical models by OpenSees.

① Numerical Model 1

$L = 100$ mm, $d = 20$ mm, $e/d = 0.6$, $L^* = 102.84$ mm, calculation results of the hysteresis curve are shown in Figure 4.

With the results of the hysteresis curve, it is not difficult to find out that the curve of the new calculation method is nearly superimposed on the curve of the original model. So, when $e/d = 0.6$, the affect of the new calculation method can be ignored.

② Numerical Model 2

In order to form a greater contrast than Example 1, this time make $e/d = 1.66$, and: $L = 100$ mm, $d = 20$ mm, $L^* = 120$ mm. The effects of the hysteresis curve are shown in Figure 5.

As with Model 1, when the value of the e/d ratio significantly increases, in addition to the little drops of the skeleton curve, the affect of the new calculation method can also be ignored.

③ Numerical Model 3

Model 3 has the same e/d condition as Model 1, while given a larger L/d value, the reinforcement geometry is as follows: $L = 100$ mm, $d = 14$ mm, $L/d = 100/14 = 7.14$, $e/d = 0.6$, $L^* = 101.40$ mm. The effects of the hysteresis curve are shown in Figure 6.

Like Model 1, the curve of the new calculation method is almost superimposed on the original model curve.

Through the three numerical examples above, the affect of the new calculation method can be ignored. The reason is that the new calculation method has just revised the geometric dimensioning, but ignored the revision of stiffness. So, simple geometric revisions can't produce the expected results.

6 CONCLUSION

This paper presented a research process to consider the lateral expansion of concrete, and recommended a new way to take the e/d ratio into account in the buckling model, and then attempted to find a simple method to use the e/d ratio to measure the lateral expansion of the concrete. A new research process has been proposed in the GA buckling model, although this is a simple and direct process with three numerical examples, it seems that the affect of the new calculation method can be ignored. This may be because of the different stiffness between a bar that possesses an initial imperfection and one that does not. So, the subsequent studies should focus more on the revision of stiffness, although not by just a simple geometric revision.

REFERENCES

Bae S, Mieses A M and Bayrak O. Inelastic Buckling of Reinforcing Bars [J]. Journal of Structural Engineering, ASCE, 2005, 131(2): 314–321.

Braga F and Gigliotti R. Analytical Stress-Strain Relationship for Concrete Confined by Steel Stirrups and/or FRP Jackets [J]. Journal of Structural Engineering, ASCE, 2006, 132: 1402–1416.n

Brown J and Kunnath S K. Low Cycle Fatigue Behavior of Longitudinal Reinforcement in Reinforced Concret Bridge Columns [R]. Tech. Rep. MCEER-00-0007, State Univ. of New York at Buffalo, N.Y. 2000.

Chang G A and Mander J B. Seismic Energy Based Fatigue Damage Analysis of Bridge Columns: Part I–Evaluation of Seismic Capacity [R]. Tech. Rep. NCEER-94-0006, State Univ. of New York at Buffalo, N.Y. 1994.

Dhakal R P and Maekawa K. Path-dependent cyclic stress-strain relationship of reinforcing bar including buckling [J]. Engineering Structures 2002, 24(11): 1383–1396.

Dhakal R P and Maekawa K. Modeling of postyield buckled of reinforcement [J]. Journal of Structural Engineering, ASCE, 2002, 128(9): 1139–1147.

Dhakal R P and Maekawa K. Modeling of Postyield Buckled of Reinforcement [J]. Journal of Structural Engineering, 2002, 128(9): 1139–1147.

Filippou F C, Popov E P and Bertero V V. Effects of Bond Deterioration on Hysteretic Behavior of Reinforced Concrete Joints. Tech. Rep. EERC-83-19, State Univ. of California at Berkeley, California. 1994.

Gomes A, Appleton J. Nonlinear Cyclic Stress-strain Relationship of Reinforcing Bars Including Buckling [J]. Engineering Structures, 1997, 19(10): 822–826.

Kunnath S K, Heo Y and Mohle J F. Nonlinear Uniaxial Material Model for Reinforcing Steel Bars [J]. Journal of Structural Engineering, ASCE, 2009, 135(4): 335–343.

Monti G and Nuti C. Nonlinear Cyclic Behavior of Reinforcing Bars Including Buckling [J]. Journal of Structural Engineering, ASCE, 1992, 118(12): 3268–3284.

Pantazopoulou S J. Detailing for Reinforcement Stability in RC Members [J]. Journal of Structural Engineering, ASCE, 1998, 124(6): 623–632.

Rodriguez M E, Botero J C and Villa J. Cyclic stress-strain behavior of reinforcing steel including effect of buckling [J]. Journal of Structural Engineering. 1999, 125(6): 605–612.

Zhang J Q, Experiment and Simulation Analysis on Cycling Behavior of Reinforcing Bar including buckling [D]. Master's degree thesis of Chongqing University, 2013 (in Chinese).

Characteristics and patterns of land development in hilly and urban fringe areas: A case study of Fangshan District in Beijing

J. Xu
School of Architecture, Tsinghua University, Beijing, China

Y. Yao
Guangdong Urban & Rural Planning and Design Institute, Guangzhou, Guangdong Province, China

ABSTRACT: As the urban land expansion accelerates, the emerging conflicts in the urban fringe areas are becoming increasingly acute. In Fangshan District of Beijing, the transfer of industries and population leads to great demand for land resources. However, the land is limited because of hilly and mountainous areas. This study separated Fangshan District into three areas – the plain area, hilly area, and mountainous area. According to the data from the TM images of 1995, 2007 and 2010, and the local official reports, the characteristics and patterns of land development in each area were analysed by GIS. In summary, there are three major problems of land development in Fangshan District, including limited land resource, unbalanced and extensive development, and the difficulty in industrial regeneration. Targeted strategies are put forward to achieve sustainability; the strategies can be applicable in other urban fringe areas with geographical restrictions.

Keywords: land development, built-up land, urban fringe; plain area; hilly area; mountainous area; GIS; Fangshan District

1 INTRODUCTION

As the land expansion of metropolis accelerates, the emerging conflicts in the urban fringe areas are becoming increasingly acute. It is because, during the fast urbanisation, industries and migrated population aggregate in the urban fringe area (Gu 1999), leading to great demand for land resources. However, the supply of land resources is limited and currently, the land development is unbalanced and in low efficiency (Zou et al. 2004). Therefore, it is urgent to analyse the patterns of land development in the urban fringe area and explore the strategies for achieving sustainable development and balancing the supply and demand.

Taking Beijing as a typical example, the built-up land keeps on expanding at a fast speed in the recent decades. Geographically restricted by the mountains in the northwest, the south area will be an essential region for future land development of Beijing (Zhao et al. 2005). Thus Fangshan District, which locates right in the southwest fringe area of Beijing, will be a major focus of development. Due to a large proportion of mountainous and hilly areas and favourable ecological conditions, the land resources are limited. It is urgent for the government to implement the policy of 'smart conservation, efficient utilisation' (Yu et al. 2009), exploring the potential of land development in the suburban hilly areas. This study focuses on the Fangshan District, analysing the major problems and patterns of land development based on GIS. Strategies are also put forward for problem-solving and decision-making in order to balance the demand and supply of land resources in Fangshan and other similar urban fringe areas with geographical restrictions. Moreover, as Fangshan District locates between urban and rural areas, the sustainable land development in Fangshan will play an important role as a bridge in the dual urban-rural structure.

2 STUDY AREA AND DATA

The study area, Fangshan District locates in southwest Beijing, spanning over 39°30′ to 39°55′N and 115°25′ to 116°15′E (Fangshan Statistical Bureau 2005). It is an outer suburban district, 38 km away from central Beijing (see Fig. 1). The total area of Fangshan is 1989 km^2 with a population of 986 thousands (2012 Census) and the area of built-up land is 380 km^2 (Fangshan Land & Resource Bureau 2007). Geographically, Fangshan lies between the North China Plain and the Taihang Mountains. Therefore the landscape is diverse and the ecological resources are various, with mountainous, hilly and plain areas taking up an average proportion. To generate targeted strategies, this study separates the whole district into three parts, which are the plain, the mountainous and the hilly areas, shown in Fig. 1.

This study used Digital Elevation Model (DEM) data and three Landsat-5 TM images in 1995, 2007 and 2010 acquired from the Chinese Science and Technology Resource. Because of the unique spectral reflectance characteristics of different types of land use, the pixels with similar colours could be regarded as the same type of land use. Accordingly, the maps shown in Fig. 2 were derived, with the grey dots representing the constructed land. Other geographic, demographic and socio-economic data, and the topographic map, were collected from the official documents such as Land Report of Fangshan District, Statistical Yearbook of Fangshan District from 2005 to 2010, Beijing Municipal Bureau of Land Resources, etc.

Because the Landsat-5 TM images are in low resolution ratio and not updated, there exist an unavoidable error in calculating the area of built-up land. Field survey was conducted to rectify the mistakes and minimise the error. Therefore the three maps above (in Fig. 2) can indicate the general trend, the spatial distribution and the patterns of land development in Fangshan.

3 RESEARCH METHODS

Based on GIS/RS, the study adopted a mixed approach with three steps as it shown in Fig. 3. The first objective was to understand how much land had been developed and how much would be available for the future. The growth rate of the area of built-up land from 1997 to 2010 was calculated and the available land was mapped according to the evaluation on ecological applicability. Meanwhile, the RS images from various locations were examined to qualitatively illustrate how the land was developed. Secondly, as there were three areas in Fangshan as shown in Fig. 1, the spatial analysis aimed to explore the specific characteristics of land development in each area. Based on GIS, the density index of the built-up land (DIBL)[1] (Jia 2009) of each town was calculated and mapped in the respective are to show the distribution of land development. Thirdly, mechanism analysis was conducted to illustrate the relationships between the built-up land and other factors, such as industrial output, population, transport infrastructure and resources, which could be essential to control the future land development. Finally, by studying on the amount, the distribution of available land, the patterns, the characteristics and the mechanism of land development, targeted strategies were recommended in order to solve the existing problems in each area and make appropriate plans for the future urban development.

4 RESULTS AND DISCUSSION

4.1 Quantity and quality of the built-up land

The land resources in Fangshan are limited but the speed of growth keeps increasing. The percentage of built-up land increased significant from 8.37% (1995) to 19.34% (2010) and the speed of growth was approximately 17 km² per year. Thus from a long-term perspective, it is urgent to calculate how much

Figure 1. Plain, mountainous & hilly areas in Fangshan District.

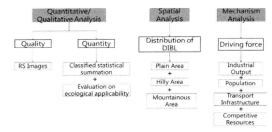

Figure 3. Diagram of research methods.

Figure 2. Maps of the built-up land in Fangshan District in 1995(a), 2007(b), and 2010(c).

---

[1] Density index of the built-up land, which refers to the ratio of the percentage of the region's built-up land to the percentage of the area of the region in Fangshan, i.e. (area of the region's built-up land/total area of built-up land)/(area of the region/total area of Fangshan).

Table 1. Categorized statistical summation of land in 2007 in Fangshan.

Types of Land Use			Area (hectare)	Proportion in Fangshan (%)	Proportion in Beijing (%)
Total			198,934.00	100.0	100.0
Built-up land			34,529.30	17.4	20.6
Un-built-up land	Total		164,405.00	82.6	66.8
	Water		5671.95	2.9	/
	Farm land	Total	115,894.00	58.3	/
		Arable land	28,275.50	14.2	/
		Garden	10,103.00	5.1	/
		Woodland	72,785.90	36.6	/
		Grassland	22.60	0.0	/
		Other	4706.60	2.4	/
	Other (unavailable land)*		42,839.00	21.5	12.6

*Unavailable land refers to wild grass ground, marshland, sand land, wasteland, bare rock land, etc.

land is available for the future development. According to the map of built up land in 2007 (Fig. 2(c)), the Local Land Report (Fangshan Land and Resource Bureau 2007) and the Comprehensive Land Use Plan (Fangshan Planning Bureau 2005), the categorized statistical summation of land is derived (see Table 1).

Among the different types of land use, 14.2% of the land was arable land, which is protected by law (The Beijing municipal people's congress 1994). 20% of land (gardens, woodland, grassland, etc.) should be conserved with another 11% including wasteland and bare rock land which cannot be developed. Considering the ecological safety, areas with slopes larger than 25 degree cannot be developed (27%) and the same with the areas within the Yongding Flood Retarding Basin (9%). Thirdly, in a social-economic perspective, large amount of regional infrastructure takes up certain amount of available land. In summary, the evaluation on the applicability of built-up land in Fangshan from these three perspectives (applicability of natural resources, ecological safety and social-economic factors) is illustrated in Fig. 4. The total area of available land is 490 km², in possession of 25% and the existing built-up land already took up 20%, which implies that the development is almost saturated and the land resources available is severely limited.

The quality of built-up land also plays an important role in achieving sustainability. Currently, the built-up land is extensively developed without adequate planning regulations.

As shown in Fig. 5, the areas around the central Fangshan New Town are crowded with illegal buildings built by farmers for owing the rents without planning permission. Moreover, some space in the plain area is in disorder and is lack of public facilities. Thirdly, the ecological/natural environment is damaged by the mining industry, even with the risks of collapse and landslide.

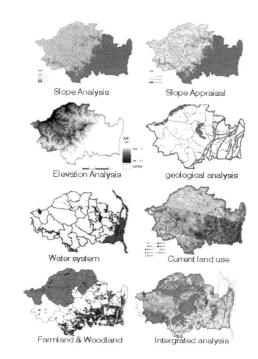

Figure 4. Evaluation on ecological applicability of built-up land in Fangshan.

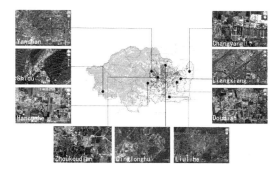

Figure 5. Quality of built-up land in Fangshan.

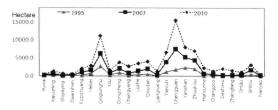

Figure 6. Areas of built-up land in 24 towns of Fangshan.

4.2 Built-up land in the plain, hilly and mountainous areas

There are 24 towns and townships in Fangshan and the areas of built-up land in different towns increase differently (see Fig. 6). According to Table 2, it is

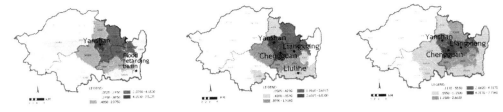

Figure 7. Distribution of the DIBL in each town in 1995(a), 2007(b), and 2010(c).

Table 2. Calculation of the DIBL in the plain, hilly and mountainous areas.

Density index of the built-up land	1995	2007	2010
The Mountainous Area	0.04	0.05*	0.04
The Hilly Area	0.14	0.24	0.27
The Plain Area	0.13	0.37	0.36

*Acceptable data error due to the low quality of the TM image.

much clear that large proportion of built-up land aggregates in the plain area and increases significantly, compared with the mountainous area.

Specifically, in 1995 (see Fig. 7(a)), the Fangshan District took the advantage of its mining resources and thus most of the hilly area developed relying on the Yanshan Industry. Considering the potential risk of the flood retarding basin, the plain area was relatively less developed. The mountainous area was less developed due to the difficulty in infrastructure construction. Then in 2007 (see Fig. 7(b)), with the industrial restructuring, high-tech industries were introduced and the focus of development shifted to Chengguan. Finally in 2010 (see Fig. 7(c)), the intensity of development in the plain area increased and gradually the formation of the Yanfang-Liangxiang developing axis became clear. The hilly area development was stimulated by some important projects such as the National Geo-park.

4.3 Driving force of the land development

The study on the driving forces is meaningful for the control of land development in the future.

(1) According to the statistics, the population and the area of built-up land are positively related. The growth range of land is significantly larger than the growth of population especially in some towns in the plain area, indicating that the land is extensively developed. Moreover, there is a positive relation between the land development and the industrial output, which implies that the towns with relatively favourable industrial conditions will have stronger motivations of land development.
(2) There are two impacts from transportation. One is the distribution by layers. Influenced by Beijing City, the built-up land aggregates within the 6th ring with a DIBL of 3.77 (similar to the intensity in Beijing City). The DIBL is 0.79 out of the 6th ring. The other impact is the agglomeration along

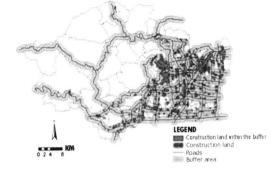

Figure 8. Built-up land within traffic buffer area.

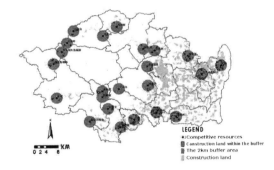

Figure 9. Built-up land within the 2 km buffer areas of competitive resources.

the main road to Beijing. In 2010, the percentage of the built-land within the 1 km buffer area of the main roads was 80.2%, a considerably large proportion (see Fig. 8). Furthermore, the figure also indicates that the density of the built-up land is relatively high where there is a high density of roads.
(3) It is indicated (see Fig. 9) that the competitive resources such as ecological and historical attractions can provide better opportunities for development, especially in mountainous and hilly areas. Based on GIS analysis, the built-up land within the 2 km buffer areas of competitive resources takes up 21%.

In conclusion, the industrial development, the population growth, the accessibility provided by transportation, and the competitive resources are the driving forces of land development.

5 SUMMARY AND RECOMMENDATIONS

In summary, the three urgent problems of land development in Fangshan are: limited land resource with the stress from the land expansion of Beijing, unbalanced and extensive land development, and the difficulty in the regeneration of mining areas. Based on the analysis of the patterns and mechanism of land development, the following strategies are recommended for problem-solving and decision-making for the sustainable land development.

Firstly, considering the limited land resources, policies should be made in order to restrict the extensive development. A dynamic monitoring system of land development is recommended for better control of the amount, the quality and the timing of land development. Secondly, with regard to the characteristics of land development in each area, the strategy of 'adjusting measures to differing conditions' (Cai, Zheng & Xiao, 2004) is provided to balance the development and solve the local problems. In terms of the plain area, land development should be strictly regulated to protect the arable land from being occupied and the development intensity of the built-up land should be adjusted in a smart way. For the hilly area, the key solution is the industrial transformation by introducing innovative and green industries to replace the essential role of mining in the area. The measures for the mountainous area are to make the best of the resources on the premise of ecological conservation and to enhance the construction of infrastructure. Thirdly, in terms of the regeneration of the mining areas, Fangshan should learn from the experiences of other places in the world, such as the Germany Ruhr Region (Couch, Sykes & Borstinghaus, 2011). Biological reclamation of the mining area is a premier step to take in order to achieve sustainability and on the other hand, innovative means of industrial transformation require exploring to stimulate the economy and optimise the land value.

Fangshan district is a typical case in the hilly and urban fringe area. Meanwhile many other regions located near metropolises have similar problems of limited land resources and unbalanced development. Similar research methods can be utilised, and targeted strategies to the specific geographic conditions can be reviewed and applied to other areas. Additionally, this analysis of the characteristics and patterns of land development can be a basis for establishing stimulation models of land development.

ACKNOWLEDGMENT

We are grateful for the supervision from Prof. WU Tinghai and Prof. LIN Wenqi at the School of Architecture, Tsinghua University and meanwhile, many thanks to the Fangshan Local Authority for the data resources and the support for the on-site investigation.

REFERENCES

Cai, Y., Zheng, W. & Xiao, L. 2004. Preliminary research on land use potential of Fangshan District in Beijing City. Territory & Natural Resources study, 2004(02), 33–34.

Couch, C., Sykes, O. & Börstinghaus, W. 2011. Thirty years of urban regeneration in Britain, Germany and France: The importance of context and path dependency. Progress in Planning, 75, 1–52.

Fangshan Land & Resource Bureau. 2007. Land Report of Fangshan District.

Fangshan Planning Bureau. 2005. Comprehensive Land Use Plan of Fangshan District (2005–2020).

Fangshan Statistical Bureau. 2005–2010. Statistical Yearbook of Fangshan District.

Gu, C. 1999. Study on phenomena and mechanism of land use/cover change in Beijing. Journal of Natural Resources, 1999(4), 307–312.

Jia, H. 2009. Study on construction land change and driving forces in Changsha City. Thesis (Master), Hunan Normal University.

The Beijing municipal people's congress. 1994. Regulation of basic farmland conservation in Beijing. [Online]. Available from: http://www.mlr.gov.cn/zwgk/flfg/dfflfg/201004/t20100429_717277.htm

Yu, K., Yuan, H. et al. 2009. Difficulties and solutions of the sustainable land use strategy in suburban hilly area in Beijing. China Land Science, 2009(11), 3–8.

Zhao, L., Chen, H., Hong, M., et al. 2005. Study on urban land use sprawling in Beijing by GIS technique. Journal of Shandong Agricultural University, 2005(4), 564–568.

Zou, D., Li, B., Zhou, J., et al. 2004. Discussions on the problems of urban land expansion. Urban Planning, 2004(7), 43–48.

Comparative analysis of CO_2 emissions of steel bar and synthetic fiber used in shield tunnel segment

Qiaosong Li & Yun Bai
Geotechnical Building, Tongji University, Shanghai, China

Andrew Ridout
Elasto Plastic Concrete, Sydney, Australia

Zhiwei Jiang
Shanghai Underground Space Architectural Design & Research Institute, Shanghai, China.

ABSTRACT: In consideration of manufacturing process, this paper consolidates the CO_2 emission calculation methods for steel bar and synthetic fiber separately. Based on the in-situ test data, the CO_2 emission factors of steel bar and synthetic fiber were obtained. The CO_2 emissions of reinforced concrete (RC) segment and synthetic fiber reinforced concrete (SFRC) segment for the same designing purpose were calculated through a FEM-based case characterized by the typical Shanghai soft soil. The result shows that in soft soil, CO_2 emissions of RC segment and SFRC segment increase with depth, and the cross joint arrangement produces more CO_2 emissions than the T joint arrangement. Furthermore, with the increase of depth, the difference between CO_2 emissions of RC segment and SFRC segment becomes significant.

Keywords: tunnel segment; reinforcement; synthetic fiber; CO_2 emission

1 INTRODUCTION

From the 'Kyoto Protocol' signed in 1997 on the UN climate change conference hosted in 2012, the greenhouse effect and CO_2 emission issues are increasingly visible in the world. In order to reduce the CO_2 emission, many technologies that help reduce the consumption of fossil energy during the steelmaking process and replace steel with materials that have less CO_2 emissions have been adopted. The syntheticfiber is one of the various materials that are used to replace the reinforcement in civil engineering.

Considering that the steel industry is a high-energy consumption and high-CO_2 emission industry, there have been many papers on the CO_2 emission of the steelmaking process. Sakamoto & Tonooka (2000) estimated the CO_2 emission from each process in the Japanese steel industry with the process analysis method, and compared the CO_2 emission factor (EFCO$_2$) of crude steel produced via the Integrated Steel Plant (ISP) route and electric arc furnace (EAF) routes. Krogh et al. (2001) analyzed the EFCO$_2$ for the production of reinforcement bar from scrap iron and the production of steel products from iron ore in Nordic countries. Sandberg et al. (2001) assessed the EFCO$_2$ of the steel produced by blast furnace (BF) route and EAF routes in Sweden by means of Life Cycle Inventory (LCI) data. Philipp (2003) introduced the status of environmental protection in the European and German steel industries. Sheinbaum et al. (2010) used logarithmic mean Divisia Index to analyze the EFCO$_2$ of Mexico's iron and steel industry. Lu et al. (2012) built a CO_2 emission model to calculate the total EFCO$_2$ of steel plants and for each production process in China. According to the studies above, the EFCO$_2$ of steel varies with country, production route, steel plant and calculation model, as shown in Table 1.

On the other hand, the existing researches on CO_2 emission of syntheticfiber are fewer than those on the emission of steel. The main component of synthetic fiber is polypropylene (PP), thus the CO_2 emission of synthetic fiber is dominated by the CO_2 emission of PP. Boustead (1993) conducted eco-profiles of the European plastic industry, and the calculated result indicated that the CO_2 emission associated with the production of PP is 1.7 kg/kg. Hammond and Jones (2008) published an inventory of carbon and energy (version 1.6a), which indicated that the CO_2 emission factor of PP in orientated film is 2.7 kg/kg and the CO_2 emission factor of PP in injection moulding is 3.9 kg/kg. Plastic Europe (2008) released an Environmental Product Declaration based on the LCI data, and reported that the CO_2 emission of the production of PP is 2.0 kg/kg. The website of Spray-All Corporation

Table 1. EFCO$_2$ of steel in different countries and routes.

Country	EFCO$_2$	
	BF (ISP in Japan) route	EAF route
Japan	2.13–2.30 kg/kg	0.51–0.70 kg/kg
Mexico	1.46 kg/kg	0.08 kg/kg
China	1.85–2.60 kg/kg	0.45 kg/kg
Sweden	1.41 kg/kg	0.32 kg/kg
Finland	1.60 kg/kg	
Denmark		0.44 kg/kg
Norway		0.24 kg/kg
German	1.30 kg/kg	

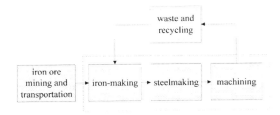

Figure 1. Life cycle of steel bar production (Calculation boundary as in the dashed box).

published a report, and it illustrated that the CO$_2$ emission of the production of PP in USA is 3.14 kg/kg. Also, the CO$_2$ emission factor of PP varies greatly in different conditions.

2 CO$_2$ EMISSION CALCULATION METHOD

2.1 Calculation theory

According to the research review of CO$_2$ emission of steel, there are many calculation methods and models used for analyzing the CO$_2$ emission of the steelmaking process. Most of the existing methods are based on the carbon balance principle, and the universal calculation formula can be expressed as:

$$EF_{totalCO2} = EF_D + EF_{ID} - EF_C$$
$$= \sum(Input\ raw\ material\ and\ energy \times EFCO_2) - \sum(Output\ product\ and\ by\text{-}product \times EFCO_2). \quad (1)$$

In Eq. (1), EF_D is the direct emission, including the emission of fossil energy, flux, etc., consumed in the production process; EF_{ID} is the indirect emission, which relates to the steel production and is not included in the direct emission, such as the purchased electric power, water, etc.; EF_C is the credit of emission permit of product or by-product, such as the metallurgical slag used for cement production.

2.2 Calculation boundary of CO$_2$ emission of steel bar

The construction industry is the final consumption of steel bar. In order to analyze the effect of CO$_2$ emission reduction due to reduced consumption of steel bar in civil engineering, it is reasonable to adopt the steel bar CO$_2$ emission factor of the whole life cycle. However, it is very difficult and complicated to calculate all the CO$_2$ emission related to the steel production, such as the CO$_2$ emission of the mining process of the iron ore and the CO$_2$ emission of the raw material transportation process.

Meanwhile, according to existing research results, the main CO$_2$ emission sources in steel production are their on-making process and the steelmaking process. In this research, the CO$_2$ emission of steel bar includes the following parts.

(1) CO$_2$ emission of the iron-making process

This part includes the CO$_2$ emission produced by the combustion of fossil energy, the oxidation process of reductants that contains carbon, and the embodied CO$_2$ emission of purchased electric power.

(2) CO$_2$ emission of the steelmaking process

This part is quite similar to the CO$_2$ emission of the iron-making process.

(3) CO$_2$ emission of the machining process

This part not only includes the CO$_2$ emission of fossil energy and purchased electric power during the machining process from crude steel to reinforcement, but also the embodied CO$_2$ produced by the loss of crude steel.

The calculation boundary of CO$_2$ emission of steel bar production is shown in Fig. 1.

2.3 Calculation boundary of CO$_2$ emission of synthetic fiber

As aforementioned, the construction is also the final consumption of synthetic fiber. It is more accurate to adopt the synthetic fiber CO$_2$ emission factor of the whole life cycle while analyzing the CO$_2$ emission of the fiber reinforced concrete. However, it is difficult to calculate the CO$_2$ emission of PP in the whole life cycle with existing statistical data.

Considering that synthetic fiber is manufactured with PP, the CO$_2$ emission of PP is the main emission source of synthetic fiber. The CO$_2$ emission of syntheticfiber can be divided into two parts: one is from the PP production process, and the other is from the manufacture process of synthetic fiber with PP. The CO$_2$ emission of synthetic fiber includes the following parts.

(1) CO$_2$ emission of the raw material used to produce the polypropylene

This part includes the CO$_2$ emission produced by the production of the main raw materials. Hydrogen gas, propylene, and ethylene are the main raw materials of PP. Large amount of energy is consumed and CO$_2$ is emitted during the production processes of those materials.

(2) CO$_2$ emission of the energy-consumption medium used to produce the polypropylene

This part includes the embodied CO$_2$ of nitrogen gas, steam, fresh water, circulating water and so on.

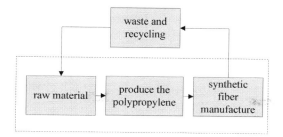

Figure 2. Life cycle of synthetic fiber (Calculation boundary as in the dashed box).

(3) CO_2 emission of the energy used to produce the polypropylene

This part includes the CO_2 emission from the energy (including the electricity and fossil fuel) consumption during the PP production process.

(4) CO_2 emission of the manufacturing process of the synthetic fiber

The CO_2 emission of this part consists of the emission of the consumed energy and the unavoidable wastage of PP during the manufacture process.

The calculation boundary of CO_2 emission of synthetic fiber is shown in Fig. 2.

3 CALCULATION OF CO_2 EMISSION FACTOR

3.1 Calculation of CO_2 emission factor of steel bar

In order to promote the application of synthetic fiber in Shanghai, it is reasonable to analyze the effect of the CO_2 emission reduction of synthetic fiber and the CO_2 emission factor of reinforcement in the steel plants in Shanghai. Considering that the main steel plant is the main producer of steel in Shanghai, the CO_2 emission factor of crude steel in Baosteel was adopted during the calculation of CO_2 emission factor of reinforcement. According to the literature (Guo et al. 2010), the CO_2 emission factor of crude steel in Baosteel is 2.06 kg/kg, while the product rate of reinforcement is 97.3% and the CO_2 emission factor of the rolling process from crude steel to reinforcement is 0.22 kg/kg. The CO_2 emission factor of reinforcement in Shanghai can be calculated with those data by the following expression,

$$e_{sbF} = \frac{2.06}{97.3\%} + 0.22 = 2.34 kg/kg \quad (2)$$

3.2 Calculation of CO_2 emission factor of synthetic fiber

The CO_2 emission factor was calculated with the data in literature and the field survey data. Considering that the synthetic fibers are manufactured in Europe, it is reasonable to adopt the data from Europe.

According to the research of Boustead (1993), the CO_2 emission of PP is 2.0 kg/kg (100 year equiv.). The CO_2 emission factor of electricity in Europe is approximately 0.5 kg/kWh. According to the field survey data, during the synthetic fiber manufacturing process, the energy consumption index is 0.482 kWh/kg, and approximately 2% of PP is wasted. The CO_2 emission factor of synthetic fiber can be calculated with those data by the formula below:

$$e_{sfF} = \frac{2.0}{(1-2\%)} + 0.482 \times 0.5 = 2.28 kg/kg \quad (3)$$

Table 2. The reinforcement ratios of RC and SFRC segments in clay.

	Reinforcement ratio	
Tunnel parameters	RC (mm²/ring)	SFRC (mm²/ring)
10 m in depth (cross joint)	1124 904	791 678
15 min depth (cross joint)	1608 1231	961 760
20 min depth (cross joint)	2010 1513	1325 1080
20 min depth (T joint)	3291 2782	2010 1714

4 ANALYSIS OF CO_2 EMISSION OF RC SEGMENT AND SFRC SEGMENT

The final analysis was to compare the CO_2 emissions of steel bar and synthetic fiber used in tunnel segment manufacturing. A hypothetical case of subway tunnel in Shanghai was chosen. The analyses below are based on a tunnel with an external diameter of 6.2 m and an inner diameter of 5.5 m. The thickness of the segment is 350 mm and the width is 1.2 m, which consist of the segment in different depth and arrangement methods. This case can represent almost all the subway tunnel in Shanghai.

In this case, the inner force was conducted with Finite Element Method (FEM), and the reinforcement ratio, controlled by crack, was calculated by the Chinese Standard CECS 38 (2004). In addition, the dosage of synthetic fiber was determined by the test results.

Table 2 presents the computed results for the reinforcement of RC and SFRC segments that meet the same design requirement (the maximum crack width was 0.2 mm).

The amount of reinforcement can be calculated with the geometric size of the tunnel (the outer diameter D = 6.2 m, the inner diameter d = 5.5 m, and the covering layer a_s = 50 mm).

On the other hand, the SFRC segment consists of synthetic fiber, reinforcement and concrete. The dosage of synthetic fiber is 8 kg/m³ and the volume of each ring is 7.72 m³, so the consumption of synthetic fiber is 61.76 kg/ring.

Table 3. The CO_2 emissions of RC segment and SFRC segment in clay.

Tunnel parameters	CO_2 emission (kg/ring)				
	RC segment		SFRC segment		
	steel bar	concrete	steel bar	synthetic fiber	concrete
10 m in depth (cross joint)	685	3273	496	141	3273
15 m in depth (cross joint)	959	3273	581	141	3273
20 m in depth (cross joint)	1190	3273	812	141	3273
20 m in depth (T joint)	2051	3273	1258	141	3273

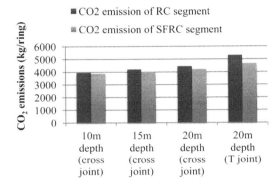

Figure 3. CO_2 emissions of RC segment and SFRC segment in clay.

The CO_2 emission of a segment includes CO_2 emitted from both types of reinforcement, when applicable, and the concrete.

According to the existing research on the CO_2 emission of concrete, the CO_2 emission of concrete is mainly from the cement, and varies with the concrete strength. Higher concrete strength leads to larger amount of CO_2 emitted from the concrete. Yu (2012) conducted an investigation of the CO_2 emission factor of concrete in Shanghai, and the investigation results indicated that the CO_2 emission factor of C55 is 424 kg/m³.

The concrete volume of each ring is 7.72 m³, so the CO_2 emission of each ring is 3273 kg. Table 3 and Fig. 3 present the CO_2 emissions of the RC segment and SFRC segment in clay.

5 CONCLUSIONS

Based on a hypothetical case of subway tunnel in Shanghai, this paper made a comparison between the CO_2 emissions of RC and SFRC shield tunnel segments. The main conclusions are as following:

(1) A method to calculate the CO_2 emission of RC and SFRC segments was consolidated, and it can be used for other projects that value the environmental effect in the design stage.
(2) Moreover, the $EFCO_2$ of steel bar and synthetic fiber were calculated. The result may be used for similar projects.
(3) Finally, the CO_2 emissions of RC and SFRC segments were calculated. The result shows that about 60%~80% of their CO_2 emissions come from concrete, and the CO_2 emission of tunnel segment in soft clay varies greatly with depth, material, and arrangement method. For tunnel segment in soft clay with deep depth and T joint, the use of synthetic fiber can reduce CO_2 emission by 32%, and the CO_2 emission of SFRC segment is 88% of the traditional segment.

ACKNOWLEDGEMENT

This project is supported by Program for Changjiang Scholars and Innovative Research Team in University (PCSIRT, IRT1029).

REFERENCES

Boustead, I. 1993. *Eco-profiles of the European Plastics Industry: Report.* PWMI, European Centre for Plastics in the Environment.
CECS 38-2004. 2004. *Technical specification for concrete structures of fiber.* Beijing: China Association for Engineering Construction Standardization. (in Chinese)
Guo, Y.C. & Li, H.X. et al. 2010. Analysis of CO_2 emission and reduction measures in Baosteel. *Energy for metallurgical industry.* 29(3): 3–7. (in Chinese)
Hammond, G. & Jones, C. 2008. *Inventory of Carbon & Energy: ICE.* Bath: Sustainable Energy Research Team, Department of Mechanical Engineering.
Krogh, H. & Myhre, L. et al. 2001. Environmental data for production of reinforcement bars from scrap iron and for production of steel products from iron ore in the Nordic countries. *Building and Environment.* 36(1): 109–119.
Lu, X. & Bai, H. et al. 2012. Relationship between the energy consumption and CO_2 emission reduction of iron and steel plants. *Journal of University of Science and Technology Beijing.* 34(012): 1445–1452. (in Chinese)
Philipp, J.A. 2003. Present status and future aspects of environmental protection in the European and German steel industry. *Proceedings of International Symposium on Global Environment and Steel Industry (ISES'03).* 15–37.
Sakamoto, Y. & Tonooka, Y. 2000. Estimation of CO_2 emission for each process in the Japanese steel industry: a process analysis. *International journal of energy research.* 24(7): 625–632.
Sandberg, H. & Lagneborg, R. et al. 2001. CO_2 emission of the Swedish steel industry. *Scandinavian Journal of Metallurgy.* 30, 420–425.
Sheinbaum, C. & Ozawa, L. et al. 2010. Using logarithmic mean Divisia index to analyze changes in energy use and carbon dioxide emissions in Mexico's iron and steel industry. *Energy Economics.* 32(6): 1337–1344.
Yu, H.Y. & Wang, Q. et al. 2012. Servicelife Period-Based Carbon Emission Computing Model for Ready-Mix Concrete. *Coal Ash China.* 23(6): 42–46. (in Chinese)

Comparative study on exploitations of Konpira Grand Theatre in Japan and Sichuan Qijian Ancient Stages in China

Li Yuan
Southwest University of Science and Technology, Mianyang, China

Masafumi Yamasaki
Ritsumeikan University, Kyoto, Japan

ABSTRACT: The ancient Konpira Grand Theatre in Japan was built in 1835. From 1972 to 1976, the theatre was reconstructed according to the original appearance. In 1985, the First Konpira Kabuki Large Show was performed publicly in the theatre. After that, the regular spring kabuki performances have been held once a year, and attracted the kabuki fans across the country. The Qijian ancient town in Sichuan, China had been the capital city of Qi State. Presently, Qijian ancient town has two well-preserved ancient stages in temples. The stages are for gods to enjoy, so the place for audiences was not considered. Instead, attention has been particularly paid to the location of the stages. Qijian ancient town deposited abundant cultural relics and intangible cultural heritage resources. Its backdrop is similar to that of the ancient Konpira Grand Theatre of Japan. So, the experience of Japan was referred to in this paper and exploitation suggestions are given for the Sichuan Qijian Ancient Stages.

To develop tourism industries and enhance the economic vitalities of remote mountainous areas, China government have decided to cultivate diversity of tourist commodities by developing sustainable rural tourism, eco-travel, and ancient town attractions. In addition to improving the infrastructure in tourist attractions, the government will strive to preserve and utilize natural environment, historic cultural resources. In Japan, with rapidly ageing population and young people going to metropolises, the recessions of small towns have been exacerbated. So, promoting tourism is not only a national growth strategy, but also a tool for developing local economy. This paper uses a successful case for Japan to preserve and utilize the ancient Konpira Grand Theatre, analyzes how to exploit local cultural resource to develop tourism attractions, and gives exploitation suggestions for the Sichuan Qijian Ancient Stage.

1 THE PRESERVATION AND UTILIZATION OF THE ANCIENT KONPIRA GRAND THEATRE IN JAPAN

The ancient Konpira Grand Theatre in Kotohira town was built in 1835. At that time, Kotohira Town, depended on Konpira Shrine development, had gathered many pilgrims and travelers. A large number of performances and shows appeared on the scene. Every March, June and August, large-scale events were held. The ancient Konpira Grand Theatre was designed according to the one in Osaka, with more modern style, and quickly became well-known in Japan. Many famous artists across the country came and appeared in this theatre. However, along with time vicissitude, it had changed hands many times, then had been utilized as cinema for some years, finally was disused (Kusanag, 1995) (Figs. 1 and 2).

To preserve the oldest theatre in Japan for coming generations, the citizens of Kotohira Town began a preservation movement. They invited constructional, dramatic experts at home and abroad to the theatre to specially investigate. In 1953, the theatre was designated as an important cultural property by the local province. In 1970, it was designated as a national important cultural property. From 1972 to 1976, the theatre was moved about 200 meters to halfway up of Atago Mountain and reconstructed according to the original appearance. In 2003, it was made a large repair and improvement, and restored the interior structure (Figs. 3 and 4).

In 1985, the First Konpira Kabuki Large Show was performed publicly in the theatre. After that, regular spring kabuki performances have been held once a year, and attracted the kabuki fans across the country. From 2001 to 2009, the annual average number of visitors to Konpira Grand Theatre reached 30,000, which is about three times the population of the town. Kotohira Town is becoming a tourist town with its own cultural characteristics.

The primary features for preserving and utilizing the ancient Konpira Grand Theatre, which is named as Kanamaruza, are as follows (Inoue, 1996).

Figure 1. The Facade (from http://www.konpirakabuki.jp/history/reform.html).

Figure 2. The Interior (from http://www.konpirakabuki.jp/history/reform.html).

Figure 3. The Facade of the theatre (from http://www.konpirakabuki.jp/history/reform.html).

Figure 4. The Interior of the theatre (from http://www.konpirakabuki.jp/history/reform.html).

1.1 The initiatives of the citizens

Before 1953, the citizens in Kotohira Town began the preservation movement for the Konpira Grand Theatre. It fully showed their consciousness. After reconstruction of the theatre, the citizens initiated the movement defending Konpira ancient town, again. Preserving traditional culture and revitalizing it make citizens remember their pride of the town, foster love to their native town, and encourage their activity toward prosperity of the local town. As people understand this, citizens made movement to preserve the theater. When the theatre was decided to hold the Konpira Kabuki Large Show every spring, the citizens actively participated in volunteering to serve for the Kabuki performance. At every performance, the young members of the Chamber of Commerce and Industry of the town take place to hold the Kabuki performance behind the stage as volunteers. Young ladies of the town also make another volunteer activity to take care and serve tee for touristic audiences. Thus, it is safe to say that the Konpira Kabuki performance is held by citizens. In the latest questionnaire survey on promoting industries of the town, 44.5% of citizens chose to promote natural environment, historic and cultural tourism, 5% higher than the second.

1.2 The protection from lows and regulations and the participation of experts and scholars

When the citizens carried out preservation movement, the theatre was designated as province-level, then national important cultural property in time. The relevant cultural properties laws and regulations steered the blindfold movement to the correct way. Through the investigation, research and reconstruction by relevant experts and scholars, the theatre restored the original appearances, which include manually operated rotating stage, trap doors for traveling around the theatre and fling device. In addition, in order to preserve and use the cultural properties of the ancient town, Kotohira Town established its own cultural properties protection regulation in 1975, the Konpira Grand Theatre's facilities management regulation in 1976, and the using rules related to Konpira Grand Theatre in 1986.

1.3 The traditional cultural education

During kabuki performances, kabuki classroom for children opens. Students in all the schools of the town and the native art societies of middle or high schools near by the town may appreciate kabuki from their teachers.

1.4 The planning to set up ancient town with its own cultural characteristics

By now and from now, the economic development of Kotohira Town was and will be based on the Konpira Shrine, so the tourism promotion planning also will focus on the Konpira Shrine and meet the needs for variety to create a tourist surrounding for travelers to like stay and return. Konpira Grand Theatre launched thanks to the Konpira Shrine. It can embody energy only in this surrounding as well. The town will optimize the opportunities for Konpira Grand Theatre to hold the regular kabuki performance, enlighten the citizens by native traditional culture, and offer them more opportunities to meet kabuki, to build up a kabuki town.

2 THE SICHUAN QIJIAN ANCIENT STAGE

The Qijian ancient town, located in Shantai county of Sichuan, had been the capital city of Qi State in the ancient time. The town has an area of 25 km^2, with a population of 13 thousands. In the 400-meter street of Qijian ancient town, there had been five ancient stages, which integrated with temples and guild halls (Fig. 5). That fully shows the prosperity of native culture in the ancient time (Li, Y. 2013).

Presently, Qijian ancient town has two well-preserved ancient stages, one is integrated with the Dizu Temple (Fig. 6), and another is Wangye Temple (Fig. 7). Both stages are gable and hip roofs with cornices and rake angles. All of the decorated brackets between beam columns are delicate wood carving. The stage tower has two parts. The front is the stage, directly facing the square, and the back is the dressing room with carving short windows. The stage is about 2.3 meters from the ground. The left door of the stage is for actors to enter and the right door is for exiting. The left and right of the stage connect with the corridor of the second floor with watch rooms. The first floor equips rooms for pilgrims or travelers to live (Fig. 8).

The stage of temple is for gods to enjoy, so it dose not include the place for audiences. Audiences can only stand or sit at will and irregularly, most of them crowd around the stage to watch. The squares between the main hall and the gate stage have a stepping up shape, which might be used as rows of watching seats when opera is performed. The designers particularly paid attention to the location of the stage. The primary spatial features of the stages integrated with temples are summarized below.

2.1 Facing the main hall

The temples generally have a main hall. It in most cases is tall, big and magnificent, for the main god. The stage of the temple is to be sacrificed to gods, so it should face the main hall.

2.2 Located on the axle wire

In the layout plan of the temple, main buildings are located on the axle wire. The left and right lay the secondary or subservient buildings. Though the stage is a service, but it is for gods to enjoy, thus it is located on the axle wire.

2.3 The relationship to the front gate of the temple

Temples generally face south and the stage is in the south of the main hall, so the relationship of the stage and the front gate is close.

The stages in Qijian Town are upper the gates. At the front of the gate, the stage cannot be seen. When entering the temple, the stage can be found upper the gate. This layout primary appeared in Dynasties of Ming and Qing. Stage and gate are interdependent. The gate because of the upper stage becomes majesty and the stage making use of the gate adds to mystery (Fig. 8).

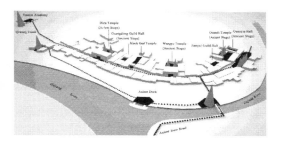

Figure 5. The space of the Qijian ancient town (drawn by the authors).

Figure 6. Dizu temple stage (took by the authors).

Figure 7. Wangye temple stage (took by the authors).

Figure 8. The plan and elevation of Dizu temple stage (took by the authors).

3 SUGGESTIONS FOR PRESERVING AND UTILIZING SICHUAN QIJIAN ANCIENT STAGES

Qijian ancient town has undergone over 2000 years of historic cultural inheritance, it deposited abundant cultural relics and intangible cultural heritage resources, especially folk Sichuan opera performance and folk events. The backdrop is similar to that of the ancient Konpira Grand Theatre of Japan. So, the experience for Japan to preserve and utilize ancient Konpira Grand Theatre is referred to give exploitation suggestions for the Sichuan Qijian Ancient Stages.

3.1 Enhancing the efforts to preserve, restore and utilize the Qijiang ancient stages

Qijiang ancient town should establish relevant regulations to protect the ancient stage and the intangible cultural heritage, and raise the preservation consciousness of citizens and encourage their initiatives. The two ancient stages should be well restored and utilized to hold the regular Sichuan opera performance. The town should create a good atmosphere of Sichuan opera, and open the Sichuan opera classroom to citizens, and set up a Sichuan opera cultural inheritance base.

3.2 Developing the Mulian plays of Sichuan Opera as its characteristic

Qijiang ancient town is close to Mulian's hometown, about 20 km away. Mulian plays are a famous series of dramas. Sichuan opera has the dramas, which had been played in Qijiang ancient town. The stories of Mulian plays derived from Buddhist scriptures. Mulian plays set operas, dances, acrobatics, and martial arts all into one. Their performance places are different from others; they may play on the stages or in the streets. Whole space of the town may be their performance stage, which more likes worship activities. Qijiang ancient town is suitable for the performance form very much.

3.3 Fully utilizing the City God Temple fair

The City God Temple Fair held in Qijiang ancient town is known far and wide. Every May 28th of lunar calendar, Qijiang town will hold a large City God Temple Fair. It is the most important traditional folk event in Sichuan. During the fair, folk troupes will be invited to play three days of Sichuan opera performance on the ancient stages. We must fully utilize the fair, increase the times of annual cultural events, connect them with the tourist resources of the town, such as Han dynasty tombs, the historic site of Qi State, ancient architectural complex of Ming and Qing dynasties, Yuntai Taoist temple, etc., which are within 3 km radius of the town, and develop the tourism as a leading industry of the town.

ACKNOWLEDGEMENT

This work was financially supported by the Ministry of Education of China on Humanities and Social Science research and the Western Frontier Regions project (11 XJAZH002).

REFERENCES

Http://www.konpirakabuki.jp/history/reform.htm

Kusanag, 1995. Kinshiro "Kanamaruza—The Oldest Thater". Kagawa Prefecture: Hoikusha.

Inoue. 1996. Shozo "Konpira Kabuki". Kagawa Prefecture: Hoikusha.

Li, Y. 2013. The Protection of Non-Material about the Ancient Town of Environmental Space. Construction of Towns 13(6):99–103.

Comparative study on street space comparative study of the Western Sichuan and Huizhou folk houses

Xukun Li, Yunzhang Li & Yihua Tang
College of Architecture and Environment, Sichuan University, Chengdu, Sichuan Province, China

Yi Cao
The Okada Architectural Firm, Fukuoka, Japan

ABSTRACT: This article selects two research objects – the Luo Mu Ancient Town in Emei city, Sichuan Province and the Guo Village in Huangshan city, Anhui Province. Some comparison analysis is carried out on the street space based on the surveying data. Western Sichuan street space gives priority to internal and external spatial penetration as well as the scale, forming the active commercial atmosphere. While the Huizhou residential street space is combined with water system, and the street interface is relatively closed due to its living function, leading to a quiet simple integral space. The formation of characteristic in both regions is not only a one-way collision of early immigrant culture, but also the integration of local and international cultures with adequate consideration of various regional factors such as natural environment, production mode and lifestyle, aesthetic taste as well as moral codes.

1 GENERAL INTRODUCTION

After the Yongjia rebellion, Han Chinese people who were on behalf of the advanced central plains culture had began migrating to south. There were three large-scale migrations that had a significant impact on the development of the southern region in history. Among those immigrants, in addition to the imperial clan, bureaucratic landlords, village clan, there were also many people who had high social status, cultural backgrounds and economic strength, including scholar-bureaucrats and scholars. After that, remarkable improvements were achieved in both economic and culture in south. Due to these large-scale migrations, northern people and aborigines went through long-term collision and integration, leading to dramatic changes in the population composition, economy, culture and production, lifestyle, custom (etiquette), language (dialect) and living patterns. For vernacular dwellings, the common points and diversities of residential types and modes result from the diversity of the geographical environment, social history, culture, customs as well as artistic standards. (Li, 2009. Sichuan Folk Houses. *Beijing: China Building Industry Press*.)

As an important part of Chinese culture, both the Bashu and Hui culture have their own features. Thus, architecture styles in these two regions vary greatly. Western Sichuan folk houses are modest and elegant, with strong mountainous characteristics and significant local landscape flavor, while Huizhou residences are calm and comely, with exquisite details. (Shan, 2009. Anhui Folk Houses. *Beijing: China Building Industry Press*.) Due to the influence of the central plains culture, coupled with the frequent exchange between the regional cultures, their formations are evolving in the coexistence with the foreign culture though thousands of miles apart. There are still a lot in common, as a result.

The past studies on western Sichuan and Huizhou folk houses were relatively rich, like "Sichuan Folk Houses" by Li Xiankui, "Value Analysis of Sichuan Ancient Town Street Spaces" by Wei Ke as well as "Anhui Folk houses" by Shan Qide, "Protection and Inheritance of Huizhou Residential Streets Space in Ming and Qing Dynasties" by Zhou Zhenxing. Those literatures did research on the development of Western Sichuan and Huizhou folk houses from the multi-angle of economy, culture, history as well as some other aspects. Throughout most of the existing researches, however, few comparative studies on Western Sichuan and Huizhou folk houses exist.

As one of academic methodologies, comparison focuses on the similarities and differences of different objects. It aims to grasp both the common point and different characteristics of objects. (Fang, 2011. *New Comparative culturally. Beijing Normal University Press*.) Research in relation with history is also referred to as "influence research". In the academic history, academics often called it historical empirical method, which is the core impact of this comparative study.

Figure 1. Street in the Luo Mu Ancient Town.

Figure 2. Square in the Guo Village.

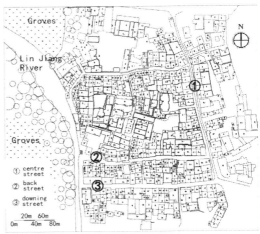

Figure 3. General Design of the Luo Mu Ancient Town.

2 THE SUMMARY

This study is a comparative study on Western Sichuan and Huizhou folk houses. The Luo Mu Ancient Town (Figure 1) is located in the southeast of Emei city, bordering on Gaoqao, Eshan, Jiuli and Guihua Bridge. In 2009, it was named "the famous historical and cultural towns in Sichuan province" by the Sichuan government. This ancient town was one of the five famed Emei towns with countless merchants gathered in dynasties and it was an important market town in southern ancient Silk Route. Nowadays, the basic street patterns of the ancient town as well as a number of folk house relics are well preserved (Baidupedia).

The Guo Village (Figure 2), with a long history, is located in the junction of Huizhou, Chizhou and Ningguo region. In the past people were used to get accommodation at Guo Village, bringing prosperity to the town. Even now, part of the ancient streets and ancient dwellings still remain relatively unscathed in Guo Village. The main reasons for choosing Luo Mu Ancient Town, Emei City in Sichuan Province and Guo Village, Huangshan City in Anhui Province are as follows.

First of all, as representatives of the typical regional culture, Luo Mu Ancient Town and Guo Village are famous historical and cultural villages with high reputation. Both of their interior space forms as well as historic buildings are well preserved. Thus the street space has its unique features. Second, there is no domestic in-depth study on Luo Mu and Guo Village yet. Finally, the whole villages have kept its original ecological landscape and have not undergone business development yet. So it provides an opportunity to get raw data. Through a comparative study on street spaces, we can summarize the common points as well as diversities between Western Sichuan and Huizhou folk houses.

3 THE LAYOUT ANALYSIS

Like most of Sichuan market towns, in site selection Luo Mu Ancient Town obeyed the principle of "terminal" site construction, namely water wharf town. It is not only a key hub which connects water and land to transport goods conveniently but also a circulation hub of goods. And it is the industrial economic principle of the site selection of market-town. There used to be a main commercial street along the Lin Jiang River. With the phasing out of waterway transport this commercial street was abandoned but people can use it to enjoy the landscape of Emei Mountain now. The Centre Street, which is parallel with the street along the river and connects external road access, has become a main commercial street instead. Along with three horizontal streets connecting the street along the river and the centre street, a pattern of 2 verticals and 3 horizontals are formed. Among them, the centre street, the backstreet as well as the downing street are relatively intact (Figure 3).

This paper chooses three main streets, the centre street, the backstreet and the downing street, as its research objects. The centre street is about 352 meters long and 9.8 meters wide on average. This street was used as a motor vehicle lane in the early stage, therefore, the road line flats out and few plans convex concave along the street. While the downing and back street are mainly for residential use now, plans along the street are irregular.

The downing street is about 266 meters long with an average width of about 8.6 meters. The back street is nearly 244 meters long with an average width of about 6.8 meters. The opening degree of these three streets' interface (except for the newly built houses) can be adjusted by disassembling the door plank, which enables the houses to be used for both commercial and residential purposes. The streets are mainly covered with cement, though it is guessed to be stone paths in old days. Beside the road are the boundaries of houses. All of them have been blocked up. It is a similar approach to pavements with stone edges. The ground floors set back from the boundary for about 1.1~1.5 meters, while eaves all extended outside the boundary, forming gallery grey space at the bottom. There are open trenches or sinks in the form of drainage in the junction of construction sites and road on both sides. And the width of the room is approximately four meters.

The general design of Guo Village (Figure 4) has distinctive characteristics of the overall arrangement of Huizhou villages. The Hen River passes through the Village, with a water gap several meters away. The Qiao Getan Square at where the etiquette worship and leisure entertainment of the villagers take place, is the center of the space sequence as well as the climax of the spiritual sense in Guo Village. It is also the commanding center node. The centre and the secondary streets go through to the Qiao Getan Square and contact foreign traffic. Other streets scattered in the village form the "core + grid" design.

In Guo Village, the Qiao Getan Square as well as the side streets is chosen as research objects. The ladder shaped square is about 60 meters long and 16 meters to 39 meters wide, covering an area of about 650 square meters. The Hen River, on which there are many ancient bridges, is 650 square meters, adding flavor to the environment. The Qiao Getan Square is paved with slab stones, where flagpoles, stone seats as well as ancient bridges are arranged to give more space levels. The Guan Yin Pavilion is the core and the most significant public building in the village and it is located in the northwest square and built across the Hen River. The east-west main streets with an average width of about four meters are relatively narrow and gather at this square. Those streets are paved with green stone boards to avoid muddy in rainy days. Besides, buildings on both sides of the streets are mainly for residential use.

4 THE VERTICAL FORM ANALYSIS

4.1 Comparison of elevation floor ratio and bay

In Luo Mu Town, two-floor buildings took a majority on the three streets, adopting the typical pattern of "living above the store". The height of each building, ranging from 6–9 meters, varies with the different purposes, but their floor vertical ratios are similar (Figure 5). The ratio of the first floor: second floor: roof height (the vertical distance from the eaves to roof) is approximately 0.35:0.3:0.35. For some single-storey buildings interspersed, proportion of storey height: roof height is about 0.55:0.45. The height of eaves of those buildings is controlled within 2.5 meters. The number of rooms in each individual building varies from a single room to seven rooms, and the width of each room is about 3–4 meters. While in Guo Village (Figure 6), the residences are mainly two-storeyed, with higher ground floor averaging about 4 meters.

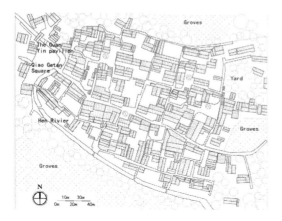

Figure 4. The general design of the Guo Village.

Figure 5. Elevation of the Luo Mu Ancient Town.

Figure 6. Elevation of the Guo Village.

Figure 7. Elevation of the Luo Mu Ancient Town.

The ratio of the first floor: second floor: roof height is around 0.35:0.3:0.35. The residences usually have three rooms, but the public buildings like the Guan Yin Pavilion have five rooms.

Through the comparative study made on facade, it is found that the room of residences in both areas is mostly 3–4 meters wide, which is reasonable for timberwork buildings. However, because Sichuan rains more and the spacious space under the eave can be used for commercial purpose, the proportion of facade on a roof is significantly larger in Luo Mu Town than in Guo Village, and the eaves extend further. In addition, most of the residences in Luo Mu Town are "living above the store" design so the second floor is often used as well. Therefore the height of two floors is usually close. On the contrary, the majority of the ground floor in Guo Village is used for primary living space, and the upper room is mainly for storage purpose. Therefore the first storey is often higher and the roof accounts less proportion. Compared with the rolling roof in Luo Mu Town, buildings in Guo Village have horse-head wall, constituting an important feature on both sides. In addition, residences in Guo Village often have three rooms, reflecting the relatively fixed way of life and building orders; while in Luo Mu Town, the number of rooms is more diverse based on business need and financial resource.

4.2 Comparison of elevation openness

Through statistics of the three streets in Luo Mu Town, the proportion of facade on holes (doors and windows) is 15% when those holes are removed. In conclusion, building facade in Luo Mu Town, which can be adjusted as needed has obvious openness variability (Figure 7).

On the other hand, the proportion of holes is relatively fixed in the Guo Village. The proportion of holes in the square, as well as inside streets, remains constantly at approximate 12%. Because there are a few small windows above the second floor and high windows on the ground floor, residences have a strong sense of closure.

The reasons for the differences of openness in the two areas are as follows. Firstly, the street space in Luo Mu Town is used for living as well as commerce, so the elevation openness is required to be variable; whereas the dwellings in the Guo Village are only for living purpose. Secondly, the western Sichuan has more rainy days and the climate is humid, leading to poor interior environment. While Huizhou home is affected by the traditional Hui business culture, thus its internal patio tends to be more open and spacious while being closed to outsiders to strengthen the construction of closure and defense.

4.3 Comparison of the skyline

Both the Luo Mu Town and the Guo Village have two-storey buildings as its majority, interspersed with a few single-storey buildings. Thus the overall skyline is smooth but each has its own characteristics and full of integrity. In Luo Mu Town, as the roofs extend outside the gable on both sides, they overlap with each other. Moreover, because of the independent construction, the skyline in the Luo Mu Town is strewn at random. Moreover the majority of buildings contain the original terrain slope, stimulating the sense of continuity of the skyline. (Wei, 2009. *The Value Analysis of Sichuan Ancient Town Street Spaces. Chengdu: Journal of Sichuan University*.)

The Guo Village, although also in the mountains, is relatively flat in terrain. Changes of the skyline are abrupt because lack of the extension of roofs. Buildings mainly use horse-head walls where the height changes. These high walls fall off in the direction of a roof and rise slightly at end, enriching the skyline of buildings. Compared with the horizontal sense of continuity in Luo Mu Town, the Guo Village lays more emphasis on the rhythm of the skyline.

4.4 Comparison of facade material and color

The facades in Luo Mu Town are mainly from wood. Most of them are natural colored fir wood, accounting for about 55% of the facade area. The grey-green roof tile, white muddy wall which have been turned tan, as well as holes of windows and doors account for 30%, 5% and 10% repectively. Overall, the facade is simple and elegant, in warm colors like deep red ochre, dark gray and taupe. In contrast, the Guo Village is typical of Hui Village, at where gray, white, and black colors are in the majority with interspersed wood. 65% of the facade area is in white walls, with gray tile roofs (25%) and deep colored bricks (5%) as well as some holes of windows and doors (5%). It leaves the elevation with a contrast but elegant impression.

5 ANALYSIS ON SECTION OF STREET SPACE

Different proportion relations in traditional dwelling houses on both sides and the width of streets will

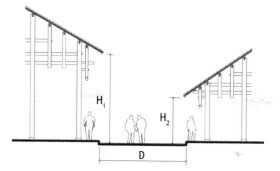

Figure 8. Section of street in the Luo mu town.

bring various space perceptions. In the street section of Luo Mu Ancient Town, (Figure 8) the centre street width (D) is about 10 meters and the height of buildings on both sides (H) is 9 m (H1)~6 m (H2). Thus, the ratio of D/H ranges from 1.63 to 1.09, in line with the street scale of the most western Sichuan folk houses. When people walk in linear space conformed by buildings on both sides, they can get a better perspective of a commercial environment in markets. For the back and down street, their average street width (D) is 6.8~8.6 meters. Because the height of both sides of residence varies, the radio of D/H is floating between 0.75 and 1.43. In this case, space is more interesting and peaceful. The gray space under eave's gallery in Luo Mu Town is also worthy of mentioning. (Ji, 2000. *Bashu Towns and Folk Houses. Chengdu: Southwest Jiaotong University Press.*) It is characteristic of western Sichuan folk houses. Buildings extend their eaves no matter single-story or multi-story, forming gray space between housing and streets as the transition between indoor and outdoor. The depth of extension is around 1.2~2 meters and the maximum is 2.4 meters. The "public space" under the eave's gallery copes with the damp and muggy climate and can be used as a shelter. Moreover, it is a socializing place where trade and business can take place.

On the other hand, the ratio of width and height (D/H) of the street section in Guo Village (Figure 9) ranges from 0.38 to 1.44, smaller than that in the Luo Mu Town. The reasons for compact design of individual building are not only because that land in this region is tight, but also because that as an immigrant clan, external defense is required to guarantee the security. Thus, the street space is deep but internal residential patio is open and convenient for activities. In addition, those deep narrow streets make it easy to prevent burning sun in summer. And the section proportion of streets in Guo Village makes walls and decoration details on both sides more purer, which conforms to the traditional artistic taste in Huizhou. (Zhou, 2012. *Protection and Inheritance of Huizhou Residential Streets Space in Ming and Qing Dynasties. Beijing: Beijing Forestry University.*)

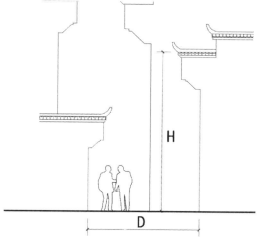

Figure 9. Section of the Guo village.

6 THE SUMMARY

Through the multi-angle comparison on street space in both the Luo Mu Ancient Town and the Guo Village, many differences between them are found. The Luo Mu Town represents the traditional home in western Sichuan while the Guo Village represents the Huizhou home. The comprehensive factors are summarized as below.

6.1 *Different natural environment*

From the comparative study, water, the primary basis of villages and towns, plays a vital role in the overall design of the two regions. Southwestern China is abundant in water resources. It can be used to drink and irrigate, and act as a transportation medium to carry out trade business. Therefore the river carries more economic sense in the town layout. Water resources in Huizhou, on the contrary, are relatively scarce. As a consequence, water diversion into the village is a necessary premise of life, and the locals lay great emphasis on the arrangement of water system in villages. Combined with the concept of feng shui, water affects the Guo Village more than to the Luomu Town.

In addition, the climate in western Sichuan is humid because of many rainy but few sunny days. Thus, residences increase the openness of the roadside facade to improve internal space comfort. Huizhou is also humid and rains a lot, but sun is blazing in the summer. So the street space is close to prevent burning sunshine and keep cool. Moreover the limited land resources in Huizhou as well as the primary lifestyle of living with clan lead to the result that the design of individual residences is intensive and the street scale is relatively tall close. While in Sichuan, the land is abundant. That is why there are stretching streets with pleasant space.

6.2 Different lifestyle

Because the market towns in Sichuan have developed commercial activities, the elevation of the ground floor is more open. So people can switch between residential and commercial use flexibly by removing and assembling moveable wooden doors. At the same time, western Sichuan dwellings pay more attention to roofs. Thus along streets there are interesting eaves gallery, which plays as an extension of the transportation and commercial space and where residences can keep off sunshine, chat with the neighborhood as well as entertaining. It creates a town with the overall environment of the communication.

On the other hand, buildings are mainly used to live because Huizhou businessmen often did business trips. Due to the security concerns and affected by Huizhou tradition, residences are relatively closed with strong defensive sense. Elevation along the street is totally closed or has a few high windows, while the internal space is open. It makes Huizhou streets more tall close and not favorable for longtime staying. Thus, the locals prefer to socialize with neighborhood in public space such as well sites and squares.

6.3 Different artistic taste and social morals

Most of the residences are plain and simple except for a few houses owned by wealthy bureaucrats because Sichuan people are easygoing. Buildings along streets are both residential and commercial. The facade is mainly wood plank in natural color. It makes street space simple and elegant and forms the modest and elegant characteristic in overall residences.

On the other hand, in Huizhou, most of the businessmen who were greatly influenced by traditional moral codes had good educational background and did business trips frequently, they did not intend to use gorgeous color to show off. So the facade is mainly in the neutral color like black, white and gray as building protection.

Huizhou street space formed its unique aesthetic characteristics by contrasting with colors in the facade along the streets and combining with the exquisite wood carvings.

In conclusion, due to different natural environment, production and lifestyle, aesthetic taste as well as ethical codes, western Sichuan and Huizhou residential street space have generated several distinctive features. Western Sichuan street space has laid more emphasis on internal and external spatial penetration as well as the scale, combined with rich space below eaves gallery, forming the active industrial atmosphere. While in Huizhou street space, it flexibly combines with the water system and is relatively closed and quiet. Finally, dwellings in both regions have improved the rationality and conveniences with local conditions. It also fully reflects that the formation of the western Sichuan and Huizhou folk houses is not only a one-way collision of early immigrant culture, but the integration of local and foreign cultures under adequate consideration of several regional factors such as nature as well as economy.

REFERENCES

Baidupedia

Fang Hanwen. 2011. New Comparative culturally. *Beijing Normal University Press*. [M]

Ji Fuzhen. 2000. Bashu Towns and Folk Houses. *Chengdu: Southwest Jiaotong University Press*. [M]

Li Xiankui. 2009. Sichuan Folk Houses. *Beijing: China Building Industry Press*. [M]

Shan Qide. 2009. Anhui Folk Houses. *Beijing: China Building Industry Press*. [M]

Wei Ke. 2009. The Value Analysis of Sichuan Ancient Town Street Spaces. *Chengdu: Journal of Sichuan University*. [J]

Zhou Zhenxing. 2012. Protection and Inheritance of Huizhou Residential Streets Space in Ming and Qing Dynasties. *Beijing: Beijing Forestry University*. [D]

Comparison and suggestions on roof structure schemes of single-storey pure basement

Haowei Kuang
Kunming University of Science and Technology, Kunming, China

Haoyi Ma
Kunming Xin Zheng Dong Yang Architectural Engineering Design Co., Ltd., Kunming, China

Wen Pan
Kunming University of Science and Technology, Kunming, China

ABSTRACT: Based on the theoretical model, this paper firstly analyzes the economy of five schemes on the structure of basement roof, which are the pure plate roof, single-direction single-beam roof, single-direction double-beams roof, crossing-beams roof and skewed-crossing-beams roof. Considering construction feasibility, design period, error probability and construction period, a roof scheme, combining pure plate roof with single-direction single-beam roof, is the best choice in most practical pure basements with irregular column grids. Finally, some useful suggestions that deal with design problems are put forward based on abundant practical experiences.

1 INTRODUCTION

In order to guarantee the use of buildings and maximally utilize the limited land resources, large-baseplate single-storey pure basement tends to be applied in many housing estates. Thus, a large area of pure basement exists. The cost of a basement is usually very high, and a rational structure of the basement can effectively save the constructive cost. In addition, the roof is a main aspect in the structure of basement, and the choice of the roof scheme is of great significance as it will affect not only the constructive cost but also the constructive period and feasibility.

At present, almost all researches focus on the roof of basement. The basement, which is a general term, does not refer in particular to a certain basement. At the same time, a large number of studies are still in theoretical stage and complicated factors in practical engineering are often ignored. Furthermore, most of those studies concern the economy. Therefore, this paper, regarding single-storey pure basement as the study object, takes more complicated factors into consideration and finally proposes a scheme combining pure plate roof with single-direction single-beam roof, which turns out to be the optimum choice in most practical projects.

2 CONTRAST OF PURE BASEMENT BETWEEN THEORETICAL MODEL AND PRACTICAL ENGINEERING

The theoretical model (as shown in Figure 1) is characteristic of regular shape, equal distance, orthogonal column grid and 100% pure basement. Actually, there are features of irregular shape, unequal distances, column grid with orthogonal and oblique crossing and 68%–80% pure basement of total basement in practical engineering (as shown in Figure 2).

3 ECONOMICAL COMPARISON AND ANALYSIS OF THEORETICAL MODELS

The calculation and analysis of theoretical models were conducted through finite element analysis and design software SATWE (2010). Theoretical models are single-storey pure basement with the floor height of 3.9 m, column grid of 5 × 5, the same column space of 8.1 m and area of 1640 m². The average thickness of covering soil of 2 m can be regarded as 16 kN/m² calculated density without the load of fire lane (it means that dead load is 32 kN/m² and live load is 4 kN/m²). Main beams are third seismic grade of

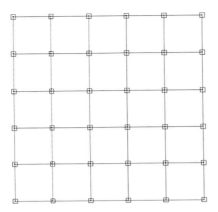

Figure 1. Column grid plan of basement of theoretical models.

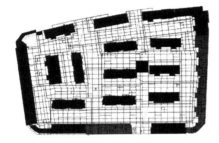

Figure 2. Column grid plan of basement of practical engineering.

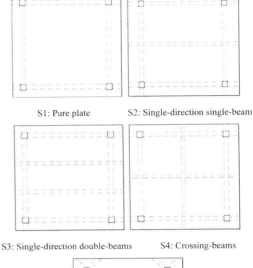

S1: Pure plate   S2: Single-direction single-beam

S3: Single-direction double-beams   S4: Crossing-beams

S5: Skewed-crossing-beams

Figure 3. Five roof schemes.

reinforced concrete structure according to "Code for design of concrete structures" (GB 50010-2010) and secondary beams are forth.

Under the same loads, this paper makes a comparison of five roof schemes (as shown in Figure 3) with regard to the economy in Table 1. The first four schemes are all applied in practical engineering. The fifth scheme is a new one that may be applied in the special engineering in the future, the purpose of which is mainly to expand ideas.

The cost in Table 2 refers to the actual cost of engineering in Kunming in 2014, where the price of reinforcement is 4000 Yuan/t, concrete is 350 Yuan/m^3, and formwork is 40 Yuan/m^2. From the Table 2, we know that single-direction double-beams and crossing-beams schemes are the most economical, and pure plate and single-direction single-beam are relatively economical, and the last one is the most expensive. There is a slight difference (no more than 8%) in the first four schemes.

4 ANALYSIS OF COMPOSITE FACTORS OF PRACTICAL ENGINEERING

When shapes of basements and column grids are all regular and there is a little difference of elevation between adjoining roofs, the economical scheme is a better choice, that is, single-direction double-beams and crossing-beams scheme. However, in most actual projects, general drawings are usually irregular for the sake of maximally utilizing of land resources. Meanwhile, the layout of carports is the top priority, which leads to irregular column grids of basement (especially column grids adjoining buildings) and a large number of oblique crossing beams that might produce irregular-large slabs. Function and usability are the main factors that should be taken into account, and naturally there will be many irregular roofs and obvious elevation differences, especially in roofs adjoining buildings. Above all, irregulars are very common in practical engineering, which inevitably lead to irregular roofs. Consequently, we should consider all kinds of factors including construction feasibility, design period, error probability and construction period when we choose the roof scheme.

When roofs are irregular or roof types are too many or elevation differences are too large, the problems that secondary beams cannot pass through and concrete materials cannot be evenly distributed are probably produced by adopting single-direction double-beams and crossing-beams schemes.

When the pure basement adjoins the buildings, it will result in oblique crossing beams and irregular-large slabs. In this case, adopting the single-direction double-beams scheme is not workable and adopting

Table 1. Project amount comparison.

Roof scheme	Reinforcement amount (kg)		Slab thickness (mm)	Section (mm×mm)
S1	Beam	36620	280	M-B 400 × 1200
	Slab	72811		
S2	Beam	59035	250	M-B 400 × 1200
	Slab	40135		S-B 300 × 1100
S3	Beam	66673	200	M-B 400 × 1200
	Slab	26350		S-B 300 × 900
S4	Beam	61450	200	M-B 400 × 1200
	Slab	32594		S-B 300 × 900
S5	Beam	77758	250	M-B 400 × 1200
	Slab	47938		

Roof scheme	Reinforcement (kg/m^2)	Concrete (m^3/m^2)	Formwork (m^2/m^2)
S1	66.7	0.386	1.59
S2	60.5	0.388	1.86
S3	56.7	0.363	2.04
S4	57.3	0.362	1.95
S5	76.6	0.412	1.84

Note: M-B, referring to the beam crossing columns, is a contraction of main beam. S-B is, referring to the beam not crossing the columns, is a contraction of secondary beam.

Table 2. Cost comparison.

Roof scheme	Yuan/m^2				
	Reinforcement	Concrete	Formwork	Total cost
S1	266.8	135.1	63.7	466
S2	242.0	135.8	74.6	452
S3	226.8	127.1	81.5	435
S4	229.2	126.7	77.9	434
S5	306.4	144.2	73.4	524

the crossing-beams scheme will lead to deep beams, which subsequently exerts adverse effects on the economy of reinforcement or causes the problem that the slab cannot be distributed evenly. The reason why that forms deep beams is that the roof of tower tends to be higher than the pure basement.

When it comes to some special positions such as fire resistance rolling shutter doors, light well, closing frame beams or courtyard closing the frame beams, the slab is hard to be distributed evenly by adopting single-direction double-beams and crossing-beams schemes.

Meanwhile, from the perspective of constructive process and period, there are many disadvantages by using single-direction double-beams and crossing-beams schemes. The specific disadvantages consist of complicated process, long period, higher error probability and hard modification.

It can be seen that single-direction double-beams and crossing-beams schemes have its limitations in irregular practical engineering.

5 CONCLUSIONS AND SUGGESTIONS

As for the basement with regular column grids and minor differences of elevation, single-direction double-beams and crossing-beams schemes are the best choice. But in most practical basements with irregular column grids or obvious elevation differences, it is better to choose the scheme combining the pure plate with single-direction double-beams scheme.

According to the past experiences in combining the pure plate with single-direction double-beams scheme, the following suggestions are proposed: 1) the types of pure plates should be as few as possible. It may be within 1–3 types as well; 2) in order to meet the standard, the roof may be as well set as double-layer reinforcement with two-way pass length and the difference of thicknesses of slabs should not be more than 30 mm; 3) considering the economy of envelope design of bottom bars and the two points above, we suggest thickness of slab should be 250 mm and 280 mm in pure basement; 4) to guarantee the space of electricity and water pipes, the elevation difference between main beams and secondary beams should be limited within 300 mm.

REFERENCES

Beijing Architectural Design Standard Office. 2005. *Beijing architectural design specifications*. Beijing: Economic Science Press. 2005.

Beijing Institute of Architectural Design and Research. 2006. *Measures of professional and technical architecture*. Beijing: China Architecture & Building Press.

Ministry of Housing and Urban-Rural Development of the People's Republic of China. 2010. *GB 50010-2010 Code for design of concrete structures*. Beijing: China Architecture & Building Press.

Ministry of Housing and Urban-Rural Development of the People's Republic of China. 2012. *GB 50009-2012 Load code for the design of building structures*. Beijing: China Architecture & Building Press.

Peilin, L. & Xuemin, W. 1994. *Structure design manual for concrete ribbed and water slab*. Beijing: China Architecture & Building Press.

Pusheng, S. 2003. *Design principle of floor structure*. Beijing: Science Press.

Wenxiang, C. & Aiqun, L. 1998. *Design of concrete floor structure*. Beijing: China Architecture & Building Press.

Concept and practice of reconstruction of European old buildings

Yanting Wang
Tianjin University of Technology, Tianjin, China

ABSTRACT: Some typical methods used in the reutilization of European old buildings from the angel of the modernization of traditional buildings are introduced in this paper. Corresponding design principles and overall thoughts from art perspective are summarized. The paper also investigates some basic attitudes and utilization methods of typical western designers through specific reconstruction plans, which give beneficial inspiration to similar projects in China.

1 HISTORY AND CONCEPT RELATED TO RECONSTRUCTION OF OLD BUILDINGS

When investigating the development history of the reutilization of buildings, two clear lines can be seen, one is functional and practical reutilization mainly aimed at economy, and the other is that to protect the heritage for the main purpose of continuation of historic culture. At the beginning, the two methods of reutilization were not closely connected in practice, but with social development and improvement of architecture, the two lines gradually combined, that is not only to seek the economic value of reutilization, but also to protect the historic culture borne by the old buildings, and to achieve the integration of social benefit, environmental benefit and economic benefit.

Actually, it is from ancient time that man had the activities of reconstruction of old buildings; it is not created by modern people, for some well-known buildings in history had been reconstructed and reutilized. For example, Parthenon was modified into a Christ church, and then a mosque after Turks occupied Athens; the King's Manor in British Yorkshire has been frequently used for different purposes, and in nearly eight hundred years from the thirteenth century to now, it has been being reconstructed into the office of a monastery, a palace, an administrative center of the parliament, an apartment, a school, a factory, etc., until now the site of a school of architecture research. During the Renaissance, master of architecture Michelangelo reconstructed the bathroom in Rome times into a church.[1] The early reconstruction and reutilization of old buildings basically resulted from the functional and economic factors.

In Italy, little changes are made to the appearance of old buildings, so the reconstruction of Italian old buildings is focused on the addition of buildings and interior modification, which mainly achieves the replacement of new functions of old buildings through modification of space, materials, lighting and facilities. Many historical buildings have been reconstructed successfully into museums, industrial zones, schools, etc., and the designers reinvigorate the old buildings by effective design. Such reconstructions give full display to the economic, cultural, art and environmental value of the old buildings in new period, thus creating good social benefit.

For Italy after the War, due to the conflict between the number of existing buildings and the demand, its concept for reconstructing old buildings was more and more radical, and the transition from the old to the new was the permanent theme of Italian architecture in the twentieth century. Works of Carlo Scarpa were favored by architects in the late twentieth century. Carlo Scarpa was born in Venice, most of his works were constructed in cities and the surroundings – he liked to add constructions to existing buildings rather than design new ones, and his core idea for his design lied in respecting history and the existing environment, which was considered incomprehensible by people of his age. In response to that, Carlo Scarpa listed the inspirational works of the past great architects: the dome of Florence Cathedral, masterpiece of Brunelleschi, was added on existing building.

2 CREATION CONCEPT AND IMPLEMENTATION FRAME OF RECONSTRUCTION AND REUTILIZATION OF EUROPEAN OLD BUILDINGS

2.1 *Reconstruction of space and epuration of functionality*

The interior space of old building loses its original function and functional replacement is the most common method in the reutilization of old buildings, so it is important to the reutilization of the building how to reconstruct the old space. Space reconstruction

---
[1] Kenneth Piwell. The Conversion and Reconstruction of Old Architecture. Translated by Yu Xin, Yang Zhimin, Si Yang. Dalian University of Technology Press, 2001.

is achieved mainly by modifying interior space, updating materials and adjusting the atmosphere of interior space, thus satisfying the new functional demand of the old buildings.

In Europe, protection of historical buildings lays emphasis on the overall images of the buildings, not on whether their original functions are continued. Besides, the space of old buildings is often in large size, and the structures are solid, providing preconditions for the reconstruction and reutilization of the spaces of the buildings.

For old buildings that have many floors, reconstruction methods of mezzanine or staggered floor can be employed, and the heights of space are rich in changes, thus creating novel and delicate space effect. Italy Milan University was reconstructed from Maggiore Hospital, which was built in the fifteenth century and had the building style of the transition from Gothic to the Renaissance. Now, the original building style remained in the appearance of the building, and the reconstruction laid emphasis on that of the interior space. Taking advantage of the large space of the original buildings, its library was reconstructed by using mezzanine in parts of the interior. Its surface is cross-shaped, and the designer added two "glass boxes" in the longitudinal directions, and connected them to the tall and spacious borrowing room by metal steps, making the interior space of the building smooth and full of levels in vision and behavior. The interior lighting was designed with combination of overall lighting and partial lighting. The design not only realized the functional change of the building, but also created the quiet and modest atmosphere of a library, and met the demand for the new functions.

2.2 *Juxtaposition and symbiosis between history and modernism*

Some historical buildings in Italy can date back to the Middle Age, even Roman Times, and with time going by and function is changing, each age has its impression on the same historical building, which is the cherishable historic value. Designers combine these memories smartly with the new space order, and display the existence and value of the old building in the combination of history and the modernism.

Museum of Castelvecchio in Italy was first built in the Middle Age, about 1354 to 1356 by Della Scala family for the purpose of preventing the invasion of enemies and rebellion. In 1925, an architect named Arnaldo Forlatio firstly reconstructed it into a museum, and later Carlo Scarpa took charge and modified it for another time.[2] It took him six years to complete the reconstruction, making it full of historical information and meanwhile mixed with modern spirit.

Scarpa refused to design according to the original style, broke all don'ts and doctrine, and used expression methods full of modern aesthetics to show the concise shape features. Modern materials including steel, glass and concrete began to enter the old building, and some characteristic of the compared sides was separated, made to stand out, strengthened and purified, enriching the meaning to be expressed by the building.[3] Here in the co-existence and conflict of the new and the old, the symbiosis of history and modernism was achieved. For the interior part, Scarpa employed door and window system completely different from the original building, so as to maintain the historical original appearance and meantime meet the demand of exhibitions. The original Gothic window style was preserved for the exterior surface of the building, and used in the corresponding positions inside were doors and windows with new De Stijl formation, which displayed the concise modernism, and juxtaposition in form was achieved in the two window styles. On the last end of the exhibition room there was a semi-circular gate arch of the original building, in front of which a steel rail was laid and the grid gate woven by iron bars was implanted on the steel, with strong comparison between the ancient and heavy arch hole and the light and thin steel sliding gate.

In the reconstruction of Manchester's Cube Gallery, the connecting joints of the new and the old are clear, showing different patterns of the building in different historical times. The original walls were remained and the plaster was stripped where the neat work was restored. Besides, some original wooden beam floor and cast iron column were preserved. Newly-added walls were clean white without any decoration. Other added elements including columns of channel steel and glass sideboard were all concise geometric bodies, whose accuracy had interesting contrast to the original unsmooth and handmade components. Plain use of building materials, such as industrial red brick, steel, stones and woods, in juxtaposition with the typical white blocks of modern sports, contrasted and strengthened the original structure.

2.3 *Reference of traditional decoration and improvement of detailed quality*

Scott once wrote in his masterwork "Humanist Architecture-Research of Sentiment History" that the past glory of history and the admiration of Romanticism to it naturally stretch to every detail where the past is preserved.[4] It is the historical deposit and improvement through times that form the delicate details of Greece and Ancient Rome columns, and the information expressed by such details takes root so deeply in people's mind that we consider it classical when seeing the details conforming to forms. Detail design is the main carrier of historical information, thus details are historical. As a result, in the design

---

[2] Chu Ruiji. Carlol Scarpa- Poetry Flowing in Space. Taiwan: Garden City Press, 2004.

[3] Sergio Los. Carlo Scarpa.an architectural guide. Arsenate Editrioe srl. Nov, 2001.

[4] Xin Huiqin, Analysis of Reconstruction and Reutilization of Italian old Buildings, [Master Dissertation], Tianjin, Tianjin University, 2004.

of the reconstruction of old buildings, the decoration and details is meaningful. Traditional decoration is an important method to achieve the continuation of the historical vein. On one hand, the formation of details is achieved by strengthening the sense of beauty in vision and structure, and on the other hand, classical or other traditional elements are used in detail decoration, and the details are signalized and integrated into the architecture. Anyhow, in the design of the whole reutilization, the architects still should give play to their talents to make the creativity that goes beyond history.

The addition of the library located in St. John's College of Cambridge University is also one of the examples where the newly-constructed buildings are harmonious with the existing historical ones. This senior library in St. John's College has been being constructed for several centuries, and its integration is damaged at a certain degree. Architects hold that the existing historical buildings were worth preserving and added a new axle on the foundation of the original architecture. The new construction adopted detail design consistent with the old buildings, including the same bricking materials, the same height of horizontal molding for each floor, and even the same triangled pediment bent winding-up of the gable walls, and moreover, a pavilion harmonious with Gothic was set on the crossing part. To be modern, deformation processing was used for the new buildings, for example, the old pavilion was built with polygon bricks, while the new one became round, glass and metal; the window of the old window was regular-shaped, while free frames were used for wall surface, virtual processing for angles of the building, and the windows are also different. In general, the old and the new blended into a harmonious whole, just because several characteristic detail processing were done well, thus achieving ideal effect.

2.4 Integration of environment and optimization of space artistry

What is also often neglected is that the aging of the environment of buildings is an important factor affecting their functions. Owing to the fading of the surroundings or the area, some buildings constructed not long before lose their functions, thus affecting the quality of interior space. As a result, improvement of the quality of the surroundings is a beneficial supplement in the practice of the reconstruction of old buildings.

For the old buildings for industry or storage, their original external surroundings including roads and facilities are designed in focus of organizations of transport, storage and production, without much consideration of man's activities, and the external space is dull and empty. Therefore, reorganization of the external surroundings is necessary for the functional change of such buildings, where it is required to establish the external open space that is person-oriented and promotes interpersonal communication.

The Media Center in Hamburg of Germany had been wrecked by the War, and the whole buildings and the surroundings suffered serious artificial damage and natural decay, conveying a sense of emptiness and coldness. When repairing and reconstructing the whole buildings, the architects, although preserved and respected the original architecture, never continued the old parts in the newly-built ones, but developed it. The design included the inner "streets" and "squares", which not only reintegrated the environment in the area, but also created more enjoyable space. The solid public and semi-public spaces preserved the full height and strengthened the true sense of city space. The combination of blocky brick-wall structure and short steel beams inside the space had a strong sense of sculpturing, creating the outstanding effect of space art, and meanwhile solving many problems related to lighting and ventilation encountered in the reconstruction and reutilization of old buildings. The design intensified the enormous change between the new and the old, and reflected the art style in the post-industrial times.

3 CONCLUSIONS

Surveying the reconstruction of European old buildings, three trends can be observed: the first is the diversification of the reconstructed objects, the second is the diversification of the reconstruction methods, including the inner reconstruction of space, addition and extension, and the application of new materials and new techniques to reconstruction of aspects and positions of the old buildings, and the third is the diversification of concepts, which not only go from the view of protecting the historical buildings, but also pay great attention to the social effects they brings about.

China also has a long and brilliant history, which leaves us a great number of old buildings. European countries provide us with much reference in this aspect, and more and more examples indicate that the practice of reconstruction and reutilization can activate the slack and depressing old space and that the revival of historical buildings will offer the space charm that can never be acquired in new architecture. Protection, reconstruction and reutilization are the main methods to protect and maintain the history of buildings and cities, and are also a piece of art that interprets the historic heritage in the modern and future life.

REFERENCES

Chu Ruiji. 2004. Carlol Scarpa- Poetry Flowing in Space. Taiwan: Garden City Press, 2004.

Kenneth Piwell. 2001. The Conversion and Reconstruction of Old Architecture. Translated by Yu Xin, Yang Zhimin, Si Yang. Dalian: Dalian University of Technology Press, 2001.

Sergio Los. 2001. Carlo Scarpa.an architectural guide. Arsenate Editrioe srl. Nov, 2001.

Xin Huiqin, 2004. Analysis of Reconstruction and Reutilization of Italian old Buildings, [Master Dissertation], Tianjin, Tianjin University, 2004.

Consolidation grouting Karst foundation treatment strategy for existing buildings

Yuanjie Xiang & Yuan He
Guizhou Construction Science Research & Design Institute Co. Ltd, Guizhou, China

ABSTRACT: Karst landform, which is common on the Yunnan-Guizhou Plateau, is famous for its complex geological situation. Due to scene conditions, ground handling is difficult for existing buildings. After many years' engaging in ground handling, the authors will describe in this paper a strategy which is beneficial to handle Karst ground handling base on an engineering practice.

Keywords: Karst; ground treatment; consolidation grouting

1 INTRODUCTION

Karst (also known as karst) is the general term of geological phenomena and morphology consist of rocks (carbonates, gypsum, rock salt, etc.) which dissolved by constant water erosion. The effects on the stability of the foundation karst come mainly from the following aspects: cave roof collapse leads to ground subsidence so that the upper structure get damaged; uneven foundation settlement; interaction of foundation under rocks along the inclined surface of the weak structure; gushing, flooding and other emergencies accidents; worsening geological conditions (*Yan Ping et al. 2012*).

2 PROJECT PROFILE

The project is built in Guiyang View Lake District and designed as a 32-story high-rise building. Its frame supports shear wall structure, the design takes artificial digging pile foundation as basis and weathered argillaceous limestone as foundation supporting layer. The single-column maximum load is 7000 kN and its geological survey suggested stratum bearing capacity is fa = 4000 kPa. The project construction has been built to 18 layers. The coring pile test from engineering department shows that: the project includes 57 hand-dug pile foundations and pile bearing stratum caverns, crevices, mud with stone and other defects exist in 17 ones of them. It means that the project does not meet the requirements and needs ground treatment (*JGJ72-2004*). Pulling detect of defective piles is shown in the Table 1 below.

In this paper, the defect of 17 piles above will be treatment via an efficient and convenient karst ground handling strategy.

3 GROUND TREATMENT STRATEGY

The underground bedrock of the building is argillaceous limestone bedrock, which has 125° dip direction and 25° dip angle. It has karst fissures and developed from caverns. And the groundwater is too abundant to be excreted. The housing construction has been built to 18 layers, but wall cracking or column settlement hasn't occurred yet. Test result by piles shows that there is rock crushing in the area three times the pile diameter (not less than 5 m) on the bottom of 17 piles. The maximum cave height is up to 3.4 m. What's more, complete rock and cave appear on the two sides of a pile respectively and the altitude difference may be up to 2.7 m. In a word, the geological condition is so complex that it's difficult to take treatment process. Considering the venue construction and engineering geological and hydrogeological conditions, it is decided to use consolidation grouting to reinforce rock bottom of the pile foundation.

Treatment for 17 piles proceeded via consolidation grouting process. Treatment scope spreads from 1 m outside the scope of the pile and 5 m under pile foundation and not less than 3d pile diameter. In the caves whose depths are greater than this value, treatment will be preceded to the cave floor (*GB50007-2011*). Each pile is arranged with 4 grouting hole (density increasing in complicated piles is acceptable) inside and around the hole with separation distance of 0.5 m~1.0 m (drilling position and separation distance can be adjusted according to the design engineering). Two sequence operations (single number such as 1, 3 is a sequence and double number such as 2, 4 is another) will set 68 grouting holes, 17 inspecting holes. Grouting consumption is calculated as 200 kg/m.

Table 1. Pulling pile Statistics of defect detection.

No.	Pile No.	Pile diameter	Pile length	Inspection Situation
1	4#	1.2 m	7.6 m	8.7 m~10.7 m Cave: clay filling
2	5#	1.2 m	7.4 m	7.4 m~8.0 m Cave: Mud and Stone filling
3	9#	0.8 m	7.9 m	10.0 m~10.8 m Cave: clay filling
4	11#	1.5 m	7.4 m	7.4~8.9 m Cave: Mud and Stone filling
5	12#	1.2 m	8.3 m	10.5~12.2 m Cave: clay filling
6	19#	1.2 m	8.2 m	11.9 m~13.5 m Cave: Mud and Stone filling
7	22#	0.9 m	7.6 m	7.9~9.7 m Cave: clay filling
8	25#	0.9 m	10.4 m	10.6~12.4 m Cave: Mud and Stone filling
9	32#	0.9 m	8.7 m	13.1 m~13.6 m Cave: clay filling
10	37#	1.2 m	7.8 m	8.0 m~8.7 m Cave: clay filling
11	38#	1.2 m	10.3 m	10.5 m~12.8 m Cave: clay filling
12	48#	1.2 m	7.3 m	9.3~10.5 m Cave: clay filling
13	51#	0.9 m	8.8 m	9.0~11.7 m Cave: clay filling
14	52#	1.2 m	8.3 m	8.4~11.8 m Cave: clay filling
15	55#	1.2 m	8.8 m	9.9~10.7 m Cave: clay filling
16	56#	0.9 m	7.9 m	8.2 m~9.9 m Cave: Mud and Stone filling
17	57#	1.0 m	10.0 m	13.0 m~14.8 m Cave: Mud and Stone filling

4 CONSTRUCTION TECHNOLOGY

Construction process: 1-Consolidation grouting hole location; 2-Drilling grouting holes; 3-Washing holes and pressurizing water; 4-Consolidation Grouting; 5-Grouting efficiency analyses; 6-Consolidation grouting quality check (JGJ123-2000).

4.1 Consolidation grouting hole positioning

Consolidation grouting holes are arranged in strict accordance with the design requirements, according to the designed position. Each pile is arranged with 4 grouting holes in two operation sequences and separation distance is 0.5 m~1.0 m.

4.2 Drilling grouting holes

According consolidation grouting hole depth and bedrock lithology conditions, a XT-100 rig is use to spirally drill geological model with carbide drill bits, initial hole diameter is 91 mm and final one is 91 mm or 75 mm. orifice tube: φ91; wall: 0.5 m to 1.5 m, in order to prevent orifice collapsing or skewing. Rigs drilling process must ensure a smooth scroll and correct orifice tube direction. Drilling records are prepared for detecting the situations about caves, drillings and collapses every 5 m (JGJ79-2002). Correction and remedy measure are launched once dip value exceed threshold. The requirement is that the maximum bias value of final pile bottom is less than 50 cm and qualified rate of pile dip reaches 100%.

4.3 Washing holes and pressurizing water

Connect the flower tube which has a few holes to the bottoms of the drilling hits so that great flow and high pressure (0.5~0.8 MPa) water can flow in. With the rises, falls and spins of drilling hits, the holes, especially the position in which caves and cracks occur are washed and cleaned in order that the soft and weathered clay can be rinsed out of holes or the areas need grouting. Then, slurry will be injected into caves and cracks and solidify on the contacting surface to reinforce foundation. The washing process will cease 10 minutes after clean water flow out. If it doesn't happen, the process must last 50 minutes at least. The water pressure test will be preceded after washing process, choose each sequence hole for a single-point water pressure test, namely, 5% water from each pile will be pumped. According to the design requirements and standard specifications, the pressure is 0.3 MPa in simple pressurizing water method. Test results reveal: average water permeability of consolidation grouting is 45.98 Lu in one-sequence and 14.3 Lu in double-sequence.

4.4 Consolidation grouting

4.4.1 Grouting materials and slurry manufacture

Ordinary silicate cement with strength value of 42.5 is selected as grouting materials. Samples from all batches must pass the quality checking and be conserved without rain or wetness to meet the requirements for grouting. The water used for grouting should also reach the level of hydraulic concrete mixing. The slurry of pure cement grout, if necessary, may be mixed with admixture and extra slurry.

Manufacture of slurry abides by "hydraulic structures cement grouting technical specifications" DL/5146-2001 and its four standard classes, Namely, 3:1, 2:1, 1:1, 0.8:1 (or 0.6:1) (DL/5146-2001). A Type-WJ200 mud mixer pulp (with capacity of 400L) is used for manufacturing slurry and a Type-100B/15 grouting pump is used for grouting. Cement slurry is stirred by a high speed mixer for more than one minute. As a result of test, the corresponding slurry gravity ratios of four classes are 1.286, 1.500, 1.589, 1.715 (or 1.800).

4.4.2 Grouting methods

Pure pressure grouting is selected as consolidation grouting method. If the bedrock length of grouting hole is less than 6 m, the full hole grouting method is used, otherwise, the bottom-up grouting method or orifice closed grouting method will be applied. The bottom circulation model is commonly used in caves while the distance of slurry pipes is kept by 30~50 cm and the length of grouting segments bases on drilling and pressurizing water test. The segment length of contact zone under bedrock is less than 4 m, generally 4~5 m, not longer than 6 m. Some individual drain holes are grouted via a low pressure, dense slurry, intermittent and repeated perfusion method for three times at least. The pressure should be controlled strictly during the whole process.

4.4.3 Grouting pressure and slurry transformation

Grouting pressure is determined by experiments, the grouting set 0.35 MPa bases on experience. Cement ratio: 3:1, 2:1, 1:1, 0.8:1 (or 0.6:1); slurry conversion: Cement ratio must not be changed when grouting pressure remains steady but the injection rate decreases constantly or injection rate is steady but pressure increases. Otherwise, it should be improved to a higher level if grouting pressure and injection rate don't get notable changes. And improvement strides levels is also acceptable according to situation when injection rate surpass 30 L/min.

4.4.4 Ceasing grouting and sealing holes

When injection rate is not higher than 0.4 L/min under predetermined pressure, grouting should last for 30 min. Grouting holes are sealed by the "pressure grouting sealing method".

4.4.5 Handles in special situation

(1) Overflow or leakage in grouting process: caulking, sealing surface, reducing pressure, thickening slurry, limiting flow, intermittent grouting can be executed for remedy.
(2) Orifice gushing: segment grouting from top to bottom with short segment length, high grouting pressure and thick slurry. Shielding and closing pulp, pure pressure grouting and quick-solidified grouting are executed until coagulation. Then pressure grouting sealing.
(3) Huge injection on grouting segments and difficulties in ceasing grouting: reducing pressure, thickening slurry, limiting flow, intermittent grouting can be executed. Accelerator may be mixed into stable slurry or compound one.
(4) Grouting caves: in caves without fillings, high flowing concrete and break stone should be put into caves depends on cave sizes, then injected with cement mortar and compound slurry until coagulation. In caves with fillings, high pressure grouting or jet grouting can be applied according to type, property and degree of fillings. At last, cement slurry is injected after caves have been cleaned.

The drilling and grouting in each sequence follows its previous one after 24 h. The succeed sequence is treated as the examination for its prior one.

Table 2. Comparison of wave velocity values before and after pile grouting.

No.	Pile No.	Average velocity before treatment (m/s)	Average velocity after treatment (m/s)
1	4#	1702	3982
2	5#	2011	3856
3	9#	2051	4345
4	11#	1836	4200
5	12#	1655	4336
6	19#	1786	3901
7	22#	2028	3860
8	25#	2100	3835
9	32#	2400	3952
10	37#	1785	4108
11	38#	1880	4273
12	48#	1899	3990
13	51#	1759	3859
14	52#	1990	3862
15	55#	2210	4007
16	56#	2108	4192
17	57#	2400	3992

4.5 Grouting efficiency analyses

4.5.1 Comparison of permeability rate and unit ash volume of injection

According to "Permeability rate curve and cumulative frequency curve in all sequences" and "ash volume of injection curve and cumulative frequency curve in all sequences" base on result materials, the grouting process is normal and efficient. The average permeability rate (permeability rate of checking holes is less than 10 Lu) and the average ash volume of injection in all sequences decrease progressively. The relationship of the average permeability rate in two sequences is: 45.98 I > 14.3 II (Lu); and the relationship of average ash volume of injection is: 981 I > 523 II (kg/m). It can be concluded that permeability rate and ash volume of injection decrease successively during the whole consolidation grouting process, especially between the two operation sequences.

It is showed that with the increase of the filling sequences, grouting efficiency becomes more and more obvious.

4.5.2 Rock compression wave changes

According to comparison of wave velocity values measured before and after grouting, the change of longitudinal wave velocity is evident from 1655 m/s~2400 m/s before grouting to 3835 m/s~4345 m/s after that. It means grouting is efficient. Comparison of wave velocity values before and after pile grouting is shown in the Table 2.

5 GROUTING QUALITY CHECK

5.1 Water pressure check

The permeable rate ranges from 45.98 Lu before grouting to 3.1~5.0 Lu after that via water pressure

Table 3. Water pressure test results of ground check holes consolidation grouting.

Hole No.	Design Requirements	Test results Water permeability $q = Q/PL$
4#	Water permeability of not more than 10 lu	4.7 lu
37#	Water permeability of not more than 10 lu	3.5 lu
5#	Water permeability of not more than 10 lu	4.2 lu
38#	Water permeability of not more than 10 lu	4.1 lu
9#	Water permeability of not more than 10 lu	4.5 lu
48#	Water permeability of not more than 10 lu	4.4 lu
11#	Water permeability of not more than 10 lu	5.0 lu
51#	Water permeability of not more than 10 lu	4.9 lu
12#	Water permeability of not more than 10 lu	3.6 lu
52#	Water permeability of not more than 10 lu	4.1 lu
19#	Water permeability of not more than 10 lu	4.4 lu
55#	Water permeability of not more than 10 lu	4.8 lu
22#	Water permeability of not more than 10 lu	3.1 lu
56#	Water permeability of not more than 10 lu	4.5 lu
25#	Water permeability of not more than 10 lu	4.4 lu
57#	Water permeability of not more than 10 lu	3.6 lu
32#	Water permeability of not more than 10 lu	5.0 lu

test through check holes. This meets requirements. Water pressure test results of ground check holes consolidation grouting is shown in the Table 3.

5.2 *Sonic wave check*

The average wave velocity rate ranges from 1655 m/s∼2400 m/s before treatment to 3835 m/s∼4345 m/s after that. It meets requirements.

5.3 *Quality check of borehole core*

Fillings are mainly clay and mud with stone before grouting and are cement stones, rocks and Cementation of cement and rocks with little mud 14 days after that. Dangerous geological phenomena such as caves, crevices, mud and stones are not found. In general, cement stones are abundant and cores are continual and integrated. Grouting replacement and filling are effective. Cement stones uniaxial pressure test shows its strength value surpasses 19 MPa and its capacity reaches 4700 KPa with reduction factor 0.3. This meets the design requirements.

6 CONCLUSION

It can be concluded from water pressure, sonic wave and borehole core tests that it is effective to apply consolidation grouting strategy to foundation treatments:

a. The weathered soft mud in caves and rock fissures in defective parts of pile bearing stratum is washed out of holes or grouting scope by high-pressure water.
b. The slurry injected into caves and fissures consolidates with caves and fissures and reinforces ground foundation.
c. Pure cement slurry is used as grouting slurry so that the uniaxial compression strength value of cement stone exceeds over design value.

In conclusion, Consolidation Grouting Karst Foundation Treatment Strategy treats existing buildings without complex platforms or huge devices so that it can be applied with just common materials in short period and is more efficient than other methods. Now, the project has been accomplished and come into service. Adverse phenomena such foundation sinking and walls cracking do not occur.

In a word, its foundation treatment performs well and is worth extending and applying.

REFERENCES

DL/5146-2001, cement grouting technology of hydraulic structures specification [S].
GB50007-2011, building foundation design specifications [S].
JGJ123-2000, the existing building foundation reinforcement of the technical specifications [S].
JGJ79-2002, building foundations processing technology specification [S].
JGJ72-2004, high-rise buildings geotechnical investigation procedures [S].
Yan Ping Tan, Caixia. Grouting in foundation treatment karst application [J]. Buildings, 2012 (13).

# Constructing the evaluation system of urban planning implement under the guideline of new urbanization—taking Tongchuan as an example

Huan Yang & Pei Zhang
Xi'an University of Architecture and Technology, Xi'an, China

ABSTRACT: The essence of the new urbanization is "human urbanization", giving full play to the function of urban and rural planning as an important public policy, putting forward higher requirements for the urban and rural planning formulation, covering a wide range of urban and rural social, economic, political, ecological, and cultural aspects. So new urbanization has a significant impact on the implementing evaluation system which focuses on the actual performance of the urban-rural planning, providing an optional path to perfect the assessment levels, rich its content, optimize the evaluation methods and enhance the quality of assessment.

1 GENERAL INTRODUCTION

In 2008, China promulgated the new *"Urban and Rural Planning Act"* as the highest legal document in the field of urban and rural planning and management, made a clear focus on the new era of China's urban and rural planning, strengthening the important public policy functions of urban and rural planning in guiding and regulating the development of urban and rural construction. The *"Urban and Rural Planning Act"* clearly put forward that the formulation authorities of provincial urban system planning, urban master planning and town master planning should organize relevant departments and experts to regularly evaluate the implementation of the plan and solicit the public opinion through hearings, discussion meetings or other means. The Commentary of the People's Republic of China on urban and rural planning et al (2008) argued that the formulation authorities should submit assessment reports and attach the case to seek opinions of the standing committees of the people's congresses at the same level, town people's congress and original examining and approving authorities[1], giving a great importance to the assessment of urban and rural planning implementation, forming a complete system of urban and rural planning based on the "preparation – approval – implementation – evaluation – revision" process, and providing a valid path to make urban and rural planning rational, scientific and sustainable.

2 THE RESEARCH REVIEW

There is a complete theoretical framework to evaluate the implementation of urban and rural planning, composed of the object, the scope, the content, the methods of assessment and so on. The types of planning evaluation range from the urban system planning, urban master planning, town planning, township or village planning to the zoning plan, controlling detailed planning. The transition from the single value oriented assessment to a diversified changes the past one which focused on assessing land size, population size, spatial structure, etc. forming an evaluation system concerning with welfare, social justice, ecological environment, the protection of natural and cultural resources, strengthening the public policy function of urban and rural planning. WeiMengkun et al (2010) argued that overall planning implementation evaluation not only deal with the degree that setting goals achieved, but also evaluate the relevant policy environment[2]. Wang Ming lee et al (2007) taking Yuyao, Zhejiang as an example, pointed out that we should evaluate planning from the three aspects of planning target, space organization and layout, the public satisfaction, and establish the system of evaluation index and its weights of the overall urban planning implementation[3]. Tian Li et al (2008) argued that the implementation evaluation mainly limited in judging the difference between the practical uses and planning[4]. Levels of planning and implementation evaluation system and urban and rural planning system being in line, are divided into urban system planning and implementation assessment, overall urban planning implementation evaluation, town planning and implementation assessment, township planning and implementation assessment, village planning and implementation evaluation. The implementation assessment has interlinkages and mutual influence among each level.

3 CONSTRUCTING EVALUATION SYSTEM

New urbanization profoundly affects the preparation of urban-rural planning, and improves the content of urban-rural planning, so exploring the evaluation system of urban-rural planning under the guideline of the new urbanization is important. From the evaluation content, the new urbanization puts forward a high content to urban planning and implementation assessment, including the assessment of planning and implementation environment, the situation that planning objectives has carried out, the direction of urban development, the performance of urban space layout, ecological and environmental protection, and historical and cultural heritage protection.

3.1 The environmental assessment of planning and implementation

The external environment (the national, regional policies, regulations, relevant planning and other macro background) that affect the planning implementation were analyzed. Implementation environment evaluation most use qualitative methods, indicating whether the external environment is conducive to the plan implementation, or provides a sufficient power for plan implement.

3.2 The assessment of planning goals

In addition to the city nature, the main economic and social indicators, population and land size, the new urbanization puts forward a high requirements for urban development goals, increases the content of basic public services, resources and the environment, changes the past situation which only emphasized on economic development, and ignored social and ecological issues. The implementation evaluation system and urban and rural planning system being in line, are divided into urban system planning and implementation assessment, overall urban planning implementation evaluation, town planning and implementation assessment, township planning and implementation assessment, village planning and implementation evaluation. The implementation assessment has interlinkages and mutual influence among each level.

3.3 The assessment of urban space layout

Assessing the community (commercial housing, affordable housing, low-rent housing), public service facilities, green space, industrial sites, transportation facilities land, municipal infrastructure land, disaster prevention and mitigation facilities has an important effect on improving the performance of urban space.

3.4 The assessment of ecological and environmental protection

It should include the situation of construction implementation in energy conservation, sewage treatment rate, waste rate, water quality, air quality, forest cover, mountain protection.

Figure 1. Tongchuan city planning implementation assessment scope.

4 AN EMPIRICAL STUDY

This study selects Tongchuan which located in Shaanxi Province, the northwest area as empirical research. Locating in the middle of Shaanxi Province, it's in the transition zone between Guanzhong Basin and Northern shaanxi plateau, with a total area of 3882 square kilometers.

4.1 Valuation area

The assessment areas: North to the boundary of new and old 210 State Road Interchange, south to Yaozhou, west to the pavilion reservoir, east to Chakpori, including Tongchuan District, Yaozhou District, Dong Town area, Huang Town area, Chakpori, Yuhuangge reservoirs, with a total area of about 150 square kilometers (Fig. 1).

4.2 The assessment content

The external environmental assessment of planning and implementation: Municipal government attaches great importance to carefully organize the urban planning, provides the organizational security for the master plan implementation to make urban planning and management system become more perfect. Besides enhancing the intensity of public participation, offering more ways for the overall planning and implementation to become more open, just and fair. planning implementation goals assessment: Under the guidance of 2005-planning overall goal. In 2011, the total production of Tongchuan was 23.453 billion yuan, and local revenues reached 1.889 billion yuan. Compared to 2010, the five main indicators of a comprehensive master plan exceeded the objectives and tasks. Actual construction land and land size scale per capita are beyond the control of planning urban space layout assessment. Urban master plan (2005) determined the spatial structure along with the main traffic arteries and distribution of resources, forming "one center, two auxiliary" (Fig. 2), and ultimately Tongchuan City's "a city double heart" led the development of groups of cities and towns. Since 2005, the actually development direction is following the planning and urban space layout. Community building performed better and living conditions are keeping improve. In 2011, Tongchuan reached 10.01 square meters per capita public green, green coverage reached 42.52% which was up to national standards. Green land was 310 hectares, accounting for 6.3% which

Figure 2. The status map of Tongchuan city (2005).

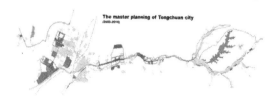

Figure 3. The recent plan of Tongchuan city (2010).

was lower than the national standard (8–15%) and thus the development of green facilities needs to be increased (Fig. 3). The industrial parks in center areas developed by leaps and bounds, planning is well implemented. With large-scale construction, new areas' sewage treatment plants and sanitation facilities had been expanding to meet the basic operation of the city, but in north area the urban sanitation facilities appeared insufficient due to excessive population density.

5 CONCLUSIONS

Overall, Tongchuan City planning (2005) had a strong guidance, forward-looking, operability and played an important role in Tongchuan development. Through assessing, we discovered that the actual development has broken the control of the original plan in aspects of scope, urban location, spatial structure, facility construction. It needs to be shifted from urban nature, industrial development, spatial development, ecological development, integrated transport, etc. to promote Tongchuan faster, better and more rational development.

REFERENCES

[1] Commentary of the People's Republic of China on urban and rural planning. Beijing: Intellectual property rights press, 2008. in Chinese.
[2] Mengkun Weg. Explore the contents of the overall urban planning implementation assessment development mode. Urban development research, Vol. 17, No. 4 (2010). in Chinese.
[3] Yuming Li. The overall urban planning implementation evaluation research. Hangzhuo: Zhejiang university press, 2007. in Chinese.
[4] Li Tian. Overall urban planning implementation evaluation theory and empirical research—the master planning of Guangzhuo (2001–2010). Journal of urban planning, No. 5 (2008): 90–96. in Chinese.

Construction situation and prospect of high-speed rail earthquake monitoring and early warning system in China

Jiang Yang, Jun Huang & Zhigao Chen
Hubei Province Key Laboratory of Earthquake Early Warning, Institute of Seismology, CEA, Wuhan, China

ABSTRACT: Earthquake disasters occurred frequently in recent years. With the great-leap-forward development of high-speed rail construction in China, the government pays more attention to the seismic safety of the high-speed rail, and has successively established high-speed rail earthquake monitoring and early warning systems in Beijing-Tianjin high-speed rail line, Beijing-Shanghai high-speed rail line and Harbin-Dalian high-speed rail line. This paper introduces the construction situation of these systems in several aspects such as observation point layout, equipment applicability and the level of monitoring and early warning. The deficiencies of existing system are analyzed and the prospect is proposed according to present situation.

1 GENERAL INSTRUCTIONS

Although earthquake is an occasional event, it is a great natural disaster for the safety of the railway. Many countries such as Japan, Taiwan, France, Germany and South Korea have established earthquake emergency control systems to mitigate the hazards of railway in earthquakes (Sun et al., 2007; Sun et al., 2011; Lu et al., 2010). China is an earthquake-prone country, thus full consideration should be given to operation risk of the railway designs in earthquake zones when promoting large-scale railway construction. The first earthquake monitoring and warning system has been established in Beijing-Tianjin high-speed rail line in 2011.

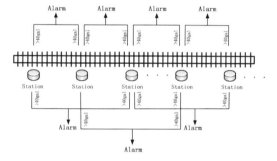

Figure 1. Principal Diagram of High-speed rail earthquake monitoring and early warning system.

2 CONSTRUCTION STATUS OF HIGH-SPEED RAIL EARTHQUAKE MONITORING AND WARNING SYSTEM IN CHINA

2.1 Earthquake monitoring station layout scheme

At present, earthquake monitoring and early warning systems have been built for Beijing-Tianjin high-speed rail line and Beijing-Shanghai high-speed rail line. The monitoring stations are designed along the railway in seismic zone within a interval of about 20∼30 km. When two of three adjacent stations monitor acceleration values that are greater than 40 gal, the alarm will be triggered. The schematic earthquake alarm diagram is shown in Fig. 1 (Yang Jiang, 2011). The alarm system is easy to operate and false alarm is rare.

2.2 Earthquake monitoring station construction scheme

Earthquake monitoring stations are designed in the railway substation. Each monitoring station consists of two earthquake monitoring points and their distance is greater than 40 m. Both of the two monitoring points need to be excavated to the original soil layer and made waterproofing to get the real seismic signals. An acceleration sensor is installed on the corresponding instrument pier. Two monitoring sensors transmit data via a cable to the earthquake disaster prevention cabinet in the equipment room. When the two monitoring points detect acceleration values that are greater than 40 gal, the seismic monitoring station reaches the alarm threshold. The seismic monitoring point layout diagram is shown in Fig. 2.

3 PROBLEMS OF HIGH-SPEED RAIL EARTHQUAKE MONITORING AND EARLY WARNING IN CHINA

Although several high-speed rail earthquake monitoring and warning systems have been established in China, they have not been tested in real earthquakes. There exist some shortcomings compared with some

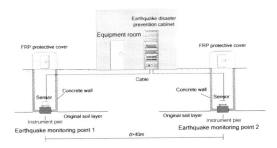

Figure 2. Earthquake monitoring point layout diagram [4].

other countries that have studies and applications earlier than China. The shortcomings mainly reflect in the following aspects.

3.1 Earthquake monitoring stations layout needs optimization

In order to achieve a better observation effect, the sites selections of the national seismic network are very strict. It is set that observation points should be laid in free field away from surrounding buildings; stations should avoid sites with large changes in local terrain, such as small ridge, valleys and so on. The background noise of station sites should be very low. The highest background noise in a station should be less than 0.0001 g (Stipulation on China Digital Strong Motion Network, 2005). High-speed rail earthquake monitoring stations are built in substations because of maintenance needs. They are generally 50 m away from the railroad, for that vibration caused by the train running and active jamming will influence seismological observation. It will be difficult to satisfy the condition of P wave early warning algorithm because of the higher background noise.

3.2 The applicability of high-speed rail earthquake monitoring equipment should be verified

Existing high-speed rail earthquake monitoring and early warning system equipments such as acceleration sensors and recorders originally were appropriate for national seismic network. However, they are not designed for high-speed rail. The environments of the high-speed rail substations are more complex, environmental factors such as site environmental interferences, strong radiation source and so on, may impact the normal operation of the instruments.

3.3 Alarm parameters and determination methods need further research

At present, the alarm algorithm of high-speed rail earthquake monitoring and warning system in China is outdated; it was used in Japan a few decades ago. It determines an earthquake alert when two stations record acceleration values that exceed a threshold value. Meanwhile, foreign scholars have been focusing on P-wave earthquake early warning technology for decades; it can estimate the original time and location as well as magnitude of the earthquake and then issue an alarm in a short time.

Currently, designs general set about 40 gal as a unified alarm threshold in all stations. But in practice, site conditions and ground motion amplification effects of viaducts are different for different stations. Therefore, a threshold should be set for each station based on the station conditions. The alarm threshold parameters need a thorough study.

In addition, the ground motion threshold alarm method also has a few new progresses, for example, Wu et al. (2005) used the initial three seconds of P-wave to calculate the displacement. They found that the earthquakes are devastating when the maximum displacement Pd > 0.5 cm (the peak values of filtered displacement) by processing data in Taiwan. Pd and τ_c (average period) (Kanamori et al., 2005) are combined to a parameter Pd*τ_c that provides a powerful destructive evaluation factor. The result shows that it is a destructive earthquake when Pd*τ_c > 1. Böse et al. (2009) also used the initial three seconds of P-wave to calculate the displacement Pd and τ_c after a high-pass filtering, and then determined the relationships between Pd, τ_c and magnitude separately and finally found the variation range of Pd and τ_c. When both of Pd and τ_c are within the range, it shows an earthquake is coming. Experiment Results show that the method can greatly reduce false triggers and miss triggers. Nakamura (2008) proposed that the inner product of the acceleration and velocity represent power. They can reflect intensity of the earthquake and they are important parameters to determine whether an alarm should be issued.

3.4 Seismic data along high-speed rail is inadequate

The first practical earthquake monitoring and early warning system was established in Japan in 1965. The first national earthquake warning system was also established by Japan and trial operation was started on August 1, 2004. The systems experienced many earthquakes; it accumulated a wealth of data and experience over the years. The seismic data along the railway is an important guarantee for the development of the Japanese earthquake early warning research.

At present, China had established several high-speed rail earthquake monitoring systems. However, none of them have record any seismic data. It limits the development of high-speed rail earthquake early warning research to some extent.

4 CONCLUSIONS AND PROSPECT

Although there are some shortcomings in the high speed rail earthquake monitoring and early warning system of China, the system's meaning is enormous. More seismic data can be accumulated in the next few years. Now, the first domestic P-wave warning earthquake monitoring equipment has been developed and will be used in the high-speed lines. A lot of manpower and material resources have been invested in the thorough research of the P-wave warning technology, equipment access standards, site selection as well as system verification and so on. It indicates that high-speed rail earthquake monitoring and early warning technology research has entered a rapid development period. Along with the accumulation of seismic data and experience, the technology will get rapid development in the next few years.

REFERENCES

Böse M., Hauksson E. & Solanki K., et al. A new trigger criterion for improved real-time performance of onsite earthquake early warning in southern California [J]. *Bulletin of the Seismological Society of America*, 2009, 99 (2A): 897.

China Earthquake Administration. JSGC-03 Stipulation on China Digital Strong Motion Network [S], Beijing: *Seismological Press*, 2005.

Kanamori H. Real-time seismology and earthquake damage mitigation[J]. *Annu. Rev. Earth Planet. Sci.*, 2005, 33: 195–214.

Lu Ruishan, Song Qilin. & Wang Fuzhang, etc. Railway earthquake disaster warning and emergency response systems and Prospect [J] *Railway Computer Application*, 2010, 19(7): 16–20.

Nakamura Y. First actual P-wave alarm systems and examples of disaster prevention by them, F, 2008 [C]. *International Association for Earthquake Engineering*.

Sun Hanwu, Wang Lan & Dai Xianchun. Automatic high-speed railway earthquake emergency disposal system research. [J]. *China Railway Science*, 2007, 28(5): 121–127.

Sun Li, Zhong Hong & Lin Gao. Speed railway earthquake early warning system status are summarized. [J]. *World Earthquake Engineering*, 2011, 27(3): 89–96.

Wu Y.M. Rapid Assessment of Damage Potential of Earthquakes in Taiwan from the Beginning of P Waves [J]. *Bulletin of the Seismological Society of America*, 2005, 95(3): 1181–1185.

Yang Jiang. The tender documents the beijing-shanghai disaster prevention system [Z]. *Wuhan, Wuhan Institute of Seismologic Instrument*. 2011.

Danger prediction system for dangerous goods transportation

Tongjie Zhang, Yan Cao & Xiangwei Mu
Dalian Maritime University, China

ABSTRACT: A system of danger prediction for dangerous goods transportation is introduced in this paper, in which LNG transportation is taken as an example. The main function of the system is to predict the state of the LNG tank – whether it will be in danger or not – based on the real-time data collected lately. The predicted accurate data of the parameters were compared with the standard data. Least square method was used to predict the accurate values of the parameters of the tank at the next moment. Three main parameters were considered in this study: level, gas concentration and pressure. Other parameters like the speed of vehicles were also considered. In addition, the relationships between the three parameters were fitted to ensure the correctness of the prediction. Regression analysis showed that the result is in accordance with the real states of the LNG tank.

1 INSTRUCTION

There has been increasing amount and varieties of chemical dangerous goods transportation (DGT), which undoubtedly raised the demand of security during the transportation. Among all the means of transporting dangerous goods, railway and road transportations are the most common ones, (Baldini, Gianmarco et al. 2012). Meanwhile, because of the physical and chemical properties dangerous goods have, like being flammable, explosive, toxic, corrosive, and radioactive etc., as well as the high potential risk and accident, guarantying the security of GDT has become a top priority task.

(Tena-Chollet et al. 2013) designed a risk value tool to evaluate hydrocarbon transportation. (Li Rongrong et al. 2013) developed a multi-objective genetic algorithm to decide the optimal transport routes. Martin Fahy & Stephen Tieman (2001) discussed the static and dynamic models of tank container using finite element analysis. Tools for making acute risk decisions with chemical process safety applications were published by Center for Chemical Process Safety (CCPS) in 1995 (Jian Chen 2010). BLEVE experiment and simulation research was applied by S.N. Chen et al. (2008) and P. Cleaver et al. (2007). However, most of the researches above are about the control or prediction before danger appeared, and studies are seldom about the prediction of accurate values of the parameters using least square method and regression analysis.

A system of danger prediction was designed in this study to get the predicted data of the main values of the parameters. Three main parameters were considered in this paper: temperature, gas concentration, pressure, and humidity. Other parameters like the speed of vehicles were also considered. In addition, the relationships between the four parameters were fitted to ensure the correctness of the prediction. Handled data was compared to the standard data in the monitor center. Once the predicted data appears abnormal, an alarm will be sent to the drivers and the monitor center, to help win more time for holding back the danger, or for rescuing.

2 FRAMEWORK OF THE DANGER PREDICTION SYSTEM

Vehicles transporting LNG are equipped with GPS, which can monitor road conditions and speed in real-time. The framework of the danger prediction system is shown in Fig. 1. The two parts of data temperature, pressure and gas concentration, and road conditions and speed are integrated by the GPS. Then the information can be displayed in the car terminal screen, and sent to the monitoring center through the GPRS communication server.

Information is shared with various monitoring terminals via LAN. Collected data is firstly handled to predict the values of the parameters of the next moment with the least square method. Then the handled data is compared with the regular range. When data is out of the regular range, an alarm will be given to the drivers and monitoring center. In addition, some information in the process of DGT can also be shared with web users, including various gas stations, managers, etc.

3 MECHANISM OF THE DANGER PREDICTION SYSTEM

Values of the three main parameters can be obtained directly from the sensors, the temperature sensor, the

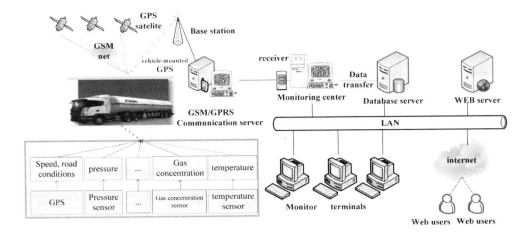

Figure 1. Framework of the danger prediction system.

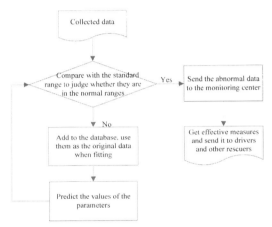

Figure 2. Flow chart of the dangerous prediction system.

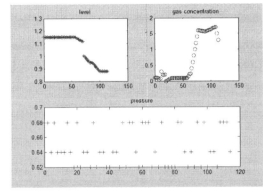

Figure 3. Level, gas concentration and pressure.

gas concentration sensor, and the pressure sensor. Speed of the tank can be obtained by the GPS equipped on the vehicles in real-time. The flow chart is as shown in Fig. 2. As the transportation begins, the collected data is firstly compared with the standard safe range to check whether the tank is in a safe state or not. The standard level of LNG in this study was 1.15 v, the standard pressure was lower than 0.3 MPa, and the standard gas concentration was 0.1 v. As the quantity of the data increased, the trend of the data was fitted with the least square method. At the same time, the relationships between the parameters were be fitted.

Combining the fitted line of each parameter and the relationships between the parameters, the value of each parameter was predicted. By comparing the values and the standard ranges, judge whether the values change in an acceptable normal ranges. If any of the values are abnormal, send alarms to the monitoring center. In this way, in the monitoring center, two groups of data are achieved each time, the real-time one and the predicated one. The difference of the two groups of data can be stored again to the database, which can be used to revise the fitted regression equation, finally reduce the error made in the former predictions. Then, analysis the value changes of each parameter synthetically and work out what triggers the changes. Proper measures are made accordingly, and sent to the driver or other rescuers to take actions. Finally, prevent the situations getting worse.

4 AN EXAMPLE OF DANGER PREDICTION

An example is used here to show the mechanism of the danger prediction system in detail. The variation curves of the three parameters are shown in Fig. 3: level, gas concentration and pressure of LNG. In Fig. 3, the x-axis stands for samples, and y-axis stands for voltage of the sensors. Suppose a group of new data is collected by the time of 100, and the values of the three parameters – level, gas concentration, and pressure – are (1.15, 1.5, 0.68).

In order to get the relationship between level, gas concentration and pressure, presume that x1 as time,

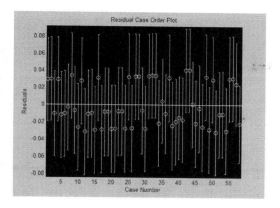

Figure 4. Residual analysis variation curve.

x2 as level, x3 as gas concentration and y as the pressure. Based on the data collected before, and the analysis of multivariate regression, the regression model is as follows:

$$\hat{y} = 0.6960 - 0.0001x_1 - 0.0387x_2 - 0.0026x_3. \quad (1)$$

The result of residual analysis is shown in Fig. 4. The exact values of the three parameters can be calculated by the regression model. Comparing the predicted values and the standard changing range, it can be seen that there is a big difference between the original value of level and the present one. And the gas concentration also accumulated in a short time. From the level of the liquid, gas concentration and pressure, it can be concluded that a leakage is going on in the tank.

5 CONCLUSIONS

A system of danger prediction for LNG transportation was designed in this study. The least square method was used to obtain the predicted data. With the predicted data, the state of danger is predicted so that prevention measures can be taken much earlier compared with the method – which judges the gas tank state by data getting in real-time. In addition, the relationships between the three parameters are also fitted to ensure the correctness of the prediction. Regression analysis showed that the result is in accordance with the real states of the LNG tank.

ACKNOWLEDGEMENT

This work is supported by Program for Liaoning Innovative Research Team in University (project code: LT2011007). We express our deep gratitude for this.

REFERENCES

Baldini, Gianmarco et al (2012). Legislative context and governance principles for dangerous goods transportation (DGT) integrated risk management. *Transport of dangerous goods: methods and tools for reducing the risks of accidents and terrorist attack*: 45–74.

Jian Chen (2010), Third International Workshop on Advanced Computational Intelligence Suzhou, Jiangsu, China: 25–29.

Li Rongrong, Leung Yee, Huang Bo (2013). A genetic algorithm for multi-objective dangerous goods route planning, *International Journal of Geographical Information Science*. Volume 27: 1073–1089.

M. Fahy, S. Tieman (2001), Finite Element Analysis of ISO Tank Containers, *Journal of Materials Processing Technology*, vol 119: 293–298.

P. Cleaver, M. Johnson, B. Ho (2007), A Summary of Some Experimental Data on LNG Safety. *Journal of Hazardous Materials*, vol 140: 429–438.

S. N. Chen, J. H. Sun and W. Wan (2008), Boiling Liquid Expanding Vapor Explosion: Experimental Research in the Evolution of the Two-phase Flow and Over-pressure, *Journal of Hazardous Materials*, vol 156: 530–537.

Tena-Chollet F. (2013) Development of a spatial risk assessment tool for the transportation of hydrocarbons: Methodology and implementation in a geographical information system. *Environmental Modelling & Software*, Volume 46: 61–74.

Deformation characteristics of a typical high rockfill dam built on a thick overburden foundation

Z.J. Zhou, J.K. Chen, W.Q. Wang, Y.L. Li & Z.Y. Wu
State Key Laboratory of Hydraulic and Mountain River Engineering, School of Water Resources and Hydropower, Sichuan University, Chengdu, China

ABSTRACT: A high rockfill dam was selected as an example, and its deformation characteristics were analyzed by finite element simulation method (FEM). The maximum deformation and deformation compatibility of the dam were analyzed based on the calculation results by FEM. In order to verify the reasonability of the calculations, the results derived from simulations and measurements were contrast with the calculation results. The conclusions of this paper can be used as an effective case to study the deformation behaviors of high rockfill dams with thick overburden foundations, or even improve the theory of the calculation model.

1 INTRODUCTION

As one of the superior dam types, an increasing number of high rockfill dams with core walls have been or will be built on thick overburden foundations (Wang, et al., 2005), including *Pubugou, Xiao langdi, Shiziping, Bikou, Nuozadu, Shuangjiangkou*, etc. There are many engineering problems that should be resolved if a high rockfill dam is determined to be appropriate on a thick overburden foundation (Wang, et al., 2001). Among these problems, the maximum deformation and deformation compatibility of the dam are some of most critical issues (Bi, et al., 2013), which deserve much attention.

Under comprehensive influences of many uncertain factors, the mechanism of dam deformation often seems to be complicated (Zhang, et al., 2003). At present, the two main methodologies adopted to study the deformation characteristics of rockfill dams are numerical simulations and in-situ measurements (Szostak-Chrzanowski, et al., 2008). Compared to the latter method, the former is more effective to forecast the stress-deformation characteristics of dams based on the geotechnical parameters obtained from laboratory tests, and is also a helpful tool for the data analysis of in-situ measurements (Yin, et al., 2009).

In this paper, the deformation characteristics of a typical high center core rockfill dam built on a thick overburden foundation was analyzed by 3-D finite element numerical method, and the numerical calculation results were compared to the in-situ measurement results.

2 GETTING STARTED

The dam discussed in this paper, shown in Figure 1, is situated on the Dadu River. The construction of the dam began in 2004 and was completed in 2009, and the filling of the reservoir was carried out in several sequential steps over the period of December 2009–October 2011. The dam is a high gravel core-wall rockfill dam with an altitude at the crest of 856 m and lowest altitude at the foundation basement of 670 m. The crest width of the dam is 14 m; while its axial length is 573 m. The dam is located on a thick overburden foundation with maximum overburden depth of 70–80 m.

The maximum cross-section of the dam and water levels of the reservoir fillings are displayed in Figure 2. The upstream dam slope is 1:2 and 1:2.25, while

Figure 1. A general view of the dam.

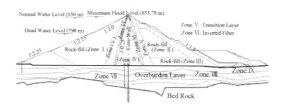

Figure 2. Maximum cross-section of the dam.

the downstream dam slope is 1:1.8. The materials of the dam and its foundation stratum can be generally divided into nine zones.

3 BRIEF INTRODUCTION OF THE CALCULATION THEORY

Constitutive models are usually applied to simulate the stress-deformation characteristics of rockfill dams, which include two categories: elastic-plastic constitutive models and non-linear elastic constitutive models. Compared to the elastic-plastic constitutive models, Duncan-Chang hyperbola models are one of most famous non-linear elastic constitutive models and extensively used to simulate the stress-deformation characteristics of rockfill dams, for the reasons that the parameters of the models can be easily obtained by laboratory tests and a large amount of engineering experience can be made reference to. The relevance theory of Duncan-Chang models is introduced briefly as follows.

The Duncan-Chang model is derived from the equation of the hyperbolic curve based on laboratory tests, shown as Equation 1:

$$\sigma_1 - \sigma_3 = \frac{\varepsilon_a}{a + b\varepsilon_a} \quad (1)$$

By taking a derivative of $\sigma_1 - \sigma_3$ with respect to ε_a, the tangent modulus of the above hyperbolic curve in the coordinate system of $\varepsilon_1 \sim (\sigma_1 - \sigma_3)$ can be deduced as Equation 2, in which SL is the stress level and can be calculated by the Mohr-Coulomb failure criterion as Equation 3.

$$E_t = K \cdot Pa \cdot (\frac{\sigma_3}{Pa})^n (1 - R_f \cdot SL)^2 \quad (2)$$

$$SL = \frac{(\sigma_1 - \sigma_3) \cdot (1 - \sin\varphi)}{2c\cos\varphi + 2\sigma_3 \sin\varphi} \quad (3)$$

In addition, Duncan-Chang models also use the data fit based on laboratory tests to obtain the bulk modulus (B) or Poisson (v). For example, the bulk modulus can be derived from Equation 4:

$$B = K_b P_a (\frac{\sigma_3}{P_a})^m \quad (4)$$

In order to reflect the irreversible deformation of soil, the modulus in unloading, calculated by Equation 5, is adopted.

$$E_{ur} = K_{ur} P_a (\frac{\sigma_3}{P_a})^n \quad (5)$$

In the above equations, $K n c \varphi R_f K_b$ and m are material parameters.

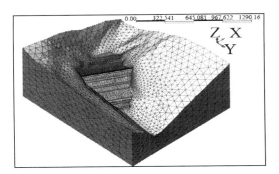

Figure 3. 3-D finite element model of the dam.

Table 1. Material parameters adopted in calculation.

Parameters	I–III	IV	V	VI	VII	VIII	XI
γ (KN/m³)	23	22	21	18	22	16	20
c (KPa)	0	40	0	0	15	11	13
Φ (°)	43	30	40	33	36	42.5	33.5
K	1000	400	800	700	780	450	750
n	0.52	0.7	0.52	0.54	0.42	0.43	0.41
R_f	0.68	0.76	0.68	0.65	0.64	0.72	0.67
K_{ur}	1800	700	1600	1500	1500	900	1300
K_b	400	400	400	400	400	350	370
m	0.12	0.18	0.18	0.18	0.18	0.18	0.18

4 THREE DIMENSIONAL MODEL OF THE DAM

4.1 The simulation range

The maximum height of the dam studied in this paper is 186 m, so with reference to this height, the range of simulation calculation can be determined appropriately as follows: along the dam axial direction, it extends 400 m towards the mountain of the left bank and 400 m towards the mountain of the right bank; along the river, it extends 400 m towards upstream, and 560 m towards downstream; along the vertical direction, it extends 400 m towards the dam foundation. The whole model, shown in Fig. 3, incorporates 19,687 grid nodes and 98,207 elements.

4.2 Preparing the new file with the correct template

In the numerical simulation calculation of this paper, the Duncan-Chang E-B model was selected to simulate the mechanical properties of the materials in Zones I–XI. Meanwhile, a linear elastic model was selected to simulate the mechanical properties of concrete and bedrock. The Young modulus of concrete is 30 GPa compared to bedrock's 27 GPa; the Poisson Ratio for concrete is 0.167 and bedrock is 0.205. The other material parameters are listed in Table 1.

Table 2. Maximum deformation in different zones.

Stage	Directions	corewall	Rockfill	Overburden
Construction	X (cm)	65	65.9	45.1
	Y (cm)	−51.6	−55.3	/
	Z (cm)	287.5	272.5	213.8
Water filling	X (cm)	83.9	84.1	46.6
	Y (cm)	−61.9	−64.0	/
	Z (cm)	295.3	282.1	214.5

5 RESULTS ANALYSIS

5.1 Maximum deformation of the dam

The processes of construction and reservoir water filling were simulated in this paper. In these two stages, the maximum settlement deformation is located in the area of the core-wall at half dam height, and the maximum deformation value in each direction of the dam is listed in Table 2. As shown in Table 2, in the construction stage, the maximum settlement value of the dam was 287.5 cm, which is approximately 1.55% of the maximum height. Compared to the construction stage, under the action of reservoir water pressure and the buoyancy force, the maximum settlement value of the dam in the reservior water filling stage increased slightly to a total 295.3 cm. The maximum settlement values of the dam in these two stages are both greater than 1.0%. It is worth mentioning that the maximum settlement is generally required to be less than 1.0% than the dam height. However, the view has been verified by engineering practices that the maximum settlement for dams with heights over 100 m with thick overburden foundations can be greater than 1.0% of the dam height (Ding et al. 2013).

The greatest horizontal deformation is in Zones I–III and V–VI. In the construction stage, the maximum deformation in X direction is 65.9 cm towards downstream; the maximum deformation in Y direction is 55.3 cm towards the right bank. In the water reservoir filling stage, under the action of water pressure, the maximum in X direction is 84.1 cm towards downstream and the value increases obviously compared to the construction stage. The maximum deformation in Y direction is 64.0 cm towards the right bank, which increases slightly from the value during construction.

The maximum settlement deformation values of overburden in the stages of construction and water reservoir filling are 213.8 cm and 214.5 cm, respectively, and its maximum X horizontal deformation values in these two stages are 45.1 cm and 46.6 cm towards downstream, respectively. Overall, the maximum deformation of the construction stage is larger than that of the water filling stage.

5.2 Settlement deformation compatibility

Deformation compatibility can directly reflect the reasonability of the fill soil sub-area design and the construction quality. Also, it can be taken as an effective index to evaluate the operational properties of the dam. As a matter of convenience, a typical cross-section was selected as an example to illustrate. Figure 4 is the contour map of the settlement deformation in the construction stage and reservoir water stage. Distributions of the settlement deformation in the whole section are continuous and deformation transitions of different zones are gradual with a relatively flat tendency. So, the settlement deformation of the dam in varied zones is harmonious, which is owing to the reasonable design of filling zones and good construction quality.

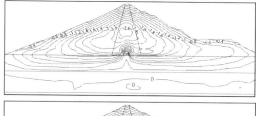
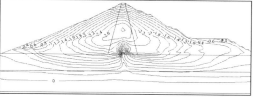

Figure 4. Contour maps of settlement deformation (Top is construction stage; Bottom is water filling stage).

5.3 Contrasts between calculation results and measurement results

In this part, the zones of core-wall and maximum section were chosen as the examples to compare calculation results with measurement results.

(1) *Deformation contrasts of construction stage*

After the completion of construction, the settlement distribution in the core-wall zone derived from numerical calculation was almost the same as the measurement results. The largest settlement was at half of the dam height, and the largest deformation calculation was 274.3 cm, which was larger by 91.4 cm than the measured 182.9 cm. Because the settlement before burying monitoring instruments is not taken into account, finite element calculation considers the whole filling procedure, so the calculation results are larger than the measurement results.

The longitudinal horizontal displacement in the core zone along the axial dam direction versus the altitude is shown in Figure 5. The displacement of the core centerline is inclined to the upstream; deformation rule calculated is consistent with the original observation. In finite element calculation, the maximum horizontal displacement along the river was 6.7 cm, and the measurement value was 6.0 cm, and both were towards upstream. Figure 6 also exhibits

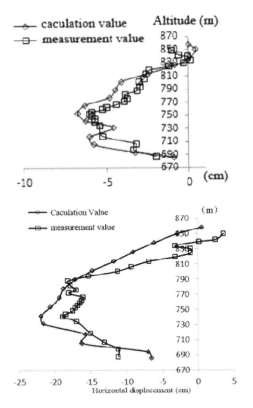

Figure 5. Horizontal displacement distribution in the core wall zone (the top is towards upstream and bottom is along the axial direction of the dam).

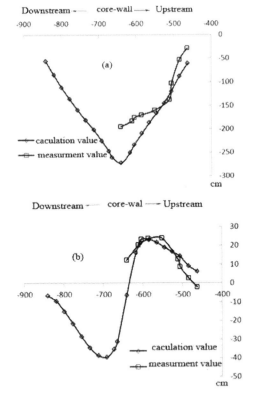

Figure 6. Settlement distribution of the maximum-section at 1/2 dam height (a); longitudinal horizontal displacement distribution of the maximum-section at 1/2 dam height (b).

the distribution of the axial horizontal displacement in the core-wall zone versus altitude. The calculation results are consistent with the measurement results.

Figure 6 is the distribution of the settlement of the typical dam sections at half dam height. Figure 6(a) shows that the settlement values calculated and measured in a typical cross-section at the halfway mark of the dam height are consistent. The maximum of the calculation settlement value was 272.5 cm, and the maximum measured was 182.2 cm, accounting for 1.47% and 0.98% of the dam height, respectively. The cause of the differential settlement is the same as what caused the settlement difference of the core-wall. Figure 6(b) shows the distribution of the longitudinal horizontal displacement of the horizontal line of the maximum section at half of the dam height. The figure shows that the horizontal displacements calculated and observed along the river present the similar law.

(1) *Deformation contrasts of water filling stage*

The deformation contrasts between the calculation results and the measurement results in the water filling stage are presented. During the process of the reservoir water filling, the settlement deformation of the core-wall decreased slightly under the action of buoyancy,

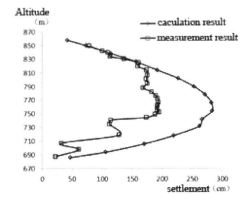

Figure 7. Distribution law of settlement deformation in the zone of core wall.

but because of the increased water pressure applied on the core-wall, the settlement deformation exhibits a slight increment compared to the construction stage. Figure 7 shows the distribution law of settlement deformation in the core-wall zone. Although parts of the calculation results are larger than the measurement results, the law of settlement deformation changes with altitude keeps a good accordance with each other. The

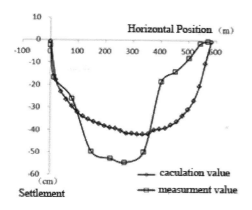

Figure 8. The settlement displacements of dam crest when the reservoir water filling is finished.

parts of the calculation results larger than the measurements may be caused by the inconformity between the installation time of measurement instruments and designed construction time.

After the reservoir was filled with water, the settlement deformation of the dam crest increased slightly compared to the construction stage. The settlement displacements of the dam crest are shown in Figure 8. The settlement displacement caused by reservoir water filling was 54.7 cm, which was much more than the measurement value of 41.90 cm. The selected Duncan-Chang EB mode cannot calculate the part of settlement deformation led by the wetting effect. This may be the main reason that the calculation values are less than the measurement values.

6 CONCLUSIONS

Finite element numerical simulation method was adopted in this paper to analyze the deformation characteristics of a typical dam with a deep overburden foundation.

The maximum deformation of the dam was larger than conventional dams located on foundations with high compressibility, and the process of reservoir water filling caused sight settlement deformation, and obvious horizontal deformation along the river. The deformation in different zones of the dam primarily showed good matching, so it can be proved that the filled soil sub-area design of the dam is reasonable and the quality of the construction is good.

Although some parts of calculation results were larger or less than the measurement results caused by some objective factors, the deformation law obtained by simulation kept good accordance with the measurement results.

ACKNOWLEDGEMENT

This research was substantially supported by the Natural Science Foundation of China (Grants No. 51109151).

REFERENCES

Bi, Q.T., Ma, S.J., Li, H., Jia, J.C. 2013. Stress Deformation Characteristics of Asphalt Concrete Core of High Dam with Thick Cover. Journal of North China Institute of Water Conservancy and Hydroelectric Power, 2013, 34(5):14–16.

Ding, Y.H., Yuan, H., Zhang, B.Y., Yu, Y.Z. 2013 Stress-deformation characteristics of super-high central core rock-fill dams. Journal of hydroelectric engineering, 32(4): 153–158.

Szostak-Chrzanowski, A., Chrzanowski, A., Massiéra, M. 2008. Study Of A Long-Term Behavior Of Large Earth Dam Combining Monitoring And Finite Element Analysis Results, Symposium on Geodesy for Geotechnical and Structural Engineering, LNEC Lisbon.

Wang, B.L., Liu, Y.Z., Wu, H.H. 2005. New development of china earth and stone dam project construction. hydroelectric power, 31(1):63–65.

Wang, Y.S., Deng, X., Luo, Y.H., etc. 2011. A study on the characteristics of deep overburden in Qizong segment of Jinsha River and its engineering effects Hydrogeology & Engineering Geology, 38(1):40–45, 64.

Yin, Z.Z. 2009. Stress and deformation of high earth and rock-fill dams. Chinese Journal of Geotechnical Engineering, 31(1):1–14.

Zhang, B.Y., Yu, Y.Z., Zhuang, J.M. 2003. Key technical problems of high embankment dams Proceedings of the 9th conference on soil mechanics and geotechnical engineering. Beijing: Tsinghua University Press, 163–186.

# Deformation monitoring and regression analysis of surrounding rock during the construction of Zhegu Mountain Tunnel

H.S. Zhang & S.G. Chen

Key Laboratory of Transportation Tunnel Engineering, Ministry of Education, School of Civil Engineering, Southwest Jiaotong University, Chengdu, China

ABSTRACT: New Austrian Tunneling Method (NATM) is a design and construction method used for dynamic monitoring of a tunnel. It is widely adopted internationally at present. Monitoring measurement plays an important role in the use of this method. By monitoring horizontal convergence and vault settlement in the construction of Zhugu Mountain Tunnel in Wenma expressway, the deformation of the surrounding rock of the tunnel was studied. Meanwhile, by using regression analysis for fitting, the ultimate deformation of surrounding rock was estimated to infer the reasonable time of secondary lining. The purpose was to prevent the occurrence of major accidents in the construction and ensure the safety and quality of the tunnel construction. In addition, this study can be taken as a reference for the construction of similar projects.

Keywords: Zhegu Mountain Tunnel; deformation of surrounding rock; monitoring measurement; regression analysis

1 INTRODUCTION

NATM, which is short for New Austrian Tunneling Method, is a reasonable construction method proposed by Professor L.V. Rabcewice et al. from Austria. Being engaged in the practice of tunnel construction for a long time, they put forward the method from the viewpoint of rock mechanics. It is a new construction method that forms a system with the theory of rock mechanics adopting spray anchor technology, construction testing and so on (Xiaohong Li 2002).

As a dynamic monitoring design and construction method widely used internationally, the method's main advantage lies in the fact that it can keep the original self-supporting capacity of surrounding rock up to the hilt and make full use of the real-time monitoring data to establish a more reasonable support system. New Austrian Tunneling Method requires making a dynamic observation of surrounding rock continuously and systematically throughout the construction process (Shanhong Liu et al. 2008). By the real-time monitoring of surrounding rock during construction, as well as analyzing and assessing the monitoring data, it further modifies and improves the original design, and guides the next phase of construction. Monitoring measurement and analysis of the results are important work to evaluate whether the design and construction is reasonable (Gaofei Dai et al. 2004).

Based on the engineering background of Zhugu Mountain Tunnel of Wenma (Wenchuan to Maerkang) expressway, deformation monitoring of surrounding rock during construction and analysis of the monitoring data were performed in this study. Which can feedback information timely to improve the design, guide the construction and guarantee the safety in construction, and in addition provide a reference for the construction of similar projects in the future.

2 PROJECT OVERVIEW

Zhegu Mountain Tunnel is a large, long and deeply buried tunnel crossing Zhegu Mountain in Wenma expressway; it is one of the key projects of the expressway from Wenchuan to Maerkang. Zhegu Mountain Tunnel is located in the juncture of Li County and Maerkang County, Aba Tibetan and Qiang Autonomous Prefecture in Sichuan Province. The tunnel entrance is in Miyaluo Town of Li County and the tunnel exit is in Wangjiazhai Gully of Maerkang County. The tunnel is a double-line tunnel with separated left and right lines, the distance between the left and right measuring lines is 35 meters. The left hole of the tunnel starts from mileage ZK179+702, ends at mileage ZK188+493, with a total length of 8791 meters. The right hole of the tunnel starts from mileage K179+730, ends at mileage K188+496, with a total length of 8766 meters.

The surrounding rock of Zhegu Mountain is featured by joint development and fragment, and the lithology is mainly composed of metamorphic feildspar-quartz sandstone, silty slate, phyllite, etc. The grade of the surrounding rock is III~V, and its stability is weak. The underground water in the tunnel zone consists of Quaternary pore water and bedrock fissure water. Rock fall, rock collapses and water seepage are likely to occur during construction.

3 MONITORING MEASUREMENT PROGRAMS

According to the *Technical Specifications for Construction of Highway Tunnel*, the project of the tunnel's monitoring and measurement must include observation of geological and supporting conditions, periphery displacement and settlement of vault and ground surface of shallow buried portal section. Among those factors, horizontal convergence and settlement of vault are the integrated embodiment of the dynamic state of the surrounding rock and support effects in tunnel excavation, the results of which can be used to judge the stability of the surrounding rock, the appropriateness of the initial support and the optimum occasion for second-layer liner. Therefore, the authors put emphasis on the measurement of the aforementioned two factors of Zhegu Mountain tunnel. Meanwhile, other parts were taken as reference.

3.1 Observation of geological and supporting conditions

Each time blasting and initial spray were finished, the tunnel face was checked by artificial observation, geological compass and hammering. Then, the geological condition of the surrounding rock, including the types of lithology, the attitude of rocks, crevices, karst caves and underground water and so on, were described and recorded. In addition, the engineering geological and hydrogeological conditions were also obtained to estimate if the grade of the surrounding rocks was in accordance with the design. At the same time, the supporting effect was observed to analyze the reliability level of the supporting structure.

3.2 The measurement of horizontal convergence and vault settlement

The measure point of vault settlement was located on the center line of the tunnel vault. In the section of preset point, the test pile with a hook was buried at the position of the point after the excavation of the tunnel hole with a depth of 30 cm. The layout of the pile is shown in Fig. 1, representing by point A. A steel tape with a suitable length was chosen when measuring. The head of the tape was hanged at point A by the hook. Then, the tape was put vertically, and when the tape stayed stable, the vault settlement was measured using a water level. The value of settlement between two adjacent measurements is expressed by $\Delta h_i = h_{i-1} - h_i$. The total subsidence after the nth measurement is

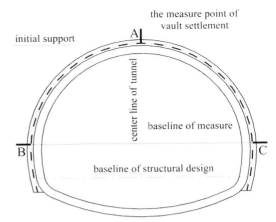

Figure 1. Measure point arrangement diagram of horizontal convergence and vault settlement.

$$h_n = \sum_{i=1}^{n} \Delta h_i \qquad (1)$$

Horizontal convergence of the tunnel means the relative displacement of the line connecting two fixed points which are in relative direction of the tunnel border. It is the most intuitive reflection of the value of the rock deformation caused by excavation of the tunnel. In the section of preset point, the test piles were buried along the sidewalls of the tunnel border with a depth of 30 cm following the blasting excavation of the tunnel. The positions of the piles are denoted by points B and C in Fig. 1. The horizontal convergence gauge of steel ruler type was used to measure horizontal convergence. The value of horizontal convergence between two adjacent measurements is expressed by $\Delta u_i = R_{i-1} - R_i$. Apparently, the value of total convergence after the nth measurement is

$$u_n = \sum_{i=1}^{n} \Delta u_i \qquad (2)$$

The measure points used for the measurement of vault settlement and horizontal convergence were arranged in the same section with the same measure frequency. On the basis of the *Technical Specifications for Construction of Highway Tunnel*, the distance between test sections for measurement of vault settlement and horizontal convergence are as shown in Table 1. Table 2 lists the frequencies.

3.3 Surface subsidence of shallow buried portal section

The settlement observation point was buried in the test range of ground surface of shallow buried section. Precision balance level was adopted to monitor

Table 1. The distance between test sections.

The level of the surrounding rock	III	IV	V
The distance between test sections / m	30~50	10~30	5~10

Table 2. The frequency of measurement.

The speed of the displacement/ (mm/d)	The distance between the section of the measurement and the excavation/m	The frequency of the measurement
>10	(0~1)B	1~2 times/d
10~5	(1~2)B	1 time/d
5~1	(2~5)B	1 time/2d
<1	>5B	1 time/week

Notes: B stands for the width of the tunnel excavation.

Table 3. The allowable relative displacement of tunnel border (%).

	The level of the surrounding rock		
Burial depth/m	III	IV	V
< 50	0.10~0.30	0.15~0.50	0.20~0.80
50~300	0.20~0.50	0.40~1.20	0.60~1.60
> 300	0.40~1.20	0.80~2.00	1.00~3.00

Notes: (1) Relative displacement refers to the ratio of cumulative value of convergence displacement to the distance between two measuring points. (2) The smaller value in the table should be taken for hard rock tunnel, and the larger value for soft rock tunnel.

the absolute sinkage of the point and calculate the settlement of the same day.

4 THE STABILITY CRITERIA OF SURROUNDING ROCK

4.1 Regression function

In the field measuring, if the deformation behavior of strata has obvious aging characteristics, the cumulative amount of deformation will increase with the excavation face going forward and the passage of time, then it will gradually stabilize and the deformation rate will gradually decrease towards zero. The above situation's normal displacement curve is similar to the shape of the hyperbolic function curve and the negative exponential function or logarithmic function curve. Thus, in the analysis of the measured data, the fitting equation of temporal curve should be chosen from those three types of equations, so that the pattern and the process of surrounding rock deformation will well agree with the actual measured results, and the error is minimized (Zhiye Li, et al. 2003).

logarithmic function: $u = a + [b/\lg(1+t)]$ (3)

exponential function: $u = a\, e^{-b/t}$ (4)

hyperbolic function: $u = t/(a+bt)$ (5)

In the above equations, a and b are regression constants, t is the time after initial reading (d), and u is the displacement (mm).

4.2 Velocity of surrounding rock displacement

According to the shape of the displacement-time curve, the criterion of the stability of surrounding rock was evaluated. The deformation curve of surrounding rock can be divided into 3 sections:

(1) Basically stable section, where $d^2u/dt^2 < 0$, deformation velocity continues to drop, meaning the surrounding rock is tending towards stability.
(2) Transition section, where $d^2u/dt^2 \approx 0$, deformation velocity keep constant for a longer period of time, indicating that the surrounding rock tends to be unstable, and a warning and strengthen support lining should be issued timely.
(3) Failure stage, where $d^2u/dt^2 > 0$, deformation velocity gradually increases, indicating that the surrounding rock is in critical condition, and excavation must be stopped immediately and measures be taken to reinforce support lining or surrounding rock rapidly.

4.3 The allowable relative displacement

The measured relative displacement at any point of tunnel border, or the final displacement calculated by regression analysis, should be less than code values. On the basis of *Code for Design of Road Tunnel*, for combined lining tunnel, the allowable relative displacement of initial support should be determined based on the analysis of the geological conditions of surrounding rock when highway tunnels are designed by bearing capacity. If in the absence of data, the allowable relative displacement should be selected according to Table 3.

4.4 Management measures of surrounding rock deformation

According to the ratio of actual displacement value of surrounding rock to the allowable value, appropriate measures should be taken in construction, as shown in Table 4. Under normal circumstances, the amount of reserved deformation will be regarded as allowable displacement in tunnel design. However, the reserved deformation of design should be revised constantly basing on the monitoring results. The reserved deformation is 6 cm when the rock grade is IV, 8 cm when the rock grade is IV strengthened (IV$_j$), 10 cm when

Table 4. Management measures of surrounding rock deformation.

The level of the measures	Deformation	Measures taken
III	$u_n < u_0/3$	can continue normal construction
II	$u_0/3 < u_n < 2u_0/3$	should strengthen the support
I	$u_n > 2u_0/3$	take special measures

Notes: u_n stands for the measured displacement, and u_0 stands for the allowable displacement.

the rock grade is V, and 12 cm when the rock grade is V strengthened (V_j).

4.5 The control of displacement velocity

When the excavation passes the test section, the displacement reaches its maximum velocity, and then gradually decreases. The stability of the surrounding rock can be judged based on the displacement velocity. Japanese design and construction guidelines for New Austrian Tunneling Method pointed out that special support is needed when the displacement velocity is greater than 20 mm/d, or it may result in instability of surrounding rock (Zhiye Li, et al. 2003). Generally, secondary lining is built after the basic stability of surrounding rock.

The construction of the latter support for tunnel using two supports should be carried out when the following three criteria are met at the same time (Qixin Yang et al. 2009).

(1) Horizontal convergence velocity of tunnel is less than 0.2 mm/d, and velocity of vault settlement is less than 0.1 mm/d.
(2) The velocities of horizontal convergence and vault settlement decline apparently.
(3) The relative displacement of surrounding rock has reached 90% of the total relative displacement.

5 ANALYSIS OF THE MONITORING DATA

5.1 Analysis of the displacement monitoring data

The data monitored in the field of typical sections of Zhegu Mountain Tunnel was organized, and the relationship curves of displacement value-time and displacement velocity-time of horizontal convergence and vault settlement for ZK188+400 section were obtained, as shown in Figs. 2 and 3.

As seen in the above figures, the changes of vault settlement and horizontal convergence have the following characteristics.

At the beginning of excavation, the displacement velocities of horizontal convergence and vault settlement were relatively large. With the increase of the distance between excavation face and the test section, the displacement velocities of horizontal convergence

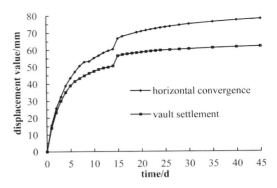

Figure 2. Measurements and records curves of displacement.

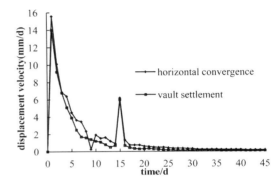

Figure 3. Rate-time curves of deformation.

and vault settlement decreased gradually. On the 15th day of excavation, for the influence of excavation disturbance of lower bench, the convergent rate increased sharply. Then, with the construction of support for lower bench, the deformation rate decreased gradually. After 16 days of excavation, the deformation rate began to stabilize to about 1 mm/d. On the 26th day of excavation, the deformation rate was basically stable and less than 0.5 mm/d.

The results of the measurement showed that the deformation of the surrounding rock develops by three stages: the rapidly-deforming stage, slowly-deforming stage and fundamentally-stable stage.

5.2 Regression analysis of deformations of surrounding rock

The data of horizontal convergence and vault settlement for ZK188+400 section were put in the kinds of regression functions separately, with the results of calculation as shown in Table 5.

5.3 Evaluation of the stability of surrounding rock

Hyperbolic function was used to fit the horizontal convergence and vault settlement for ZK188+400 section separately, and the calculation results and the measured data are as shown in Fig. 4.

Table 5. Regression equations of deformation for ZK188+400 section.

Regression function	Regression equation	Correlation coefficient
Horizontal convergence		
Logarithmic function	$u = 87.3 - 26.93/\lg(1+t)$	0.9087
Exponential function	$u = 80.48e^{-2.954/t}$	0.9781
Hyperbolic function	$u = t/(0.06063 + 0.01154t)$	0.9956
Vault settlement		
Logarithmic function	$u = 71.3 - 20.82/\lg(1+t)$	0.9306
Exponential function	$u = 64.27e^{-2.376/t}$	0.9789
Hyperbolic function	$u = t/(0.05104 + 0.01498t)$	0.9931

Conclusion: The values of the correlation coefficient showed that the use of hyperbolic function for fitting horizontal convergence and vault settlement has the highes precision.

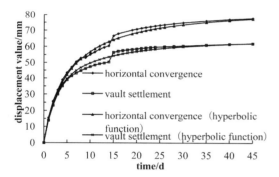

Figure 4. Displacement comparison of measurement and calculated results for ZK188+400 section.

As seen in Fig. 4, the measured displacements of horizontal convergence and vault settlement well coordinate with the regression analysis results. Therefore, regression analysis by using hyperbolic function is more reasonable in this example (the level of the surrounding rock is V_j).

By the regression functions $u = t/(0.06063 + 0.01154t)$ and $u = t/(0.05104 + 0.01498t)$, the final displacement values and the displacement rates of horizontal convergence and vault settlement for ZK188+400 section were obtained. The final displacement values were 86.655 mm and 66.756 mm, and the displacement rates on the 45th day were $du/dt = 0.1803 < 0.2$ mm/d and $du/dt = 0.0971 < 0.1$ mm/d.

According to geologic examination, the depth of ZK188+400 section is 57.21 m, and the level of the surrounding rock is V_j. Table 3 shows that the allowable relative displacement of horizontal convergence for ZK188+400 section is 1.60%, and the final measured relative displacement of horizontal convergence was $86.655/11625 = 0.75\% < 1.60\%$, conforming to the requirement of code. The final measured convergence of tunnel inner perimeter was 43.328 mm, and the final displacement of vault settlement was 66.756 mm, both in the range of $u_0/3$ to $2u_0/3$ (the rock level is V_j, $u_0 = 120$ mm), thus the support should be strengthened in the construction.

Analysis of Fig. 4 shows that for ZK188+400 section, 90.07% of the deformation of horizontal convergence and 92.88% of the deformation of vault settlement completed in about 45 days.

It was determined that the horizontal convergence and vault settlement of the surrounding rock and initial support were basically steady after 45 days of excavation. Moreover, the acceleration of the deformation of the surrounding rock was $d^2u/dt^2 < 0$. The results show that the support parameter was reasonable for guaranteeing construction safety.

5.4 The reasonable time of secondary lining

According to the analysis results of monitoring measurement data and stability criterion of the surrounding rock, after 45 days of excavation, for ZK188+400 section, the velocities of the horizontal convergence and the vault settlement were below 0.2 mm/d and 0.1 mm/d, respectively. Meanwhile, the amounts of their deformations both reached 90% of the total deformation. At this point, the surrounding rock tends to be stable and its bearing capacity is at the maximum level. In addition, the resistance of initial support is fully utilized. The stress redistribution of the surrounding rock leads to the dynamic balance between its deformation stress and support resistance. It is the best time for secondary lining form the perspective of making sure it is under the least stress. If the secondary lining is not constructed at this time, it cannot provide enough restraint stress for the rheology of the surrounding rock, probably resulting in the cracks of initial support on account of excessive force.

6 CONCLUSIONS

By the real-time monitoring of the construction site in Zhegu Mountain Tunnel, this paper analyzed the deformation of surrounding rock from two aspects: horizontal convergence and vault settlement, and the following conclusions were drawn.

(1) From the horizontal convergence and vault settlement data monitored in the field, it can be seen that the deformation of the surrounding rock develops by three stages: the rapidly-deforming stage, slowly-deforming stage and fundamentally-stable stage. The value of deformation is related to the level of the surrounding rock and the excavation method. The reliability of the monitored data is related to the time of test point arrangement and the precision of measuring instrument.

(2) The displacement velocity of surrounding rock is very large at the moment of excavation. The measuring points should be arranged and measurement started immediately after excavation during construction.

(3) Because of the soft rock influenced by the excavation disturbance of lower bench, the velocities of horizontal convergence and vault settlement will suddenly increase when the lower benching is excavated. At this point, the monitoring measurement should be strengthened and supporting parameters adjusted in real time to ensure the stability of surrounding rock.

(4) Based on the regression analysis of the monitoring data, the deformation data of surrounding rock meets the regression equation, the acceleration of the deformation for surrounding rock and the time required for the stabilization of surrounding rock were inferred, then the stability of surrounding rock and initial support was judged. In addition, the ultimate deformation of surrounding rock was estimated, and the reasonable time of secondary lining was inferred by combining the measured data.

REFERENCES

Gaofei Dai, Song Ying, Caichu Xia, et al. 2004. The Monitor Measuring of Highway Tunnel in NATM. *Journal of Chongqing University* 27(2): 132~135. (in chinese)

Qixin Yang, Mingnian Wang. 2009. *Construction and management of underground engineering*. Chengdu: Southwest Jiaotong University Press. (in chinese)

Shanhong Liu, Yi Liu, Fang Li. 2008. Application of NATM Principle to Deformation Monitoring of Shilongshan Tunnel Surrounding Rock. *Journal of Chongqing Jiaotong University (Natural Science)* 27(1): 44~48. (in chinese)

Xiaohong Li. 2002. *New Austrian Tunneling Method and Measurement Technology*. Beijing: Science Press. (in chinese)

Zhiye Li, Yanhua Zeng. 2003. *Design theory and method of underground structure*. Chengdu: Southwest Jiaotong University Press. (in chinese)

Details & decoration—an interpretation of Carlo Scarpa's architecture work

Chao Huang
Tianjin University of Technology, Tianjin, China

ABSTRACT: Carlo Scarpa (1906–1978), a well-known Italian architect, is honored as the master of light application, mater of details and connoisseur of material. Based on the simple narration of the cultural creation background of Carlo Scarpa, this paper analyzes the decoration characteristics and basic methods of Scapra's architecture and concludes some commonly used mother themes of decoration to point out that all the connotations of Scarpa's architecture are hidden in the details.

1 INTRODUCTION

Carlo Scarpa is hailed as one of the most famous architect designer in the 20th century. He graduated from Architecture Institute in Academy of Fine Arts of Venice, and acquired teaching post of Architectural Drawing. Then, he worked as teaching assistant in Higher Education Institute of Venice which fixed himself in the architecture education career. Scarpa is a master who pays attention todetail construction. Although he didn't design large-scale buildings, every piece of his work is extremely fine and smooth. "All connotations of architecture are contained in the details of it" and "the overall is harmonious with details" are the absolute characteristics of Scarpa's works.

2 THE CULTURAL BACKGROUND OF SCARPA ARCHITECTURE CREATION

The influence of Scarpa's personal experience and cultural background is undeniable on his description and preference of decoration. Scarpa was born in Venice, Italy, the intersection for eastern and western cultures in the history and whose culture is not only characterized by western culture, but also infused with eastern contents. Venice has genetic capacity for exotic cultures and is good at absorbing exotic cultures. In addition, its liberal political system forms cultural landscape equipped with both eastern and western cultures. Glass artware is a traditional crafts in Venice, and Scarpa once was engaged in the design of glassware and decoration in world famous Murano Island. Scarpa grew in the period when the modern architecture was emerging and developing. He was deeply influenced by masters of modernism, especially Wright. Scarpa visited America and experienced Wright's works in person. Wright's architectures with modern decoration style enlightened him a lot. In 1969, Scarpa visited Japan. Simple but elegant and exquisite traditional Japanese architectures left a strong impression on him, and those art works which were totally different from western paintings, such as Japanese ukiyoe and ancient Chinese paintings gave him new enlightenments to enable him to discover the expression of lines and integrate some eastern contents in his works. There are two aspect of eastern contents manifested in Scarpa's works: one is the similarity in spiritual condition, which is particularly similar to traditional Japanese aesthetic, and his architectures are always presented in quiet, poetic and elegant temperament; the other is the similar specific architecture processing technique, such as the application of cross-type separation and process of light.

3 REASONABILITY AND MODERNITY OF DECORATIVE NODE

Scarpa's architectures have strong decorative function. Unlike those modern architectures in which decoration is totally abandoned, his architectures inherit decorative function of classic architectures. He held an opinion that modern architectures should also achieve such elegant and graceful effects like classic architectures. He discovered the node construction as a proper reason for the decorative function of modern architecture.[1] It is obvious that Scarpa accepted

---

[1] Chu Ruiji, Carlo Scarpa: The Poetic of Spatial Flow [M] Taipei: The Gardencity Press of Taiwan, 2004.

Louis Kahn's view that nodes are the origination of decoration, so node construction plays an important role in the former's designing. Just such meticulous construction of node builds the fine and exquisite appearance of Scarpa's architectures.

In node construction, Scarpa expressed his constructional logic clearly. For example, in the transformation of CASTLEVECCHIO Museum, the weight of floor firstly is loaded on the two perpendicular concrete girders, and then the weight is transmitted to the large steel girder in the center by a special steel bracket. As was said by Scarpa, "to reserve the facility of all exhibition rooms, I can't repair the main girder roughly". A large steel girder supports at the cross point of two perpendicular concrete girders, which suggests that invisible pillar and emphasis the characteristics of square exhibition room. In addition, different materials, different stories that, when brought close together though kept rigorously separate, begin a sort of dialogue.

The node construction of Scarpa is equipped with decorative characteristics. Scarpa regarded the nodes as the essence of the design of overall architecture, which is also the language of modern architecture. He tried to present the elegant aesthetic effect in his modern architectures that are the same as classic architectures. The point of his opinion is similar to Huffman's that he uses modern language to enable the decorative effect of architecture, rather than additional decoration.

4 BASIC TECHNIQUE OF NODE CONSTRUCTION

The creation techniques used in Scarpa's architectures are diverse and affected by many aspects. They not only have some of Wright's ideas, but also are equipped with some traditional Japanese architecture styles. However, his most basic creation ideas originated form de Stijl, and his most basic creation technique is new modeling doctrine.

New modeling doctrine is the modeling technique of de Stijl, which was firstly proposed by Mondrian. In his series of paintings whose themes are "image composition with yellow", and "the rhythm of lines", such style is expressed. Those paintings mainly are constituted by some separated horizontal and vertical lines. Scarpa introduced this technique into his designs. He liked vertical and horizontal picture composition, liked pure application of geometric shapes and asymmetric forms. Scarpa's technique extended the experiment and exploration of neo-platicism in visual effects of shaping and picture composition, enabling his architectures to be listed in the true modern architecture.[2]

5 THE WORLD OF IMAGES

Scarpa's works not only take abstract composition language in de Stijl as reference, but also apparently present complicated images, which consist of an abundant image system. Scarpa's image world reflects the close relationship of his architecture creation with traditional image art. Like other modern architecture, Scarpa's architecture adopts simple geometric shapes and fully presents abstract composition. For example, in the entrance porch of Briant family cemetery, two circles are piled and the end of wall is removed. His image world was influenced by eastern culture deeply, especially the Japanese culture. In the corner of the wall of Briant family cemetery, a pierced image is designed, which is similar to the Chinese character twin " 喜 "; The small temple in the cemetery has a round doorway, which remind people of those kinds of doorways in classic Chinese gardens.[3] Such elements enrich the formal characteristic of Scarpa's architectures, and enhance the expression of the architectures' cultural connotation, which is totally against the modernism that pursues absolute geometric shapes but abandons formal expression and decorative function. Scarpa's form does not abstract Purism, but is full of the meaning of "symbols".

To soften the architecture and ease the sense of weight, Scarpa always adopted corbelled decorative lines on the cornice. Such decorative lines are formed by corbelling, which is equal to the decorative pictures on the cornice of classic architectures. In the design of the Bank of Verona, Scarpa did it.

The newly built bank is the extension part of an old one, and is located between two old buildings. The newly built part is a building with three layers and two facades, one facing the square and the other facing the garden. Both ends of the building are completely the same with those of the old buildings. Basically, Scarpa adopted the classical three-stage facade pattern with new changes. The thin and continuously repeated corbelled decorative lines separate the base and the central part; the eave part is composed of the steel girder supported by twin steel round tube pillars and projected marble overhangs; the cornice is also corbelled. Although this architecture is built with modern steel materials, everywhere of it manifests the characteristics of classic architectures, making it harmonious.

6 CONCLUSIONS

In the history of modern architecture, Scarpa is a unique architect. His works always are free from the main stream of modern architecture. Even when the modern advocates rationality, Scarpa still persisted in his detail exploration. His fine and exquisite details

---

[2] Sergio Los, Carlo Scarpa an architectural guide.

[3] Pan Zhaihui The Architectural Idea and Woks of Italian Architecture Carlo Scarpa [C] Collection of academic dissertation for master degree in Tongji University.

and poetic expression deeply influence subsequent designers. Just as is said by Tadao Ando:

"… He notices form problem and the material usage, and want to discover the things deep concealed in everything to express the power for inner side to outside. It can be said that things that other architects don't concern are what I'm interested in."

REFERENCES

Chu Ruiji, Carlo Scarpa: The Poetic of Spatial Flow [M] Taipei: The Gardencity Press of Taiwan, 2004.

Pan Zhaihui The Architectural Idea and Woks of Italian Architecture Carlo Scarpa [C] Collection of academic dissertation for master degree in Tongji University.

Sergio Los, Carlo Scarpa an architectural guide.

Detecting the interface defects of steel-bonded reinforcement concrete structure by infrared thermography techniques

Y.W. Yang, J.Y. Pan & X. Zhang
Department of Civil Engineering, Zhejiang A&F University, Hangzhou, China

ABSTRACT: The new nondestructive testing method based on infrared thermography techniques is introduced to evaluate the interface defects of steel-bonded RC structures. A review of various detecting test methods for interface defects is given. Comparing with traditional test methods, the interface defect positions and size can be measured exactly by infrared thermography technique. Then, the testing environmental conditions, equipments requirements and testing procedures of this method are discussed to conform to the code requirements for acceptance construction quality.

1 INTRODUCTION

There are several traditional methods mainly including visual observation, coin-tap, and ultrasonic nondestructive test (NDT) for detecting the interface defects of steel-bonded RC structures. Visual observation method is used to speculate the bonding quality by observing the structural adhesive overflow from the air vents. Then coin-tap method is used by knocking the outside surface of bonded steel plate to determine the bond quality. The crisp sound means it is dense and dull sound means that there are some hollows. These two methods need experienced examiners and can not give exact positions and degrees of bond defects. The ultrasonic method is not imprecise for the defects size. So these three methods can not reflect the defects position, size completely for detecting the qualities of interface defects of steel-bonded RC structures.

The infrared thermography techniques are used for detecting the bonded defects of the composite material layers of the aircraft shells in civil and military aerospace originally. Heating the workpieces, the surface thermal wave radiations of workpieces are measured and infrared scanning images are obtained. Then the bond interface defects are obtained according to the thermal radiation temperature gradient.

From 90's of the last century, the infrared detection technology is widely used in the detection of internal bonding defects of components and layers of aircraft's skin structures which are made of composite material in developed countries. Government research institutions and multinational corporation of USA and other countries have carried out relevant research and established technical standards (Xavier M. 2001, Favro L.D. & Newaz G.M. et al. 2002). The WTI (Western Telematic Inc) of USA has developed the thermal wave imaging detection instruments which applied to the NDT, and developed a more advanced analysis software for business.

After 2003, the Capital Normal University and Beijing Institute of Aerial Materials *etc.* carried out more and more research projects related to thermal infrared detections. Then they established the first infrared thermal wave nondestructive testing laboratory of China. The infrared thermal wave detection of rocket engine was used to carry out related research about defect detections of composite layer materials (Jiang S.F. & Guo X.W., et al. 2005, Xu Y.G. 2008). Also Research on infrared thermography testing method was carried out in 2005, which is presented to detect defects that related to the detection time of honeycomb aluminum composite material (Huang S.L. & Li L.M., et al. 2005).

2 INFRARED TECHNIQUES APPLICATION IN STEEL-BONDED RC DEFECTS MEASUREMENT

For being applied to the field of civil engineering with the infrared thermal technology, this method is proposed to defect measuring of airports, freeways, bridges, and building surface qualities. (Gary J. & Wei L. 1998a,b,c). Then, infrared thermal wave detection technology applications are introduced to the field of domestic civil engineering (Zhang R.Y. 2001), and the interface defects detection of steel-bonded RC structure by the infrared thermal wave technology are also studied (Huang P. & Xie H.C. 2004), the application of this method to leak inspections of house, the interface defects of steel-bonded RC structure, the cracking and hollowing inspections of wall heat preservation were discussed in 2011 (Xie C.X. 2011).

Table 1. The main material thermal conductivity.

Number	Material	Thermal conductivity (w/m.k)
1	concrete	1.8
2	Steel-bounded plate	55
3	adhesive	1.3–1.63
4	air	0.025

3 DEFECT DETECTION MECHANISMS

The interlayer between the steel interface and concrete surface will be filled with air when the adhesive interface of steel-bonded RC structure is hollowing. According to the Table 1, the thermal conductivity of sealed air is smaller than steel, concrete and adhesive layer. Therefore, when the heat transfers along the steel plate surface to concrete, the temperature of the steel-bonded layer will change and a high temperature zone at the hollowing place by the air layer's thermal insulation effect called "heat island" is formed. Then the infrared camera is used to record the temperature distribution of steel plate surface and the infrared thermography temperature distributions are obtained. According to the thermal gradient analysis of isothermal air layer on the steel back, ultimately the plate hollowing region is determined.

On the thermal distribution images, the infrared isotherms can be shown at different temperature, as shown in the Figure 1, where the temperature difference is equal in each adjacent isotherms. Also shown that, the closer to the hollowing edge, the temperature gradient mode is larger. With the normal coordinate $\tau - r$ isotherm system, the tangential direction is set as τ axis, the normal direction as r axis, so the thermographic temperature distribution is expressed as $T(r, \tau)$. The module of temperature gradient can be expressed as:

$$|\Delta T| = \sqrt{(\frac{\partial T}{\partial r})^2 + (\frac{\partial T}{\partial \tau})^2} \quad (1)$$

at the isotherm position: $\frac{\partial T}{\partial \tau} = 0$, then $|\Delta T| = |\frac{\partial T}{\partial r}|$. So the maximum of $|\Delta T|$ is in the location of the interface defect edge of adhesives hollow, and shown as the intensive isotherm zone.

At present, there are two problems of applying infrared thermal wave method for detecting the steel-bonded RC structure bonding quality mainly:

1) The first is steel surface infrared radiation heating control. The surface temperature of steel structure is nearly equal to the surrounding environment under normal Circumstances. So there is no isotherm on the thermal distribution images. For the interface detection, the steel-bonded RC structure is needed to be heated. In the heating or cooling process, Heat will transfer into its inner or radiate to the outside atmosphere of structure. The structural surface temperature will be different on the intact or

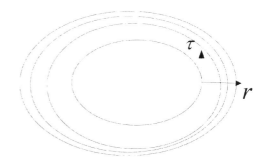

Figure 1. The infrared isotherms distribution images.

defection of adhesion interfaces. This thermal gradient can be detected by infrared images. Usually, two kinds of method for heat transfer or radiation. One is cooling method, the steel-bonded RC structural surface is covered with liquid nitrogen. Liquid nitrogen is gasified, diffuses, absorbs heat of steel plate, then steel surface temperature decreases and causes heat conduction of steel-bonded RC structure from inside out. However, this detection accuracy is not stable for uneven nitrogen spraying.

The other is heating method, inappropriate heating also can influence the detection precision. There often have continuous heating and pulse heating method. So the heat power and frequency and the absorption ability of the structure surface are needed to be determined.

2) The second is infrared image recognition. For the condition of steel plate rough surface, uneven cooling or heating process, the infrared thermal image of steel surface contains non defect information inevitably. Therefore, the defect recognition of infrared thermal image is crucial.

This paper focuses on the active heating method, and finds proper heating method, infrared thermal image recognition. The infrared heating and detecting system are expressed as Figure 2.

4 RESEARCH STEPS

According to the two main problems of bonding quality detection of infrared thermal wave application in steel-bonded RC structure, the detection procedure can be carried out according to the following steps:

1) Numerical model analysis: using Ansys and Comsol software to set up a numerical model for thermal analysis in accordance with test conditions.
2) The structure surface treatment: in connection with the steel surface roughness factors that influence on the infrared thermal absorption, polishing plate surface and brushing the better endothermic material for thermal efficiency and uniform absorption before testing.
3) Heating control during the experiment: using pulse infrared excitation, the infrared radiation lamp can

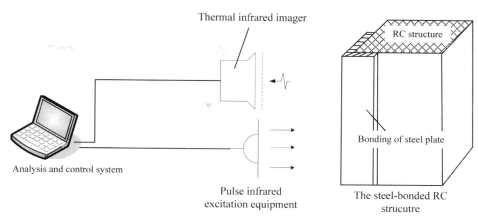

Figure 2. The infrared heating and detecting system.

adjust the transmission power and transmission frequency. The infrared thermal imager will take the temperature distribution picture of the steel outside surface during the infrared heating gap.

4) Infrared image recognition of experimental results: for infrared image fuzzy edge of defects, the fuzzy recognition method based on multi-source is carried out (Wang B.S. 2007).

REFERENCES

Favro L.D., Newaz G.M. *et al*. 2002. *Progress in Thermosonic Crack Detection for Nondestructive Evaluation*. DARPA Prognosis Bidder's Conference.

Gary J., Wei L. 1998. *Nondestructive Testing of Airport Concrete Structures: Runways, Taxiways, Roads, Bridges, and Building Walls and Roofs*. Proceedings of SPIE-The International Society for Optical Engineering. v3397.

Gary J., Wei L. 1998. *Remote Sensing of Voids in Large Concrete Structures: Runways, Taxiways, Roads, Bridges, and Building Walls and Roofs*. Proceedings of SPIE-The International Society for Optical Engineering. v3436.

Gary J., Wei L. 1998. *Infrared Thermographie Pipeline Leak Detection Systems for Pipeline Rehabilitation Programs*. Proceedings of SPIE-The International Society for Optical Engineering. v3398.

Huang S.L., Li L.M., Yang H.Q. 2002. *Evaluation of Composites Bonding Defects by Infrared Imaging Testing*. Aerospace Materials and Technology 6:43–46.

Huang P., Xie H.C., Yuan X. 2004. *Detecting of Defects in Bonding Interface Between an Armor Plate and a Concrete Component by Infrared Thermography*. Laser and Infrared 34(5):350–353.

Wang B.S. 2007. *Compound-source Fuzzy Numbers and Their Operations*. Fuzzy System and Mathematics 21(3): 24–28.

Xavier P.V. 2001. *Infrared and Thermal Testing, Nondestructive Testing Handbook, third edition: Volume 3:* the American Society for Nondestructive Testing.

Xie C.X. 2001. *Study on the Infrared Thermography in the Steel-bonded Reinforcement Concrete Structure*. Journal of Sichuan Architecture 31(3):170–172.

Xu Y.G. 2008. *the Application of composite Load Technique Infrared Wave Nondestructive Testing*. Beijing: Capital Normal University of China.

Jiang S.F., Guo X.W., Shen J.L. 2005. *Infrared Thermal Wave NDT on the Disbonds of the Heat Insulated Layer in Solid Propellant Rocket Motors*. Laser and Infrared 35(8):584–586.

Zhang R.Y. 2001. *New Technique for the Detection of Construction Engineering Quality*. Beijing: China Planning Press.

Detection, appraisal and strengthening of the support structure of a boiler fan in part II of a power plant

Mengxi Tan
China University of Geosciences, Beijing, China

ABSTRACT: This paper is about the safety appraisal and strengthening of a support frame of blower in a power plant. According to the appraisal conclusion of the structure, the denitration devices are applied on the upside of the original boiler fan. The bearing capacity is calculated. On the basis of the results of calculation, the corresponding reinforcement schemes are put forward in order to ensure the safety of the structure in its subsequent service life.

Keywords: supports of blower, detection, appraisal, strengthening, bearing capacity

1 ENGINEERING SITUATION

The support structures for the second stage 3#, 4# boiler fan are both two-storey reinforced concrete frame structures. There are two transverse spans, 3.5 m and 11 m long respectively, and four longitudinal spans, 7 m, 12.23 m, 12.23 m, and 7 m long respectively. The total length of transverse spans is 14.5 m and that of the longitudinal spans is 38.46 m. The height of the first storey is 9 m, with no floor slab; and the height of the second story is 10.4 m; therefore the total height is 19.4 m. In the original design, the frame beam columns of the upper part all adopt C40 concrete. Concrete cover for beam and column is 25 mm thick. Cross-sections of the frame columns are all rectangular sections in the dimensions of 700 mm*800 mm. The longitudinal and transverse cross-sections of the main frame beam are in the dimensions of 500 mm*1300 mm, and the cross-section of beam in secondary floor is in the dimensions of 450 mm*700 mm and 500 mm*1200 mm. There's no maintenance structure. The foundation is of pile foundation with six piles rectangular bearing platform, and the pile is precast concrete square pile with the section size of 450 mm*450 mm and vertical bearing capacity of 2000 kN for a single pile. The section size of bearing platform is 2600 mm*4200 mm and 2800 mm*4000 mm, the height of bearing platform is 1500 mm. In the original design, the strength grade of concrete for the pile foundation bearing platform is C35, while the that for cushion is C10; the bearing platform foundation cover is 35 mm thick, and the bearing platform foundation is buried 3.0 m deep under the outdoor floor. In this area, the seismic intensity scale of structure is VII, design earthquake group is group II, and site category is category III. So it is considered as a seriously liquefied site, and vibro-replacement stone pile is used to avoid liquefaction.

2 DETECTION AND IDENTIFICATION OF ENGINEERING STRUCTURES

2.1 Contents and conclusion of field detection (GB/T50344-2004)

2.1.1 The result of structural arrangement and member size check

After checking against the original design drawings, it is found that the plant 3#, 4# boiler fan structure arrangement are in accordance with the original design, and the section size of the boiler fan components are also in accordance with the original design.

2.1.2 Inspection results of apparent damage of structures components

Inspection was conducted onsite concerning the appearance damage on 3#, 4# blower support structures, including the appearance of foundation, frame column, and beam frame.

i) The foundation

It's a frame structure building with two floors. According to the on-site inspection, the building's ground foundation shows no significant slop or differential settlement. Beams and columns of upper structure frame show no new diagonal cracks caused by differential settlement of the ground foundation. The structure has no significant inclination, and no static loads defect on the building foundation and pile foundation. The present situation of the ground foundation is basically well.

ii) The upper structure

According to the on-site inspection, 3# and 4# blower bracket have no floated coat; and due to concrete peeling, there are exposed steel bars and hooping corrosive cracking in the root of columns. Beams at the elevation of 9 m and 19.4 m have no floated coat, too. The bottom, side and top surface of part of beams show exposed steel bars and corrosive hoopings.

2.1.3 Test results of concrete sampling strength (CECS220:2007, GB/T50107-2010, JGJ/T23-2011)

Concrete strength of beams and columns in fan support frame did not meet the requirements of the original design of C40, but can only meet the requirements of current national standard C20 concrete strength.

2.1.4 Test results of reinforcement arrangement sampling (JGJ/T152-2008)

The test results show that the frame column is unilaterally equipped with 8 reinforcements, and the bottom of beam is equipped with 7 reinforcements. Stirrup spacing basically meets the requirements of the original design and the standard deviation.

2.1.5 Test results of durability performance (CECS220:2007)

Tests on the thickness of reinforcing bar protection cover, depth of concrete carbonation, content of CL- and the status of the reinforcement corrosion of beams and columns were carried out respectively on site.

The thickness of bar protection cover in the test area is mostly in accordance with the original design and standard deviation requirements. Some individual components are a little thin and don't meet the requirements of national standard, which may cause the problem of durability. If the protection cover is too thick, it will reduce the calculated height of the section and lower component bearing capacity.

According to the result of concrete carbonation depth test, due to lack of plaster layer, the carbonation depth of columns and beams is generally deeper, at average of 15 mm.

The carbonation depth of the beam is larger. For the structures with thin protective cover, the carbonation depth is close to or reaches the minimum thickness of concrete cover, and the concrete can no longer provide alkaline protection to the steel bar. Deep carbonation will affect test values of the concrete strength. Cl- is one of the most common reasons, which cause problems of durability (especially steel bar corrosion) in the reinforced concrete structure. In order to learn the current durability of reinforced concrete column and provide basis for repair and maintenance, tests are made especially to measure the content of Cl in the reinforced concrete column. Testing on the content of Cl- for 4/6-4/6 beams in 3# Fan support at the elevation of 19.4 m and 3/14-1/13/D beam in 4# fan support shows that, the Cl content accounted for the largest percentage of cement dosage is 0.105%, less than the maximum content of Cl 0.15% as stipulated in the codes. So it meets the specification requirements.

The results of corrosion detection combined with exterior damage are mainly about the corrosion level of steel bars in frame columns. The electrochemical non-destructive testing method combining with small damage detection method is used to test the corrosion of reinforcement. From the test results, it can be considered that the reinforcements are basically unstable with corrosion rate not more than 5%. By chiseling parts being suspected of rust, it is found that the steel does not corrode, and there are only slightly floating rust in several places where the concrete cover is thin.

2.1.6 Measurement of slope deformation for structures (GB50144-2008)

Inspections on the whole slope deformation of structures were carried out. Measuring of column verticality mainly aims at checking the condition of tilt for columns in air supply unit support frame, and it is an important means to test the performance of upper structure component. Total station (HTS – 362 RL) is used to measure tilt deformation of the frame column. Based on the top and bottom deviation test results on all directions of ten columns in 3#, 4# fan support, it is found that tilt value for each column varies in different directions.

According to the Article 7.3.9 in Industrial Building Reliability Evaluation Standard (GB50144-2008), if the top tilt value of concrete structure is less than 40 mm, it must not consider the effects of the additional internal forces on the bearing capacity of structure, which are caused by the horizontal tilt. The measuring results show that the deformation of the individual components are more tilted than the limit, but the tilt value is less, which does not affect the normal use of the upper structure. Additional internal forces caused by tilted structure are small and can be ignored during the structure calculation.

2.2 Structures details of seismic design analysis (GB50023-2009)

Evaluation of seismic measures focuses on macroscopic control and construction evaluation, which comprehensively evaluates the seismic factors of buildings from the overall structure and construction. The 3# and 4# boiler fan support frames were designed and built in 2003 as reinforced concrete frame structures. According to the national standard and evaluation standards, the boiler fan support structure should be evaluated on the basis of the requirements of a building whose service life is 50 years, seismic intensity scale is VII, design earthquake classification is group II and seismic fortification is class b, type C. According to the specification (2001), the original seismic fortification is class b. Based on the results of tests for various components, the strength grade of concrete for some beams does not meet the requirements, and other various seismic construction measures are under requirements of evaluation standard. The influence of structure that does not meet the requirements should be considered when it comes to seismic comprehensive ability evaluation.

2.3 Calculation of bearing capacity of the structures

The site where the support frame is located falls into class III, where the seismic fortification category is B, seismic intensity scale is VII, basic seismic acceleration value is 0.1 g, seismic group is group II, and subsequent service life is 50 years. The basic wind pressure value is 0.48 kN/m², the ground roughness is Class B.

2.3.1 Calculation of bearing capacity of support frame under the original loads

Because the test results show that the concrete strength of 3# and 4# blower bracket frame is low, shear and torsion bearing capacity is calculated based on C20. Frame beams for elevation of 9 m and 19.4 m can't provide the requirements of bearing capacity with a difference of 30%. Bearing capacities of the remaining components in 3# and 4# blower bracket structure, ground and foundation satisfy the specifications and standards. Elastic displacements between the layers meet specification requirements.

2.3.2 Bearing capacity checking after increasing the denitration loading

After applying the denitration loading, bearing capacity of pile foundation in 3#, 4# boiler fan support structure is calculated. Bearing capacity of pile caps is far below the requirements of specification and standard, but elastic displacements between stories basically meet the requirements of specification. The maximum axial compression ratio of columns in the first floor is 1.46, higher than 0.75, the limit of secondary frame axial compression ratio. The amount of steel bars in most of the beams and columns do not meet the current standard specification requirements. The biggest difference between the standard requirement and the actual usage of steel bars is 72%, indicating the need for reinforcement. Bearing capacity of pile foundation meet the requirements, but bearing capacity of pile caps in the pile foundation does not. The ratio between cutting resistance of the pile to pile caps and load effect is 0.74. The maximum elastic interstory displacement angle of the north-south frame and east-west frame is 1/677 and 1/633 respectively (storey two), within the specified limit of 1/550.

On the basis of the current specification, after on-site inspection, testing, calculation and analysis of the blower bracket structure, the following conclusions are obtained:

Based on the inspection and test results of the blower bracket structure, under the condition of existing structural system and the status quo, except the shear and torsion bearing capacity of lateral frame beam does not meet the requirements, bearing capacity for the rest of the component all meet the requirements.

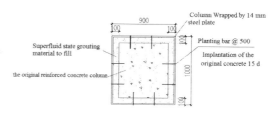

Figure 1. The reinforced chart of concrete column.

After increasing denitration process loads, bearing capacity of 3# and 4# blower frame seriously doesn't meet the requirements of current national standards, which affect the overall safety, so reinforcement must be made to the blower bracket structure.

3# and 4# blower bracket structure mainly has the following problems:

i) Concrete strength of columns and beams in 3#, 4# fan support frame do not meet the requirements of the original design of C40, but can only meet the requirements of current national standard, concrete strength of C20 (conditions permit shall also core validation).

ii) Most of the frame column roots are exposed, where reinforcements are corrosive; there are exposed and corrosive reinforcements at the bottom of some beams, as well as on lateral and top side of individual beams.

iii) According to the durability test, some parts show durability damage in different degrees. The surface has slight cracking phenomenon. Construction quality of some components is bad. Concrete cover thickness does not meet the requirements. The places of the thin layer have problems such as large areas of rusted exposed stirrup.

3 THE ADVICE FOR REINFORCEMENT

In order to achieve the goal of safe production for the 50 years of use period, developing a reasonable reinforcement and reconstruction scheme for the building is very important in terms of reducing capital investment. In consideration of site conditions and with reference to the experiences of existing engineering reinforcement practices, the following reinforcement measures are to be taken (Figure 1):

i) Ground and foundation: The original pile caps shall be strengthened by the method of enlarged cross section.

ii) Concrete frame column: The method of outsourcing steel plate sleeve shall be adopted to strengthen the frame columns, which can not only greatly improve the bearing capacity of the pillars, but also help to meet the requirements that the column section will not increase too much. The column reinforcement scheme is as follows:

Prior to reinforcement of frame columns, voids, pits and porous concrete should be cut firstly; then

strengthening can be carried after dense and fresh concrete is exposed.
iii) Concrete frame beam and beam-column joints: New steel column foot beam-column joints and the connected beam end are main places to be reinforced. It is recommended that method of enlarged section and node outsourcing plate be adopted for reinforcement.
iv) In order to strengthen the safety and seismic performance of the whole structure, a certain amount of steel braces shall be vertically and horizontally placed in the upper and lower floors respectively.
v) For all concrete members, it is recommended that concrete protection liquid be applied on the surface coat of the concrete to prevent it from erosion by harmful medium. This liquid can immerse the concrete and protect the steel from the influence of further carbonization, thus guaranteeing the durability of the concrete. It is better to apply the liquid in every 3~5 years.

REFERENCES

CECS220:2007, Reinforced Concrete Structure Durability Evaluation Standard. *Beijing, China Building Industry Press*, 2007.

GB/T50344-2004, The Building Structure Detection Technology Standard. Beijing: *China Building Industry Press*, 2002.

GB/T50107-2010, The Concrete Strength Inspection Evaluation Standard. Beijing, *China Building Industry Press*, 2010.

GB50144-2008, Industrial Building Reliability Evaluation Standard. Beijing, *China Building Industry Press*, 2008.

GB50023-2009, Building Aseismic Accrediting Standards. Beijing, *China Building Industry Press*, 2009.

JGJ/T23-2011, Technical Specification for Concrete Compressive Strength of the Rebound Method. Beijing, *China Building Industry Press*, 2002

JGJ/T152-2008, The Detection Rate of Reinforcement in Concrete Technology Discipline. Beijing, *China Building Industry Press*, 2008.

Developing an Artificial Filter Bank (AFB) for measuring the dynamic response of civil structures

Gwanghee Heo & Joonryong Jeon
Department of Civil and Environmental Engineering, Konyang University, Nonsan, South Korea

ABSTRACT: In order to efficiently attain dynamic responses of civil structures in real time, an artificial filter bank (AFB) was developed using a band-pass filter optimizing algorithm and peak-picking algorithm. The optimization of the AFB was performed by using the E1-Centro earthquake wave, and the reconstruction and compressive effects were quantitatively compared to the original signal using the reconstruction error and the compressive ratio. The artificial filter bank designed, developed, and evaluated in this study sufficiently captured the original signal in both time and frequency domain, and at the same time acquired the necessary dynamic (acceleration) responses of the compressed size in the frequency range of interest.

1 INTRODUCTION

For structural health monitoring (SHM) of civil infrastructures, it is important to measure dynamic (acceleration) responses of the structure in real-time. Conventional dynamic response measurements of civil infrastructures in the past were mostly based on wired systems, but current research of SHM operate on wireless sensor networks (WSN) that integrate rapidly developing information technologies and RF communication technologies. Although a WSN lacks the power consumption, efficiency, and robustness of wired networks, they are expected to expand in development and application due to their mobility, easy placement of sensors, and low cost of system configuration and maintenance [1]. However, dynamic responses from WSNs require more data handling loads than static responses do, which may cause data loss due to bottlenecks when wirelessly transmitting dynamic data, and the costs increase when maintaining a large database. Therefore, in order to efficiently obtain, transmit, and maintain only the necessary data, an alternative technology needs to be proposed to filter significant dynamic responses [2].

In this study, an artificial filter bank (AFB) using a band-pass filter optimizing algorithm and peak-picking algorithm has been developed. This ABF filters and compresses response data in order to efficiently acquire dynamic responses. The AFB was optimized by using the random E1-Centro earthquake wave often used in structural dynamic studies, and the reconstruction error (RE) and compressive ratio (CR) between the filtered signal and the original signal were calculated in order to quantitatively evaluate the effectiveness of reconstruction and compression, respectively. Finally, the AFB developed in this study was able to obtain effective dynamic responses whose size was compressed around the frequency range of interest while still sufficiently capturing the necessary time and frequency data of the original signal.

2 CONCEPT OF ARTIFICIAL FILTER BANK

A filter bank as defined in the field of signal processing as an array of band-pass filters designed to filter and output only a particular range of the input signal. A filter bank performing such decomposing and composing of a signal can be designed according to the designer's purpose and interest. The more band-pass filters and the closer their spacing, the closer the output signals to the input signal. However, greater numbers of band-pass filters mean less efficiency in computation and processing, which eventually requires further optimization of the filter-bank. For optimal numbers, the bandwidth and spacing of band-pass filters are considered main design factors. Basically, filter bank optimization is to satisfy the three design factors for a predetermined input signal via repetitive and numerical computations.

The filtering seems effective because the decomposed and composed output signal contain only the dynamic characteristics of the input signals of particular interest. However, the output signal is still the same size as the input signal due to the attributes of the band-pass filter. Therefore, a data compression technique is necessary to efficiently attain the dynamic response of a structure based on WSNs with limited wireless bandwidth. A key point in achieving a compressive technique is that dynamic characteristics of the original signal must be well reflected in the compressed signal as it was the same for the filter bank. Based on many design factors and conditions, this study

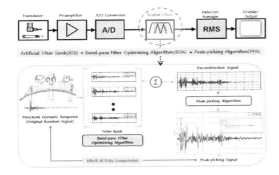

Figure 1. Concept of artificial filter bank (AFB).

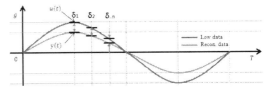

Figure 2. Concept of reconstruction error (RE).

developed an AFB using a band-pass filter optimizing algorithm (BOA) for filter bank configuration and a peak-picking algorithm (PPA) for data compression. Figure 1 is the conceptual diagram of the AFB developed in this study for obtaining dynamic responses of a structure.

2.1 Band-pass filter optimizing algorithm (BOA)

The AFB developed in this study requires a paralleled configuration of multiple band-pass filters to make it appropriate for evaluating the target mode to effectively attain the structural dynamic (acceleration) response of a structure. Such a filter bank selects the signal components of a specific frequency range from the input signal or the raw data with a larger spectrum of frequency, and filters them into separate bands. For the configuration of a filter bank, the number, bandwidth, and spacing of band-pass filters need to be determined. In this section, a band-pass filter optimizing algorithm (BOA) is developed for optimizing the filter bank and determining the three design factors. The computation processes of developing a BOA are as follows: First, choose the frequency range necessary for evaluating the target mode. Second, assume the number of band-pass filters based on the selected frequency range. Third, determine the filter bandwidth and center frequency spacing based on the assumed number of band-pass filters. Fourth, calculate the reconstruction error (RE) and compressive ratio (CR) based on the determined bandwidth and spacing of the band-pass filter. Last, adjust the number of filters based on the calculated RE and CR. The reconstruction error is the comparative difference between the composition of filtered signals and the original signal as expressed in Equation (1) and the graphs in Figure 2

$$RE = \frac{\int_0^T |u(t) - y(t)|/|u(t)|}{T} \quad (1)$$

where u(t) is the raw data (acceleration) against response time, y(t) is the reconstructed signal (acceleration) against response time, and T is the total period of response time in seconds. The closer the RE to zero, the better the reconstruction effectiveness will be. Lynch [2] and others apply a statistical theory to use the residual sum of squares (RSS) for calculating the RE, but this shows only a relative difference between the reconstructed signal and the raw data. Therefore, in order to accurately portray the reconstruction effectiveness, the difference between the reconstructed signal and the raw data is expressed in Equation (1) as the absolute difference.

2.2 Peak-picking algorithm (PPA)

As mentioned earlier, both selective filtering and compressive techniques are required to efficiently obtain the dynamic response of structures based on WSNs. In this section, a peak-picking algorithm (PPA) was developed in order to reduce the size of attained data, while still including the dynamic characteristics within the frequency range of interest. The developed PPA extracts peak values from the reconstructed signal, but not the raw data where peak values are extracted, and not from each decomposed signal, but from the reconstructed signal that combines the decomposed signals. This significantly reduces the computation process. The computation processes of the developed PPA are as follows: First, read the reconstructed signals determined during the first step of the AFB. Second, divide the reconstructed signals into three data sets. Third, calculate the derivative of each data set. Fourth, extract the peak values by evaluating the slopes of data set derivatives. Last, resample the extracted signals with the time information of peak values.

In this study, a central difference method is used to extract peak values during the computation process of PPA. Figure 3 shows the concept of central difference while the derivative is calculated using Equation (2).

$$f'(x(i)) = \frac{f(x(i+1)) - f(x(i-1))}{x(i+1) - x(i-1)} \quad (2)$$

The compressive effectiveness can be determined by comparing the size of the resampled compressed sinal based on peak values extracted using Equation (2) and the reconstructed signal determined by Equation (1). The compressive effectiveness can be evaluated using the compressive ratio (CR) in Equation (3)

$$CR = \frac{NS_C}{NS_O} \quad (3)$$

where, NS_C is the number of samples of the compressed signal, and NS_O is the number of samples of the reconstructed signal. The closer the CR is to zero, the better the compressive effectiveness.

3 NUMERICAL STUDY OF AFB

An AFB was developed in Section 2 to efficiently obtain the dynamic responses of structures. In this section, the developed AFB is optimally designed for random signals, and its performance and validation are evaluated from a numerical simulation of optimum AFB. Earthquake signals such as El-Centro, Kobe, and Northridge are major random signals used in the field of construction. Such random signals are considered and applied as sudden shock events during design, construction, and maintenance of a structure. This study chose El-Centro earthquake wave as the input signal to numerically evaluate the performance of the developed AFB. Figure 4 shows the time and frequency responses of El-Centro used as the input signal.

If the earthquake in Figure 4 is assumed, the dynamic response of the structure is measured over the frequency spectrum of the random earthquake, including the natural frequency of the structure. For structural health monitoring, it is necessary to find the natural frequency of the structure included in the random frequency. The target mode required for structural health monitoring can be limited to a specified frequency bandwidth below 10 Hz because the seismic behaviors of large-scale structures are usually low frequency. Therefore, the frequency range of interest is selected below 10 Hz for BOA of the AFB where the number of band-pass filters is also assumed to be 10. The frequency range of interest selected here applies to existing large-scale structures in general, and the number of band-pass filters is arbitrarily selected because it is to be optimized during the computation process of BOA later. Table 1 and Figure 5 are reconstruction errors (RE) calculated using Equation (1) while considering the assumed band-pass filter number is 10 and the selected frequency range of interest is below 10 Hz. Reconstruction errors are calculated for 100 cases by increasing the bandwidth and spacing of band-pass filters by 0.1 Hz between 0 and 1 Hz.

From Table 1 and Figure 5, the reconstruction error is minimized at a bandwidth of 0.6 Hz and spacing of 1.0 Hz for the filter bank with an initial filter bank number of 10. Selected bandwidth and spacing are determined as the optimal condition for BOA. Next, Figure 6 is the comparison between the reconstructed signals and the original signals for the optimal condition determined in Table 1 and Figure 5. Figure 6 shows that the reconstructed signal sufficiently describes the original signal. The above result shows that the initially selected frequency range of interest and the number of band-pass filters are sufficient for the description of

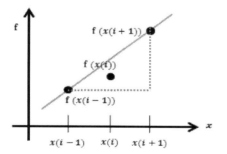

Figure 3. Concept of central difference.

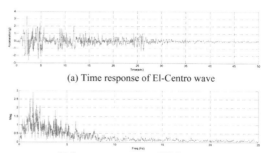

(a) Time response of El-Centro wave

(b) Frequency response of El-Centro wave

Figure 4. El-Centro wave as input signal of AFB.

Table 1. Reconstruction error (RE) for a filter bank with 10 filters.

		Bandwidth of filters (Hz)										
		0.0	0.1	0.2	0.3	0.4	0.5	0.6	0.7	0.8	0.9	1.0
Filter Spacing (Hz)	1.0	–	0.19930	0.16497	0.13378	0.10931	0.09347	0.08713	0.09036	0.10278	0.12373	0.15250
	0.9	–	0.19515	0.15896	0.12816	0.10692	0.09705	0.09893	0.11214	0.13588	0.16920	0.21118
	0.8	–	0.19374	0.15711	0.12810	0.11125	0.10870	0.12073	0.14650	0.18474	0.23406	0.29320
	0.7	–	0.18970	0.15631	0.13387	0.12719	0.13762	0.16462	0.20688	0.26290	0.33124	0.41053
	0.6	–	0.18396	0.14814	0.13360	0.14171	0.17114	0.21986	0.28583	0.36719	0.46229	0.56957
	0.5	–	0.18357	0.14861	0.14443	0.17345	0.23217	0.31612	0.42141	0.54484	0.68378	x
	0.4	–	0.19240	0.17890	0.21053	0.28428	0.39277	0.52962	0.68971	x	x	x
	0.3	–	0.19745	0.21661	0.29817	0.42797	0.59387	0.78741	x	x	x	x
	0.2	–	0.23026	0.29689	0.42203	x	x	x	x	x	x	x
	0.1	–	0.25073	x	x	x	x	x	x	x	x	x
	0.0	–	0	–	–	–	–	–	–	–	–	–

the original signal, and that the determined bandwidth and spacing are optimal.

This study also developed a PPA that extracts peak values of the reconstructed signal in order to reduce the size of the measurement data in real time. Next, Figure 7 shows the compressed signal of the reconstructed signal using Equation (2).

Figure 7 shows that the compressed signal extracts the peak values of the reconstructed signal. The sample data size of the compressive signal is 424, giving the compressive effectiveness of approximately 84% from the reconstructed signal, with a sample data size of 2500.

As a last step in the optimum design of the AFB, this study determined the number of band-pass filters based on the optimally selected bandwidth (0.6 Hz) and spacing (1.0 Hz) of the band-pass filters. The reconstruction error and compressive ratio were calculated against the different number of band-pass filters. The calculated RE and CR are shown in Figure 8, and the results are tabulated in Table 2 along with peak-picked values.

Figure 8 shows that RE and CR have an inverse relationship, and the optimum number of band-pass filters are determined to be 6, where the relative difference between RE and CR is at a minimum. Also Table 2 shows that the size of the compressed data with band-pass filter number 6 is 312, which gives more efficient compression effectiveness compared to compressed data size 424 of the initially assumed band-pass filter number 10. Conversely, the reconstruction effectiveness becomes slightly decreased.

Next, Figures 9 and 10 are the time and frequency responses of the filter-bank of the determined optimum conditions where the number of band-pass filter is 6, bandwidth is 0.6 Hz, and spacing is 1.0 Hz. Figures 9 and 10 show that the filter bank is optimally designed using the determined conditions.

The reconstructed signals produced under the optimum conditions are shown in Figure 11. The reconstruction error increases by 48% from that of the

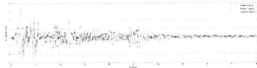

Figure 7. Original vs. Recons. vs. Comp. signal.

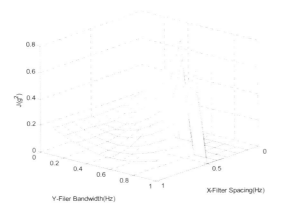

Figure 5. RE results for filter bank with 10 filters.

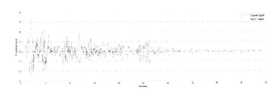

Figure 6. Original vs. Reconstruction Signal.

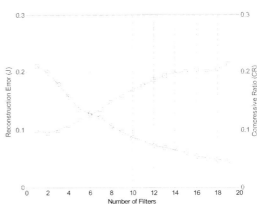

Figure 8. Optimization of number of filters.

Table 2. Optimization of number of filters (Reconstruction error (RE) vs. Compressive ratio (CR)).

No. of Filters	1	2	3	4	5	6	7	8	9	10
	11	12	13	14	15	16	17	18	19	20
RE	0.2108	0.1997	0.1792	0.1571	0.1366	0.1289	0.1222	0.1072	0.0965	0.0871
	0.0808	0.0732	0.0694	0.0650	0.0605	0.0545	0.0516	0.0489	0.0473	—
CR	0.1000	0.0952	0.0992	0.1080	0.1160	0.1248	0.1368	0.1480	0.1600	0.1696
	0.1768	0.1872	0.1944	0.2000	0.2008	0.2040	0.2008	0.2056	0.2136	—
No. of P-p's	250	238	248	270	290	312	342	370	400	424
	442	468	486	500	502	510	502	514	534	—

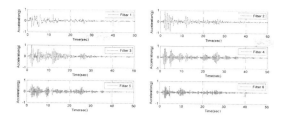

Figure 9. Time domain of optimal condition.

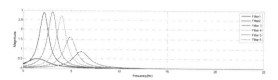

Figure 10. Frequency domain optimal condition.

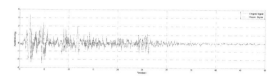

Figure 11. Original vs. Recon. Signal.

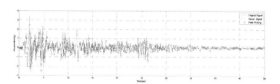

Figure 12. Original vs. Recon. vs. Comp. signal.

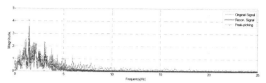

Figure 13. Original vs. Recon. vs. Comp. signal.

band-pass filter number 10 as seen in Table 2, but Figure 11 shows that the reconstructed signal of the optimum conditions still satisfactorily capture the original signal. Utilizing the validated PPA, the data is compressed for optimal condition and illustrated in Figure 12, which shows that the compressed signal is accurately extracting only the peak values of the reconstructed signal.

For structural health monitoring, the capturing of the original signal by the reconstructed signal and compressed signal in frequency domain must also be considered along with the responses in time domain. In other words, the compressed signal by the PPA must contain the data from the target frequency range while compressing the measurement data by extracting peak values from the reconstructed signal. To validate this, Figure 13 plots the frequency response of the reconstructed signal from the BOA and compressed signal from the PPA. It shows that the reconstructed signal sufficiently captures the data of the original signal within the target range of frequency. It also shows that the compressed signal sufficiently captures the original signal in the frequency domain of interest. From this, it can be concluded that the optimally designed AFB using a BOA and PPA can efficiently attain the dynamic responses of structures, while it is also numerically validated that the reconstructed signal and compressed signal all contain the necessary original signal data in both time and frequency domain.

4 CONCLUSION

This study developed an artificial filter bank using a band-pass filter optimizing algorithm and peak-picking algorithm for efficient attainment of significant dynamic (acceleration) responses of structures, and the AFB was further designed and validated by numerical simulations. In order to evaluate the performance and validity of the AFB, the obtained reconstructed signal and compressed signal were quantitatively compared in time and frequency domain with the original data. Finally, from these numerical simulation results, the following could be concluded:

1) The band-pass filter optimizing algorithm developed here was proven useful in selectively displaying the data of the frequency range of interest from a wide range of frequencies of random signals.
2) The peak-picking algorithm developed in this study was proven effective in compressing the data by selecting peak-values and containing only significant modal information of the dynamic response.
3) Finally, the AFB developed in this study is expected to provide a new paradigm in wireless measurement, transmission, and management of dynamic structural response in real time because a variety of optimization skills for the AFB can be applied to obtain dynamic responses not only of construction structures, but of other structures as well.

ACKNOWLEDGMENT

This research was supported by Basic Science Research Program through the National Research Foundation of Korea (NRF) funded by the Ministry of Education, Science and Technology (grant number: NRF- 2013R1A2A1A01016192), and funded by the Ministry of Science, ICT & Future Planning (grant number: NRF- 2013R1A1A1063540).

REFERENCES

[1] Heo, G. and Jeon, J. 2009. A Smart Monitoring System Based on Ubiquitous Computing Technique for Infra-structural System: Centering on Identification of Dynamic Characteristics of Self-Anchored Suspension Bridge, KSCE J. Civil Engineering, 13(5), pp. 333–337.

[2] Peckens, C.A. & Lynch, J.P. 2013. Utilizing the Cochlea as a Bio-inspired Compressive Sensing Technique, Smart Materials and Structures, 22(10), 105027.

Development and prospects of structural vibration control

Long Yue
College of Civil Engineering and Architecture, Southwest Petroleum University, Chengdu, China

ABSTRACT: Structural vibration control is a systematic project. It combines the structure and control systems together to resist the external dynamic loads to control the structure and morphology, reducing the structural dynamic response requirements. Its development is currently focused on the theory of algorithms, control instruments and the selection of structural materials. Its specific application is embodied in bridge structures and high-rise buildings and other areas of seismic vibrations wind. Currently, the project is the development of hot spots will study structural vibration control and seismic. There is a gradual trend of transition from previous passive controlling to artificial intelligence controlling. The current trends in the project can provide some advices as well as directions for the future development of structural vibration control studies.

Keywords: structural vibration control; technology; system; material; prospects

1 INTRODUCTION

Structural vibration control is an interdisciplinary discipline that includes automatic control, structural mechanics and mechanics. In 1972, J.T.P. Yao proposed the concept of structure control. Then, scholars have introduced control theory and control systems only in structural engineering, so that the structure and control systems work together to resist the external dynamic loads, so as to control morphology and reduce the structural dynamic response requirements. With the further development of this discipline, we can study structural vibration control from different angles. In respect of the system, structural vibration control can be divided into active control, passive control, semi-active control, hybrid control, etc. Based on the structures of the control algorithms, they can be divided into classical algorithms (Root locus method point, Laikuisite map, Figure omen, Laichiersitu map, etc.) and modern algorithms (modal control method, optimal control method, pole configuration, etc.). Based on the materials used for the structures, they can be divided into non-smart structures and smart structures.

Recently, the Earth has become increasingly active, for example frequent earthquakes. That is bound to cause a huge impact on the safe use of housing constructions, bridges and other large structures, thus promoting the development of structural vibration control technology, which becomes essential, especially in the civil engineering seismic, wind and other fields.

2 DEVELOPMENT OF STRUCTURAL VIBRATION CONTROL TECHNOLOGY

2.1 Development of structural vibration control algorithm theory

In 1972, Yao Zhiping (Yao J.T.P.) described the control of civil engineering structures under wind and earthquake conditions. Then, with the mode control theory proposed by Porter B., Martin C.R. et al. analyzed the control problem of multilayer structures. Yang J.N. et al. derived the transfer matrix method of civil structures active vibration control. Abde-l Rohman M. et al. researched the control optimization problems of civil structures. Chang J.C.H. et al. applied active mass damper for civil structures to optimize the feedback controlling. Yang J.N. et al. proposed open-loop control method for optimizing critical modal high-rise buildings under earthquake excitation. Meirovitch L. et al. optimized independent modal space control (IMSC) method to realize the building structures under seismic control. Meirovitch L. et al. discussed to suppress unstable suspension bridge with IMSC method Flutter Modal. Pu J.P. et al. researched IMSC method to optimize the use of discrete-time linear output feedback control for multi-storey structures. After being investigated by scholars from many countries in recent decades, structural vibration control has evolved into sets of vibration control systems, control technology, algorithm theory, nonlinear, multi-scale, multi-coupled system dynamics theory and simulation technology in one of the macroscopic theory.

2.2 Development of structural vibration control systems and apparatus

Because of the random loads (wind load and earthquake), scholars use some dampers to control structural vibration. This method also promotes the active control, passive control, semi-active and hybrid systems under the control of the various instruments used in vibration control development.

2.2.1 Active control

Active Control Technology is a high-tech method that applies modern control theory and automatic control technology into the structural seismic (Takeda Hisaichi, 1997); it is used in civil engineering since late 1960s. Meanwhile, any structure can be controlled in theory and the damping effect is better than that of passive control systems. Currently, the active control system consists of (Guangyuan Wang, 1980) sensors, computing and applied force actuator, the main system consists of (1) Active tuned mass damper vibration control system (ATMD), (2) Active Mass Damper vibration control system (AMD), (3) Active cables vibration control system (AST), (4) Active support vibration control system (ABS), and (5) Active vibration control system of variable rigidity (AVSS). Japanese scholars Kobori et al. (Kobori T. 1990) first proposed active variable stiffness system (AVS). Nasu et al. (Nasu T. et al. 1996) studied the towering structure of a device equipped with Active Variable Stiffness. At present, Japan has more than 20 buildings using active controlling. In the State University of New York, United States of Buffalo, T.T. Professor Song used vibration table to perform a experiment on steel frame model fitted Afs, and the damping effect is obvious (Jing Cai, 2001).

2.2.2 Passive control

Currently, passive control consists of base isolation, energy dissipation damping and shock absorption. The latest research base isolation technology are mainly laminated rubber bearing friction sliding isolation, ball and roller isolation, braced swing isolation, lead rubber isolation pads, and mixing isolation (Qiaobin Peng et al. 2007). The technology of energy dissipation control is developing rapidly. According to incomplete statistics, the countries launching structural damping technology research are over 20 at present, the practical application of engineering has more than 300 items, and has achieved remarkable results (Yun Zhou et al. 2002). The study of structural damping technology produced steel consumption transducer, lead dampers, friction dampers, viscous dampers and viscoelastic dampers, etc.

2.2.3 Semi-active control

Semi-active control is the control of a small amount of additional energy. Although it is also controlled by the motion of control that the device itself generates passively, the control device can use external energy initiative to adjust its parameters to play a regulatory control effect in the control process. Semi-active control system was first produced in the 1920s. In 1983, Hrovat et al. (Kamopp D. et al. 1974) first proposed semi-active control technology in the field of structural engineering. After a series of development, currently, semi-active control system has generated the following devices: semi-active variable stiffness device, semi-active variable damping device, semi-active damping fluid control device, variable friction control device, semi-active tuned mass (liquid) damper control device, electromagnetic fluid dampers control device, and semi-active isolation device (Yun Zhou, 1996 and Guiqing Li et al. 1992).

2.2.4 Hybrid control

Hybrid control system is the one that put passive control and active control as a whole to analyze the structure in order to overcome the limitations of passive vibration control. In the analysis of the vibration control, control is reduced while reducing the power of the external control device, the volume of energy and maintenance costs, increase system reliability will be active control and passive control in combination, can give full play to the advantages of the two control systems to overcome their shortcomings, only a small energy input to achieve better control effect. Currently, there are three main types of control devices, active control with the combination of base isolation (Yang J.N. et al. 1993), active control (including artificial neural network control) with the combination of energy dissipation devices (Nagashima I. et al. 1997), hybrid control system of the main mass damper (AMD) with tuned mass damper (TMD) system or tuned liquid damper (TMD) system, hybrid control of damping energy dissipation and active support system (ABS), and blended mass damper HMD et al. (Yang J.N. et al. 1995).

2.3 Development of intelligent structural vibration control

In early 1980s, Swigert used piezoelectric ceramic element to study the controlling problem of cylindrical antenna model, creating smart structures for structural vibration control precedent. Subsequently, many scholars have studied a variety of flexible smart material structural vibration control problems, such as Bailey and Hubbard, who used PVDF piezoelectric film to control the vibration of the beam successfully, and Chen et al. who studied the vibration control problems of the beam with upper and lower surfaces attached by a thin layer of SMA. At present, smart materials have been widely used in military, machinery, medicine, and civilian products in the field of structural vibration control. In the U.S., the Naval Surface Weapons Center clark invented a ship magnetostrictive material Terfenol-D, which is expected for the vibration control actuators. NASA used electroluminescent material to improve the Hubble Space Telescope. Intelligent material systems and structures Center, Virginia Tech University and the State University prepared a dense fiber network by Ss, while put it on a polymer matrix composite and made this composite material have high impact resistance. In the near future, intelligent structural vibration control system will replace the traditional bulky mechanical or electromechanical vibration control device. Meanwhile, the vibration control device tends to miniaturization and intelligence.

3 STRUCTURAL VIBRATION CONTROL TECHNOLOGY IN CIVIL ENGINEERING

Structural vibration control of civil engineering can effectively reduce the structural response and damage accumulation in the dynamic action of the vehicle, wind surf, water, earthquakes, etc., and improve the capacity and resilience vibration performance of the structure. It is an effective method and technique in structure vibration mitigation and disaster prevention and mitigation positive. Currently, the research and application is generally divided into three areas, base isolation, passive energy dissipation and active, semi-active and intelligent control.

After nearly 30 years of research, structural vibration control in civil engineering applications has been reflected in various fields. In the field of seismic isolation: Engineers built the world's first lead-rubber mat isolated building-William Clayton government office buildings in 1981. In 1985, the world's first high-damping rubber pad construction was built in California. In the field of bridge seismic wind, Gu et al. (Gu M. et al. 2002) proposed a leveraged tuned mass damper and simulated large-span bridges under wind-induced vibrations. Chen et al. (Chen S.D. et al. 2008) studied the problem that TMD inhibits long-span highway bridge vibration under wind loads and car loads. Wang et al. (Lin C.C. et al. 2005) studied vibration control of Taiwan High Speed Railway Bridge about MTMD. Evidently, the theory is based on the principle of multiple tuned mass dampers. Considering the deck roughness factor, Jo et al. (Jo B.W. et al. 2001) studied the effect of TMD on the three-span continuous steel box girder bridge of civil action arising under vehicle loads. Similarly, structural vibration control is also applied in some high-rise buildings. For example, the World Trade Center towers in New York were installed with 360 tons of TMD damping system at the top, the Lohn Hancock Tower in Boston was installed with two 300-ton TMD systems at the top, the Sydney Tower had TMD system installed at the top and in the center (Xiaoqin Yuan et al. 2012).

4 CONCLUSIONS

In summary, the concept of structural vibration control (Guangheng Zhao et al. 1999, Yunjie Shen et al. 1999 and Dichun Hua et al. 2008), from a proposal to the establishment of a vibration control theory, after years of engineering applications, has made remarkable progress. Nevertheless, the following aspects of structural vibration control require further study (Fei Hu et al. 2001 and Jian Mao et al. 2001).

(1) To use new types of seismic vibration control materials.
(2) To establish energy component of ductile structures, new mechanical model architecture and structure of the population containing rational thought control invention, the preparation of the corresponding calculations, and analysis software.
(3) To improve the vibration control mechanics, structural mechanics, computer-controlled areas such as methods and theoretical techniques, and explore the mechanical properties of structural vibration clearly associated with the material.
(4) To absorb new principles and new technologies in other disciplines, develop the theory of intelligent control structure, determine the structural development plan of intelligent control, and promote the development of intelligent structural vibration control to accelerate the practical process of shape memory alloy in the field of structural vibration control. (Sakai F. et al. 1989).

Thus, structural vibration control has a broad prospect; it will give civil engineering a revolution, especially in the field of seismic resistant design.

REFERENCES

Chen S.D. & Wu J. 2008. Performance enhancement of bridge infrastructure systems: Long-span bridge, moving trucks and wind with tuned mass damper [J], Engineering structures, 30:3316–3324.

Dichun Hua & Qiting Fan. 2008. Vibration Control in Civil Engineering Application of [J]. Chinese waterway, (8):191–192.

Fei Hu & Zhigang Zhou. 2012. New method for structural vibration control [J]. Shanxi Architecture, 38(4): 56–57.

Guangheng Zhao. 1999. Advances in civil structural vibration control [J]. Seismology Journal, (3):35–42.

Guangyuan Wang. 1980. Towering wind vibration control [C]. Towering structure Symposium.

GB50011-2001. 2001. Seismic Design of Buildings [s]. Beijing: China Building Industry Press.

Guiqing Li & Hongwei Zhou. 1992. Progress in structural control [M]. Beijing: Earthquake Press.

Gu M., Chen S.R. & Chang C.C. 2002. Control of wind-induced vibrations of long-span bridges by semi-active lever-type TMD [J]. Journal of Wind Engineering and Industrial Aerodynamics 90:111–126.

Jing Cai. 2001. Span cable-stayed bridge seismic response of active control [D]. Chengdu: Southwest JiaoTong University (PhD thesis).

Jo B.W., Tae G.H. & Lee D.W. 2001. Structural vibration of tuned mass damper-installed three-span steel box bridge [J]. International Journal of Pressure Vessels and Piping, 78:667–675.

Jian Mao, Qingzhong Qin & Jie Zhang. 2001. New progress of structural vibration control [J]. Control Theory & Applications, 18(5):647–652.

Kamopp D., Crosby M.J. & Hal-wood R.A. 1974. Vibration control using semi-active force generators [J]. Journal of Engineering for industry, 96(2):619–626.

Kobori T. 1990. Technology development and forecast of dynamical intelligent building (DIB) [J]. Journal of Intelligent Material systems and Structures, 1(4):391–407.

Lin C.C., Wang J.F. & Chen B.L. 2005. Train-induced vibration control of high-speed railway bridges equipped with multiple tuned mass dampers [J]. Journal of Bridge Engineering, (4):398–414. October 2005.

Manlin Wu, Ping Tan & Mao Ye [J]. 2009. Water Resources and Architectural Engineering, 7(4):19–26.

Nasu T., Kobori T., Takahashi M. et al. 1996. Analytical study on the active variable stiffness system applied to a high-rise building subjected to the records in osaka plain during the 1995 hyogo-ken nanbu earthquake [J]. Journal of Structural Engineering, 42(8):1–8.

Nagashima I. & Shinozaki Y. 1997. Variable gain feedback control technique of active mass damper and its application to hybrid structural control [J]. Earth Engrg & Struct Dyn, (26):815–838.

Proceeding of Seminar on Seismic Isolation. 1993. Passive Energy Dissipation and Active Control, ACT17-1 California: Redwood.

Qiaobin Peng, Xiaolei Han & Jing Ji. 2007. Application Status passive control techniques in civil engineering [c]. Seventh National Symposium on Modern structural engineering [A]. Tianjin.

Sakai F., Takaeda S. & Tamaki T. 1989. Tuned liquid column damper-new type device for suppression of building vibrations. In Proceedings of International Conference on High-rise Buildings: 926–931. China: Nanjing.

Takeda Hisaichi. 1997. Building isolation, vibration and vibration control [M]. Beijing: China Building Industry Press.

Xiaoqin Yuan, Xijun Liu & Suxia Zhang. 2012. MR-TMD damping system for dynamic control of Continuous Box Girder Bridge [J]. Vibration and Shock, 31(20):153–157.

Yun Zhou & Tong Xu. 2002. Retrospect and Prospect of Energy Dissipation Technology [J]. Mechanics in Engineering, 22(5):1–7.

Yun Zhou. 1996. Energy dissipation theory and design [M]. Heilongjiang: Harbin University of Architecture.

Yang J.N., Danielians A. & Liu S.C. 1995. Aseismic hybrid control systems for building structures [J]. J. Eng. Mech., ASCE, 12(4):555–567.

Yang J.N., Li Z. et al. 1993. Nonlinear control buildings using hybrid systems [C]. Proc. of ATC-17-1 Seminar on seismic isolation, passive energy dissipation, and active control. vol. 2: 11–12 Mar 1993. San Francisco California.

Discussion on construction technology of asphalt pavement on expressway

Xiang-Ping Ou, Kui Ding, Wei-Lu Xu, Fei Gao & Lei Jiang
Wuhan University of Technology, Wuhan, Hubei, China

ABSTRACT: In recent years, with the rapid development of highway construction and the dramatical increase in the number of vehicles in China, there has been more early damage of asphalt pavement on expressway. Therefore, improving controls of the quality in asphalt pavement construction process is imminent. Aiming at problems of the quality occurred in China's expressway asphalt pavement construction, this article put forward some main points of quality control in the process of asphalt pavement construction, such as the selection of raw material of asphalt pavement, the mixing and transporting of asphalt mixture, and the paving and rolling in asphalt mixture.

Keywords: expressway; asphalt pavement; asphalt mixture; pavement construction; paving

1 INTRODUCTION

Asphalt pavement is a mixture pavement structure composed of various substrates and cushions and surface layer, which is constructed with mixing materials including mineral aggregate and asphalt that are used as binder material. The construction process of it is as follows: firstly, design the proportion of asphalt mixture; secondly, mix and transport asphalt mixture; finally, pave and roll asphalt mixture. Asphalt pavement has merits of smooth and seamless surface, comfortable driving, wear resistance, small vibration, low noise, short construction period, easily maintaining and repairing, construction by stages and so on. Therefore, it has been widely used in China and has made an important contribution to the economic development.

However, in the operation process, there will be some damages, such as longitudinal cracks, lateral cracks, reticular crack, rutting, loosing, peeling off and surface polishing, etc., resulting in reduction in the strength of pavement. Under the vehicle load, pavement becomes distorting, cracking, loosing and boiling. Especially in sections that have poor climatic, hydrographic and geological conditions, the damage phenomena are more serious. These diseases directly cause decreases in pavement performance, therefore, some sections within a short time after the completion have to be overhauled; some highways even have early severe damage in the opening two or three years. These damages not only reduces the efficiency in the use of the maintenance funds for highway construction, also greatly affects the social image of highway construction.

2 REQUIREMENTS OF ASPHALT MIXTURE

2.1 Characteristics of asphalt mixture

Asphalt mixture is a senior pavement material mixed with mineral mixture that has passed through an artificial, rational selection of gradation and the right amount of asphalt materials at given temperature. As the main pavement material in high grade highway, it has the superiority to which other many building materials are unable to compare (Zhi-Qing Jang, 2004). Its specific performances are as follows:

(1) Asphalt mixture is a kind of elastic-plastic viscous material, so it has a certain high temperature stability and low temperature crack resistance. The pavement is flat and flexible, and does not need to set expansion joints. Driving is also quite comfortable on it.
(2) Asphalt mixture pavement is a little rough, providing a good resistance to sliding on rainy days. The pavement can guarantee certain smoothness too; for example, the smoothness of highway pavement is 1.0 mm. Furthermore, asphalt mixture pavement is so black that it has no strong reflection; this can protect the safety of driving.
(3) Asphalt mixture pavement construction is convenient, fast, and has a short curing period, so it can open to traffic in a timely manner.
(4) Asphalt mixture pavement can be transformed by stages and recycled. With the increase of road traffic, widening and thickening the original road pavement are feasible. For old asphalt mixture, modern technology can to used to recycle it in order to save materials.

Of course, the asphalt mixture pavement also has some problems, such as aging, which will make the road surface loose, causing the pavement damage. In addition, it has a poor thermal stability. The pavement is easy to soften under high temperature in summer, causing rut, hug, and wave phenomenon; and it is easy to generate brittle crack under low temperature in winter, causing cracks under repeat load of vehicles (Xue-Jun Deng, 2006).

2.2 Proportion between asphalt and mineral powder

Asphalt is too little to be a film bonding surface of mineral aggregate particles. With the increase in the amount of asphalt, the mineral aggregate surface can be completely packed, and asphalt gradually forms a structure of asphalt, so that the binding power between asphalt and mineral aggregate increases with the increase of the dosage of asphalt. When the asphalt content is the most suitable, the bonding power of asphalt cement is the best. Subsequently, if the asphalt content continues to increase, the asphalt would gradually push away the mineral aggregate particles and change into "free asphalt" between the particles without interaction with the mineral powder. At the moment, the bonding power of asphalt cement decreases with the increase of "free asphalt". When asphalt dosage increases to a certain amount, the cementing force of the asphalt mixture depends mainly on free asphalt, so the shear strength is almost unchanged. With the increase of asphalt content, at this point, the pitch is not only a binder, but also works as a lubricant. It damages the dense structure of coarse aggregate and reduces the friction within the asphalt mixture.

In the case of a fixed amount of asphalt, the amount of mineral powder directly affects the dense degree and the cohesive force of asphalt mixture. There would not be too much mineral powder, especially that the content of mineral powder since its particle size of less than 0.075 mm is unfavorably overmuch. Otherwise, the asphalt mixture clusters into pieces and is not easy for construction (Industry Standard of PRC, China Communications Press, 2004).

2.3 Effects of temperature and loading rate on the shear strength of asphalt mixture

With the increase of temperature, asphalt cohesion decreases, while the deformation capacity enhances. When the temperature decreases, the mixture cohesion and strength increase, while the deformation capacity weakens. However, overly low temperature would cause the cracking of asphalt mixture pavement. Loading at high frequency could make the asphalt mixture produce excessive stress, plastic deformation and slow elastic recovery, even generate permanent deformation, which could not be restored (De-Gang Xin, et al. 2006).

Table 1. Main Material of Asphalt Concrete Pavement on Jiangnan Highway.

Number	Material Name	Specification
1	Asphalt	Petroleum asphalts for heavy traffic road pavement AH-90
2	Coarse aggregate	3~5 mm limestone mass of stone 5~10 mm limestone mass of stone 10~15 mm limestone mass of stone 15~20 mm limestone mass of stone
3	Fine aggregate	0~5 mm, 0~4.75 mm washed-out sand
4	Mineral power	0~0.6 mm limestone mass of stone
5	Anti-stripping agent	TA-1

3 CONTROL OF ASPHALT MIXTURE RATIO

3.1 Control requirements of material quality

Material factors have absolute importance in cost and quality in construction of asphalt concrete pavement. In a larger sense, material quality control is to control the quality of pavement structure layer. At present, the materials used in the Jiangnan project pavement engineering are mostly specified by the owners, thus the material factors are at a disadvantage. Monitoring, reclaiming and strengthening inspection are able to reduce the impact of adverse factors.

Materials supply departments should carry out reclaiming inspection in the material field that assigned by Party A instead in the mixing field, and make sure the material is relatively uniform and contains small amount of water in the first step in reclaiming. The test departments should reinforce standard testing on materials in order to put an end to unqualified materials involved in the construction and ensure good quality materials used in each contract section.

3.2 Proportion design of asphalt mixture

Bitumen mix design is to determine varieties and proportion of different raw materials, mineral aggregate gradation and the best bituminous content. The mix proportion design of asphalt mixture includes target mix design, production mix design and production mix verification (Xue-Jun Deng, 2006). The Chinese *Technical Specifications for Construction of Highway Asphalt Pavement* (JTGF40-2004) clearly stipulates that ratio design of hot mix asphalt mixture should use the Marshall stability method. First of all, choose mixture types, basic properties of raw materials, primary mix range and asphalt content. Then, form specimens according to the Marshall test method and determine its physical indicators after 12 hours, such as bulk volume density, void ratio, asphalt saturation and voids in mineral aggregate. At last, measure the Marshall stability and flow values. On the premise of assurance specification, the asphalt concrete mix ratio design in Jiangnan project highlights the cost. On the one hand, it reasonably uses low asphalt aggregate ratio. On the other hand, it reduces the more expensive materials in the graded aggregate ratio appropriately under the same condition of technical standards.

4 QUALITY CONTROL OF PAVING AND COMPACTION TECHNOLOGY

4.1 Asphalt mixture paving

Before paving the bituminous mixture, it is necessary to make preparations on the through layer. It is better to clear the through layer thoroughly by manual work and air compressor ($9 m^3$). The value of the evenness must be controlled in the permissible range when retesting technical indices, otherwise, it should be remedied. Use the milling machine to mill the bulge at the grassroots, and dig a pit to backfill the sunken at the grassroots. Although the distributing layer is mainly used to connect each structure layer closely to sustain vehicle load as an entirety, if distribution is too much, it will easily lead to bleeding. Bituminous concrete paver is a special equipment that can pave bituminous mixture to the grass roots or with the lower level homogeneity as the technical requirements and tamp preliminary (Shao-Zhong Zhao, 2004). A paver consists of drive system, feed equipment (receiving hopper, flight conveyers, spiral distributor), and working device (beam vibrator, screed unit, automatic leveling device).

4.2 Technology of asphalt mixture paving

It is hard to be informed of the reasonable compaction technology before a project. After selecting the type and combination of the road roller, the rolling speed, frequency, amplitude, rolling operation length, effective compaction time, etc. can be preliminarily determined according to experience. Finally, the compaction technology is determined by experiment on test road.

The compaction is divided into three processes: initial pressure, re-pressing and terminal pressure. Initial pressure is to equalize and stabilize mixture, and make a benefit to the re-pressing, it is the basic of compaction, thus the smoothness of the pressure should be noticed. The effect of the re-pressing is to make the mixture dense, stabilized and formed, and the compaction rate of the mixture depends on this process. This is the reason to choose the type of road roller and frequency and amplitude of the road roller reasonably. Terminal pressure is to eliminate the wheel track to format a compaction surface, thus heavy road roller cannot be used on this process (Hai-Xin Yao 2005).

The driving wheel should face to the paver while rolling. The path and direction of the rolling cannot be changed suddenly since it will lead to the mixture displacement. The starting and stopping of the road roller must proceed slowly. To decrease or eliminate part upheaval at the joint of rolling, the rolling direction of the road roller can be adjusted to diagonal lines, but the adjustment must be done smoothly, or new upheaval will appear.

5 ASPHALT CONCRETE PAVEMENT CRACK TREATMENT

Some generation of the grassroots crack will influence the asphalt concrete pavement, bringing some quality problems to the pavement and reducing the road durability. The prevention of the crack should be focused on, and the design of the embankment and pavement layer and all aspects of construction should be paid attention to. Once taking effective measures to regulate the management, organizing construction process rationally, and preventing the occurrence of some small failure factors, the crack problems could be reduced. The pavement cracks should be solved promptly after the asphalt pavement crack appears, in order to prevent intrusion of the water and other harmful substances and to decrease the impact on the road use life. Before pouring, the seam, seam side debris and garbage must be cleared, in order to ensure that the seam is desiccant. After pouring, the surface should be sprinkled by coarse cloth or 0~5 mm stone chips.

(1) The pouring cracks of the irrigation pouring sealant asphalt pavement are affected by factors such as grooving depth, dryness of the groove walls, cleanliness, construction temperature, and pouring materials, etc. Before pouring, the crack must be dug, and the seam must be cleared and dried.

(2) Construction process. Firstly, close the traffic, put safety signs in the right places, appoint someone to direct the traffic, and move the traffic signs in accordance with the progress at any time. Secondly, slot (expand the seam). Generally, if the seam is smaller than the width of 1.5 cm, it should be expanded along the crack force direction to required size (normally slit width 1 cm, deep 2 cm), and the groove edges must stay straight and tidy. Then, clean up the crack. Using compressed gases and blower, thoroughly remove the particles and debris inside the crack. For crack that is full of water, use hot air gun, ensuring an environment that is clean and free of ash, to integrate closely the crack wall and filling materials (Chang-Shun Hu, 2003). Preheat the groove before pouring with ordinary liquefied gas external spitfire device, till the temperature reaches 80–100°C, which is good for firmly bonding between pouring glue and asphalt concrete. When the pouring sealant heating temperature reaches 188°C, the oven dish stops heating automatically, and then turns into the insulating state; after that, the pressure nozzle, which works as a calibrating device along with the irrigation sewing machine, pours the sealant evenly into the crack. The crack should be filled in two times; at the first time, pour the groove crack in of 4/5 depth, and at the second time, fill the crack and pull a T-seal affixed layers, whose

width is 50 mm, and thickness is 2∼3 mm (above the road surface). Finally, sprinkle some conservation material. After finish the pouring, spreading powder or sand should be sprinkled on the sealed plastic surface. The traffic should not be opened until the pouring glue has cooled down to normal temperature. Generally, the cooling time is about 15 minutes.

6 TECHNOLOGY OF CURVE PAVING

Curve paving is an important part of the construction technology and a technical difficulty in construction of asphalt pavement. Currently, there is not any kind of construction technology and force based on strict scientific method in China. The interchange of highway between the upper and lower usually needs a ramp to connect each other, and the interchange has small turning radius, big superelevation between the inner and outer parts and short superelevation runoff, leading to severe change of cross slope.

The underlying cause that influences the accuracy and smoothness of cross slope on curve paving is a time lag inevitably exists between the given thickness control signals and the achieved corresponding thickness of asphalt mixture paving layer, both of which are determined by the characteristic of floating screed (Ming-Jian Shao, 2001). The control measures are as follows:

(1) When setting paving benchmark, on the basis of referring to the balance beam standard and manually controlling input signal of cross slope, the variable slope was divided into several parts along a high side according to the arc length, and the cross slope values that were required to change of every arc length were calculated according to the changed values of the total horizontal slope.
(2) Layer thinning caused by the error adjusting signals of thickness can be solved by appropriately thickening the layer of the ultra high side in advance and always checking the layer thickness of the ultra high side in the process of changing slope paving. When necessary, continuously fine-tune longitudinal sensor can be used to control the imperative thickness.
(3) Paver steering is required to conduct slowly and in balance in accordance with the direction line drew ahead of time to avoid substantiation.
(4) To avoid confusion and error in construction and guarantee the quality of curve paving, the construction technology plan shall be worked out in detail before paving test road, from the preparatory work of construction to the specific operation of cross slope controller.

7 CONCLUSIONS

With the further promotion of asphalt pavement in China, it will certainly play an important role in Chinese road construction. In order that pavement roughness can ensure a comfortable driving, it requires more attention paid to the roadbed from construction preparation stage. Through the research of the process quality control of construction process on surface course of asphalt pavement on Jiangnan highway, this article drew the following conclusions:

(1) Based on test and control of raw materials, fine-tuning mixture ratio according to the specific requirements of engineering in the mix the allowable scope of design can not only ensure the construction quality, but also reduce the engineering cost.
(2) The specific selection of mixing equipment should be based on the engineering quantity and optimum combination required by period. At the same time, comprehensive debugging on asphalt mixing station should be done before extensive mixing in order to select the mixing parameter.
(3) Controlling cold silo flow, size and inclination angle of vibrating screen, uniformity of receiving hopper on hot silo and mixing temperature and time can effectively reduce the production of unqualified material and put forward corresponding solving measures directed at the abnormal phenomena during the mixing process.
(4) In the transportation of mixture, strengthening the reasonable matching of transport vehicles, controlling the transport speed and avoiding transporting for long distance, can effectively reduce the segregation of mixture and the heat aging process phenomenon generated in transport.
(5) Precisely setting on reference line and strictly controlling of paving speed, paving thickness, paver running direction, through layer flatness, as well as compacting in accordance with the rolling procedure, are able to improve the pavement smoothness and durability effectively.

REFERENCES

Deng, X.J. 2006. *Subgrade and Pavement Engineering*. Beijing: China Communications Press.
Hu, C.S & Huang, H.H. 2003. *Construction Technique of Subgrade and Pavement of high grade highway*. Beijing: China Communications Press.
Industry Standard of PRC. 2004. *Technical Specifications for Construction of Highway Asphalt Pavement* (JTG D50-2006). Beijing: China Communications Press.
Jang, Z.Q. 2004. *Road Construction Materials*. Beijing: China Communications Press.
Shao, M.J. 2001. Construction Technique of Mechanization and Quality Control on Asphalt Pavement. Beijing: China Communications Press.
Xin, D.G. & Wang, Z.R. & Zhou, X.L. 2006. *Materials and Structure of Expressway Asphalt Pavement*. Beijing: China Communications Press.
Yao, H.X. 2005. *Discussion of Paving Technology and Development Trends for High-grade Highway [J]*. Construction Machinery Technology & Management 2005(08): 41–44.
Zhao, S.Z. 2004. *Study on Quality Control of Asphalt Pavement of Advanced Highway [D]*. Hunan: Civil Engineering of Hunan University.

Effects of elevation difference on GPS monitoring accuracy

W. Qiu, T. Huang & X.M. Wang
Earth Science and Engineering College, Hohai University, Nanjing, Jiangsu, China

ABSTRACT: Several commonly used models for GPS tropospheric delay correction are introduced, and the effects of the three main meteorological elements, temperature, pressure and relative humidity, on the results of correction are analyzed. The results show that meteorological differences caused by elevation will have some influences on GPS calculating results in dam projects, which have characteristics of short baseline and large elevation difference. Using average meteorological data between stations to process those large elevation difference baselines in control network may reduce the impacts to a certain extent.

1 GENERAL INSTRUCTIONS

With the development of technology, GPS has been widely used in various types of construction engineering measurement. For general short baselines, tropospheric delay is highly correlated and can be eliminated by double-difference (Li and Huang, 2005). However, in hydraulic engineering, elevation difference between the two ends of the baseline is usually large, so is the meteorological differences. This may cause that the tropospheric delay cannot be eliminated effectively by double-difference or cause the failure of integer ambiguty operation even when the baselines are short (Ding, 2009). Researches have shown that when the satellite's elevation is 20°, relative tropospheric delay of 1 mm will lead to elevation error of 3 mm (Dai, 2011), which becomes the main constraint of GPS elevation accuracy. This short baseline and large elevation difference situation sometimes may be hard to avoid especially in dams, buildings and bridges, therefore the study of meteorological parameters in tropospheric delay correction is particularly important. In this paper, the effects of elevation difference on GPS monitoring results will be analyzed and some useful conclusions will be given.

2 COMMON TROPOSPHERIC MODELS AND ACCURACY ANALYSIS

Using empirical models for correction is the most commonly used method to deal with tropospheric delay. For general engineering surveying, the baselines are usually calculated with software that the manufacturer offers and contain Saastamoinen and Hopfield models.

Hopfield model (Hopfield. 1969):

$$D_{trop} = D_{dry} + D_{wet}, \quad (1)$$

$$D_{dry} = 1.552 \times 10^{-5} \times \frac{P_0}{T_0} \times (H_d - h), \quad (2)$$

$$D_{wet} = 7.46512 \times 10^{-2} \times \frac{e_w}{T_0^2} \times (H_w - h), \quad (3)$$

$$H_d = 40136 + 148.72(T_0 - 273.16), \quad (4)$$

$$H_w = 11000 m, \quad (5)$$

$$e_w = rh \times 6.11 \times 10^{\frac{7.5 T_0}{T_0 + 273.3}} \quad (6)$$

where P_0 is the ground-surface pressure in mbar; T_0 is the ground-surface temperature in K; e_w is the water vapor pressure in mbar; h is the station elevation in m; rh is the relative humidity; D_{dry} is the dry zenith delay; D_{wet} is the wet zenith delay; and D_{trop} is the zenith totally delay.

Saastamoinen model (Saastamoinen, 1972):

$$D_{dry} = \frac{0.002277}{f(\varphi, H)} P_0, \quad (7)$$

$$D_{wet} = \frac{0.002277}{f(\varphi, H)} [0.05 + \frac{1255}{T_0}] e_w, \quad (8)$$

$$f(\varphi, H) = 1 - 0.0026 \cos(2\varphi) - 0.00028 H \quad (9)$$

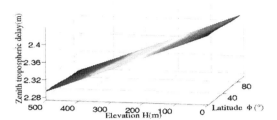

Figure 1. Zenith tropospheric delay with different elevation and latitude of Saastamoinen model.

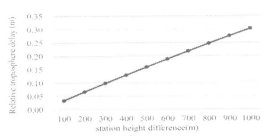

Figure 2. Relationship between relative tropospheric delay and elevation difference.

where $f(\varphi, H)$ is a function close to unity and takes into account the variation of the gravity field with latitude φ and ellipsoidal altitude H. Other variables are the same with those of Hopfield model.

Hopfield error formula:

$$\sigma_{D_{dry}} = \sqrt{0.0023^2 \sigma_P^2 + 0.0001^2 \sigma_T^2}, \quad (10)$$

$$\sigma_{D_{wet}} = \sqrt{0.1461^2 \sigma_{rh}^2 + 0.0036^2 \sigma_T^2}, \quad (11)$$

Saastamoinen error formula:

$$\sigma_{D_{dry}} = \sqrt{0.0023^2 \sigma_P^2}, \quad (12)$$

$$\sigma_{D_{wet}} = \sqrt{0.1496^2 \sigma_{rh}^2 + 0.004^2 \sigma_T^2}, \quad (13)$$

Since only the meteorological differences are analyzed, latitude and elevation errors are not considered. It can be concluded from the formulae that pressure deviation is the main error to dry delay as well as water vapor pressure deviation to wet delay. In tropospheric zenith delay, dry component accounts for 90% while wet only for 10% (Fan et al. 2009). However, there is no temperature error in Saastamoinen dry delay error formula, which makes Saastamoinen model better than Hopfield in practice. In this paper, Saastamoinen model is discussed.

Firstly the impact of meteorological error on station's zenith tropospheric delay is discussed and then its influence on relative positioning is analyzed.

Figure 1 shows the zenith tropospheric delay of Saas with different elevations and latitudes. It can be seen that the higher the station altitude, the smaller the tropospheric delay correction (Zhou, 2012). In addition, the correction is no less than 2.3 m. However, the impact of latitude is much smaller, so experience correction models have general applicability all over the word.

It is assumed that the reference station elevation is 80 m (quietly accords with the actual situation) and the height difference increases from 100 m.

Figure 2 shows the relationship between relative tropospheric delay and station elevation difference. As can be seen from the results, when the elevation difference is large, even to a short baseline, the relative tropospheric delay cannot only be eliminated by double-difference. Larger elevation difference would lead to larger relative tropospheric delay. In the ideal meteorological state, the baseline relative error caused by zenith delay is as follow (Ou, 1998):

$$\frac{\Delta l}{l} = \frac{d_{trop}}{R \sin E_{min}}, \quad (14)$$

where d_{trop} is the zenith tropospheric delay; R is the station geometric distance; and E_{min} is the satellite's minimum elevation angle. When zenith delay is 2.3 m and main elevation angle is 15°, the relative error of baseline is 1.4 ppm. The elevation error caused by relative tropospheric delay is as follow:

$$\Delta h = \frac{\Delta d_{trop}}{\sin E_{min}}, \quad (15)$$

where Δd_{trop} is the relative tropospheric delay of both ends of baseline. When $E_{min} = 15°$, 1 mm relative tropospheric delay will cause 4 mm error in vertical component. Thus it can be seen that the impact of meteorological error on relative positioning is mainly in vertical component of baseline. Since this error has nothing to do with length, the effect of short baseline is particularly prominent.

3 DATA ANALYSIS

3.1 Steps of experiment

In order to study the influence of meteorological difference caused by elevation on hydraulic engineering GPS data processing, this paper takes a dam's observation data as example. The steps of the experiment are as follows.

Firstly, since this article focuses on practical application of engineering rather than scientific research, normal GPS software the manufacturer offer for processing was chosen. Saastamoinen and Hopfield are the most widely used models and the software usually

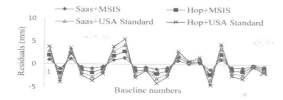

Figure 3. North residuals.

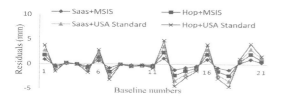

Figure 4. East residuals.

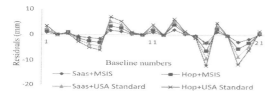

Figure 5. Elevation residuals.

contains MSIS and USA standard models. Two of them were selected randomly to calculate the baselines.

Secondly, usually when processing baselines, the meteorological parameters are default values. This leads to the condition that when we use tropospheric models at any time of a year, the correction is a fixed value, which is clearly inconsistent with reality. For Saastamoinen, standard meteorological model and measured values were chosen to calculate separately.

And lastly, for those large elevation difference baselines, average meteorological values between stations were applied in tropospheric models.

3.2 *Results and analysis*

In order to demonstrate the difference between groups of the first experiment, Figs. 3–5 show the four models' baseline residuals in east, north and elevation directions.

As can be seen from the figures, no matter in what direction, the group of Saastamoinen and MSIS is better than others. The standard deviations of this group in north and east directions are 2.37 mm and 2.22 mm while in elevation is up to 6.20 mm, which also verifies the result derived from the theory above.

Since meteorological factors have influence on the result of empirical tropospheric models, why do we usually use default values in general engineering? Based on this question, this paper attempts to use measured meteorological data and average values instead of default values in baseline calculation. The mean square errors of positions are shown in Table 1.

Table 1. Comparison of point accuracy between default and measured meteorological values.

Point number	Default			Measured			Average		
	X	Y	Z	X	Y	Z	X	Y	Z
XC01	1.33	1.74	4.45	3.11	2.08	6.13	1.79	1.83	1.64
XC02	2.45	1.80	4.98	3.20	2.82	7.22	1.88	1.91	1.91
XC03	1.97	1.65	3.18	1.08	1.16	1.57	1.26	1.23	1.48
XC04	2.04	1.63	3.29	1.05	1.12	1.54	1.06	1.49	1.60
XC05	1.89	1.51	2.95	1.03	1.06	1.49	0.97	0.94	1.90
XC06	1.96	1.90	2.32	1.20	1.46	1.85	1.23	1.41	1.83
XC07	2.13	1.71	3.35	1.08	1.25	1.60	1.05	1.29	1.04
Average	1.97	1.71	3.50	1.68	1.56	3.06	1.32	1.44	1.49

As can be seen from Table 1, adding measured meteorological data will improve the point accuracy wholly. However, for XC01 and XC02, the accuracies decrease instead. On a field trip, it was found that these two points are on the tops of different mountains. They have large elevation differences with other monitoring points. The maximum is up to 120 m. Due to sun exposure and distance away from reservoir, the max temperature, pressure and relative humidity differences are 4°C, 17 mbar and 30%, respectively. The reason that using measured data will decrease accuracy is consistent with the responses to local representative errors. The ground meteorological data cannot reflect the actual signal path atmospheric condition effectively. The larger the elevation differences, the larger the difference of meteorological elements, so is the meteorological correction.

However, when we take the same average meteorological values of both stations in processing, the points' accuracies increase wholly. This can be understood as an adding of "virtual reference station" in the middle of large elevation difference baseline. The station's meteorological parameters are average values, thus reducing the difference between stations in disguised form. Meanwhile, taking average values can reduce local representative errors caused by vegetation in a certain degree.

4 CONCLUSIONS

Elevation differences between stations have a certain influence on GPS baseline calculation. The effect on plane is little while on elevation is considerably large. In addition, the larger the elevation difference, the larger the elevation errors.

For general structural monitoring, taking measured meteorological parameters might bring ground factors errors. For those small difference baselines, double-difference can eliminate most tropospheric delay or use Saastamoinen and default meteorological values.

The tropospheric delay caused by elevation difference should be carefully considered if the difference is large or there is certain requirement in elevation accuracy. Taking average meteorological values for those large elevation difference baselines in control network can weaken the influence on elevation direction to some extent.

To those condition-allowed projects, large elevation difference baselines should be avoided. An extra reference station can be added in the middle of baselines so as to close the gap.

REFERENCES

Dai, W.J et al. 2011. Modeling regional precise tropospheric delay. *Geomatics and information Science of Wuhan University* 36.4: 392–396.

Ding, X.J. 2009. Research of tropospheric delay model and applications based on the GPS data processing. *Xian: Changan University*.

Fan, G.Q et al. 2009. Study of meteorological parameters' effect on tropospheric delay. *Science of surveying and mapping* 34(6): 37–39.

Hopfield, H.S. 1969. Two-quartic tropospheric refractivity profile for correcting satellite data. *Journal of Geophysical research* 74(18): 4487–4499.

Li, Z.H & Huang, J.H. 2005. *GPS surveying and data processing*. Wuhan: Wuhan university press.

Ou, J.K. 1998. Research on the correction for the neutral atmospheric delay in GPS surveying. *Acta Geodaetica et Cartographica Sinica* 27(1): 31–36.

Saastamoinen, J. 1972. Contributions to the theory of atmospheric refraction. *Bulletin Geodesique* 105(1): 279–298.

Zhou, S. 2012. The analysis of some tropospheric correction model and application in LGO and pinnacle. *GNSS word of China* 37(2): 48–52.

Estimation of fault rupture length of Wenchuan earthquake using strong-motion envelopes

Deyu Yin, Qifang Liu & Maosheng Gong
Institute of Engineering Mechanics, China Earthquake Administration, Harbin, Heilongjiang, China

ABSTRACT: This paper presents a new method to estimate the rupture length of large earthquakes by utilizing the envelopes, which are calculated through the data of aftershock. Using the method, the length of Wenchuan earthquake was calculated by using the records of Lushan earthquake. The strong-motion data from Lushan earthquake were used to obtain the envelope parameters relationship with epicentral distance, which are utilized to get the envelopes of Wenchuan earthquake. Then, the fault rupture length of Wenchuan earthquake was calculated by the envelopes through two steps, using the genetic algorithm, based on the finite-fault model and sub source model. The first step showed that the northeast was the main rupture direction with the rupture length of 280 km, and the total rupture length was 300 km. The second step indicated that the Yingxiu and Beichuan areas had larger slip. The simulation results agree with the actual situation well. Meanwhile, the method requires an adequate distribution of near-source stations.

1 INTRODUCTION

On 12 May 2008, the great Wenchuan earthquake with the magnitude of $M_w 7.9$ occurred in the Longmen Shan fault zone, and caused great loss of lives and properties. In the same region, in the southern part of Longmen Shan fault zone, the Lushan earthquake with the magnitude of $M_w 6.7$ occurred on 20 April 2013. After these earthquakes, several studies using different datasets have investigated the rupture processes of the two earthquakes (Feng et al. 2010, Hartzell et al. 2013, Shen et al. 2009, Wang et al. 2008, 2013, Zhang et al. 2008).

The distance between the epicentres of these two earthquakes is about 80 km (Fig. 2). In macroscopic, the Lushan earthquake can be regarded as an aftershock of Wenchuan earthquake (Wang et al. 2013). In the near fault zone, the large earthquakes can be synthesized by the small earthquakes. Based on this idea, a new method is presented in this paper to estimate the lengths of large earthquakes by using the envelopes of aftershocks. Yamada & Heaton (2008) estimated the fault rupture extent by using envelopes of acceleration. The dataset used to predict waveform envelopes included 30,000 seismograms in a region (Cua 2005). The method applied in this study adopts the idea proposed by Yamada & Heaton (2008) with the variation that the data of the aftershock was utilized to obtain the envelopes and the intervals of subsources were inequal.

First, the envelope parameters attenuation relationship was obtained by using the strong-motion data from the Lushan earthquake. Then, the attenuation relationship was used to calculate the envelopes of Wenchuan earthquake, and the genetic algorithm was used to estimate the fault rupture length through two steps.

2 THE ENVELOPE FUNCTION OF LUSHAN EARTHQUAKE

The data for analysis was the strong-motion dataset from the 2013 Lushan earthquake. The envelope function was divided into arising, stationary, and descending parts (Huo et al. 1991, Qu et al. 1994). The three-part envelope function model was adopted, using genetic algorithm to fit peak envelopes of the dataset. Then, the model parameters were expressed as the function of epicentral distance.

2.1 *Envelope function model*

Huo et al. (1991) utilized the three-part envelope model to calculate model parameters and their attenuation relationship. The envelop function f(t) is composed of arising (t < t1), stationary (t1 < t ≤ t2) and descending (t > t2) parts (Fig. 1a). The envelope function model is expressed as follows:

$$f(t) = \begin{cases} I_0 (t/t_1)^2 & (t \le t_1) \\ I_0 & (t_1 < t \le t_2) \\ I_0 e^{-c(t-t_2)} & (t > t_2) \end{cases}, \quad (1)$$

where f(t) is the envelope function, I_0 is the amplitude of the stationary part, t_1 is the end time of the arising

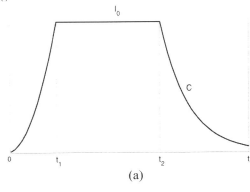

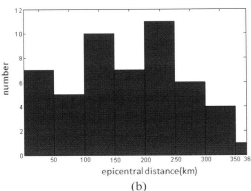

Figure 1. (a) The three-part envelope function model. The original point represents the P wave arrival time. (b) The number of records for this analysis. The epicentral distance is from 11 km to 368 km.

Table 1. Coefficients of UD component.

Parameter	C_1	C_2	C_3	R_0	ε
t_1	−2.30	0.31	0.55	10.00	0.17
I_0	0.03	0.65	−1.58	10.00	0.32
Ts	−1.95	0.36	0.42	10.00	0.09
C	1.90	−0.37	−0.11	10.00	0.17

Table 2. Coefficients of EW component.

Parameter	C_1	C_2	C_3	R_0	ε
t_1	−2.38	0.30	0.67	10.00	0.24
I_0	0.04	0.65	−1.50	10.00	0.31
Ts	−1.88	0.37	0.33	10.00	0.09
C	1.82	−0.38	−0.02	10.00	0.20

Table 3. Coefficients of NS component.

Parameter	C_1	C_2	C_3	R_0	ε
t_1	−2.33	0.31	0.64	10.00	0.19
I_0	0.04	0.65	−1.50	10.00	0.25
Ts	−1.89	0.37	0.35	10.00	0.08
C	1.81	−0.38	−0.01	10.00	0.13

part, and t_2 is the stop time of the stationary part. C represents the attenuation coefficient and the length of the plate region T_s is $t_2 - t_1$.

Eq. (1) contains four variables, I_0, t_1, t_2, C. To calculate the variables, it was stipulated that the energy of the stationary part accounts for 70 percent of the total seismic record (Huo et al. 1991, Qu et al. 1994). The formula is signified as follows:

$$\int_{t_1}^{t_2} a^2(t)dt = 0.7 E(\infty), \qquad (2)$$

where a(t) represents the strong-motion record, and $E(\infty)$ is the total energy of the record. Then Eq. (1) only includes three independent variables I_0, t_1, C.

2.2 Envelope parameters attenuation relationship

The strong-motion records of the Lushan earthquake were collected. The dataset contained 51 group records and each group included three components: one vertical and two horizontal (Fig. 1b).

The records of every station were baseline corrected by removing the mean. Acceleration envelopes were obtained over a 1-sec window by taking the maximum absolute amplitude of the strong-motion dataset. To get stable envelopes, Hanning-window was selected to smooth the envelopes. The genetic algorithm was chosen to get the envelope parameters I_0, t_1, C and Ts of each record. Then the attenuation function (Eq. (3)) was used to get the relationship between envelope parameters and epicentral distance.

$$\log Y = C_1 + C_2 M + C_3 \log(R + R_0) - \varepsilon, \qquad (3)$$

where Y represents the envelope parameters I_0, t_1, C or Ts, and M is the magnitude (assumed as 6.9 in this study). The coefficients C_1, C_2, C_3 and R_0 were calculated by fitting Eq. (3). R in the equation is the epicentral distance, and ε is the deviation. Tables 1 to 3 present the coefficients of each component.

3 INVERSION OF FAULT RUPTURE LENGTH OF WENCHUAN EARTHQUAKE

Based on the finite-fault model and subsource model, the fault rupture length of Wenchuan earthquake was calculated by using the genetic algorithm. The attenuation relationship obtained from the Lushan earthquake was used to calculate the envelopes of Wenchuan earthquake. Then, the fault extent was estimated by using the envelopes (Yamada & Heaton 2008). Here, the intervals of subsources were inequal.

3.1 The subsource model

The fault surface was divided into some subfaults, and each subfault was expressed by a point source, which is called sub source. In order to simplify the problem, it was stipulated that all sub sources had uniform magnitude and the P waves were radiated when the rupture front arrived at the sub source (Yamada & Heaton 2008).

The final envelope at a station was the sum of envelopes generated by each sub source. It is expressed as Eq. (4) below.

$$E_{Itotal}(R) = \sqrt{\sum_{i=1}^{N_1} E_i^2(R) + \sum_{i=1}^{N_2} E_i^2(R)}, \quad (4)$$

where $E_{itotal}(R)$ is the final envelope of station I, R is the epicentral distance, N_1 and N_2 are the numbers of the sub sources to the northeast and southwest, N_1 includes the sub source in the epicentre, and $E_i^2(R)$ is the envelope produced by the sub source i. In the calculation, it was assumed that the magnitudes of all sub sources were 6.9, which was the same dimension as the Lushan earthquake. The envelope parameters at each station were calculated by Eq. (3), using the coefficients in Tables 1 to 3. Then, the envelope E(R) was obtained using the three-part envelope model. When the direction of the rupture was determined, the envelope E(R) at station I was only related to the epicentral distance R. When calculating the total envelope at station I by using Eq. (4), the time delay due to the rupture propagation was considered.

The best estimate of the model parameters can be obtained by minimizing the residual sum of the squares between observed envelopes and predicted envelopes. The following equation represents the best fit.

$$RSS = \sum_{i=1}^{ns} \sum_{j=1}^{3} (A_{observed} - A_{predicted})^2, \quad (5)$$

where RSS is the sum of squares, ns is the number of stations, j represents the three components (one vertical UD and two horizontal EW, NS), and $A_{observed}$ and $A_{predicted}$ are the observed and predicted envelopes. The time is in 1-sec interval.

3.2 The fault rupture length of Wenchuan earthquake

To get the fault rupture length of Wenchuan earthquake, some parameters were assumed according other investigators. The velocity of P wave was 5.85 km/s (Zhao & Zhang 1987), the rupture velocity was 2.7 km/s (Wang et al. 2008), the direction of rupture was 225° (Zhang et al. 2008), the magnitude of all sub sources were 6.9 and the space were 20 km and 10 km. Using the attenuation relationship from the Lushan earthquake, the envelopes at stations near the Wenchuan earthquake epicenter were calculated. Here, each data record was normalized to peak amplitude of

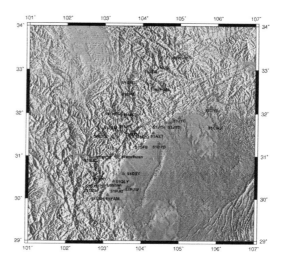

Figure 2. Locations of the stations and epicenters of Wenchuan earthquake, and Lushan earthquake, the stars mark the epicenters, and the triangles represent the stations.

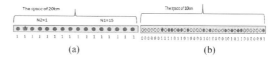

Figure 3. (a): The results of the first step. (b) The results of the second step. The epicenters are expressed as a star. The solid circles and the number 1 uder the solid circles represent the subsources. The hollow circles and the number 0 indicate that the location does not have subsources.

1.0. Thus, each record had nearly equal weight in the inversion. 32 stations were utilized. The amplitude of ambient noise was a constant 0.0174 gal, which was the mean value of the noise parts of the 32 stations. The location of the stations and epicenters are showed in Fig. 2.

Here, two steps were taken using the genetic algorithm to inverse the length of fault rupture.

First, it was assumed that the interval of the subsources was 20 km and the sub sources were in uniform distribution. The result was $N_1 = 15$, $N_2 = 1$ (Figs. 3a, 4a). The rupture lengths to the northeast and southwest were 280 km and 20 km. The total length was 300 km. The results indicated that northeast is the main rupture direction.

Then, it was presumed that the value range was larger. The N_1 was 17 and N_2 was 3. The rupture lengths to the northeast and southwest were 320 km and 60 km. In wider range, the part that has more severe rupture was searched for. The interval of the sub sources was 10 km. The results are shown in Figs. 3b, 4b.

Fig. 5 presents a comparison of the observed envelopes and predicted envelopes from best-fit source model.

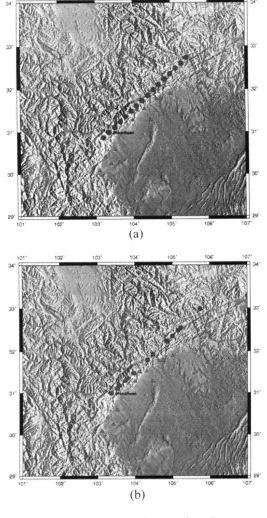

Figure 4. Locations of the subsources from the two-step inversion method, (a) The results of the first step. (b) The results of the second step. The stars mark the epicenters and the blue circles mark the subsources.

4 RESULTS AND DISCUSSION

Figs. 3a, 4a show that the length rupture of Wenchuan earthquake is 300 km and the main rupture direction is northeast at which the rupture length is 280 km. The simulation results indicate that the evaluation was effective (Zhang et al. 2008, Zhang et al. 2008), though the assumptions simplified the problem. In this way, whether the fault rupture is unilateral or bilateral and the fault rupture extent can be concluded.

After an earthquake, what concerns researchers the most is which areas have the larger slip on the fault. The second step tries to solve the problem. From the second results, it can be obtained that the points to northeast with epicentral distances of 0 km, 20 km, 30 km, 50 km, 60 km, 80 km, 130 km, 150 km, 200 km, 220 km, 240 km, 250 km, 320 km had subsources (Fig. 3b).

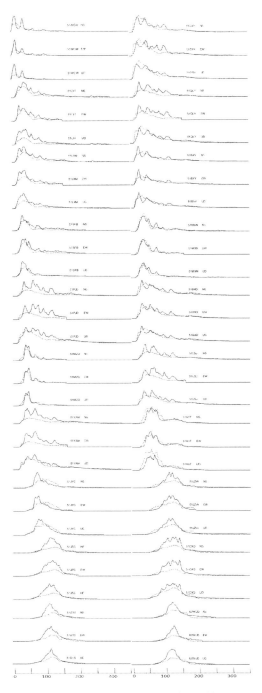

Figure 5. Comparison of observed envelopes (blue curves) and predicted envelopes (red dashed cures). The left characters are the station codes, and the right characters are the components of seismic record. Time scale is in seconds.

Figs. 3b, 4b indicate that three areas had larger slip. The first area has the epicentral distance from 0 km to 80 km, which represents the Wenchuan-Yingxiu fault. The area has 6 subsources. The second area included 2 subsources and the epicentral distance from 130 km to 150 km, which signifies the Beichuan fault. The results are in line with the outcome of other studies that used different datasets to get the rupture process of Wenchuan earthquake (Wang et al. 2008, Zhang et al. 2008, Zhang et al. 2008). The earthquake damage also indicated that the Yingxiu and Beichuan counties were severely damaged. Form Fig. 5, it can obtained that the predicted envelopes of the stations to the southwest agree well with the observed envelopes except for the stations 51WCW, 51PJD, 51LSJ, 51YAM. The observed envelopes and predicted envelopes of the stations 51PJD, 51LSJ, 51YAM had a common feature: the vertical results had small differences and the horizontal results had significant differences. The peak amplitudes for UD, EW, NS of the three stations were 47.65gal, 125.99gal, 147.95gal, 41.37gal, 81.31gal, 84.22gal and 38.34gal, 117.55gal, 115.87gal, respectively. The amplitudes for horizontal were much larger than those for vertical. Possibly, the records themselves lead to the differences. It should be noted that near the two areas there were adequate near-source stations. The third area with the epicentral distance from 200 km to 320 km included 5 subsources. Fig. 5 shows that the differences between observed envelopes and predicted envelopes of the stations near the third area, 51JYC, 51JZW, 51JZG, 51CXQ, 51GYS, 62WUD, were significant. It may be the inadequate station coverage that caused the phenomenon (Fig. 4b). To obtain reliable results, the station coverage should be adequate.

5 CONCLUSIONS

In this paper, a new strategy was proposed to estimate the rupture lengths of large earthquakes through the data of aftershocks. The rupture length of Wenchuan earthquake was obtained by utilizing the strong-motion envelopes of Lushan earthquake. Two steps were taken. From the first step, the main rupture direction and the fault rupture extent were obtained. Then, the areas with the larger slip were found. However, the method requires adequate near-source station coverage.

ACKNOWLEDGEMENT

This research is supported by the National Natural Science Foundation of China under Grant No. 51078337; 91315301-10.

REFERENCES

Cua, G. 2005. Creating the virtual seismologist: developments in earthquake early warning and ground motion characterization. *Ph.D.Thesis*. Department of Civil Engineering, California Institute of Technology.

Feng G.C, Hetland E.A, Ding X.L. et al. 2010. Coseismic fault slip of the 2008 M_w 7.9 Wenchuan earthquake estimated from InSAR and GPS measurements. *Geophys. Res. Lett*. 37, L01302.

Hartzell S, Mendoza C, Ramirez-Guzman L, et al. 2013. Rupture History of the 2008 Mw 7.9 Wenchuan, China, Earthquake: Evaluation of Separate and Joint Inversion of Geodetic, Teleseismic, and Strong-Motion Data. *Bull. Seismol. Soc. Am*. 103: 353–370.

Huo J.R, Hu Y.X, Feng Q.M. 1991. Study on envelope function of acceleration thim history. *Earthquake Engineering and Engineering Eibration*. 11(1): 1–12.

Qu T.J, Wang J.J, Wang Q.X. 1994. Study on characteristics of envelopes function of ground motion in local site. Earthquake Engineering and Engineering Eibration. 14(3): 71–80.

Shen Z.K, Sun JB, Zhang PZ, et al. 2009. Slip maxima at fault junctions and rupturing of barriers during the 2008 Wenchuan earthquake. *Nat. Geosci*. 2: 718–724.

Wang W.M, Zhao Lian. Feng, Li J. et al. 2008. Rupture process of the Ms 8.0 Wenchuan earthquake of Sichuan, China. *Chinese Journal of Geophysics*. 51(5): 1403–1410.

Wang W.M, Hao J.L, Yao Z.X. 2013. Preliminary result for rupture process of Apr. 20, 2013, Lushan Earthquake, Sichuan, China. *Chinese Journal of Geophysics*. 56(4): 1412–1417.

Yamada M, Heaton T. 2008. Real-time Estimation of Fault Rupture Extent Using Envelopes of Acceleration. *Bull. Seismol. Soc. Am*. 98(2): 607–619.

Zhang P.Z, Xu X.W, Wen X.Z, et al. 2008. Slip rates and recurrence intervals of the Longmen Shan active fault zone, and tectonic implications for the mechanism of the May 12 Wenchuan earthquake, 2008, Sichuan, China. *Chinese Journal of Geophysics*. 51(4): 1066–1073.

Zhang Y, Feng W.P, Xu L.S, et al. 2008. The rupture process of the great Wenchuan earthquake. *Science China Earthquake Sciences*. 38(10): 1186–1194.

Zhao Z, Zhang R.S. 1987. Primary study of crustal and upper mantle velocity structure of Sichuan province. *Acta seismologica sinica*. 9(2): 154–166.

Evaluation and analysis of bridges with encased filler beams

V. Kvocak, D. Dubecky, R. Kocurova, P. Beke & M. Al Ali
Faculty of Civil Engineering, Technical University in Košice, Slovak republic

ABSTRACT: The paper presented focuses on the research into progressive bridges with encased filler beams of modified steel sections designed in the way to minimize steel consumption without affecting essentially the overall structure resistance. The design is based on the latest findings of national and international research in the field of composite structures. Special attention is paid to the current theoretical and experimental behavioural analysis of such deck bridges employing various sections and using the principle of composite action between steel and concrete ensured through shear connecting strips, classical rolled-steel sections, and welded sections.

1 INTRODUCTION

Composite structures are more and more frequently used both in theory and practice. They make it possible to take advantage of good mechanical properties of concrete in compression and steel in tension. One of the commonest types of composite structures is deck bridges with encased filler beams. These types of construction have been employed in Slovakia and all over Europe without any major change since the beginning of the 19th century. Among their advantages is their quite clear static action, simple structural design, a short period of construction and low maintenance costs. Their main disadvantage is their economical inefficiency. The design of deck bridges is currently based on STN EN 1994-2, which allows only the utilization of rolled or welded sections of constant cross-sections. The verification and calculation of deck bridges with modified sections is not specified in the standards currently in force. Some detailed design and construction methodology for a deck bridge is virtually absent from the current standards. Therefore, new research into deck bridges with various steel sections and methods of composite action is particularly desirable, particularly the one experimentally focused. The appropriate design of rigid reinforcing members in deck bridges (their appropriate type, shape, number and arrangement) as well as the appropriate method of composite action can bring great savings in steel consumption [2, 4].

2 TYPES OF LOADED BEAMS

The programme of experimental research includes experiments on five different types of beam with steel reinforcement of varying rigidity at the Faculty of Civil Engineering.

Beams N1 are made of encased hollow sections. The hollow section is made by welding its upper flange

Figure 1. Shape and geometry of Specimens N1.

and walls made of a 6 mm thick U-shaped steel sheet to another 6 mm thick steel plate creating the lower flange with overhanging ends. There are holes 50 mm in diameter cut by flame in an axial distance of 100 mm in the beam. Reinforcement bars 12 mm in diameter are threaded through every third hole in the beam. In addition, holes 50 mm in diameter are cut in the upper flange in an axial distance of 100 mm. The holes are arranged in such a manner that there is one hole either in the flange or the wall in each section [3].

Beams N2 are composed of encased T-sections. They are made by cutting rolled IPE 220-sections straight in the longitudinal direction. Transverse reinforcement 12 mm in diameter is located at a distance of 100 mm and is threaded in the wall through the holes 20 mm in diameter. The holes are situated 55 mm above the bottom edge of the beam.

Another type of specimen is beams N3. Similar to the previous specimens, they consist of encased T-sections made of cut rolled IPE 220-sections, where the edge of the wall is comb-shaped. The shape and dimensions of the section correspond to the comb-like strip previously tested at the laboratories of the Civil

Figure 2. Shape and geometry of Specimens N2.

Figure 3. Shape and geometry of Specimens N3.

Figure 4. Shape and geometry of Specimens N4.

Engineering Faculty in Košice [2], which made it possible to calculate the bearing capacity of the composite structure theoretically. Again, there are reinforcement bars running 55 mm high in an axial distance of 105 mm.

Beams N4 are also made of steel IPE 220-sections, while the edges of the walls are perforated in such a special manner that the resulting section provides stronger composite action between the two materials – steel and concrete. The edge of the wall is modified – holes at two levels 90 mm apart are flame-cut in each beam. Equally, transverse reinforcement runs at two levels at a distance of 135 mm.

The length of all specimens was 6 m and the width was 900 mm. The depth of the individual beams resulted from the layout dimensions of the specimen

Figure 5. Shape and geometry of Specimens N5.

so as to make the proportional conditions for the placement of steel members in a real construction, i.e. 270 mm.

The last type of specimen is Beams N5. Similar to N2 beams, these specimens are composed of encased T-beams formed of rolled IPE 220-sections. Transverse reinforcement for concrete, 12 mm in diameter is placed 40 mm high at an axial distance of 300 mm. Composite action is secured by means of 50 × 100 mm loops of reinforcement steel welded horizontally to the wall of the steel section 60 mm above the lower edge.

3 STATIC LOADING TESTS

During the concreting phase the specimens were placed on a solid base, so the zero loading state corresponded to the dead weight of the beams. The moment caused by the self-weight was $M_g = 27.33$ kNm. The specimen was loaded by vertical forces applied at a distance of 2000 mm from both edges; the axial distance between the forces being 1800 mm and a free end overhanging the support 100 mm.

All the laboratory tests were carried out and the specimens manufactured in the laboratories of the Institute of Structural Engineering. The specimens were loaded by two symmetrically arranged hydraulic presses so that in the section between the presses simple bending occurred. The zero loading state was identical with the self-weight of the beam. The following loading procedure was gradual, incremental, while the compression in the hydraulic presses was increased by 10 bars at a time, this corresponding to approximately 15 kN. The specimens were unloaded twice, the first time from 60 kN per cylinder to 15 kN, and the second time from 75 kN to 30 kN. Hairline cracking occurred in the concrete on the stretched side of the beam under a load of 20 kN when the concrete tensile strength was exceeded. These cracks opened out later and developed until they reached a length of approximately 230 mm, which was the anticipated position of the neutral axis. The tests finished when it was impossible to increase the load transmitted by the specimens any more. Deflections in the specimens started do rise considerably without any increase in loading.

Table 1. Test results.

	F_{exp} (kN)	M_{exp} (kNm)	$M_{exp,ave}$ (kNm)	M_{theor} (kNm)	Difference %
N1-1	154.0	335.33	339.37	318.69	+6.48%
N1-2	155.5	338.33			
N1-3	158.5	344.44			

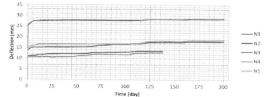

Figure 7. Time-dependent deflection of the beam.

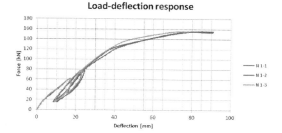

Figure 6. Correlation between deflection and load (force) applied.

Based on the equation (1) it was possible to calculate the moment of resistance of the deck bridge specimens reached in the laboratory conditions under static short-term load:

$$M_{exp} = F_{exp} \cdot 2{,}0 \cdot r_a + M_g \qquad (1)$$

Table 1 shows the maximum forces obtained from the hydraulic presses on selected measurement N1 beam, by which the specimens were loaded just before the completion of the test, the moments M_{exp} corresponding to the maximum load, and the resulting average moment of resistance at the point of ultimate strength $M_{exp,ave}$. Besides, the resulting moment was compared with the numerical calculations and the difference expressed in percentages.

Strain (relative deformation) was measured and recorded by means of strain gauges situated at the points most subjected to bending and those in the areas around holes. Inductive sensors detected the midspan deflection and the sinking in the supports. The correlation between the maximum deflection and load (force) applied is shown in Figure 6.

The graph above represents the correlation between the deflection and loading force in the specimen of a composite beam. The beam was unloaded at the loads of 60 kN and 80 kN and even though there was no further external load applied, severe permanent deformation in the beam remained. There was sufficient composite action/shear connection between the steel and concrete elements of the section, which is illustrated by the graph itself. There is no substantial increase in deflection that would otherwise have occurred if the composite action between the elements had failed.

4 LONG-TERM LOADING TESTS

During the long-term loading tests the specimens are placed on their sides. There are always two specimens making a pair of beams supported and loaded simultaneously. The beams are turned with the upper flanges facing each other at a distance of 50 mm. The support structure consists of frames which compress the beams against each other at their theoretical supports. Compression load is exerted using air pillows located in the gap between the beams. The load activated in such a manner is continuously uniform all over the top surface of the beam. The constant pressure in the pillow is maintained with an air compressor connected through valves to air pressure gauges. It is possible to set a specific pressure for each pillow. The specimens have been loaded by small incremental advances of 5 kPa up to a pressure of 30 kPa in N1 beams and up to 25 kPa in other beams. The long-term pressure that will be applied to the specimens corresponds to as much as 40% of their bending resistance.

The graph in Figure 7 shows the relationship between deflection and time. It represents the gradually increasing permanent deformation of the beam under continuous load (i.e. creep) exerted by the rubber pillow. Deflection increased sharply at the initial stage of loading. When the beam stabilised over time, the increase in deflection became very modest. The test will continue by adding more load to the beams and, after the consolidation, the load will be increased for the third time. It is possible to observe time-and-load-dependent rheological changes in the beam.

5 DYNAMIC TESTS

During the process of concreting the specimens were placed on a frame, so that the zero loading state corresponded to the loading of the beams by their own weight. The magnitude of the moment caused by the own weight of the specimen was $M_g = 27.33$ kNm. The test specimen was loaded by two vertical forces applied in a distance of 2 000 mm from the edge, with the axial distance between the forces 1 800 mm, and the free end of the beam protruding from the support by 100 mm.

Fatigue tests were performed on ten specimens. Variable load with peak-to-peak amplitude per one

Figure 8. Data collected from the control position sensors during the dynamic loading test.

actuating cylinder from $-10\,kN$ to $-90\,kN$ was applied to half of the specimens, corresponding to approximately 66 per cent of their theoretical resistance. Another half of the beams were loaded with amplitude between $-10\,kN$ and $-50\,kN$, corresponding to approximately 33 per cent of the theoretical resistance of the experimental beams.

The following diagram shows the position of control position sensors during the dynamic loading test. Positions of the individual actuating cylinders were scanned, as well as the position of mid-span deflection of the beam. The frequency of data recording was 10 Hz. The diagram also depicts the unloading moment at the end of the test. Dynamic loading was actuated by means of INOVA – a hydrodynamic pulsator – transmitting the load through its actuating cylinders on the composite beam.

6 CONCLUSION

The paper presented pertained to the description of the experimental research programme currently conducted at the Institute of Structural Engineering at the Civil Engineering Faculty of the Technical University in Košice.

All presented results of experimental and theoretical analyses as well as those currently in progress provide clear and distinct information on the overall behaviour of deck bridges with modified types of steel members with respect to the examined effects of support, load, geometric and material properties and also types of composite action.

Some partial results confirm that in view of bending resistance, the application of T-sections in deck bridges seems as a suitable solution. Nevertheless, the method of composite action in the directly cut T-beams appears to be very problematic as for the effectiveness of the composite action between the two materials. Moreover, the research proved that the effective composite action between the steel and concrete parts can be ensured by means of the polygonal cutting of the steal section, or, possibly, by the application of a box section in the steel member.

The continuation of the experimental measurements followed by the theoretical analysis within the adequate software environment will enable the researchers to optimize the method of composite action and formulate conclusions and recommendations for professionals and the general public.

ACKNOWLEDGMENT

The paper was supported by the project ITMS: 26220220124 "Development of Bridges with Encased Beams of Modified Sections."

REFERENCES

Gramblička Š., Bujňák, J., Kvočák, V., Lapos, J. 2007: *Zavádzanie Eurokódov do praxe: Navrhovanie spriahnutých oceľobetónových konštrukcií STN EN 1994*, Bratislava, Inžinierske konzultačné stredisko Slovenskej komory stavebných inžinierov, ISBN 978-80-89113-36-1.

Kvočák V., Karmazinová M., Kožlejová V. 2012, *Verification of the Behaviour of Deck Bridges with Encased Filler Beams*. In: International Journal of Systems Applications, Engineering and Development. Vol. 6, No. 1, pp. 163–170. ISSN 2074-1308.

Kvočák V., Kožlejová V. 2013. *First Load Phase of Long-Term Tests on Composite Beams*, In: Procedia Engineering: Concrete and Concrete Structures 2013: 6th International Conference, Slovakia. Vol. 65 (2013), pp. 423–427.

STN EN 1994-2 Design of Composite Steel and Concrete Structures, Part 2: General Rules and Rules for Bridges.

Vargová R., Beke P., Chupayeva K. 2013: *Príprava experimentálneho programu mostov so zabetónovanými oceľovými nosníkmi modifikovaných tvarov*, In: Vedecko-výskumná činnosť ž ÚIS: Prezentácia vedeckých výsledkov projektov ÚIS za rok 2012: 5. ročník: zborník príspevkov: Herľany, 4.–5. december 2012. – Košice: TU, 2013 S. 19–24. – ISBN 978-80-553-1300-9.

Evaluation of cultural heritage resources of Yanhecheng in Beijing

Run-Lian Miao, Mei Li, Yan-Ling Xu & Hui-Na Zhang
Beijing Municipal Institute of Science and Technology Information, Beijing, China

ABSTRACT: A new classification system for the cultural heritage of Yanhecheng region is provided based on field surveys and literature research. Evaluation of distribution pattern and heritage resources was done through a detailed analysis. The cultural heritage of the Yanhecheng region is divided into 16 types based on their existing form, distribution and function. Military cultural heritage dominates the region, which has military buildings related to the Great Wall. These buildings were built mainly in the middle of the Ming Dynasty. The Yanhecheng region is regarded as the standard for research on military, political and cultural events of the Ming Dynasty because of its relatively aggregated, distinctive and high level of cultural heritage.

1 INTRODUCTION

Cultural heritages inherit and carry the memory of human civilization. They are irreplaceable resources representing the history context and cultural impression of the human being. The Convention Concerning the Protection of the World Cultural and Natural Heritage and the Convention for the Safeguarding of Intangible Cultural Heritage defined cultural heritage. The Notice of the State Council on Strengthening Protection of Cultural Heritages defined the nature of cultural heritage in legal terms. The definition and classification of cultural heritage in these treaties have great significance in guiding the protection and utilization of cultural heritage resources. However, with the exploration of research and protective measures, the diversity and complexity of cultural heritage resources often exceed those defined in the treaties. A more practical standardization system is needed for general investigation of cultural heritages. In this study, the ancient Yanhecheng Pass area, which has relatively concentrated cultural heritage resources, was investigated by conducting field survey and literature review, to explore a macro-scale cultural heritage classification system. In addition, the cultural heritages were qualitatively evaluated, with a view to helping with the protection and utilization of them.

2 GENERAL SITUATION OF CULTURAL HERITAGES IN YANHECHENG

Yanhecheng is located in the mountains on the south bank of the Yongding River, andon the northwest of Mentougou district of Beijing. It is an ancient village with a long history, administered by the Zhaitang town of Mentougou District. The village was once a town of great military importance in the Chinese history. It spans an area of about 108 km^2. Yanhecheng was a fort on the border in the Ming Dynasty, with the army stationed. Since the town was near the Yongding River, it was named Yanhecheng, which literally means the city along the river.

Yanhecheng was one of the most import passes of the Great Wall in western Beijing; defense was first setup in Yanhecheng in the 33rd year of Emperor Jiajin (1554), and the town was established in the 6th year of Emperor Wanli (1578) [1]. Various important historic cultural properties and natural heritages are preserved in this area, mostly military sites. Ancient village culture and religious culture are also important elements of the historic cultural properties. The military sites include three important passes, Hegucheng, Tianjinguan (also called Huangcaoliang) and Hongshuikou, as well as their surrounding watch towers. The ancient village culture is mainly represented by the ancient city, temples, and former residences of famous people. The religious sites include the Baishan Temple, Laojun Hall and other temples, as well as monuments and stone carvings.

3 CLASSIFICATION OF CULTURAL HERITAGES IN YANHECHENG

3.1 Classification

In the history, Yanhecheng served for a long term as a military fort, while it was also a rural settlement. According to the cultural heritage classification standards currently in effect, the cultural heritages in Yanhecheng can be classified as tangible and intangible heritages. In this study, based on the actual requirements of cultural heritage preservation and development, a three-level classification standard was proposed in view of the existence form, spatial location

Table 1. Classification of Cultural Heritages in Yanhecheng.

Level I	Level II	Level III	Type	Quantity
Tangible heritage	Inner-settlement heritage	Military heritage	Fort	1
			The Great Wall and its attached facilities	6
		Civilian heritage	Civilian building	9
			Old tree	5
	Outer-settlement heritage	Military heritage	The Great Wall	1
			Watch tower	19
			Beacon tower	8
			Pass	17
			Cemetery and tomb	2
		Politic heritage	Government office and stone carving	1
		Civilian heritage	Archaeological ruin	5
			Road and bridge	4
		Religious heritage	Temple, tower, tomb engraving	7
Intangible heritage			Folklore	2
			Poem	4
			Quyi and play	1
Total				92

and function of heritage sites. The cultural heritages in Yanhecheng were classified accordingly (Table 1). Level I of the standard classifies cultural heritages into tangible and intangible heritages. Level II further classifies tangible heritage sites according to their spatial location. With the wall of Yanhecheng, the heritage sites in Yanhecheng were classified into inner-settlement sites and outer-settlement sites. Level III divides heritage sites, according to their functions, into military heritages, civilian heritages, religious heritages and politic heritages. Military heritages include the Great Wall and its attached passes, watch towers, beacon towers and cemeteries. Civilian heritages include civilian buildings, archaeological ruins, old trees, roads, bridges, etc. Religious heritages include temples, tomb engravings, tomb towers, etc. Politic heritages include government office, stone carvings and so on. As for the intangible heritages, they are mostly folklores, poems, Quyi and plays, and so on [2–4].

As the field survey and the literature [5] show, Yanhecheng hosts in total 92 cultural heritages. On average there is one cultural heritage within each square kilometers of area, indicating the high concentration of cultural heritages in Yanhecheng. The 92 cultural heritages include 85 (92%) tangible heritages and 7 (8%) intangible ones. There are 54 (59%) military sites, including seven inside the settlement and 47 outside the settlement. This high percentage of military sites is in accordance with the function and importance of Yanhecheng as a military fort. In the Ming and Qing Dynasties, Yanhecheng was an important fort protecting the capital. According to the literature, there were once 15 temples in Yanhecheng, worshiping the Budda, Zhenwu, Guandi, the fire god, the Dragon King, the Horse King, etc. Most of these gods are believed guardian of the ranks. Unfortunately, most of these temples no longer survive nowadays.

3.2 *Spatial-temporal distribution*

Temporally, the cultural heritage distribution in Yanhecheng is not consecutive. Besides the five archaeological ruins from the Neolithic times and the Han Dynasty and the two tombs from modern times, the other 78 tangible cultural heritages are all from the Ming and Qing Dynasties. Among these 78 heritages, only nine civilian buildings inside the settlement are from the Qing Dynasty, the others are all from the Ming Dynasty, accounting for 81% of the tangible cultural heritages. These 69 cultural heritages from the Ming Dynasty include 52 (75%) military sites. The defense system in Yanhecheng was first established in the 2nd year of Emperor Jingtai (1451), the same period when the passes of Yanhekou, Shigangkou, Dongxiaolong, Tianjinguan, Donglongmen and Tianqiaoguan were built [5]. From the 2nd year of Emperor Jingtai to the 2nd year of Emperor Longqing (1568), 17 passes were built. From the 5th year of Emperor Longqing (1571) to the 2nd year of Emperor Wanli (1574), the construction of the Great Wall, watch towers and beacon towers was gradually finished [5]. As of the 6th year of Emperor Wanli, the town of Yanhecheng was established, forming a complete defense system. Therefore, Yanhecheng is a place with concentrated military sites from the Ming Dynasty, and the values of these sites await further investigation.

Spatially, the 85 tangible heritages include 21 (25%) heritages inside the settlement and 65 (75%) outside the settlement. The heritages are scattered around a single center in a loose distribution. This distribution is related to the construction order of the defense system in Yanhecheng; the passes, watch towers, and the Great Wall were built before the construction of the facilities inside the town. Therefore, the preservation and utilization of these cultural heritages should focus on the defense system as a whole.

4 EVALUATION OF CULTURAL HERITAGES IN YANHECHENG

4.1 Distinctive characteristics

Yanhecheng has various types and relative large amount of cultural heritages. It has a long history of military culture, and has been once a town of great military importance. The military sites include the Great Wall, forts, beacon towers and ancient tomb engravings. The defense system of Yanhecheng and its attached Great Wall is well preserved. In fact, the only pass of the Great Wall that still survives in Beijing area is the Great Wall pass of Yanhecheng. Moreover, the town witnessed the military defense and the relationship between different ethnic groups in the Ming and Qing Dynasties, and now has the only well-preserved inner-Great Wall defense system in Beijing area. These characteristics endowed the town with apparent military heritage feature. Meanwhile, there are several temples inside the town, worshiping a variety of gods. Among similar military forts, even comparing with other cultural settlements, such diversification of religions and their peaceful coexistence are rare.

4.2 Centralized distribution

The cultural heritage resources in Yanhecheng exhibit a clearly centralized spatial-temporal distribution. Spatially, the well-preserved ancient fort of Yanhecheng is the center and cultural heritages scatter around it. Temporally, the cultural heritages are mainly elements of the Great Wall constructions from the mid Ming Dynasty, and mostly military sites. This area is of great importance in investigating the military affairs, traditions and national relationships of the Ming Dynasty.

4.3 High ranks

In terms of protection level and protection order, the cultural heritage resources in Yanhecheng are of national level. In 1984, Yanhecheng and the watch towers were listed as officially protected sites of Beijing. In 2006, Yanhecheng was announced an officially protected site of China. In 1987, the Great Wall in Yanhecheng and its attached facilities were listed in the world cultural heritage directory. Yanhecheng is a part of the Great Wall, and it is of important historic and cultural values. On the other hand, the town itself has great architectural and artistic values. The wall is well-preserved, and there are many historic sites; including ancient residences, roads, temples and trees and archaeological ruins. These cultural sites represent the age-old history of Yanhecheng and its profound cultural foundation, carrying great historic importance.

5 CONCLUSIONS

In this study, by referencing related standards and regulations, combining studies of many scholars, and conducting a field survey, a classification system consisting of two major classes, two subclasses and four secondary classes was proposed. With this classification system, the 92 cultural heritage resources of Yanhecheng were classified and organized. It was summarized that the cultural heritages in Yanhecheng are mainly military sites, accompanied by ancient village culture and religious culture. With the town as the center, the heritage resources are scattered around in a loose manner, and this is a characteristic of the heritage distribution in Yanhecheng. The values of these heritages were also evaluated. This study will provide helpful reference for the protection and utilization of the cultural heritage resources of Yanhecheng, as well as the resource planning and development of other ancient villages.

REFERENCES

Gu Jin-fu & Wang Xian-cheng. Research on Evaluation System of Exploitation Value of Intangible Culture Heritage Resource. Resource Development and Market. 2008, 24(9): 793–795.

Lei Xing-chang. The classification research on world heritage resources. Gansu social science. 2010(5): 230–233.

The village culture editorial board of Beijing Mentougou district. Village culture records of Beijing Mentougou district (second). Beijing: Yanshan press.

Wang Chang Song. The research on military history geography of Yanhecheng in Beijing. China Local Chronicles. 2009, (10): 59–63.

Zheng Yuedan. The index system building research on intangible cultural heritage resources value. Cultural Heritage. 2010, (1): 6–10

Evaluation on cultural heritage resources of Gubeikou fort in Beijing

Yan-Ling Xu, Run-Lian Miao, Min Zhang & Mei Li
Beijing Municipal Institute of Science and Technology Information, Beijing, China

ABSTRACT: On the basis of literature review and field survey, the cultural heritage resources of Gubeikou were systematically organized. The characteristics of these resources were comprehensively evaluated and it was shown that (1) Gubeikou is a typical area for studying ancient military affairs, diplomacy and national relationship; (2) The cultural heritages within the territory of Gubeikou are abundant, various, characterized, and with high quality, endowing them with obvious advantages in being industrialized and promoted; and (3) Strengthening the rational development and utilization of the cultural heritages in Gubeikou is imminent, in order to facilitate the protection and inheritance of the resources.

1 INTRODUCTION

Gubeikou is a town of Miyun county in the northeastern Beijing. It is a critical place in war, known as the Key to the Capital. Once a place of strategic importance on the border for both military affairs and trading activities, Gubeikou is now known for its cultural heritages, after a history of 2400 years. The town has various cultural heritages and historic sites. Within the town, there are more than 130 historic buildings with documented history, including well-preserved remains of the ancient Great Wall, tombs of famous people, revolutionary sites, remains of ancient royal roads, ancient temples, etc. Some of these historic sites are the famous Simatai Great Wall, Linggong Temple, Tomb of Qilang, Yibusanyan Wells, and Liangbusanzuo Temples. From the Northern Song Dynasty, batches of literates have created abundant poems and songs here, leaving our generation valuable intangible cultural heritages. All these cultural heritage resources are precious cultural wealth of Gubeikou. More and more studies are focusing on the cultural heritages in Gubeikou along with the development of China's culture industry and the high attention paid to the cultural soft power by the government in these years. However, these studies mostly focus on the hydrological geology [1–5], vegetation [6], historical and cultural functions [7–12], cultural heritage preservation [13, 14], etc. of Gubeikou. There has rarely been any study systematically organized and evaluated the various cultural heritages in Gubeikou. On the basis of literature review and field survey, this paper comprehensively evaluates the characteristics of these resources, in order to provide reference and basis for the protection and utilization of the cultural heritage resources of Gubeikou.

2 CLASSIFICATION

This study focuses on the cultural heritages within the territory of Gubeikou town of Miyun, Beijing. Through literature review and field survey, tangible cultural heritages including settlements, passes, watch towers, temporary imperial palaces, government offices in feudal China, folk houses, tombs and tomb engravings, and intangible cultural heritages including poems, folklores, customs and folk arts, were investigated. According to the existence form, spatial location and function of these heritage resources, a three-level classification standard was adopted to categorize them (Table 1). Level I classifies the heritages into tangible and intangible heritages according to their existence form. Level II further classifies the tangible heritages into inter-settlement heritages and outer-settlement heritages according to their spatial distribution, with the pass of Gubeikou as the boundary. Level III classifies the heritages as politic heritages, military heritages, civilian heritages and religious heritages. Political heritages include the temporary imperial palaces, royal roads, government offices, etc. Military heritages include the Great Wall and its attached passes, watch towers, beacon towers, barracks, granaries and cemeteries. Civilian heritages include civilian buildings, archaeological ruins, and old trees, roads and bridges. Religious heritages include temples, tomb engravings, tomb towers, etc. As for the intangible heritages, they are mostly folklores, poems, Quyi and plays, and so on.

The cultural heritages of Gubeikou were divided into two major classes, three subclasses, six categories and twenty types. The 448 cultural heritages include 437 (98%) tangible cultural heritages and 11 (2%) intangible ones. Among the 437 tangible heritages,

Table 1. Classification of cultural heritages of Gubeikou.

Level I	Level II	Level III	Type	Quantity
Tangible heritage	Inner-settlement heritage	Politic heritage	Temporary imperial palaces	1
			Government office	12
			Ancient royal road	1
		Military heritage	Fort	1
			Barrack	1
			Granary	1
			Drill ground	2
			Cemetery and monument	10
		Civilian heritage	Old well	1
			Ancient building	1
		Religious heritage	Temple	37
	Outer-settlement heritage	Military heritage	The Great Wall	6
			Watch tower	331
			Beacon tower	3
			Pass	19
			Cemetery and monument	10
Intangible heritage			Traditional technique	3
			Poem and traditional opera	5
			Tradition	2
			Surname culture	1
Total				448

68 (16%) are located inside the Gubeikou settlement, and the other 369 (84%) are located outside. Most of the tangible heritages are distributed inside the administrative range of Gubeikou. The 384 military heritages include 15 heritages inside the settlement and 369 outside the settlement, accounting for 88% of the total tangible heritage resources. Apparently, large amount of the cultural heritages in Gubeikou are military heritages, and most of which are located outside the settlement. This composition is closely related to the strategic importance of Gubeikou as a military fort in the history.

3 SPATIAL DISTRIBUTION AND VALUES

3.1 *Spatial distribution*

Among the 437 items of tangible cultural heritages in Gubeikou, 18 (4%) of them were built before the Ming Dynasty, 360 (82%) were built in the Ming Dynasty, and 59 (14%) were built in the Qing Dynasty or the modern times. Figure 1 presents an illustration of such temporal distribution. The earliest military facilities in Gubeikou were built during the reign of the King Yanzhao. In 283 BC, the King Yanzhao started the construction of the North Great Wall; beacon towers were built along the way from the Great Wall to Yuyang county (south of the Tongjun Village of Miyun nowadays) [7, 15]. In the 2nd year of Emperor Yuanshuo in the Western Han Dynasty (127 BC), a general Li Xi started to build the Tixi city (the current Hexi Village of Gubeikou) [15]. In the 6th year of Emperor Tianbao in the Northern Qi Dynasty (555), the Great Wall was built to defend against the Xianbei nation; it was the first section of Great Wall that was built in Gubeikou

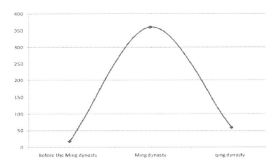

Figure 1. Spatial distribution of cultural heritages of Gubeikou.

[7]. The construction of the Great Wall defense system in Gubeikou peaked in the Ming Dynasty. In the 1st year of Emperor Hongwu (1368), general Xu Da conquered the capital of Empire Yuan, and took the order from his emperor to start repairing the passes of Juyong Guan, Gubeikou, and Xifengkou, etc. In the 11th year of Emperor Hongwu (1378), the city of Gubeikou was established. In the 1st year of Emperor Yongle (1403), the capital of Ming was migrated to Beijing by Emperor Cheng. Thereafter the defense matters in Gubeikou became an important job of the government. After Emperor Cheng, the emperors of the Ming Dynasty all prioritized the construction of the Great Wall as an important matter of border defense [7]. During the years of Emperors Longqing and Wanli, the construction led by the commanding officer Qi Jiguang of the Ji town was the largest project among all the projects these emperors have ordered. Eventually, a defense system consisting of towns, roads, passes, forts, watch towers, walls, cities and beacon towers,

Table 2. Values of cultural heritage resources in Gubeikou.

Value	World heritage site	National heritage site	Officially protected site of Beijing	Officially protected site of district or county	Total
Quantity	1 (Simatai section of the Great Wall)	1 (Simatai section of the Great Wall)	5	7	13

Note: from the list of officially protected sites of national and city levels.

which were all tightly and reasonably distributed, was established. In the Qing Dynasty, the defense strategy against the Mongolians was changed. Justice and mercy were tempered against the Mongolian royals. This new strategy formed a new defense system that was said to be "stronger than the Great Wall". Therefore, no more sections of Great Wall were being built, yet armies were still stationed in Gubeikou. In the Qing Dynasty, the emperor would connect with the Mongolian royals through an activity called "Mu Lan Qiu Mi" in the Mulan hunting field in Chengde, Hebei, and Gubeikou was the way that must be passed. During the 19th to 21st year of Emperor Kangxi (1680 to 1682), the Wanshou palace was built in Gubeikou. This palace is the only palace that was built at the location of a pass of the Great Wall.

3.2 Values

Table 2 presents the values of the cultural heritage resources in Gubeikou. Among the 437 tangible cultural heritages of Gubeikou, 13 are listed as world cultural heritage or officially protected sites of different levels. There are one world heritage site, one national heritage site, five officially protected sites of Beijing, and seven officially protected sites of Miyun.

4 EVALUATIONS

4.1 Abundant and various

In total, there are 448 cultural heritages in Gubeikou, including politic, military, civilian and religious heritages. Within the territory of the town, there are more than 20 types of heritages, including palace, government office, ancient royal road, fort, barrack, granary, drill ground, cemetery, monument, well, ancient structure, temple, the Great Wall, watch tower, beacon tower, gateway, and traditional technique, poem and traditional opera, culture, surname culture, etc.

4.2 Distinctive characteristics and obvious advantages in industrialization

Most of the cultural heritages in Gubeikou are military heritages located within the territory of the town, especially outside the settlement. The military heritages are mainly elements of the Great Wall defense system constructed in the Ming Dynasty. This feature is closely related to the strategic importance of Gubeikou as a military fort. The Great Wall, and the military fort are the two characteristics that facilitated the regional culture of Gubeikou. During the development and utilization of these cultural heritages, they are easier to be industrialized.

4.3 High quality and easy promotion

There are 13 cultural heritages in Gubeikou that have been listed as world, national and other levels of heritage sites, and most of them are military heritages. The high quality of the Gubeikou cultural heritage resources is apparent. Meanwhile, Gubeikou is on the way of the Qing emperor to the royal "Mu Lan Qiu Mi" activity to connect with the Mongolian royals. It is the only place that has a palace and a Great Wall pass; its special importance is prominent. Therefore, after industrialization of the cultural heritages in Gubeikou, the promotion would be easy and effective.

5 CONCLUSIONS

The unique geographical environment of Gubeikou has facilitated its special military importance in the Chinese history. Gubeikou was first a military fort, then royal road, diplomatic passageway and trading passageway; it played an important role in the politic, military affairs and diplomacy of ancient China. It is a typical area that must be studied in order to investigate the ancient military affairs, diplomacy and national relationship. Within the territory of the town, there are various cultural heritage resources. These resources are abundant, various, strongly characterized, high-quality, and are easy to be industrialized and promoted. The town awaits reasonable development and utilization of its cultural heritage resources, and the tasks of preservation and inheritance of these resources are imminent.

REFERENCES

Bai Tian. Gubeikou's past. Beijing: China city press. 1997. 1–83.

Hang Zhi-ying, Yu Xin-xiao, Jiao Yi-zhi. Analysis on the Stand Structure of Larix principis-rupprechtii in the Chaoguanxigou Basin, Beijing. Journal of Shanxi Agricultural Sciences, 2011, 39(5): 443–445.

Li Jian. Yanshanian Tectonic Evolution of the Chengde Basin and the adjacent area in the Eastern Segment of the Yanshan fold-and-thrust belt. Chinese Academy of Geological Sciences, 2006.

Li Huanqiang, etc. Some Suggestions about repair "gubeikou". China's Great Wall museum, 2009, (1): 43–44.

Long Long. Gubeikou being the north-west fortress. The front. 2007, (9): 60–61.

Ma Yanjun, Li Xianwen, Hu Chunhui. The Preservation and the Renovation of the Gubeikou Historic Village in Beijing. Modern Urban Research. 2005, 20(11): 41–43.

Qu Hongjie, Zhang Yingli. Characterization of Tuchengzi formation in chengde area and its structural significance. Geotectonica et Metallogenia. 2005, 29(4): 462–477.

Wang Yongdang. The Comparison and analysis on water quality monitoring results on Chaohe Gubeikou section. Hebei province water conservancy. 2009, (11): 36–37.

Wu Jinyu. What is the role of progress of the Beijing coup. historical study, 2003, (11): 9.

Yu Jianhua, Fu Huiqin, Zhang Fenglan, etc. The Chieheng—Gubeikou plutonic rock belt of rapakivi suite and proterozoic rifting. Jour Geol & Min Res North China. 1994, 9(1): 34–45.

Zhang Baoxiu. Historical Development of the Gubei Pass. Journal of Beijing Union University, 1998, 12(3):5–10.

Zhang Jidong, Li Xiang, Li Guangdong. Gubeikou Thrust Nappe Structure and Brittle-ductile Deformation Belt of the Tuchengzi Formation. Geology in china. 2002, 29(4): 392–396.

Zheng Jianbang. The Gubeikou battle in the Great Wall resistance. Unity. 2005, (Supplement): 28–32.

Zheng Yunshan. visiting the ancient at Gubeikou. Beijing chronicle. 2005, (12): 62–63.

Zhou Feng. Gubeikou in the period of Liao and Jin. Journal of Liaoning Educational Institute. 2000, 17(4): 38–41.

Experimental study on the deformation characteristics and strength parameters of rock-soil of landslide of levee's bank slope

Rongyong Ma & Hui Liang
Institute of Civil Engineering, Guangxi University, Nanning, China

Siqi Dai
Zhu Wei Hydraulic Design and Research Institute, Nanning, China

ABSTRACT: With geotechnical material of typical landslide of levee bank in Nanning as research background, geotechnical material of the three landslides of levee bank slope was researched. In order to eliminate the impact of stress structure, stress history and other complex factors of the undisturbed soil sample on the test results, remoulded soil was used in the test. They were the representative of the levee. Soil samples were prepared in strict accordance with the specification. Some test work were completed with the help of the strain controlled triaxial test system. Through the analysis and comparison of experimental results, deformation characteristics and strength parameters of the geotechnical material of landslide of the levee bank slope were obtained. The study provides a scientific basis for the governance of typical landslides and safety management of levee.

Keywords: levee bank; landslide; deformation characteristics of rock and soil; strength parameters

1 INTRODUCTION

Strict analysis methods for geotechnical problems include three parts: Firstly, the model describing the geotechnical properties is set up, as a linear elastic body, nonlinear elastic body, or elastoplastic body must be clear. Secondly, the required parameters of the model are confirmed. Thirdly, the calculation method is found for the model to solve the engineering practical problem. Geotechnical stress-strain relationship is complex in practical engineering, having the characteristics of nonlinearity, elastic-plastic and viscoplasticity, dilatancy, anisotropy, etc., at the same time influenced by stress path and strength development degree and geotechnical condition, composition, structure and temperature and so on (Shen Zhujiang 1980, Shen Zhujiang et al. 1982). At present the studies on rock and soil mechanics are mostly for a certain type of rock and soil, and have rarely been used in specific project. To estimate the stress and displacement caused by external load in the soil, the stress-strain relationship should firstly be understood and then analyzed according to the boundary conditions. But for a long time, limited by computational difficulties and inadequate understanding of rock-soil, deformation and strength have been artificially separated during calculation. That is one problem of calculating the deformation and displacement, another is the problem of calculating maximum load or maximum stress that causes soil damage. In fact, the deformation of soil continuously happens in the initial stress state until destruction. As a result of artificial simplified to only consider deformation without effect of yielding in local, and the only damage and no matter how deformation, making the analysis results inconsistent with the actual situation (*Mechanics teaching and research section of water conservancy institute in east China* 1984). Because elastic-plastic model is practical in the description of soil properties, domestic and foreign researches of elastic-plastic model have made certain achievements and elastic-plastic model is made a useful try in engineering application. However, because of the complexity of the model itself and the material of soil, whether in theory or in practice, elastic-plastic model remains constantly being improved and perfected. Especially, applied research of the model is worth further exploration in combination with regional characteristics of soil.

2 THE GENERAL SITUATION OF FLOODWALL ENGINEERING AND THE SELECTION OF TYPICAL LANDSLIDE IN NANNING

2.1 *Project summary*

Nanning is located in the middle and lower reaches of the Yujiang, which is the largest tributary of Xijiang river system of Pearl River Basin. Nanning is the capital of Guangxi, it is also one of the first batch of national key cities for flood control in 1987. The total length of levees in Nanning is 59.930 km. The crest level of the levees is about 82.70 m. The designed standard of the levees is for 50-year flood. The designed flow of the levees is 19100 m^3/s. The designed level of the levees is 80.38 m.

2.2 The selection of typical landslide

Geotechnical materials of natural typical bank slope landslides of three areas were chosen to do three groups of triaxial tests, that three areas including Henan West Dike near the village of Weicun, Jiangbei Minsheng Wharf and Jiangbei East Dike near Yongjiang Bridge.

(1) Bank slope of Henan West Dike near the village of Weicun

The bank slope's height of that reach of Yongjiang is 16~18 m and bank slope is steep. Slope is composed of Quaternary alluvium yellow silty clay, grey silty clay, round gravel and the upper part of the soil is hard ~ hard plastic shape. The bank slope of dry state is steep, prone to weathering and collapsing, water softened. The lower part of yellow soil, gray soil are plastic shape, and part of it is in soft plastic condition, lower shear strength, easily to produce small collapse and sliding. The erosion and scour of flood to the river bank made the bank slope instable, sliding.

(2) Bank slope near Jiangbei Minsheng Wharf

The river reach is concave bank of Yongjiang river and erosion on one side of the river is serious. Some sliding repeatedly happened in levee of Minsheng Wharf. Its reason are: stratum in holocene GuiPing group is generally weak with poor properties; The groundwater is very rich, high underground water level in dry season, and dynamic water pressure of groundwater is lager; Banks are seriously scoured and 45° steep bank slope forms in underwater; Urban sewage discharge is seriously impeded; Two underground drainage channel are punched in building of Jiangbin Road, and a large amount of water seepage make dynamic water pressure increase quickly, triggering landslides.

(3) Bank slope in Jiangbei East Dike near Yongjiang Bridge

The river reach is concave bank of Yongjiang. In the long-term's scour, under the action of lateral erosion, bank slope often has little damage. Damage types are mainly scour -collapse, sliding, etc. But its scale is small and its distance to levees is 20~50 m, little impact on embankment.

3 THE TEST EQUIPMENT AND TEST METHOD

3.1 The test equipment

Test equipment is triaxial apparatus SJ-1A of strain controlled that was produced by Nanjing electric power automation equipment factory, and it mainly includes the following several parts (Yang Xizhang 1993).

1) Triaxial pressure chamber; 2) The axial loading system; 3) Axial compressive force measurement system; 4) The surrounding pressure regulator system; 5) Pore water pressure force measurement system; 6) The axial deformation measurement system; 7) Backpressure bulk system and so on.

Test process was continuously monitored in the process of consolidation undrained shear, and axial force, strain and pore water pressure were recorded and storied in the process of test. Before each test, parameters and coefficient of load sensor, displacement sensor, pore pressure sensor should be checked and calibrated.

3.2 Test method

In this paper, test all used test method of the conventional static triaxial consolidated undrained.

3.3 The physical properties of soil sample in test

Original soil samples were from some typical bank of Nanning levee, including soil depth of about 2.2 m in Henan West Dike, soil depth of about 3.6 m in the middle-embankment of Jiangbei, soil depth of about 3.8 m in Jiangbei East Dike. Take a small amount of soil samples to do soil particle-size analysis experiment (2 in each group) to determine soil property. The method used was the hydrometer method, the results of particle-size analysis are shown in Table 1.

Test results of particle-size analysis show that soil property of the soil samples is silty clay, can be used as the research object of the model.

3.4 Preparation and saturation of soil samples

The stand or fall of sample preparation is the key to the success of experiments, moreover different sample molding ways will influence the result of the test. To reduce or eliminate the impact of the different methods of sample molding on the result of the experiment, experiment adopted hierarchical dry loading. Test strictly used geotechnical test method standard GB/T50123-1999, and made every effort to minimize bias.

Representative soil samples of a certain quality are selected, after drying and sieving, according to the dry density of sample loading, weighing the quality of soil sample required. The samples were 3.91 cm in diameter, and solid cylinder specimens of height of 8 cm were divided into five copies. All samples were divided into 5 layer compaction, and the quality of the soil of each layer was equal, paying attention to control the density and uniformity of soil at the same time. After the compaction to the height of the demand on each layer, making surface dehair, then the second layer soil was filled in. Like that, until the last layer.

Table 1. Test results of particle-size analysis of soil sample in test.

Soil sample number	Soil depth	Names and size of particles (mm)							nonuniform coefficient Cu	nonuniform coefficient Cu	coefficient of curvature Cc	coefficient of curvature Cc	Soil sample name
		gravel		sand			silt	clay					According to the Ip or particle-size analysis
		middle >20	thin 20~2	thick 2~0.5	middle 0.5~0.25	thin 0.25~0.075	0.075~0.005	<0.005	(d70)	(d60)	(d70)	(d60)	
		Content (%)											
1-1	2.0–2.2	0.0	0.0	0.0	1.4	21.6	62.8	14.3	25.2	18.3	1.9	2.7	Silty clay
1-2	2.0–2.2	0.0	0.0	0.0	0.0	21.1	56.0	22.9	30.0	14.0	1.1	2.3	Silty clay
2-1	3.4–3.6	0.0	0.0	0.0	0.0	22.5	65.3	12.2	20.0	15.7	1.4	1.8	Silty clay
2-2	3.4–3.6	0.0	0.0	0.0	0.0	11.9	53.0	35.1	30.8	21.8	0.8	1.1	Silty clay
3-1	3.6–3.8	0.0	0.0	0.0	4.1	25.1	52.8	18.0	46.4	27.9	1.9	3.1	Silty clay
3-2	3.6–3.8	0.0	0.0	0.0	0.0	23.5	63.9	12.6	27.3	21.8	2.5	3.1	Silty clay

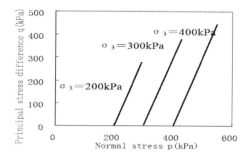

Figure 1. Consolidated undrained stress path.

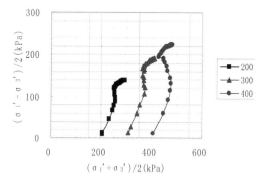

Figure 2. Actual stress path curves.

3.5 Test steps

1) Checks of instrument performance;
2) Sample installation;
3) Sample's drainage and consolidation;
4) Sample shear.

3.6 Test plan

We took remoulded soil as test object and test into three groups according to the different sections of the levee. Stress path was path of p increased. σ_3 was equal to the constant of consolidated-undrained triaxial test in the isotropic consolidated conditions. Loading, unloading and compression test of loading again under the condition of three consolidated confining pressure that $\sigma_3 = 200, 300, 400$ kPa (for example) were successively conducted. Stress path is shown in Figure 1.

4 TEST RESULTS

4.1 Results of triaxial compression test

The effective stress path curve of triaxial compression test is shown in Figure 2, the relationship of axial strain and the principal stress difference is shown in Figure 3.

4.2 Isobaric experiment results of each direction (hydrostatic pressure test)

Height of test sample was 80 mm, and the drain readings, confining pressure, water discharge, volumetric strain ε_v, lg p etc. were tested.

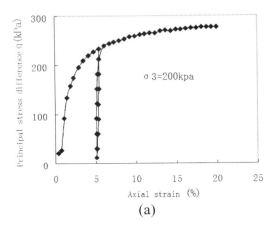

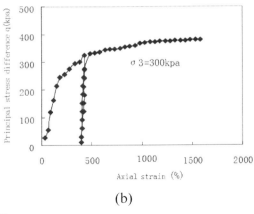

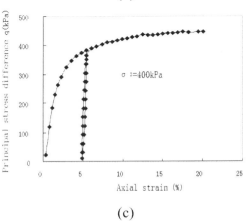

Figure 3. $\varepsilon_a \sim q$ curves.

5 GEOTECHNICAL DEFORMATION CHARACTERISTICS AND STRENGTH PARAMETERS OF LANDSLIDE

5.1 Geotechnical deformation characteristics of landslides

Through the above conventional triaxial consolidated-undrained compression test and static water pressure test of different confining pressure, test's stress and strain characteristic of each soil sample is obtained as follows.

Under the condition of the three-way isobaric springback test, change relation of average stress p and the volume strain ε_v of the geotechnical materials are shown in Figure 4. Under the condition of constant pressure test, it can be seen from the diagram that volume strain changes with the increasing of average pressure and slope of hysteresis is loop of rebound test decreases with increasing of rebound initial pressure.

The soil's stress-strain relationship in triaxial consolidated-undrained compression test is nonlinear, its deformation characteristics relate to the initial and final stress level. Test shows that stress-strain relationship curves of test under different confining pressure are different.

Space limited, for example of soil sample test of the third group, curves of stress-strain relationship in the triaxial compression test under different confining pressure are shown in Figure 5.

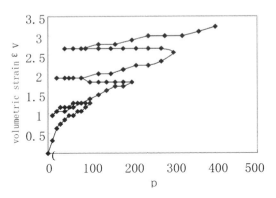

Figure 4. Hydrostatic pressure test curves.

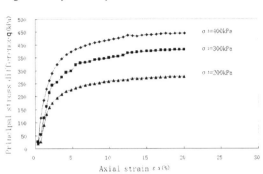

Figure 5. Hyperbola stress-strain curve.

Relation curves $\varepsilon_a \sim (\sigma_1 - \sigma_3)$ of conventional triaxial compression test show better hyperbolic form, and the initial period of curves are relatively steep. Their slope obviously largen with the increasing of ambient pressure; Back end of curves are relatively flat with no obvious peak value, but value of $(\sigma_1 - \sigma_3)_f$ increases with the increase of ambient pressure.

Because rock-soil is multiphase materials, and pore is contained in rock-soil particles. Thus in the action of all directions' pressure, the water and gas in the rock-soil particles discharge. It can produce plastic deformation. With the increase of axial compression, the soil samples at both ends up and down are the more pressured and more dense, and the central of it bulges gradually. The soil samples are eventually destroyed.

5.2 Geotechnical strength parameters of landslide

According to fracture strength got by the conventional triaxial shear test under different confining pressure (or the corresponding peak intensity when strain reaches 15%), the relationship between breaking strength and confining pressure is drawn to a damaged stress circle, and the strength envelope is made, namely the shear strength parameters of the soil sample is got, as shown in Figure 6.

For undrained test, due to pore water pressure existing, it often takes effective stress as a design value of shear strength in engineering. It is shown in Figure 7.

Engineering excavation has a significant influence on soft rock-soil mass. It often leads to cracks open, rock-soil mass loose, thus it leads to rock-soil mass's strength weakening and the change of deformation properties.

As shown in Figure 8, the breaking strength of silty clay in Henan West Dike near the village of Weicun is 472.20 kPa, 624.04 kPa and 734.10 kPa when confining pressure is 200 kPa, 300 kPa and 400 kPa. The breaking strength of silty clay in Jiangbei Minsheng Wharf is 476.10 kPa, 630.81 kPa and 742.23 kPa when confining pressure is 200 kPa, 300 kPa and 400 kPa. The breaking strength of silty clay in Jiangbei East Dike near Yongjiang Bridge is 472.66 kPa, 675.76 kPa and 839.38 kPa when confining pressure is 200 kPa, 300 kPa and 400 kPa.

According to the test results of destructive strength on specimens under different confining pressure, it can be seen that the breaking strength of rock-soil materials significantly reduces and it shows a good linear relationship between breaking strength and confining pressure with the decrease of the confining pressure. Breaking strength of soil sample of some different soil's depth of different embankment changes little corresponding to soil depth in low confining pressure; After the increase of confining pressure, breaking strength of the deep soil samples is significantly higher than the shallow soil samples. Attenuation law of breaking strength can better reflect the excavation unloading's effect on the engineering's stability.

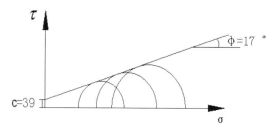

Figure 6. Stress Mohr circle and the strength envelope.

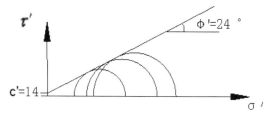

Figure 7. Effective stress Mohr circle and the strength envelope.

6 CONCLUSIONS

Based on the analysis of geotechnical material of typical landslides of levee bank slope in Nanning, soil's physical properties of bank slope is qualitatively analyzed by conventional experimental results;

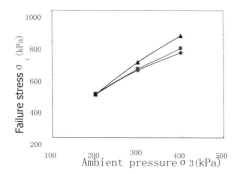

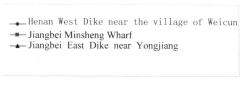

Figure 8. Stress Mohr circle and the strength envelope.

Geotechnical deformation characteristics and strength parameters of typical landslide in bank slope are quantitatively analyzed by the triaxial test data. It is of great significance that study on mechanical properties of geotechnical materials to give full play to the material strength, improve the design level and reduce the cost of city levee engineering. It also can provide some valuable reference to engineering design, evaluation of stability analysis and landslide treatment of city levee bank slope in the future.

REFERENCES

Mechanics teaching and research section of water conservancy institute in east China. 1984. 6. Volume 1 of geotechnical principle and calculation: water resources and electric Power Press.

Xizhang, Yang. 1993. 6. Geotechnical test and principle. Tongji university press.

Zhujiang, Shen. 1980. the soil's reasonable form of elastic-plastic stress-strain relationship. Journal of geotechnical engineering, 80, 2 (2): 10–17.

Zhujiang, Shen. ShuXin, Cheng. 1982. Uniqueness assumption of the theory of stress and strain of soil. Waterway transportation science, 82 (2): 11–19.

Fabric-formed concrete: A novel method for forming concrete structures

R.P. Schmitz
RPS Structural Engineering, LLC, Brookfield, Wisconsin, USA

ABSTRACT: Concrete members have traditionally been cast using a rigid formwork. However, casting concrete in a flexible formwork may in fact be used nearly anywhere a rigid formwork is used and is beginning to attract attention as a method of construction. Straightforward methods of analysis and design are available for the traditionally cast concrete member – be it a concrete floor, beam, wall or column member. This is not the case for the concrete member cast in a flexible fabric formwork.

This paper provides a state-of-the-art update on the use of fabric as a formwork for concrete construction and introduces an analytical modeling and design technique that will offer the design community, architects and engineers, 1) an alternative method for expressing themselves using flexible fabric formwork 2) the ability to optimize concrete members and 3) realize economies of construction leading to a conservation of construction materials and a *greener* more *sustainable* planet.

1 STATE-OF-THE-ART

1.1 First formworks

Since its invention by the Romans, concrete has been cast into all manner of formworks whether temporary or permanent. All-rigid formworks including rubble, brick and wood have become the containment form of choice for our modern concretes and an industry standard practice ever since humankind first sought to contain these early forms of mortar and "concrete" in their structures.

Historically both civil engineering and architectural projects have benefited by the use of fabric as a formwork for concrete containment. This versatile means of containing concrete saw some of its first use in civil engineering works such as erosion control. Developed and patented by Construction Techiques, Inc. in the mid-1960's Fabriform® is the original fabric-formed concrete system. Their products include Articulated Block, Filterpoint, Unimat, Concrete Bags and Pile Jackets. Engineers who have reported on the use of fabric-formed concrete lining used for slope protection include Phildysh & Wilson (1983) and Lamberton (1989).

This paper will highlight a few of those engineers, architects, designers and researchers worldwide who have made use of this unique way of forming concrete and focuses on fabric formworks for use in forming concrete members used in architectural works.

1.2 Modern-day formworks

One of the first architects to use a flexible formwork in an architectural application was the late Spanish

Figure 1. Juan Zurita residence (Studio Miguel Fisac), to an otherwise cold and hard substance.

architect Miguel Fisac with his 1970's design of the Juan Zurita residence in Madrid, Spain, (Fig. 1). His use of rope and plastic sheeting to create these precast panels imparts a sense of "warmth and softness".

Another architect whose work has softened up concrete is Japanese architect Kenzo Unno. Working independently of Fisac he has developed several cast-in-place (CIP) fabric-formed wall systems since the mid-1990's. The Kobe earthquake on January 17, 1995 provided the motivation for Unno to create residential designs that are intended to provide safe housing

Figure 2. Eiji Hoshino Residence (Mark West photo).

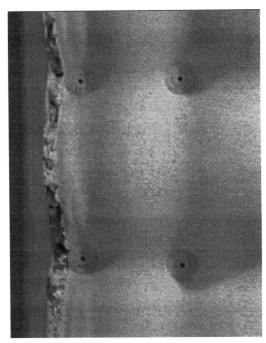

Figure 3. Quilt-like pattern detail for Eiji Hoshino Residence (Mark West photo).

using simple methods of construction with as little construction waste as possible. Using standard wall ties and the wall's reinforcement for support of the fabric membrane his quilt-point restraint method, for example, creates a pattern reminiscent of a quilt for the Eiji Hoshino Residence (Figs. 2 & 3).

For the Susae Nakashima "Stone Renaissance" house a "frame" restraint method was employed using pipes at a slight angle to restrain the fabric and give these walls their own distinct character (Fig. 4).

Two other practitioners that come to mind are Sandy Lawton, a Vermont, USA design-builder, and Byoung Soo Cho, a Seoul, South Korea architect. Lawton used geotextiles to form the columns, walls and floors for a nontraditional "treehouse" which was completed in 2007 and Cho crafted a Korean visitor center and guesthouse completed in 2009 using geotextiles to form its walls. See 'FURTHER INFORMATION' for links to these designers' websites.

Industries are sometimes slow to embrace new technologies and industries utilizing fabric formworks are few. Several industries that have benefited by using flexible formworks are; Fab-Form Industries, Ltd. based in Vancouver, British Columbia, Canada, Monolithic (air inflated domes) based in Italy, Texas, USA and Concrete Canvas Ltd. based in Pontypridd, UK.

It has been said "The beautiful rests on the foundation of the necessary. – Ralph Waldo Emerson". This quote aptly applies to fabric-formed structures as well beginning with the foundations. Since 1993 Richard Fearn, owner and founder of Fab-Form Industries, Ltd., has developed and marketed several fabric forming products including; Fastfoot® for continuous and spread footings; Fastbag® for spread footings

Figure 4. Susae Nakashima "Stone Renaissance" house (Kenzo Unno photo).

and Fast-TubeTM for piers and columns. See Fab-Form Industries' website listed under 'FURTHER INFORMATION'.

Several methods of construction using inflated forms have been available since the early 1940's but it was only recently that ACI (American Concrete Institute) Committee 334 (2005) introduced a standard guide for the construction of thin-shells using inflated forms.

David South, president and founder of Monolithic is the co-inventor of the Monolithic Dome and has been constructing thin-shell domes for more than 40 years. Monolithic's basic steps for constructing a dome are inflating an airform fixed to a foundation, applying a layer of polyurethane foam, hanging reinforcement and applying up to five layers of shotcrete. The inherent tensile strength of the PVC-coated or polyester fabric used for the airform allows it to be inflated to a sufficient strength to support all the applied construction materials until the concrete has cured to the point where the dome is self-supporting. Monolithic's use of fabric allowed the construction of thin-shell domes to once again be done economically. See 'Monolithics' website listed under 'FURTHER INFORMATION'.

William Crawford and Peter Brewin are directors and co-founders of Concrete Canvas Ltd., UK. Their approach to creating a concrete structure is similar to Monolithic's by using inflation to support the PVC form temporarily. However, that is where the similarity ends. The structures, which can be used as emergency shelters has a PVC form impregnated with concrete that hardens upon hydration leaving a self-supporting structure in place. The companies' concrete impregnated canvas may also be used in civil engineering projects for erosion control. See 'Concrete Canvas' website listed under 'FURTHER INFORMATION'.

1.3 Formwork applications

These examples highlight where flexible fabric formwork has been used forming architectural applications. Fabric forming applications include:

– Walls
– Cast-in-place
– Precast
– Shotcrete thin-shell curtain wall systems
– Beam and floor systems
– Trusses
– Columns
– Vaults
– Prefabrication of thin-shell funicular compression vaults
– Molds for stay-in-place concrete formwork pans
– Foundations
– Continuous and spread footings
– Civil engineering works
– Revetments, underwater pile jackets
– Coastal and river structures

While it is true that a flexible fabric formwork may be used nearly anywhere a rigid formwork is used, a significant amount of research remains to be done to bring these systems into everyday practical use by the construction industry. Standards and guidelines for using flexible fabric formworks need to be developed for the design community to take full advantage of this unique method of forming concrete members.

Countries with architectural and engineering students conducting most of the current research include Belgium, Canada, Chile, Denmark, England, the Netherlands, Switzerland and Scotland. The most prolific research currently being conducted is under the

Figure 5. Model wall panel formwork (C.A.S.T. photo).

direction of Professor Mark West, Director of the Centre for Architectural Structures and Technology (C.A.S.T.) at the University of Manitoba, Canada.

2 BASIC PRINCIPLES

2.1 An introduction to flexible formwork

The author's first introduction to flexible formwork came from reading an article entitled "Fabric-formed concrete members" published in *Concrete International* by Professor West, West (2003). A visit to C.A.S.T. in June of 2004 exposed the author to this unique method of forming concrete members. Professor West and his architectural students at C.A.S.T. first began exploring the use of flexible formwork for precasting concrete wall panels in 2002, West (2002, 2004). The shape a wall panel could take was first explored using a plaster model with various interior support and perimeter boundary conditions (Fig. 5). The cloth fabric, when draped over interior supports and secured at the perimeter, deforms as gravity forms the shape of the panel with the fluid plaster as shown in the completed plaster casts (Fig. 6). Once a satisfactory design has been obtained, a full-scale cast with concrete can be made.

The casting of a full-scale panel using concrete requires finding a fabric capable of supporting the weight of the wet concrete. For this purpose, a geotextile fabric made of woven polypropylene fibers was utilized. Assorted interior supports were added to the formwork (Fig. 7) and the flexible fabric material was

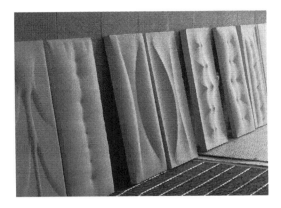

Figure 6. Completed plaster cast wall panels (C.A.S.T. photo).

Figure 7. Placing blockouts and interior supports prior to stretching in fabric in full-scale wall panel formwork (C.A.S.T. photo).

Figure 8. Securing fabric and placing reinforcement in full-scale wall panel formwork (C.A.S.T. photo).

pretensioned at the perimeter (Fig. 8). Depending upon the configuration of these interior support conditions, three dimensional funicular tension curves are produced in the fabric as it deforms under the weight of the wet concrete (Fig. 9). The completed panel is shown in Figure 10.

Figure 9. Placing concrete in full-scale wall panel formwork (C.A.S.T. photo).

Figure 10. Completed concrete wall panels (C.A.S.T. photo).

2.2 *Supporting elements*

Geotextile fabric as a formwork has a number of distinct advantages including:

- The forming of very complex shapes is possible.
- It is strong, lightweight, inexpensive, will not propagate a tear and is reusable.
- Less concrete and reinforcing are required leading to a conservation of materials.
- Filtering action of the fabric improves the surface finish and durability of the concrete member (Fig. 3 & 11).

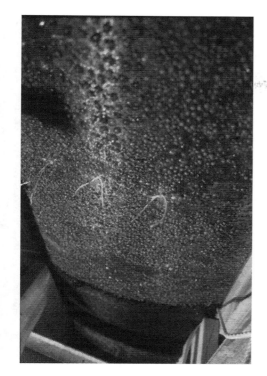

Figure 11. Filtration of excess water and air bubbles through geotextile fabric (C.A.S.T. photo).

It also has several disadvantages including:

– Relaxation can occur due to the prestress forces in the membrane.
– There is the potential for creep in the geotextile material, which can be accelerated by an increase in temperature as might occur during hydration of the concrete as it cures.
– The concrete must be placed carefully and the fabric formwork must not be jostled while the concrete is in a plastic state.

The author believes however, until new fabrics are developed the benefits of using geotextiles far outweighs any disadvantages.

3 STRUCTURAL MODELING AND ANALYSIS

3.1 *An FEA procedure for a flat cast wall panel*

The design of a fabric formed concrete panel may be approached in several ways. Each approach must take into account the panel's anchor locations to the backup framing system. One approach might be to locate the anchor points based on the most efficient panel design. Another approach could be to locate the anchor points based on the most pleasing appearance the panel takes due to the deformed fabric shape, and still another could be to consider both efficiency and appearance as a basis for the anchor locations.

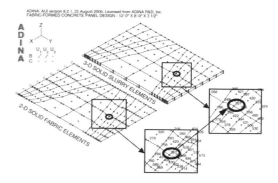

Figure 12. Form-finding concrete panel shape using finite element analysis (FEA).

How might a precast wall panel system, for example, be engineered? Straightforward methods of analysis and design are available for the traditionally cast concrete wall or floor panel. This is not so for the panel cast in a flexible fabric formwork. Shapes as complex as these require the use of finite element analysis (FEA) software. A procedure to "form-find" and analyze the complex panel shape is required (Fig. 12). Prior to a thesis and a paper by the author to introduce a design procedure that allows one to design a fabric cast concrete panel, no design procedures or methods to predict the deflected shape of a fabric cast panel had been developed, Schmitz (2004, 2006).

Briefly, the steps in this procedure are as follows:

1. Determine the paths the lateral loads take to the wall panel's anchored points.
2. Use the load paths, defined in Step 1, to model the fabric and plastic concrete material as 2-D and 3-D Solid elements, respectively. Arrangement of these elements defines the panel's lines of support.
3. "Form-find" the shape of the panel by incrementally increasing the thickness of the 3-D Solid elements until the supporting fabric formwork reaches equilibrium. The process is iterative and equivalent to achieving a flat surface in the actual concrete panel – similar to a ponding problem.
4. Analyze and design the panel for strength requirements to resist the lateral live load and self-weight dead load.

If, after a completed analysis of the panel in Step 4, it is found that the panel is either "under-strength" or too far "over-strength", adjustments to the model in Step 2 will be required and Steps 3 and 4 repeated. With this iterative process, it should be possible to obtain an optimal wall panel design. It should be noted this procedure was developed using a plain concrete model.

The structural analysis program ADINA was employed to analyze the formwork and the concrete panel cast in it. The final panel form, function and performance of the fabric membrane and the reinforcement of the panel for design loads all add to the complexities of the panel's analysis and design (Fig. 13).

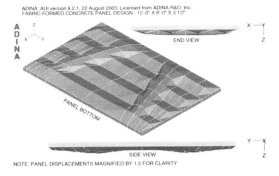

Figure 13. Final concrete panel shape using FEA.

A precast fabric-formed concrete wall panel, poured flat in a bed, may be one of the simpler concrete members to analyze, but when it comes to other concrete member shapes, one thing is clear, the system will undoubtedly be very complex and a procedure using finite element methods will be required.

3.2 Engineering procedures for more complex forms

Most recently, C.A.S.T. research has focused on thin-shell concrete vaults formed from fabric molds. These vaults can themselves serve as molds for stay-in-place formwork pans or glass fiber reinforced concrete (GFRC) applications. Another vault option being explored is a direct-cast fabric-formed thin-shell vault that can span between abutments in a beamlike fashion. These members are formed using a single flat rectangular sheet of fabric simply hung from a perimeter frame and used as a mold to form a double curvature vault (Fig. 14). Using a carbon fiber grid in lieu of conventional reinforcing steel allows for a creation of a very thin section – only 3 cm thick as shown in the completed vault (Fig. 15).

C.A.S.T.'s innovative work closely follows methods of funicular shell formation first pioneered by Heinz Isler. Isler used small-scale funicular models to determine full-scale geometry and structural behavior of reinforced concrete thin-shell structures.

The engineering of complex and exotic vaults and thin-shell panel shapes will require an approach different than the form-finding approach described above for a horizontal fabric formed precast panel. Whereas Isler tested small scale models of his shell structures and then scaled them up to full size these vaults and thin-shell wall panels may already be at full-scale before they are put to use. Two approaches to the engineering analysis of these thin-shell panel shapes might be considered. One is a photographic method using a commercially available software program called PhotoModeler® Scanner. This program imports images from a digital camera to create a dense point cloud and mesh data, which can be exported to FEA software. Another method might be to use High Definition Laser Scanning which also creates a dense point

Figure 14. Fabric mold stretched lengthwise in frame to form double curvature funicular thin-shell vault (C.A.S.T. photo).

Figure 15. Completed double curvature funicular thin-shell vault (C.A.S.T. photo).

cloud and mesh data which can be exported to FEA software.

Both approaches will involve an iterative process where one would first image the basic member shape and then analyze it for the superimposed design loads. Results of the first analysis would show where weak points in the member occur. Further analysis would suggest to what degree the member needs to be built-up using additional textile reinforcement and concrete materials.

4 CONCLUSIONS AND FURTHER RESEARCH

4.1 Conclusions

By utilizing a flexible fabric formwork, such as a geotextile, several advantages have been noted:

– The forming of very complex shapes is possible.

- Improved surface finish and durability – due to its filtering action.
- A more efficient and sustainable design is possible since material is placed only where it is needed – "form follows function".
- Flexible fabric formwork increases freedom of design expression and can spark the imagination of architects and designers to think beyond the simple prismatic shape.
- The development of a fabric formwork system has the potential to significantly reduce man's impact on the environment in terms of materials and energy usage.

4.2 *Further research needs*

The advancement of *Fabric-formed Concrete* would be furthered by:

- Design and modeling verification for research work being done on precast concrete wall panels.
- Investigating reinforcement options:
  - Fiberglass rebar
  - Alkali resistant (AR) glass textile
  - Carbon-fiber grids
- Finding the most advantageous reinforcing textiles for the reinforcement of all fabric-formed members including thin-shell shapes.
- The development of new fabrics, with improved properties over those of geotextile fabrics, for use as flexible formworks.
- The development of standards and guidelines for use in precast and cast-in-place forming systems are needed for this method of forming to be of practical use to the design community.

5 FURTHER INFORMATION

Readers interested in additional information are encouraged to visit the following websites especially, the C.A.S.T. website at the University of Manitoba where numerous examples and literature on this topic may be found.

- Author's research dedicated website:
  http://www.fabwiki.fabric-formedconcrete.com/
- The Centre for Architectural Structures and Technology (C.A.S.T.) at the University of Manitoba, Canada:
  http://www.umanitoba.ca/cast_building/
- The International Society of Fabric Forming (ISOFF):
  http://www.fabricforming.org/
- Byoung Soo Cho Architects, South Korea:
  http://www.bchoarchitecs.com/
- Sandy Lawton ARRODESIGN, Vermont, USA:
  http://www.arrodesign.org/
- Fab-Form Industries, Ltd., BC, Canada:
  http://www.fab-form.com/
- Monolithic (air inflated domes), Texas, USA:
  http://www.monolithic.com/
- Concrete Canvas Ltd., Pontypridd, UK:
  http://www.concretecanvas.co.uk/

REFERENCES

ACI Committee 334. 2005. *Construction of Concrete Shells Using Inflated Forms.* Farmington Hills, Michigan: American Concrete Institute, (ACI 334.3R-05).

ADINA R & D, Inc. 2008. *ADINA,* [Software], version 8.5. Watertown, Massachusetts, USA.

EOS Systems, Inc. 2009. *PhotoModeler® Scanner* [Software], version 6.3.3. Vancouver BC, Canada.

Lamberton, B.A. 1989. Fabric Forms for Concrete. *Concrete International* 11(12): 58–67.

Pildysh, M. & Wilson, K. 1983. Cooling ponds lined with fabric-formed concrete. *Concrete International* 5 (9): 32–35.

Schmitz, R. 2004. *Fabric-formed concrete panel design.* Thesis (Masters), Milwaukee School of Engineering, Milwaukee, Wisconsin, USA.

Schmitz, R. 2006. Fabric-formed concrete panel design. In: F.A. Charney, D.E. Grierson, M. Hoit, eds. *Proceedings of the 17th Analysis and Computation Specialty Conference, St. Louis, MO, USA, May 2006*, Structures Congress CDROM published by ASCE, ISBN 0-7844-0878-5.

West, M. 2002. Prestressed fabric formworks for precast concrete panels. Materials Technology Workshop, Department of Architecture, University of Manitoba, Canada. Web address: http://www.umanitoba.ca/cast_building/resources.html

West, M. 2003. Fabric-formed concrete members. *Concrete International* 25(10): 55–60.

West, M. 2004. Prestressed fabric formwork for precast concrete panels. *Concrete International* 26(4): 60–62.

Feasibility of discharging tailing in the open pit of Fengshan copper mine using transforming open-pit into underground mining

Liang Xia
Wuhan University of Technology, China

Shi-you Xia
Dongfeng Design Institute Limited, China

ABSTRACT: Aiming at the insufficient cubage of the tailing base due to growth of transforming open-pit into underground mining, this paper proposed a technique approach of dumped open pit for discharging ore tailing, and applied computational software, ANSYS and FLAC3D, to perform numerical simulation analysis on stress and strain of underground rock after discharging tailing. This study also analyzed the potential safety problems and the treatment.

1 INTRODUCTION

With the rapid development of mining industry, the mining mode of many mines in China as well as oversea mines has been switched from open-pit to underground. When the underground extension of deposit is relatively large and the covering layer is thin, the upper layer of the mine is operated using typical open-pit method, whereas the remained deposit (primarily underground part) is mined with underground method. This mining method is so called transforming open-pit into underground mining. This method can save tremendous manpower and material resources. Meanwhile, using this mining method also reduces the risk of high and steep slope during open-pit. Therefore, transforming open-pit into underground mining is the optimal method of mining.

Waste dump and tailings dam have been big issues for mining engineering. Particularly, for mines after mining for decades, the lease of these industrial lands such as waster dump and tailings dam faces issues of expiration. To solve these issues, the mining industries have to re-process land confiscation, re-build waster dump and tailings dam. However, with the increase of global population, land resources are significantly limited and no appropriate land can be used for this purpose. Even with available land, the economic pressure would be a huge burden for these miming companies. Therefore, some mining companies have to increase the height and volume of the pond dam of tailings dam to extend the age of usage, which arises significant safety issue. Even worse, some mining companies directly emit the tailing to the environments, which lead to very bad consequences to the ecological environments. Under this severe situation, understanding using open pit to solve insufficient cubage of tailing base is essential to theoretical values and meanings for protecting ecological environments [1~2].

2 ENGINEERING BACKGROUND OF MINE

Fengshan mine located in the east of Fuchi zhen, Yangxin xian, Hubei Province, which is in the middle of valley. The surface water is not abundant and weather is wet and warm, with clear distinguished four seasons. It is close to Yangzi River, and connected to Fuchi zhen by road, with well-developed transportation.

The strata in the mine area is simple: distributed from top to bottom as quaternary eluvial, talus, alluvium, triassic system, low series Daye formation, the third to seven dolomitic limestone strata in Daye formation, lower series of Triassic system, and limestone. Due to thermal metamorphic effect of magma, most of them have been metamorphosed into marbles. This copper mine consists of numerous copper and molybdenum ore bodies with different sizes, which are distributed as zonal pattern surrounding Fengshan granodiorite porphyry body. According to the spatial distribution of ore body, ore is mainly distributed in the southern and northern margins and the contact zone of Daye formation limestone (metamorphosed into marble). Hence, it can be decomposed into two Ore belts, so called ore belt of southern edge and northern edge, with approximately 500 m apart in the middle. The two belts converge at 20-line of the western edge of the rock body.

The Rock structure in the mine is mainly composed of tight linear fold, with the direction of structure line of 285°. The fold structure has Litou overturned syncline and secondary fold Dalong overturned syncline, Sianrendong overturned anticline, Kejiatang overturned anticline, and Zhulintang overturned syncline. The fault structure has not been well developed, with relatively small scale. The total fault structure is 22 with different sizes, which can be categorized into three groups according to their strike tendency: northwest direction, nearly east-west direction, and northeast direction.

3 TREATMENT METHOD OF DISCHARGING TAILING IN THE OPEN PIT

Tailing problem has always been issue for mining companies. Traditional treatment of tailing is to requisite land and build tailing dam close to the mine. The tailings are then directly disposed onto this land. This method is simple and has advantage of low cost. But it occupies large area of land, which wastes the land resources and severely containments environments. With the development of society and growing environmental awareness, this method has been proved to be unscientific with many problems. Therefore, researchers have developed a new tailing treatment method, i.e., pressure filtration and dry heaping [3]. This technology has significant advantage compared to the traditional method. Firstly, it is environmental friendly, which utilizes and recycles industrial water. It reduces the contamination of ground water, and avoids the contamination to downstream residents in terms of production and domestic water usage. The pre-dewatering of tailing before discharging enables the feasibility of discharging tailing in the open pit during transforming open-pit into underground mining. Secondly, the procedure of dry heaping of tailing mine is quite simple. The tailing slurry discharged after ore dressing entered stirred tank for buffering, which would pumped into pressure filtration workshop by slag slurry pump. It was then compressed into dry flake-like tailing slag cake by filter press [4]. Finally, it was transported by belt conveyor to and layered-placed into open pit. Plate and frame filter press was typically used in mining. The content of tailing slag cake was about 80% and the water content was about 20%.

Based on above analysis, it was decided to use pressure filtration and dry heaping as the current method to place tailing in the open pit.

4 ANALYSIS OF SIMULATION RESULT OF DISCHARGE TAILING IN THE OPEN PIT

The computation model was established according to Fengshan copper mine #17 cross-section profile and Fengshan mine map. The block group is shown in figure 1. The open pit bottom elevation is −65.18 m, the maximum elevation of left pit top reaches 83.5 m, and the maximum elevation of right top is 38.7 m. Furthermore, a main shaft and an auxiliary shaft were

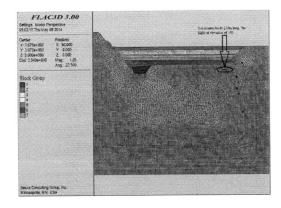

Figure 1. Block group of cross section of exploration #17.

Table 1. Parameters of rock mechanics.

Rock body	Weight (kg/m³)	Tensile strength (Mpa)	Cohesive force (Mpa)	Internal friction angle (°)	Modulus of elasticity (Gpa)	Poisson ratio
Surrounding rock	2600	0.55	1.8	35	10	0.16
Industrial orebody	2740	0.55	1.5	45	14.5	0.31
Heaped load body	1780	—	—	—	0.03	0.25

on the original cross-section profile (for simplifying the computation, only four points were drawn. 1 and 2 show the vertical location of the main shaft, 3 and 4 show the vertical location of the auxiliary shaft).

To further simulate the effect of open pit heaped load on the underground mining, a 5-meter high, 30-meter long mined-out area was selected in the industrial orebody. The classification of the heaped load was then processed. The elevation of total five ordered heaped load was 35.14 m, 22.63 m, 25.33 m, 17.14 m, 15.64 m, respectively.

According to the characteristics of elastic-plastic model, displacement constraint was set for horizontal direction at both sides of the model and the vertical direction of the horizontal bottom. The top of the model was set as free constraint. The initial rock stress field was defined as self-weight stress field. The lithology of the rock body was simplified several major ones, shown in the table 1.

The maximum principal stress of surrounding rocks under natural non-heaped load condition was 13.05 MPa shown in figure 2. With the mining of the lower part of the rock body, the maximum stress of the surrounding rocks changed to 2.06 MPa shown in figure 3. The surrounding rocks appeared obvious stress relaxation. When stating heaped load, the stress change was not large at the mining area and the bottom of the open pit. The maximum stress of surrounding rocks increased with increase of the height

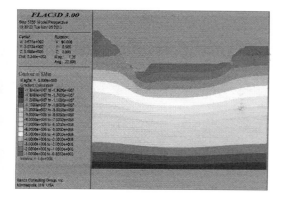

Figure 2. Contour of maximum principal stress under natural non-heaped load.

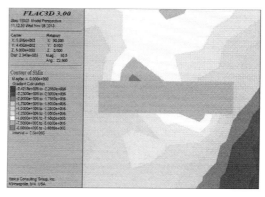

Figure 5. Contour of maximum principal stress under second order heaped load.

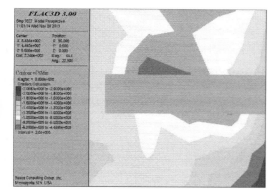

Figure 3. Contour of maximum principal stress of underground orebody during mining.

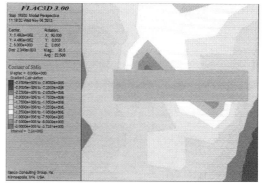

Figure 6. Contour of maximum principal stress under third order heaped load.

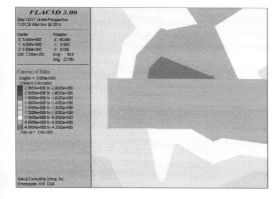

Figure 4. Contour of maximum principal stress under first order heaped load.

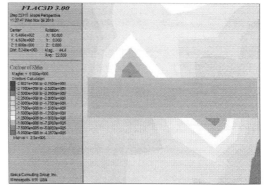

Figure 7. Contour of maximum principal stress under fourth order heaped load.

of heaped load, but with decreasing amplitude of increment. The stress is shown in figure 4 to figure 8 was 2.09 MPa, 2.42 MPa, 2.65 MPa, 2.86 MPa, 3.22 MPa, respectively.

Despite increasing of stress of surrounding rocks, bottom displacement of open pit, and top displacement of underground mining area, the surrounding rocks were still in stable condition, because the area of the plastic breaking of surrounding rocks was relatively small.

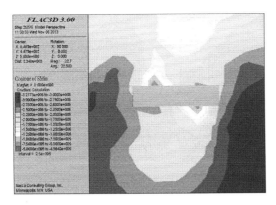

Figure 8. Contour of maximum principal stress under fifth order heaped load.

5 POST-ENGINEERING CONTROL APPROACH OF DISCHARGING TAILING IN THE OPEN PIT

Despite treatment of dry slag of tailing, it still contained fair amount of water. When applying blasting excavation to underground mines, this amount of water would infiltrated into underground, which contaminates groundwater, even threatens the underground excavation. Under the effect of water, the mechanical character of rock body could be further deteriorated. The seepage of tailing water would act normal filtration pressure force and tangential drag force on the fissure wall, which resulted in seepage field that affected the stability of the mining field [5]. Therefore, to ensure the safe mining, the discharge of tailing must be carefully treated. Firstly, according to the real situation of the open pit, an aquifuge need to be arranged at the bottom of the open pit. The aquifuge was made of 30 cm thick clay layer. After paving, rolling and examined for acceptance, a geomembrane was placed on the top of the clay. The geomembrane must reach a certain strength requirement, to avoid breaking during the operation. Secondly, seepage prevention and reinforcement must be done on the side slope. Considering severe weathering of rock body of side slope of Fengshan copper mine and existence of dangerous rock in some local area, the reinforcement and seepage prevention were performed. The decayed rock and protruding rock body were cleaned to make a smooth rock surface. Thereafter, appropriate reinforce measures were applied to strengthen the unstable rocks. After the reinforcement of the side slope, the gaps and cracks were eliminated. However, the requirement of prevention from seepage was not achieved. Therefore, a layer of soft waterproof material was necessarily applied on the surface of the side slope to prevent from seepage.

6 CONCLUSION

Aiming to provide reliable base theory of discharging tailing open pit of mine, this paper studied the stress field variation of underground rock body during discharge of tailing of open pit. This study comprehensively considered geology, hydrology, rock mechanics, and the real situation of the mine, used ANSYS to establish three-dimensional model of the tailing discharge of open pit. The result was introduced into FLAC3D to process analysis. The possible safety issues were analyzed and treated. The results showed that discharging tailing in the open pit is feasible for Fengshan copper mine. This study provided new insight and valuable information for waste open pit during transforming open-pit into underground mining.

REFERENCES

[1] Wu C, Zeng Y, Qin Y. Present Situation, Application, and Development of Simultaneous Extraction of Coal and Gas [J]. Journal of China University of Mining & Technology, 2004, 33(2): 137–140
[2] Chen Y, Gan D, Sun W. The Feasibility Study of Discharging Tailing in the Open pit of Chengjiagou Iron Mine from the Open-pit to Underground [J], Journal of Hebei Polytechnic University, 2008
[3] Li Z, Zhang Y, Gan D, Chen C. Study on the Stability Mechanism of the Underground Stope after Discharging Tailings into the Open Pit in chengjiagou Iron Mine [J]. Ming Research and Development, 2007, (4): 70–72
[4] Liu G. Construction Measures of Antiseepage Work When Using Open Pit as Tailing Storage [J], Nonferrous Metals Engineering & Research, 2010
[5] Cai M, He M, Liu Y. Rock Mechanics and Engineering [M]. Bejing: Science Press. 2002

Flow around the 3D square cylinder and interference effects on double square cylinders

Lijun Meng & Xiangwang Ma
College of Civil Engineering, Shijiazhuang Tiedao University, Shijiazhuang, Hebei, PR China

Wenyong Ma & Liqian Hou
Wind Engineering Research Center, Shijiazhuang Tiedao University, Shijiazhuang, Hebei, PR China

ABSTRACT: Using the Cobra probe, the flow around the single three-dimensional square cylinder was tested in wind tunnel, and the effects of cylinders on disturbed flow were discussed. The mean wind loads distributions on cylinder were studied with rigid model pressure testing in uniform flow. As the main concern, the interference effects of double 3D square cylinders were described based on the comparison of wind pressure distributions on interference cylinder and single cylinder, and the relationship between flow characteristics and wind pressure distribution was discussed to prove the feasibility of revealing the interference effects by the flow around the single 3D square cylinders. The results presented in this paper can be used for accessory structure design, evaluations of pedestrian level winds and square-section high-rise building design.

Keywords: three dimensional square cylinder, flow characteristics, wind tunnel test, interference effects

1 INTRODUCTION

Three-dimensional (3D) square cylinders and similar structures, which are one of the common shapes in wind engineering, are widely used in high-rise buildings and other structures (Quan, Chen et al. 2010). The flow around the 3D square cylinders and the wind pressure distribution are worthy to be discussed further (Hui, Tamura et al. 2012, Chen, Xiao et al. 2012, Huang, Zhu et al. 2012, Quan, Zhang et al. 2012, Shen, Wang et al. 2012, Tang, Jin et al. 2012, Wang, Shi et al. 2012, Zou, Liang et al. 2012, Hui, Tamura et al. 2013, Hui, Yoshida et al. 2013, Gu, Zhang et al. 2013, Zhang, Quan et al. 2013). Considering as the hot issues in recent years, the flow characteristics and wind loads distributions are essential to wind engineering, and the interference effects of double square cylinders remain unclear so far (Xie and Gu 2004, Xie and Gu 2007).

The flow around the single 3D square cylinder was tested, by using Cobra probe in wind tunnel, and the wind velocities and turbulence intensities were analyzed. The rigid model pressure test was conducted to obtain the mean wind pressure distribution on 3D cylinder in uniform flow. Based on the comparison of the wind pressure distributions on interference cylinder and single cylinder, the interference effects of double 3D square cylinders were described. The relationship between flow characteristics and wind pressure distribution was discussed to prove the feasibility of revealing the interference effects by the flow around the single 3D square cylinders. The results presented in this paper can be used for accessory structure design, evaluations of pedestrian level winds and square-section high-rise building design.

2 METHODS

The wind speed profile and turbulence intensity in the empty tunnel show the flow development along the test section from the end of the first contraction section to the end of the low speed test section 24 m in total length (see Figure 1). The mean wind velocity was $U = 16.2$ m/s, and the background turbulence intensity was less than 0.5 %.

The 3D square cylinder model, as shown in Fig. 2, was made of organic glass, the width B was 10 cm and the height H was 60 cm. The model was fixed on the origin of coordinates and defined the wind direction as the positive x axis. The coordinates system was used to test the flow around the single cylinder and the interference effects.

The 220 pressure taps were evenly arranged at the heights of 10 cm, 20 cm, 30 cm, 35 cm, 40 cm, 45 cm, 50 cm, 54 cm, 56 cm, 58 cm and 59.5 cm.

3 THE FLOW AROUND THE SINGLE 3D SQUARE CYLINDER

The model was tested at the wind velocity of 16.2 m/s and position of $L = 6$ m. The wind velocity ratio and turbulence intensity along the wind were given by

$$r_u = \frac{\overline{U}_x(x,y,z)}{\overline{U}} \quad TI_x = \frac{\hat{U}_x(x,y,z)}{\overline{U}_x(x,y,z)} \times 100 \quad (1)$$

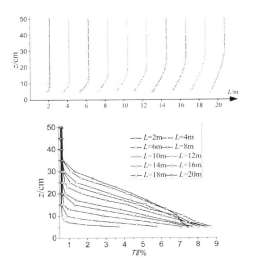

Figure 1. Wind Speed Profile and Turbulence Intensity in Empty Wind Tunnel.

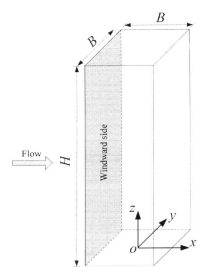

Figure 2. The 3D square cylinder model.

where r_u is the down wind velocity ratio, TI_x is at down wind turbulence intensity, $U_x(x,y,z)$ is down wind velocity (x direction), $\overline{U}_x(x,y,z)$, $\hat{U}_x(x,y,z)$ and $\overline{U}$ are the mean down wind velocity, the root mean square value of fluctuating wind and mean wind velocity.

The down wind velocity ratio around the single square cylinder and the turbulence intensity are plotted in Fig. 3.

For wind velocity, neglecting the wind velocity variation at range of 5%, the disturbed region is about H in front of the cylinder (negative x), about $2B$ at the top of the cylinder (z), three times as wide as B at both sides (negative and positive y) and more than $5H$ behind the cylinder. The wind velocities at the top and both sides are higher than the undisturbed wind velocity.

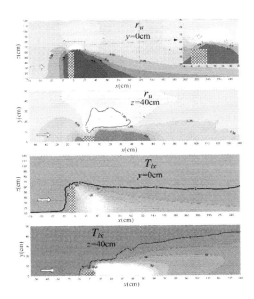

Figure 3. Flow Characteristics around the Single Cylinder.

For the turbulence intensity, the disturbed region is about $1/5H$ in front of the cylinder and smaller behind the cylinder. The turbulence intensity is 10% at $x = 4H$, $z = 0.5H$.

Obtained from the flow characteristics, the disturbed flow region is about H in front of the cylinder, 2B at the top, 3B at the sides and 5H behind. Based on the wind velocity distribution, the interference effects can be neglected as long as the distance between the interfering cylinder and the principal cylinder is five times longer than the width of the cylinder in a side by side arrangement, and the effect of the interfering cylinder located further than the width of the cylinder at leeward side on the principal cylinder can be neglected in tandem arrangement. The strong wind near the top and the corner should be taken into account in the accessory structure design, and the evaluations of pedestrian level winds.

The results of the rigid model pressure test are plotted in Fig. 4. The wind pressure distribution in Figure 4 was used as the wind pressure distribution on the undisturbed cylinder to compare with that on the interference cylinder, and it agrees well with the exist results and the free end effects are obvious.

4 THE INTERFERENCE EFFECT OF DOUBLE CYLINDERS

The wind pressure coefficients used in the paper is given by

$$Cp_i(t) = \frac{P_i(t) - P_0}{0.5\rho U_0^2} \qquad (2)$$

where $Cp_i(t)$ is the wind pressure coefficient at tap i, $P_i(t)$ is the wind pressure at tap i, P_0 is the static

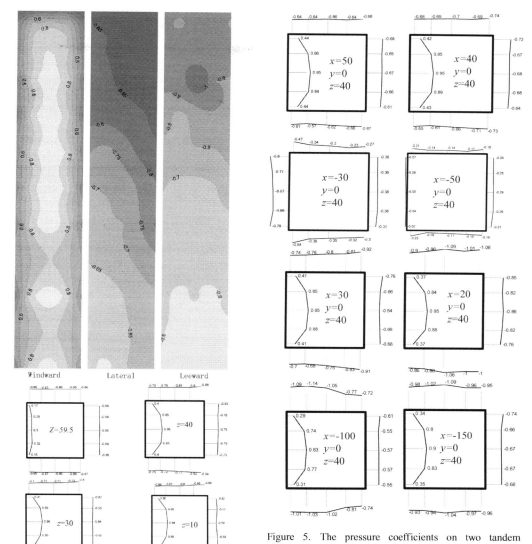

Figure 4. Wind Pressure Distribution on Single Cylinder.

Figure 5. The pressure coefficients on two tandem cylinders.

pressure at reference point, ρ is the air density, and $U_0 = 16.2$ m/s is the wind velocity at reference point. Cp_i, which used to describe the wind load distribution is the mean of $Cp_i(t)$.

The interfering cylinder is the same as the principal cylinder which is located at the origin of the coordinates system. Then x and y can describe the location of the interfering cylinder, and z is used to describe the height.

4.1 Tandem arrangement

The mean wind pressure coefficients around the cross-section at the height of 40 cm are plotted in Fig. 5 For the interfering cylinder placed on the leeward of the principal cylinder in the along-wind direction, comparing with the undisturbed cylinder, the negative pressures on the leeward side and both lateral faces are stronger within the range of 50 cm (the same as the wind velocity ratio influencing range in front of the single cylinder).

For the interfering cylinder placed in front of the principal cylinder in the along-wind direction, the mean interference effects generally present as 'shielding effects' where the interfering buildings tend to decrease the mean wind load on the principal building. The shielding effects decrease with the increase of the spacing between the two cylinders. The pressure on the windward face is negative at the distance of 50 cm (center to center distance). As one of the results of shielding effects, the mean drag force may be negative when the distance is small.

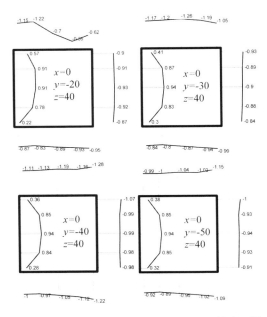

Figure 6. The pressure coefficients on two side-by-side cylinders.

4.2 Side-by-side arrangement

The pressure coefficients distributions on the cross section at the height of 40 cm are shown in Fig. 6. The side-by-side arrangement produces the channeling effect which induces mean lift force that may produce the adverse static amplification effects on the principal cylinder The unsymmetrical pressure reaches the maximum at the clear distance 2B, and the pressures on the two lateral sides are almost same at the clear distance 4B.

Figure 6 the pressure on windward surface close to interfering cylinder is stronger than the value away from interfering cylinder.

5 CONCLUSIONS

The scope of the influence of wind velocity was estimated by neglecting the wind velocity variation at range of 5%. The influenced region is about H in front of the cylinder (negative x), about $2B$ at the top of the cylinder (z) and three times as wide as B at both sides (negative and positive y); for the positive x the influenced region is more than $5H$. As one of the results of shielding effect, the mean drag force may be negative when the distance is small in the tandem arrangement. The side-by-side arrangement produces the channeling effect, which induces mean lift force that may produce the adverse static amplification effects on the principal building.

REFERENCES

Chen, K., Xiao, C. Z., Jin, X. Y., et al. 2012. Time-domain analysis method for the wind-induced 3D responses of super high-rise buildings. *China Civil Engineering Journal (In Chinese)* 45(7): 1–9.

Gu, M., Zhang, Z. W. and Quan, Y. 2013. Aerodynamic Measures for Mitigation of Across-wind Responses of Super Tall Buildings: State of the Art. *Journal of Tongji University (Nature Science)(In Chinese)* 41(3): 317–323.

Huang, D. M., Zhu, L. D. and Chen, W. 2012. Amplitude characteristics of wind loads on high-rise buildings with interference from surrounding buildings. *China Civil Engineering Journal (In Chinese)* 45(9): 1–10.

Hui, Y., A. Yoshida and Y. Tamura. 2013. Interference effects between two rectangular-section high-rise buildings on local peak pressure coefficients. *Journal of Fluids and Structures* 37(0): 120–133.

Hui, Y., Y. Tamura and A. Yoshida. 2012. Mutual interference effects between two high-rise building models with different shapes on local peak pressure coefficients. *Journal of Wind Engineering and Industrial Aerodynamics* 104–106(0): 98–108.

Hui, Y., Y. Tamura, A. Yoshida and H. Kikuchi. 2013. Pressure and flow field investigation of interference effects on external pressures between high-rise buildings. *Journal of Wind Engineering and Industrial Aerodynamics* 115(0): 150–161.

Quan, Y., Chen, B. and Gu, M. 2010. Acrosswind Loads and Responses of Square Highrise Buildings with Large Aspect Ratios. *Building Structure (In Chinese)* 40(2): 89–92.

Quan, Y., Zhang, Z. W., et al. 2012. Study of the RMS values of across-wind aerodynamic base moment coefficients of high-rise buildings with square or rectangular sections. *China Civil Engineering Journal (In Chinese)* 45(4): 63–70.

Shen, G. H., Wang, N. B., et al. 2012. Calculation of Wind-induced Responses and Equivalent Static Wind loads of High-rise Buildings Based on Wind Tunnel Test. *Journal of Zhejiang University (Enginering Science)(In Chinese)* 46(3): 448–453.

Tang, Y., Jin, X. Y. and Yang, L. G. 2012. Study of the interference effects of wind loads on tall buildings in staggered arrangement. *China Civil Engineering Journal (In Chinese)* 45(8): 97–103.

Wang, Q. H., Shi, B. Q. and Zhang, L. L. 2012. Influence of Wind Direction on Equivalent Static Wind Loads of a Super High-Rise Building. *Journal of Shantou University (Nature Science) (In Chinese)* 27(2): 48–52.

Xie, Z. N. and M. Gu. 2004. Mean interference effects among tall buildings. *Engineering Structures* 26(9): 1173–1183.

Xie, Z. N. and M. Gu. 2007. Simplified formulas for evaluation of wind-induced interference effects among three tall buildings. *Journal of Wind Engineering and Industrial Aerodynamics* 95(1): 31–52.

Zhang, Z. W., Quan, Y., et al. 2013. Effects of corner recession modification on aerodynamic coefficients of square high-rise buildings. *Civil Engineering Journal (In Chinese)* 46(7): 58–65.

Zou, L. H., Liang, S. G., et al. 2012. Analysis of three dimensional equivalent static wind loads on symmetrical high-rise buildings based on wind tunnel tests. *Journal of Building Structure (In Chinese)* 33(11): 27–35.

Civil Engineering and Urban Planning III – Mohammadian, Goulias, Cicek, Wang & Maraveas (Eds)
© 2014 Taylor & Francis Group, London, ISBN 978-1-138-00125-1

Identification of highway dangerous driving behavior based on Hidden Markov Model

Jian Wang, Jun Li & Xiaowei Hu
School of Transportation Science and Engineering; Harbin Institute of Technology, Harbin, Heilongjiang, China

ABSTRACT: This paper extracts four dangerous driving behavior characteristic factors (vehicle speed, acceleration, time headway, distance from the right edge of the line to vehicle). Combined with Hidden Markov Model, they are applied in the study of highway identification of dangerous driving behavior. Taking recognition accuracy as evaluation index, proved the feasibility of Hidden Markov Model. Results show that the model can effectively identify the highway dangerous drivng behavior.

1 INSTRUCTIONS

2011 Chinese traffic accident statistics shows that in 97000 road accidents, factors associated with driving behavior accounted for 91.85%, road traffic accidents were closely relevant to driving behavior. Therefore it is necessary to establish a suitable method which can identify highway dangerous driving behaviour to explore highway dangerous driving behaviour and its factors, providing a basis for decision-making related to accident prevention and management departments.

Hidden Markov Model is developed on the basis of the Markov chain and became a widely used statistical models, the classic theory put forward by the Baum et al, It was firstly applied to speech recognition systems, and had become the most successful way in the voice recognition system after years of development, HMM is widely used in computer Science Electrical Engineering information Sciences, Computational Mathematics, Statistics and other fields, it has become an important tool in the field of the voice recognition gene identification crypt analysis target tracking image processing and in the field of signal processing (Wang Xinmin 2003)

In foreign countries, Alex P. Pentland applied HMM to the driver's behavior and intention recognition applications, and its forecast rate reached over 95%. Gautam B. Singh (2009) used HMM model to study the vehicle collision prediction methods, and the predication proved that this method has a good timeliness to prevent collisions. In China, Yin Ming (2007) proposed a congestion detection algorithm based on HMM, this algorithm has a good ability to predict and detect a sequence of images of the vehicle blocking the vehicle; Wu Zhou (2013) extracted seven features of state insurance driving factors from the bus driver's driving state insurance behavior, which was combined with a hidden Markov statistical model, used to drive hazard identification study of behavior. Lu An (2010) shored combined with Gaussian mixture hidden Markov model (GM-HMM), carried identification of highway overtaking phenomenon and the analysis of the driver's intentions and overtake phenomenon normality, results showed that this method could effectively identify a variety of driving conditions in normal overtaking phenomenon.

2 CHARACTERISTIC FACTORS EXTRACTION AND ANALYSIS

Based on the video data collection, this paper carried out whole day's video and used these videos to get the vehicle speed, acceleration, time headway, distance from the right edge of the line as dangerous driving behavior characteristic factors.

(1) Vehicle Speed (X_1), When the vehicle speed exceed the speed limit of the highway, prone to collision, overtaking collision accidents. It is necessary to deal with overspeed behavior of dangerous driving behavior characteristics analysis.

The vehicle speed probability distribution (figure 1) statistics shows that the speed of the probability distribution is close to normal distribution, The vehicle speed average is 73.11 km/h, this paper select $V_5 = 41.8$ km/h as low vehicle speed threshold and $V_{95} = 91.3$ km/h as vehicles speeding threshold.

(2) Vehicle Acceleration (X_2), The speed of acceleration change directly influences the rate of vehicle's speed, the acceleration is an important indicator reflecting moderate degree of change of speed and the controllability of driver to the vehicle.

Through the vehicle acceleration probability distribution (figure 2) statistics, this paper finds that the vehicle acceleration appears similar to the normal distribution as the vehicle speed, and

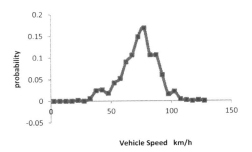

Figure 1. Probability distribution of vehicle speed.

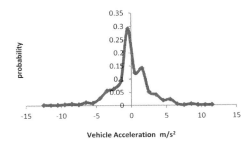

Figure 2. Probability distribution of vehicle acceleration.

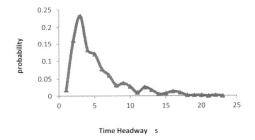

Figure 3. Probability distribution of time headway.

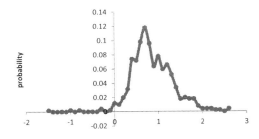

Figure 4. Probability distribution of distance from the right edge of the line to vehicle.

concentrates between $-5\sim5$ m/s^2, so this paper chooses $(-5, 5)$ as the acceleration index in dangerous driving behavior of the threshold.

(3) Time Headway (X_3), Keeping safe time headway in the process of vehicle driving on the highway is one of the major means of reducing accidents.

Through the headway probability distribution (figure 3), it's obvious that he headway is roughly negative binomial distribution, and the peak appears in the position of the headway about 2.5 s. The smaller of time headway, the bigger of the vehicle collision probability, according to the conception of V_{85} and the corresponding distribution of cumulative probability, this paper selects the 5th headway $X_3 = 1.2$ s as vehicle spacing less than the threshold.

(4) Distance from the right edge of the line to vehicle (X_4), When the vehicles running on the highway near the right side, it is easy to collide with guardrail or fell into the ditch, causing traffic accidents.

Through the probability distribution of distance from the right edge of the line to vehicle (figure 4), it shows that when a car travels in the normal condition, the distance from the right edge of the line to vehicle is $X_4 \geq 0$, when $X_4 < 0$ vehicles had entered the emergency parking area, this situation belongs to the dangerous driving behavior, when $X_4 \geq 2$, it indicates that vehicle overtaking behaviors in the test section, which not belongs to dangerous driving behavior, thus this paper selected $X_4 = 0$ as the dangerous driving behavior judgment threshold.

3 HIDDEN MARKOV MODEL AND ALGORITHM

3.1 Hidden Markov Model construction

The driving behavior on the highway is uncertain and invisible, to some extent, it can be expressed by the random process of the controlled vehicles. The random process of vehicles and driving behavior which we observed is not corresponding but related through a group of probability distribution HMM has strong time series, by driving behavior totransiteprobability of vehicleswe can make a prediction of the driving behavior of the after a period of time. When using hidden Markov model to predict driving behaviour, the parameters of the righteous are as follows:

(1) S represents the driving behavior state, $S = \{S_1, S_2\}$, S_1 represent Normal driving behavior, S_2 represent dangerous driving behavior;
(2) O represents Collection of an observed sequence $O = \{O_1, O_2, ..., O_M\}$, composition by the four dimensional vector X_1, X_2, X_3, X_4, through the video monitoring data, M from X_1, X_2, X_3, X_4 can output the number of different observations;
(3) Driving behavior matrix of transition probability $A = \{a_{ij}\}_{2\times 2}$, a_{ij} represents dangerous driving behavior transition probability;
(4) Driving behavior matrix of observation probability $B = \{b_j(k)\}$, it represents the probability of the observed state O under different driving S;
(5) Initial state distribution $\pi = \{\pi_i\}$, it represents the initial probability of each driving behavior.

3.2 The HMM algorithm

For calculating the output probability $P(O|\lambda)$ must solve the problem of evaluation, this can be achieved by "forward–backward algorithm". Define forward variable $\alpha(i)$ for a given HMM parameters, output part of the observed sequence $\{o_1, o_2, \ldots, o_T\}$ The probability of state S_i in the t moment $\alpha_t(i) = P(o_1 \ldots o_t, q_t = s_i|)$; Define backward variable (i)for a given HMM parameters, output part of the observed sequence $\{o_{t+1}, o_{t+2}, \ldots, o_T\}$, The probability of state S_i in the t moment $\beta_t(i) = P(o_{t+1}, o_{t+2}, \ldots, o_{T+q_t} = s_i|)$.

Through the forward and backward variables to get the output probability:

$$P(O|\lambda) = \sum_{i=1}^{N} \alpha_t(i)\beta_t(i)$$

$$= \sum_{i=1}^{N}\sum_{j=1}^{N} \alpha_t(i)\alpha_{ij}b_j(O_{t+1})\beta_{t+1}(j) \quad (1)$$

$$= \sum_{i=1}^{N} \alpha_T(i)$$

Using Baum-Welch algorithm to train the HMM train, this paper define two other variables $\varepsilon_t(i,j)$ and $\gamma_t(i)$. $\varepsilon_t(i,j)$ is Given model λ and observation sequence O in state S_i, in t moment, and the probability of state S_i in the $t+1$ moment:

$$\varepsilon_t(i,j) = P(q_t = S_i, q_{t+1} = S_j, O|\lambda)$$
$$= \frac{q_t = S_i, q_{t+1} = S_j, O|\lambda}{P(O|\lambda)} \quad (2)$$

$\gamma_t(i)$ is Given model K and observation sequence O in state S_i, in t moment:

$$\gamma_t(i) = (q_t = S_j, O|\lambda) = \frac{\alpha_t(i)\beta_t(i)}{P(O|\lambda)} \quad (3)$$

From the definitions of $\varepsilon_t(i,j)$ and $\gamma_t(i)$, define E_{ij} to expected frequency from state S_i to S_j, define E_{if} to expected frequency from state S_i, define E_i to expected frequency in state S_i:

$$E_{ij} = \sum_{t=1}^{T-1} \varepsilon_t(i,j) \quad (4)$$

$$E_{if} = \sum_{t=1}^{T-1} \gamma_t(i) \quad (5)$$

$$E_i = \sum_{t=1}^{T} \gamma_t(i) \quad (6)$$

Using formulas (4)~(6) can get the HMM parameters. This paper defines to expected frequency of observed value V_k in state S_i, a new revaluation model $\bar{\lambda} = (\bar{\pi}, \bar{A}, \bar{B})$ can be obtained by the following equation:

$$\bar{\pi}_t = \gamma_1(i) \quad (7)$$

$$\bar{\alpha}_{1j} = \frac{E_{ij}}{E_{ij}} = \frac{\sum_{t=1}^{T-1} \varepsilon_t(i,j)}{\sum_{t=1}^{T-1} \gamma_t(i)} \quad (8)$$

$$\bar{b}_j(k) = \frac{E_{jv_k}}{E_j} = \frac{\sum_{t=1}^{T-1} s.t O_t = V_k \gamma_t(i)}{\sum_{t=1}^{T} \gamma_t(j)} \quad (9)$$

According to the given A, and randomly given B, gets the initial model $\lambda = (\pi, A, B)$. Using training traffic data sequence O, according to formulas (7)~(9) calculate the model after the revaluation. Thus, this model will be more closer than the actual model, $P(O|\bar{\lambda}) > P(O|\lambda)$. Using $\bar{\lambda}$ instead of λ, repeating the above process of revaluation, until a limiting condition is satisfied, for example $P(O|\bar{\lambda}) - P(O|\lambda)$ less than a given threshold Repeat this calculating, it can improve the output of the training sequence O probability. The final result of this revaluation program called the HMM maximum likelihood estimate. After the training, getting the $\lambda = (\pi, A, B)$ is the final HMM model.

4 INSTANCE ANALYSIS

4.1 Data sources

This paper takes the video vehicle detector to collect the required information data, the location is located in Hu-Ning highway ($K219 + 040S$). Video vehicle detector was mainly composed of the camera (head) and the image recognition processing system. Its working principle: The camera was taken to road traffic flow in the two consecutive frames, carried on the comparison to all or part of its area, such as difference shows detection area had a moving object (e.g., a running vehicle). Then based on the principle of using $C++$ programming language to develop a video processor Vehicle Tracking Demo Upgrade.

By tracking the vehicles can get the position information of each car in the video, it could be further calculated on the basis of these information including vehicle speed, acceleration, time headway, distance from the right edge of the line tovehicle. The video processing program is shown in figure 5.

4.2 Identification of highway dangerous driving behavior

The process of solving the HMM model contains HMM model training and HMM model identification.

(1) HMM model training: first, to differentiate various parameters of dangerous driving behavior, by using the seating arrangement ratio method, then

Figure 5. Demo video processing.

Table 1. Dangerous driving behavior identify.

	Probability Value	result
HMM$_1$	0.98	normal driving
HMM$_2$	0.26	behavior

Table 2. Dangerous driving behavior identify.

	Vehicle Speed	Vehicle Acceleration	Time Headway	Distance from the Right Edge of the Line to Vehicle
Precision Rate	93%	91%	94%	90.5%

to use the literature Baum-Welch training algorithm get the final $\lambda = (\pi, A, B)$, and it's initial value in training is set up according to the corresponding percentage of statistics of sample data.

(2) HMM model identification: through the transformation of the conditional probability, and by using trained HMM model to find the probability $P(O|\lambda)$ at the hidden state corresponding to the sequence, comparing the probability about normal driving behavior of the HMM model λ_1 with dangerous driving behavior of the HMM model of λ_2, calculates the probability of a larger model to determine whether the state of driving behavior belongs to the normal driving condition or the state of dangerous driving behavior. Identification of the state of driving behavior process is shown in table 1.

In this paper, experimental conditions are as follows: randomly selected 200 normal driving behavior data and 50 dangerous driving behavior data as the training data to train the HMM model; Randomly selected 200 data from the normal driving behavior data and dangerous driving behavior data to identify them; the total number of identified data is divided by the number of properly identified data and the result is the recognition probability. Different dangerous driving behavior characteristic factors based on the HMM experiment result are shown in table 2.

5 CONCLUSIONS

This paper uses hidden markov model, combining with the video recording data, and achieves the modeling and identification of dangerous driving behavior on the highway, the main conclusionsare as follows:

(1) Analyzing thehighway dangerous driving behavior and temporal characteristics, putting forward four dangerous driving behavior characteristic factors (vehicle speed X_1, average acceleration X_2, time headway X_3, distance of the vehicles from the right edge of the line X_4), then applying it to hidden markov models.

(2) Hidden markov model identification is very accurate, it's recognition accuracy rate reachedover 90%.

REFERENCES

Alex P. Pentland, Graphical Model for Driver Behavior Recognition in a Smart Car. Proceedings of the IEEE Intelligent Vehicles Symposium University of Parma, Parma, Italy June 14–17, 2004.

Gautam B.singh. Using Hidden Markov Models in Vehicular Crash Detection, IEEE Transactions on Vehicular Technology, 58(3), March 2009.

Lv An. Recognition and Analysis on Highway Overtaking Behavior Based on Gaussian Mixture-Hidden Markov Model. Automotive Engineering, 2010, 32(7): 630–633.

Ming Yin, Hao Zhang. An HMM-Based Algorithm for Vechile Detection in Congested Traffic Situations, Proceedings of the 2007 IEEE Intelligent Transportation Systems Conference, 2007.

Wang Xin-min, Huang Xin-tang, Yao Tian-ren. A HMM training algorithm based on grouping multiple observations by multiple correlation coefficient. Journal of Central China Normal University, 2003, 37(3):179–182.

Wu Zhizhou. Reckless Status Identification of Bus Driving Behavior based on Hidden Markov Model. The eighth Chinese intelligent transport annual outstanding papers, 2013.

Impact of the climate conditions on waterproofing materials in long-term use in industrially polluted regions

V. Bartošová, D. Katunský, M. Labovský & M. Lopušniak
Faculty of Civil Engineering TUKE, Vysokoškolská, Košice, Slovakia

ABSTRACT: When choosing a hydro-insulation roofing sheets to consider his conduct throughout the period of exploitation of the building. For the purpose of examining external influences on the qualitative change in the behavior of the waterproofing materials was established corrosion station in polluted area. The goal was to observe the physical and chemical processes, supply of materials, processes of aging and predisposition of the individual components of which insulation is made. The basic criterion was to assess the climate resilience of waterproofing sheets based on visual examinations and laboratory tests. Acquired knowledge allows estimating the expected lifetime and hence the climate resistance of the types of materials. Results show protection of the waterproof sheets with a layer of gravel has not significant influence on preservation of their mechanical properties.

1 INTRODUCTION

The impact of external climate to the lifetime and quality of hydro-insulation roofing sheets is a content of many research projects. For example, research of roof pollution impact to increase of heat gains of materials, which leads to the fastest degradation or overheating of construction and also building (Ronnen Levinson) [1]. The further is research of impact of heat ageing process to the performance of PVC membranes (Paroli) [2]. The description of results of 7-year-long exposure to TPO membranes in different climate areas (Delgado) [3]. Some works are focused on fire-safety resistance and UV stability (Yang) [4]. Also results of comparison of research methods of shorten lifetime to performances of TPO membranes are published (Xing) [5]. As well as, works comparing performances of various roof membranes (Cash) [6].

In a corrosion station in Košice waterproofing materials placed directly to a heat insulation made of foam polystyrene and mineral wool were tested so that the all research process and comparison copies real conditions by which waterproofing insulations are placed on a concrete flat roof. The station was established in the year 1994, particular samples were placed in stages to the year 1997. Not all observant places were filled. Trial cut holes – elements of strips were taken in the years 2000 and 2010, sizes of approximately 20 × 20 cm from each holder were taken and sent to an accredited laboratory. Particular results, mainly organisation assurance of research and measurement methodology have been published in domestic, as well as foreign publications [7, 8, 9, 10].

2 CORROSION LAB

The role of the corrosion center of Civil Engineering Faculty was to create a station, which will record an influence of weather conditions on performances of selected materials. This station is the second of its kind in Slovakia, there was another one at the time when it started to work, where the samples of building materials were exposed for a long-term to extreme weather conditions. The corrosion station in Košice records the variation of performances caused by industrial air pollution in contrast to the station in the High Tatras. This station is focused on the determination of time-process variation of waterproofing coating performances under natural climate. There were monitored selected exterior parameters, as well as data of visually evaluated samples and results of laboratory tests. Among these exterior parameters, recorded during the year belong: exterior temperature, relative humidity, rainfall quantity, sunshine duration and wind velocity.

The monitoring and consideration of air pollution has also an important role. Freely waving dust and dust falling on rainwater samples were considered. The following visual changes were as follows: surface cracks, surface crumpling, spreading flow away and infiltration of penetration material.

The defined area of the climatic corrosion station is located on the area of scientific and research laboratory. Experimental samples with dimensions of 1 × 1 m were placed on various horizontal, vertical and sloped bases (Fig. 1). The tested materials were placed on stands directly on a thermal insulation layer made from expanded polystyrene or mineral wool. These

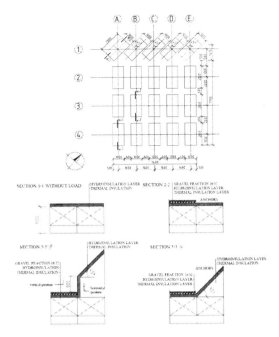

Figure 1. Schematic representation of the position stands the test specimens in floor plan and sections.

alternatives served as a background for monitoring of waterproofing coatings as roof fragments, and their behaviour in real conditions. The scheme pictures of stands and arrangement of samples are shown in the Fig. 1. The stands are made from steel with dimensions of 2.0×1.0 m. Two types of samples are placed on each stand.

There are totally 20 monitored points. Five representative materials were selected for the paper. It is also because that 15 years is a long time, during which it was impossible to backwardly get the measured values of some data (closed laboratory, where measurements were done).

3 ENVIRONMENT

So called "rain" containers were used for consideration of industrially polluted air and for determining the quantity of each element investigated. The containers allowed monitoring of particulates from the air and their observation and analysis of their chemical composition. The containers were complied with distilled water up to 1–2 cm high in summer season. They were complied with methanol in winter season. Subsequently, after the removal samples were transferred to the drying-plant and they were dried at 105°C. Spectrochemical analysis allowed us to determine 17 secondary and trace elements. When comparing samples exposed to climatic and industrial impacts those in corrosion station are less polluted than those from Main TUKE Building. Professor Florian [11] claims even ten times higher pollution for the same climatic

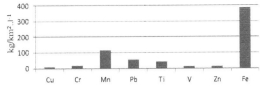

Figure 2. Metallic pollution load in February 2000.

influences and trends. The periodical repetitive trend occurs in amount of dust particles and Fe. Other particles show no periodical repetitions. The changes and repetitions in cycles are not only influenced by industrial conditions changes but by impact of climatic conditions too [11]. In 2000, air pollution by dust particles was highest in February with amount of $75.3 \text{ t/km}^2.\text{J}^{-1}$. Air pollution by metals in this month is shown in Fig. 2.

4 METHODOLOGY

Testing was performed after five and fifteen years of exploitation. Laboratory tests were carried out according to the applicable regulations and standards and conducted by thermomechanical analysis and spectrochemical analysis in selected laboratories based on standard requirements.

5 WATERPROOFING MATERIALS

According to material basis they are classified as: bitumen, plastic (film), silicate (special mastics based on cement), clay materials (based on clays and bentonite) and special (chemical grouting agents). Insulation based on the plastics is used as protection against water and moisture. By the source materials they are divided into: thermoplastics, elastomers and thermoplastic elastomers. Thermoplastics: the key feature is the activation of the surface by hot air-softening. They are divided into: polyvinyl chloride (PVC), polyethylene (PE), polypropylene (PP), polyisobutylene (PIB). Samples were monitored on the basis of plastic materials, more specifically the following thermoplastics:

PVC homogeneous, thickness 1.5 mm without protective gravel (M1); PVC homogeneous, thickness 1.2 mm loaded with protective gravel (M2); PVC fabric reinforced, thickness 1.5 mm without protective gravel (M3); PVC fabric reinforced, thickness 1.2 mm, loaded with protective gravel (M4); PIB thickness 1.5 mm reinforced with fiberglass without protective gravel (M5).

6 RESULTS

Measurements show, that homogeneous PVC-P sheet 1.5 mm thick, without protective gravel shows after 15 years, a significant decrease in tensile

Table 1. PVC-P homogenous M1 and M2.

Properties		Material without gravel		Material loaded with gravel	
Mean values	Units	after 5 years	after 15 years	after 5 years	after 15 years
Nominal thickness	mm	1.5	1.5	1.2	1.2
Tensibility	%	18	19	11	14.5
Tensile strength	N/50 mm	380	265	200	30
Flexibility at low temperatures	°C	−45	−25	−35	−10
Perforation at low temperatures	°C	–	−20	–	−10

Table 2. PVC-P fabric reinforced M3 and M4.

Properties		Material without gravel		Material loaded with gravel	
Mean values	Units	after 5 years	after 15 years	after 5 years	after 15 years
Nominal thickness	mm	1.5	1.5	1.2	1.2
Tensibility	%	25 (24–25)	19 (16–25)	25	18
Tensile strength	N/50 mm	1375	1410	1250	1110
Flexibility at low temperatures	°C	−30 (−30)	−20 (−10 −30)	–	−30
Perforation at low temperatures	°C	–	−20 to ±0	–	−20 to ±0

strength (Table 1). The decrease is from the value >1000 N/50 mm for a new sheet until the value 265 N/50 mm after fifteen years (Table 1). The sheet shows no loss of tensibility. Also we do not notice a loss of flexibility at low temperatures as well as a loss of thickness is not notice. Measurements also show that homogeneous PVC-P sheet 1.2 mm thick, with protective gravel shows after 15 years, a more significant decrease in tensile strength. The decrease is from value >1000 N/50 mm for a new sheet until the value 30 N/50 mm after fifteen years (Table 1). The sheet shows no loss of tensibility. Though, sheet shows a loss of flexibility at low temperatures (−10°C, for a new sheet −25°C). A loss of thickness is not notice.

Homogeneous PVC-P sheets 1.5 and 1.2 mm thick reinforced with fabric do not show loss of tensile properties (Table 2). Tensibility and tensile strength are insignificant for evaluation of these sheets due to reinforcement. The measured properties reflect properties of the reinforcement and not properties of PVC. Negative results are noticed in a test of a perforation at a low temperature. The results show on loss of a frost resistance (−20°C to 0°C). It shows that sheets are prone to perforate at low temperatures.

PIB sheets show after 15 years a significant decrease in tensibility. After five years it was 310% (the same as for a new sheet). But after fifteen years it was only 145% (Table 3). PIB sheets also show after fifteen years a loss of flexibility at low temperatures (−20°C, for a new sheet −25°C). Negative results are noticed in a test of a perforation at a low temperature. The results show on a loss of frost resistance (−10°C).

Presented results show, that results are similar for the sheets with the gravel protection and for the sheets without the gravel protection. A gravel protection has no positive impact on physical and mechanical properties of roof sheets or membranes. A statement, that a gravel layer protects roof sheets against aging is wrong.

Figure 3, illustrates the changes in tensile strengths for materials after 5 and 15 years.

Table 3. PIB.

Properties		Material without load	
Mean values	Units	after 5 years	after 15 years
Nominal thickness	mm	1.5	1.5
Tensibility	%	310	145
Tensile strength	N/50 mm	320	300
Flexibility at low temperatures	°C	–	−20
Perforation at low temperatures	°C	–	−10

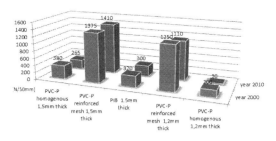

Figure 3. Tensile strength.

Figure 4, depicts changes in flexibility at low temperatures for materials after 5 and 15 years.

Figure 5, shows the perforation of materials at low temperatures after 15 years.

From point of sheets implementation on roofs a gravel protection layer is more convenient not to use. For example, a gravel layer makes complication in a case of defects diagnostic, like water leaks or unwanted perforations, etc.

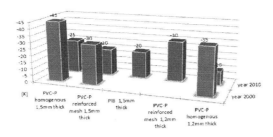

Figure 4. Flexibility at low temperatures.

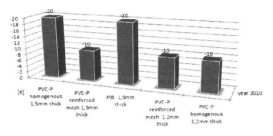

Figure 5. Perforation at low temperatures.

The rate of changes after aging of all sheets could not be fully measured. It is due to limited space options for large number of observed samples exposed to the weather conditions, as well as due to an occurrence of some unplanned damages during their exposure. It led to changes in dimensions and quality of samples and it is inappropriate for quality analysis. So we analyzed only samples with an appropriate level of quality.

In general, an evaluation of roof hydroinsulation sheets in real climatic conditions for five to eight years is inadequate. It is a too short time period to show all changes of material properties. Results show, that the fifteen year period is more adequate.

7 CONCLUSION

Based on the measured data, it can be deduced that hydro-insulation roofing sheets are adversely affected by industrially polluted climatic conditions and degrade significantly over time. Significant changes were observed in the physical and mechanical properties of all the elements that were subjected to laboratory testing, i.e. homogenous PVC, fabric reinforced PVC and fiberglass grid reinforced PIB. In addition, the protection of the hydro-insulation roofing sheets with a layer of gravel did not significantly influence the preservation of their physical and mechanical properties.

ACKNOWLEDGMENT

This paper was created and realized through the project ITMS 26220120037 "Centre for excellent research of progressive building construction, materials, and technologies", with support of the research and development program financed by the European Regional Development Fund.

This contribution was created and realized through collaboration with the research project VEGA 1/1060/11 "Monitoring changes in physical parameters of envelope constructions in quasi-stationary conditions".

REFERENCES

C. G. Cash, Comparative testing and rating of thirteen thermoplastic single ply roofing membranes, Interface (Raleigh, North Carolina) 17(10) (1999) 28–33.

A. H. Delgado, M. Ludwig, S. Elliot, K. C. Barnhardt, C. Chapman, R. Ober, R. M. Paroli, Thermoplastic polyolefin roof membranes exposed in the western united states for seven years, Paper presented at the ASTM Special Technical Publication 1538 (2011) 1–41.

K. Flórián, Schlussbericht 1999 über die Forschungsprojekte von WGD No.012: Messungen von Staubniederschlägen am Freibewitterungstand WGD/TU in Košice, WGD-Tagungsbuch (2000) 71–87.

D. Katunský, R. Schoepe, Untersuchung des Alterungsverhaltens, Dachdecker-Handwerk Zentralverband des Dachdeckerhandwerks 10 (2000) 2.

D. Katunský, Pozorovanie prirodzeného starnutia strešných hydroizolačných, Rekonštrukcia a sanácia striech (2000) 154–159.

D. Katunský, Beobachtung der Veränderungen der physicalisch-mechanischen Eigenschaften von Dichtungsbahnen in realen Bedingungen, WGD-Tagungsbuch (2000) 43–45.

D. Katunský, R. Schope, Untersuchung des Alterungsverhaltens von Dachabdichtungen am Freibewitterungsstand der WGD an der Technischen Universität in Košice (Slowakische Republik), WGD-Tagungsbuch (2000) 57–61.

R. Levinson, P. Berdahl, A. Asefaw Berhe, H. Akbari, Effects of soiling and cleaning on the reflectance and solar heat gain of a light-colored roofing membrane, Atmospheric Environment 39 (40) (2005) 7807–7824.

R. M. Paroli, O. Dutt, A. H. Delgado, H. K. Stenman, Ranking PVC roofing membranes using thermal analysis, Journal of Materials in Civil Engineering 5(1) (1993) 83–95.

L. Xing, T. J. Taylor, Correlating accelerated laboratory, field, and thermal aging TPO membranes, Paper presented at the ASTM Special Technical Publication 1538 (2011) 50–70.

L. T. Yang, L. Xing, T. Taylor, A bright future – single ply thermoplastic polyolefin roofing, Paper presented at the Annual Technical Conference – ANTEC, Conference Proceedings 3 (2009) 1509–1513.

Implementation of PMS at a local level-case study based on StreetSaver®

G. Wang
Quality Engineering Solutions, Inc., Gainesville, FL, USA

D. Frith
Quality Engineering Solutions, Inc., Reno, NV, USA

D. Morian
Quality Engineering Solutions, Inc., Conneaut Lake, PA, USA

ABSTRACT: In this study, a case study based on pavement management system (PMS) program StreetSaver® was conducted on a west costal city in USA to evaluate the overall condition of the city street network and highlighted the impacts of various funding levels on the network pavement condition and deferred maintenance backlog. Through the case study, the PMS program StreetSaver® demonstrates that it is an effective tool for local pavement network. Local governments can predict the future condition of their pavement for different levels of funding and show the effects of under-funded road programs. With the progress of urbanization in China, the study may help China local agencies to gain the knowledge of pavement management practice on a local level and thus give some meaningful implication for the future application of PMS in China.

1 INTRODUCTION

Pavement management system (PMS) is defined as "a set of tools or methods that (can) assist decision makers in finding cost-effective strategies for providing, evaluating and maintaining pavements in a serviceable condition" (FHWA, 1989). The goal of a PMS is to optimize the value of the maintenance funds and to provide the highest possible pavement quality with the allocated resources (Wells, 1984). In order to effectively and efficiently manage urban roadway network, the United States (US) Congress approved the Intermodal Surface Transportation Act (ISTEA) in 1991, which called for the involvement of Metropolitan Planning Organizations (MPOs) in the development and implementation of PMS (Sohail, et al., 1996). Since then, many PMS programs such as MicroPaver and StreetSaver® have been implemented in many cities and counties in the US. For instance, StreetSaver® has more than 100 users in the San Francisco Bay Area and more than 250 users nationwide and internationally (Schattler et al., 2011), and users are predominantly cities and counties.

The importance of PMS to local agencies can be attributed to several reasons (Vasquez, 2011):

- Provide a means to objectively deal with volatility in pavement treatment and repair costs and variation in annual tax-based revenues
- Sophisticated modeling methods employed at the state and federal levels are inappropriate for local agencies because local agencies lack the staff and money to manage them.
- Use of pavement management systems supports better allocation of the local agencies' money by giving them a means to choose cost-effective ways of treating the pavement network system, and reduce work zone visual and operational impacts.

PMS has been proved as a cost effective tool in planning budget, allocating the funding and maintaining the pavement network in a best possible serviceable condition (Sohail, et al., 1996; Schattler et al., 2011; Vasquez, 2011; Chen et al., 2011; Romell et al., 2010).

On the other hand, with the progress of historical urbanization in China, more and more small and medium sized cities have been developed in China. Roadway and streets, as a critical part of urbanization, play a significant role in public service. Billions of dollars have been invested on roadways in urban areas. How to protect those huge investments on urban roadways, preserve and maintain city roadway networks in a cost effective way has been a big challenge to local agencies. Although PMS has been proved an effective for attacking such a challenge, the implementation of PMS on a local level is almost blank in China. Therefore, the purpose of this paper is to introduce a US based successful implemented PMS program StreetSaver® through a case study to gain the knowledge of pavement management practice on a local level and thus give some meaningful implication for the future application of PMS in China.

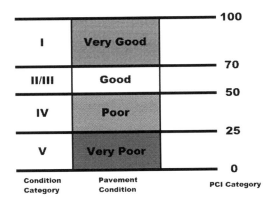

Figure 1. Pavement condition categories.

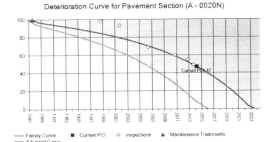

Figure 2. An example of adjustment of pavement performance curve.

2 STREETSAVER®

In this paper, PMS program StreetSaver® developed by the Metropolitan Transportation Commission (MTC) of the San Francisco Bay Area is used for the case study. The implementation of the StreetSaver® began with six local jurisdictions in the bay area in 1980s (MTC, 2013) and has developed more than 400 users nationwide and internationally after more than two decades. The success of its implementation resides in user-friendly interface, easy to operate, customer-defined decision tree and the technical support provided by MTC which includes training, on-site assistance to address special cases, assistance in the budgeting process, and continuous feedback (MTC, 2014).

StreetSaver® considers five surface types in its decision support tools, namely asphalt concrete (AC), portland concrete cement (PCC), asphalt concrete over asphalt concrete (AC/AC), asphalt concrete over portland concrete cement (AC/PCC), and surface treatment (ST) (MTC 2013). All pavement sections are grouped into four functional classes: arterial (A), collector (C), residential/local (R/L), and other (O) (MTC, 2013).

Pavement condition in the StreetSaver® is expressed in terms of pavement condition index (PCI) ranging from 0 to 100, with 100 representing a newly constructed or overlaid pavement while 0 representing a completely failed pavement. A typical MTC definition of pavement condition categories based upon the PCI value is shown in Figure 1.

3 PAVEMENT PERFORMANCE MODELS

Pavement performance models are used to predict pavement deterioration over time due to factors such as traffic loads, climatic conditions, age, material durability, poor drainage and construction quality. Pavement performance prediction models are among the most critical parts of a well-defined PMS since they affect the timing for applying treatments, budget planning, inspection scheduling and work planning (Silva et al., 2000). In StreetSaver®, pavement performance prediction models are based on a sigmoid equation (Deshmukh, 2009):

$$PCI = 100 - \frac{CHI \times \rho}{\left[\ln\left(\frac{\alpha}{AGE - SHIFT}\right)\right]^{\frac{1}{\beta}}} \quad (1)$$

where, AGE is the age in time from the construction to the point at which the PCI is to be calculated; PCI is the projected PCI value at some value of AGE; α, β and ρ are regression constants that control the shape of the performance curve; CHI and SHIFT are the projection modifiers that are adjusted to force the PCI projected value to match the observed value for each management section (Deshmukh, 2009). The initial CHI and SHIFT values are set to be 1 and 0 respectively, and they are modified based on the following equations to make sure the inspected PCI value at a certain point is equal to the projected PCI value (Deshmukh, 2009).

$$CHI = \frac{100 - PCI_{Obs}}{100 - PCI_{Fam}} \quad (2)$$

$$SHIFT = AGE_{Fam} - AGE_{Insp} \quad (3)$$

where, PCI_{Obs} is the observed PCI at the time of inspection; PCI_{Fam} is the value from the family curve based on the eq. 1 with current CHI and SHIFT values. AGE_{Insp} is the age of the section at the time of inspection, which is the time difference between the date of construction and the date of inspection. AGE is calculated based on the PCI value from the inspection with current CHI and SHIFT values as follows (Deshmukh, 2009):

$$AGE_{Fam} = SHIFT + \alpha e^{-\left(\frac{CHI \times \rho}{100 - PCI_{Insp}}\right)^{\beta}} \quad (4)$$

An example of adjustment of performance curve is shown in Figure 2.

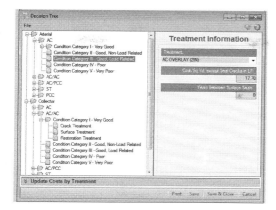

Figure 3. Typical decision tree in StreetSaver®.

Table 1. Example of PP strategies incorporated in the decision tree of StreetSaver®

Functional Class	Surface	Condition Category	Treatment Type	Treatment
Residential/Local	AC	I-Very GOOD	Crack Treatment	Seal Cracks
			Surface Treatment	Slurry Seal
			Restoration Treatment	Thin Overlay (1.5 inches)
Treatment	Cost per Sq Yd except Crack Seal in LF	Yrs between Crack Seals	Yrs between Surface Seals	# of Surface Seals before Overlay
Seal Cracks	0.80	3		
Slurry Seal	2.7		5	
Thin Overlay (1.5 inches)	11.4			3

4 DECISION MAKING PROCESS

The decision making process is modeled based on the decision tree in the StreetSaver®. The decision tree contains "branches" for each functional classification, surface type, and condition category as shown in Figure 3. The decision tree screen allows agencies to modify costs associated with treatments on specific types of pavement according to their jurisdiction.

Integrating pavement preservation (PP) strategies in the decision tree is another key feature of StreetSaver®. As shown in Table 1, a PP strategy can be easily set up for a residential/local street with AC surface. Under "Very Good" condition category, seal cracks will be applied at $0.80/LF at a 3-year interval between crack treatments, and slurry seal at $2.7/SY every 5 years between surface seals. The pavement will receive a 1.5" thin overlay as a surface restoration at unit cost of $11.4/SY.

Table 2. Statistic summary of street network.

Functional Class	Total Sections	Total Center Line Miles	Total Lane Miles	PCI
Arterial	183	27.4	89.3	71
Collector	285	21.6	46.6	69
Residential/Local	734	65.1	130.2	72
Other	67	14.2	32.1	63
Total	1269	128.3	298.2	
Overall Network PCI as of December 2013:				70

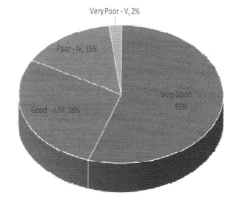

Figure 4. Pavement condition summary by condition categories (2013).

5 CASE STUDY

A StreetSaver® based PMS from a west coast city in California was used as the case study. The case study evaluated the overall condition of the street network and highlighted the impacts of various funding levels on the network pavement condition and deferred maintenance funding shortfalls. Detail description about this study is given in the following sections.

5.1 Network description and existing pavement condition

The City is responsible for the repair and maintenance of approximately 128 centerline miles of streets, or 1269 pavement sections. The City's street network replacement value is estimated at $150.8 million. This represents a significant asset for City officials to manage. This asset valuation is assessed by the assumption of replacing the entire street network at today's dollar.

Based upon the field condition surveys completed, the average overall network PCI of the City's street network is 70, which indicates that the street network is between 'Good' and 'Very Good' condition. Table 2 summarizes the number of sections, length, and average PCI of the network by functional class. Figure 4 presents the pavement condition categories of the network. As shown, 83% of network falls into the

Figure 5. GIS-based map for current network PCI conditions.

'Good' or 'Very Good' condition category, while 17% of the network falls into the 'Poor' or 'Very Poor' condition category. Illustrated in Figure 5 is a GIS-based map of the current network PCI conditions.

5.2 Budget needs

StreetSaver® develops a maintenance strategy targeting at improving overall network condition to an optimal PCI at around 80 and sustaining it at that level based on the principle of less cost to maintain streets in good condition than those in bad condition.

The overall PCI for the City street network is 70, which is between the 'Very Good' and 'Good' condition categories. However, load-related distresses are shown in a large area of the network and $14.9 million deferred maintenance backlog will be accumulated in the first year of the scenario. If these problems are not addressed, the condition of the street network will inevitably decrease. Therefore, a cost-effective strategy for funding and maintenance and rehabilitation (M&R) must be implemented. And the analysis of budget needs for maintaining the whole street network should be first performed.

In determining relative budget scenarios over a five year period, a representative interest rate of 5% and inflation rate of 1% is selected in the analysis. The interest rate is used to describe an annual percentage increase in invested funds that would be realized if it were not instead spent on rehabilitation and maintenance activities. The inflation rate describes the rate of change of prices especially in relation to the construction cost index where a positive inflation rate indicates a loss in purchasing power over time and a negative inflation rate indicates an increase in purchasing power. Purchasing power simply describes the number of goods or services that can be purchased with a unit of currency.

An estimation of $34.5 million over the next five years is required based on the budget need analysis. Of the $34.5 million in maintenance and rehabilitation needs shown, approximately $2.3 million or

Table 3. Summary of budget needs analysis.

Year	PCI Treated	PCI Untreated	PM Cost ($1000)	Rehab Cost ($1000)	Total Cost ($1000)
2014	79	69	$1,284	$13,638	$14,922
2015	78	67	$71	$4,983	$5,054
2016	79	65	$368	$5,260	$5,628
2017	80	63	$161	$5,724	$5,885
2018	80	61	$395	$2,646	$3,041
		%PM	PM Total Cost ($1000)	Rehab Total Cost ($1000)	Total Cost ($1000)
		6.60%	$2,279	$32,251	$34,530

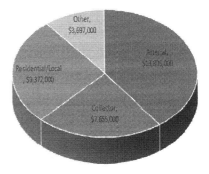

Figure 6. Budget distribution by functional classification.

6.60% is earmarked for preventive maintenance or life-extending treatments, while the remaining $32.3 million or 93.40% is allocated for more costly rehabilitation and reconstruction treatments. If the City follows the strategy recommended by the program, the average network PCI will increase to 80 and eliminate the current maintenance and rehabilitation backlog. If, however, there is no funding for pavement maintenance and there is little or no maintenance is applied to the street network over the next five years, the network PCI will decrease to 61 due to the continue deterioration of streets. The results of the Budget Needs analysis are summarized in Table 3 below. Figure 6 illustrates funding distribution by street functional classification.

5.3 Budget scenarios

Once the budget needs for the street network are determined, the next step in developing a cost-effective M&R strategy is to conduct "what-if" analyses. the effects of different budget scenarios on pavement conditions in terms of PCI and deferred maintenance (backlog) were evaluated. By comparing PCI and deferred maintenance, the advantages and disadvantages of different funding levels and M&R strategies become clear. In this study, the following five scenarios were run for a five-year period.

1. Unlimited Budget — Under this scenario, the budget is not constrained and the total budget amount

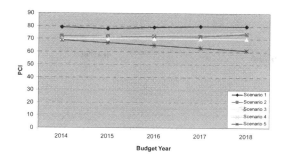

Figure 7. Comparison of PCI over time using different budget scenarios.

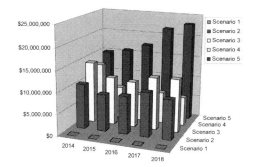

Figure 8. Comparison of deferred maintenance over time using different budget scenarios.

per year is defined as "needs". A total of $34.5 million, as identified in the Budget Needs analysis, were input into the Budget Scenarios module. This scenario will improve the condition of the City's street network to a PCI of 80 and eliminate deferred maintenance backlog in the first year,
2. Current Investment Level — Under this scenario, the City would limit annual budget to $4.5 million per year. A 5% budget was split to the preventive maintenance. Stop gap costs are taken from preventative maintenance funds.
3. Maintain Current PCI — In order to maintain the current PCI level at 70, a Target-Driven Scenario model was used to determine the required budget. The result indicated that a five year total of $15.9 million is needed, with $13.6 million for rehabilitation and $2.3 million for preventive maintenance.
4. Increase Current PCI by 5 points — In order to increase the current overall PCI by 5 points, to 75, by the end of the fifth year, a Target-Driven Scenario model was used to determine the required budget. The results indicate that a five year total of $23.3 million is needed, with $21.0 million for rehabilitation and $2.3 million for preventive maintenance.
5. Do Nothing — If no maintenance or rehabilitation is applied over the next five years, the condition of network will deteriorate to an overall PCI of 61. The maintenance backlog will increase to $23.1 million.

Figures 7 and 8 graphically illustrate the comparison of the five scenarios in terms of PCI and deferred maintenance. Figure 7 shows the comparison of the change of PCI over time using different budget scenarios. As shown, Scenario 1 (Unlimited Budget) will ultimately reach a PCI of 80, while Scenario 5 (Do Nothing) will decrease to a PCI of 61 after five years. Under Scenario 2 (Current Investment Level), the PCI will increase to 74 in year 2018. Figure 8 illustrates the change in deferred maintenance over time for each scenario. As expected, Scenario 1 will completely eliminate the deferred maintenance, while the amount of deferred maintenance will continuously increase and reach $23.1 million in 2018 if no monies are spent on maintenance and rehabilitation.

6 DISCUSSION AND RECOMMENDATIONS

As demonstrated in the budget scenario analysis, the idea strategy for the City is Scenario 1, which not only increases the network PCI to an optimal level of 80, but also eliminates the entire deferred maintenance backlog in the first year. However, nearly $14.9 million funds are required from the City in the first year, which makes this strategy unrealistic for the City. But, this scenario can be served as a base line for comparing other scenarios.

Under the Scenario 2, namely current City budget level, the network PCI is anticipated to increase to 74 after five years. And the maintenance backlog will stay between $8 million and $10 million. The City's current funding level is clearly insufficient to maintain the whole of the street network in 'Very Good' or 'Good' condition.

The high maintenance backlog will increase future cost as more expensive treatments such as reconstruction are needed in order to complete deferred street repair. A significant amount of funds are needed on expensive rehabilitation and reconstruction projects so that deferred maintenance backlog will be reduced. And thus, the pavement condition of the whole street network will be improved and more money can spent on less expensive treatments such as seal coat and cracking seals in the future.

7 CONCLUSIONS

A Citywide Pavement Management Program will assist City personnel by providing current technical data to maintain a desirable level of pavement performance, while optimizing the expenditure of limited fiscal resources. Specifically, the program will provide administrators and maintenance personnel of local agencies with:

– A current inventory of all public roadways.
– The current pavement condition for all public roadways.
– A project listing of all pavement needing maintenance, rehabilitation, or replacement.

– A forecast of budget needs for maintenance, rehabilitation, or replacement of deficient sections of pavement for a multiple year Capital Improvement Program.

Through a case study, the PMS program StreetSaver® demonstrates that it is an effective tool for local pavement network. Local governments can predict the future condition of their pavement for different levels of funding and show the effects of under-funded road programs. With the progress of urbanization in China, an effective PMS program like StreetSaver® is needed in order to protect huge investments on urban roadways, preserve and maintain city roadway networks in a cost effective way.

REFERENCES

De Melo e Silva, F., Van Dam, T.J., Bulleit, W.M. &Ylitalo, R. *Proposed Pavement Performance Models for Local Government Agencies in Michigan.*" Transportation Research Record: Journal of the Transportation Research Board, No. 1699, Transportation Research Board of the National Academies, Washington, D.C., 2000, pp. 81–86.

Ding Xin, C., Sui, T., & Hicks, R.G. *Optimizing Pavement Preservation into Pavement Management System.* 8th International Conference on Managing Pavement Assets. Santiago, Chile, 2011.

Federal-Aid Highway Program Manual. Federal Highway Administration (FHWA). Washington DC, 1989.

Maithilee Mukund Deshmukh. Development of Equations to Determine the Increase in Pavement Condition due to Treatment and the Rate of Decrease in Condition After Treatment for a Local Agency Pavement Network. Master's Thesis Submitted to Texas A&M University, Texas A&M University, 2009.

Metropolitan Transportation Commission (MTC) Homepage. Metropolitan Transportation Commission, CA. http://www.mtc.ca.gov/services/pmp/. Accessed March 10, 2014.

MTC Pavement Management Software 9.0: Computer User Guider. Metropolitan Transportation Commission, California, 2013.

Romell, R. & Sui, T. Regional Asset Management Efforts and a Performance-Based Approach to Local Streets and Roads Funding Allocation.Compendium of Papers from the First International Conference on Pavement Preservation, 2010.

Sohail, F., Dossey, T. & Hudson, W.R. "Implementation of the Urban Roadway Management System". Report No. SWUTC/96/465550-1, Southwest Region University Transportation Center, College Station, Texas, 1996.

Schattler, Rietgraf, Wolters, & Zimmerman. "Implementing Pavement Management Systems for Local Agencies." Report No. FHWA-ICT-11-094, Illinois Center for Transportation, Urbana, IL, 2011.

Vasquez, C.A. *Pavement Management Systems on a Local Level.* Master's Thesis submitted to Utah State University, Utah State University, Logan. Utah, 2011.

Wells, W. *Developing Pavement Management Systems at County and City Levels.* Transportation Research Record: Journal of the Transportation Research Board, No. 997, Transportation Research Board of the National Academies, Washington, DC, 1984, pp. 41–46.

Inheritance and creation of regional architecture features in urbanization process

Yi Han & Qiong Wang
Xi'an University of Architecture and Technology, Xi'an, China

ABSTRACT: Traditional regions in China have distinct architecture characteristics, which have been facing or is about to face significant changes in urbanization process. On the basis of theoretical research, this article highlights the combination of regional architectural culture and urbanization practices and emphasizes the integration of traditional architecture features and modern construction techniques. By analyzing the causes and performance of regional architectural features, and investigating its survival advantages in special natural environmental and resource context, this article explored complex patterns of residential buildings in the southern of Maowusu Desert with modern life adaptation, and on the basis of inheriting, some of the design principles for regional architecture in the context of urbanization are proposed.

Keywords: urbanization, regional, architectural features

1 INTRODUCTION

1.1 Waning of regional architecture features

The main background of our times is the construction of modern civilization along with the inheritance of nation's outstanding traditional culture; the development of regional economy and the creation of a better living environment; as well as the minimized waste of resources and the ecological balance of natural environment. These are the main problems that we have to face inevitably in the process of urbanization.

In such context, it is quite necessary and important to combine modern architecture elements with traditional architecture factors, as already shown by some regional architectural theories and recent worldwide studies and practices. However, with accelerated urbanization, regional architecture identity associated with local natural environment, traditional culture, and social organizations and so on has been rapidly weakening and even disappeared. The construction in China's large and medium cities provides a good proof, in which distinct regional characteristics are lost – constructions are gradually becoming more consistent not only between cities in China, but also between cities in China and that in other countries.

Seen from the perspective of historical development, the formation and evolution of local traditional architecture is restricted by natural resources and regional social culture. Following the progress of globalization, when multiple values and modern building technology become a common phenomenon, local architecture will break the existing constraints of local resources and ideology to choose more economical, more convenient and more efficient construction mode. In such context, modernist architecture takes the advantages of standardized construction, which replaced traditional architecture in many aspects such as materials, construction technology, morphology, and so on.

Seen from the perspective of social development, most studies have confirmed that local traditional architecture also has various adaptive advantages of sustainable development in the inevitable trend of urbanization and modernization. The key lies in how to inherit its characteristics correctly, how to combine it with modern building technology creatively, and how to ring it in line with the objective laws of social development and cultural inheritance.

Specific research work to be done is as follows:

1. Analyze the formation factors of local architecture features including natural and geographical factors, socio-cultural factors, economic factors, etc., using holistic research methods
2. Summarize the characteristic system of architecture features under the guidance of system theory and sum up different types of cultural elements based on the relationship between space activities and architecture environment;
3. Follow the dynamic laws and developing trends, integrate the characteristics of local architecture and useful modern elements including multiple layers of combination, such as building materials, construction methods, organizational space integration, etc.

The aim is to form a rational, beneficial, unique pattern of regional architecture form in urbanization, which belongs to our times and suits our living environment, with scientific theory system, economical construction methods, and minimum environment interference characteristics and resource conservation being properly incorporated.

1.2 *Urbanization is an integration of modernization and regionalism*

Urbanization is the only way for developing a country from an agricultural one to an industrial one. In China's 2010 Central Economic Work Conference, Central Rural Work Conference and Central Document No. 1., it had been pointed out that urbanization shall be promoted in an active and steady manner. The 18th CCP large conference also emphasized the importance and necessity of urban modernization and rural urbanization.

Urbanization is a harmonious developing process of various social systems. The field of urban and rural planning and construction is to create a physical environment to adapt to it. For nearly 30 years, the development of urban planning and architectural design has caused great changes to China's large and medium-sized cities. Nowadays international metropolis has become a rush goal, while regional architecture features which were once the externalization performance of local culture disappeared day by day. Urban construction brought us more efficient and more convenient city life, as well as rapid development of urban society, economy and culture. However, the cultural diversity of the region has been severely challenged. In Lewis Mumford's book "History of Urban Development – the Origin, Evolution and Prospects," it is pointed out that the future task of a city is the full diversified development of various regions, cultures, and individuals (Lewis, M. 2005).

Thus urbanization does not exactly equal to modernization. In fact, the combination of regional culture and modern civilization is the dominant approach to urbanization. Modern civilization brings the transformation of nature, usage of resources, more efficient methods and techniques of creating social value, while regional culture particularly reflects the values of local life and values. Einstein characterized our times as complete means and confusion goals, which is the fundamental objective of our research in cultural heritage area. In the interdisciplinary field of architecture, it can be materialized as an integration of traditional architectural characteristics and modern construction techniques.

2 FORMATION AND REPRESENTATION

2.1 *Regional architectural theory – theoretical basis*

Mr. Wu Liangyong had clearly pointed out in his book "Broad Architecture" that regional architecture is an objective reality, which cannot be ignored by architectural discipline. The regional features are produced by geography, economic development and social culture. Regional architecture inevitably reflects the changes of architectural forms and styles. The revaluation of regional architecture showed an effort to rescue architectural culture poverty from international architecture ideological restraints (Liangyong, W. 2011). In academic sense, provincialism is one of the essential presences of architecture, especially in the historical background of the conflict between regional cultural diversity and international monotony.

Shan jun in his book "Regional Cities and Architecture – the Concept of the Living Environment of a Region of Architectural Research" (Jun, S. 2010), summarized the regional theories of architecture since the 20th century, from the 1960s vernacular architecture research, to 1980s critical regional architectural theory, place theory, and the "Beijing Charter" in which regional architecture theory in the context of globalization are widely identified as an important identity. It not only clarified the general context of regional development theory, but also touched related important research area – vernacular architecture and theory, and this will undoubtedly expand the scope of theoretical study and practical areas of architectural theory.

These works played an important role in the formation of regional architectural theory. However, there are still problems to be solved as follows for the development of regional architecture theory:

1. In-depth studies are lacked. For example extensive contacts of theoretical systems in the architecture phenomenology theories and environmental behavior theories are lacked;
2. Specific study object in architectural sense has to be established. For example architecture materials, construction methods, forms, spatial and other factors on regional architecture characteristics have to analyzed and summarized;
3. By taking spatial level as classification and regional characteristics as the core, architectural heritage conservation and development, vernacular settlement study, preservation and renewal of historic towns and other areas shall be included into the research category to form a holistic space research system;
4. Dynamic development shall be studied in combination with urbanization of social practice; specific issues in China shall be researched to solve the problems for ordinary people.
5. Problems in specific practices shall be identified to seek solutions; and regional development experiences from local construction shall be summarized.

2.2 *The influential factors in formation of architectural feature*

According to the interpretation of "cultural identity" from Amos Lapp Bute (Amos, R. B. 2004), cultural

Figure 1. Traditional residential building – Willow house.

Figure 2. Wide steps of Baicao Temple in Qingjian County.

Figure 3. People at Baicao temple fair.

identity can be considered not only as an important cultural manifestation affected by building and its environment, but also as an concept and method to study the relationship between regional building and its environment. Even the Culture-specific design study is culturally an important practical direction emphasized by regional urbanization.

Regional architectural characteristics are deeply related with geographical and ethnic regional culture, reflecting the human-environment interaction mechanism, which is relatively stable in a certain historical period and varies when a particular element or mechanism changes. Formation factors of regional architectural characteristics include natural geographic features, resources, economics, construction, social activities, and folk culture.

Regional natural environment and resource conditions formed different survival pattern, as well as different architecture characteristics. Laomaoji Village is located in Yulin area in Shaanxi Province in the southern edge of Maowusu Desert, a weak ecological agro-pastoral zone. The main plant – Salix and the surface gravel layer – white mud as the main source of building material, formed traditional residential buildings – Willow house (Figure 1). In order to adapt to the special climate (windy, bitter cold or tarried heat, drought), the thickness of the white mud wall is up to 0.8–1 m to provide good thermal insulation.

Socio-cultural is another very important formation factor for regional architectural characteristics, which, in turn, is the material system of socio-culture. Most of Chinese traditional settlements have typical public activity areas, which hold community gatherings, such as religious festivals, Opera, fair trade in the New Year and so on in fixed time Festival. Temples of the village and the surrounding environment are fixed locations for these activities; trade, entertainment, worship and other various social and economic activities are also occurring here. Take Bai Cao Temple in Qingjian County as an example, there are theatre space at the entrance, composed of multiple steps about 550 cm wide and 400 cm high as the outdoor seats. Therefore, the architecture of the village temple is to meet the acting and entertainment needs (Figure 2 and Figure 3).

It should be noted that, in specific regional architecture and its environment, the unique and distinctive regional architectural characteristics are the combined affection of multiple factors (Kingston W. H. 2009).

2.3 *Expressions of regional architecture characteristics*

Characteristics of regional architecture are the essence of architecture, which is one of the main expressions of the diversity of regional culture. The nature of architecture is not only about the composition of the architecture itself, but also about the functional component. Architecture characteristics are reflected in the selection and processing of building materials, building technologies, building construction, architecture and decorative arts and so on; The function features of architecture cultural include: regional natural environment, location, architectural group, the relationships between neighborhoods and other space, different life scene (such as live events, production and other economic, social, leisure and religious activities, and so on).

Therefore, it can be seen that architectural identities are associated with two factors: One is regional restrictions such as regional resources and economic sources; the other is the thought of ideal human settlement. For example in willow house, with high toughness and strong plasticity, Salix can be tied and bent to about

Figure 4. Facades and roofs of willow house – made from white mud.

Figure 5. Willow roof bearing structure.

20 centimeters bundles used as rafters in roof load-bearing structure of cave system (Figure 4). Wall's thickness can be up to 0.9 m by using white mud as main material, which has good adhesion, high density. These walls not only have the physical properties of thermal insulation, but also can resist lateral thrust from arched roof (Figure 5). There is a high degree of unity of building materials, reasonable building structure, and natural appearance in willow house. Generally speaking, these two materials are the expression of local architectural characteristics limited by natural resources.

As another example, Chinese traditional residential environment is a significant cultural habitat. By the spatial organization of residential buildings and faiths worship buildings such as temples and shrines, people in Mutouyu Village in northern Shaanxi Province established an ideal habitat with symbiosis among nature Gods, clan gods and the occupants.

The four temples on the beach and hilly mountain of Mutouyu Village: Heshen Temple, Kuixing Temple, Sanhuang Temple and Guiyun Temple surround the central living space, embody the imagination about the natural world and the dependent of the nature God, reflecting the order of nature and human settlement. In the village center, Wenchang Temple and Guanyin Temple were built on a raised platform, reflecting

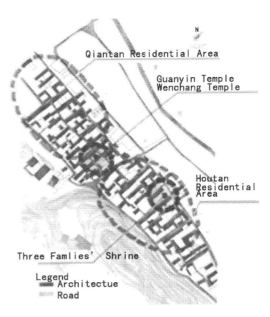

Figure 6. Relationships between residential buildings and ritual construction in Mutouyu Village, Jia County.

the faith of the village; Sanxing memorial temple in the village meets the needs of blood clan worship activities (Figure 6).

These spaces as the combination of material and spirit are one of the important manifestations of local architecture identity.

3 ADVANTAGES

3.1 *Effective use of regional resources*

There are great contradictions between resources and population in China. The scale and speed of current urbanization is bound to make this conflict more prominent. Yue Bangrui in his book "Oasis Architectural Theory – To Create Patterns Oasis Settlement Area under Resource Constraints" raised the issue of efficient use of regional resources on the basis of "Suitable construction", which is not a new set of values, and pointed out adaptive survival and development advantage of traditional architecture in the new era (Bangrui, Y. 2011).

This adaptive advantage did exist in regional architecture. Taking the formal example of Laomaoji Village in the north of Shaanxi Province, the regional use of mixing Salix and white mud as building materials has the following characteristics:

1. Anti-seismic building structures: with light willow roof, the wall is not easy to collapse in earthquakes;
2. Adaptability to local climate: willow house is an earth sheltered building, which is warm in winter and cool in summer, very suitable for the harsh local

Figure 7. Salix-local plant in Maowusu Desert.

climate of long period of cold and windy winter and badly hot summer.
3. Affordable building materials: Salix and white mud is local resource, whose processing is not complicated.

It can be seen that unique forms of local architecture and distinctive regional architectural culture comes from the unique natural conditions, and the remarkable survival wisdom under resource constraints. Architecture identity is a reflection of people's understanding, utilization of and adaptation to local natural environment.

3.2 Symbiosis mode with natural environment:

What must be emphasized is that the main material of willow house – Salix is a kind of windbreak trees in desert, which can effectively prevent desertification (Figure 7). It has fast growth rate and fantastic growth characteristics, whose long branches must be cut in a certain period. It will wither up if you do not cut, and the more cut the more prosperous it will become.

After simple treatment, it can be used as construction materials, widely used in all aspects of building environment: the load bearing structures, the roof of storage warehouses, fences, livestock storages and so on. So in this ecologically fragile area of Laomaoji village, Salix can achieve win-win results for not only preventing ecological degradation, but also acting as low-carbon green building materials (Table 1).

Regional natural environment, lifestyle, architectural forms have internal and external unity. In the changing times, we should not change or abandon the traditional way of living eagerly; on the contrary, we should fully aware of its suitability and ecological advantages. Through appropriate reformation it still can meet the specific needs of production and changing lifestyle. Only in this way the symbiotic of living environment and ecological environment can be achieved as we expected.

By planting Salix, villages on the edge of the Maowusu Desert not only protect local environment, but also meet the needs of construction material, which is the symbiotic mode worthy to inherit and carry forward.

Table 1. Features of Salix building material in traditional architecture in Laomaoji village.

Classification	Location	Structure	Appearance
Courtyard border – the fence	Courtyard wall	Salix branches are arranged in vertical rows, using their thicker side as the bottom, and using thicker strips of wood as their horizontal joint.	Salix branches are neatly arranged in double or single row, according to the length and intensity.
Residential building	Roof	Interconnect dozens of wickers to 3 m long and 20 cm thick bundles, bent them into a bow to make a willow shed, then place the shed between willow fence and wheat straw. A good thermal insulation dust cover is formed.	Horizontal wood girders and vertical willow rafters constitute square mesh smallpox.
Storage	Roof	Firstly, willow branches with tenacity are brined so as to have insect prevention and to be bent into different size circles easily, and then they are closely packed to form a pavilion roof.	Short conical roofs are covered by slim wickers.
Stables railings	Envelope	The same as the fence structure.	Walls are surrounded on three sides and the top is covered by willow branches.

3.3 Composite residential buildings – innovation and creation of regional architectural characteristics

Affected by many factors, regional architectural characteristics change in the new era. The traditional residential building – willow house in the south of Maowusu Desert is no exception, as the roof scale is restricted by the length of willow, making the indoor space relatively small, which can not meet the growing needs of living standards, not to mention its small windows, thick walls, poorly indoor lighting and ventilation effect. Along with the living conditions improvement and lifestyle changes, more alternative factors such as more convenient materials – brick, precast concrete panels, etc., as well as the acceptance of foreign cultures, have produced new residential patterns in the village. The new flat roof brick structure has improved the original residential living conditions in this area. So old traditional buildings and new residential buildings coexist in Laomaoji village.

As the result of social development, it is possible for regional architecture to break through the limitations of local resources to have more choices on the architecture materials, techniques and forms. Changes in regional architectural characteristics should be compatible with existing lifestyle and current social production. New architecture creations should take into account of local geographical and climatic characteristics, and should be the fusion of traditional architecture characteristics and modern technology. For example, in a residential unit of Laomaoji Village, we can create different space with different functions, separated and effectively linked in a complete interpenetration way. Specifically, this complex residential building model taking the advantages of the old and new architectural features is to make new building in a courtyard form enclosed shape with old willow house. Willow building has good thermo-physical properties, with relatively small space and window proportions very suitable for bedroom; while new building can be used as hall to accommodate living, cooking, dining, working and other daily functions. The courtyard between two buildings is the space effectively linking them, not only an extension of the indoor space, but also a convenient space for other courtyard activities. Such spatial patterns not only take advantage of the original building, waste no resources, but also create a courtyard to resist sand attacks effectively.

4 CONCLUSION

As an important carrier of local material and spiritual sense, local architecture plays an important role in cultural diversity region, which shows its significance in the context of urbanization. In conclusion, the design principles have been preliminarily formed as follows.

Firstly, theoretical system – Taking regional architectural features as the main research object, this article studies the sciences of human settlements, combines it with the living environment science, which integrated Regional architectural theory, Architectural culture research, Environment-behavior studies, Environmental psychology, and summarizes the formation and expression of the architectural characteristics in specific area.

Secondly, suitability and economic rationality – This article pays attention to the integration, transformation and utilization of the regional resources and modern construction technology, such as effective use of regional resources, inheritance of local economic construction technology, renovation of local unreasonable factors, integration of modern valuable elements, such as building materials, construction methods, spatial organization, and so on.

Thirdly, controllability – This article emphasizes on the resource conservation and environmental protection, which not only strike a balance between economic development and environmental protection, but also demands more symbiotic relationships between human habitat construction and natural environment. Ecological living wisdom of local dwellings is worthy of our in-depth research, comprehensive inheritance, and more effective use of it.

REFERENCES

Amos, R. B. 2004. Culture, Architecture and Design. Beijing. China Architecture & Building Press.

Bangrui, Y. 2011. Building on Oasis – Oasis Settlement under the Constraint of Local Resources to Create a Model, Shang Hai. Tongji University Press.

Jun, S. 2010. Regional of Architecture and Cities – A Research on Regional Architecture with Human Settlements Concept. Beijing. China Architecture & Building Press.

Kingston W. H. 2009. Vernacular Architecture and Regional Design: Cultural Process and Environmental Response. Architectural Press.

Lewis, M. 2005. The City in History – A Powerfully Incisive and Influential Look at the Development of the Urban Form through the Ages. Beijing. China Architecture & Building Press.

Liangyong, W. 2001. Introduction to Science of Human Settlements. Beijing. China Architecture & Building Press.

Liangyong, W. 2011. Integrated Architecture. Beijing. Tsinghua University Press

Tong, Z. 2003. The Overall Regional Architecture. Nanjing. Southeast University Press

Inhibitory effect from house price rising expectation on adverse selection: Evidences from resale house trading experiments

H. Zhang, X. Sun & Y. Zhang
Department of Construction Management, Tsinghua University, Beijing, China

ABSTRACT: This article focuses on the inhibitory effect from house price rising expectation on adverse selection in resale house market. Firstly, the formulae of bullish sentiment index (BSI) and adverse selection level (ASL) were derived. Then, economic experiments were designed and implemented based on resale house market's characteristics, and trading processes of resale houses in different bullish sentiment were simulated by experiment. Experimental data were collected. Finally, BSI and ASL were calculated with the experimental data and the influence from bullish sentiment on adverse selection was tested. The research shows that severe information asymmetry is the main cause of adverse selection in resale housing market. Currently, bullish sentiment of traders in the market temporarily restrains the phenomenon of adverse selection to some extent. Reducing the degree of information asymmetry is the basic method of preventing and solving adverse selection problem in resale housing market.

Adverse selection is a market misallocation phenomenon caused by information asymmetry. The phenomenon often exists in secondary market (Akerlof, 1970) in which bad products drive out good ones and trade activity atrophies. Without original quality certification, buyers can't get actual quality information of second-hand products in the market. It becomes hard to eliminate information asymmetry and these market features will finally lead to adverse selection.

In China's resale housing market, information asymmetry is severe; it is hard for buyers to search for actual quality information of their target houses (Li & Guo, 2003). It is in line with the basic condition of adverse selection market. But the phenomena including bad products driving out good ones and trade activity atrophy do not exist in China's resale housing market. On the contrary, from the year 2000 to 2013, the resale house price in this market increased by 133% and the trading volume increased by 78% based on the China economic census yearbook written by China's National Bureau of Statistics (2013). There are plenty of resale houses with high quality in this market. This article focuses on the reason of the paradox fact stated above.

Resale house has dual properties of consumption and investment (Liu & He, 2009). Traders' evaluation of resale house's using function (quality) and investing function (price expectation) are two key factors influencing trading decisions. In fact, Buyers with house price bullish sentiment will still make purchasing decisions even without actual house quality information, which will weaken the negative influence from information asymmetry of house quality and avoid adverse selection phenomena like trading atrophy.

Few existing studies focus on the influence exerted by house price expectation on adverse selection. Most of the researches discuss the cause and control method of adverse selection from the perspective of information transfer. Heal (1976), Bond (1982), Hendel & Lizzei (1997) and Levin (2001) separately discussed the influences on adverse selection exerted by long-term interest, market offset system and search cost, non-market system and socialization of private information. To study the relationship between house price bullish sentiment and adverse selection, it is essential to collect numerous data of trader's house price sentiment and trading activity, which cannot be obtained with traditional methods like market research. In this circumstance, it is hard for researchers to measure house price bullish sentiment and adverse selection level. Further analysis becomes even impossible.

Therefore, this article tries to collect interrelated research data by the method of economic experiment. In controlled environment of experiment, researchers can generate house price bullish sentiment by delivering specific information to traders; then researchers can record mass experimental data by observing the traders' behavior and adverse selection phenomenon (Zhang & Zhang, 2012). The bullish sentiment index and adverse selection level can be calculated precisely based on experimental data. The influence from bullish sentiment on adverse selection can be analyzed finally.

1 RESEARCH ROADMAP

1) The formulae of bullish sentiment index (BSI) and adverse selection level (ASL) were designed. This

study firstly defined the bullish sentiment index of house price. Then, the relationship of house price, quality and adverse selection was illustrated. Finally, the calculation formula of adverse selection level was designed.
2) Economic experiment was designed and implemented and experimental data were collected. This study designed the resale housing market experiment to build up market environments with different house price bullish sentiments. Then, the experiment was implemented in laboratory. Meanwhile, experimental data were collected.
3) The influence from bullish sentiment on adverse selection was analyzed based on the experimental data. With the data collected and formulae designed, the bullish sentiment index and adverse selection level were calculated. Then, this study analyzed the influence from bullish sentiment on adverse selection in resale housing market. At last, the final conclusion was drawn.

2 INDEX DEFINITION AND FORMULA DESIGN

2.1 *Formula of bullish sentiment index*

House price bullish sentiment is a kind of investor's expectations. In resale housing market, buyers and sellers all propose to gain profit via trading activities. They can be viewed as investors to some certain extent. Seller's expectation is reflected by the lowest price charged, which is known as seller's reserved price. Buyer's expectation is reflected by the highest price offered, which is known as buyers' reserved price. The seller's and buyer's expectations will change with the house price bullish sentiment.

House price bullish sentiment is related with the difference between buyer's and seller's reserved prices. It can be illustrated from the following three dimensions. In time dimension, seller's reserved price cannot be lower than the buying price in the past while buyer's reserved price cannot be higher than the possible selling price in the future. When the price is expected to rise, the difference between the two reserved prices is more obvious. In risk dimension, sellers mainly focus on the past while buyers focus on the future. The uncertainty that buyers faced is bigger than that of sellers. When the price is rising, the gap between buyer's and seller's reserved prices will be widen. In psychological dimension, the resale houses offered by sellers are often the second or third house of the family, while buyers often buy resale houses as their family's first house. It means that the expected profits of sellers and buyers are different. Sellers have less expected profit while buyers' purchase willing is strong. When expectation changes, buyer's and seller's reserved prices will not change in the same way.

Therefore, this article uses the ratio of reservation price difference index (RPDI) in different market stages to measure bullish sentiment of house price and designs BSI (bullish sentiment index) of resale housing market.

$$BSI = \frac{RPDI}{RPDI_0} = \frac{\sum_{i=1}^{n_1}(V_i - U_i)/n_1}{\sum_{j=1}^{n_2}(V_{0j} - U_{0j})/n_2}, \quad (1)$$

where, V_i represents the buyer's reserved price of house "i" in the current stage, V_{0j} represents the buyer's reserved price of house "j" in the past stage, U_i represents the seller's reserved price of house "i" in the current stage, U_{0j} represents seller's reserved price of house "j" in the past stage, and n_1 and n_2 represent the numbers of buyers and sellers houses, respectively.

2.2 *Formula of adverse selection level*

Price and quality are the most obvious characters of adverse selection market. In this article, price of resale house represents the sold price of resale house, while quality is the actual value of the resale house's composite characters including floor, house type, construction quality, location, etc. To measure the influence exerted by house price bullish sentiment on adverse selection, the variable ASL (adverse selection level) is introduced to reflect the degree of adverse selection with the deviation of market price and quality.

The article uses Price & Quality index (PQI) to represent house price and quality in resale housing market. Meanwhile, the article uses ASL to represent adverse selection level. In a market of continuous trading, ASL can be calculated with the following formula.

$$ASL = \frac{PQI}{PQI_0} = \frac{\left\{\sum_{i=1}^{n_1}\left|\frac{[P_{\max}(\theta_i)-P_i][\theta_{\max}-\theta_i]}{[P_{\max}(\theta_i)-P_{\min}(\theta_i)][\theta_{\max}-\theta_{\min}]}\right|\right\}/n_1}{\left\{\sum_{j=1}^{n_2}\left|\frac{[P_{\max}(\theta_j)-P_j][\theta_{\max}-\theta_j]}{[P_{\max}(\theta_j)-P_{\min}(\theta_j)][\theta_{\max}-\theta_{\min}]}\right|\right\}/n_2}, \quad (2)$$

where n_1 and n_2 represent the total numbers of transactions in test market and standard market, respectively, P_i represents the transaction price of house "i" in the market, θ_i represents the quality of house "i", $P_{\min}(\theta_i)$ and $P_{\max}(\theta_i)$ represent the lowest price sellers charge and highest price buyers offer for houses with quality "θ_i", respectively, $\theta_{\min}$ and $\theta_{\max}$ represent the lowest and highest quality of houses, respectively. PQI_0 is the PQI of standard market, $PQI \in (0, 1)$, and it is related with the quality and price level of the market.

3 EXPERIMENT DESIGN AND IMPLEMENT

3.1 *Simplification hypotheses*

To increase the operability and clear the direction of the research, several hypothesizes were proposed to simplify the experiment.

Table 1. Experimental parameters.

Parameters $\theta/1$	Control group Experiment 1			Experimental group Experiment 2			Experiment 3		
	1	2	3	1	2	3	1	2	3
$V(\theta_i)/¥$	15000	33000	51000	15000	33000	51000	22500	49500	76500
$U_1(\theta_i)/¥$	5250	17250	41250	5250	17250	41250	5250	17250	41250
$U_2(\theta_i)/¥$	9000	21000	45000	9000	21000	45000	9000	21000	45000

1) To guarantee the market consistency of house characters and highlight the difference in house quality, experimental house quality was divided into three levels including "low", "middle" and "high" to simplify the experiment (Zhang & Zhang, 2013).
2) To guarantee the market consistency of house price and keep the unity of reserved price and market reality, transaction price of the second house was assumed higher than the first one, which also fits the law of diminishing marginal utility.
3) To guarantee the market consistency of trading system and simulate the situation of seller's market, the price was decided by sellers in experimental market, which means the seller quotation system was adopted in the experiment.
4) To guarantee the market consistency of transaction subject and simulate the present situation of market where supply is less than demand, more buyers than sellers were set in the experiment.

3.2 Basic experimental parameters

As is listed in Table 1, the basic parameters of the formulae were designed in accordance with the simplification hypothesizes. θ represents house quality, which is decided by house's composite characters. $\theta \in (0.5, 3.5)$, $V(\theta_i)$ represents buyer's reserved price, and $U_1(\theta_i)$ and $U_2(\theta_i)$ represent seller's reserved prices for the first and second houses transferred. The reserved prices above are all functions of θ.

3.3 Experiment design

Based on the purpose of the study, this experiment was designed from four aspects, including experimental purpose, experimental contents, experimental settings and experimental procedure (Zhang, Zhang & Li, 2013).

3.3.1 Experimental purpose and contents

The purpose of the experiment was to observe the adverse selection of markets with different degrees of information asymmetry and house price bullish sentiments, by collecting price and quality data of those markets. The experiment contents consist of three parts. The first part is constructing parallel market of the resale housing market. The second part is recruiting experimental participants and implementing experiments in different market conditions. The third part is observing experimental transaction activities and collecting data of house price and quality.

3.3.2 Experimental settings

The experimental settings consist of market structure settings, experimental mechanism settings and experimental group settings.

The market structure settings include market entity setting and market object setting. Market entities of the experiment are the traders of resale houses, who need to know the basics of real estate field and have decision-making ability. Meanwhile, the entities' purpose should be getting payments from experiment. There are 3 three buyers and four sellers in the experiment, who are senior undergraduates from the department of real estate in Tsinghua University. The market objects are the tradable resale houses. To meet the purchasing needs of buyers and simulate oversupply environment, the market objects consist of six houses for each period. The house price floated between buyer's and seller's reserved price. The house quality is divided into three levels, including "low", "middle" and "high", to reflect house difference and simplify the experiment.

The experimental mechanism settings consist of trading mechanism settings and incentive settings.

In the trading mechanism settings, auction is the transaction method in the experiment. The house price is decided by sellers. The trading market is controllable experimental market. In the information disclosure settings, transaction price and house id are open to the public, while reserved prices were kept by buyers or sellers themselves. The disclosure of quality information is related to different experiment groups to simulate different market information conditions.

In the stimulation settings, payments of participants, which are material stimulations, are decided by participants' experimental profits. The experimental payments are approved by participants and are directly related to the experimental performance. With the purpose of stimulating participants and avoiding irrational behavior, the payments are moderate. The performances of participants are listed after the experiment. The best participant will get praised and souvenir, which is mental stimulation of the experiment.

Table 2. Descriptive statistics of experimental data.

Sub-Experiments	Index	Minimum	Maximum	Average	Standard deviation
Experiment 1	Transaction price (¥/m^2)	6000	48000	19526	9099
	House quality (1)	1.00	3.00	1.79	0.52
Experiment 2	Transaction price (¥/m^2)	9500	22000	15053	4831
	House quality (1)	1.00	2.00	1.21	0.41
Experiment 3	Transaction price (¥/m^2)	12000	43500	23158	9585
	House quality (1)	1.00	3.00	1.79	0.69

In the experimental group settings, the experiment is divided into three sub-experiments: experiment 1, 2 and 3. Experiment 1 is the control experiment, where the quality information is disclosed before transaction and house price rising expectation is relatively low. In experiments 2 and 3, the quality information is disclosed after the transaction period is finished, so buyers cannot make purchase decisions by referring to quality information. Reserved price in experiment 3 is 50% higher than that of experiments 1 and 2, which means the house price rising expectation in experiment 3 is higher than those in experiments 1 and 2.

3.3.3 Experimental procedure

Each transaction of the experiment is divided into 4 steps. Firstly, each seller will make selling decisions to determine the price and quality of his/her two houses that will be transacted. Organizers of the experiment will collect the decisions and display them on the projector. Secondly, each seller will make purchasing decision with the information displayed. Each buyer can only buy one house from one seller at a time. Thirdly, each seller will make decision on whether to sell their second house if a buyer wants to buy it. At last, the experiment organizers will collect the transaction information, calculate the profit of each participant and grant experimental payments.

3.4 Experimental data

The experiment is divided into three sub-experiments and each sub-experiment includes five periods. There are four transactions in each period. Therefore, the experiment includes 60 groups of transaction information. Each group of information includes transaction price, house quality and traders ID. The transaction price and house quality information are shown in Table 2.

4 DATA ANALYSIS

4.1 Index calculation and comparison

The influence from house price bullish sentiment on adverse selection was analyzed by the following procedure. Firstly, experiment 1, which represents the information symmetry market, was regarded as the baseline market. The bullish sentiment indexes (BSI) of experiments 2 and 3 were calculated with Eq. (1). Then, with the experimental data and Eq. (2), the adverse selection levels (ASL) of experiments 2 and 3 were calculated, as listed in Table 3.

Table 3. Results of bullish sentiment index and adverse selection level calculation.

Index	Experiment 1	Experiment 2	Experiment 3
Bullish sentiment index (BSI)	1	0.78	2.09
Adverse selection level (ASL)	1	1.34	1.06

As shown in Table 3, based on the standard market of experiment 1, the BSI and ASL of experiment 2 are 0.78 and 1.3, while the BSI and ASL of experiment 3 are 2.09 and 1.06. This means that compared to the information symmetry market (experiment 1), both the information asymmetry markets (experiment 2 and 3) show adverse selection in different levels. Information asymmetry is the main cause of adverse selection. In the two information asymmetry markets, the BSI rise from 0.78 to 2.09, and ASL falls from 1.34 to 1.06. The adverse selection level almost recovers to that of the information symmetry market, which means bullish sentiment will inhibit adverse selection in resale housing market.

4.2 Market data analysis

To perform further and intuitive analysis of the inhibition by house price bullish sentiment on adverse selection in the market, the diagram of market transaction price (Figure 1) and the diagram of house quality distribution (Figure 2) were drawn with the data collected from the three sub-experiments. By comparing the transaction price data and house quality data from different markets, the inhibition and the phenomenon of "bad products driving out good ones" were analyzed exhaustively.

Firstly, comparing the transaction prices and house qualities from experiment 1 and experiment 2, Figure 1 shows that the transaction price in experiment 2 is

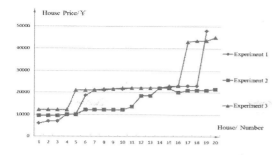

Figure 1. Diagram of market transaction price.

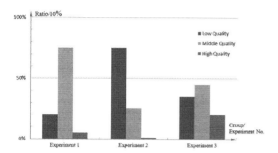

Figure 2. Diagram of house quality distribution.

Table 4. Results of Wilcoxon-Mann-Whitney rank test.

The null hypothesis	P value
H1: There is no difference in resale house price between experiment 2 and 3	0.001
H2: There is no difference in resale house quality between experiment 2 and 3	0.006

obviously lower than that of experiment 1, while Figure 2 shows that the house quality level in experiment 2 is also lower than that of experiment 1. This means that obvious phenomenon of adverse selection has emerged in the information asymmetry market of experiment 2 compared to the information symmetry market of experiment 1. It is information asymmetry that causes adverse selection in resale housing market.

Secondly, by comparing the transaction prices and house qualities from experiment 1 and experiment 3, Figure 1 and Figure 2 show that price and quality level barely change in the two markets with different information conditions. This means that there is no obvious adverse selection in the information asymmetry market of experiment 3.

Thirdly, the difference between the transaction prices and house quality levels in experiments 2 and 3 were analyzed quantitatively. By normal distribution test and variance homogeneity test of the price and quality data in the two experiments, the study shows that they do not obey normal distribution and there is no variance homogeneity. However, as the sample data were collected independently, the method of Wilcoxon-Mann-Whitney rank test can be used to compare the difference of the two sets of data. The test results are shown in Table 4.

As is shown in Table 4, the null hypothesizes of H1 and H2 are rejected at significant level of 1%. It means that there are obvious differences in the house prices and qualities of experiments 2 and 3. The markets of experiments 2 and 3, in which exists the same information asymmetry, show different adverse selection results. The main difference between the two markets is the traders' bullish sentiment. Therefore, it is bullish sentiment of house price that finally inhibits the adverse selection in the information asymmetry market of experiment 3.

5 CONCLUSIONS

This article tries to simulate resale house trading procedure in different bullish sentiment conditions by implementing economic experiment and collecting experimental data. Then, bullish sentiment index (BSI) and adverse selection level (ASL) were calculated based on the data collected. The influence of bullish sentiment and adverse selection was tested at last. The result shows that the severe information asymmetry is the main reason for adverse selection in resale house market, but bullish sentiment of traders can inhibit adverse selection to some extent. Reducing information asymmetry is still the fundamental way to prevent and solve the problem of adverse selection in resale housing market.

The house price expectations include not only optimistic expectations, but pessimistic expectations as well. In the follow-up study, the influence from pessimistic expectations on adverse selection will be focused on by changing the experiment design. Therefore, more exhaustive conclusion of the relationship between house price expectations and adverse selection will be drawn.

REFERENCES

Akerlof G A. 1970. The market for "lemons": Quality uncertainty and the market mechanism. *The Quarterly Journal of Economics* 84(3): 488–500.

Bond, E W. 1982. A direct test of the "lemons" model: the market for used pickup trucks. *The American Economic Review* 72(4): 836–840.

China's National Bureau of Statistics. 2013. *China economic census yearbook 2013*. Beijing: China Statistics Press.

Heal G. 1976. Do bad products drive out good? *The Quarterly Journal of Economics* 90(3): 499–502.

Hendel I. & Lizzeri A. 1997. Adverse selection in durable goods markets. *National Bureau of Economic Research* 89(5): 1097–1115.

Levin J. 2001. Information and the Market for Lemons. *RAND Journal of Economics* 32(4): 657–666.

Li Jianhua & Guo Xiaoling. 2003. Information asymmetric and real estate market efficiency. *Economy and Management* 2003(2): 59–60.

Liu Haiyan & He Yufu. 2009, The nature of investment should be taken full account in making macro-control policies for real estate. *Revolution and Strategy* 2009(3): 138–140.

Zhang Hong & Zhang Yang. 2012. Incorporation of experimental methods into teaching real estate economics: Process, practice and development. *The New Educational Review* 29(3): 121–133.

Zhang Hong & Zhang Yang. 2013. Effects of buyer demand information transmission on housing search: An experimental study. *Journal of Hunan University (Science and Technology)* 39(7): 99–103.

Zhang Hong, Zhang Yang & Li Rui. 2013. Effect of information disclosure on resale housing trading price based on experimental method. *Journal of Hefei University of Technology (Science and Technology)* 38(5): 621–624.

Instability warning model of open-pit mine slope based on BP neural network

Zhenhua Xie
Department of Safety Engineering, China Institute of Industrial Relations, Beijing, China

Shasha Liang & Tingting Luan
Civil and Environment Engineering School, University of Science and Technology Beijing, Beijing, China

ABSTRACT: The study of slope instability has always been a key technical issue for the safe production in open-pit mine. In order to establish an early warning model based on back propagation (BP) neural network, adhesion, internal friction angle, slope angle, slope height, ratio of pore water pressure and bulk density were selected as system input units. Twenty sets of sample data selected were used for completing the training of BP neural network. Lastly, the early warning model was used for two working slopes of a certain Iiron mine for early warning analysis on slope instability. The conclusions were accordant with the current actual situation and the methods discussed were worth being applied and spread.

1 INTRODUCTION

Slope instability warning is a complicated and comprehensive work. Slope stability is influenced by geological factors and engineering factors. The main methods to analyze slope stability are numerical analysis, physical simulation and intelligent analysis forecasting design. Compared with the former two, intelligent analysis can study multiple factors. The slope is used as the uncertainty and unbalanced complex system to consider, analyzing slope stability from several aspects. At present, fuzzy analysis (Wang, Y.M. 2011), grey theory (Zhou, N. et al. 2006), and extension theory (He, Z.M. et al. 2011) are widely used in intelligent analysis count model. There are many factors influencing the stability of open-pit mine slope. Most of the factors are fuzzy and random, and influence each other, which makes the relationship between variables very complex. It is difficult to use a mathematical equation to describe the relationship accurately (Qi, Z.F. et al. 2012). Therefore, it requires a process of slope instability warning that is dynamic and nonlinear and can handle both certain and uncertain information. Also, the process should objectively identify the stability state of the slope based on a number of existing engineering examples.

Artificial neural network (ANN) is also called neural network. It is linked together by a number of simple neurons, and it can realize the learning, judgment and reasoning in artificial intelligence. Neural network evaluation experience is mainly a result of its function approximation property. Namely, it can be expressed with the knowledge of any nonlinear relationship, and provides a nonlinear relationship for many involved in the evaluation of new solutions (Wang, W.X. et al. 2006). Using ANN study method and theory, the various factors affecting the slope stability can be used as the input variables as much as possible. A highly nonlinear mapping model could be established between those quantitative and qualitative factors and the slope forewarning grades, and then used for slope failure warning.

2 BASIC PRINCIPLE OF BP NEURAL NETWORK

Back propagation (BP) neural network was proposed by the scientists led by Rumelhart and McCelland in 1986. It is a multilayer feedforward neural network trained by error back propagation algorithm, and it is one of the most widely used neural network models. BP neural network can learn and store a large amount of information – output model mapping, without prior to reveal the mathematical equations describing the mapping relationship. Its learning rule is to use the steepest descent method constantly to adjust the weights and thresholds through the back-propagation, so that the square error of the network is minimum (Qing, Y.L. et al. 2009).

The typical network topology of BP network model includes the input layer, hidden layer and output layer. The network structure is as shown in Figure 1. Each layer contains a number of neurons. The neurons in the same layer have no contact, and the upper and lower layers have full contact. BP network learning process is divided into two parts, forward propagation and back-propagation (Wang, Z. et al. 2009). After obtaining the learning samples, input information transmits to the output layer through each layer. Meanwhile, learning results are compared with expectations mapping through output layer. With the direction of reducing the error, the connection weights are corrected layer by layer. Finally, output information goes back to the input layer. This process is repeated until the system output error is less than a predetermined value, and then the training is completed.

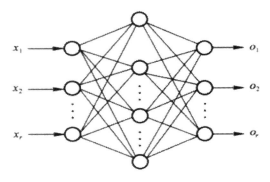

Figure 1. BP neural network structure.

3 SLOPE INSTABILITY WARNING MODEL BASED ON BP NEURAL NETWORK

BP neural network was applied to the instability warning process of open-pit slope, and the process was mainly the structural design of the system. It included determining the number of layers, determining the number of neurons in each layer and selecting the excitation function. As for the training samples, the MATLAB software toolbox was used for the training set to get trained system structure. Finally, the instances of samples were selected into the calculations to verify the correctness of system structure.

3.1 Selection and processing of data

Selecting the appropriate indicators is an important task related to the accurateness and reliability of slope disaster warning. On the basis of the study of the existing research results, six indicators of high significance to the study were selected, according to the actual situation of open-pit mine slope. The indicators are adhesion, internal friction angle, slope angle, slope height, ratio of pore water pressure and bulk density. At the same time, the stability of the slope was classified into five grades as shown in Table 1.

Based on the literature review and on-site survey, 20 sets of sample data similar to the studied slope were collected, as shown in Table 2.

Table 1. Classification of mine slope stability.

Grade	I	II	III	IV	V
State	Extremely stable	Stable	Basically stable	Unstable	Extremely unstable
Output value	[1 0 0 0 0]	[0 1 0 0 0]	[0 0 1 0 0]	[0 0 0 1 0]	[0 0 0 0 1]

Table 2. Sample data.

Serial number	Adhesion (MPa)	Internal friction angle (°)	Slope angle (°)	Slope height (m)	Ratio (–)	Bulk density (kN·m^{-3})
1	0.015	30	25	10.7	0.38	18.8
2	0.055	36	45	239	0.25	25
3	0.025	13	22	10.7	0.35	20.4
4	0.033	11	16	45.7	0.2	20.4
5	0.063	32	44.5	239	0.25	25
6	0.063	32	46	300	0.25	25
7	0.014	25	20	30.5	0.45	18.8
8	0.007	30	31	76.8	0.38	21.5
9	0.048	40	45	330	0.25	25
10	0.069	37	47.5	263	0.25	31.3
11	0.012	26	30	88	0.45	14
12	0.024	30	45	20	0.12	18
13	0.069	37	47	270	0.25	31.3
14	0.059	35.5	47.5	438	0.25	31.3
15	0.1	40	45	15	0.25	22.4
16	0.02	36	45	50	0.5	20
17	0.068	37	47	360	0.25	31.3
18	0.068	37	8	305.5	0.25	31.3
19	0.005	30	20	8	0.3	18
20	0.035	35	42	359	0.25	27

3.2 Determining the network topology structure

1) Determining the number of BP network layers

When the number of hidden layers increases, it can enhance the experience of the system to solve complex and nonlinear problems. However, too many hidden layers will prolong the system learning time, and it is not conducive to getting results. At the same time, Kolmogorov's theorem states that, for any continuous functions in a closed interval, BP system with a single hidden layer can achieve arbitrary mapping. Therefore, three-layer network with a single hidden layer was selected for training (Feng, H.M. et al. 2008).

2) Determining the number of neurons in each layer

The numbers of neurons in the input layer and the output layer of neural network were determined by the problem's input vector and output vector. As the input vector was 6 and the output vector was 5, so the number of neurons in the input layer was 6, and that in the output layer was 5.

The formula recognized internationally for determining the number of neurons in the hidden layer is as follows (Yang, T. 2012):

$$s = \sqrt{0.43mn + 0.12n^2 + 2.54m + 0.77n + 0.35} + 0.51 \quad (1)$$

In the formula, m and n respectively represent the numbers of the input and output layer neurons, $m = 6$, and $n = 5$.

The result of the calculation is $s = 6.45 \approx 6$. Therefore, the number of neurons in the hidden layer was set to 6.

3) Selecting the excitation function

Empirically, the transfer function of neurons in the hidden layer can be S-type tangent function 'tansig'. As the output function is in the interval [0, 1], the transfer function of the output layer neurons can be S-type logarithmic function 'logsig'.

4) Selecting the network training function

Considering the limited number of samples and the particular limitation of the traditional algorithm, the improved BP network training algorithm was selected. The function 'traingdx' contains additional momentum and adaptive learning rate, and it can avoid falling into local minimum and improve the accuracy. Therefore, the function 'traingdx' was chosen as the training function.

3.3 Training the BP neural network model

1) Selection of initial weight value and threshold value

If the output value of each neuron after the original weighting is close to zero, it is ensured that each neuron weight can be adjusted at the point of greatest change of excitation function. The initial value of weights and thresholds take a random number between 0 and 1.

2) Selection of learning rate

The learning rate selection range is 0.01~0.07. Considering that the system structure was complex and the number of neuron was large, the learning rate was selected as 0.05.

Table 3. BP network training parameters.

Network structure	6-6-5
The number of samples	20
Training times	10000
Training function	traingdx
The learning rate	0.05
Target error	1e−4

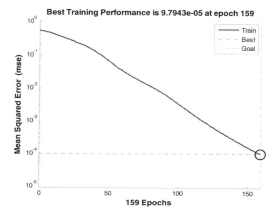

Figure 2. Identified system training results.

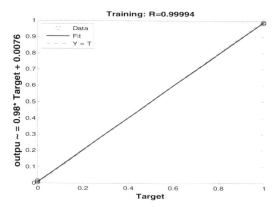

Figure 3. Regression figure.

3) Standard processing of input data

The excitation function used in this study is the Sigmoid (S-type and bipolar). When the variables are in the saturated zone, the convergence is slow. Because of the diversity of index values in the original data, therefore, for the convenience of calculation and to prevent part of the neurons to reach saturation state, the primary data was standardized by MATLAB to [0, 1].

4) BP network training

The parameters selected are shown in Table 3.

The neural network toolbox in MATLAB was employed to proceed the specific calculation. The target error as set to 10-4, as shown in Figure 2. After 159 times of training, the error was reduced to the target range, and the regression diagram is shown in Figure 3.

By comparing the training results with the expectations, the system accurately identified the training samples. A nonlinear relationship between disaster factors and warning levels of slope instability was established.

4 APPLICATION OF THE SLOPE INSTABILITY WARNING MODEL

A set iron mine in Anhui province is the main iron ore mining production base. The geological structure of the mine is featured by volcanic structure and fracture structure in the late metallogenic period. The fractured structure is well developed, and appears in shear joint form. In space and orientation shear joint is on the close relationship with the regional fracture structure. The two sets of shear joints in northwest and northeast formed an intensive X fissure zone, and it often makes the rock and ore body ruptured and medium structures. Each step height of the upper stope from $+115$ m to $+45$ m is 14 m, and there are a total of 5 steps. Each step height of the lower stope from $+45$ m to -165 m is 15 m, and there are a total of 15 steps. Each step height from -165 m to 201 m is 12 m, and there are a total of 3 steps. The east side, south side and west side of the stope are transport corridors. The fixed step width equals to or is greater than 7 m, the producing step width equals to or is greater than 20 m, the slope angle of the step is 60° and the overall slope angle is 35°~43°. Because of the sound rock alteration at the north side, the rock is loose and soft. The slope angle of the steps is 45°~50°, and the overall slope angle is 38°~42°. Because the slope is high, the slope is steep and the rock is loose and soft. Affected by gravity and airport inclined, the slope is prone to deformation and instability. Therefore, early warning of instability is needed.

Through the investigation of the mine stope, relevant data was collected from the Institute of Mining Research, and the data was confirmed by the mine management personnel. A group of data was selected from side I and II, respectively, for early warning, and the data is shown in Table 4.

The output results of the slope stability rating for stope sides I and II were 1 and 3, respectively, namely extremely stable and basically stable. This is consistent with the current status of the mine, indicating that the design has a certain practicality. By the field research, the mine studied and implemented framework beams and prestressed the anchor and applied other control measures for the slope stability of side I. There are some areas of side II with the state of flaking and loose rock. If this phenomenon further develops, corresponding treatment should be taken.

5 CONCLUSIONS

1) The BP neural network prediction model for slope instability was established by using a large number of engineering examples as sample data for prediction. Meanwhile, the model can give a timely and intelligent warning on open-pit slope instability.
2) A series of system training parameters were defined, and the input data were standardized by choosing a suitable rate. The learning and test of the system model were completed using the MATLAB toolbox. The result was reliable and it showed that the established BP neural network prediction model is reasonable.
3) Two open pit slopes of an iron ore mine were analyzed by trained warning design, and the result was consistent with the field situation. It shows that the prediction model established in this study is reasonable and has a particular popularization value.

ACKNOWLEDGMENTS

The research is supported by the national "Twelfth Five-Year Plan" Science and technology support program project of China (project No. 2012BAK09B05).

REFERENCES

Feng, H.M. et al. 2008. Spontaneous combustion of coal forecast and the realizing in MATLAB based on neural network. *China Coal* 34(5): 82–84.

He, Z.M. et al. 2011. Numerical analysis for stratified rock slope stability reinforced by bolt. *Journal of Central South University: Science and Technology* 42(7): 2115–2119.

Qi, Z.F. et al. 2012. Stability analysis of abutment slope at left bank of Jinping-I hydropower project during construction. *Rock and Soil Mechanics* 33(2): 531–538.

Qing, Y.L. & Li, H.Y. 2009. Third-party logistics resource integration risk early warning model based on BP neural network. *Statistics and Decision* 19(7): 31–33.

Wang, W.X. et al. 2006. BP neural network of MATLAB in the evaluation of slope stability. *West-China Exploration Engineering* 11(2): 273–274.

Wang, Y.M. 2011. *Stability of dump and its disaster prevention and control*. Beijing: Metallurgical Industry Press.

Wang, Z. & Guo, Y. 2009. Safety assessment model of underground non-coal mine based on BP neural network. *China Safety Science Journal* 19(2): 17–21.

Yang, T. 2012. *Research of mine safety evaluation system based on BP neural network*. Taiyuan: Taiyuan University of Technology.

Zhou, N. et al. 2006. Correcting factor of gray with prediction model unequal interval time-varying parameters. *Chinese Journal of Geotechnical Engineering* 28(6): 756–760.

Table 4. Early warning data.

Serial number	Adhesion (MPa)	Internal friction angle (°)	Slope angle (°)	Slope height (m)	Ratio (–)	Bulk density (kN·m^{-3})
1	0.074	31.1	24	115	0.25	18.8
2	0.012	25	37.5	145	0.25	14

Investigating urban transportation planning using IC card trip data

Shuzhi Zhao & Yuefeng Gao
College of Transportation, Jilin University Acceptable, Changchun, China

Qingfei Tian
Jilin Energy Saving Evaluation Center, Department One, Changchun, China

Fang Zong
College of Transportation, Jilin University Acceptable, Changchun, China

ABSTRACT: This study investigated the collection of trip data from IC card systemsfor the purpose of obtaining multi-mode trip data. This contributes to understanding the change regulation of bus trips, the development of effective traffic management policies, the promotion of public transport strategies, and the alleviation of traffic congestion in urban areas. Using a trip chain analysis method, this paper discusses the basic concept of a trip, methods of trip chain construction, an acquisition method of trip information, and the advantages and disadvantages of investigating these issues using IC card data.

Keywords: IC card, trip investigation, trip chain, public transit

1 INTRODUCTION

The IC (Integrated Circuit) card was created in the 1970s and first applied to the financial and transportation sectors. The development of IC card technology led to more convenient passenger travel and promoted public transportation systems in urban areas. Many researchers have applied the data from IC cards to investigate trip data and transportation planning. Zhou Chonghua provided a method of generating Origin-Destination (OD) travel data from subway IC cards [1]. Shi Fumin discussed bus trip OD based on IC card data [2]. Guo Jifu conducted a network reliability evaluation using taxi IC card data [3]. Cao Yeke analyzed operating conditions of bus lines in Shenzhen using IC card data [4]. Bagchi M, White P R, et al. developed a trip investigation project based on IC card data.

The majority of past studies concentrate on the simple transport mode of IC card data where researchers focused on the mode where on and off data can be obtained, such as the subway. Smart card data applications of multi-mode analysis and prediction of travel behavior research is very limited. This paper analyzes the construct method of trip chains, obtains travel information for bus IC card data platforms, and summarizes the advantages and disadvantages of our investigation method based on bus trip data of IC cards.

2 INVESTIGATION METHOD BASED ON BUS TRIP DATA FROM IC CARDS

2.1 Basic definition of trip unit

The IC card data comes from the part of theIC card where payment records are kept; not the area that focuses on traffic system planning. Therefore, it is necessary to find the relationship between the traditional variables of transport planning and link them with this data source to achieve the data needed for travel behavior analyses and prediction.

2.1.1 Transfer
The data can be divided into multi mode transfer and same mode transfer variables which limit the IC card payment platform that can be used for research.

- Multi mode transfer: Transfer between different transport methods. For the data from the smart card system, this refers to transfers between subway, light rail, bus and taxi;
- Same mode transfer: Transfer between the same modes. For the data from the smart card system, this refers to bus-bus (limited to between different bus lines) or subway-subway transfers;
- Same bus line transfer: Aspecial kind of bus transfer. In general, IC card data cannot reveal the trip

Table 1. Daily trip sheet based on smart card system.

Line	License plate No.	Card Type	Card No.	Balance	Consumption	Date	Time	Station Number
6	1489	01	0000313691	99.10	−0.90	2007/9/6	07:30	0002
270	1690	01	0000313691	98.20	−0.90	2007/9/6	07:56	0005
270	1510	01	0000313691	97.30	−0.90	2007/9/6	16:50	0008
6	1728	01	0000313691	96.40	−0.90	2007/9/6	17:10	0006
Subway 1	–	01	0000313691	94.40	−2.00	2007/9/6	18:05	12
Subway 1	–	01	0000313691	94.40	0.00	2007/9/6	18:25	15
Subway 1	–	01	0000313691	92.40	−2.00	2007/9/6	19:13	15
Subway 1	–	01	0000313691	92.40	0.00	2007/9/6	19:35	12

directions. Thus, for the bus trip in the same line between certain intervals, we assume that the return bus is in the same line as the departure.

2.1.2 *Trip section*

In the data obtained from the current IC card system, each trip section, either a multi-mode transfer or a same-mode transfer, can be identified as an on-off from a bus or in-out of a station.

2.1.3 *Trip*

Basic element of transport system and made up of one or several trip sections. It is the process of spatial movement from one place to another [5].

2.1.4 *Trip frequency*

The number of trips per time period (generally use "day" as a unit in transport planning). It is a macroscopic variable of network and transport planning measuring the overall demand of the system.

Travelers may have several trips per day where each trip is made up of one or several trip sections. An important step in defining a trip is solving the association between trip and trip sections. That is, how to recognize one trip section in this time period or the next time period must be determined. However, the investigation method based on IC cards cannot obtain the trip purpose data. Therefore, this study can determine the trip sections a trip is made up of according to the intervals between trip sections. According to related surveys, if the IC card system is not including off-bus departure data, the interval is 30 minutes. That means if the interval is less than 30 minutes, the interfacing trip section can be looked at as one trip. If the IC card system is including off-bus data, the interval is 15 minutes.

2.2 *Trip chain structure with case study*

The case study explains how to use IC card data to form a basic trip chain. Table 1 gives the trip data of the mode a traveler uses in one day under the smart card database.

According to the data from Table 1, it can be concluded that the daily trip sheet includes eight items, the first four of which are relating to bus trips while the rest refer to subway. All of the bus trip data includes only on-bus data while the subway trip data includes in-station and out-station data. First, the closely-connected on-bus or out-station intervals were selected and labeled as trips. Since there are only on-bus data in the table, only the relationship according to the definite 30 minute intervals is found. The 1st and 2nd items refer to the same mode transfer, which is made up of two trip sections, becoming the 1st trip when combined. The 3rd and 4th items are the same mode transfer, which is made up of two trip sections, constructing the 2nd trip. The 5th and 6th items construct the 3rd trip and the 7th and 8th items make up the 4th trip. Analyzing the time relationship of each trip, the 1st and 2nd trip chose the same bus line combination but the sequence is in reverse. The time period is in accordance with the commuting trip time period. Therefore, the two trips can be considered as inbound and outbound parts of the journey. While the 3rd and 4th trips choose subway as the trip mode, including data of entrance and exit of stations, which provides directional data. The two trips end in the opposite direction and can also be considered as an inbound and outbound journey.

From the analyses above, the trip of this person can be connected by using a trip chain, as shown in Figure 1. The trip chain includes six trip sections, four trips, and two same-mode transfers (Other transfers can be changed into an outbound journey or activity, such as working, eating at home, meeting friends, etc.). If the trip purpose is added into the chain (obtained by supplementary investigation), the trip chain can then be analyzed, including a morning peak trip to work, evening peak commuting trip, evening meeting friends trip and outbound trip. So, according to the data obtained by the smart card payment platform, the same trip items can be obtained and analyzed to investigate trip behavior.

2.3 *Trip information obtained by platform based IC card*

According to the case study, the data based on IC cards can provide researchers with many items for analyzing bus trip behaviors. Furthermore, with this data, the urban bus system can be evaluated to provide the basis for route planning and operational management. Items

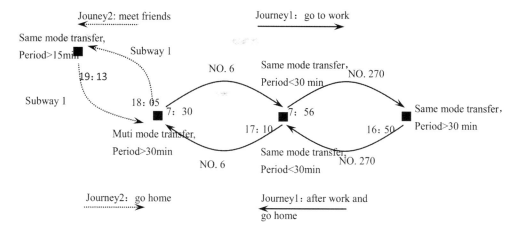

Figure 1. Daily trip chain analysis.

Table 2. Passenger flow of Changchun No. 5 bus.

Stop	Number of People Getting on the Bus
Children's Hospital	8
Municipal Government	7
Department Store	10
Square Primary School	4
4th Street	1
3rd Street	1
2nd Street	0
Nanguan	7
...	...
Waterworks	0
Total	44

of trip information obtained by platform based on IC cards are shown below.

2.3.1 Bus mode sharing rate and daily rate frequency

Urban bus trip frequencies can be obtained by the payment platforms on IC cards. Then it can be combined with full-day trip rate modes obtained by traditional trip surveys. City bus sharing rates can also be obtained to be incorporated as an index for evaluating urban traffic structure and the level of transit service. According to the investigation data of the bus IC cards and residents' trip surveys in Changchun in 2011, the bus trip sharing rate was 22.21% and the average bus daily trip rate was 0.57 times per day. If the payment platform is based on IC cards, the daily trip frequency and the overall trip sharing rate of the subway, light rail and taxi can be found. For example, the light rail trip sharing rate of Changchun in 2011 was 0.19% [6].

2.3.2 Passenger flow by line

Using the number of people who got on the bus for each stop on a line, the traffic distribution in that line can be analyzed. Table 2 shows the number of people who got on the bus for each stop on Line 5. By observing the daily passenger flow data, the variation in passenger flow in a day can be obtained. Figure 2 shows the Changchun No. 306 bus passenger flow distribution.

2.3.3 Passenger flow for each station

Incorporating the number of on or off bus transfers for each station on a line, the passenger flow data of each station can be yielded. Based on the smart card system and according to a data query, the passenger flow data of a large hub can also be attained. For example, the passenger flow for subway, bus, and taxi hubs at each time point can be learned. This data is then used as reference for determining the scale of stops or transfer hubs, distribution of stops, distance between stops, etc.

2.3.4 Transfer message

Using the definition of transfer based on IC cards mentioned above, the transfer properties of travelers including transfer time, transfer route, transfer stops, etc. can be acquired. Transfer data between different lines and different trip modes or stops (transfer hubs) can also be found. Additionally, the daily transfer rate, stop distribution, and time distribution of the transfer volume can be tabulated. For example, the bus transfer rate was 12.42% [6] in Changchun in 2011.

2.3.5 Trip behavior analysis data

Using the daily trip chain method and IC card data, the trip frequency, trip locations, time of trip and transfer mode can be analyzed. When combined with basic data of travelers, how personal attributes influence trip behavior can be studied and then a trip behavior model to predict urban bus trip behavior can be established.

For IC card data, only registered cards and non-registered cards exist. For registered cards, age, occupation, and address of cardholder is available in the system log.

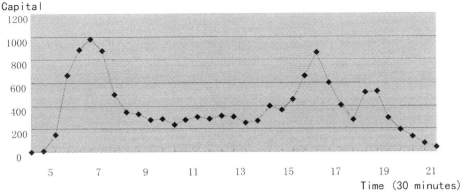

Figure 2. Variation of residents travel time distribution of Changchun No. 306 bus with time.

2.3.6 *Bus trip OD*

For the subway and light rail modes that have on-bus and off-bus data, the trip OD data between stops can be obtained by assembling on-bus and off-bus data to predict passenger flow. For the IC card system that charges when going on and off the bus, it can use this to obtain an OD matrix. When the data lacks off-bus data, based on the principle that the inbound journey and outbound journey are the head and tail corresponding with each other, the off-bus stop can be estimated. Combining this with the off-bus time obtained by the planning service time, the trip OD matrix can be obtained [2].

2.4 *Supplemental investigation and calculation*

When analyzing traveler classification regarding how personal attributes may affect trip behavior and trip OD data, a supplemental investigation must be conducted.

2.4.1 *The percentage of population using IC cards as payment option for public transportation*

Because there are still many travelers paying by cash in the transport system, a sample is required to confirm the percentage of travelers that are cardholders before any large scale bus planning can take place. In some cases, it may also be necessary to estimate regular travelers to adjust for the collection scope not being comprehensive.

2.4.2 *Actual start point and end point investigation*

IC card systems can only provide start, stop, and end stop data. However, as travelers generally need to walk to a stop, the actual start/end points of a trip do not spatially overlap with the start/end points of the transit stops. It is essential to investigate the actual start and end points of travel. Likewise, the IC card data can also be used for trip access analysis.

2.4.3 *Off-bus investigation*

For IC card systems that only record on-bus information, model calibration is essential for estimating the data of off-bus time and off-bus stop. Changes can be made to permit the swiping of the card a second time when passengers get off the bus or some other way that records the off-bus data.

2.4.4 *Trip message and traveler personal attributes investigation*

Currently, the residents' trip purpose cannot be obtained by investigating bus trip data from IC cards. Thus, in order to conduct large scale bus planning, supplementary investigations including trip purpose, age, vehicle occupation of travelers, etc. are needed.

2.4.5 *Bus service data investigation*

Based on the IC card data from swiping the card a second time or from segmented pricing, the average speed of bus travel between stops can be known. However, the data describing the bus service, such as convenience and comfort of the trip, waiting time at stop, accessibility of stop, etc. cannot be known, making bus service data research necessary.

3 CONCLUSIONS

The proposed investigation method based on bus trip data of IC cards can make significant contributions towards understanding bus trip behavior and bus system planning.

Using the data from the IC cards, the trip information like bus mode sharing rate with daily frequency, passenger flow of every line, passenger flow of every station, transfer message, trip behavior analysis data, bus trip origin-destination, etc. can be obtained.

Investigation methods based on bus trip data of IC cards has advantages and disadvantages. Compared to traditional residents' trip analysis, this method allows for a long observation timeline. Also, the sample size is larger than traditional residents' trip investigations. And this method requires less money and time compared to traditional data collection methods while at the same time providing us with data that are more accurate. However, the method still has limitations, such as subjective errors and the inability to cover all trip modes. Also, the data are different because IC cards are not uniform across different cities.

ACKNOWLEDGMENTS

Project Fund: National Natural Science Foundation of China (51078168).

REFERENCES

M. Bagchi P.R. White. The potential of public transport smart card data[J]. Transport Policy, 2005, 12, 464–474.
Cao Geke. Developing intelligent transportation scheduling system using electronic ticketing system [J]. China's traffic information industry, 2004, 6: 72–74.
GuoJifu, Wen Huimin, Gao Yong, Yu Lei, Chen Kun. Beijing road network reliability evaluation based on taxi IC card data [J]. City Intelligent Transportation, 2007: 28–130.
Shi Fumin. Construction method study of bus OD matrix based on IC card data [D]. Master's degree thesis of Jilin University, 2004.
Zhou Chonghua. Information processing of railway OD based on IC card data [J]. The modern city track traffic, 2007, 2: 47–49.
Zong Fang. Research on the selection model based on travel time and mode of activity [D]. Master's degree thesis of Jilin University, 2005.

ISM-based identification of factors influencing pedestrian violations at signalized intersections

Ming-ming Zheng & Zhong-yi Zuo
Department of Traffic Engineering, Dalian Jiaotong University, Dalian, Liaoning, China

ABSTRACT: Signalized intersections serve as important nodes of city traffic. Pedestrian violations at such places will not only reduce the efficiency of them but may also cause traffic accidents. This paper studied the main factors that induce pedestrian violations by constructing a model based on ISM. The research shows that the key causes of pedestrian violations are as follows: unreasonable design of pedestrian crossing facilities, lacking of traffic safety awareness and traffic law awareness; in addition, the poor traffic condition is also a mediating factor influencing pedestrian violations. In the end, this paper proposes some effective measures to correct the pedestrians' improper behaviors.

1 INTRODUCTION

Pedestrian traffic is an important part of urban traffic. Among all the trips made in China, 20% are made by pedestrians. As the travel demand in urban city continues to grow, the contradiction between the fast moving motorized vehicles and the pedestrian will tend to deepen. According to the statistics, about 1/3 of traffic accidents are related directly with pedestrians, and almost 1/4 of those who die in road traffic crashes are pedestrians. One of the most major causes behind this is pedestrian violation. The purpose of this paper is to find out the regularity and reason of pedestrian violations at signalized intersections, and provides measures to prevent and correct pedestrian violations. All of these have significant meaning in reality.

Sun Shi-jun (2007) analyzed the cause of traffic offences, psychological characteristics of pedestrians and their psychological action by using the expectation theory. Liu Guang-xin (2008) introduced the development of study on the latency time and speed of pedestrian crossing intersections, the acceptable clearance, pedestrian violation of traffic rules, the influence of traffic circumstance to pedestrian psychology and behavior. Yuan Hong-wei (2008) analyzed pedestrian traffic behaviors based on both domestic and international experience. Wang Yang-feng (2009) established an influencing factor index system for pedestrian violation behavior, and a choice model of pedestrian violation behavior was constructed by adopting Logistics regression model. Pan Han-zhong (2010) examined the group psychology of pedestrians when violating traffic laws to cross road at signalized crossings. Meng Hu (2012) built a Binary Logit model to predict whether the pedestrian will obey the traffic rule at signalized intersections or not. Based on Interpretive Structure Modeling of system engineering, this paper analyzed the factors influencing pedestrian violations. A 3-layer hierarchal interpretative structure model is put forward and measures to reduce pedestrian violations at signalized intersections are suggested.

2 FACTORS IDENTIFICATION

Before a pedestrian crosses the street at signalized intersection, he will give the utility an evaluation based on the real-time traffic conditions, pedestrian crossing facilities nearby and pedestrian characteristics. Such evaluation will determine his behavior – violation or compliance. Figure 1 shows the decision-making

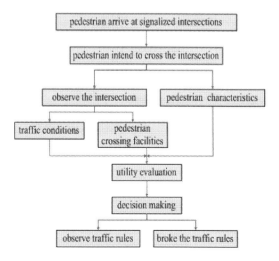

Figure 1. Decision-making process of pedestrians when crossing the intersection.

process when a pedestrian decides to cross a signalized intersection.

According to the decision-making process in Figure 1, three kinds of factors, including pedestrian characters, traffic conditions and pedestrian crossing facilities, will influence the pedestrian's decision-making process. Based on the systematic principle, scientific principle, comprehensive principle, level principle and simplicity principle, these three factors are further subdivided into 18 key elements, which are shown in Table 1 in detail.

3 THE BASIC PRINCIPLE OF ISM MODEL

ISM belongs to conceptual model, whose basic idea is to transform complex system problems into a visible model that has clear logic structure (Wu Biao et al., 2010 Rajesh Attri et al. 2013).

The specific steps to achieve this end are as below:

(1) Identify the problem to be studied.
(2) Collect and select related elements that influence the problem. Establish an adjacency matrix according to the binary relation between the elements.
(3) Derive a reachability matrix by calculating the adjacency matrix.
(4) Identify the level of each element by analyzing the reachability matrix.
(5) Set up a multi-level hierarchical graph based on the reachability matrix and the level partitions.
(6) By using the multi-level hierarchical graph, discern and analyze the essential factors of the complicated problem.

Table 1. Factors influencing pedestrian violations at signalized intersections.

Category	Factors	Code
Pedestrian characters	The long latency time	S1
	Following others behavior	S2
	Lucky-me attitude	S3
	Impatience mood	S4
	Lacking traffic safety awareness	S5
	Lacking traffic law awareness	S6
Pedestrian crossing facilities	Unreasonable design of pedestrian overpass/underpass	S7
	Unreasonable distance between bus stop and crosswalk	S8
	No pedestrian barrier	S9
	Unreasonable design of pedestrian crossing	S10
	Unreasonable pedestrian signal timing	S11
	Lack of management	S12
Traffic conditions	Low speed of traffic stream	S13
	Large distance headway	S14
	Traffic congestion	S15
	Illegal occupation of pedestrians' facilities	S16
	Vehicle violations	S17
	Pedestrian violations	S18

4 AN ISM APPROACH FOR MODELING OF FACTORS

4.1 Adjacency matrix

On the basis of causal relationship, this paper establishes a direct relationship between elements influencing pedestrian violations. The relationship was obtained by Delphi, in which a group of experts' opinions were considered. The adjacency matrix is presented in matrix A, in which 1s indicates direct relationship between elements and 0s indicates indirect relationship between elements.

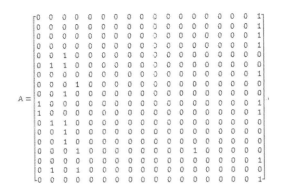

4.2 Reachablity matrix

Because of the transitivity of the relationship between elements, a reachability matrix is proposed based on the adjacency matrix. Reachablility matrix indicates whether or not a column element can be reached from a row element along a continuous directed path. It can be obtained by solving the following equation:

$$R=(A+I)^r \quad (1)$$

where: I = an identity matrix having the same order as matrix A. r = maximum number of delivery.

The value of r is computed as follows:

$$(A+I) \neq (A+I)^2 \neq (A+I)^3 \neq \cdots \neq (A+I)^{r-1}$$
$$\neq (A+I)^r = (A+I)^{r+1} = \cdots = (A+I)^n \quad (2)$$

According to the above equations, the reachability matrix R is obtained by using MATLAB.

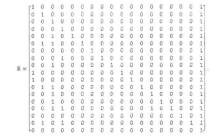

4.3 Level partitions

First of all, the reachability set and advanced set are found based on the reachability matrix R, reachability set includes the element itself and the elements that are impacted by S_i, and advanced set includes the element itself and the elements which impact S_j. Secondly, this paper gets the intersection set of reachability set and advanced set for all the elements. The top level L_1 includes the elements whose reachability set is exactly the same as the intersection set. Then the elements belonging to the top level from 18 elements above are removed. For the other elements the same step is repeated until all the elements are classified. The specific steps are as follows:

$$L_0 = \Phi$$
$$L_1 = \{S_i/S_i \in P - L_0, C_0(S_i) = R_0(S_i), i = 1,2,\cdots,n\}$$
$$L_2 = \{S_i/S_i \in P - L_0 - L_1, C_1(S_i) = R_1(S_i), i < n\}$$
$$\vdots$$
$$L_k = \{S_i/S_i \in P - L_0 - L_1 - \cdots - L_{k-1}, C_{k-1}(S_i) = R_{k-1}(S_i), i < n\} \quad (3)$$

The final results are presented in Table 2.

Table 2. Levels of each factors.

Level	Sets
1	$L_1 = \{S18\}$
2	$L_2 = \{S1, S2, S3, S4, S7, S16\}$
3	$L_3 = \{S5, S6, S8, S9, S10, S11, S12, S13, S14, S17\}$
4	$L_4 = \{S15\}$

4.4 Developing the digraph based on the partitions

The structural model proposed in the form of the hierarchical pyramid is represented as a digraph (Fig. 2). In this digraph, the links between elements are shown. If there is a relationship between the elements S_i and S_j, it will be shown by an arrow pointing from S_i to S_j. The first level element (S1) is placed at the top of the figure, while the fourth level element (S15) is placed on the bottom of the figure.

By working out the hierarchical arrangement of system elements, it is easy to identify the significant elements that influence pedestrian violations at signalized intersections. The conclusion is very useful to propose measures that may significantly reduce pedestrian violations.

The direct acting factors that influence pedestrian violations at signalized intersections are the long latency time (S1), pedestrian's psychology (S2, S3, S4), unreasonable design of pedestrian overpass/underpass (S7) and illegal occupation of pedestrian crossing facilities (S16). The indirect acting factors are unreasonable design of pedestrian crossing (S10), unreasonable pedestrian signal timing (S11), unreasonable distance between bus stop and crosswalk (S8), vehicle violations (S17), lacking traffic law awareness (S6), lack management (S12), low speed of traffic stream (S13), lacking of traffic safety awareness (S5), no pedestrian barrier (S9) and large distance headway (S14).

The unreasonable design of pedestrian crossing facilities is an important factor that both directly and indirectly affects pedestrian violations, thus is a key factor that causes pedestrian violations. Lacking of traffic safety awareness and traffic law awareness are the underlying causes of pedestrian violations,

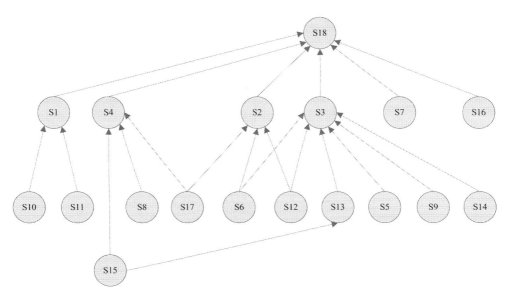

Figure 2. Interpretive structural model showing levels of factors influencing pedestrian violations.

while the poor traffic condition is a mediating factor influencing pedestrian violations.

5 MEASURES TO REDUCE PEDESTRIAN VIOLATIONS

1. Optimize the designs of pedestrian crossing/overpass/underpass

   The pedestrian crossing/overpass/underpass should be such designed that their quantity, location and width are fully considered, the pedestrians and the vehicles enjoy the same rights at signalized intersections, and the goal of both safety and efficiency are achieved.

2. Work out an optimal signal timing plan

   Both the pedestrian green time and the waiting time are the key factors that influence pedestrian violations, for which the real-time sign-controlling scheme for intersection traffic can be considered. In order to ensure that the pedestrians do obey the signal timing plan, a few policemen can be arranged to render supervision.

3. Strengthen publicity and education of traffic safety/rules.

   The underlying reasons behind the pedestrian violations are that pedestrians have a weak awareness of traffic safety and traffic rules. So, enhancing the pedestrian's traffic safety awareness and traffic rule awareness can play a decisive role.

6 CONCLUSION

This paper proposed a pedestrian violation model based on ISM. The research shows that the key cause of pedestrian violations is the unreasonable design of pedestrian crossing facilities. Lacking traffic safety awareness and traffic law awareness are the underlying causes of pedestrian violations, and the poor traffic condition is the mediating factor influencing pedestrian violations. Based on these results some effective measures are proposed, i.e. optimize the designs of pedestrian crossing/overpass/underpass, work out an optimal signal timing plan, and strengthen publicity and education of traffic safety/rules.

REFERENCES

Pan Han-zhong, Chen Peng, Ma Jing-jie (2010). Pedestrians' Group Psychology When Violating Traffic Laws to Cross Road at Signalized Crossing. *Transport Standardization*, 234(12):150–156.

Liu Guang-xin, Li Ke-ping, Ni Ying (2008). An Overview on Pedestrian Psychology and Behavior When Crossing Intersections. *Technology & Economy in Areas of Communications*, 49(5):58–61.

Yuan Hong-wei, Xiao Gui-ping (2008). Study on the Unsafe Behaviors of Pedestrians Based on Traffic Psychology, *China Safety Science Journal*, 18(1):20–26.

Sun Shi-jun, Wang Chi, Zhang Ying (2007), Psychological Research on Pedestrian Violations at Urban Road Intersections, *Urban Transport of China*, 5(4):91–96.

Meng Hu, Chen Yan-yan, Liu Hua (2012). Prediction Model for Pedestrian Obeying Traffic Rule at Signalized Intersections. *Journal of Highway and Transportation Research and Development*, 29(12):114–118.

Wu Biao, Xu Hong-guo, Dai Tong-yan (2010). Identifying Safety Factors on Expressway Work Zone Based on DEMATEL and ISM, *Journal of Transportation Systems Engineering and Information Technology*, 10(5):130–136.

Rajesh Attri, Sandeep Grover, Nikhil Dev, Deepak Kumar (2013), An ISM Approach for Modelling the Enablers in the Implementation of Total Productive Maintenance (TPM). *Int J Syst Assur Eng Manag (Oct–Dec 2013)* 4(4):313–326.

Mechanical behavior of water diversion tunnel lining before and after consolidation grouting

Yansong Li, Shougen Chen & Zelin Zhou
Key Laboratory of Transportation Tunnel Engineering, Ministry of Education, Southwest Jiaotong University, Chengdu, China

ABSTRACT: Based on engineering background and finite element simulation of the diversion tunnel of Dong Song hydropower, this paper represent a mechanical behavior analysis of diversion tunnel lining between and after consolidation grouting respectively. The results present that consolidation grouting can reinforce rock effectively, modify crack propagation around tunnel, improve stress condition about lining, and enhance the anti-deformation and anti-fracture ability of lining. However, consolidation grouting will cause significant additional load in lining. The simulation results show that consolidation grouting caused 30% augment of the maximum stress in lining, and in turn caused secondary deformation. Thus, inappropriate consolidation will cause the increase of compressive stress, even the possibility of the thorough overwhelm of lining. Finally, the paper proposes that segmental consolidation grouting causes little additional stress in lining than that of full hole grouting at a time, thus, different construction methods will decrease the additional stress effectively.

Keywords: the diversion tunnel; consolidation grouting; lining mechanical properties; numerical simulation

1 INTRODUCTION

China's water resources rank first in the world, however the extent of the development stage of Chinese water resources is only 31%, far below the average level of developed countries, having great potential for the development of the water resources. With the development of China's water conservancy and hydropower construction projects engineering more and more difficult and complex geological conditions, and blasting rock tunnel excavation disturbance and damage the rock. The purpose of consolidation grouting is to reinforce the tunnel rock, close tunnel surrounding rock fracture, improve rock integrity and deformation resistance, and enhance the ability to resist penetration of surrounding rock to prevent water penetration from occurring in destroyed adjacent hydraulic structures. Therefore, consolidation grouting is an effective means of dealing with poor geological conditions and rock damage. Consolidation grouting is widely used in water conservancy and hydropower projects to reinforce the rock in China.

According to the *Code for Design of Road Tunnel* (The Professional Standards Compilation Group of Peoples Republic of China, China Communications Press, 2004), lining structure is required to fit closely with the rock. However, tunnel project is hidden works, the engineering geology cannot be accurately understood and construction problems are often caused by over-excavation phenomena, where lining structure can not to fit closely with the rock.

Hydraulic tunnel backfilling and consolidation grouting behind the lining can prevent cavities, thus analysis and evaluation of the mechanical behavior of water diversion tunnel lining before and after consolidation grouting have important theoretical significance and application value.

Based on engineering background and finite element simulation of the diversion tunnel of Dong Song hydropower, numerical simulation software was employed to simulate cavern excavation, shoring, filling and other processes, the mechanical behavior of diversion tunnel lining was analyzed, in order to carry out a comprehensive evaluation of realistic.

2 PROJECT DESCRIPTION

The Dong Song Hydroelectric Project is located downstream of the Shuo Qu river in Ni Si and Dong Song townships of the Xiang Cheng county. It is the hydropower planning "one library level six" to the fourth power plant in development plan of the Shuo Qu river trunk stream in Xiang Cheng and De Rong section.

Because the engineering geological structure of the area is complex, as it is in the north-south Dege – Xiangcheng fault zone, the tectonic stress field in the spatial distribution is very uneven. According to hydraulic fracturing method in situ gravity test data, the maximum horizontal principal stress of the diversion tunnel is 18.60 MPa, with the direction of NE15°

(North east 15°). In this section vertical stress is the main stress and the lateral pressure coefficient is 0.6~0.8. According to the *Standard for Engineering Classification of Rock Masses* (The Ministry of Water Resources of the People's Republic of China, China Planning Press, 1995), this section of tunnel surrounding rock is in high ground stress zone.

Whole section reinforced concrete lining support scheme was used in this diversion tunnel. The class III surrounding rock reinforced concrete lining thickness is 40 cm, the class IV 1 surrounding rock reinforced concrete lining thickness is 60 cm, the class IV 2 surrounding rock reinforced concrete lining thickness is 70 cm, and the class V surrounding rock reinforced concrete lining thickness is 80 cm. Meanwhile, class IV and V surrounding rocks need consolidation grouting.

3 ANALYSIS PARAMETERS AND MODEL

3.1 Premise and assumptions to calculate

(1) It was assumed that the soil is isotropic materials, with no joints, no cracks, and horizontal distribution.
(2) In calculating steel arch and shotcrete, what composed of temporary support, as homogeneous continuous reinforced concrete lining, without regard to longitudinal connection concrete and steel arch.
(3) To simulate the consolidation grouting reinforcement of surrounding rock, it is needed to modify the tunnel rock parameters. In order to reflect the effect of consolidation grouting on diversion tunnel lining, this paper assumes that the strength of grouting material gradually enhances in the calculation.
(4) Soil, lining and filling materials were set to obey the Mohr-Coulomb yield criterion.
(5) By applying vertical pressure on the top of the model, diversion tunnel buried depth of 600 m was simulated.
(6) In order to calculate diversion tunnel consolidation grouting, this paper uses numerical simulation software to simulate in situ stress balance, cavern excavation, tunnel support filling and other processes.
(7) To study mechanical behavior of water diversion tunnel lining before and after consolidation grouting, it to simulate without consolidation grouting, one-time consolidation grouting and break consolidation grouting three conditions.
(8) The external grouting pressure of the lining is 2 MPa. This pressure was retreated at the end of the grouting.
(9) To study the mechanical behavior of the tunnel lining, it is needed to modify internal water pressure of the diversion tunnel to 1 MPa, 2 MPa, 3 MPa, 4 MPa, 5 MPa, 6 MPa, 7 MPa, 8 MPa, 9 MPa, or 10 MPa.

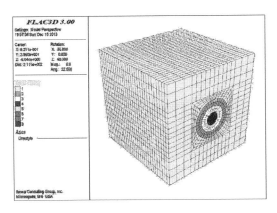

Figure 1. The computing tunnel model.

Table 1. Parameters of soils and construction materials.

Material name	E MPa	μ	ρ kN/m³	C kPa	φ (°)
Surrounding rock	1400	0.3	21	180	24
Primary lining	23000	0.2	25	–	–
Secondary lining	23000	0.2	25	–	–
Solid grouting body	20000	0.2	20	–	–
Liquid grout body	2180	0.25	17	–	–

3.2 Finite element model

The FLAC finite difference software was used for 3D simulation calculation (Peng Wenbin, China Machine Press, 2011). In order to reduce the impact of simulation results of the model boundary and accelerate computing, the model was a cube with a side length of 60 m (Diversion tunnel inside diameter is 6 m and outside diameter is 7 m. The distance model boundary to model center is about 4 times the tunnel diameter). According to the calculation requirement, the calculation model is shown in Figure 1.

3.3 Calculation parameters

According to the geological report, indoor test and rock point load test data of Dong Song hydropower station diversion tunnel were obtained. The formation and material parameters used in the calculation are shown in Table 1.

4 RESULTS ANALYSIS

Consolidation grouting is important for the diversion tunnel; for example, the purpose of consolidation grouting is to reinforce the tunnel rock, close tunnel surrounding rock fracture, improve rock integrity and deformation resistance, and enhance the ability to resist penetration of surrounding rock to prevent water penetration occurs adjacent hydraulic structures destroyed. Therefore, consolidation grouting is

Table 2. Lining deformations of key locations.

Conditions	Vault	Left spandrel	Right spandrel	Left wall	Right wall	Bottom arch
Condition 1	0	0	0	0	0	0
Condition 2	−12.3	−9.27	−9.36	−6.94	−6.95	2.13
Condition 3	−9.1	−7.97	−8.06	−6.42	−6.42	−4.02

Table 3. Lining influence in key positions consolidation grouting.

Vertical deformation values   mm

Conditions	Vault	Left spandrel	Right spandrel	Left wall	Right wall	Bottom arch
Condition 1	0	0	0	0	0	0
Condition 2	−12.3	−9.27	−9.36	−6.94	−6.95	2.13
Condition 3	−9.1	−7.97	−8.06	−6.42	−6.42	−4.02

Table 4. Lining influence in key positions consolidation grouting.

Horizontal deformation values   mm

Conditions	Vault	Left spandrel	Right spandrel	Left wall	Right wall	Bottom arch
Condition 1	0	0	0	0	0	0
Condition 2	0	0.9	−0.91	−0.52	0.52	0
Condition 3	0	1.16	−1.16	1.18	−1.21	0

an effective means of dealing with poor geological conditions and rock damage. The analysis results also confirmed the above points of view (Zhen Jiquan et al. 2001, Su Lihai et al. 2008, Yang Xuexiang et al. 2010). Consolidation grouting has a significant influence on surrounding rock, and also has a significant influence on the mechanical behavior of the tunnel lining. In order to reflect the effect of consolidation grouting on diversion tunnel lining, this paper uses the FLAC finite difference software to simulate the consolidation grouting of the diversion tunnel and the diversion tunnel under the internal water pressure after operation.

Diversion tunnel's deformation of the key points is shown in Table 2, which shows that the down and right displacements are positive, and up and left displacements are negative. In this table, the following results of numerical simulation can be observed: Vault sink 115.4 mm, bottom arch uplift 72.8 mm, and the left and right wall displacements are 63 mm. Because this section of tunnel surrounding rock is in high ground stress zone, the actual displacement is larger than the simulation results. According to the actual data, deformation of the part of the tunnel reached 1000 mm, and the maximum deformation was 2000 mm, which led diversion tunnel into expanding excavation. Because the stability of the diversion tunnel of weak surrounding rock under high in-situ stress environment has little influence on mechanical behavior analysis of diversion tunnel lining between and after consolidation grouting, no further research was done.

4.1 Lining deformation analysis

Condition one, which regards the casting of primary support and secondary support diversion tunnel as the initial state diversion tunnel, and set its deformation value as 0, is the diversion tunnel without consolidation grouting. Condition two is the diversion tunnel of one-time consolidation grouting. Condition three is the diversion tunnel of break consolidation grouting, which constructs 0 m to 4 m of consolidation grouting first, and then constructs 4 m to 8 m of consolidation grouting during the cement slurry condensation.

Table 3 shows the vertical displacement values of diversion tunnel lining under the influence of consolidation grouting. Table 4 shows the horizontal displacement values of diversion tunnel lining under the influence of the consolidation grouting. From Tables 3 and 4, it can be seen obviously that under the consolidation grouting pressure and cement slurry weight, lining deformation occurred. This shows that consolidation grouting produces additional stress on the lining, which forces lining deformation to occur. It can also be seen that the displacement of condition three is smaller than that of condition two, indicating that displacement can be controlled by changing the construction method. By changing the construction method, diversion tunnel lining can escape from damage caused by the additional stress and cement slurry weight.

The internal water pressure of the diversion tunnel has a significant influence on lining. To study mechanical behavior of the tunnel lining in this case, it is needed to modify internal water pressure of the diversion tunnel to 1 MPa, 2 MPa, 3 MPa, 4 MPa, 5 MPa, 6 MPa, 7 MPa, 8 MPa, 9 MPa, or 10 MPa. With the internal water pressure increases, the lining without consolidation grouting deformation reaches a dozen millimeters or even dozens of millimeters. From Figures 2 and 3, it can be seen that with the internal water pressure increases, the lining, without consolidation grouting deformation reaches a dozen millimeters or even dozens of millimeters, indicating that deformation is growing fast and in a non-linear fashion. However, with consolidation grouting, there is only little deformation on the lining and the deformation grows slowly and in a linear fashion, and almost unchanged with the increasing internal water pressure. These results were calculated in the case where the lining structure close to the rock was considered. If the cavity behind the lining and the lining of uneven are considered, the lining deformation, without consolidation grouting, will be greater and the stress will be more negative. The consolidation grouting can effectively reduce the impact of the internal water pressure on the diversion tunnel lining. From the figures it can also be seen that the deformation of condition two is very close to the deformation of condition three, indicating that changing the construction method has little influence on the diversion tunnel lining of operating stage.

4.2 Lining stress analysis

Due to limitation of the article length, it is not possible to give all the stress clouds of numerical simulation. This paper only gives the stress nephogram of condition one and condition three and the lining stress nephogram under 2 MPa of internal water pressure of condition one and condition three. Figure 4(a) shows the lining stress nephogram without consolidation grouting (condition one). From Figure 4(a), it can be seen that under the surrounding rock pressure, the lining compressive stress at the wall is greater than that at the vault under pressure, proving that the surrounding rock pressure at the vault is greater than that at the wall. The maximum compressive stress is about 49.1 MPa at the left and right wall and the minimum compressive stress is about 1.4 MPa at the vault and bottom arch. Figure 4(b) shows the lining stress nephogram with consolidation grouting (condition three), and the location of the maximum and minimum pressure stresses did not change. The maximum compressive stress is about 63.7 MPa and the minimum compressive stress is about 4.8 MPa. In Figure 4(b), the maximum

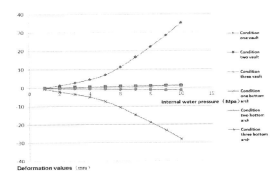

Figure 2. Vaults and inverted arch bottom vertical deformation curve.

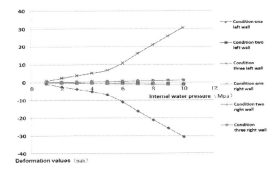

Figure 3. Left and right lateral wall deformation curves.

compressive stress increases by 29.7% as compared with that in Figure 4(a). The stress state of condition two is very close to that of condition three, where the maximum compressive stress is about 64 MPa and the minimum compressive stress is about 5.4 MPa.

Figure 4(c) shows the condition one stress nephogram under 2 Mpa of internal water pressure, and the maximum compressive stress is about 35.0 MPa and the minimum compressive stress is about 2 MPa. Figure 4(d) shows the condition three stress nephogram under 2 Mpa of internal water pressure, and the maximum compressive stress is about 62.3 MPa and the minimum compressive stress is about 6.1 MPa. The stress state of condition two is very close to that of condition three, where the maximum compressive stress is about 62.7 MPa and the minimum compressive stress is about 6.6 MPa.

From Figures 4(a) and 4(b), it can be seen obviously that consolidation grouting has a significant influence on lining, producing additional stress on the lining. It can improve the integrity and impermeability of the lining and close tunnel surrounding rock fracture when the grouting pressure is suitable or cement slurry injecting science. If the consolidation grouting pressure is too large or cement slurry injecting is too much, it will produce an excessive burden on lining.

As well known, the compression properties of reinforced concrete are better than its tensile properties. If the concrete cracks, it will cause huge harm to the tunnel. From Figures 4(c) and 4(d), it can be seen that

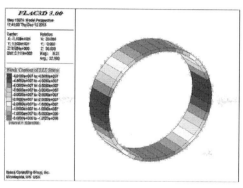
(a) Condition one without consolidation grouting

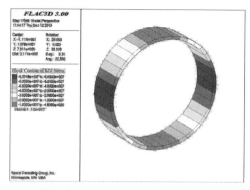
(b) Condition three with consolidation grouting

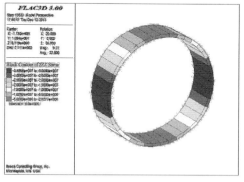
(c) Condition one under 2 Mpa of internal water pressure

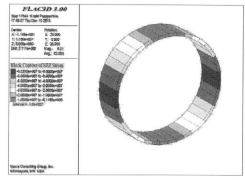
(d) Condition three under 2 Mpa of internal water pressure

Figure 4. Part of the diversion tunnel lining stress cloud.

the consolidation grouting produces additional stress on the lining, preventing the tensile stress of the lining under the internal water pressure. The stress state of condition two is very close to that of the condition three, and the effect of one-time consolidation grouting is stronger than break consolidation grouting, but the effect is not great.

5 CONCLUSIONS

Based on engineering background and finite element simulation of the diversion tunnel of Dong Song hydropower, this paper represents a mechanical behavior analysis of diversion tunnel lining between and after consolidation grouting. The following conclusions can be drawn.

(1) Consolidation grouting can reinforce rock effectively, modify crack propagation around tunnel, improve integrity and deformation resistance of surrounding rock, and enhance the ability to resist penetration of surrounding rock to prevent water penetration from occurring and adjacent hydraulic structures from being destroyed.
(2) Because of consolidation grouting, lining structure can fit closely with the rock, make surrounding rock and lining work together, improve stress condition of lining, enhance the anti-deformation and anti-fracture ability of lining, and prevent internal water pressure from occurring and diversion tunnel lining from being destroyed.
(3) The consolidation grouting produces additional stress on the lining, which forces lining deformation to occur. It can improve the integrity and impermeability of the lining and close surrounding rock fracture of the tunnel when grouting pressure is suitable or cement slurry injecting science. If the consolidation grouting pressure is too large or cement slurry injecting is too much, it will produce an excessive burden on lining or even damage the lining.
(4) The additional stress on the lining caused by consolidation grouting can be controlled by changing the construction method, and the effect of one-time consolidation grouting is stronger than break consolidation grouting, but the effect is not great.

REFERENCES

Peng Wenbin. FLAC 3D practical tutorial [M]. Beijing: China Machine Press, 2011. (in chinese))
Su Lihai, Qu Xing, Li Ning. Numerical Simulation of Consolidation Grouting in Side Slope at Left Abutment of Jinping I Hydropower Station [J]. Journal of Water Resources

and Architectural Engineering, 2008, 6(2): 118–120. (in Chinese)

The Ministry of Water Resources of the People's Republic of China. GB 50218-94 Standard for Engineering Classification of Rock Masses [S]. Beijing: China Planning Press, 1995. (in Chinese)

The Professional Standards Compilation Group of Peoples Republic of China. JTG D70-2004 Code for Design of Road Tunnel [S]. Beijing: China Communications Press, 2004. (in Chinese)

Yang Xuexiang, Li Yan. Consolidation Grouting Experiment for Foundation of High Arch Dam of Dagangshan Hydropower Station [J]. Journal of Water Resources and Architectural Engineering, 2010, 8(1): 35–38. (in Chinese)

Zhen Jiquan, Lai Jiehe, Quan Hai. Effect Inspection of Grouting Test for Weak Batholith in Xiluodu Hydroelectric Power Station [J]. Chinese Journal of Rock Mechanics and Engineering, 2001, 20(Supp): 1851–1857. (in Chinese)

Numerical analysis of Hefei metro shield tunnel construction crossing pile foundation of building

Kewei Ding & Dawei Man
School of Civil Engineering, Anhui Jianzhu University, Hefei, PR China

ABSTRACT: By using the finite-difference software FLAC^{3D} to conduct a numerical simulation of shield tunnel crossing pile foundation of buildings in Hefei, which has a strata that is soft in the upper part and hard in the lower part, an analysis of the law of ground subsidence and the deformation characteristics of pile foundations was analyzed in this study. Results show that the surface settlement of cross-section is close to the normal distribution, and the maximum subsidence occurs right above the tunnel. Before the shield reaches the bottom of the pile, the vertical displacement of the pile increases while its distance from the shield decreases, and the maximum appears in the shield leaving. In addition, the maximum horizontal displacement occurs when the shield reaches the bottom of the pile, and the impact will be reduced in accordance to the distance it stays away from the shield.

Keywords: upper soft and lower hard, shield construction, FLAC^{3D}, ground settlement, pile deformation

1 INTRODUCTION

Shield, as a common method of tunnel construction that has made great improvement, still causes strata displacement and ground subsidence (L.H. Zeng. 2010). The Hefei Metro Line1 project is located in an area that is crowded with buildings, and most of the buildings are in the form of pile foundation. Sometimes the shield tunnel will drive through adjacent to the pile foundations, the shield construction will certainly affect the internal force and deformation of the pile foundations, and a certain degree of loss will happen to the stability and safety of the upper structure (M.W. Li. 2011). Therefore, in order to protect the upper structure and to ensure the smooth tunnel construction, this study aims to unveil the effects of tunneling on adjacent pile foundations, which will lead to acknowledgement of the serious practical significance of this constructing method.

At present, domestic and foreign scholars have done a lot of research on the influence of shield tunnel excavation on the pile foundations. Ruan and Li (1997) have analyzed the shield-ground-structure system; they believe that the pile deflection is mainly affected by being thrust forward. Mrough H. and Shahrour (2002) took advantage of finite element numerical analysis software to establish 3D models, in order to simulate the effects of tunneling on adjacent pile foundations. Liu and Tao (2006) used finite difference software FLAC^{3D} to study the spatial effects of ground deformation induced by shield-driven tunnel crossing building foundations (X.G. Tian. 2010).

Based on the shield tunneling crossing building foundation in Hefei Metro Line1 porject, this paper will use infinite-difference software FLAC^{3D} to simulate the effects of shield-driven tunnel crossing pile foundations. Analyses of the ground settlement pattern and deformation characteristics of pile foundations are also presented (M.J. Zhang et al. 2013).

2 THE PROCESS OF FINITE-DIFFERENCE SOFTWARE FLAC3D SIMULATION

(1) The related parameters

The related parameters are shown in table 1, table 2, table 3.

(2) Establish a 3D model (S.Q. Li et al. 2009)

As figure 1 shows, the shield tunnel has a outside diameter of 6.0 m, lining thickness of 0.3 m, grouting thickness of 0.15 m, and buried depth of 18 m. Considering the area that shield tunneling might affect, the distance of 19.5 m was selected as the left and right

Table 1. Lining segment parameters.

Density (kg/m³)	Elastic modulus (GPa)	Poisson ratio	Bulk modulus (GPa)	Shear modulus (GPa)
2500	33	0.17	16.67	12.50

Table 2. Layer parameters.

Layer	Density (kg/m³)	Thickness (m)	Prop bulk (MPa)	Shear (MPa)	Fric (°)	Cohesion (KPa)	Tension (KPa)
Plain fill	1850	4	5	2.3	14	18	2.5
Clay (2)	1920	6	9.83	4.53	15	38	3.2
Clay (3)	1940	10	9.83	4.53	15	42	3.2
Strong weathered muddy sandstone	2100	4	16.6	7.96	33	52	4.0
Moderately weathered muddy sandstone	2200	16	20.83	9.61	35	56	4.2

Table 3. Pile parameters.

Density (kg/m³)	Elastic modulus (MPa)	Poisson ratio	Unit linear shear coupling		Unit linear normal coupling	
			Cohesive force (N/m)	Fric (°)	Cohesive force (N/m)	Fric (°)
2250	200	0.18	42.8	26.5	0.0	0.0

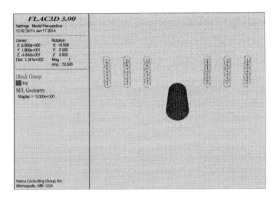

Figure 1. The model of shield tunnel.

margins, and the distance from the center of the tunnel to the bottom of the model was 19 m. Six piles were put at both sides of shield tunnel, stayed in the same horizontal line and the interval between the bottom and the top of tunnel was 2 m. As figure 1 shows, the width of the final model is 45 m, height is 40 m, and length is 36 m. In order to facilitate the simulative calculation, the longitudinal length of each tunnel excavation is 2.0 m.

(3) The numerical simulation process (C. P et al. 2008)

(1) Establish the model and apply gravity load to it, let the model began to consolidate until it arrives a stable stage under the effect of its own gravity stress, and get the initial stress field.
(2) Set the layer displacement value to 0 and activate the pile structure unit. Calculate the deformation of the pile–soil system under the effect of building load.
(3) Set the pile – soil displacement value to 0 after the model reaches balance.
(4) Simulate shield excavation and impose 0.3 MPa of force to the shield's heading face. Apply 0.3 MPa of grouting pressure and install the lining.
(5) Calculate a certain step to complete a full working condition of excavation.
(6) For the next excavating condition, cycling, calculates until the excavation complete and the surface subsidence value appears to be stable.

3 ANALYSIS OF THE RESULTS OF NUMERICAL SIMULATION

3.1 *Analysis of the law of ground settlement*

The vertical displacement of the strata when shield tunneling to different diatance are shown in figure 2, figure 3, figure 4, figure 5, figure 6. After the shield tunneling started, as the figures indicate, the tunnel arch subsides while the tunnel bottom upsides. Accompanied with the continuous advance of shield, the deformation of assembled segments becomes more and more obvious. The surface settlement increases suddenly and settlement rate reaches its maximum when the segment comes out of the shield tail. Therefore, the deformation control when the segment comes out of the shield tail is the key to control ground settlement in shield tunneling. The maximum settlement of tunnel arch reaches 8.9 mm. Since the buried depth of tunnel reaches 18 m, the soil will consume part of the settlement, so the settlement of the surface just above the tunnel axis is only 5.5 mm. This sedimentation value meets the urban subway shield construction requirements. In addition, as figure 7 shows, the surface settlement of cross-section is close to the normal distribution and the surface subsidence of the area,

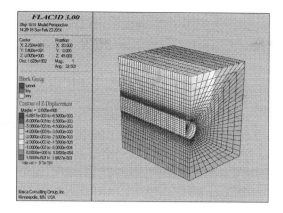

Figure 2. The vertical displacement of the strata when shield tunneling to 2 m.

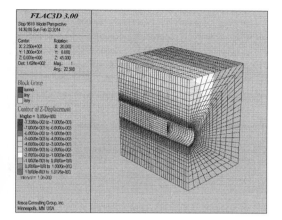

Figure 3. The vertical displacement of the strata when shield tunneling to 10 m.

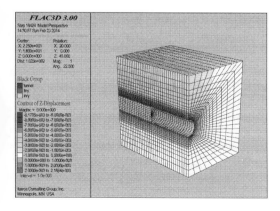

Figure 4. The vertical displacement of the strata when shield tunneling to 18 m.

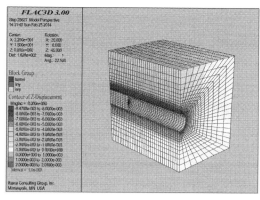

Figure 5. The vertical displacement of the strata when shield tunneling to 26 m.

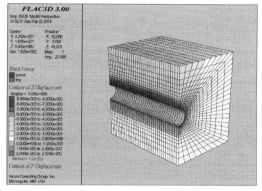

Figure 6. The vertical displacement of the strata when shield tunneling to 36 m.

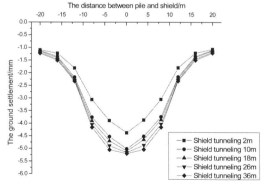

Figure 7. The curve of ground settlement.

where existence of pile foundation is greater than the open ground subsidence. The reason is that the presence of the building had an impact on surface subsidence.

3.2 Analysis of pile deformation

(1) Vertical displacement of the pile

This paper simulates the changes of the piles' vertical displacement in the process of shield tunneling at

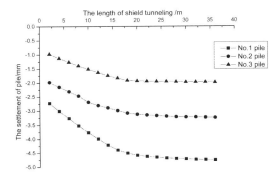

Figure 8. Vertical displacement of the pile.

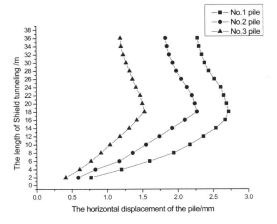

Figure 9. Horizontal displacement of the pile.

different locations of the tunnel. As figure 8 shows, before the shield reaches the bottom of the pile, the vertical displacement of the pile increases as its distance to the shield reduces. The maximum appears in the shield leaving. In addition, the vertical displacement of the pile is inversely proportional to the distance from the pile to the center of the tunnel, and the changing rule of each pile stays in line with normal distribution. As the figures show, when the shield comes out of the tunnel, No. 1 pile, which is the closest to the center of the tunnel, has the largest vertical displacement (4.75 mm). This value meets the urban subway shield construction requirements.

(2) Horizontal displacement of pile foundation
This paper simulates the changes of the piles' horizontal displacement at different locations around the tunnel in the process of shield tunneling. As figure 9 shows, in the process of shield tunneling, the horizontal displacement of the pile is inversely proportional to the distance from the pile to the center of the tunnel and reaches its maximum when the shield excavates to the bottom of the pile. As the figures indicate that when the shield reaches the bottom of the tunnel, No.1 pile, which is the closest to the center of the tunnel, has the largest horizontal displacement (4.75 mm). This value meets the urban subway shield construction requirements.

4 CONCLUSIONS

Based on the upper soft and lower hard strata of Hefei region, this paper uses finite-difference software FLAC^{3D} to simulate the effects of shield-driven tunnel crossing pile foundations. Analyses of the ground settlement rule and the deformation characteristics of pile foundations were conducted. The conclusions are as follows:

(1) The surface settlement of cross-section is close to the normal distribution, the maximum subsidence occurs right above the tunnel.
(2) The vertical displacement of the pile is inversely proportional to the distance from the pile to the center of the tunnel. Before the shield reaches the bottom of the pile, the vertical displacement of the pile increases as it distance to the shield reduces, and the maximum appears in the shield leaving.
(3) In the process of shield tunneling, the horizontal displacement of the pile is inversely proportional to the distance from the pile to the center of the tunnel and reaches its maximum when the shield excavates to the bottom of the pile. In addition, the horizontal displacement of the pile increases as the distance from it to the shield reduces.

ACKNOWLEDGEMENTS

This project is supported by Anhui Provincial Laboratory Project Funding through grant No. 1106c0805024.

REFERENCES

C. P, Y.L. Ji, H.B. Luo, et al. 2008. Numerical Simulation of Effects of Double-tube Parallel Shield Tunnelling on Neighboring Building [J]. Chinese Journal of Rock Mechanics and Engineering: 3868–3874

L.H. Zeng. 2010. Study on Influence of Shiled Tunnel Construction on Pile Foundation-Frame Structure [D]. Guangzhou: South China University of Technology: 12–23

M.J. Zhang, X.J. Liu, Y.X. Du. 2013. Building Settlements Induced by Shield Tunneling in Close Proximity [J]. Journal of Beijing University of Technology, 39(2):215–217

M.W. Li. 2011. A Study on Construction Technologies of Shield Tunnel Crossing Complex Building [D]. Chengdu: Southwest Iiaoton University: 10–21

S.Q. Li, Y.H Cao. 2009. Numerical Analysis on Influence of Shield Tunneling to Neighboring Piles [J]. China Harbour Engineering. 8(4):1–3

X.G. Tian. 2010. Shield Tunneling Effect on the Building and its Control Technology [J]. Railway Engineering, 05(3):34–36

Numerical analysis of soil deformation in shielding tunneling considering the effect of shield gap and grouting patterns

Zelin Zhou & Shougen Chen
The College of Civil Engineering, Southwest Jiaotong University, Chengdu, China

ABSTRACT: The distribution forms of shield gap and grouting patterns are important factors influencing the ground settlement in shield tunneling. It is necessary to study the impact of these factors on ground deformation. In this study, the distribution forms of shield gap and grouting patterns were researched. Furthermore, a nonlinear numerical model was established by Flac3D to study the influence of shield gap and grouting on ground settlement. It was found that: (1) The patterns of back-filled grouting distribute differently in different soil layers; (2) The effects of grouting fill-back determine the values of ground settlement directly; (3) The distribution range of back-fill grouting has an important influence on the ground settlement, when the grouting angle is equal to 120°, the ground settlement reaches the maximum value. The conclusions obtained through this research can offer a reference for the prediction and control of the ground settlement in shield tunneling.

Keywords: shield tunneling; shield gap; grouting parents; ground settlement; numerical calculation

1 INTRODUCTION

Under the influence of shield excavation and soil conditions, disturbance on the soil layers in shield tunneling is unavailable, thus it is a key issue to predict and control the deformation of ground settlement at the stage of design and construction. The process of shield tunneling is a three-dimensional problem. On the one hand, the shield tail gap formed after the last assembled segment will induce ground loss, which has a significant influence on the ground displacement at the tunnel' cross section. On the other hand, the deformation of tunnel' excavation face, as well as the friction between shield machine and soil, will have an important influence on the ground displacement at the tunnel' vertical section.

The main methods that have been used to study the ground settlement are empirical formula method, analytic method and numerical simulation method. Perk (Perk et al. 1969) proposed the famous Perk theoretical formula, which assumes that the curves of ground settlement meet the distribution of Gaussian curve, to predict the ground settlement. Rowe (Rowe et al. 1983) proposed the concept of *gap parameter* and simplified the ground loss into an equivalent two-dimensional gap to calculate the displacement of soil layer. Loganathan (Loganathan et al. 1998) derived an analytical solution to predict ground movements around the tunnel in clays. Wei Gang (Gang, Wei et al. 2005) derived a ground deformation formula which considers the influence of excavation face' supporting force and shield machine' friction based on the Mindlin solution in elastic mechanics. Wang Minqiang (Minqiang Wang et al. 2002) proposed a three-dimensional finite element numerical model to simulate the excavation of shield tunnel and studied the deformation pattern of soil layer. Zhu Hehua (Hehua Zhu. 2000) carried out a finite element numerical simulation to study the characteristics of grouting materials and tubes sheet in construction. Zhang Yun (Yun Zhang. 2002) proposed a concept of *equivalent circle zone* to reflect the shield tail void, the degree of filling grout and the disturbance of soil when using numerical simulation to predict the ground settlement of shield tunnel.

The forms and distributions of shield gap and diffusion model of grouting, which have a great influence on the deformation of soil layer, are not the same in different soil layers. However, current research concern this is very few. It is necessary to study the impact of shield gap and grouting forms to the ground settlement. Compared with theoretical and experimental studies, numerical modeling provides a convenient, economical approach to study the deformation of soil layer. This paper aimed at the distribution of shield tail gaps and diffusion model of grout, by using finite difference software Flac3D to perform numerical modeling and analyze the characteristic of ground settlement on the tunnel' cross section. Some useful conclusions obtained in this study can provide a reference for the ground settlement control of shield tunneling in soft soil layer.

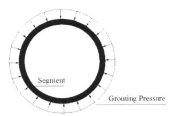

Figure 1. Even distribution.

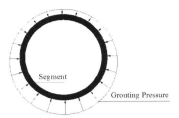

Figure 2. Uneven distribution.

2 DISTRIBUTION OF SHIELD GAP AND GROUT

After assessment of the last segment in the shield tail, there will be a temporarily physical gap between the segment and soil layer, this gap is called shield gap Δ_0. Δ_0 consists as the flowing parts: (1) Gap Δ_1 (2~3 cm) caused by overbreak of shield cutter; (2) Gap Δ_2 (3~4 cm) generated by the thickness of shield shell; (3) Gap Δ_3 (2~3 cm) reserved for assessment space of segments in the shield machine; (4) Gap Δ_4 (1~6 cm) formed by the friction, which will take away the clay adhered to the shield shell and result in soil loss, between shield shell and soil layer. Therefore, the value of Δ_0 ($\Delta_0 = \Delta_1 + \Delta_2 + \Delta_3 + \Delta_4$) is in the range of 8 cm to 16 cm. Because there will be a time interval between the formation of shield gap Δ_0 and synchronous grouting, the gap Δ_0 is the most important factor to cause the ground loss and induce a ground deformation.

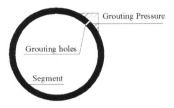

Figure 3. Triangular distribution.

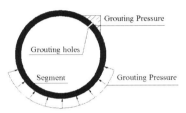

Figure 4. Distribution in soft soil layer.

The distribution forms of shield gap are closely related to soil layer conditions. In stiff soil layer, where the soil condition is good, the shield gap distributes evenly around the segment back. However, in soft soil layer, when the last segment is departing from the shield machine, one end of the segment is connected by bolt with adjacent segment, the other end of the segment is put on the shield machine, forming a temporary beam-structure to bear force. In this case, the soil located at the upper and sidewall part of the segment will collapse and extrude to fill the shield gap naturally under the action of gravity and soil pressure. Therefore, the distribution of shield gap is uneven at the bottom of the segment and mainly distribute below the segment.

The patterns of back-filled grouting are different with different distribution forms of shield gap. In stiff soil layer or porous, permeable and sandy soil layer, the grout slurry will wrap around the segment. In this case, the grouting distribute through the ring evenly or unevenly, as shown in Figures 1 and 2. If the shield gap is filled by soil, the grouting slurry will extrude or diffuse into soil and form a mixture of 'soil–grouting–water'. In this case, the grouting effect is bad and the grouting distribution pattern through the ring, recommended by International Tunnelling Association (2000), is triangular, as shown in Figure 3. However, in soft soil layer, as analyzed above, the gap located at upper and sidewall of the settlement has been filled by soil, and the gap is mainly distributed below the segment actually. Thus, the distribution pattern of grouting is a combination of the above two cases, as shown in Figure 4. The different distribution patterns of back-fill grouting will cause different effects on the ground settlement.

3 NUMERICAL MODELING

3.1 Calculation model

When analyzing the influence of shield gaps and grouting' distribution pattern on the ground settlement, the following several factors, including filling the gap by soil' naturally moving, elimination the gap by synchronous grouting, as well as the distribution forms and range of gaps and grouting, are closely related with the soil condition and workmanship in construction, and those factors are hard to be analyzed quantitatively. In this paper, the above three factors were generalized as an arc distributed "Mixture of soil and slurry" in numerical simulation. The "Mixture of soil and slurry" is a circle of material with elastoplastic properties, and its mechanical parameters are between the mechanical parameters of soil and grouting. To simplify the calculation, the alerting of mechanical properties of "Mixture of soil and slurry" can be reflected by the modulus of elasticity 'E_1', and the distribution ranges

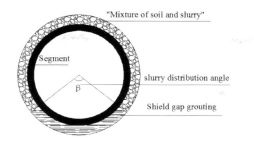

Figure 5. Calculation model.

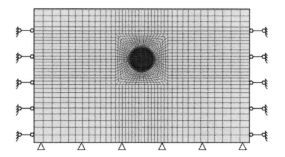

Figure 6. Model meshing and boundary conditions.

of shield gaps and grouting filling are reflected by the parameter of slurry distribution angle 'β'. The calculation model is shown in Figure 5.

3.2 Numerical model

The values of parameter E_1 and β, with highly nonlinear characteristics, are closely related to soil condition and workmanship in construction, and can be determined by the method of numerical back analysis. To discuss the influences of parameter E_1 and β on the ground settlement, in this paper, these two parameters were analyzed univariatly to quantify the value.

Analyses were conducted using the finite element method by the software Flac3D, and two-dimension numerical modeling was carried out. The numerical model is in the length of 50 m and width of 25 m, with meshing and boundary conditions as shown in Figure 6. Considering the boundary condition, the displacement is constrained in both bottom and direction, while the zero displacement is imposed to the upper boundary. The soil behavior is described using an elastoplastic constitutive relation based on the nonlinear Mohr-Coulomb criterion, and the behavior of lining is described as linear-elastic. The coefficient of the lateral stress $\gamma = 0.75$, the thickness of the lining is equal to 0.5 m, and the shield gap is equal to 9 cm.

In this paper, the modulus of elasticity of soil layer is assumed to increase with depth according to the following expression:

$$E(z) = E_0 (P_m / P_0)^{0.5}. \qquad (1)$$

Table 1. Material parameters.

Materials	E_0 (MPa)	u	c (kPa)	φ (°)	ρ (KN/m³)
Soil	8	0.3	3.0	30	18.0
Mixture of soil and slurry	30	0.25	6.0	33	19.1
Grouting material	300	0.2	100	38	24.0
Segment	35000	0.2	–	–	28.0

In the above expression, E_0 corresponds to the modulus of elasticity at mean pressure $P_m = P_0$; P_m denotes the mean stress at depth z, and P_0 denotes the mean pressure of soil layer. This expression takes into account the variation of modulus of elasticity with the mean pressure, which increases with depth due to the soil gravity.

The mechanical parameters of lining, soil, shield gap and grouting materials are shown in Table 1.

In Table 1, parameter c means the cohesion of soil, φ means the friction of soil, and ρ means the heavy of soil.

3.3 Results of analysis

The results of the numerical analysis are as flows.

(1) Influence of parameter E_1

To discuss the influence of parameter E_1, this paper keeps the parameter of $\beta = 60°$. Analyses were conducted for four different values of E_1, including 0.5 MPa, 5.0 MPa, 20 MPa and 50 MPa. The ground settlement curves of shield tunnel on cross-section are shown in Figure 7. As shown in Fig. 7, the values of ground settlement decrease with the increasing values of E_1. Greater value of E_1 means better effect of grouting, greater stiffness of 'Mixture of soil and slurry' and greater supporting force to the upper soil layer, thereby the deformation of soil layer is small. However, if the value of E_1 keeps increasing, the decreasing gradient of ground settlement becomes smaller; when $E_1 = 50$ MPa, the maximum settlement is equal to 1.78 cm. Because of the constrains by other factors, the grouting fill rate of shield gaps could only amount to 80%. Analysis shows that the effects of grouting fill-back determine the values of ground settlement directly.

(2) Influence of parameter β

To discuss the influence of parameter β, this paper keeps the parameter of $E_1 = 20$ MPa, and the numerical calculation was conducted for a set of values of β with an interval of 15°, including 15°, 30°, 45°, ... 180°. Figure 8 shows the variation of maximum ground settlement with the distribution angle β. As shown in Figure 8, the distribution range of back-fill grouting has an important influence on the deformation of ground, the variation of maximum ground settlement with β is not simple increase or decrease. If the

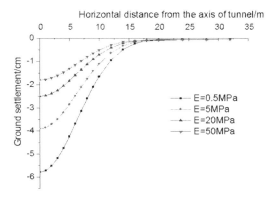

Figure 7. Ground settlement curves with different Young's moduli of hybrid.

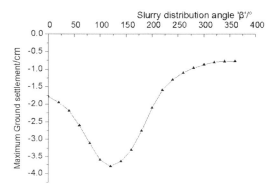

Figure 8. Maximum value of ground settlement with the change of grouting scope.

value of β is greater or smaller, the maximum ground settlement is smaller. When $\beta = 360°$, which means full-face grouting, the maximum settlement reaches a minimum value of 0.76 cm. However, if β is an intermediate value, the maximum ground settlement is greater. For instance, when $\beta = 122°$, the maximum settlement reaches a maximum value of 3.79 cm. If the distribution range of grouting is large, the grouting filling range is large, indicating better grouting and reinforcement effect, thus the loss of soil layer is small. Reversely, the grouting filling range is small, and the soil at the bottom of arch will move upward in the effect of soil pressure, which will reduce a part of ground settlement. Though the maximum settlement is small in this case, it is disadvantageous for the lining structure to bear force.

4 CONCLUSIONS

In shield tunneling, the distribution forms of shield gap and grouting patterns are important factors influencing the ground settlement. In this paper, the distribution forms of shield gap and its relation with back-filled grouting patterns were studied in detail. To analyze the influence of shield tail gaps forms and diffusion model of grout on ground settlement, numerical modeling was carried out, and the parameters of E_1 and β were introduced to reflect the mechanic parameter and distribution range of "Mixture of soil and slurry" in numerical simulation. Some useful conclusions are drawn as follows.

1) The patterns of back-filled grouting are different in different distribution forms of shield gap back segment. In stiff soil layer, the grout slurry will wrap around the segment. In soft soil layer, the gap and grouting are mainly distributed at the bottom of the segment.
2) The values of ground settlement decrease with increasing value of E_1, which means that the effects of grouting fill-back determine the values of ground settlement directly.
3) The distribution range of back-fill grouting has an important influence on the deformation of ground; when the grouting angle β is equal to 120°, the ground settlement reaches the maximum value.

The conclusions obtained by this research can offer a reference for the prediction and control of the ground settlement in shield tunneling.

REFERENCES

Gang, Wei, Riqing, Xu. 2005. Prediction of longitudinal ground deformation due to tunnel construction with shield in soft soil. Chinese Journal of Geotechnical Engineering 27(9): 1077–1081. (In Chinese)

Hehua Zhu, Wenjun Ding, Xiaojun Li. 2000. Construction Simulation for the Mechanical Behavior of Shield Tunnel and its Application. China Civil Engineering Journal 33(3): 98–103. (In Chinese)

International Tunneling Association. 2000. Working Group 2. Guidelines for the design of shield tunnel lining. Tunneling and Underground Space Technology 15(3): 303.

Loganathan N, Poulos H G. 1998. Analytical prediction for tunneling-induced ground movements in clays. Journal of Geotechnical and Geoenvironmental Engineering, ASCE 124(9): 846–856.

Minqiang Wang, Shenghong Chen. 2002. 3-Dimensional Non-linear Finite Element Simulation of Tunnel Structure for Moving-forward Shield. Chinese Journal of Rock Mechanics and Engineering 21(2): 228–232. (In Chinese)

Perk R B. 1969. Deep excavations and tunneling in soft ground [C] State of the Art Report. Proceedings of 7th International Conference on Soil Mechanics and Foundation Engineering . Mexico City (s.n.): 225–290.

Rowe R K, Lo K Y, Kack G J. 1983. A method of estimating surface settlement above tunnels constructed in soft ground. Canadian Geotechnical Journal 20(8): 11–22.

Yun Zhang, Zongze Ying, Yongfu Xu. 2002. Analysis on Three-Dimensional Ground Surface Deformations Due to Shield Tunnel. Chinese Journal of Rock Mechanics and Engineering 21(3): 288–392. (In Chinese)

Numerical analysis of soil nailing's instability in deep foundation pit

Hongguo Dong & Daikun Chen
China University of Mining and Technology (Beijing), Beijing, China

ABSTRACT: This paper applies FLAC^{3D} software to simulate and analysis deep foundation pit design based on some unstable situation such as the top soil body of side slope dehiscence and the base soil body of side slope collapse appeared in substep excavate and support by soil nailing in deep foundation pit located on Beijing of fang shan magmatic university city building design. Apply Mohr Coulomb elastic-plastic model in numerical simulation to compute side slope's displacement deformation indifference steps, then acquire displacement deformation cloud atlas of the whole deep foundation pit and the maximal displacement cross-section through post-processing method, comparing numerical simulation result with reality measured data to determine the approximate state of potential sliding surface and collapse in deep foundation pit. Furthermore, it could offer some important reference for improving support pattern of major displacement area in design programme.

1 INTRODUCTION

With the rapid development of society and increase of urban population intensive of the large and medium-sized city, in order to improve the utilization of urban land, construction develop to high-rise buildings and underground space day by day, all these things make the deep foundation pit become an important research direction[1]. Deep foundation pit is the foundation pit engineering whose excavated depth is more than 5 meters, when reach a particular deep excavation, only relying on the soil itself is difficult to ensure the deformation within the scope of the security, stability, therefore, all kinds of foundation pit support modes such as soil nails and anchor pile is generally used[2]. The geological features of the different regions decide the degree of difficulty or ease of the support, so support process should be selected flexibly according to the specific soil and the depth of excavation of foundation pit[3]. The feature of temporary support of foundation pit want to be considered at the same time, so that avoid to cause the consumption of material and artificial[4].

Due to the complexity of geological structure, diversity, the water content increasing variability and uncertainty of the original ground stress, Current research on the theoretical calculation of the deep foundation pit lead to limitations a single mechanics calculation of internal force of the structure of the soil and the retaining wall of foundation pit analysis has set[5]. FLAC^{3D} which is based on finite difference method can not only establish a consistent with the actual of retaining the structure with 3d-model, considering the soil distribution, ground stress, pore-water pressure and seepage flow, but also can analysis and

Figure 1. Soil body cracking in top of side slope neighbouring area.

calculation elastic-plastic deformation, stress distribution of cloud, to predict the maximum stress and displacement of concrete numerical value and its corresponding site under the action of the supporting structure in foundation pit excavation supporting, so the FLAC^{3D} numerical simulation has been widely applied in geotechnical engineering analysis.

Foundation pit project of Beijing Fang Shan magmatic university town adopted soil nailing structure, stratified excavation and layered support (a total of five layers). After fulfilling the first three layer excavation support, because of failing to follow the design plan, when continuously excavating layer of soil four and five to 7.6 meters deep underground, the soil produced larger banded cracking in the long side about 3.9 meters away from the top position, the cracking width is about 50 millimetres (As shown in Fig. 1). There are also smaller ribbon craze on the top of the slope, collapse occurs (As shown in Fig. 2) on the

Figure 2. Part of soil body collapse in bottom of side slope.

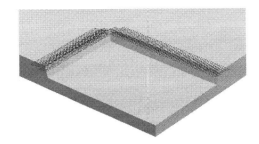

Figure 3. The excavate and support model of foundation pit.

bottom of the slope. The analysis of the causes of instability: 1, The over break and failed to timely support; 2, At the bottom of the foundation pit, soil poor mechanical properties. This paper uses FLAC^{3D} to simulate pit supporting which was excavated according to working method and gets the distribution of the maximum deformation in different excavate and support stages by post-processing (the location where the most likely emerge sliding face). It can provide a reference for rearrange design plan and ensure safety construction by contrast and combined with the settlement and horizontal displacement of foundation pit deformation of the measured data.

2 GENERAL SITUATION OF THE PROJECT

The total land area of Fang Shan magmatic university city building projects is approximately 30000 square meters, including excavation site length × width = 150 m × 110 m. This area is relatively flat terrain, soil mainly insilty clay and sandy clay, underground depth of 7.30∼8.80 metres mainly containing the layer of medium fine sand which aquifer size range between 0.075 millimetres and 0.5 millimetres, its water permeability and compressibility is heavy, cohesive force and the shear modulus can be ignored. The methods of using the tube well drain off water can reduce the moisture content of soil, but still cannot change the condition fine that the medium is sand soft interlayer for the whole foundation pit after drainage, often due to adjacent excavation of medium-fine sand soil layer stress redistribution and produce larger deformation and instability[6].

The length of the soil nails that used in the first layer of foundation pit is 6.8 meters, the length of Steel wire that used in the second layer is 15.0 meters and the prestress is 100 kN, the length of the soil nails that used in the third layer is 7.8 meters. The slope gradient is 1:0.3, the layout of soil nails is plum blossom shape. Inject M20 cement slurry into the hole after the placement of the soil nails, the water cement ratio of the cement slurry is 1:0.5. The Soil nailing wall layer consists of Φ6.5@ 200 mm × 200 mm Steel mesh and C20 fine stone concrete (Cement:Sand:Stone = 1:2:2) of which the thickness is 100 cm.

3 PROGRAMME DESIGN OF NUMERICAL SIMULATION

Considering the symmetry of the foundation pit shape, establish 1/4 the actual excavation model in the FLAC^{3D} simulation with length × width × height = 225 m × 165 m × 15.2 m, and the excavation length × width × height = 75 m × 55 m × 7.6 m, the slope gradient of the model is 1:0.3. The simulation of excavation process is divided into four steps, the depth of excavation of the first three times is 1.5 metres, support after Stratified excavation, the excavation depth of the fourth time is 3.1 metres and no support after excavation. In consideration of the actual conditions of sand aquifer 7.3∼8.8 metres underground, set the initial pore water pressure in the depth range, the supporting model of foundation pit is shown in Fig. 3.

4 ANALYSIS OF MONITORING DATA WITH NUMERICAL SIMULATION RESULT

Horizontal and vertical displacement are two important factors that determine the stability of foundation pit, through the analysis of the results of numerical simulation and the measured data, accurately comprehend the situation and development trend of soil deformation[7]. The simulation analysis focuses on getting the displacement cloud atlas that directly reflects the deformation of foundation pit of all parts by the post processing program, then determine the approximate location of maximum deformation region, providing an important reference for the reform in the soft soil support scheme[8].

4.1 Analysis of horizontal displacement in long and short side slope

The horizontal displacement of long and short side of foundation pit in the displacement step excavation and support cloud conditions is shown in Fig. 4 and Fig. 5, the long and short side of the horizontal displacement is roughly same. Gradually increased with the depth of excavation and support, the horizontal displacement of side slope wall increase and its range spread gradually. Before the first three layers excavation were supported completely, the maximum horizontal displacement of the position still have a certain distance

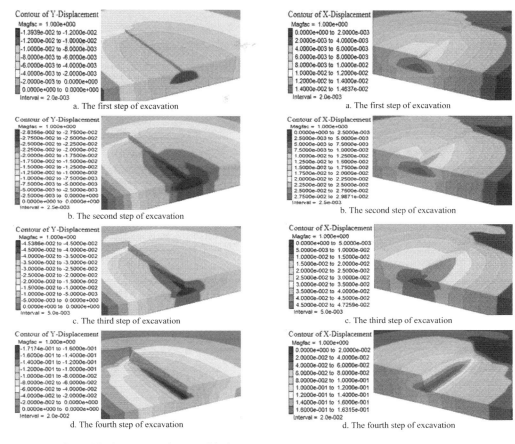

Figure 4. Horizontal displacement nephogram of the longer side slope.

Figure 5. Horizontal displacement cloud atlas of the shorter side slope.

from the bottom, because of the foundation pit about 7.3~8.8 metres depth of medium-fine sand aquifer is weak in terms of mechanical property, caused by upper excavation supporting ground stress, redistribution of the perturbation is larger, therefore produce obvious horizontal displacement tends to excavation position, the displacement cloud atlas distribution is similar to the "mushroom" shape. Excavating but not supporting of foundation pit in finished fourth and fifth layer, each side in the center of the foundation pit slope bottom along both directions respectively spreading around 50 metres within the scope of the horizontal displacement of more than 140 millimetres, the top center to expand on each side about 30 metres within the scope of the horizontal displacement of more than 100 millimetres, belongs to the serious deformation in the deep foundation pit engineering. Only six hours after the fourth and fifth layer completelybeing excavated (4.8~7.6 metres deep underground layer), banded cracking appeared where keep 3.9 metres from the top of the slope, and the continuation of crack width has increased gradually, over time achieves the stability after eight days.

The horizontal displacement data obtained by post-processing program show that the horizontal displacement increment combine with every step of excavation and support, and as the excavation support gradually, the horizontal displacement growth increasing. The top of the "mushroom" closer to a layer of excavation soil nailing supporting position, thus, through the passive stress to a certain range, soil nail achieve a more balanced distribution of soil displacement, play an important role in the foundation pit slope support. In the first three layers excavation completed that two larger horizontal displacement contour area are smaller, a preliminary analysis is due to the uniqueness of the regional soil makes the third layer after the excavation of soil stress distribution produced by changes of its inside the foundation pit horizontal displacement, the effect similar to unloading.

4.2 Contrast numerical simulation result with monitoring data

In order to grasp the deformation of foundation pit in different phases of the excavation support more accurately, getting relevant data from monitoring sites in a periodicity time[9]. This section focuses on analyzing the whole cracking and collapse in the excavation of

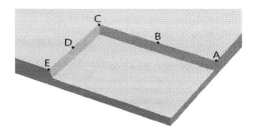

Figure 6. The distribution map of several monitoring sites.

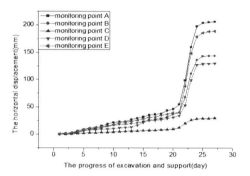

Figure 7. The horizontal displacement chart of monitoring sites.

foundation pit slope stability, selected representative five points from a host of monitoring stations: A and E points are the center of long and short side, B and D points are 1/4 of long and short side, C point for foundation pit slope around the corner, As shown in Fig. 6.

From figure 7 can see that with the extension of time horizontal displacement finally reached stable in each substep excavation. Due to the adjacent fine medium sand aquifer and without timely supporting, the horizontal displacement value of both point A and B (after the fourth excavation) almost four times of the value caused by the formerly three excavation that resulted in emerging instability. To suppress the soil collapse on the surface of the slope and the cracking degree of soil near the slope top, the fourth and fifth excavated layer had been backfilled along the foundation pit slope so that the monitoring data show that the displacement tends to stable after the following 5 days.

5 CONCLUSION

(1) Numerical simulation and monitoring data show that it firstly formed a plastic zone at the slope toe before instability generates widely, then a tension crack and shear plastic belt appeared at a certain distance from the slope top, both plastic zones extend to expand to the middle part of the slope, the overall sliding failure of pit occurred when they ran through from the bottom to the top of the slope to form a sliding surface of slope.

(2) The displacement cloud atlas of foundation pit partial excavation in the long and short edge shows that the medium-fine sand aquifer affected by the excavation produces a large deformation, which becomes the main reason for instability of foundation pit. The fourth and fifth excavation steps occur simultaneously and it fails to get timely support, which is also an important factor leading to instability of the pit.

(3) Based on Mohr-Coulomb elastic-plastic model, FLAC^{3D} simulates partial excavation and support of the base pit and reflects the effect of soil deformation, which is closer to the real situation and has reference value and guidance significance to engineering.

REFERENCES

[1] Shu LI, Dingli ZHANG, Qian FANG, Zhijia LI. 2012. Research on characteristics of retaining wall deformation due to deep excavation in Beijing [J]. Chinese Journal of Rock Mechanics and Engineering, 2012, 31(11): 2344–2353.

[2] Zhaoyuan CHEN, Jinghao CUI. 2000. Application of soil nailing in foundation pit engineering (The second edition) [M]. China Architecture & Building Press, 2000. (in Chinese).

[3] Mingmin ZHANG. 2011. FLAC^{3D} Simulation and Field Testing of Deep Excavation Supported by Soil Nail Walls [M.S. Thesis] [D]. Beijing: China University of Geosciences, 2011. (in Chinese)

[4] FU Wenguang, YANG Zhiyin. Formulae for overall stability of composite soil nailing walls [J]. Chinese Journal of Geotechnical Engineering, 2012, 34(4):742–747.

[5] Yong FENG, Ping LI, Chunyan JIANG 2013 Application of FLAC^{3D} in Numerical Simulation Analysis for Deep Foundation Pit Excavation [J]. Journal of Water Resources and Architectural Engineering, 2013,11(4):17–20

[6] Denghua YU, Ji YIN. 2010. Finite element analysis of the displacement and stability of composite soil nailed foundation pits [J]. Chinese Journal of Geotechnical Engineering, 2010, 32 (supplement 1):161–165.

[7] Yanlin ZHAO, Weiguang AN. 2011. Reliability-based stability analysis of a foundation pit supported by composite soil nailing [J]. Journal of Harbin Engineering University, 2011, 32(10):1300–1304.

[8] Hongxin WANG, Song ZHOU. 2013. Several issues on deformation and stability analysis of Bar-load-spring model of foundation pit design and recommendations [J]. Chinese Journal of Rock Mechanics and Engineering, 2013, 32(11):2349–2358.

[9] Xuelin YANG. 2012. Several Issues in Design, Construction and Monitoring of Foundation Pits [J]. Chinese Journal of Rock Mechanics and Engineering, 2012, 31(11):2327–2333.

Numerical simulation of the characteristic of air motion in the spillway tunnel with aerators*

S.B. Yue & M.J. Diao
State Key Laboratory of Hydraulics and Mountain River Engineering, Sichuan University, Chengdu, China

C. Qiu
Sichuan College of Architecture Technology, Deyang, China

ABSTRACT: The characteristic of the air motion in the spillway tunnel with aerators was numerically simulated, by combining the VOF model and the RNG $k-\varepsilon$ turbulence model. The calculated results agreed well with the experimental results. Visualization of the patterns of air motion in flood discharge tunnel was achieved by means of numerical simulation. Three partial air rotary regions were formed in the spillway tunnel. The first region located at the reverse curve section in flood discharge tunnel. The second was on the slope between two aerators. And the third was seated on the slope between the second aerator and the outlet. They became smaller when the flood discharge became larger.

Keywords: numerical simulation; RNG $k-\varepsilon$ turbulence model; Volume of Fluid (VOF) method; characteristic of air motion

1 INTRODUCTION

High-speed flows will bring many special hydraulic problems, such as strong pressure fluctuation, wave rolling phenomenon, water aerated flow, cavitations of hydraulic structures, and so on.

Spillway tunnel is one of the most easily damaged flood discharge structures because of cavitations. For this, our predecessors have done a lot of researches and found that air entrainment is one of efficient and inexpensive methods to prevent cavitations in a lot of hydropower projects (Russell 1974 and Rutschmannn & Volkart 1988). Therefore, various kinds of aeration facility will be taken into account on almost all large flood discharge spillway tunnels, such as aerator with airshaft, sudden expansion and fall aerator and so on (Boes & Hager 2003; Pegram 1999; Liu et al. 2008; Xu et al. 2008 and Wang et al. 2009). So it is of great significance to study the characteristic of the air motion in the spillway tunnel with aerators.

The characteristic of the air motion in the spillway tunnel with aerators can be estimated by project method, dimensional analysis and numerical simulations. As far as the flow of a physical model is concerned, the air-water two-phase turbulent flow patterns in the mid-gate chamber with ventilation are complicated, and it is very difficult to measure the hydraulic parameters in physical model tests. Numerical simulation is an efficient way to simulate the movement of air-water two-phase turbulent flows of this kind.

In this paper, hydraulic characteristics of the aeration flows are analyzed. Especially, the characteristic of the air motion in the spillway tunnel with aerators was simulated by combining the VOF model and the RNG $k-\varepsilon$ turbulent model in an actual project.

2 MATHEMATICAL MODEL

The numerical simulation adopted the VOF model and the RNG $k-\varepsilon$ turbulence model. The Volume of Fluid (VOF) model is usually used to track the free water surface, such as in simulating the water surface profile of a spillway, and the flows in stilling basin, shaft and many other discharge structures. Studies (Constantinescu et al., 2003; Karim & Mali, 2000; Wang et al., 2009; Wu & Ai, 2010 and Chen et al.,2010) show that the $k-\varepsilon$ two-equation turbulence model is a useful tool in simulating this kind of complex water flows. The flows in a stepped spillway have been simulated by using the VOF model combined with the standard $k-\varepsilon$ turbulence model (Chen et al. 2002). The VOF model and the Mixture model, respectively, together with the RNG $k-\varepsilon$ turbulence model have been used to simulate the air entrainment of flows in a stepped spillway (Cheng et al. 2006). The Mixture

---

*Project supported by the Academic and Technical Leaders Training Foundation of China' Sichuan Province (NO:2012DTPY020)

model and the realizable $k-\varepsilon$ turbulence model have been adopted to simulate the air entrainment of flows (Qian et al. 2009) and it was shown that the realizable $k-\varepsilon$ turbulence model is better than the RNG $k-\varepsilon$ model with respect to the air entrainment of flows. The energy dissipation characteristics of the hydraulic jump in a stilling basin of multi-horizontal submerged jets have been studied by using the VOF RNG $k-\varepsilon$ and Mixture RNG $k-\varepsilon$ turbulence model (Chen et al. 2010).

Considering that the traditional $k-\varepsilon$ turbulent model failed to simulate efficiently the strong anisotropy in high speed jet, the Renormalization Group (RNG), which was introduced by Charles, Yakhot and Orszag (Charles et al. 1991 and Yakhot & Orszag, 1992), was employed in the turbulent model. The continuity equation, momentum equation, k (turbulent kinetic energy) equation and the ε (dissipation rate of turbulent kinetic energy) equation are elaborated as follows:

the continuity equation:

$$\frac{\partial \rho}{\partial t} + \frac{\partial \rho u_i}{\partial x_i} = 0 \qquad (1)$$

the momentum equation:

$$\frac{\partial \rho u_i}{\partial x_i} + \frac{\partial}{\partial x_i}(\rho u_i u_j)$$
$$= -\frac{\partial p}{\partial x_i} + \frac{\partial \rho}{\partial x_j}[(v + v_t)(\frac{\partial u_i}{\partial x_j} + \frac{\partial u_j}{\partial x_i})] \qquad (2)$$

the k equation:

$$\frac{\partial(\rho k)}{\partial t} + \frac{\partial(\rho u_i k)}{\partial x_i}$$
$$= \frac{\partial}{\partial x_i}[\frac{v+v_t}{\sigma_k}\frac{\partial k}{\partial x_i}] + G_k - \rho\varepsilon \qquad (3)$$

the ε equation:

$$\frac{\partial(\rho\varepsilon)}{\partial t} + \frac{\partial(\rho u_i \varepsilon)}{\partial x_i}$$
$$= \frac{\partial}{\partial x_i}[\frac{v+v_t}{\sigma_\varepsilon}\frac{\partial \varepsilon}{\partial x_i}] + C_{1\varepsilon}\rho\frac{\varepsilon}{k}G_k - C_{2\varepsilon}\rho\frac{\varepsilon^2}{k} \qquad (4)$$

where ρ and μ are the average densities of the volume fraction and the molecule viscous coefficient, respectively, p is the modified pressure, and v_t is the turbulent viscous coefficient which can be derived from turbulent kinetic energy (k) and dissipation rate of turbulent kinetic energy (ε) as below:

$$v_t = C_u \frac{k^2}{\varepsilon}, \quad C_{1\varepsilon} = 1.42 - \frac{\eta(1-\eta/\eta_0)}{1+\beta\eta^3},$$
$$\eta = \frac{Sk}{\varepsilon}, \quad S = \sqrt{2\overline{S_{ij}}\overline{S_{ij}}},$$

Figure 1. 2-D View of the Numerically Simulated Region.

where $i = 1, 2, 3$ in each equation, namely, $\{x_i = x, y, z\}$, $\{u_i = u, v, w\}$, and j is the sum suffix. The values of η_0, β, C_u, $C_{1\varepsilon}$, $C_{2\varepsilon}$, σ_k, and σ_ε are 4.38, 0.012, 0.0845, 1.42, 1.68, 0.7179, and 0.7179, respectively.

The control-volume method (Patankar 1980) was introduced to discretize the partial differential equation, and the SIMPLER method, which has good convergence characteristics, was employed for numerical simulation (Qian et al. 2009). The non-slip boundary condition was used on the wall, the near-wall condition was given by the standard wall functions, and the normal velocity was set as zero on the wall (Charles et al. 1991). The VOF method was introduced to simulate the free surface, and α_q represented the volume fraction of q phase fluid: when $\alpha_q = 0$, it meant that there is no q in the controlling body; when $\alpha_q = 1$, full of q; and when $0 < \alpha_q < 1$, part of q. The total fraction was 1 in the controlling body, that is $\sum \alpha_q = 1$.

The simulated object in the mathematical model was combined with a practical project. The numerically simulated region is shown in Figure 1, where the number symbols respectively represent circle pipe inlet, transition sections, radial gate, transportation tunnel, wind complement tunnel, mid-gate chamber, anti-arc section, first aerator with airshafts, second aerator with airshafts, and outlet.

3 ANALYSES AND RESULTS

Four cases (see Table 1) were simulated, and Table 2 shows a comparison between the simulated and the experimental values at typical sections. It can be seen that the simulation gave consistent results to the actual experiment. Note that the air motion in flood discharge tunnel is invisible in experiment, but the visualization of the patterns of air motion in flood discharge tunnel was achieved by means of the simulation. So, the pattern of air motion in flood discharge tunnel was obtained.

Air moves to the downstream along with high-speed water flow in the flood discharge tunnel while air gets in from tunnel top remainder of outlet. The typical airflow characteristic in flood discharge tunnel is given in Fig. 2.

When the gate opening degree e < 0.5 (see Figure 2.a,b,c), the water depth is small, so the tunnel top remainder of the outlet is broad. Air would get in directly from the tunnel top remainder of the outlet to the junction of flood discharge tunnel and mid-gate chamber, forming three partial air rotary regions while

Table 1. Cases of Calculation and Experiment of the Water Discharge Tunnel.

Case	Orifice	Water level (m)	Water discharge (m³/s)
1	e = 0.33	1835.00	1144
2	e = 0.50	1835.00	1644
3	e = 0.75	1835.00	2275
4	e = 1.00	1835.00	2757

Table 2. Comparison between Simulated and Experimental Values of the Four Cases.

Case		Water level (m)		Water velocity (m/s)		Cavity length (m)	
		Gate	Outlet	Gate	Outlet	Shaft 1	Shaft 2
1	cal	2.63	2.24	36.25	42.56	44.62	20.80
	exp	2.56	2.19	37.24	43.53	45.00	20.63
2	cal	4.00	3.15	34.25	43.49	40.51	21.85
	exp	3.95	3.29	34.68	41.64	41.25	21.38
3	cal	6.41	5.32	29.58	35.64	35.78	22.94
	exp	6.38	5.24	29.72	36.18	36.25	22.50
4	cal	8.52	7.45	26.97	30.84	33.87	26.55
	exp	8.63	7.38	26.62	31.13	34.05	26.25

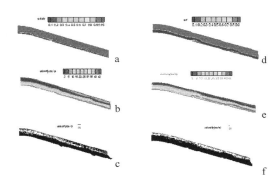

Figure 2. The typical airflow characteristics in flood discharge tunnel when e = 0.33 (a,b,c) and e = 0.75 (d,e,f).

the air on the tunnel top remainder moves against the air along with the high-speed water flowing to the downstream. The first region locates at reverse curve section in the flood discharge tunnel. The second is on the slope between the two aerators. And the third is seated on the slope between the second aerator and the outlet.

Moreover, when e > 0.75 (see Figure 2.d,e,f), there are still three partial air rotary regions. But the origin of the first region is different. Due to the larger discharge, the remainder of tunnel top in the outlet is relatively narrower. Air getting in from outlet cannot reach reverse curve, therefore, the direction of the airflow in reverse curve section is consistent with the direction of the water flow, forming a smooth area. After the smooth airflow reaches the crown, smaller partial air rotary region is formed in the junction of the flood discharge tunnel and mid-gate chamber.

From the experimental and simulated results, it can be known that the three rotary regions would be smaller when the flood discharge is larger.

4 CONCLUSIONS

In this article, combining the VOF model and the RNG $k-\varepsilon$ turbulent model, the characteristic of the air motion in the spillway tunnel with aerators was simulated. The simulated results agreed well with the experimental results in terms of water flow, length of cavity, water velocity and air velocity in the shafts. So, the visualization of the characteristics of air motion in flood discharge tunnel was achieved by means of the numerical simulation. Air moved to the downstream with high speed in the flood discharge tunnel from the mid-gate chamber while air got in from tunnel top remainder of outlet. Three partial air rotary regions formed in the spillway tunnel. The first region located at the reverse curve section in the flood discharge tunnel. The second was on the slope between the two aerators. And the third was seated on the slope between the second aerator and the outlet. They became smaller when the flood discharge became larger. On the other hand, the pattern of the air motion is instructive and meaningful to set up the ventilating shaft in spillway tunnel with aerators.

REFERENCES

Boes, R. M., Hager, W.H. 2003. Hydraulic design of stepped spillway. *J. Journal of Hydraulic Engineering, ASCE*, 129(9): 671–679.

Charles, S.G., Thomas, G.B., Neassan, F. 1991. An analysis of RNG base turbulent models for homogeneous shear flow. *J. Phys Fluids*, 1991, 3(9): 2278–2281.

Chen, J.G., Zhang, J.M. and Xu, W.L. et al. 2010. Numerical simulation investigation on the energy dissipation characteristics in stilling basin of multi-horizontal submerged jets. *J. Journal of Hydrodynamics*, 22(5): 732–741.

Cheng, X.J., Chen, Y.C., Luo, L. 2006. Numerical simulation of air-water flow on stepped spillway. *J. Science in China Series E: Technological Sciences*, 36(11): 1355–1364.

Constantinescu, G., Chapelet, M., Squires, K., 2003. Turbulence modeling applied to flow over a sphere. *J. AIAA Journal*, 41(9): 1733–1742.

Karim, O. A. K., Mali, K. H. M., 2000. Prediction of flow patterns in local scour holes caused by turbulent water jets. *J. Journal of Hydraulic Research*, 38(4): 279–287.

Liu, C., Zhang, G.K., Li N.W., et al. 2008. Influence of aerator on the side cavity length. *J. Journal of Sichuan University: Engineering Science Edition*, 40(1):1–4.

Patankar, S.V. 1980. Numerical Heat Transfer and Fluid Flow. New York: *Hemisphere Publishing Corporation and McGraw Hill Book Company*, 1980.

Pegram, G. S. 1999. Hydraulic of skimming flow on modeled stepped spillway. *J. Journal of Hydraulic Engineering, ASCE*, 125(5): 500–509.

Qian, Z.D., Hu, X.Q., Huai W.X. 2009. Numerical simulation and analysis of turbulent flow over stepped spillways. *J. Science in China Series E: Technological Sciences*, 39(6): 1104–1111.

Russell, S.O., Sheenan G.J. 1974. Effect of entrained air on cavition damage. *J. Canadian Journal of Civil Engineer*, 1(1): 68–86.

Rutschmannn, P., Volkart, P. 1988. Spillway chute aeration. *J. Water Power&Dam Construction*, 40(1): 10–15.

Wang, K., Jin, S., Liu, G. 2009. Numerical modeling of free-surface flows with bottom and surface-layer pressure treatment. *J. Journal of Hydro-dynamics*, 21(3): 352–359.

Wu, J.H., Ai, W.Z. 2010. Flows through energy dissipaters with sudden reduction and sudden enlargement forms. *J. Journal of Hydrodynamics*, 22(3): 360–365.

Xu, Y.M., Wang, W., Xu, W.L., et al. 2008. Calculation of the cavity length of jet flow from chute aerators. *J. Progress in Natural Science*, 2008, 18(5): 591–598.

Yakhot, V., Orszag, S.A. 1992. Development of turbulent models for shear flows by a double expansion technique. *J. Phys Fluids*, 4(7): 1510–1520.

One-dimensional consolidation settlement numerical analysis of a large-scale land reclamation project

Jie Zhao, Yuhao Jin & Jian Yi
Dalian University R&D Center of the Civil Engineering Technology, Dalian, China

ABSTRACT: For a large-scale land reclamation project, the numerical analysis of consolidation settlement is conducted to examine the soft soil foundation settlement based on Terzaghi one-dimensional consolidation theory. The post-construction residual settlement graph of the entire site is drawn to predict the development tendency of settlement. In addition, as there is measurement accuracy error in the domestic laboratory apparatus, necessary sensitivity analysis is conducted concerning the mechanical parameters that have a significant impact on one dimensional settlement, with the aim to obtain the influence rule of main parameters in one-dimensional calculation on settlement results. The results of the analysis show that the distribution of post-construction residual settlement of the entire site can be predicted well by using one-dimensional rock-soil consolidation settlement design based on Terzaghi consolidation theory. This calculation and analysis approach can also provide some reference for forecasting the settlement of other land reclamation project areas.

Keywords: e-p curve; coefficient of consolidation; soft soil; one dimensional settlement; numerical analysis

1 INTRODUCTION

Soft soil foundation is widely distributed in our country's coastal areas, such as Dalian, Qingdao and other places where coastal sedimentary facies prevail. The soft soil is mainly composed of silt and silty soil in terms of its engineering properties. Due to its characteristics of high water content, low strength, low bearing capacity, low permeability and long duration of deformation, settlement is very likely to appear in soft soil regions.

The foundation settlement can be divided into initial settlement and post-construction settlement. The magnitude of settlement, on one hand, depends on the dead load and load distribution of structure itself; and on the other hand, depends on the type and thickness of each foundation soil layer, the compressibility of the soil and the depth and distribution of groundwater, etc. The post-construction settlement of soft soil foundation, as an important standard to measure construction quality of land reclamation project, will affect the normal operation and service life of the whole project.

At present, due to the complexity of finite element program calculation model and theory, part parameters required by the 2-D Biot consolidation theory[one] cannot be determined by conventional experimental methods, which requires support from high quality soil samples and relevant measuring means. Moreover, compared to the traditional one-dimensional settlement analysis, it is more difficult to accurately select calculation model and calculation parameters for 2-D and 3-D finite element analysis, making them hard to be accepted by the general designers, thus limiting their use in engineering practice. Furthermore, for a large-scale land reclamation project, to draw a macro settlement graph for the entire site requires a great deal of computing cell, which is hardly bearable for any general computer. However, by using the traditional geotechnical theory for calculating multipoint one-dimensional settlement and the TECPOOT software for drawing, the macro settlement diagram of the entire site can be obtained.

In this paper, the numerical analysis of consolidation settlement based on Terzaghi one-dimensional consolidation theory is conducted for a large-scale land reclamation project. Analysis concerning the soft soil foundation settlement is made and the post-construction residual settlement graph of the entire site is drawn to predict the development tendency of the project's settlement.

2 BASIC PRINCIPLES OF CONSOLIDATION

According to the basic principles of Terzaghi one-dimensional consolidation theory[two], the seepage of pore water follows Darcy's law. The permeability coefficient and compression coefficient is constant. In the soil only the vertical compression deformation and vertical seepage of pore water occur. The load is applied instantaneously. In Terzaghi's theory, the residual pore water pressure effect on the dissipation and the diffusion of pore water pressure is considered, and it is assumed that the total stress and void ratio does not vary with time during the process of consolidation of soil without considering the seepage consolidation. The analysis results based on Terzaghi's consolidation theory for 2-D or 3-D is an approximate solution, which assumes that the total stress is constant in the counting process of one-dimensional consolidation settlement. While in the one-dimensional settlement analysis, the solution is accurate.

In the process of one-dimensional deformation settlement, the average total stress σ_z of soil equals to the ground load q and varies with q. It is independent of the deformation of the soil skeleton and the coupling effect of seepage of pore water. The equation of Terzaghi one-dimensional consolidation under the condition of gradually increasing load can be obtained as follows.

$$C_v \frac{\partial^2 p}{\partial z^2} = \frac{\partial p}{\partial t} - \frac{\partial q}{\partial t} \quad (1)$$

In the formula: $C_v = k(1+e)/\rho_w a_v$, where C_v is one-dimensional coefficient of consolidation of soil, e is the initial void ratio; k is the permeability coefficient; ρ_w is the water bulk density; a_v is the compression coefficient; and p is the pore water pressure.

The load q is applied almost instantaneously and invariable, so equation of Terzaghi one-dimensional consolidation can be simplified as

$$C_v \frac{\partial^2 p}{\partial z^2} = \frac{\partial p}{\partial t} \quad (2)$$

The total settlement of foundation consists of three parts: instantaneous settlement, primary consolidation settlement and secondary consolidation settlement [three]. For the primary consolidation settlement, the layered summation method [four, five] mainly include e-p curve method [2] and e-log p curve method [six]. The e-log p curve method can better reflect the characteristics of the soil itself as it considers the impact of soft soil stress history.

In the settlement calculation, when the clay is under the action of load, the formula of calculating primary consolidation settlement using e-p curve method is [7]:

$$S_c = \sum_{i=1}^{n} \frac{e_{0i} - e_{1i}}{1 + e_{0i}} \Delta h_i \quad (3)$$

where S_c is the primary consolidation settlement; n is the number of foundation layers in the calculation of foundation settlement; e_{0i} is the firm void ratio of layered midpoint on the layer ith without being applied the next load in soil ground; e_{1i} is the stable void ratio of layered midpoint on the layer ith with the load in soil foundation being applied; Δh_i is the thickness of layer i in the calculation of foundation settlement, which shall be 0.5–1.0 m preferably, and in this project 0.5 m is taken. In case of encountering natural stratigraphic boundary, the natural one will be taken as the boundary layer.

3 A PROJECT CASE

3.1 Project profile

It is a land reclamation project for an airport industrial park. The site to be reclaimed is a rectangular artificial island with the length of 6621.1 m, the width of 3328.3 m, and the area of about 20.87 km². It mainly consists of revetment area and backfilling land area. During site investigation 389 boreholes were dug with depth ranging between 35 m and 80 m. According to the survey data, the rock and soil in the site show a macro trend of relatively soft in upper part, plastic and hard plastic in central part, dense in lower part, and hard in the bottom.

This large-scale land reclamation project is divided into the following areas: revetment wave-walls area, the runway and apron area, terminal area, construction area, soil surface area and construction accesses. Except for the soil surface area and construction area, all the other areas will undergo shallow layer foundation treatment by using the method of mud replacement by blasting.

3.2 Calculation parameters and models

According to the borehole logs and rock-soil's physical-mechanical parameters provided in the geotechnical investigation report, one-dimensional consolidation settlement numerical analysis model is established (see Figure 1), through which the one-dimensional settlement volume is obtained.

The average physical and mechanical parameters of the rock in the area are used when the physical and mechanical properties of some boreholes are absent.

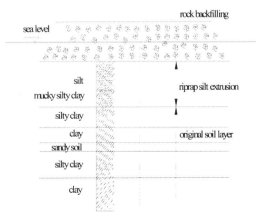

Figure 1. Calculation Model of Borehole Settlement.

The average physical and mechanical properties of land backfilling area are shown in Table 1. As the physical and mechanical parameters of the sand layer are missed in the report, they are taken by referring to similar projects in Dalian or other domestic areas.

The original elevation of soil surface of the construction area is between −6.0∼−4.6 m, and the water level in calculation adopts design low water level of −1.39 m. In the calculation, the sandy soil layer and the backfilled aggregate are used as drainage layer. Plan layout of areas and main boreholes are shown in Figure 2.

The geotechnical investigation report provides detailed experimental data of e-p curves, including void ratios e_i, compression modulus E_s, and compression coefficients a_v of the same soil layer at different depths of each borehole and at different stress stages. The following Figure 3 and 4 shows the profile of typical backfilling area and the profile of typical revetment area, respectively.

4 ANALYSIS OF COMPUTING RESULTS

The one-dimensional geotechnical settlement calculation model is established to compute the multipoint residual settlement of the entire site of 2 years, 5 years, 10 years, 20 years, 50 years and 100 years after construction respectively. The residual settlement graph of the site (see Figure 5) is drawn by putting the settlement data into TECPLOT software Meanwhile, two representative cross-sections: X4-X4 and Y3-Y3, are selected, for which the residual settlement graphs of different years are obtained (see Figure 6).

According to the results of Figure 5, it can be seen that the settlement extent of south construction area and the soil surface area are the biggest, while the runway and apron area are the smallest. The residual settlement of the soil surface and construction area of different years after construction is more significant than other areas. After a period after construction, the soil layers close to the ground surface will enter the stage of secondary consolidation, while the soil layer

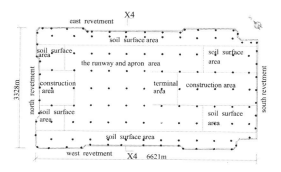

Figure 2. Plan Layout of Areas and Main Boreholes.

Figure 3. Profile of Typical Backfilling Area.

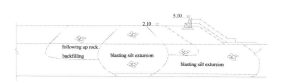

Figure 4. Profile of Typical Revetment Areas.

Table 1. Average Physical-Mechanical Property of Soil in Land Backfilling Area.

Area	Number of soil layer	Soil name	Moisture content ω (%)	Natural weight-specific density γ kN/m³	Vertical coefficient of consolidation C_v cm²/s	Horizontal coefficient of consolidation C_h cm²/s	Coefficient of secondary consolidation cm²/s	Compression modulus $E_{s0.1-0.2}$/ MPa
Land backfilling area	①	Silt	28.1	19.1	0.01	10	0.0005	12.10
	①₁	Silty clay mixed with sand	31.2	18.7	0.00447	0.00460	0.0003	5.70
	①₂	Mucky silty clay	42.2	17.6	0.00129	0.00169	0.0350	2.90
	①₃	Mud	63.7	16.0	0.00025	0.00025	0.0500	1.51
	①₄	Silt clay	47.8	17.2	0.00068	0.00080	0.0350	2.44
	①	Clay	39.1	17.9	0.00063	0.00083	0.0030	4.72
	②₁	Silty clay	25.1	19.6	0.00180	0.00238	0.0020	6.04
	②₂	Silt	20.2	20.8	0.01000	0.01000	0.0005	12.76
	③₁	Clay	28.7	19.1	0.00139	0.00175	0.0030	6.90
	③₂	Silty clay	24.6	19.7	0.00253	0.00300	0.0020	7.39
	③₃	Clay	27.2	19.3	0.00124	0.00161	0.0030	8.05

Residual settlement of 2 years after construction

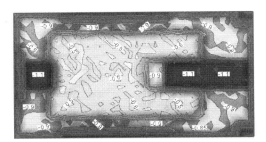
Residual settlement of 20 years after construction

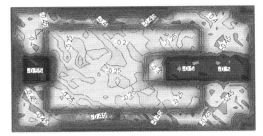
Residual settlement of 5 years after construction

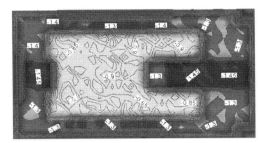
Residual settlement of 50 years after construction

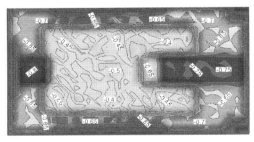
Residual settlement of 10 years after construction

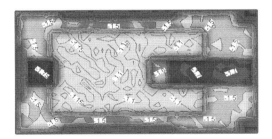
Residual settlement of 100 years after construction

Figure 5. Residual Settlement of the Site at Different Years after Construction.

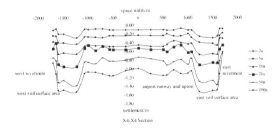

Figure 6. Settlement of X4-X4 Section after Construction.

below will still stay at the primary consolidation stage. Along with development of consolidation, after 100 years after construction, the maximum residual settlement of the soil surface area, building area, southern terminal area, runway and apron area, and revetment area will be 1.502 m, 1.752 m, 1.648 m, 1.150 m, and 0.912 m, respectively. The main reasons behind the different settlements in the site are as below: different areas of the site have different upper loads; the compression modulus and coefficient of consolidation of different soil layers vary greatly; the thicknesses of the same soil type in each area is different; in the areas where the sandy soil layer is thick or there are many layers of sandy soil, the settlement will be very small due to large compression modulus of sand layer. In addition, the way of foundation treatment is different: the shallow foundation treatment using mud replacement by blasting is applied in the revetment wave-walls area, runway and apron area and terminal area, but not in the construction area and the soil surface area. The coefficient of primary consolidation of backfilled aggregate after shallow treatment will be two or three orders of magnitude higher than that of muddy soil's coefficient of consolidation before treatment.

As shown by Figure 6, the post-construction settlement of the same cross section in different areas is significantly different; the settlement of revetment, runway and apron area are relatively small, while the terminal area, construction area and soil surface area is larger. Along with the growth of time the trend

Table 2. Post-construction Settlement of Various Years after Multiple Increase of Coefficients of Primary Consolidation.

Borehole No.	Location	Coefficient of primary consolidation	Post-construction Settlement (unit: m)					
			2 years	5 years	10 years	20 years	50 years	100 years
P3	Revetment area	The original data	0.103	0.235	0.411	0.729	1.293	1.752
		2 times	0.453	0.660	0.902	1.269	1.861	2.315
		3 times	0.562	0.816	1.087	1.469	2.069	2.467
T31	Runway area	The original data	0.074	0.155	0.256	0.410	0.674	0.849
		2 times	0.417	0.537	0.675	0.861	1.098	1.179
		3 times	0.474	0.608	0.752	0.937	1.135	1.187
H11	Backfilling area	The original data	0.070	0.162	0.295	0.549	1.090	1.603
		2 times	0.577	0.735	0.951	1.312	1.968	2.395
		3 times	0.661	0.872	1.146	1.570	2.215	2.507

Table 3. Consolidation Degree at Various Depths after 100 Years after Multiple Increase of Coefficients of Primary Consolidation.

Borehole No.	Location	Coefficient of primary consolidation	Degree of consolidation				
			5 m	10 m	20 m	30 m	40 m
P3	Revetment area	The original data	0.852	0.724	0.527	0.417	0.379
		2 times	0.936	0.879	0.792	0.741	0.724
		3 times	0.972	0.947	0.908	0.886	0.878
T31	Runway area	The original data	0.970	0.939	0.895	0.903	0.949
		2 times	0.997	0.994	0.990	0.991	0.995
		3 times	0.999	0.999	0.997	0.998	0.999
H11	Backfilling area	The original data	0.807	0.689	0.642	0.723	0.870
		2 times	0.945	0.912	0.898	0.921	0.963
		3 times	0.984	0.975	0.971	0.977	0.989

becomes more and more prominent. The main reasons are as follows: different areas have different backfilling heights, leading to relative great difference in their upper loads; the shallow foundation treatment using mud replacement by blasting is applied in the revetment, runway and apron area. In the late period of consolidation settlement, as the pore water pressure begins to decrease and the settlement rate becomes slow, the secondary consolidation settlement begins to play a leading role. As the dimension stones in the revetment, runway and apron areas subject to mud replacement and the muddy soil in the soil surface area and construction area have different coefficient of secondary consolidation, the secondary consolidation settlement produced will also vary greatly.

5 SENSITIVITY ANALYSIS OF PARAMETERS

Because the accuracy of settlement analysis will be affected by the complexity of rock body's mechanical property change, the selection way of soil sample, and the error of geotechnical experiment, sensitivity analysis is conducted for the parameters that have relatively large impact on the soil body's settlement.

The coefficients of primary consolidation of representative boreholes are selected for parameters sensitivity analysis. According to the practical engineering experience, the horizontal coefficient of consolidation and the vertical coefficient of consolidation are enlarged by double and triple times respectively. The quantity of consolidation settlement (see Table 2) in different years after construction is obtained. Meanwhile the consolidation level of backfilled foundation at different depth after 100 years of construction is outputted (see Table 3). From Table 2 and Table 3, it can be seen that the coefficient of primary consolidation has significant effect on consolidation degree and primary consolidation settlement. As the depth of the soil layer increases, the level of consolidation will become smaller and smaller, and the time required for secondary consolidation will become longer and longer. When the compression modulus and other parameters remain constant, the increasing of coefficient of primary consolidation will lead to more complete consolidation of soil layer in different depths. Consequently, the primary consolidation settlement of different soil layers will increase, too, leading to the total increase of settlement after construction.

6 CONCLUSION

The soft soil is treated by shallow foundation treatment method. In the early stage of consolidation settlement of soft soil foundation, the coefficient of consolidation directly affects the size of the initial consolidation settlement; while in the later stage of consolidation settlement, as the pore water pressure begins to decrease and the settlement rate becomes slow, the secondary consolidation settlement begins to play a leading role. As the various soil types have different coefficient of secondary consolidation, the secondary consolidation settlement produced will vary greatly over time.

Under the condition of domestic existing technology, the finite element program cannot completely replace the classical analysis method. To obtain the macro settlement graph of the entire site of a large-scale land reclamation project, and meanwhile avoid the uncertainty of parameters in finite element program and finite difference program as well as the computational burden, the rock-soil consolidation settlement analysis method on the basis of Terzaghi consolidation theory is adopted in this paper to obtain the residual settlement after construction, and the settlement graph of the entire site is achieved by using TECPLOT drawing software.

REFERENCES

[1] Xu Shuping, Liu Zude. The Use of Biot's Consolidation Theory in Surcharge Preloading Engineering Field [J]. *Geotechnical Mechanics*, 2003, 24(2):307–310.

[2] Gong Xiaonan. Geotechnical Computer Analysis [M]. *Beijing: China Building Industry Press*, 2000.

[3] Zhang Mengxi. The Principles of Soil Mechanics (Second Edition). *Wuhan: Huazhong University of Science and Technology Press*, 2010:111–112.

[4] Wang Yong. The Analysis of Calculating Foundation Settlement by Using Layer-wise Summation Method [J]. *Shanxi Architecture,* 2010, 36(13):84–87.

[5] Xu Jinming, Tang Yongjing. Some Improvements in Calculating Settlement by Using Layer-wise Summation Method [J]. *Geotechnical Mechanics*, 2003, 24(4):518–521.

[6] Hao YuLong, Jiang Dongqi, Chen Yunmin. The Surcharge Preloading Consolidation Theory and Application of Double-layered Foundation

[7] JGJ79-2002. Code for Technology of Building Foundation Treatment [S]. *Beijing: China Building Press*, 2002.

Optimization of clapboard with wind box type construction ventilation in super-long railway tunnel

Zhengmao Cao, Qixin Yang & Chun Guo
Key Laboratory of Transportation Tunnel Engineering, Ministry of Education, School of Civil Engineering, Southwest Jiaotong University, Chengdu, China

ABSTRACT: For the construction ventilation problems of double tunnels district in single-track super-long railway tunnel, the different working conditions of construction ventilation schemes were calculated by three-dimensional numerical simulation method, and a reasonable ventilation scheme was determined to provide technical support for the construction ventilation of similar projects. With the new Guanjiao tunnel located on Xining-Golmud railway for case study, numerical simulation of flow field in wind box under different working conditions was conducted via fluid calculation software FLUENT, the ventilation efficiencies of different working conditions were obtained, the influence length of jet fans was calculated and analyzed, and a reasonable arrangement of jet fans was determined. It was shown that the ventilation efficiency of the existing scheme is lower than those of other improved schemes, due to the short length of the wind box and the strong mutual influence of axial fans. The average ventilation efficiency improvement of scheme 2 was 18.35%, which is the highest of all schemes. The jet fans on the clapboard in incline shaft should be at least 50 m away from the tunnel portal, and the spacing of jet fans should be more than 150 m. According to existing equipment condition, the spacing was determined as 400 m.

Keywords: railway tunnel, wind box, construction ventilation, ventilation efficiency

The construction ventilation is an important factor that influences the progress of long tunnel construction. Pressed ventilation in heading face can meet the ventilation requirements in shorter tunnels, while for longer tunnels, tunnel ventilation is conducted by excavating auxiliary channel or ventilation vertical shaft, or increasing inclined shafts to shorten the construction length to conduct auxiliary ventilation. In some cases, because of the terrain conditions, the ventilation inclined shaft is too long to meet the construction ventilation requirements of several working faces simply by single head ventilation (Xi, Z. 2007). In order to solve this problem, a new method of inclined auxiliary ventilation, which is inclined shaft and separated tunnel with wind box joint ventilation technology, was adopted (Haifeng, C. & Qixin, Y. 2011). Based on the actual situation of the new Guanjiao tunnel, which is located on the Xining-Golmud railway, the construction ventilation scheme of inclined shaft 7 construction section was optimized (Zhengmao, C. 2011).

1 PROJECT OVERVIEW

The new Guanjiao tunnel is 32.645 km in length, at an altitude of 3300 m on the Qinghai-Tibet railway from Xining to Golmud, across the south mountain of Qinghai. It is a rare high-altitude super-long railway tunnel in the world at present. The spacing between tunnel I and tunnel II is 40 m. From DK 280 + 650, tunnel is in the continuous uphill with the slope of 8‰, and from DK 295 + 700, with the slope of 9.5‰ in the downhill, there are 10 inclined shafts to auxiliary construction for main tunnel. The area of the tunnel is in special natural conditions, which is semi-arid Tibetan Plateau sub-arctic climate. Unlike plains regions, there are many special features such as high altitude and coldness, hypoxia, sparse population, sandstorm, drought and so on.

Inclined shaft 7 is to the right of Guanjiao tunnel line II, the point of intersection with the main tunnel mileage is DyK 300 + 210, the angle is 44°04′27″, and the elevation of rail surface is 3446.77 m. The length of inclined shaft is 2248 m, with the slope of 10.56%. There are 6658 m of construction task for inclined shaft 7, including 3278 m for tunnel I, and 3380 m for tunnel II. The distance of single head ventilation in the entrance is 4515 m, and the air volume in the excavation face of trackless transport is large.

2 EXISTING VENTILATION SCHEME

The interior space of inclined shaft is divided into two parts by clapboard. The space above the clapboard is for fresh air to flow into the tunnel, and the blow is

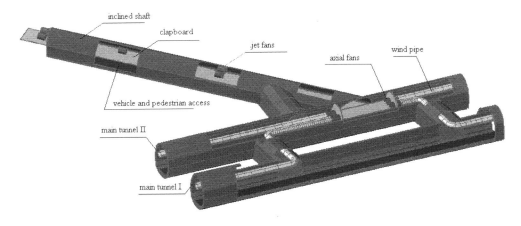

Figure 1. Existing construction ventilation scheme.

Table 1. Parameters of ventilation equipment.

Type	Specifications	Technical Parameters			Quantity
		Speed type	Wind pressure Pa	Air volume m³/min	Power kW	
Axial fan	SDF(C)-No12.5	High-speed	1378~5355	1550~2912	110 × 2	4
		Medium-speed	629~2445	1052~1968	34 × 2	
		Low-speed	355~1375	840~1475	16 × 2	
	SDF(B)-No14		1078~6860	2113~4116	185 × 2	2
Wind pipe	PVCΦ1800 mm	The average one hundred meters leakage rate is 0.15%, and frictional resistance coefficient is 0.02				6658 m
Jet fan	SSF-No12.5	55 kW				8

for exhausting polluted air, pedestrian, and vehicular access. Jet fans are installed in the above space to increase the fresh air pressure.

Six jet fans are set on the clapboard at 0 m, 600 m, 1200 m, and 1800 m of the inclined shaft, which is SSF-No12.5, 55 kW. There are 4 SDF(C)-No12.5 axial fans in the main tunnel, with 4 ventilation pipes to the working faces. Polluted air is discharged through the channel of inclined shaft under the clapboard Two axial fans are set at the intersection of inclined shaft and main tunnel, and two jet fans are set in the inclined shaft to accelerate the exhaust gas. The existing ventilation scheme is illustrated in Figure 1. The construction ventilation equipment is shown in Table 1.

According to the on-site construction ventilation test, there is little fresh wind around the working face of the tunnel; the construction environment is bad in the tunnel. Analysis show the reason is that due to defects in the form of wind box set, the length of the wind box is too short, thus each axial influences greatly because of the short distance, and there are several bending of the wind pipe in tunnel II (Lei, H. 2005). The wind resistance is increased, which leads to low ventilation efficiency. According to the existing ventilation system deficiencies improvements of the wind box should be done to improve the ventilation efficiency.

3 NUMERICAL SIMULATION MODEL

3.1 Model building

The computational fluid dynamics software FLUENT was used to conduct model building; the width of the inclined shaft is 8m, the width of each main tunnel is 5.9 m, the width of the cross-channel is 6.6 m, the spacing between the two main tunnels is 40 m, and the height of the wind box is 2.5 m. According to the actual situation and the results of the theoretical analysis, three schemes were proposed and compared with the existing scheme. For the existing ventilation scheme, the length is 20 m, the width is 5.8 m, the height is 2.5 m, the power of axial fan 1 and axial fan 2 is 110 × 2 kW, and that of axial fan 3 and axial fan 4 is 185 × 2 kW. Figure 2 below shows the model views.

3.2 Meshing

Considering the computational stability (Xiujuan, Z. 2009), convergence and computing scale, three-dimensional unstructured grid was adopted. Compare with the structured grid, the calculation process of unstructured grid is more complex, but the local encryption is easier (Zhanzhong, H. 2004). There is

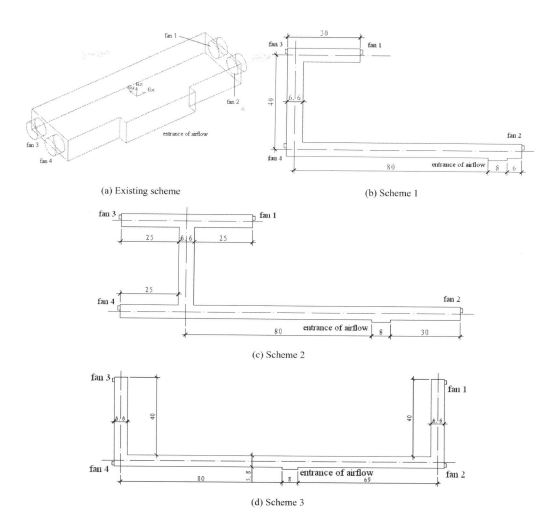

Figure 2. Various ventilation schemes.

high air flow rate and pressure gradient near the axial fans, its nearby grid was encrypted. When meshing, the most important thing is to determine the length of the unit, as the size of the unit will directly affect the number of mesh and the mesh density; the number of grid computing is nearly 300,000.

3.3 Boundary conditions and parameter settings

Before numerical simulation, assuming the air is an incompressible fluid, using $k-\varepsilon$ two-equation model of turbulence in an incompressible constant fluid appropriate simplification of airflow was done, and rational calculation parameters were designed (Yaohua, Z. & Jiapeng, H. & Hongming, F. 2007).

(1) Airflow directions of inlet and outlet in the axial fans are in the axial direction.
(2) The airflow is incompressible Newtonian fluid.
(3) The affect of gravity on the flow field is ignored in calculation.

The boundary conditions are set as follows:

(1) The entrance of air flow in inclined shaft 7 is pressure inlet. The inlet pressure is defined as 0, which is relative atmospheric pressure.
(2) The outlet of the axial fans is set as velocity inlet; according to the fan parameters, the wind velocity of fan 1 and fan 2 is -25.8 m/s, and that of fan 3 and fan 4 is -36.4 m/s.
(3) The surface of the wind box is the wall. The wall roughness constant of wind box is $C_K = 0.5$, and roughness thickness is $K_s = 0.01$.

4 RESULTS AND ANALYSIS

Fluid calculation software FLUENT was used to conduct three-dimensional numerical simulation. The calculation results of effective wind pressure and axial velocity at the center of the end face of the fan were obtained under every working condition.

Under a certain rotational speed, when the air volume and pressure changes, the ventilation efficiency of the fan also changes. There must be a value of maximum efficiency, which is called the best condition. Delivering a quantity of air at a certain pressure, the fan needs to consume the energy provided by electric motor (Liying, X. & Mengli, G. 2008). Energy consumption per unit time is called the power N, which is expressed in kW.

The ratio of effective power to the shaft power is the ventilation efficiency of the fan.

$$\eta = \frac{N_y}{N} \times 100\%, \quad (1)$$

where η is the ventilation efficiency of the fan; N is the shaft power of the fan, kW; N_y is the effective power of the fan, kW.

Under a certain rotational speed, when the air volume and pressure change, the effective power of the fan also changes (Shuxun, Z. & Qing, Z. 2010).

$$N_y = \frac{QH}{102}, \quad (2)$$

where Q is the effective air volume, m³/s; H is the effective wind pressure generated by the fan, mmHg.

4.1 Calculation results

CFD computational analysis of the model was conducted. When the iteration convergence, the calculation results of effective wind pressure and axial velocity at the center of the end face of the fan in the wind box were obtained under every working condition. Figure 3 shows part of the velocity contour and pressure contour at the sections in the wind box.

The length of the wind box is only 20 m in the existing ventilation scheme, and the power of the four axial fans is high. When the fans are working, both ends of the ventilator draw in fresh air in opposite directions. As the fans' longitudinal spacing is too short, the influence of each fan lead to the low ventilation efficiency, and the effective wind pressure of the fans reduces. Although the power of two parallel fans is the same, the influence exists for the fresh air enters from the sides.

4.2 Analysis of calculation result

The relative pressure of airflow entrance of each fan was read under every working condition, as shown in Table 2.

Figure 4 shows the intake airflow pressure of each fan under every working condition.

The calculated intake airflow pressures of the three schemes are less than that of the existing one. Table 3 and Figure 5 show the ventilation efficiency improvement of each fan comparing with the existing scheme under every working condition.

(a) Pressure contours of existing scheme

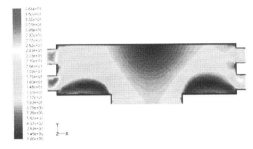

(b) Velocity contours of existing scheme

(c) Pressure contours of scheme 2

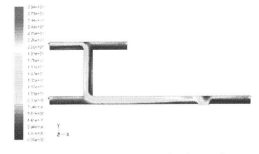

(d) Velocity contours of scheme 2

Figure 3. Flow fields in the wind box.

Figure 5 shows that the ventilation efficiencies of fan 1 and fan 3 improved more than fan 2 and fan 4. It is suggested that with scheme 2 adopted, the power of fan 1 and fan 2 is 110 × 2 kW, and that of fan 3 and fan 4 is 185 × 2 kW.

Table 2. Relative pressure of each fan at the airflow inlet.

	Existing scheme	Scheme 1	Scheme 2	Scheme 3
Fan 1	−7.52	−299.52	−267.91	−143.22
Fan 2	−53.21	−73.16	−99.65	−87.69
Fan 3	−6.16	−237.96	−326.98	−263.60
Fan 4	−70.88	−158.60	−192.69	−100.78

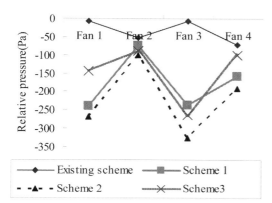

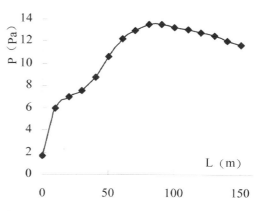

Figure 6. Average static pressure at the longitudinal cross-section.

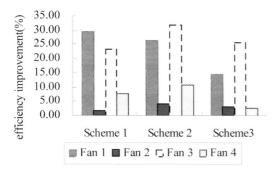

Figure 4. Airflow pressure of each fan under every working condition.

Table 3. Efficiency improvement of each fan.

	Scheme 1	Scheme 2	Scheme 3
Fan 1	29.44	26.41	14.45
Fan 2	1.81	4.21	3.12
Fan 3	23.37	31.90	25.83
Fan 4	7.82	10.86	2.67
Average value	15.61	18.35	11.52

Figure 5. Efficiency improvement of each fan compared with the existing scheme.

5 ANALYSIS OF JET FANS INFLUENCE

In order to optimize the arrangement of the jet fans on the clapboard, numerical simulation of the flow field in the inclined shaft was conducted.

The jet pressurized effect of fans play an important role in the jet, and the effect is closely associated with the jet energy conversion process. In the segment of jet induction, induced effects and pressurized effect exist simultaneously (Zhen, Z. & Yimin, X. 2010). The cross-section pressure reaches a maximum in the phase where the induction segment is completed. Figure 6 shows the changing process of average static pressure at the cross-section along the jet direction from the induction segment.

The figure shows that the minimum pressure is at the starting cross section of induction, and with the gradual increase of jet pressure, the maximum pressure is generated in the stage of the completion of the induction. The air flow in this segment is in reverse pressure flow state, which is one of the most significant characteristics of jet in the induction segment. Then the airflow flow enters the pressured ventilation section, where the pressure distribution declines slowly. The cross-section average static reaches the maximum when $L = 90$ m along the jet direction. Figure 7 shows the velocity vector graphics of cross-section in inhalation segment.

The sphere of jet fan inlet influence is small in the tunnel. When $L = 1$ m away from the inlet of fan, strong suction effect occurs near the inlet of fan. With the distance increases, the wind velocity distribution at the section affected by the inhalation effects is significantly reduced (Yong, F., Yu, L., Yanhua, Z., Yurong, Y. & Yuchun, Z. 2009).

Based on previous field measurement, model test and numerical simulation method were used to analyze the length of the jet fan suction, which is generally not more than 10 m.

Therefore, when using pressurized jet fan during the construction ventilation, considering 40∼50 m airstream affected at the entrance of the tunnel (Xiujun, Y., Xiaowen, W. & Jianzhong, C. 2008), the jet fans should be installed at least 50 m away from the entrance of tunnel, and the spacing of each jet fan should be more than 150 m, in this case the jet fan can produce a better guide, boost effect.

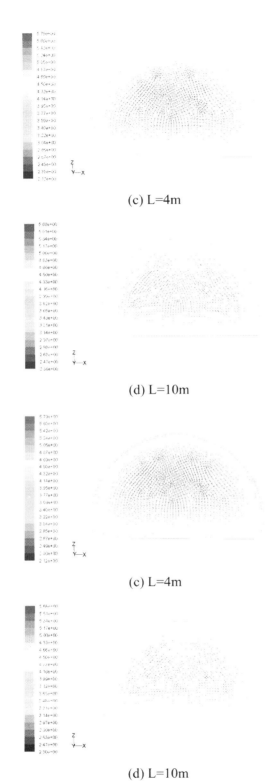

(c) L=4m

(d) L=10m

(c) L=4m

(d) L=10m

Figure 7. Velocity vector graphics of cross-section in inhalation segment.

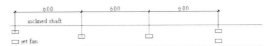

Figure 8. Jet fans layout of existing scheme.

Figure 9. Suggested jet fans layout.

Taking the length of inclined shaft 7 (2248 m) into account, Figure 8 shows the existing layout of jet fans above the clapboard.

According to the results of numerical simulation and theoretical analysis, Figure 9 shows the changed layout of jet fans above the clapboard in the inclined shaft 7.

6 CONCLUSIONS

Through numerical simulation and analysis of calculation result, with new Guanjiao tunnel which is located on Xining-Golmud railway for case study, research was conducted on the construction ventilation scheme of the construction zone in inclined shaft 7. The regularity of axial fan ventilation efficiency varying with each factor and the influence length of jet fan in the tunnel were obtained. The adjusted scheme improved the ventilation efficiency of axial fan and the construction ventilation effect in the tunnel. The following conclusions were made.

(1) The ventilation efficiency of the existing scheme is lower than those of other improved schemes, because of the short length of the wind box and the strong mutual influence of axial fans.
(2) The ventilation efficiencies of the three improved schemes are higher than that of the existing scheme, and the ventilation efficiency of fan 1 and fan 3 are improved more significantly than fan 2 and fan 4.
(3) The average ventilation efficiency improvement of scheme 2 is 18.35%, which is the highest of all schemes. It is suggested to adopt the wind box structural form of scheme 2 to optimize the construction ventilation.
(4) In order to produce a better guide, boost effect, jet fans above the clapboard should be installed at least 50 m away from the entrance of tunnel, and the longitudinal spacing of each jet fan should be more than 150 m. According to existing equipment condition, the longitudinal spacing of each jet fans above the clapboard in the inclined shaft was as 400 m.

ACKNOWLEDGEMENT

This research was supported by the National Natural Science Foundation of China (No. 51108384).

REFERENCES

Haifeng, C. & Qixin, Y. 2011. Probe into Air Cabin Ventilation in Long and Large Tunnels Construction, *Technology of Highway and Transport, 2011(5):124–128.*

Lei, H. 2005. Analysis on flow characteristic of key basic unit in ventilation network of the underground excavation cluster of hydropower station, *ChongQing: ChongQing University.*

Liying, X. & Mengli, G. 2008. Experimental Study on the Efficiency of Jet Flow Ventilation, *Modern Tunnelling Technology 45(3):50–53.*

Shuxun, Z. & Qing, Z. 2010. Performance evaluation based on the overall efficiency of the main mine fan, *Coal Mine Safety (12): 102–104.*

Xi, Z. 2007. Structure Optimization of Axial-flow Fan In Air-Condition Based on CFD, *Shang Hai: Dong Hua University.*

Xiujuan, Z. 2009. The key skills of meshing finite analysis elements, *Mechanical engineering and automation, 185–186.*

Xiujun, Y. & Xiaowen, W. & Jianzhong, C. 2008. Study on minimum longtitudinal distance of jet fan in ventilation of highway tunnel, *Journal of Chongqing Jiaotong University (Natural Science) 27(1): 40–164.*

Yaohua, Z. & Jiapeng, H. & Hongming, F. 2007. The study of subway fire CFD simulation boundary conditions, *Journal of engineering thermal physics, 28(2):310–312.*

Yong, F. & Yu, L. & Yanhua, Z. & Yurong, Y. & Yuchun, Z. 2009. Study on the Installation of Jet Fans in a Three-lane Highway Tunnel, *Modern Tunnelling Technology 46(2): 90–98.*

Zhanzhong, H. 2004. FLUENT engineering simulation examples and application, *Beijing: Beijing Institute of Technology Press.*

Zhen, Z. & Yimin, X. 2010. Discussion of the optimal longitudinal spacing for jet fans in subway, *Railway Standard Desing. S(2): 76–79.*

Zhengmao, C. 2011. Study on the Technology of Operation Ventilation and Disaster Prevention of Guanjiao Railway Tunnel. *Chengdu: Master's thesis of Southwest Jiaotong University.*

Participation in urban planning

Y.Y. Liu, Z.C. Zhang & Dashan Gao
Southwest University of Science and Technology, Mianyang, Sichuan, China

Mingyan Wang
Huaneng ZhaoJiao Wind Power Co. Ltd., Chengdu, Sichuan, China

ABSTRACT: When the Chinese translation of Jane Jacob's classic "The Death and Life of Great American Cities" was published after nearly half a century, a vigorous reading classic movement began. An increasing number of domestic scholars begin to review the modern urban planning in Jacob's people-oriented "civilian perspective". In addition, China is undergoing in the turn of the economic transition, the industrial restructing, the urban expansion and urbanization accelerated. The traditional urban planning featuring "elite decision-making" is facing an unprecedented challenge. The term "public participation" is gradually gaining popularity in urban planning. Yet the current comments to that are overwhelmingly positive and its authenticity needs to be investigated. This paper gives the investigation and analysis about the current public participation in its implementation and puts forward some constructive suggestions.

Keywords: public participation; urban planning; dialectical analysis; China

1 INTRODUCTION

1.1 *The definition of public participation*

Public participation is a coordinated strategy in cases of social stratification, diversity of public demands and intervention of interests groups. It emphasizes the public participation, decision-making and management in the process of urban planning and management[1]. The aim of public participation in urban planning is to encourage citizens, especially those who are influenced by the content of the planning to participate in compilation and discussion in the process of planning. Meanwhile, the planning department should hear public opinions of all kinds and take them into consideration in decision-making, making them an integral part of planning decisions[2]. To put it in a plain way, public participation in urban planning is the direct involvement of citizens in urban planning formulation and implementation process through certain methods and procedures.

1.2 *The implementation of public participation*

Public participation mainly includes two aspects. One is policy orientation. The other is for people to participate in. Public participation involves the citizens' awareness of the urban planning and their individual quality. As policy makers of city construction, the government should guide planning properly and orderly, improve the public awareness of their own rights, and advocate people's awareness of effective feedback to planning through a variety of ways. Finally, a "bottom-up" planning will be formulated after the planning departments communicate with the public to improve urban planning scheme.

2 THE CURRENT SITUATION AND EXISTING PROBLEMS OF PUBLIC PARTICIPATION IN CHINA

2.1 *The current situation of public participation in China*

Project planning includes several stages: the early stage planning, planning, examination and approval, modification, and implementation. At present, public participation in China mainly exists in the planning process and focuses on the presentation and evaluation of design scheme while in other stages public participation is limited and incomplete. In the early stage, planners usually carry on certain investigations to understand the status quo using questionnaire, interviews and so on. But this is only a kind of "passive" participation which has not really aroused the enthusiasm of people; therefore, it cannot reflect the true idea of the public effectively. As to public participation in planning approval stage, it undertakes mainly by expert argumentation and examination and approval of local people's congress, which confines the participants solely to those from academic research institutions and local political circle. Public participation in planning implementation stage targets more serious problems which are caused by certain urban

Table 1.

Planning stage		Public participation
Planning establishing stage	Start working	Completely without
	The early draft of the proposed	Completely without
	The consultation of the draft plan	Involved; Revised draft stage there is no guarantee of participation
	The consultation of the draft plan environment report	Some special planning
Planning determining stages		No; Only some expert
Planning implementation stage	The publication of the planning	No; only the query
	The consultation in the process of planning emission	Part of the project
	The consultation of the draft plan of the project environment report	Part of the project
Planning revision plans	Planning in assessment phase	Possible, the specific location of interested party
	Modify the startup phase	–
	Modified scheme establishment phase	–

construction activities which affect people's work, life and study[3].

2.2 The problem of public participation in China

2.2.1 The one-way and passive public participation

On the one hand, most of our urban planning enacted from top to bottom is controlled by government. Public participation activities are launched by the planning department which owns the dominant power. For example, visiting the masses, giving out questionnaire survey, interviewing, and other forms of public participation are launched by the planning department. The interviewed residents simply answer the questions and the independent participation consciousness is still weak. On the other hand, owing to the lack of effective mechanism for information collection, subsequent adoption screening, and tracking, public participation information feedback is insufficient, not to mention "proposed–feedback–and then put forward" mechanism, the information flow system is hindered. With only a little information gathering, it is hard for the public to directly participate in policy establishment and implementation.

2.2.2 The discontinuity of public participation in the later implementation

The document – CHARTER OF MIACHU PICCHU declared that: "Planning is a dynamic process, including not only planning formulation, but also the implementation of the plan." [4] The same process goes to public participation in urban planning, which should also be a dynamic and continuous process. But the current situation is that the public participation only exists in the preparation of planning and the final part. Participation in the intermediate process is almost blank and public participation is in coherent (just as the Table 1).

3 THE CASE ANALYSIS

3.1 Description

In 1959, Boston's "Big Dig" project cross over the city's central artery buried, reshaped the city ground space and formed nearly 1 square kilometers of urban green space. This project not only solved the long-standing disturbing traffic problem in the city, but also combined the residential, commercial and green space with urban corridor perfectly. While the construction of this 12.6 km central artery cost nearly $15.9 billion, it is a vivid case in the whole process of public involvement.

At the initial stage of the project, there were 66 social organizations including different groups participated in the design, such as students, social labor groups, and experts. They came up with a variety of forms to express their ideas. Some use painting to give their design ideas while others focus on public space network and the housing problem. During the planning permission, public participation was mainly about public review and hearings. Three versions of "environmental impact assessment" were distributed among supermarkets, libraries, post offices and so on for a month in order to obtain the public assessment. Finally, in "the environmental impact assessment amendment" absorbed 264 public comments[5]. The last stage of the overall design was that the expert scheme combined with public opinion and four schemes were formed.

Public review is still effective in the process of project implementation. When the 3-D model revealed before constructing, the plan immediately prompt public opposition. The result is to modify the plan to satisfy all concerned parties. As a result, the project lost as much as $23 million because of delay. When the underground engineering project was coming to an end, the 12.6 kilometers project was divided into 23 small plots, each plot in the planning of use was posted on line in detail.

In spite of the large budget, the Massachusetts Turnpike Authority chairman Matthew J. Amorello said: What Boston gets is the bridge which boasts the landmark of the beautiful city, rather than a few decades of regret.

The "BIG DIG" cost more than 200% of the budget. On the other hand it shows that the public participation is not the more the better, it should guarantee proper participation and effective coherence.

3.2 Case study

From the "tunnel" project, we can conclude the following characteristics for further reference:

1. Sound services and guidance institutions. American governmental organizations play a great role in guidance and organization. Government agencies'

guidance in the planning promote the smooth and effective follow-up of public participation, and the adoption of public opinion and public participation in subsequent planning are also an important factor to promote the process.

2. Good awareness of public participation. Americans have strong awareness of safeguarding the legal rights and active initiative in public participation. The Non-government organizations (NGOs) are often founded by the masses spontaneously and these spontaneous formations of the organization system have clear functions and are also highly-professional. And they also adopt a bottom-up mechanism, thus enjoy high participation and comprehensive.

3. Continuous tracking feedback mechanism. Despite the high degree of public participation in the previous work, it will not have any effect if the advices are not adopted in the follow-up plan, modification and implementation process. Therefore, that's why the continuous tracking feedback is so indispensable in achieving the integrity of public participation.

4 SUGGESTIONS FOR IMPROVEMENTS

Despite the differences in political background and cultural tradition in different countries, the content depths, breadth of public participation are correspondingly different. But on the whole, the public participation should run through the whole process of urban planning.

4.1 *The establishment of community organization*

Generally speaking, communities are the representative of the grassroots where the possibility of public participation in planning and decision-making is often higher than that in the higher level. The establishment of community organization is a good solution. The counterpart of this kind of organization in china is the sub district office. Because everybody wants to maximize their own interests, screening public opinions and feedback effectively are crucial.

4.2 *To achieve a variety of values*

The decisions made by municipal government and experts are not always correct and cordial with the interests of the people. Once assured that the experts' judgment is the core, we cannot avoid sacrificing some interests of other groups of the society. In this way, however the urban planning cannot get the support of the community from all walks of the society. On the other hand, the urban planning is operated on the basis of the expert's choice and judgment, and cannot forecast the public utility; sometimes perfectly planned new areas are not as satisfactory as the old ones.

The theory put forward by Davidoff provides a good foundation for public participation. From the point of views of different people and different groups with different values, he thinks that the plan should not suppress other values but should be supportive for realizing a variety of value. Urban planners are playing the role of providing technical support for the multiple values and public participation in urban planning is an important way to realize the values by absorbing all views of different people.

In collecting all kinds of information, the planning department should care more about the opinions of the majority, consider vulnerable groups, and avoid the "out of system" activities (such as private negotiation between business leaders and government), thus ultimately achieve the fairness.

4.3 *Weighing the good and bad and improving the legal system*

Applying a dialectical view of public participation which is upheld now, it is generally believed that the more people participate, the better it helps to make it more distinct and comprehensive. Yet, with the increase in number, size, and matters involved; it will also bring negative effect – the operating costs increases while the operating efficiency decreases. For instance, the Boston tunnel case costs more than twice of the original budget which is consistent with the marginal benefit and the law of diminishing in economics. As the project operation time increases, the degree of public participation reduces, the marginal utility diminishes, and the operation cost surges. When planning and public expectations are inconsistent, public confidence in the government is lower, leading to the subsequent "indifference" in public participation. This is similar to the public price hearings in China, which emerges into the price-rising conference. It does not fulfill the expected function of the hearing.

In addition, the increase of the number of people participating in the city planning and the expanding scope pose a greater challenge to information integration and feedback capacity of planning departments.

In view of this phenomenon, while ensuring interactive public participation, efficiency should be improved. On the other hand, while ensuring the continuity of public participation, the public participation in the different items should be limited in appropriate range and depth. Hence, it is necessary to build corresponding legal norms. However, the existing laws and regulations which involve the public participation are theoretical, and do not clearly define the rights and interests of the public, with no refinement of the corresponding content of each city planning stage. Without the refinement of the law, public participation will be impractical.

5 CONCLUSIONS

Public participation in urban planning is given much expectation since the beginning, yet the practice in china is lagging behind. With the deepening of public participation in China, our concerted efforts are required to make a better use of public participation in urban planning. Citizens should enhance their awareness of participation, improve their initiative, enrich their knowledge and participate in urban planning with proper organization and guidance. Community institutions should strengthen the organization and improve the citizens' ability of information integration. Urban planners should change their roles and communicate with the public in a more effective way. Government departments should improve the legal system and carry out the independent and rational public participation under regulations. The government should establish a network system of public participation that is not just the lower level communication in the system or blind horizontal public communication.

REFERENCES

[1] Chen Zhi. 2010. Urban planning and public participation in research. technology and science and life (5): 197.
[2] Huang Hong. 2008. Analyses the public participation in China today's urban and rural planning. Engineering construction 40 (4): 17–20.
[3] Tang Fenfang. 2011. The application aboutthe landscape planning and design of public participation in China. Knowledge economy (1): 81–82.
[4] Lu Chunyang. 2008. Land use planning implementation evaluation study. Agricultural mechanization research (6): 237–239.
[5] An Rongquan.2009. How to keep the government and the society harmonious interaction some thought from Boston. The CPC Hangzhou municipal committee party school (6): 25 to 31.

AUTHOR'S BRIEF INTRODUCTION

Liu Yiyun, Chengdu, Sichuan. Female, Southwest university of science and technology, urban planning, bachelor degree in reading.
Address: fucheng district in mianyang city, Sichuan , tsing lung road middle west Kowloon 59, southwest university of science and technology A – 423 Phone number: 18281517887
E-mail: liuyiyunOcean@163.com

Performance investigation of steel slag porous asphalt treated mixture

Canhua Li & Xiaodong Xiang
Hubei Key Laboratory for Efficient Utilization and Agglomeration of Metallurgic Mineral Resources, Wuhan University of Science & Technology, Wuhan, China

Quan Guo
Jiangsu Liding New Construction Materials Technology Co. Ltd., Jintan, China

Lixin Jiao
The Research Center of the Metal Resources Co. Ltd., of WISCO, Wuhan, China

ABSTRACT: Steel slag is a byproduct making up a portion of 15–20% of steel output in an integrated steel mill. Most of them are deposited in slag storing yards. Then, many serious environment problems have been resulted in China. Therefore, a new preparation method of open-graded friction course (OGFC), a permeable and porous surface asphalt mixture by using SBS modified asphalt when the coarse aggregate steel slag and the fine aggregate are selected respectively, was developed in this study. Marshall Stability test was used to evaluate the properties of the asphalt mixture. The results indicated that OGFC mixture is still stable with high percentages of steel slag. It was found that in the permeability coefficient test, OGFC mixture had a good performance of water seepage. These experiments showed that the utilization of steel slag as aggregates for porous asphalt mixture presented excellent pavement performance.

Keywords: steel slag; open-graded friction course (OGFC); asphalt mixture

1 INTRODUCTION

In recent years, the asphalt pavement has been developed rapidly in China due to its excellent road performance and comfortable driving performance. By the end of 2012, the total mileage of asphalt pavement has been reached 4×10^6 km, while the mileage of highway has reached 9.7×10^4 km, of which the vast majority is asphalt pavement. More than 90% of completed asphalt pavement was designed based on multilevel dense built-in gradation, making the pavement dense graded asphalt concrete pavement with small void ratio. The pavement structure has dense, impermeable characteristics with high strength, water stability, low temperature crack resistance, and good durability. However, the high temperature stability of the pavement structure is very poor. And considering its antiskid surface, it is hard for surface water to drain away. It is easy to cause mist, glare and water skip at high running speed, which affect the pavement skid resistance and road safety. Therefore, Euramerican in the developed countries such as Germany, France, Britain, Holland, the United States, etc., launched researches on permeable asphalt pavement.

Permeable asphalt pavement is a porous structure skeleton embedded crowded pavement, and it has many characteristics. This kind of pavement can not only reduce surface water effectively, but also provide adequate surface roughness. In addition, it can reduce traffic noise along the lane, while its aggregate skeleton structure enhances the ability to resist deformation rut. Along with the national economy development and people's lives being improved, the functional requirement for road transport increases gradually. Now, pavements should not only meet the basic demand for transport but also have comfortable driveability that water splash should be avoided and good wear skid resistance should be guaranteed to ensure driving safety. In addition, permeable asphalt pavement is a potential type of new pavement structures.

Researches indicated that traditional aggregate permeable asphalt pavement has functional defects such as aging, loosing and seed shattering, which cause early pavement problems, therefore stronger adhesion between asphalt and aggregate is needed. Steel slags have good angularity and wear resistance, which make them suitable for asphalt mixture. Both the test section of steel slag concrete built in WISCO, 2002 and that of steel slag asphalt mastic built in Wu-Huang Highway Heavy-repairing Project, 2003, showed good pavement performance and durability. It is very important that road cost can be reduced, natural stone saved and ecological environment protected if steel slag is used as permeable asphalt concrete aggregate.

Table 1. Chemical composition analysis of slag.

Name	SiO$_2$	MgO	Al$_2$O$_3$	P$_2$O$_5$	MnO	Fe$_2$O$_3$	CaO	Loss	others
Content/%	19.24	5.19	3.25	1.41	1.77	24.55	42.77	0.32	1.50

Table 2. Slag aggregate performance test results.

Pilot project		Test results	Requirements
19–13.2 mm	Apparent R.D	3.26	≥2.6
	Water absorption/%	1.2	≤3
13.2–9.5 mm	Apparent R.D	3.27	≥2.6
	Water absorption/%	1.9	≤3
9.5–4.75 mm	Apparent R.D	2.684	≥2.6
	Water absorption/%	2.7	≤3
Crushing value/%		17.6	≤26
Los Angeles abrasion/%		12.9	≤28
Polished value		67	≥42
Rugged/%		1.5	≤12

2 MATERIAL AND METHODS

2.1 Asphalt

PG76-22 SBS modified asphalt provided by Wuhan Materials Co. Ltd., with high viscosity, was used in this experiment. The property tests of the asphalt were carried out in reference to Chinese norm (JTG E20-2011). Experimental results suggested that all pavement performances of PG76-22 SBS modified asphalt met the industry standard named *Technical specification for construction of highway asphalt pavement* (JTGF40-2004), so it can be used as a good cementing material in making porous asphalt mixtures.

2.2 Aggregate

The chemical composition of slag is listed in Table 1. The classification of coarse aggregates is formed based on particle sizes as follows: 1 # is steel slag aggregate of 19−13.2 mm, 2 # is steel slag aggregate of 13.2−9.5 mm, 3 # is steel slag aggregate of 9.5−4.75 mm, and 4 # is limestone of 4.75−0 mm as fine aggregate. The technical performances of coarse and fine aggregates were tested strictly in accordance with *Test Methods of Aggregate for Highway Engineering* (JTG E42-2005).

Slag's road performance tests were carried out based on *Test Methods of Aggregate for Highway Engineering* (JTG E42-2005). The main technical parameters include the apparent relative density, water absorption, crushing value, Los Angeles abrasion value, etc. Test results are shown in Table 2.

The results from Table 2 indicate that the apparent relative densities of slag 1 # and 2 # are large (>3.2%), but the water absorptions of them are small (<2.0%), indicating that slag 1# and slag 2# are suitable as excellent asphalt concrete aggregates. Meanwhile, slag 3 # has a lower apparent relative density and a higher water absorption, which meet regulatory requirements. Slag aggregates road performance test results showed that the slag road performances meet regulatory requirements. Further analysis showed that its outstanding road performances include its good adhesion and excellent abrasion resistance, especially when applying to permeable asphalt pavement structure, which can effectively enhance water resistance, dispersion resistance and wear resistance of the permeable asphalt concrete. Therefore, the unique material properties of steel slag make it more suitable as a permeable asphalt concrete aggregate.

2.3 Test methods

2.3.1 Preparation of test pieces

Three rut samples and several marshall standard samples whose specification is 101.6 × 63.5 mm were prepared based on test projects. The preparations of samples were carried out with the following procedure. Asphalt mixtures with certain mixture ratios were packed in rut test models directly, and then the rut samples were roller-compacted through rutting compaction apparatus. Mixture was put into modeling apparatus then knocked by Marshall compaction apparatus to gain marshall standard samples. Besides, both of form removals were carried out after 24 h of placement.

There exits strict requirement on marshall standard samples that forming height error has the ranges of ±1.5 mm at most. The ratio of mixture, the section of asphalt-aggregate ratio, packing temperature, knocking time of marshall compaction apparatus and knocking power have influence on forming height, so the packing mass should be gained through many experiments. Finally, repeated experiments proved that the height could meet experimental requirements when the packing mass was determined at 1120 g.

2.3.2 Asphalt mixture test

Two different test methods, immersion Marshall test (T 0709-2011 standard) and Lott man test (AASHTO T 283) were used to evaluate the water sensibility of asphalt mixture (Wu S.P., 1997). Permeability performance of materials is often tested by its water permeability coefficient through pavement seepage meter. In this study, friction coefficients of OGFC specimens were tested with pendulum apparatus.

3 MIXTURE DESIGN

According to Chinese standard *Technical specifications for construction of highway asphalt pavements* (JTGF40-2004), two types of OGFC are OGFC-16 and OGFC-13 respectively. In this research, the steel slags of 10–19 mm were put in a 19 mm square mesh sieve. After more than 19 mm of the steel slag particles

Table 3. The mix and OAC of various aggregates.

Type	1#	2#	3#	4#	5#	OAC/%	Slag content/%
OGFC 13-1	35%, slag	51%, slag	13%, limestone	1%, powder		4.5	86
OGFC 13-2	35%, slag	49%, limestone	15%, limestone	1%, powder		4.4	35
OGFC 16-1		51%, slag	13%, limestone	1%, powder	35%, slag	4.6	86
OGFC 16-2		40%, limestone	14%, limestone	1%, powder	45%, slag	4.4	45

Table 4. RMS and RST results of OGFC.

	OGFC 13-1	OGFC 13-2	OGFC 16-1	OGFC 16-2	Basalt
RMS/%	87.9	89.8	85.7	91.9	83.7
TSR/%	83.5	82.4	87.3	85.1	80.9

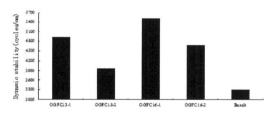

Different types of OGFC

Figure 1. Dynamic stability of steel slag based OGFC.

were removed, the remaining was used for coarse aggregates of OGFC-16 Similarly, the steel slags of 10–16 mm were put in a 16 mm square mesh sieve. After more than 16 mm of the steel slag particles were removed, the remaining was used for coarse aggregates of OGFC-13. The mix and the optimal asphalt content of various aggregates in this study are shown in Table 3.

4 RESULTS AND DISCUSSION

4.1 Water sensibility

The residual marshall stability (RMS) and tensile strength ratio (TSR) test results of different steel slag based OGFC are shown in Table 4. The RMS and TSR of steel slag based OGFC meet the Chinese code. The steel slag based OGFC presented a good performance of water sensibility.

Dynamic stability reacts the resistance deformation capacity of asphalt concrete at high temperature. Dynamic stability test results of the four types of steel slag based OGFC are shown in Fig. 1. The

Table 5. The results of water seepage coefficient.

	OGFC 13-1	OGFC 13-2	OGFC 16-1	OGFC 16-2	Basalt
WSC/ml/min	3750	4250	3960	4680	3560

results in Fig. 1 indicate that dynamic stabilities of all types of steel slag based OGFC are greater than 3000 cycles/mm, and meet the Chinese code. This confirms the strong resistance deformation performance of steel slag based OGFC at high temperature.

4.2 Seepage test

Water seepage coefficient (WSC) is not only an important index to evaluate the drainage performance of asphalt mixtures, but also an indirect index to show the mixture gradation composition, which is the macrography index of the internal porosity (Shen D.H., Wu C.M., Du J.C., 2008). The permeability performances of rutting specimens were tested (see Table 5). That the water permeability coefficient is very large suggests that its large inner structural porosity results in good water permeability performance.

5 CONCLUSIONS

1) The optimum asphalt contents of OGFC-13 and OGFC-16 using SBS modified asphalt were 4.5% and 4.6%, respectively. Compared with traditional permeability asphalt mixtures, un-improved dosage of asphalt could help to cut cost.
2) The gradation project of slag permeability asphalt mixtures made slag as coarse aggregate (larger than 4.75 mm) while the fine aggregate was limestone. Slags accounted for more than 85% of the mixtures, and the utilization of slag increased. Due to the good cohesion between slag and asphalt, the road performances such as the water stability, the water damage resistant of slag permeability asphalt mixtures were much higher than those defined in the correlation Chinese code.
3) The extensive use of slag not only decreases the reliance of traditional stone in road construction and saves resource, but also relieves the damages of ecological environment caused by bombing mountain etc., which tallied with the strategy of sustainable development.

ACKNOWLEDGEMENT

This work was supported by grants from the major scientific and technological achievements into the project, Wuhan Science and Technology Bureau (code: 2013010803010378).

REFERENCES

JTG E20-2011.2011. Test specification for asphalt and asphalt mixture in highway engineering. (in Chinese)

JTG E42-2005.2005. Test Methods of Aggregate for Highway Engineering (in Chinese)

JTG F40-2004.2004. Technical specifications for construction of highway asphalt pavements. (in Chinese)

Shen, D.H. Wu, C.M. Du, J.C. 2008. Performance evaluation of porous asphalt with granulated synthetic lightweight aggregate. Construction and Building Materials. 22(5): 902.

Wu, S.P. 1997. Study on Flame-retardant Modified Asphalt Membrane. Chinese J. of New Building Materials. 24(8): 34–35. (in Chinese)

Probe into the landscape planning of Chaoshou Village, Chengbei Town, Giange County, Guangyuan City

Chenxi Ma, Aimei Liu & Li Yuan
School of Construction Engineering, Southwest University of Science and Technology, Mianyang, Sichuan, China

ABSTRACT: This paper analyzes the landscape planning of Chaoshou Village, which is located in Chengbei Town, Jiange County, Guangyuan City, Sichuan Province. By combing and summarizing the ideas and methods of this village's landscape planning, the paper aims to provide a reference for building new countryside with its own scenic features.

1 OVERVIEW OF THE CASE

1.1 Background and significance of planning

Jiange County is a large agricultural county located in northern Sichuan. In recent years, the Government of the County has attached great importance to the construction of New Countryside. They have determined and implemented a series of agricultural structure readjustment measures which can be summarized as "southern pears, northern peaches and mulberry all the county; southern birds, northern cattles (sheeps), and pigs (rabbits) all the county". As a result of these measures, plenty of well-off families and villages are generated. According to the requirement of the Province which can be summarized as the "five new things and one good thing (developing new industries, constructing new houses, shaping new landscape, establishing new mechanism, cultivating new farmers, creating a good team)", Jiange County has been striving to build a new socialist countryside through their experienced practice. And its achievement has won approval from the leaders at all levels. However, to build a pilot new countryside in the undeveloped villages located in the low mountains of northern Sichuan, there still exist many problems, including the shortage of money, low level of the scale and standardization of the leading industries, the imbalance of development, and the insufficiency of facilities. So it is pretty urgent to explore development ways for construction of new socialist villages by promoting rural infrastructures, accelerating development of modern agricultural industry, increasing the peasant's income, and establishing and perfecting the system and mechanism of new countryside construction.

1.2 Location and geomorphology

The Minzhu Village is located in about 10 kilometers north of Chengbei Town, east longitude 105°09′ to 105°49′ and north latitude 31°31′ to 32°21′. It is next to Dongya Village, Feifeng Village in the west, adjacent to the Beiliang Village in the south, close to Xinhua Village in the west and join the center village of Hanyang Town in the north, covering an area of 8.5 square kilometers. The village is located between the old and new town, and the National Road 108 runs through the whole territory which improves the traffic of the village fundamentally. The Chaoshou Village to be planned is located in the Minzhu Village, starting from Mengziliang in the north and ends at Beibeiliang in the south, covering an area of 2.6 square kilometers. With convenient traffic, it enjoys apparent location advantages.

Minzhu Village is of hilly land, high in the north and low in the south. The terrain of the planned plot is relatively flat, with the highest elevation being 715.02 meter and the lowest being 750.40 meter. Most slopes are less than 8%, making it suitable for construction. But there are 1000 square meters in the southwest of the plot that are subject to geological disaster, thus is not unsuitable for house construction.

2 PRINCIPLES OF LANDSCAPE PLANNING

The landscape planning takes full advantage of the local terrain, geology, and hydrology. The layout of the buildings and landscape is mainly in natural and regular hybrid layout. Highlight the diagonal and included angle landscape effect and the landscape periphery's nodes take full advantage of the skill of building with borrowing, leakage, comparing, and avoidance, isolation which leads to well-combination with mountains, forests and houses, creating a natural harmonious village scene. On the residential layout level, paying attention to the combination of the shops and residence, the traditional houses and new built houses, forms a activity space where living and industry

South street facade before modification

South street facade after transforming

Figure 1. The contrast of south street.

Figure 2. The contrast of north street.

integrate; on the production of cultural landscape level, the landscape based on farming culture as the core focuses on the economical courtyard, highlighting the farmer residence's combination with plant scenery and creating a rural poetic imagery; In respect of enhancing cultural scene, exploring autochthonous culture, enriching the culture deposits, and establishing an environment with history and culture of Xinshou Village. Constructing a cultural landscape's plaza, an orchard with fruits in an appropriate and prominent place where the villagers enjoy themselves and exercise after eating. Forging a landscape called "green mountains and clean water, fragrant fruits and delicious vegetable, deep cultural details, expression at all times.

3 CONCRETE STEPS IN PLANNING THE LANDSCAPE

3.1 Architectural style

The principle of "respecting tradition and strengthening unity" is adopted, which means that the present buildings won't be demolished to a large degree, but will only be adjusted in some details to unify building elements and achieve nice overall effect.

In terms of building colors, large area of white wall will be used to match with the surrounding environment; the roofing will use blue-gray colors as bright spots.

In terms of building style, townhouses and detached houses are in the majority, with sloping roofs adopted to echo with the stretching mountains; the economic plants such as pears, loquats, and vegetables will be planted in the courtyard to create pleasant architecture space.

In terms of building height, most of which will be 2–3 storey ones and the elevation difference of the terrain will be made use of to achieve height disorder sense.

In terms of decoration of facades, the imitation wood coating will be applied on the wall with joints in brown. The strip window, composite window, and local wooden artifacts will be used to form layering sense in the façade (Figure 1 and 2) [1].

Figure 3. Logo.

3.2 Style of landscape

A culture logo is specially designed for Chaoshou Village in combination with Minzhu Village's history and culture and based on the legend of "Chaoshou" and "phoenix". The logo can be engraved on the memorial arch at the entrance of the village or sculpture, or portrayed on the cultural and residential wall, or incorporated into the lamp ornament or the plaza's pavement.

The logo encompasses the following themes: green mountains and clear waters, fragrant fruits and delicious vegetables, and profound Shu-Han culture.

Its detailed interpretation is as below: there are endless mountains from the east and blue waters from the west; It has Feeing Mountain of the imperial cypress' ancient post road on the top from the east and the style of the phoenix means that phoenix is homesick when it flies, and the flower on the phoenix's head means the ancient post road. The phoenix looks back at the village at the Feifeng mountain, and then, it flies to the Fengquan temple on the west to live, and the phoenix brings the water out from now on; the style of the central buildings means that the new construction, the residents and the orchard with the loquat form the synthesis; The whole pattern reflects a comprehensive impression of the Minzhu Village's various historical and cultural legends (Figure 3).

In the planning a great deal of attention has been paid to the details of village landscape. A lot of landscape art pieces are designed to promote the overall cultural atmosphere, including architectural ones like sculptures, murals, pavilions, pavilions, and memorial archways; living facility ones like seats, phone booths,

Figure 4. Sculpture.

Figure 5. Aerial view.

mailboxes, and trash cans; and road-related ones like stop boards, streetlights, guard rails, and road signs (Figure 4).

3.3 Construction of the landscape nodes

3.3.1 Entrance of the new village

The entrance of the new village, as a buffer zone leading to village complex, is a channel to link the village complex with the outside towns and a carrier to exchange social materials and information between the village complex and the outside.

Mark monument, located on one side of the road into the village, is set at the entrance of the village. The main purpose of its establishment is to show the culture of the new village. Meaningful patterns and shapes will be designed to symbolize the flourishing of the new village. The mark monument is made up of natural stone as the base body which coordinates with the circumjacent environment, and it shines beautifully under the flowers, bamboos, and trees, perfectly displaying the overall image of the new village.

3.3.2 The central plaza

The central plaza, being a popular place where people can gather, is a key area requiring special attention during the planning. It will be an important center of arts and culture, entertainment and sporting integrating fitness, fair, scenery, leisure, and recreation.

Minzhu Village, with a long history, is rich in both historical and cultural allusions, myths and legends. The plaza, located opposite to the village committee and covering an area of 2090 square meters, is composed of four plots with different elevations. It will be planned based on the culture of three kingdoms of Minzhu Village and the phoenix legend of Feifeng Mountain. As the plaza has a large area, it will be sliced into dynamic area and static area and detailed design will be made for each of them. The plaza is divided into three areas: the left, right, and middle areas. There is a large space as the main body in the middle which based on the simple pavement design in wide vision, and the central theme 'pattern and sculpture oddments form the plaza's center. Go straight to the viewing booth, and then walk down the stone stairs along the path through the trees which makes people direct to a pool to enjoy themselves; the beautiful flowers and green trees on the left side constitute a place where people can relax and exercise; The landscapes with loquat trees show the village's industry culture. The path along the trees associate industry with landscape well which forms a kind of plaza scene. Because of the beautiful scene, combining with the construction of zhanfeng observation Gallery, xinyi pavilion, luofeng parterre, realize the harmonious of learning, viewing, entertaining, exercising.

3.3.3 The rural tourism areas

The rural tourism area 1 will be made by reconstructing the original buildings featuring typical northern Sichuan style. Guideboards will be erected along one side of the road in the rural tourism area, and flowers, shrubs will be planted in the front and back yards of the farmhouse, which, together with the crape myrtle flower nursery in front of the farmhouse, will form pleasant and beautiful scenery. The rural tourism area 2 will retain their original front-yard ponds where lotus will be planted. Flowers and loquat trees will be planted along the paths leading to the farmhouses, so as to improve the overall environment taste of the tourism area.

3.3.4 The relax garden along the street

There is a land parcel of 1000 square meters in the southwest of the planned plot, which is unsuitable for house construction due to the hazard of geological disaster. It is thus designed to be a leisure green space to offer the functions like leisure, sightseeing, disaster prevention and refuge. A petty street garden will be created in combination with the terrain, which will make full use of the local raw materials based on the practical and economic principal and meanwhile fully embody the local culture. A sculpture of Chaoshou will be installed at the entrance and crape myrtle pavilions will be planted in two eye-catching places to fully embody the ecology landscape. The pavement will use stones in harmony with the nature and will be distributed in the grass naturally. The plants are mainly typical local trees with plenty of loquats as the basic tone at the outer ring, the grass and local bushes base at the middle as the echoing tone, and the beautiful loquat trees as the main tone in the front. Walnut trees will be planted in both the dynamic and static areas (Figure 5) [2].

4 CONCLUSIONS

It is easy to become ticky-tacky or disorderly for the construction of new countryside. The landscape planning of the Chaoshou new village, Chengbei town, Jiange County, by combining with the architectural style, landscape art pieces style, landscape nodes, has found a best balance between the integrality and the diversity.

REFERENCES

Cui Heng, Huang Zhe. The Multi-dimensional Analysis and Reflection on the Traditional Architectural Landscape of Sichuan, [J]. Scientific Research on Architecture in Sichuan, 2010, 36(3):221–225

Qiu Lianfeng, Zou Nini. The Connotation and Practice on the Features of a City Landscape — Taking Sanjiang City's Landscape as An Example, Planners, 2009, 25(12):26–32

Rational performance tests for permanent deformation evaluation of asphalt mixtures

Q. Li & F. Ni
Southeast University, Nanjing, China

ABSTRACT: The objective of this study is to develop a simple laboratorial test that can identify the permanent deformation characteristics of asphalt mixtures during the mix design process. Four types of performance tests were conducted on three different types of asphalt mixtures with various volumetric properties. It was observed that the dynamic modulus indicator $|E^*|/\sin\delta$ was not accurate enough to identify the rutting performance. Similarly, the indirect tensile strength of asphalt mixtures did not have a good correlation with the permanent deformation. From the triaxial compressive strength test, it was discovered that the shear strength represented in combination of friction angle and cohesion was strongly correlated with the permanent deformation. Finally, the triaxial compressive strength is recommended to be a rational tool for evaluating rutting performance during a mix design process.

1 INTRODUCTION

Permanent deformation or rutting is one of the most important distress modes in asphalt concrete pavements. Thus, a rational laboratory performance test for identifying this distress is urgently needed to complement the volumetric mix design method. The focus of the test is to simply measure the fundamental engineering property that can be correlated with the pavement performance.

Many laboratory tests have been developed over forty years in an attempt to capture the permanent deformation characteristics of asphalt mixtures in field, such as the Marshall stability test, wheel tracking (WT) tests, shear tests, creep tests, strength tests, static creep permanent deformation (SCPD) test, repeated load permanent deformation (RLPD) test, and dynamic modulus $|E^*|$ test (Witczak et al. 2002, Christensen & Bonaquist 2002, Gardete et al. 2008, Zhu et al. 2010, Li et al. 2013, Szydlo & Mackiewicz 2005, Collop et al. 1995, Ossa et al. 2010). However, most of the existing tests have both strengths and weaknesses. Mohammad et al. (2006) observed that the dynamic modulus $|E^*|$ could not differentiate the permanent deformation characteristics for various asphalt mixtures at high temperature. The SCPD test with static loading mode cannot offer a better simulation of field traffic conditions. WT test has a limitation that it is incapable of providing an accurate simulation of field stress states (Li et al. 2010). Although the RLPD test was proposed as a good tool to identify the permanent deformation performance, the complexity in performing it and the equipment requirements may limit its application for a mix design.

2 EXPERIMENTAL PROGRAM

2.1 Materials

Three asphalt mixtures were designed in the laboratory. Two of them, AC-19C and AC-19M, were 19 mm nominal maximum aggregate size (NMAS) dense graded mixtures with conventional PG 64-22 and SBS modified PG 76-22, respectively. The other one, SMA-13, was 13 mm NMAS SMA mixture with PG 64-22 binder. Five asphalt contents were designed for each mixture to simulate the mix design procedure. The details are as provided by Li et al. (2010).

2.2 Specimen preparation

The specimens of 150 mm in diameter and 175 mm in height for the $|E^*|$, TCS and RLPD tests were compacted using a Superpave gyratory compactor (SGC). The SGC could supply a consolidation pressure of 600 kPa with a gyration angle of 1.25. The gyration speed was 30 gpm. The height controlled mode rather that gyration number controlled mode was selected for compacting. Then, they were cored from the center and sawed from each end to obtain test specimens of 100 mm in diameter and 150 mm in height. The IDT strength tests were performed on specimens of 150 mm in diameter and 50 mm in height also compacted using SGC.

2.3 $|E^*|$ test

All the tests were conducted on MTS testing system. An appropriate haversine compressive stress was

Table 1. Test results for the AC-19C mixture.

Parameters	C-4.0	C-4.5	C-5.0	C-5.5	C-6.0
$|E^*|/\sin\delta$ (MPa)	6184	5242	4438	3385	2595
σ_{IDT} (MPa)	0.066	0.092	0.109	0.106	0.098
C (kPa)	367	377	378	376	353
ϕ (°)	46.61	42.89	41.70	41.47	41.53
τ_f (kPa)	756	765	737	685	664
ε_{pf} (%)	4.64	5.91	7.06	9.74	15.14

Table 2. Test results for the AC-19M mixture.

Parameters	M-4.0	M-4.5	M-5.0	M-5.5	M-6.0
$|E^*|/\sin\delta$ (MPa)	7105	4379	3859	2967	2320
σ_{IDT} (MPa)	0.203	0.270	0.233	0.213	0.189
C (kPa)	558	626	624	611	573
ϕ (°)	47.94	40.98	38.77	38.51	38.50
τ_f (kPa)	1070	1132	1107	1086	1028
ε_{pf} (%)	1.48	1.60	1.67	1.73	2.44

applied on samples with the frequency of 5, 1, 0.5 and 0.1 Hz at the temperature of −10°C, 5°C, 20°C, 35°C and 50°C to achieve a target vertical strain level of about 100 microns. The dynamic modulus $|E^*|$ and phase angle δ of all cylindrical specimens were measured before operating the TCS and RLPD tests.

2.4 IDT strength test

The IDT strength test was performed at a constant displacement rate of 50 mm/min and a temperature of 50°C. The bulging effects were corrected using the existing procedure (Roque & Buttlar 1992). The average values of the IDT strength σ_{IDT} on two replicates were used for analysis.

2.5 TCS test

The triaxial cell was set up on the MTS system for the TCS and RLPD tests. Thin and fully lubricated membranes at the test specimen ends were used to minimize end effects. The TCS tests were run at a load rate of 50 mm/min and temperature of 50°C under the confining pressures of 0, 69 and 138 kPa, respectively. An average value of the deviatoric stresses at failure obtained from two replicated tests was used to calculate shear properties (cohesion C and friction angle ϕ) using the Mohr-Coulomb failure theory.

2.6 RLPD test

The RLPD test was performed under the deviatoric stress of 690 kPa and the confining pressure of 69 kPa at the temperature of 50°C. A specimen was subjected to a repeated haversine axial compressive load pulse, a 0.1 second of loading followed by a 0.9 second of unloading. The cumulative permanent strain ε_p was recorded as a function of the number of load cycles over the full test period until 40,000 cycles where the tests stopped. The axial displacement ε_{pf} at 40,000 of loading cycles of two replicates was measured as a comparison basis.

3 TEST RESULTS

A brief summary of the results from various performance tests is provided in Tables 1–3. The mixtures shown in the table are named as mixture type-binder content (%). For example, the C-4.0 represents the AC-19C mixture with binder content of 4%.

Table 3. Test results for the SMA-13 mixture.

Parameters	S-5.0	S-5.5	S-6.0	S-6.5	S-7.0
$|E^*|/\sin\delta$ (MPa)	6208	5582	4820	3252	2508
σ_{IDT} (MPa)	0.134	0.159	0.127	0.102	0.081
C (kPa)	367	381	367	348	326
ϕ (°)	48.39	47.24	46.78	45.54	45.47
τ_f (kPa)	768	808	774	691	664
ε_{pf} (%)	3.88	4.56	4.63	5.71	7.38

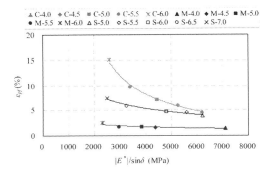

Figure 1. Relation of the permanent strain and dynamic modulus indicator.

4 EVALUATION OF VARIOUS TESTS

4.1 $|E^*|$ test

The $|E^*|/\sin\delta$ value measured at a frequency of 5 Hz and a temperature of 50°C, as recommended by Witczak et al. (2002), was correlated with the permanent strain ε_{pf} at the loading cycles of 40,000 measured in RLPD test, as shown in Fig. 1. Generally, it is discovered that ε_{pf} decreases with the increase of $|E^*|/\sin\delta$ value for the same type of mixtures with various volumetric properties. Thus, the $|E^*|/\sin\delta$ can distinguish the rutting susceptibility for a specific type of mixtures. However, it cannot take care of mixture type dependency. At the same binder content, the AC-19M mixtures with SBS modified binder show much better rutting performance than the AC-19C mixtures with conventional binder in the RLPD tests, as shown in Tables 1–3. However, from Figure 1 it is observed that the $|E^*|/\sin\delta$ values of the AC-19M mixtures are even a little lower than those of the AC-19C mixtures at the

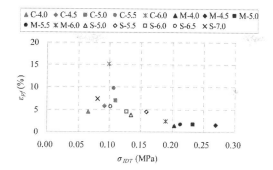

Figure 2. Relation of the permanent strain and IDT strength.

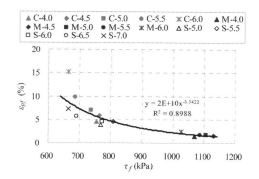

Figure 3. Relation of the permanent strain and shear strength.

same binder content for most cases except for the pair of M-4.0 mixture and C-4.0 mixture. The same phenomenon is also found in SMA mixtures. The findings confirm that the dynamic modulus $|E^*|$ alone is not accurate enough to predict the rutting performance of different mixtures (Mohammad et al. 2006).

4.2 IDT strength test

Plot of the permanent stain ε_{pf} versus IDT strength σ_{IDT} measured on all mixtures at a temperature of 50°C is presented in Figure 2. It is seen in the figure that the σ_{IDT} value does not have a good agreement with the ε_{pf} value in this study. From this figure, it is easy to find that the σ_{IDT} peak values appear at the binder content of 5.0, 4.5 and 5.5% for AC-19C, AC-19M and SMA-13, respectively. The mixture with the highest σ_{IDT} does not correspond to that with the highest rutting resistance. Thus, it is not a good measure of the rutting resistance even though it is widely used as an important criterion in Marshall mix design method. IDT strength even cannot identify the rutting performance of the same type of mixtures with various volumetric properties, not to mention taking care of mixture type dependency. It is not a good measure of the rutting resistance. Both of the two important criteria adopted in existing mix design methods, air void and IDT strength, do not provide a good correspondence with the permanent deformation. It emphasizes once again the need of a performance-based mix design method.

4.3 TCS test

The classic Mohr-Coulomb failure theory was employed to calculate the cohesion and friction angle of asphalt mixtures based on the TCS tests. It is observed from Tables 1–3 that for a given type of mixtures, ϕ decreases as binder content increases. However, similar to $|E^*|/\sin\delta$, it cannot take care of mixture type dependency. The AC-19M mixtures exhibit the best rutting performance but the worst interlocking capability of the aggregate matrix among the three types of mixtures. Thus, friction angle is not a rational indicator of permanent deformation.

For a given type of mixtures, the maximum c occurs at approximate optimum binder content obtained from volumetric design criteria. The AC-19M mixtures with SBS modified binder show much higher cohesion values than the other two types of mixtures with unmodified binder. However, the c value does not associate well with the ε_{pf} value obtained from the RLPD tests. For example, similar c values are observed between the AC-19C and SMA-13 mixtures since the same binder type is used. However, it is obvious that the SMA-13 mixtures have much better rutting resistance than the AC-19C mixtures. Thus, cohesion is not a good measure of permanent deformation either.

It can be concluded that friction angle represents the interlocking capability of the aggregate matrix from the applied loads, and it can only identify the effects of aggregate gradations on the permanent deformation. Similarly, cohesion can only identify the effects of the adhesion or bonding mechanism of the binder. It cannot consider other factors such as volumetric properties and aggregate gradations. Thus, neither friction angle nor cohesion alone can take care of all mixture properties dependencies, including aggregate gradation type, binder type and content. Fortunately, shear strength represented in combination of these two parameters can represent both advantages on the permanent deformation characterization. Moreover, it can be conveniently obtained from the same test. Thus, it is a potential rational indicator of permanent deformation.

The curve of the permanent deformation obtained from the RLPD test at the confining pressure of 69 kPa versus the shear strength τ_f of asphalt mixtures calculated based on the TCS test at the same confinement level is shown in Fig. 3. A good correlation with a coefficient of determination R^2 value of about 0.9 represented using power law function was established between the τ_f value and the ε_{pf} value on the data of all mixtures. The mixture ranking of the τ_f value nicely corresponds to that of the ε_{pf} value for most cases. Moreover, the correlation is independent of mix types and volumetric properties. It proves that the shear strength is a rational indicator to differentiate the rutting performance for different asphalt mixtures and could be used as rutting criteria for a performance-based mix design.

5 DISCUSSIONS

In this study, the TCS test is finally proposed as a rational laboratory performance test for evaluating the permanent deformation in asphalt mix design process since the shear strength obtained from this test is able to identify the rutting performance irrespective of mixture types and volumetric properties. Moreover, the shear strength is a function of normal stress and shear properties according to the Mohr-Coulomb failure theory. It means that this indicator can also possibly reflect the effects of pavement mechanical responses. As a result, it could be a bridge between mix design and pavement structure design. It could be used as a performance-based mix design tool to determine the selection of aggregate gradations, binder types and contents with respect to the permanent deformation. For the places where the complicated equipment for the triaxial test is unavailable, an abbreviated procedure, a combination of unconfined compressive strength (UCS) test and IDT strength test is also recommended to calculate the shear strength.

6 CONCLUSIONS

(1) The dynamic modulus indicator $|E^*|/\sin\delta$ is not accurate enough to identify the rutting performance because it does not consider mixture type dependencies.
(2) The IDT strength of asphalt mixtures does not have a good correlation with laboratorial permanent deformation characteristics and is not a reliable indicator to mix design.
(3) Shear strength is a rational indicator to differentiate the rutting performance for asphalt mixtures with different materials and volumetric properties. Thus, the TCS test or a combination of the UCS test and IDT test is recommended to be the simple laboratorial performance tests for permanent deformation evaluation in a mix design process.

ACKNOWLEDGEMENTS

This work was sponsored by the Project supported by the National Natural Science Foundation of China (Grant No.51308303), China Postdoctoral Science Foundation funded project (Grant No.2013M531252), the Project Supported by Basic Research Plan (Natural Science Foundation) in Jiangsu Province of China (Grant No.BK20130980), Specialized Research Fund for the Doctoral Program of Higher Education (Grant No.20123204120011), Jiangsu Planned Projects for Postdoctoral Research Funds (Grant No.1202010C). The authors gratefully acknowledge the support.

REFERENCES

Christensen, D.W. & Bonaquist. R. 2002. Use of strength tests for evaluating the rut resistance of asphalt concrete *J. Assoc. Asphalt. Paving. Tech* 71: 692–711.

Collop, A. et al. 1995. Viscoelastic approach to rutting in flexible pavements *J. Transp. Eng* 121(1): 82–93.

Gardete, D. et al. 2008. Permanent deformation characterization of bituminous mixtures using laboratory tests. *Road. Mater. Pavement* 9(3): 537–547.

Li, L. et al. 2013. Integrated experimental and numerical study on permanent deformation of asphalt pavement at intersections *J. Mater. Civ. Eng* 25(7): 907–912.

Li, Q. et al. 2010. Characterization of permanent deformation of asphalt mixtures based on shear properties. *Transport. Res. Rec* 2181: 1–10.

Mohammad, L.N. et al. 2006. Permanent deformation analysis of Hot-Mix Asphalt mixtures with Simple Performance Tests and 2002 Mechanistic-Empirical Pavement Design software *Transport. Res. Rec* 1970: 133–142.

Ossa, E. et al. 2010. Triaxial deformation behavior of bituminous mixes *J. Mater. Civ. Eng* 22(2): 124–135.

Roque, R. & Buttlar. W.G. 1992. The development of a measurement and analysis system to accurately determine asphalt concrete properties using the indirect tensile mode *J. Assoc. Asphalt. Paving. Tech* 61: 304–332.

Szydlo, A. & Mackiewicz, P. 2005. Asphalt mixes deformation sensitivity to change in rheological parameters. *J. Mater. Civ. Eng* 17(1): 1–9.

Witczak, M.W. et al. 2002. *National Cooperative Highway Research Program (NCHRP) Report 465: simple performance test for Superpave mix design.* Washington, D.C.: Transportation Research Board, National Research Council.

Zhu, H. et al. 2010. Relationship between repeated triaxial test and Hamburg wheel tracking test on asphalt mixtures. *Journal of Southeast University (English Edition)* 26(1): 117–121.

Reliability analysis of asphalt pavement considering two failure modes

Huili Liu & Xu Xu
Civil Engineering Department of Shanghai University, Shanghai, China

ABSTRACT: Asphalt pavement has multiple failure modes, which are usually considered as independent modes. In this paper, reliability analysis of asphalt pavement considering both the fatigue cracking and rutting failure modes based on the Response Surface method and the Monte Carlo method was carried out. Then, sensitivity analysis was conducted to find that the most important parameter affecting the reliability for both failure modes was the surface layer thickness. The calculation results showed that the probability of simultaneous occurrence of these two failure modes can be as high as the probability of either mode individually. Therefore, to improve the asphalt pavement performance, simultaneous failure modes should be taken into account instead of the individual failure mode. In addition, the simultaneous failure probability was found to increase considerably with small increments in the traffic loads.

1 INTRODUCTION

Due to the vast difference in climate conditions, economic levels, and traffic in different regions of our country, technical problems with asphalt pavements (pavement cracking and mud pumping) inevitably occur. In the present design method of asphalt pavements, the parameters are all identified as deterministic values. However, these parameters are variable to some extent. Similarly, because of non-uniformity of the material properties and the effects of construction deviation, environmental and loading condition variations, the design parameters of pavements vary as well. Thus, the concept of reliability needs to be applied in the design of the asphalt pavements. The introduction of the concept of reliability dates back to 1996, when design procedures considering material variability were first formulated by America California Highway Bureau and the Asphalt Institute (Kim, 1999). Afterwards, the reliability-based models were widely used in studying the performance of asphalt pavements (Chou, 1990). However, the only country that employs this reliability-based method in designing asphalt pavements is America.

In asphalt pavement design, the damage and failure of the pavement can be induced by fatigue cracking, rutting (permanent deformation), thermal cracking, and traffic overloading. Although there are multiple failure modes, pavement design methods are intended to only guard against fatigue cracking on the surface and rutting within the subgrade (Sanchez, et al., 2005). Despite the fact that fatigue cracking and rutting have been found in the empirical design method as the most important pavement failure modes, they are usually considered as independent failure events. However, reliability analysis of asphalt pavement design is most critical if it can be applied to the combination of all failure modes in pavement design (Retherford and McDonald, 2010).

The objectives of this study are to build a reliability analysis framework based on fatigue cracking and rutting models, and to determine the reliability of the pavement. Through sensitivity analysis, the influence of the pavement design parameters on the reliability is to be established. Furthermore, the importance of the simultaneous failure probability in view of the fatigue cracking and rutting failure is to be highlighted. The investigation also attempts to determine the influence of varying traffic demands on the reliability and the joint probability of failure.

2 TWO FAILURE MODELS

2.1 Fatigue cracking model

In general, transfer functions are used to calculate the number of load repetitions that lead to the failure of the pavement. In this paper, the transfer function for fatigue cracking takes the following form (Huang, 1998):

$$N_{fc} = c_1 \cdot \varepsilon_t^{-c_2} \cdot E_1^{-c_3} \qquad (1)$$

where N_{fc} is the number of standard axles to produce 20% cracking area; ε_t is the tensile strain at the bottom of the asphalt surface layer (in microstrain); E_1 is the elastic modulus of the asphalt surface layer (MPa); c_1, c_2, and c_3 are the regression coefficients of the fatigue cracking equation.

2.2 Rutting model

In this paper, the relationship between the number of load repetitions that lead to the failure of the pavement and the vertical compressive strain at the top of the subgrade is described by the following transfer function (Huang, 1998):

$$N_r = c_4 \cdot \varepsilon_c^{c_5} \quad (2)$$

where N_r is the number of standard axles to produce 12.7 mm of rutting; ε_c is the vertical compressive strain at the top of the subgrade (in microstrain); c_4 and c_5 are the regression coefficients of the rutting equation.

3 RELIABILITY OF ASPHALT PAVEMENT

3.1 System reliability

In reality, for most practical engineering structures, there are multiple failure modes. Thus, the reliability of the asphalt pavement should be defined as the system reliability. For asphalt pavement structures, the system fails if either fatigue cracking or rutting has occurred. Therefore, the probability of two failure modes for the pavement structure can be represented in a series relationship as follows:

$$p_f = \Pr\left(\bigcup_{i=1}^{2} Z_i \leq 0\right) = \int_{\bigcup_{i=1}^{2} Z_i \leq 0} f_X(x) dx \quad (3)$$

where Z_i are the functional equations ($i = 1, 2$ represents fatigue cracking and rutting event).

Due to the difficulty in obtaining the solutions of the integration in Equation (3), approximate approaches have been used in this paper. One of the methods is the wide bound method for the failure probability of the series system (Cornell, 1967):

$$\max_{1 \leq i \leq 2} p_{fi} \leq p_f \leq 1 - \prod_{i=1}^{2}(1 - p_{fi}) \quad (4)$$

where P_{fi} is the failure probability of the ith failure mode. Obviously, the correlation of each failure mode is neglected so that the wide bound method is simple in order to evaluate the system failure probability.

The other approach is the narrow bound method for the failure probability of the series system given as follows (Ditlevsen, 1979):

$$p_{f1} + \sum_{i=2}^{2}\max(p_{f2} - \sum_{j=1}^{1} p_{fij}, 0) \leq p_f \leq \sum_{i=1}^{2} p_{fi} - \sum_{i=2}^{2}\max_{j<i} p_{fij} \quad (5)$$

where p_{fij} is the joint probability of the occurrence of both failure modes. In the narrow bound method, it is not easy to calculate the joint probability of the simultaneous occurrence of the two failure modes, since the bivariate normal integral is difficult to solve. Thus, approximate methods are usually employed in engineering.

In 1989, Feng presented an estimate for the joint probability (Feng, 1989):

$$p_{fij} \approx \left(\sum_{i=1}^{2} p_{fi}\right)\left(\frac{\pi - \arccos(\rho_{ij})}{\pi}\right) \quad (6)$$

where ρ_{ij} is the correlation coefficient between the ith and jth failure mode.

In 1990, Dong gave a more accurate estimate prediction for the joint probability (Dong, 1990):

$$p_{fij} \approx \begin{cases} \max[p_{f1} + p_{f2}] + \min[p_{f1} + p_{f2}]\left(\frac{\pi - 2\arccos(\rho_{ij})}{\pi}\right), & \rho_{ij} \geq 0 \\ \min[p_{f1} + p_{f2}]\frac{2(\pi - 2\arccos(\rho_{ij}))}{\pi}, & \rho_{ij} < 0 \end{cases} \quad (7)$$

3.2 Monte Carlo method

The Monte Carlo method is widely used for reliability analysis of engineering structures. In this method, a function $M(x)$ is defined and denotes the failure state ($M(x) = 1$), or the safe state ($M(x) = 0$) for a set of input random variables x. Using a number generation algorithm, a sequence of N_s input variable x_i is obtained according to their joint statistical distribution. Then the failure probability of the asphalt pavement structure can be estimated by:

$$p_f = \frac{\sum_{i=1}^{N_s} M(x_i)}{N_s} \quad (8)$$

3.3 The implementation of technical methods

Firstly, in order to determine the tensile strain ε_t at the bottom of the asphalt surface layer and the vertical compressive strain ε_c at the top of subgrade for fatigue cracking and rutting, respectively, the pavement design and analysis software KENPAVE (developed by Huang at the University of Kentucky) was used. Then, the reliability analysis was conducted implementing STRUREL software, which was able to calculate both reliability of the individual failure mode and system reliability of the two modes using the Monte Carlo method. In addition, for simplicity in calculation, the Response Surface method was used for predicting the critical strains of the pavement responses.

4 EXAMPLE

For this paper, a three-layer asphalt pavement structure consisting of an asphalt surface layer, base course, and subgrade, was subjected to dual wheel loads, and is shown in Figure 1. The design parameters that were adopted for the asphalt pavement are depicted in Table 1, where all the variables are assumed to

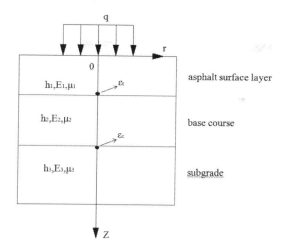

Figure 1. Pavement structure.

Table 1. Design parameters.

Parameter	Average value	COV (%)
h_1 (cm)	20	5
h_2 (cm)	25	10
E_1 (MPa)	3000	15
E_2 (MPa)	555	20
E_3 (MPa)	110	20

Table 2. Correlation coefficient matrix of design parameters.

Parameter	h_1	h_2	E_1	E_2	E_3
h_1	1	−0.25	0	0	0
h_2	−0.25	1	0	0	0
E_1	0	0	1	0	0
E_2	0	0	0	1	0.35
E_3	0	0	0	0.35	1

be normally distributed and the coefficients of variation (COV) are appropriately assumed. The correlation coefficient matrix of the design parameters is shown in Table 2, according to the practical case.

According to the design parameters in Table 1 and their COV values, the tensile strain and the compressive strain were obtained using KENPAVE software, and then the Response Surface method was used to determine the relationship between the critical strains and the design parameters, given as follows:

$$\varepsilon_t = 205.3 - 9.27h_1 - 0.502h_2 - 0.048E_1 - 0.348E_2 - 0.092E_3 \quad (9)$$

$$\varepsilon_c = 571.14 - 8.92h_1 - 4.62h_2 - 0.029E_1 - 0.263E_2 - 1.582E_3 \quad (10)$$

In this paper, the transfer functions for the fatigue cracking and rutting were adopted from the Asphalt Institute and presented by:

$$N_{fc} = 0.0796(\varepsilon_t)^{-3.291}(E_1)^{-0.854} \quad (11)$$

Table 3. Results comparison based on RSM and MCM.

Failure mode	Reliability index β		Failure probability p_f (%)	
	RSM	MCM	RSM	MCM
Fatigue cracking	0.52	0.56	31	29
Rutting	0.61	0.67	28	26

$$N_r = 1.365 \times 10^{-9}(\varepsilon_c)^{4.477} \quad (12)$$

Before performing the reliability analysis, the allowable traffic repetitions for the fatigue and rutting failure modes were evaluated. In this paper, the average value of the traffic demand N_d was assumed to be 20 msa with a log normally distributed with a COV of 45%. Therefore, the performance functions are given as follows:

$$Z_1(x) = N_{fc} - N_d; \quad Z_2(x) = N_r - N_d \quad (13)$$

Based on the performance functions, the reliability analysis was then carried out using the Response Surface method. The sensitivity analysis was performed via STRUREL structural reliability software, and the system probability and the joint probability for the two failure modes were estimated through Eq. (4), Eq. (5), Eq. (6), and Eq. (7), and the results were compared to those calculated with the Monte Carlo method.

5 RESULTS AND DISCUSSIONS

5.1 Reliability for individual failure modes and sensitivity analysis

Based on the design parameters depicted in Table 1, the reliability analysis was performed using the Response Surface method (RSM) and the Monte Carlo method (MCM) to obtain the results shown in Table 3. It can be seen that the differences between RSM and MCM results were small, being no more than 7% for both failure modes. Thus, any method is available and accurate enough to obtain the results.

Using STRUREL, the sensitivity analysis was performed to obtain the index of elasticity for evaluating the sensitivity of the design parameters. These results are shown in Figure 2. It can be observed that the most critical parameter for both failure modes was the thickness of the asphalt surface layer.

A mere 1% increase in the asphalt layer thickness can lead to 9.5% and 8.3% increase in the reliability index for the fatigue cracking and rutting modes, respectively. The second most critical parameter for the fatigue cracking mode is the base modulus, whereas for the rutting mode, it is the thickness of the base layer.

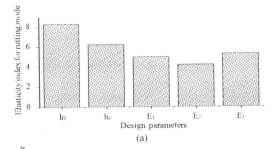
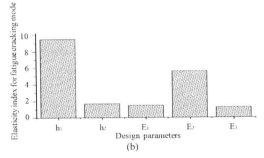

Figure 2. Elasticity index for (a) fatigue cracking mode and (b) rutting mode.

Table 4. Failure probability for both failure modes.

P_f	WB	NB	MCM
Lower bound (%)	22.45	25.60	28.75
Upper bound (%)	36.71	32.14	28.75

5.2 Reliability for both failure modes and estimation of joint probability

The failure probability for both failure modes was predicted with the wide bound method and narrow bound method. The results were compared with those from the MCM, shown in Table 4. The two methods correspond to the events when the modes are fully uncorrelated and correlated. The results demonstrate that the MCM solutions are 28.75%, which is within the range of both the wide bound (WB) and narrow bound (NB) methods. In addition, the narrow bound method gives a more accurate prediction for reliability than the wide bound method.

The joint probability for both the fatigue cracking and rutting modes was estimated with Eq. (6), Eq. (7), and the Monte Carlo method, as depicted in Table 5. The results estimated with Eq. (6) are approximately 20.04%, 22.25% with Eq. (7), and 24.67% with the MCM. Also, Dong's prediction equation was found to be more accurate than Feng's. Furthermore, by comparing the joint probabilities with the probabilities for the individual modes in Table 3, it is reflected that the joint probability of both modes is close to the probability of either mode alone. Thus, the improvement of asphalt pavement performance should not only consider an individual failure mode, but also focus

Table 5. Joint probability of different methods with different traffic demands.

Traffic demand (msa)	Joint probability (%)		
	Feng	Dong	MCM
16	4.86	5.36	5.81
18	11.56	12.43	13.05
20	20.04	22.25	24.67
22	28.88	30.13	32.34
24	37.43	39.56	40.12

on the simultaneous failures of multiple modes. The correlation of the two failure modes should not be ignored.

Owing to varying daily transportation circumstances, traffic demands increase and decrease. Hence, there is a need to investigate the effects of uncertainties in traffic loads on pavement performance by changing the average value of the traffic load between 16 and 24 msa. The results are shown in Table 5. The joint probability for all three methods increase as the traffic demand grows. And, as the traffic demand grows from 16 to 24 msa, the joint probability increases considerably, to nearly nine times of the original values.

6 CONCLUSIONS

In this paper, a reliability analysis of asphalt pavement for two failure modes, including fatigue cracking and rutting was performed to obtain the system failure probability and joint probability based on the Response Surface method and the Monte Carlo method. Some conclusions were drawn as follows:

(1) The reliability for individual failure modes was obtained based on the Response Surface method and the Monte Carlo method using the reliability analysis software STRUREL, and it was found that the results of the two methods are nearly the same. As such, both methods can be used to obtain the reliability.
(2) The sensitivity analysis was performed using STRUREL software based on the elasticity index. The design parameter that affects reliability the most for both failure modes is the asphalt surface layer thickness.
(3) The correlation between the fatigue cracking mode and the rutting mode was considered. The probability of simultaneous occurrence of the two modes is as high as those of either failure mode alone. Therefore, the joint probability of failure is crucial and worth paying more attention to in reliability analysis for improving the performance of asphalt pavement structures.
(4) With small increments in traffic loads, the joint probability grows considerably. The influence of uncertainties in traffic loads on the pavement performance should be given more focus.

REFERENCES

Chou Y. T. 1990. Reliability design procedures for flexible pavements. *Transp. ETransP. Engrg. ASCE* 116(5): 165–172.
Cornell C. A. 1967. Bounds on the reliability of structural system. *Journal of Structural Division, ASCE* 93: 151–158.
Ditlevsen O. 1979. Narrow reliability bounds for system. *Journal of Structural Mechanics* 7(4): 453–472.
Dong C. 1990. *Reliability analysis and optimization of structural system*. Xi'an: NPU.
Feng Y. S. 1989. A method of computing structural system reliability with high accuracy. *Computers and Structures* 33(1): 1–5.
Huang Y. X. 1998. *Analysis and design of pavement*. Beijing: People Transportation Publisher.
Kim H. B. 1999. *Framework for incorporating rutting prediction model in the reliability-based design of flexible pavements*. Miehigan: Miehigan State University.
Retherford J. Q., McDonald M. 2010. Reliability methods applicable to mechanistic-empirical pavement design method. *Transportation Research Record* 25(14): 130–137.
Sanchez S. M. et al. 2005. Reliability based design optimization of asphalt pavements. *Int. J. Pavement Eng* 6(4): 281–294.

Reliability analysis of geotechnical engineering problems based on an RBF metamodeling technique

Q. Wang & J. Lin
Department of Civil and Environmental Engineering, Manhattan College, Riverdale, NY, USA

J. Ji
China Harbour Engineering Co., Ltd., USA Office, San Ramon, CA, USA

H. Fang
Department of Mechanical Engineering and Engineering Science, The University of North Carolina at Charlotte, Charlotte, NC, USA

ABSTRACT: In this paper, a reliability analysis method is presented for geotechnical engineering problems using a meta-modeling technique based on augmented Radial Basis Functions (RBFs). The RBFs are able to create accurate approximate models for linear and highly nonlinear functions. In this work, reliability analysis was studied using augmented RBFs that were identified from a previous study to be highly accurate models. An RBF-based metamodel was created to approximate a limit state/performance function. Once the metamodel was constructed, the failure probability was calculated using a sampling method such as the Monte Carlo simulation (MCS) method. As for numerical examples, two practical geotechnical engineering problems were considered. Various sample sizes were tested and the failure probability obtained using the RBF metamodeling technique provided good accuracy with a reasonable number of sample points. Compared with other conventional methods and the direct application of MCS method, the proposed reliability analysis method was very efficient, which required fewer evaluations of the original implicit limit state/performance function.

Keywords: reliability analysis; geotechnical engineering; failure probability; metamodeling technique; Radial Basis Function (RBF); Monte Carlo Simulation (MCS)

1 INTRODUCTION

Reliability analysis is commonly used to assess the failure probability of an engineering system, such as a geotechnical engineering system. Calculation of probability of failure, denoted by P_F, is the computation of multidimensional probability integrals (Ditlevsen & Madsen 1996):

$$P_F \equiv P(g(\mathbf{x}) \leq 0) = \int_{g(\mathbf{x}) \leq 0} p_x(\mathbf{x}) d\mathbf{x} \quad (1)$$

where $\mathbf{x}$ is a m-dimensional random variable vector; $g(\mathbf{x})$ is the limit state/performance function, and a system is deemed as failed if $g(\mathbf{x}) \leq 0$; and $p_x(\mathbf{x})$ is the joint probability density function (PDF) of the variable vector $\mathbf{x}$. In most practical engineering applications, Eq. (1) is difficult to be obtained due to the nonlinear and implicit nature of $g(\mathbf{x})$. A detailed simulation model, such as a finite element (FE) analysis model of the system is required in conjunction with a reliability analysis approach. The most probably point (MPP)-based methods and the sampling methods are the two kinds of reliability analysis methods commonly used in practice. The MPP-based methods include the first-order reliability method (FORM) or second-order reliability method (SORM), which are widely used for linear or nonlinear engineering systems (Kiureghian et al. 1987, Ditlevsen & Madsen 1996). Since the calculation of the first-order derivatives of system responses (simulation outputs) with respect to random variables (simulation inputs) is required, the integration of FORM/SORM with an existing analysis program is not an easy task. The sampling methods, such as MCS, require the sampling of random variables and calculation of the limit state/performance function in a repeated manner (Rubinstein 1981). The advantage of the sampling methods is that they do not require sensitivity analysis, i.e., derivatives of the limit state/performance function; therefore its application is much simpler. However the direct use of a sampling method is not computationally efficient, especially when expensive FE simulations are required.

To improve the computational efficiency, various metamodeling methods have been developed for reliability analysis of engineering systems involving expensive simulations. In the metamodeling method, the simulation or analysis tool is replaced by an approximate metamodel, which is an explicit function in terms of random variables. Therefore it is very efficient to calculate its function values. The most popular metamodel is the least-square polynomial model which is typically referred to as the response surface method (RSM). Due to its simplicity and efficiency, the RSM has been applied to reliability analysis of engineering problems (Kim & Na 1997). However a global RSM model is not appropriate to approximate highly nonlinear functions, since only a single polynomial is used for the entire input space. To improve the accuracy of the RSM, different techniques were proposed in the literature (Kang et al. 2010). Apart from the conventional polynomial-based RSM, other types of approximation techniques were developed, for example, kriging (Jin et al. 2001), artificial neural networks (Dai et al. 2011), high-dimensional model representation (HDMR) (Chowdhury & Rao 2009), and radial basis functions (RBFs) (Fang & Wang 2008). There is a need to develop metamodels to achieve the computational efficiency without compromising on the accuracy of results.

Recent studies demonstrated that the RBFs performed better than the global RSM for highly nonlinear functions (Fang & Horstemeyer 2006). One major advantage of RBFs over other approximation methods is that the RBF models pass through the sample points; therefore there are no errors at sample points. Some augmented RBF models were developed and shown to perform better than the commonly used non-augmented RBF models for all test functions. Little research, however, has been conducted to study the relatively new augmented RBF models, including the compactly supported functions for reliability analysis of geotechnical engineering problems. The present study is intended to fill the gap. The augmented RBF model created by a compactly supported RBF, which was identified as an accurate RBF model, was used in the reliability analysis in this work.

This paper presents a study of geotechnical engineering reliability analysis using a metamodeling technique based on augmented RBFs. RBFs were employed to create explicit approximations of the original limit state/performance function. Once the limit state/performance function was written in an explicit form, the MCS method was employed to evaluate the failure probability and reliability index. The results were compared with those obtained using the direct application of MCS method (without using metamodels) to assess the efficiency and accuracy of the augmented RBF-based method.

2 AN RBF META-MODELING TECHNIQUE FOR RELIABILITY ANALYSIS

For most engineering applications, a limit state/performance function $g(\mathbf{x})$ is implicit but can be numerically calculated for any given input random variable vector $\mathbf{x}$. Therefore, it is desirable to create an explicit and accurate approximate function of $g(\mathbf{x})$ for the entire input space of $\mathbf{x}$.

After some sample points are generated based on a design of experiments (DOE) method, an approximate function of $g(\mathbf{x})$ can be constructed using the values of $g(\mathbf{x})$ at these sample points that shall be well distributed and can represent the entire input space. Assuming there are totally n sample points used, the approximate function based on RBFs has the following general form:

$$\tilde{g}(\mathbf{x}) = \sum_{i=1}^{n} \lambda_i \phi(\|\mathbf{x} - \mathbf{x}_i\|) \qquad (2)$$

where $\mathbf{x}_i$ is the random variable vector for the ith sample point, $\|\mathbf{x} - \mathbf{x}\|$ is the Euclidean norm, ϕ is a basis function, and λ_i is the unknown coefficient, respectively. In Eq. (2), $\tilde{g}(\mathbf{x})$ represents a linear combination of RBFs with weighted coefficients that need to be determined. The most commonly used basis functions were studied by Fang & Horstemeyer (2006) and Fang & Wang (2008), including Gaussian, multiquadric, and compactly supported functions, etc.

It was found that the RBF models in Eq. (2) were less accurate to approximate linear responses. To solve this issue when linear responses are considered, augmented RBF models can be defined as follows:

$$\tilde{g}(\mathbf{x}) = \sum_{i=1}^{n} \lambda_i \phi(\|\mathbf{x} - \mathbf{x}_i\|) + \sum_{j=1}^{p} c_j f_j(\mathbf{x}) \qquad (3)$$

where $f(\mathbf{x})$ is a linear polynomial function, and p is the total number of terms in the polynomial. The unknown coefficients c_j ($j = 1, \ldots p$) need to be determined. Since the total number of unknown parameters is more than the number of equations available, the coefficients λ and c given in Eq. (3) need to be solved using the orthogonality conditions (Fang & Horstemeyer 2006).

To represent an augmented RBF model, a suffix '-LP' was used together with the symbol for its corresponding non-augmented model. For instance, an augmented RBF model created with the compactly supported function CS20 used the symbol 'RBF-CS20-LP'. The augmented RBF models were shown to be more accurate than the corresponding non-augmented models. The RBF and augmented RBF models are explicit in terms of the random variables; therefore they can be evaluated very efficiently. It is important to note that the Analysis of Variance (ANOVA) is not useful to evaluate the accuracy of RBF metamodels, since there are no model errors at sample points. Therefore, the accuracy of RBF metamodels is typically evaluated by calculating the RMSE values at randomly generated off-sample points (Fang & Wang 2008).

Eq. (2) and Eq. (3) provide RBF and augmented RBF approximation $\tilde{g}(\mathbf{x})$ of function $g(\mathbf{x})$, respectively. Once an explicit RBF or augmented RBF approximation is obtained, the MCS method can be employed to calculate the failure probability P_F in an efficient manner for any sample size. Note that the total cost of original implicit response simulations is determined by the sample size used in an RBF approximation. Therefore the majority of the computational cost is from the original implicit response simulations used for creating an RBF metamodel.

3 NUMERICAL EXAMPLES

Two geotechnical engineering examples from literature were adopted for evaluating the proposed reliability analysis method based on one augmented RBF model, i.e., RBF-CS20-LP. Latin hypercube sampling (Montgomery 2001) was applied to create different samples that were used to construct RBF metamodels. In the subsequent MCS method, a sample size $N = 10^6$ was adopted to calculate the failure probability P_F. The reliability analysis results using augmented RBFs were compared to those obtained by other researchers. When evaluating the computational effort, the original limit state/performance function was considered, and its required number of evaluations was compared. In this study, the direct MCS method was used to represent the application of the MCS method without using metamodels, and in this case the number of original limit state/performance function evaluations was governed by the MCS sample size. However, in the proposed RBF metamodel-based method, the number of implicit function evaluations was much smaller, which was the same as the sample size used to create the RBF metamodels.

3.1 Example 1: A tunnel problem

This example is a circular tunnel excavated in a homogeneous and isotropic rock, as shown in Figure 1. The hydrostatic far field stress is denoted as p_0 and the uniform internal support pressure in the tunnel as p_i. If p_i is less than the threshold value, i.e., the critical pressure p_{cr}, a plastic zone will be generated around the tunnel. The radius r_p of such plastic zone can be calculated using the Mohr–Coulomb failure criterion (Li & Low 2010). The limit state/performance function of the circular tunnel is defined as:

$$g(\mathbf{x}) = L - \frac{r_p}{r_0} \qquad (4)$$

where $L = 3$ is the permissible threshold about the plastic zone, and $r_0 = 1$ is the tunnel radius. In this study the two pressures were regarded as deterministic variables; $p_0 = 2.5$ MPa and $p_i = 0$ MPa. The elastic modulus E, cohesion c, and friction angle ϕ were random variables and their statistical properties are given in Table 1. The derivation of $g(\mathbf{x})$ in terms of E, c, and ϕ can be found in the literature (Li & Low 2010); therefore the details are not presented here.

Table 2 contains the numbers of original limit state/performance function evaluations associated with various sample sizes, i.e., 21, 51, and 81. A total of 2,197 off-sample points were randomly generated and used to study the accuracy of the RBF models. RMSE values were calculated and listed in Table 2. The RMSE values and thus the failure probability estimation

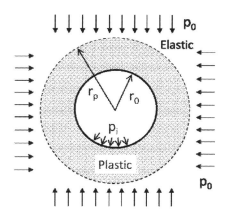

Figure 1. A Tunnel Problem in Example 1.

Table 1. Three Random Variables for Example 1 (Li & Low 2010).

Parameter	Mean value	Standard deviation
E (Mpa)	373	48
c (Mpa)	0.23	0.068
$\pi(°)$	22.85°	1.13°

Table 2. Numerical Results for Example 1.

Method	Failure probability	Error (%)	No. of function evaluation	No. of off-sample points	RMSE
RBF-CS20-LP	0.2587	5.3%	21		0.113
	0.2445	−0.5%	51	2,197	0.029
	0.2459	0.1%	81		0.022

errors using augmented RBFs became smaller when more sample points were used. The failure probability P_F based on the direct MCS method was 0.2457, as reported by Li & Low (2010). In this work, the errors estimating the failure probability became very small when 51 and 81 sample points were used. With 21 sample points, the failure probability P_F value was 0.2587, representing a 5.3% error. When the sample size was increased to 51, the error was reduced to 0.5%.

3.2 Example 2: A geotechnical settlement problem

This example is a geotechnical settlement problem (Ang & Tang 1975, Chowdhury & Rao 2009). The settlement of a point on the ground caused by the building construction is governed by the consolidation of the clay layer. The settlement S can be calculated for normally loaded clay, assuming the secondary soil consolidation and the induced settlement is negligible (Chowdhury & Rao 2009). If the required settlement shall be less than 2.5 in., the limit state/performance function can be written as:

$$g(\mathbf{x}) = 2.5 - S = 2.5 - \frac{C_c}{1+e_0} H \log \frac{p_0 + \Delta p}{p_0} \qquad (5)$$

Table 3. Five Random Variables for Example 2 (Chowdhury & Rao 2009).

Random variable	Mean	COV	Distribution type
C_c	0.396	0.25	Gaussian
e_0	1.19	0.15	Gaussian
H	168 (in)	0.05	Gaussian
p_0	3.72 (ksf)	0.05	Gaussian
Δp	0.5	0.2	Gaussian

Table 4. Numerical Results for Example 2.

Method	Failure probability	Error (%)	No. of function evaluation	No. of off-sample points	RMSE
RBF-	0.08335	2.9%	21	3,125	0.044
CS20-	0.08111	0.2%	51		0.012
LP	0.08089	−0.1%	81		0.005

where C_c, H, and e_0 are the compression index, thickness, and void ratio of the clay layer, respectively; p_0 is the original effective pressure before construction; and Δp is the pressure increase due to the construction of the structure. The five random variables and their statistical properties are given in Table 3.

Table 4 summarizes the numerical results for 21, 51, and 81 sample points. A total of 3,125 off-sample points were randomly generated and RMSE values were calculated to study the accuracy of the RBF models. It is obvious that the RMSE values and thus the failure probability estimation errors became smaller as the number of sample points increased. Compared with $P_F = 0.08096$, which was calculated based on the direct MCS method (Chowdhury & Rao 2009), the errors for estimating failure probability became very small when 51 and 81 sample points were used. With 21 sample points, the failure probability P_F value was 0.08335, representing a 2.9% error.

4 CONCLUDING REMARKS

The reliability analysis of practical engineering problems poses a challenge to the traditional MPP-based and direct sampling methods, due to the nonlinear and implicit limit state/performance functions involved. A study on accurate and efficient metamodels for reliability analysis is needed before they can be applied to practical large-scale problems. In this paper, a study of metamodeling techniques based on augmented RBFs was conducted which enabled a reliability analysis in an efficient manner. An RBF metamodel using compactly supported basis functions was used, which was identified as a highly accurate model from a previous study. Different numbers of sample points were tested in the augmented RBF model and the effects on the estimated failure probability were investigated. The RMSE values were calculated at off-sample points to evaluate RBF model errors. Two geotechnical engineering examples were selected from literature for assessment of the new method. Numerical examples showed that with a reasonable sample size, the proposed method worked well and accurate estimation of the failure probability was obtained. Compared with the direct MCS method, the proposed method was computationally efficient, as the number of original limit state/performance function evaluations was greatly reduced. A very small sample size should be avoided to create RBF models as the models may not provide accurate approximations of the true responses. Metamodels based on augmented RBFs have great potential for practical reliability analysis of engineering problems, especially when expensive response calculations are required, such as nonlinear transient FE simulations.

REFERENCES

Ang AHS & Tang WH. 1975. *Probability Concepts in Engineering Planning and Design, Vol. 1 Basic Principles*. New York: Wiley.

Chowdhury R & Rao BN. 2009. Assessment of High Dimensional Model Representation Techniques for Reliability Analysis. *Probabilistic Engineering Mechanics* 24:100–115.

Dai HZ, Zhao W, Wang W, Cao ZG. 2011. An Improved Radial Basis Function Network for Structural Reliability Analysis. *Journal of Mechanical Science and Technology* 25(9):2151–2159.

Ditlevsen, O & Madsen, HO. 1996. *Structural Reliability Methods*. Chichester: Wiley.

Fang, H & Horstemeyer MF. 2006. Global Response Approximation with Radial Basis Functions. *Engineering Optimization* 38(04):407–424.

Fang H & Wang Q. 2008. On the Effectiveness of Assessing Model Accuracy at Design Points for Radial Basis Functions. *Communications in Numerical Methods in Engineering* 24(3):219–235.

Jin R, Chen W, Simpson TW. 2001. Comparative Studies of Metamodeling Techniques under Multiple Modeling Criteria. *Structural and Multidisciplinary Optimization* 23:1–13.

Kang S-C, Koh H-M, Choo JF. 2010. An Efficient Response Surface Method Using Moving Least Square Approximation for Structural Reliability Analysis. *Probabilistic Engineering Mechanics* 25:365–371.

Kim SH & Na SW. 1997. Response Surface Method Using Vector Projected Sampling Points. *Structural Safety* 19(1):3–19.

Kiureghian D, Lin H-Z, Hwang S-J. 1987. Second Order Reliability Approximations. *Journal of Engineering Mechanics, ASCE* 113(8):1208–1225.

Li H-Z & Low BK. 2010. Reliability Analysis of Circular Tunnel under Hydrostatic Stress Field. *Computers and Geotechnics* 37: 50–58.

Montgomery DC. 2001. *Design and Analysis of Experiments*. New York: John Wiley & Sons, Inc.

Rubinstein RY. 1981. *Simulation and the Monte Carlo Method*. New York: Wiley.

Research of Chinese healthy city construction practice

Fei Lv, Yu Zhang & Xinyan Song
Department of Architecture, Harbin Institute of Technology, Harbin, Hei Longjiang, China

ABSTRACT: With the development of urbanization in China, many cities face a series of so called "city diseases" such as overcrowding, housing shortage, traffic jam and environmental pollution, which brought more and more attention to the healthy urban development in recent years. As a guiding theory of harmonious development between city and human being, healthy city has a great significance for creating a good living environment and promoting sustainable urban development. Based on the summary of healthy city assessment indicator system and building practices of five cites including Suzhou, Shanghai, Hangzhou, Beijing and Hong Kong, this paper provides reference and guidance for the improvement of the healthy city construction and index system.

1 THEORY OF HEALTHY CITY

1.1 Origin of healthy city

"Healthy Cities" came forth in the 2000 International Healthy Toronto Conference which was carried out in Canada in 1984 (NiyIA, 2003). This paper, which triggered a strong reaction, broke out the outdated connotation of outdated health, medical assistance and other concept intensions. It suggests that people living in healthy cities should enjoy the good natural environment and harmonious life. Subsequently, the city's health problem has been focused on as one of the most important problems. The World Health Organization began to intervene "Healthy Cities Project" in 1986. By formulating healthy urban planning, appropriate hygiene regulation and organizing citizens to participate in construction of healthy city project, WHO achieved praiseworthy achievement. In October 2003, WHO convened a healthy city international conference in the UK. In this meeting, the World Health Organization predictted the future on the foundation of summarizing the past experience and lessons, and announced that Europe entered the fourth stage of healthy cities construction. (International healthy cities conference, 2003) During the same period, active healthy cities movement had being rolled out in a large number of countries and making a hot trend in Canada, the U.S, New Zealand, Australia, Japan which became an international movement that varioous cities participate in. (Zhou Xianghong, 2008)

1.2 The concept of healthy city

Originally Healthy City is defined by WHO as: Where all citizens live in harmony, committed to sustainable development, respectful of diversity, reaching the highest possible quality of life and equitable distribution of health, by promoting and protecting health in all settings. (Hancock, T. & Duhl, L. 1988) After the first phase of European healthy city project, the definition added: Healthy City is a process rather than a result of definition, it is not only a city that has reached the special level of health condition, but a city that has a clear understanding of healthy situation and tried to improve it. Therefore, regardless of their current health level, any city can be a healthy city. What we really need is to make a commitment to improve the health level and set up appropriate structures and procedures to implement the commitment. (Tsours, A. 1990)

2 THE OVERVIEW OF CHINA HEALTHY CITY CONSTRUCTION

In China, national healthy city activities founded in 1989 have made the foundation for the construction of healthy city. In 1994, officials came to investigate China. They think China has the condition and it is essential for China to launch healthy city movement construction. In August 1994, the ministry of health, together with the leaders of Beijing, Shanghai municipal government made Beijing Dongcheng and Shanghai Jiading district as "healthy city project testing in China". In June 1995, Haikou and Chongqing Yuzhong district joined in the movements of making a healthy city. In 1996 Dalian, Suzhou, Baoding, have also carried out the activities of the healthy city. China health city construction stepped into the stage of all-round progress.

On December 28t, PHCCO held a kick-off meeting for the pilot work of the construction of healthy cities. They officially launched the movements of construction of healthy city, district (town), such as

Shanghai, Suzhou, Dalian, Karamay, Zhangjiagang, ten cities. Beijing Dongcheng district and west district, were determined as the nationwide first batch of pilot construction of healthy cities. The conference is held as an important milestone of our country's healthy city movement. It promotes further development of healthy city construction all over China. Since then, in order to enhance the urban environment and the health and quality of life, many cities create healthy city construction voluntarily. (Lv Fei & Shun Cheng, 2007) By the time, cities in our country which join in the healthy city alliance are Shanhai, Suzhou, Kunshan, Zhangjiagang, Changshu Naing, Jiangning, District, Wujiang, Huai'an, Tongzhou, Qian'an, Rugao, Tacang, Jilin, Tonghua, Luohu District, Shenzhen, MacaoSAR, Sai Kung District, K-wai Tsing District, The Kowloon City District, Kwun Tong District, Tai Po District, North District, Southern District, Wan Chai District, Wong Tai Sin District, YauTsim Mong District, TsuenWan, Sha Tin District, Islands District. (http://www.alliancehealthycities.com/htmls/members/index_members.html) There are 35 cities all together, 10 cities were accepted as pilot town of healthy town construction. According to preliminary statistics, six provinces had made plans of healthy city activities, 25 cities have been performed the activities.

This article selects five cities like Suzhou, Hangzhou, Shanghai, Beijing, Hong Kong to discuss, since they started the healthy city construction earlier. Their experience and system are relatively mature. Their economic, social development, healthy city construction was walking in the forefront. It is more helpful to analyze their healthy city index system.

2.1 Example of healthy city construction cities

2.1.1 Healthy city construction of Suzhou

Suzhou which was declared to WHO Western Pacific Area by the PHCCO, is the first pilot of healthy city in China, it began the healthy city project in 2002. The leading group of Suzhou construction healthy city organized the related departments to participate in formulating a healthy community, health schools and other 11 kinds of project standards. In 2004, it held a meeting of the construction of healthy city leading group members, researching the job of planning for the construction of healthy city, Suzhou's healthy city construction began to enter the stage of implementation. Suzhou focus on combining with its own feature and obtained the remarkable result. The number of international healthy city awards Suzhou has been regarded up to 25 by 2012. In recent years, the construction of healthy city in Suzhou even more combines urban and rural water environment management and protection, the ancient city of protective reconstruction and the improvement of residents' living environment. Try its best to build the first city group which takes Suzhou town as the center, with a healthy environment, healthy social, health services and healthy citizen in China.

2.1.2 Healthy city construction of Hangzhou

Hangzhou is the only pilot city listed as "national construction of healthy city" among the capital of China city. Activity of building healthy city in Hangzhou has experienced three stages: the early researching, pilot projects and officially launch. In 2005, Hangzhou city carried out the feasibility of the construction of healthy city project research, then started the pilot projects in the uptown, downtown and Gongshu district. In 2007, Hangzhou put forward seven "Everyone Share" as the goal during the process of promoting healthy city movement. In July 2007, Hangzhou made the plan "Hangzhou action of healthy city in , three years (2008 to 2011)". In the same year, Hangzhou authorized by national PHCCO to set the standard for "national healthy city" index system and evaluation system. In 2013, Hangzhou was voted as the most healthy deputy provincial cities in China during the assessment of healthy city in China.

2.1.3 Healthy city construction of Shanghai

History of health-city-construction in Shanghai dated from 1994 at Jiading district is the first phase of healthy experiment in China. Shanghai opened the first three years' healthy city from 2003 to 2005, ever since it has taken the plans for the three years' twice building healthy city. According to the principle of the whole advancement and personal development, the health city construction in Shanghai promoting various of items, such as the healthy building of Jingan district, healthy markets of Minhang district, healthy lanes of Huangpu district and so on. As the same time, the construction had been fully combined with the key work such as the 2010 Shanghai Expo preparation, the new rural construction and so on. WHO Health Cooperation Center was set up in Shanghai in 2011, which makes the hygienic and patriotic movement joined with the international further. As the first megacity who built healthy city in China, Shanghai provide theory and practical experience for other big cities and megalopolises.

2.1.4 Healthy city construction of Beijing

Beijing, as the center of political and cultural and international communication hub in China', walked in the forefront of the construction of the national health city when the concept of "Healthy City" introduced in China in the 1990s. Then, In 1993 WHO formally established cooperative relations with Beijing Dongcheng district. In the following year, Beijing Dongcheng district launched the project pilot of federal primary batch of healthy city. In 2001, Beijing patriotic health campaign committee took healthy urban activity on the agenda; In September 2004, Beijing started a three-year health community activity. After the district (town) started the construction of healthy city activities, the patriotic health campaign office took the Dongcheng district and the two district of Beijing as the construction of a healthy city. During the 2008 Olympic Games. Beijing carried out in the national health series of activities which the theme is 'Health Olympics, healthy Beijing throughout the city. In August 2011, Beijing health association for the advancement of urban construction created, the advancement focused on '12th five-year plan' of health Beijing, mainly involved in the construction of healthy city and the national health activities, taught the new concept of healthy city, the new model to the masses, emphasized on public participation. Meanwhile, it established the plan "Health from Beijing – the national health promotion action plan (2009–2018) for ten years", "the health Beijing 12th five-year development plan" and other relevant laws, regulations and special planning. These promoted the health of the people as the center concept throughout all aspects of the urban construction and development.

2.1.5 *Healthy city construction of Hong Kong*

The movement of Hong Kong healthy city began in the late 1990s. After SARS in 2003, it got the substantial advance. At the end of 2013, the western-central district, the Jiulong district, Xigong district and more than a dozen areas had promoted the healthy city activities. According to their own characteristic, districts operated in different development phases. The department of health also compiled Health-City-Construction, promoting healthy urban planning guidance in Hong Kong'. And taking the Jiangjunao as the first city to promote health activities in Hong Kong. Scientifically and reasonably promoting the construction of Healthy City. At the end of 2012, the Guardian in the UK held the vote of the most healthy city in the world, Hong Kong won the second place.

2.2 *Analysis and evaluation index system of Chinese healthy cities*

Beijing, as a modern metropolis of historic values, is heading towards a magnificent goal of creating a world city with Chinese characteristics. So it is important for Beijing to continue building healthy city construction. There are 35 indicators in healthy city indicator system of Beijing, including 10 healthy levels indicators, 10 health services indicators, 15 healthy environments indicators. The feature of healthy city indicator system can be observed from the selection of indicators. In the course of selecting crowd indicators, it especially selected age-specific mortality rates of heart disease and other four indicators which need several years or even decades to evaluate. Moreover, Beijing also highlighted the equality of health in the selection of health city indicators, which includes the expectation life gap between urban and rural areas indicator, obesity rate of primary and secondary school students indicator and other indicators that related to health environmental protection, transportation, education, agriculture and other sectors.

Shanghai whose overall healthy level standing in the front ranks of the global has a high health level. Therefore, the selection of healthy city indicator system of Shanghai focused on the action plan. Shanghai healthy city indicator system includes 31 indicators, which including 4 everyone eating healthy diet actions, 3 everyone limiting alcohol and tobacco actions, 2 everyone scientific exercising actions, 2 everyone feeling enjoyable actions, 7 everyone cleaning home actions, 13 major actions. (Shanghai Municipal People's Government, 2008) Hangzhou, as one of the original pilot healthy city, has achieved astonishing results through years of active exploration. Its construction of healthy city is already in the domestic advanced level. Healthy city evaluation index system of Hangzhou includes 44 indicators, including 7 environmental indicators, 10 population indicators, 10 service indicators, 10 social indicators, 7 political polling indicators. (Hangzhou Municipal People's Government, 2011) Indicators involved are more conspicuous in citizens demand and life needs, such as the superior air days throughout the day, the number of pension institutions beds for per 100 elderly and others which it is relevant to our lives; the number of medical beds per thousand people, the number of practicing nurses per thousand and other indicators. That has been concerned about expensive medical bills and difficult access to quality medical services.

Suzhou City health cities index system is total of 95 items, which is divided into three items that include core indicators, fundamental indicators and development indicators. Specifically, there are 12 core indicators and 81 basic indicators. (Suzhou Municipal People's Government, 2008) Health Indicator System of Suzhou is relatively delicate, which set out more specific demands of the scope and the number of healthy scenario building, and has closely coupled with all walks of life and reflect wide adaptability. The healthy city indicator system of Suzhou has been revising every three years, which reflects the step-by-step of the construction of health city. Furthermore, every indicator is specific to the responsibility and collaboration sectors of the health index system of Suzhou city, which reflected the characteristics of government promoting, department heading and community involving. (Xie Jiangfeng, 2005)

Healthy city indicator system of Hong Kong is divided into seven major themes, 38 indicators which including 5 population problems, 6 health status, 7 lifestyle, 7 living environment, 5 social and economic conditions, 1 inequality, 7 physical and social basis. (Department of Health Hong Kong SAR, 2007) The phenomenon of Hong Kong's increasing population and decreasing land is very prominent. Although Hong Kong's land is very tense and the living area is small, the government opened up large tracts of land which used for greening. What's more, the environment and the developmental level of Hong Kong are at the forefront of the world, there are colorful public areas within the residential. Therefore, the indicator system of Hong Kong give a more prominent concern to public health and life, including stress management, regular physical examination, the phenomenon of inequality, and other specific indicators (Table 1).

Based on the work experience of the healthy cities pilot construction work. At the end of 2013,

Table 1. Comparison of Suzhou, Hangzhou, Shanghai, Beijing, Hong Kong healthy city indicator system.

	Senior indicators
Suzhou	Core Indicators
	Basic indicators
	Development Indicators
Hangzhou	Environmental Indicators
	Population indicators
	Service indicators
	Social indicators
	Public opinion indicators
Shanghai	Everybody healthy diet
	Everybody limit tobacco and alcohol
	Everybody scientific fitness
	Everybody mental and physical pleasure
	Everybody clean their homes
	Key Project
Beijing	Healthy level
	Healthy service
	Healthy environment
Hong Kong	Demographic indicators
	Health status indicators
	Lifestyle indicators
	Living Environment indicators
	Social and economic indicators
	Inequalities
	Tangible and social infrastructure

Table 2. Healthy cities index system of China.

Senior indicators (weight)	Secondary indicators (weight)
Healthy environment [40%]	The number of days air quality meets standards [30%]
	Urban sewage treatment rate [10%]
	Life garbage treatment rate [30%]
	Built-up area greening rate [20%]
	Urban population density [10%]
Healthy culture [15%]	Number of theaters per square kilometer in the town [33.3%]
	The total reserves of the public library per thousand citizen [33.3%]
	Internet penetration [33.4%]
Health Conditions [15%]	Number of medical institutions in the city per square kilometer [25%]
	Number of beds in hospitals and health centers of per thousand people [20%]
	Number of medical practitioner of per thousand [25%]
	Number of the people attending basic medical insurance [30%]
Healthy society [30%]	Food safety attention level [25%]
	Fitness attention level [20%]
	Health attention level [15%]
	Fog haze attention level [40%]

China Urban Development Research Task Force set up an evaluation system for Chinese healthy city. In reference to the WHO experience and other relevant index system, the relevant ministries, researching institutions and leading experts considering the new development concepts and practical experience abroad, initially constructed the system. (http://www.chinacity.org.cn/cstj/zxg-g/127303.html) The indicator system contents 4 premier indicators – healthy environment, healthy culture, healthy condition, healthy community, which is subdivided into 16 secondary indicators including: related the greening rate of built-up area, food safety attention, haze attention and so on (Table 2). Through the assessment of two hundred cities and towns all over the country, the actual circumstances in various cities. The index concludes survey the results of city healthy levels. After the results released, it triggered a wide-ranging discussion and more and more cities' government began to focus on healthful city construction.

3 CONCLUSIONS AND RECOMMENDATIONS OF HEALTHY CITY ASSESSMENT AND INDEX SYSTEM

The above analysis shows that, the current understanding of healthy city has become more and more thorough. Healthy city are no longer limited to the health sector, also involves education, psychology, transportation, civil organizations and many other departments and organizations. More than just a concept of healthy population, healthy city is a comprehensive one that includes various aspects of urban construction and development. At the same time, although the introduction of healthy city in China has been nearly the 20th anniversary, the number of cities which practice healthy city construction is not enough. There is not any clear indicator system in most cities' construction and the construction is still at the planning level. Cities which selected in this paper are the ones that have a specific indicator system and reflect the goals and philosophy of healthy city perfectly. However, these existing indicator systems are mainly resulting indicators, lacking of the action and supervision indicators. So, in order to propel the construction work of a healthy city, it has yet to be refined and promoted. (Yu Haining et al. 2012)

3.1 Improving management levels and complete evaluation system

The building of healthy city index system needs further standardized and improved, it should increase specific actions, monitoring and management indicators and put the health cities building into practical action. Moreover, the assessment of indicators should form a definite periodicity. In order to facilitate the comparative analysis of the latter data, we should focus on the data collection and preservation of diverse construction period. Such as Suzhou, which has modified every three years to eliminate unjustified targets and increase demand indicators to makes the indicator system improved constantly updated.

3.2 *Increasing featured indicators and maintaining the vitality of the index system*

The index system should continue to improve itself by combining with the cities' own level of economic development and other urban characteristics in different stages and environment. For example, the health indicators system of Tainan added the number of nut stands and stray dogs and other indicators which are full of local characteristics. Nevertheless, indicators for mental health are insufficient. The pace of modern life becomes increasingly fast, urban stress and psychological problems grow with each passing day. It is necessary to add the psychological indicators into the system. (Chen Zhaojiao & Xv Liangwen, 2013).

3.3 *Establishing awards selection mechanisms and relevant departmental cooperation*

Indicator system should improve the incentive mechanism to encourage different sectors of the society to participate in building healthy cities. It should promote the development of healthy cities by improving recognition reward system, implementing the responsibility assessment, building healthy cities and communities demonstration area. The construction work of healthy city is a wide-ranging and long-term movement. Therefore, for the smooth implementation of projects in various fields from the policy, it should urge relevant departments to cooperate and make healthy cities project into departmental planning.

REFERENCES

Beijing Municipal Health Bureau. Health Beijing "Twelfth Five Year Plan" development and construction planning (2011–2015) [Z]. 2011.

Chen Zhaojiao, Xv Liangwen. 2013. Research on Evaluation and Index System Healthy Cities [J]. Health Research: 5–9.

China healthy cities evaluation index system and evaluation results for the year 2013 [EB/OL] http://www.chinacity.org.cn/cstj/zxgg/127303.html

Department of Health Hong Kong SAR. Building Healthy Cities—Promoting healthy urban planning guidelines in Hong Kong [Z] 2007.

Hancock, T & Duhl L. 1988. Promoting health in the urban. context[M]. Copenhagen: FADL.

Hangzhou Municipal Government, Health Hangzhou "Twelfth Five Year Plan" (2011) [Z] 2011.

International Healthy Cities Conference. 2003, Belfast. 19–22.

Lv Fei & Shun Cheng. 2007. Strategies for the implementation of the overall construction of Healthy Cities [A] Urban Planning Society of China. Harmony Urban Planning—2007 China Urban Planning Annual Conference [C]. Urban Planning Society of China.

NiyIA. 2003. The Healthy cities approach-reflections on framework for improving global health, Bull WHO. 81(3):222.

Shanghai Municipal People's Government. Shanghai Urban Construction Health Action Plan (2012–2014) [Z] 2012.

Suzhou Municipal People's Government. Suzhou Building Healthy Cities Index System (2008–2010) [Z] 2008.

Tsouros, A. 1990. *World Health Organization healthy cities project: a project becomes a movement-review of progress 1987 to 1990* [M]. Copenhagen: FADL and Milan. Sogess.

The members of alliances for healthy cities. [EB/OL] http://www.alliancehealthycities.com/htmls/members/index_members.html

Xie Jiangfeng. 2005. Suzhou Urban Health Indicator System [D]. Suzhou University.

Yu Haining, Cheng Gang, Xv Jin Wang Haipeng, Chang Jie, Meng Qingyue. 2012. Comparative analysis of the health index system of urban construction [J]. China Health Policy Research: 30–33.

Zhou Xianghong. 2008. *Healthy Cities: International Experience and China strategy* [M]. Beijing. China Building Industry Press.

Research on the function hierarchies of MA rail transit network system based on the time goal

Ming Yang & Xiucheng Guo
Southeast University, Nanjing, Jiangsu, China

Xiaojing Ling & Xiaoying Bi
Nanjing Institute of City and Transport Planning Co., Ltd, Nanjing, Jiangsu, China

ABSTRACT: In order to guide rail transit network planning in metropolitan areas, the article firstly concludes the basic structure, function hierarchies, and technical features of rail transit network in international mature metropolitan areas, analyzes the commute time patterns in these areas, and specifies the commute time goals of rail transit according to their types. Based on these control factors, the article then proposes the function hierarchies of metropolitan area (MA) rail transit network.

1 OVERVIEW

A metropolitan area (MA) is a highly integrated combination of a densely populated urban core and less-populated surrounding territories. These territories are socio-economically tied to the urban core. An MA is a city space structure that takes form in the advanced stage of urbanization and is the basic geographical unit for city statistics collection and research (Xu 2006). Based on the strength of impact from urban core, a metropolitan area can be divided into three circle layers from inside out: central city, fringe region and countryside, among which the central city is most urbanized and has the most concentrated functions. The three circle layers decrease in population density and employment rate. An MA may have several communities. And the formation of such communities requires a comprehensive city transportation system, especially a system that takes high-capacity rail transit system for the backbone. As an intensive public transportation method, rail transit will display more and more advantages in a situation where energy and environmental issues deteriorate day by day.

With the rapid development of MAs in China, the construction of regional or city rail transit network also enters into a high-speed stage. However, due to the lack of mature MAs and the incomplete understanding of rail transit in MAs, current researches on rail transit planning are largely restricted to the traditional administrative districts of a city, and rail transit network planning explores only the morphology and layout of subways. In fact, MAs have broken the administrative districts of a city, and thus are characterized by large size, circle-layered structure, and long commute distance. Besides subway, high-speed railway, inter-city railway and suburban railway all play significant roles in meeting the long- and medium-distance commute needs in an MA. Therefore, we need to understand the differentiated commute needs that vary with circle layers and the function hierarchies of rail transit network so that we can build one with well-defined functions. These understandings are of great importance to the sustainable development of rail transit in MAs and make it possible for an MA to fully display the advantages of rail transit network system.

2 SUCCESSFUL EXPERIENCE OF RAIL TRANSIT NETWORK IN INTERNATIONAL MAS

2.1 Function hierarchies and technical features of rail transit network in international MAs

After hundreds of years' evolution, international mature MAs have developed complete rail transit network with clear function hierarchies. Table 1 shows the function hierarchies and technical features of rail transit network in Tokyo, Paris, London, and Moscow.

These MAs may differ in city size, space structure and rail transit names, but their function hierarchies of rail transit network are essentially the same, i.e., regional railway, suburban railway, and urban railway. Urban railway, such as subway and light rail, serves the highly urbanized central city within the radius of 15–20 km. Urban railway features high capacity, high service frequency, short stop spacing, and low travelling speed. The average line length is about 40 km and the highest driving speed is 80 km/h. Suburban railway serves commuters from fringe region to central city, within the radius of 50–60 km. Suburban railway features longer stop spacing and faster travelling speed.

Table 1. Function hierarchies and technical features of rail transit network in international MAs.

Circle Layer		Central city	Fringe region	Coutryside	Nation-wide
Metropolitan Areas	Tokyo	Subway	JR commuter	JR commuter/JR intercity	Shinkansen
	Paris	Subway	RER	RER/Rgional railway	TGV
	London	Subway	Suburban railway		National railway
	Moscow	Subway	Suburban railway		National railway
Speed (km/h)		25–35	40–70	40–80	200–320
Stop spacing (km)		0.5–1.2	2–3	2–5	30–50

Table 2. Functional hierarchies and technical features of rail transit network in China's major cities.

		Central city	Administrative region	Nation-wide
Beijing	Circle Layer	R: 20 km S: 1085 km^2	R: 80 km S: 16808 km^2	High-speed railway, inter-city railway, and common railway
	Rail transit	Subway Light rail	Suburban railway	
Shanghai	Circle Layer	R: 15 km S: 660 km^2	R: 45 km S: 6340 km^2	
	Rail transit	Subway Light rail	Suburban railway Express suburban railway	
Wuhan	Circle Layer	R: 20 km S: 678 km^2	R: 50 km S: 8549 km^2	
	Rail transit	Subway Light rail	Suburban railway Express suburban railway	

Table 3. Technical features and matching relationship between China's and international rail transit network systems.

	International			
China	National railway	Intercity railway	Suburban railway/Community express railway	Urban railway
Shinkansen /High-speed railway/ Eurocity high-speed railway	○	○	–	–
JR/National railway	○	○	○	–
Privately owned railway/Regional railway	–	–	○	○
Urban railway	–	–	–	○

Note: ○ indicates the matching relationship.

The highest driving speed is 120–160 km/h. Regional railway such as high-speed railway and inter-city railway, meets the national and inter-city commute needs. Based on the transit layout of an MA, some regional railway may also serve the long- and medium-distance commute needs of travelling between MA circle layers, such as between countryside and central city.

2.2 Function hierarchies and technical features of rail transit network in China's major cities

Compared with developed countries, the urbanization and rail transit construction have started later in China. However, MAs around some megacities are growing fast and the coverage of rail transit network is also extended. Learning from the successful experience of international MA planning, China has fostered a relatively complete rail transit network system. In most Chinese cities, their rail transit systems are mainly composed of urban railways. In only a few cities such as Beijing, Shanghai and Wuhan, they have also built several suburban railways. Table 2 shows the function hierarchies and features of rail transit network in China's major cities.

2.3 Comparison between international and China's rail transit network systems

The function hierarchies and technical features of rail transit network vary with countries. In order to understand and absorb the successful experience of international MAs, the article compares China's rail transit network systems with those of international cities. Table 3 shows their technical features and the matching relationship in between.

China has not fostered a mature MA by now. A big proportion of commuters live in the central city. There are no strong commute needs of travelling from suburbs to central city. So, the city will not pay sufficient attention to suburban railways during rail transit network planning. In fact, suburban railway plays a quite important role in the transportation network of a mature MA. Connecting the vast suburban area to the central city, suburban railway has the ability to serve heavy commuting flows. With the further development of MAs in China, city space will keep expanding, leading to more residents travelling from city periphery to central city and in turn growing needs for suburban railways. In addition, according to the required travelling speed, international MAs further divide urban railways into express railway and common railway by adjusting stop spacing. Currently in China, only Shenzhen and Nanjing are taking stop spacing into consideration and set express railway and higher-frequency railway in planning the rail transit network. In fact, there has been no real distinction between express railway and common railway in China by now.

3 COMMUTE TIME OF MAS

3.1 Average commute time

In the early stage of urbanization when city size is expanding, the commute time of the city will increase. But commute time will finally stabilize in the latter stage of urbanization (Szalai 1972, Tanner 1996). Theoretically, longer distance between the place of residence and central city leads to longer commute time and lower employment rate in the central city. When the commute time exceeds residents' accepted range, they will give up long-distance travels. Under the impact of these factors, the commute time of an MA usually shows a rising and then falling pattern, as shown in Figure 1.

Compared with international MAs, China's city residents mainly live within the central city and the average commute time is about 30 minutes (Zhu 2010), as shown in Table 4. With the progress of urbanization, commute time will rise. The commute time of residents living in the highly urbanized central city is about 30–35 minutes.

3.2 Centripetal commute time

Centripetal commute time reflects the relationship between city size and residents' travelling status. It also indicates the maximum commute time that can be tolerated by residents. Taking Tokyo as an example, as shown in Figure 2, the average centripetal commute time is 68.7 minutes. In particular, the commute time of travelling within central city is 30 minutes, the commute time of travelling from fringe region (within the radius of 20–40 km) to central city is 30–45 minutes, and the commute time of travelling from countryside to central city is 45–60 minutes.

Based on the analysis above, the article concludes the following characteristics of commute time of

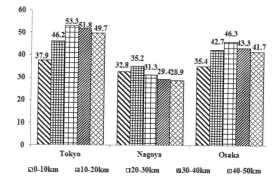

Figure 1. Average commute time by distance in Tokyo, Nagoya, and Osaka (min).

Table 4. Average commute time of China's major cities.

City	Beijing	Chengdu	Nanchang	Fuzhou
Average commute time	33.31	32	28	22.43

City	Hangzhou	Xiamen	Suzhou	Shijiazhuang
Average commute time	32.9	24.31	22.74	26.11

City	Hefei	Changsha	Guiyang	Wuhan
Average commute time	28.3	30	28.3	33

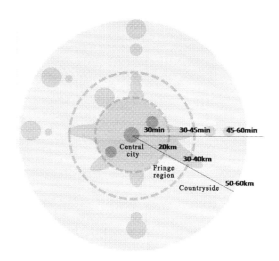

Figure 2. Average centripetal commute time of Tokyo.

MAs: (1) generally, the average commute time in an MA is about 45 minutes and the maximum centripetal commute time does not exceed 60 minutes; (2) the commute time of residents living in the central city is less than 30 minutes; (3) the commute time of residents living in the fringe region ranges from 30 minutes to 45 minutes; (4) the commute time of residents living in the countryside does not exceed 60 minutes, which is also the maximum commute time that can be tolerated by residents.

4 FUNCTION HIERARCHIES OF RAIL TRANSIT NETWORK BASED ON THE TIME GOAL

Based on the commute time of international and China's MAs, we can set goals for the rail transit system in each MA circle layer: The maximum commute time of residents living in the central city (within the radium of 5 km) is 30 minutes, the maximum commute time of residents travelling from fringe region to central city is 45 minutes, and the maximum commute time of residents travelling from country side to central city is 60 minutes.

4.1 Travelling speed requirements for each circle layer

A typical MA consists of central city, fringe region and countryside. Central city covers an area with a radius of 15–20 km, fringe region covers an area with a radius of 30–40 km, and countryside covers an area with a radius of 50–60 km, as shown in Figure 3. Therefore, the distance between city core and central city, fringe region, and countryside are respectively 10–15 km, 25–35 km, and 45–55 km.

Given the radius and commute time goals of each circle layer, we can determine the travelling speed requirements of rail transit system. For central city, the maximum commute time is 30 minutes and average shuttle time is 10 minutes, so the travelling speed must reach 30–45 km/h. For fringe region, the maximum commute time is 45 minutes and average shuttle time is 15 minutes, so the travelling speed must reach 50–70 km/h. For countryside, the maximum commute time is 60 minutes and the average shuttle time is 15 minutes, so the travelling speed must reach 60–75 km/h.

4.2 Stop spacing requirement for each circle layer

Given fixed transportation infrastructure and train model, rail travelling speed is mainly determined by stop spacing. Longer stop spacing leads to higher travelling speed. Assume the travelling distance of a train is L, the total travelling time T_travel is the sum of total driving time T_drive, total acceleration and deceleration time in n stations T_change, and the total stop time T_stop. In Figure 4, d indicates the stop spacing.

Assume the acceleration when a train decelerates or accelerates is both a, t_a is the acceleration and deceleration time at a single stop, s_a is the sum of driving distance during acceleration and deceleration, V_h is the highest driving speed, and t is the time that the train stays in a station. Then, the total travelling time of a train that has passed n stations is calculated as follows:

$$T_{tra} = T_v + T_{stop} + T_h = n(d-s_a)/v_h + (n-1)t + nt_a$$

$$= \frac{L}{d}(d - \frac{v_h^2}{a})/v_h + (\frac{L}{d}-1)t + \frac{L}{d}\frac{2v_h}{a} \quad (1)$$

The travelling speed is calculated as follows:

$$v_{tra} = L/T_{tra} = \frac{1}{\frac{1}{d}(d - \frac{v_h^2}{a})/v_h + (\frac{t}{d} - \frac{t}{L}) + \frac{2v_h}{d \cdot a}} \quad (2)$$

According to the rail transit statistics in China (Huang 2013), the time that the train stays in a station $t = 45$ s, the acceleration $a = 1.0$ m/s^2, the highest driving speed v_h has three values, respectively, 80 km/h, 120 km/h, and 160 km/h, to calculate the travelling speed, as shown in Figure 5.

Judging from the figure above, stop spacing has strong correlation with travelling speed. When stop spacing is between 0.5 km and 1.5 km, the highest driving speed has little influence on travelling speed. However, when stop spacing rises to between 2 km and 5 km, the highest driving speed can dramatically elevate travelling speed. Thus, we can adjust stop spacing to cater for the travelling speed requirements of each circle layer. For example, urban railway has many stops, so the highest driving speed of trains

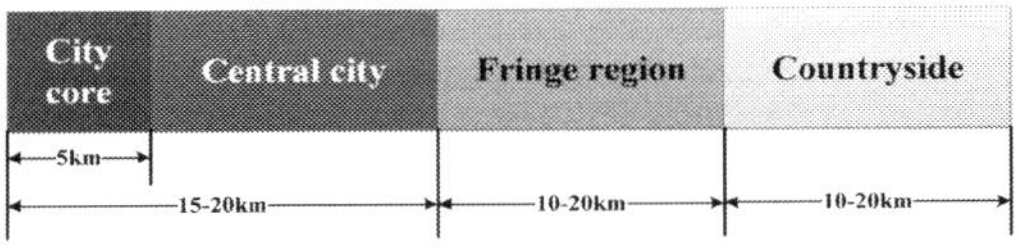

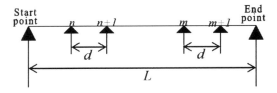

Figure 3. Radius of each MA circle layer.

Figure 4. Railway travelling process.

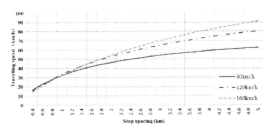

Figure 5. Relationship between station spacing and travelling speed.

should reach 80 km/h; suburban railway has few stops, so the highest driving speed of trains should reach 120–160 km/h. Based on the travelling speed requirements analyzed in the previous section, the article can determine the stop spacing for each MA circle layer.

Central city, this circle layer is urbanized with concentrated functions. Railway service for this region highlights high coverage and availability. So, the suggested stop spacing ranges from 0.8 km to 1.2 km.

Fringe region, this circle layer boasts the most frequent connection with central city. Railway service for this region must provide higher travelling speed to ensure that residents can reach the central city promptly. To provide such service, we suggest two modes: (1) outreached urban railways (express railway in central city). In this mode, we need to increase the stop spacing in urban railways or set overtaking tracks. The suggested stop spacing ranges from 1.5 km to 2 km. (2) Outreached suburban railways. In this mode, suburban railways serve both the fringe region and countryside. The suggested stop spacing ranges from 2 km to 5 km.

Countryside, this circle layer features long- and medium-distance railways that are connected in series. Such railway service leads to independent peripheral communities. Compared with fringe region, countryside requires more express railway service to meet the commute needs, so the stop spacing should be longer, ranging from 3 km to 8 km. In particular, for small-sized communities with commute needs, we should consider whether national railways and inter-city highways can meet their needs to reduce construction investment.

Table 5 describes the stop spacing suggestion for rail transit.

4.3 Function hierarchies and technical features of rail transit network

To achieve the commute time goals of each circle layer and promote the coverage of rail transit network and travelling efficiency, this article suggests that the rail transit network in MAs be divided into three hierarchies according to their functions: (1) regional railway, serving commuting flow travelling between central city and suburban areas. According to the service scope, regional railways can be further divided into national railways and inter-city railways. Inter-city railways provide services for residents in both the countryside and central city. (2) Suburban railway, serving commute needs of residents living in fringe region and countryside. (3) Urban railway, serving commuting flow within the central city and diverting commuting flows from regional and suburban railways. According to the service scope and travelling speed, urban railways can be further divided into express railway and common railway. Figure 6 shows the function hierarchies and service scope of MA rail transit network. Table 6 describes their technical features.

Table 5. Stop spacing for rail transit.

Circle	Central city	Fringe region	Countryside
Time goal	30 minutes	45 minutes	60 minutes
Speed requirement	30–40 km/h	40–55 km/h	50–65 km/h
Station spacing	0.8–1.2 km	Outreached urban railway: 1.5–3.5 km Outreached suburban railway: 3–5 km	3–8 km

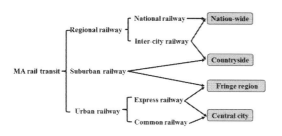

Figure 6. Function hierarchies and service scope of MA rail transit network.

5 CONCLUSION

With the development of urbanization in China, metropolitan areas are becoming more and more important. As an essential factor in the MA space structure, rail transit system faces new challenges either in network planning or operation and management. The article drew references on the successful experience of international MAs, analyzed the commute time of MA residents, and explored the relationship

Table 6. Technical features of MA rail transit network.

Item		Service scope (km)	Stop spacing (km)	Travelling speed (km/h)	Highest driving speed (km/h)
Urban railway	Express railway	About 20	1.5–2	35–40	80–100
	Common railway	Below 20	0.8–1.2	25–35	70–80
Suburban railway		20–100	2–5	45–60	80–350
Regional railway	Inter-city railway	50–300	10–20	160–250	250
	National railway	Above 200	50	80–350	350

between travelling speed and stop spacing based on commute time goals. Moreover, the article considered the differentiated commute requirements of MA circle layers on the rail transit system, and then proposed the basic structure, service scope, function hierarchies and technical features of the rail transit system, which boasted referential significance to the rail transit network planning for China's MAs.

REFERENCES

Huang, Rongsong. 2013. Research on Stop Spacing in Suburban Railway based on the Shortest Commute Time. *Transport Information and Safety* 5: 74–81.

Szalai. A. 1972. *The Use of Time* (ed.), The Hague: Mouton, 113–144.

Tanner, J.C. 1961. *Factors Affecting the Amount of Travel* (ed.), London: Road Research Tech: Paper No. 51, H.M. Stationary Office.

Wang, Xu. 2006. *American City Development Mode: Urbanization to Metropolitan Area* (ed.), Beijing: Tsinghua University Press.

Zhu, Weiguo. 2010. Rail Transit Function Positioning based on Commute Time. *Urban Rapid Rail Transit*. 6: 60–63.

Research on the spatial morphology and holistic structural preservation of East Asian city moat area—sample: Haohe river historic area of Nantong city

Leilei Sun
School of Architecture and Urban Environment, Soochow University, Suzhou, China

Hua Gu
General Manager, JoMo Architectural Design Co. Ltd, Shanghai, China

ABSTRACT: As a special historical district which carries the history and culture of a city, the city moat basin in East Asia makes the unique urban images as well as the urban features. The old city core is just approximately overlapped with the city moat basin, and it is exact the reason why the research on the theme is vital.

This thesis takes the typical city moat basin—the Haohe River historical district in Nantong city for example, from the view of the holistic preservation and renovation, use the specific urban morphology of city moat area and architecture typology analysis as tools to deduce the practically and methodologically effective discussion and experiment.

Keywords: city moat area; spatial morphology; historical heritage; holistic; structural preservation

FOREWORD

River, the cradle of human civilization, has been the most important natural condition considered for people to settle down. The survival of original inhabitants and the rivers were inextricable. The river has very long-standing ties with the city, which was an important factor to determine the location of a city in ancient times, and the moat was the basis of building the original city. "Cheng" (city) and "Chi" (moat) in Chinese history were associated. The literal meaning of "Cheng Chi" was city walls and moat, refers to a city in general.

The moat is carrying the city's historical memory. It is not just a river, but rather a carrier of urban culture. The core area of the old town coincides with the area of the moat basin considering the logic of the development of the urban form and spatial structure, which is exactly the significance of the research on the protection and innovation of the moat basin.

1 AN OVERVIEW OF THE MOAT BASIN'S SPATIAL FORM

As part of a defense system of ancient cities, a moat was a ditch artificially excavated and filled with water to provide it with a line of defense. As an obstacle outside the walls, moats made access to the walls difficult for siege. In the process of historical evolution, the moat has a variety of functions: urban flood control, fishing and hunting, salt shipment, water transport, irrigation, water supply, farming, shipping, recreation, tourism, etc. Moats in many cities in China are well preserved with a complete pattern, such as Suzhou, Beijing, Xi'an, Nantong, however, the functions of the moat has been changed. In a sense, moats had accelerated the growth of cities as well as carried the historical memories of cities. The moat is the most primitive core in the pattern of urban space, serving as a container of the city's historical precipitation. As the initial framework of the spatial structure, the moat basin is a spatial element of special urban districts, as well as an important consideration of the development of urban form and the evolution of spatial pattern. The original outline of the city can be found at the moat basin, where retains the historical relics of the original city, blended with the urban spatial structure in the historical process.

The moat basin, the study subject in this thesis, is defined as the one in a space-time framework consisting of the space of the moat waters, the boundary between land and water, and the waterfronts in certain areas. The moat basin in this thesis will be referred to as mentioned above as a type of urban spatial morphology, which is derived from the original meaning of the moat basin.

In recent years the focus of study has gradually shifted toward emphasizing the historical significance, the protection of local districts, the coordination of the features of historical districts and the proper disposal of historical relics. Even so, there are still a lot of problems because of the complexity of the protection and development of urban historical districts, as well as the particularity of the spatial form of the moat basin. On a practical level, lack of integrity in vision and innovation in methods are the outstanding problems.

2 METHODS

Nantong, as many other traditional cities in China, is on its way to rapid modernization. To protect and update the historic districts of the moat in Nantong city, known as the Haohe River, represents the practical needs of inheriting in modern cities. Planners and architects are required to apply theory and strategy to practice when facing problems of protecting the material forms in their designs, consequently, the discussion on methodology has the significance of reflecting both the theoretical perspectives and the guidance strategy. Therefore, the method of protection needs to return to the understanding of the city and to be developed from the basic theory of urban structure.

2.1 Structure

"One won't be able to understand a building, a group of buildings, particularly the urban space if the concept of structure is not introduced."[1]

The theory of city structure can be regarded as a deduction from the structuralism in philosophy. The structuralist method has widely been used in the analysis of language, culture and social context since the second half of the twentieth century. This thesis does not discuss structuralism, but rather focus on the holistic principle based on the methodology – namely, a whole has preferential important in logic than a part; and the nature of the structural elements lies in the contact with the others. Therefore, the urban structure, various in its form, is a result of the relationship and interaction between various elements that constitute the structural system. Each element is part of the overall urban spatial structure; however, the overall urban structure is much larger than the simple sum of each element. With regard to the moat basin, the structural elements are mainly of the spatial structure of the moat, traffic structure, network of roads and bridges and the city context.

2.2 Morphology and typology

The form refers to the description of "How" is the form of completion, location, perimeter, and internal and external relations.[2] Morphology is a comprehensive study of the composition law of the form and structure of things. It has a profound connection with the theory of structuralist within the context of architecture. The inter-relationship and the way of composition of element have been the main ideas of structuralism. The change of elements is to be dependent on the overall structure, but it can maintain its own significance. "Structuralists believe that structure is an entirety that contains a variety of relationships. The conversion of the elements will not change the overall structure; but the change of the relationship between the elements will lead to the change of structure."[3]

The concept of type is the basis of typology, which was originally a generic term of a methodology for grouping and classifying the natural science. "The term of type as to the field of design means an intent to make a copy of or imitate one thing, instead of things having common characteristics."[4] Thus the concept of type in architecture is an idea that determines the rule of formation, which is dynamic and differs from the static concept or icon symbols. The "diachronic" of type is associated closely with the historical change and development of architectural form, while the "synchronic" of type is related to the formation mechanism and the continuity of urban space. The development of the architecture typology has a long history; however, it is revolved around the type, the core concept of the basic architectural significance. The analysis of urban problems is no other than the study of urban form system and urban building types. The protection and renovation of the historic districts requires an understanding of the internal syntax of the urban form, and continue the urban structure and architectural vocabulary that are in line with the collective memory of the city.

3 ANALYSIS ON THE SPATIAL STRUCTURE OF THE MOAT IN NANTONG

3.1 The evolvement

Historic sites prove that original tribes dwelled at Qingdun District during the Neolithic Age 5000 years ago. The downtown area of Nantong, where a county established in Five Dynasties (907–960 AD), was called Jinghai. In the 5th year of Xiande of the Later Zhou Dynasty (958 AD), the city walls were built and named Tongzhou. It was renamed Chongzhou (Chongchuan) in the 1st year of Tiansheng of the Song Dynasty (1023 AD). The walls were demolished in the early years of Republic of China (around 1928 AD) with only three gate towers left on the east, south, west and the North Pavilion. The Haohe River in history was bigger; however, the shape of it is almost the same as what it was.[5] [Figure 1]

The old town of Nantong was surrounded by the Haohe River in the shape of a gourd. It flows for 14.96 kilometers and drains a land area of 64.53 ha. The maximum width of it is 215 meters and the minimum width is only 10 meters. The old town of Nantong before mordent time was a city built under the typical feudal town pattern enclosed by the moat, which featured square city walls, T-crossing and symmetrical. The moat basin contains the cultural extension of the city's invisible structure, and nowadays it is still a comprehensive and complicate area with traditional features.

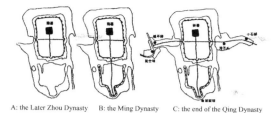

A: the Later Zhou Dynasty   B: the Ming Dynasty   C: the end of the Qing Dynasty

Figure 1. The evolvement of the spatial form of the moat basin (from the Later Zhou Dynasty to the Ming Dynasty to the end of the Qing Dynasty). Source by: *The development of cities in modern China*, China City Press, Beijing, 1998.

3.2 The relationship between the spatial form of the moat and the urban space of Nantong

Seeing from the relationship between the moat and the urban spatial structure, a typical feudal town pattern had formed along with the moat in Nantong before modern times. From 1895, Zhang Jian, an industrialist, established his spinnery at Tang Jia Zha and built a port for cargo ship. The urban pattern of "one city three towns" had been made in initial shape. Later on, Lu Jing Gang- a passenger port and Wushan-a Buddhist scenic spots were finished, forming a multi-functional district.

"The morphological evolution of Nantong city is basically moving from the center to the periphery, especially to the north and south. Although a variety of space expansion has alternately taken place to the city center, the overall form turns out a zonal expansion and asteroidal structure. The old town, surrounded by the Haohe River, has been the core and the most important central area of Nantong and also the place where historic sceneries concentrated."[6] The core area of the old town coincides with the area of the moat basin considering the logic of the development of the urban form and spatial structure, which is exactly the significance of the research on the protection and renewal of the moat basin. Therefore, today we can find the renovation and infiltration of new commerce, services and finance.

The urban planning in early Nantong was to set its city functions by zoning. Some functions were particularly emphasized but enjoyed mutual benefit. The linkages in geography, economy and society between urban groups were emphasized with a reasonable topology. Thus the city's multi-core group as well as the core of group level had been formed on the logic of the spatial structure. [Figure 2]

3.3 Historic relics in the moat area

The following three aspects show the physical manifestation of specific historical remains of the moat in Nantong. 1) The physical environment of the moat; 2) Traditional residential settlements; 3) Heritages that spread all over. Preliminary research shows that the traditional spatial form of the historic districts of the Haohe River depends on the structure of the river, street direction and the change in its width, and the

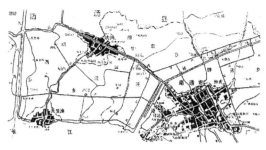

Figure 2. The structural grade of Nantong—a pattern of "one city three towns". Source by: Nantong Planning Bureau.

Figure 3. Historical relics around the Haohe River area. Source by: Photos.

settlement pattern of residential buildings. The structure of the whole river basin basically remained in its primitive form after many years, which is now the dominating spatial form framework of the core area of Nantong. The network of roads and bridges around the river basin has formed a linear structure network. Residential settlements are combined by the traffic skeleton of streets and lanes. Houses are mostly with timber structure and walled-district system, the courtyards of which are linked in series with different sizes. Numerous historic spots, modern historical sites and historic buildings are dotted about the moat basin. Therefore, the historic districts of the Haohe River are able to reflect the traditional style and local characteristics of Nantong City within a certain historical period, which is more rich, complete and real. [Figure 3]

The ancient moat and the ruins of ancient cities since the Song and the Yuan Dynasty as its core, the heritage conservation district of the Haohe River covers an area of 326.2 ha, where has numerous of well-preserved historic sites—1 national cultural heritage conservation board, 5 provincial cultural Heritage conservation units, 20 municipal cultural heritage conservation units and 52 excellent historic buildings. The Haohe River historical and cultural protection areas are the highest concentration of regional historical and cultural relics in the city, of which Si Jie and the Xi Nan Ying block are relatively well preserved. In this block, there are many fine mansions and courtyards of different historical periods, which have been an important historical witness for the development of the city of Nantong.

In view of the complexity of the protection and the innovation of urban historical districts, moreover, the particularity of the special form of the moat basin, common problems still lie in protecting and developing the moat basin: 1) overdeveloped; 2) functions are too dense, and the structure of land use is not reasonable; 3) bad living conditions at the historical block with poor infrastructure and the eco-environmental cycle disorders; 4) lack of integrity of historic preservation and environment update.

4 STRATEGY AND METHOD OF THE STRUCTURAL PROTECTION OF THE MOAT BASIN

The preservation of historic districts of the moat, unlike to protect historic buildings as well as its surroundings, cannot rely on the protection strategy as done in the museum. When the protection extended to a specific urban space, including where has a function of living; it should be considered under the vision of urban design and the thinking of integrity to resolve the conflict between the protection of historical relics and the development of the city. In most cases, people are in need of continuing their lives in the old town, and the activities of automobile conflict with the old structure most of the time. The protection and update focus on the real carrier of history. Consideration should be given to the transmission of internal grammar of integral structure, together with the preservation of the proper character of the block, to resuscitate the culture life in the historic districts when the historic memories go on.

From functionalism to structuralism, strategies and methods to solve urban issues have gone through different stages. The structural protection strategy is proposed in this thesis upon the protection of the historic districts, which is from the perspectives of integral protection and renovation, uses urban morphology and building typology and benefited by the approaches of urban design.

Figure 4. Plan of the Haohe River. Source by: Nantong Planning Bureau.

Figure 5. The structure of streets. Source by: The Haohe Tour Map.

4.1 *Structural protection*

The concept of structure can be traced to the same origin, which has been the commonness of urban issues. The state of urban structure is made up of streets with specific forms, watercourses and city lands. The structural protection, however, should focus on the integrity of the spatial structure of the historic districts in the beginning to protect and continue the internal grammar of the city language. Rational protection methods can be available through clarifying the inner logic of the spatial structure. Structural protection is not so much a specific way, but a guiding concept and strategy. The specific methods, upon the structural elements of different types and levels, can be devised through inheriting the city fragments by means of reorganizing and making them up in a new context without destroying the integrality of the city structure. The analysis of material remains of the Haohe River can be carried out correspondingly in accordance with the hierarchical division of urban structure.

4.2 *Case study—the integral structure of the moat basin*

The integral structure of the Haohe River contains the shape of the moat, the trend of waterway, the water network and the transportation. Protection of the integral structure of the river basin should focus on the following aspects.

1) To protect the Haohe River waters, the boundary between land and water, and the waterfronts in certain areas as a whole. Continue the natural gourd shape, maintain the "日"-shaped water system and preserve the diversity of forms at key nodes. [Figure 4]
2) To protect the street structure and the network of roads and bridges. As the city's branches, the street is an important part of the structural system. The trunk transportation network in the Haohe River districts is "日"-shaped while the sub-network is "田"-shaped, which tightly fit the structure of the river network. To extract a rational structural framework of the moat form by means of layering and replacement to make a controlling guide for conservation planning. [Figure 5]

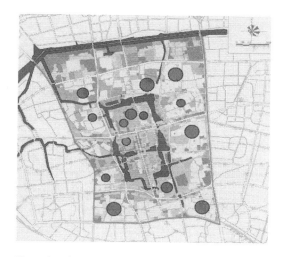

Figure 6. The texture analysis. Source by: Drawing.

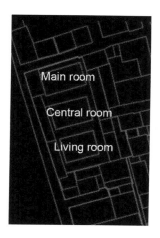

Figure 7. The texture of No. 95, Xi Nan Ying.

3) To classify the lots in three grades as below: Class A land use; Related land use and Peripheral control land use.[7]
4) To explore the potential edge effects. Integrate the diverse functions of the river basin, connect the urban ecological axes, and enhance the structural integrity of the river basin.
5) To reduce the destruction made by landfills and island development, and maintain the ecological environment and natural ecological communities. To control the infiltration on the edge of the river made by the expansion of urban space.

4.3 Case study—the city texture and surface

"The urban interface" of the historical sections of Haohe River shows regularity. When overlooking the roof texture, walking through the traditional streets and feeling the walls that enclosed, one can have an unforgettable and unique experience. Analyzing the interface can be done from the horizontal and vertical dimensions, while the characteristics and regular patterns of urban structure in respect of texture and surface can be analyzed in a manner of plane. [Figure 6]

A large number of residential buildings constitute the main "matrix" part of the city, and the analysis of urban structure in building groups as well as single building is to start with typology. Regardless of the dialogue between the building structure and building function, buildings and walls or building clusters combined by courtyard can be analyzed in the category of typology. The basic types of houses are similar to the root in grammar, and how the courtyards and communities are organized, to some extent, are similar to different syntax. The urban form can be further developed to provide reasonable and extensible methods on the analogy of the structure of vocabulary.

Shi Jie and Xi Nan Ying are currently the most complete retained residential communities in Nantong.

Figure 8. The texture of the historical blocks, Xi Nan Ying.

Traditional houses situated on the main axis in the old town are mostly dual pitched roofs with clear skylines stepped tiled in dark gray with delicate texture. Bungalows dominate at Shi Jie and Xi Nan Ying, showing the plan layout of one-courtyard-three-halls. Take the house at No. 95 Xi Nan Ying for example, it was built in the Qing Dynasty with a layout of one line three halls, of which the south part was Xu's and the north part was Chen's. Each entrance gate has a door lock and was well-decorated, which are the typical residential features in Nantong. [Figure 7] The layout of the courtyards is characteristic of a crisscross of lanes, vertically from door to hall then hallway and finally principle room, and surrounded by concrete walls. The spatial structure of the residential cluster can be clearly embodied when the roof texture of them seen from the sky. [Figure 8] The architectural space in residential buildings that features a gradual combination of "unit", "cluster" and "courtyard" can be used from the perspective of typology.

Figure 9. Aerial view of the Haohe River (part). Source by: Haobin Forum.

4.4 *Differentiated renewal—a supplement to the structural protection*

Jacobs Jane calls for a diversity of city uses and identifies the preconditions for the creation of diverse cities: mixtures of primary uses; small-scale, pedestrian-friendly blocks and streetscapes, and high densities of population and activities. The differentiated renewal of the historic lots within the moat basin can be understood as, within the framework of protecting the environment of urban physical space, to concern over the diversity of various elements as well as the topological relations between each other in the structure and to establish a flexible structure; avoiding the protection means to be stiff or rigid. The differentiated renovation is either a respect for the existing complicate urban spatial forms, or a logical addition to the structural protection.

Specifically, the physical space of the historic section is complicate where functional activities are mixed. River, neighborhood, commercial facilities, public green space, and new lot to be developed are involved in the land use. Rational multi-function should be protected while the structure is clear and encourages integrated land use. [Figure 9]

5 CONCLUSION

As a special historical district where succeeds the history and culture of the city, hand down updating information of the city, the moat basin has composed unique city image and characteristic of it. This thesis carries out a geographical empirical research on the Haohe River in Nantong which exemplifies the moat basin, from the view of the holistic preservation and renovation, use the specific urban morphology of city moat area and architecture typology analysis as tools to deduce the practically and methodologically effective discussion and experiment. This subject can hopefully have the significance for others to offer their valuable ideas on the protection and innovation of the core areas in traditional cities, and inspire more experiments on rebuilding the cities in harmony.

REFERENCES

[1] Tange Kenzo's speech Quoted from The Design Theory and Method of Modern Cities, By Wang Jianguo Southeast University Press, Nanjing, 2001. P96
[2] The Innovation of Modern Cities By Yang Jianqiang, Wu Mingwei Southeast University Press, Nanjing, 1995. P51
[3] Theories of Modern Architecture By Liu Xianjue, China Architectural Industry Press, Beijing, 1998. P377
[4] Theories of Modern Architecture By Liu Xianjue, China Architectural Industry Press, Beijing, 1998. P303
[5] According to the documents from the Nantong Records and Haohe River museum
[6] Protecting the Scene of the First City in Modern Times—Research on the Overall Urban Design and Framework System of Nantong, By Xue Peihua, City Development and Research, 2010.10
[7] River System and Urban Space By Xing Zhong, Chen Cheng, City Development and Research, 2007.01

Research on urban ecological water system construction: A study case of Wulijie, China

S. Ye

School of Architecture and Urban Planning, Huazhong University of Science and Technology, Wuhan, Hubei, China

ABSTRACT: In China, although the contradiction between water supply and demand is increasingly prominent, some areas are still over-exploiting water resources, leading to serious damage to the water environment. This paper uses Wulijie as a study case, which is a small town in Wuhan and will be built into an ecological city in the future plan, to find out the ways of protecting and reusing water resources during the construction of an ecological city. It firstly introduces the present conditions of water system in Wulijie according to the site investigation, analyzes present problems, and predicts problems that may occur during the developing process. Then it makes specific designs from three aspects, which are water supply system, water drainage system, and water resources reservation. Finally, it proposes some suggestions to guarantee the sustainability of water resources and improve the living standard in Wulijie.

1 INTRODUCTION

The total amount of freshwater resources in China ranks sixth in the world. However due to the large population base, the per capita water capacity in China is only a quarter of the world's average level. Though the contradiction between water supply and demand has become increasingly prominent, some areas are still over-exploiting water resources, causing serious damage to the water environment. The current situation of rivers and lakes is quite worrying, 80% of untreated sewage is poured into rivers and lakes directly, more than one third of the rivers are polluted, 90% of urban water is polluted seriously, and 70% of the urban lakes are not suitable for drinking (Sun, 2010). Industrial pollution in the city and widespread pollution from the overuse of fertilizers and pesticides in the country have undermined the sustainable use of water resources dramatically, which are huge threats to both human health and living environment.

This paper uses Wulijie, a small town abuts the central city in Wuhan, as a study case, to find out the ways of protecting and reusing water resources during the construction of an ecological city. Wulijie is a complexity with both the urban area and rural area. As it has abundant water area, farm land and green open space, it will be built into an ecological city in the future plan. Hence, how to develop water resources sustainably is the priority of the plan. This paper firstly introduces the present conditions of water system in Wulijie according to the site investigation, analyzes present problems, and predicts problems that may occur during the developing process. Then combining with experience at home and abroad, it makes specific designs from three aspects, which are water supply system, water drainage system, and water resources reservation respectively. Finally, it proposes some suggestions to guarantee the sustainability of water resources and to improve the living standard in Wulijie.

2 DEFINITIONS

Urban water system is a system changing with the dynamic spatial-temporal process (Song, 2005). It consists of the social circulation system and the natural circulation system. The social circulation system mainly has four elements including water resources, water supply, water consumption and water drainage (Li & Liu, 2007). The natural circulation system is composed of the processes like precipitation, runoff, infiltration, evaporation and others, which relate to the area of rivers, groundwater, wetlands, estuaries and so on (Gui, et al. 2009). In a word, urban water system is made up of all water affairs in the city.

Urban ecological water system refers to the continuous water in both natural facilities and artificial control facilities, with the comprehensive consideration of various factors, like water quantity, water quality, water ecology, aesthetics, etc. It is a complexity with various functions like water conservation, water supply, rain and flood storage, self purification, biological habitat, recreation and education. It should be easy to manage, consume less energy, and help to reduce the impact of one region's surface water drainage on other regions.

The study is supported by State Natural Sciences Fund: No. 51178200.

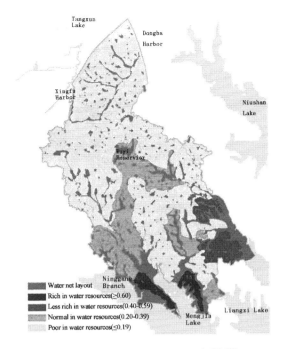

Figure 1. The Layout of Water Resources in Wulijie.

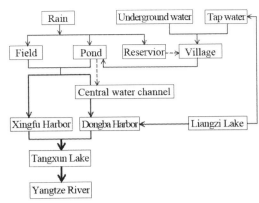

Figure 2. Present Self-circulating Water System in Wulijie.

3 CONDITIONS AND PROBLEMS

3.1 Present conditions

The total area of Wulijie is 4743.33 hectares, among which the water area covers 705.46 hectares, taking 14.87% of the total space, so the water resources are very abundant in Wulijie, as shown in Figure 1. The river system of Wulijie belongs to the basin of Tangxun Lake and Liangzi Lake, and is the potential recharge area of these two lakes. Liangzi Lake is the spare drinking reservoir in Wuhan, while Tangxun Lake directly flows into the Yangtze River. The water quantity and quality of these two lakes directly influence Wulijie's water condition and water ecology. Before the construction of Wutai Watergate, Zhifang Lake in the site is connected to Tangxun Lake by the central harbor. After the construction of Wutai Watergate, Xingfu Harbor, Dongba Harbor, Wuyi Reservoir, and other irrigation canals, some stream segments have no running water perennial, so the floods are not severe.

The distribution of surface water in the site is very even. Large water area mainly locates in the southern area, which is the rural part of Wulijie. The surface water resources mainly have three types – ponds, linear flows, and reservoirs. Connected by surface runoff, irrigation ditches and natural river courses, they form a stable self-circulating water system, as shown in Figure 2.

The whole site drainage direction is from south to north, except for the center port, which is from the southwest to northeast. In the northern township, water for living and production is provided by a water supply company, which induces water from Liangzi River. In the southern countryside, tap water and ponds are the main resources for living, while Liangzi River supplies water for agriculture.

3.2 Present problems

The treatment of rain and sewage has not been divided yet. They all flow arbitrarily by the force of gravity. Ground pollution and fertilizer pollution flow into harbors and then river courses along with the rainfall. Villager's domestic sewage and wastes are poured out without treatment, directly polluting the farmlands.

Current flood control and water storage are very poor, resulting in the risks of waterlogging and droughts. The site's drainage mainly relies on gravity and other natural forces. The human intervention is very limited, except for building water pumps and valves, and digging irrigation ditches. The villages are facing the threat of being submerged by heavy rain. Current water storage facilities are only ground ponds, which are affected hugely by the temperature, humidity and other natural factors, and the water in which can only stay for half a month during long period of drought. The reservoirs can also provide some water for irrigation.

3.3 Probable developing problems

The construction of the new city will decline the diversity of original water functions, for some ladscape projects only focus on the creation of beautiful water scenery. With the intervention of water consolidation and canalization, some functions will fade away, such as water supply, self-purification, and providing habitats for living beings, etc.

The increase of rigid pavements will lead to an increase of surface runoff and the risks of waterlogging. As the terrain is relatively flat, the construction of roads and municipal facilities would easily affect the original surface runoff system. Thus, an adjustment on the current storage system is needed.

There will be more types and amount of pollution after the urban construction, as the completion of roads and shopping center will leave suspended sentiment, heavy metal and many other poisonous pollutants on the site; also the amount of sewage will increase a lot with more people and industries moving in the city.

The increase of human activity, pavements, landscape isolation, and the change of microclimate and habitats will all lead to the migration or disappearance of species, which may cause a fracture of the food chain and a series of chain reactions in the ecosystem.

4 CONSTRUCTION IN WULIJIE

According to the General Planning of Wulijie, the urban built area will be 1257.78 hectares, covering 26.52% of the total town, so the rural area will cover the left 73.48%. The farmland and the forest occupy 60.11% of the town, and 635 hectares of ecological water area will be kept, taking up 13.38% of the whole area. Thus, the maintained environmental layout is very conducivefor Wulijie to become an ecological city. In accordance with the content of water system, new water supply system, water drainage system, and water resources reservation project will be built in Wulijie.

4.1 Water supply system

Two kinds of water supply systems will be built in Wulijie, as shown in Figure 3. One is living and production water supply system for urban and rural built area, the other is agricultural production water supply system in the southern country. The first system will still use the same company to supply water to residential areas, including township, three new rural central communities, and ten featured rural communities. The second system pumps water from Liangzi Lake directly for agricultural production. And the water pipe system covers 100% of cultivation area.

4.2 Water drainage system

Water drainage system can be divided into three parts, rain drainage, sewage drainage, and flood control. The separation of rain and sewage is the key to reduce water pollution diffusion and reuse rainfall. Flood control should also be improved to avoid loss of properties, especially the damage to agricultural crops.

4.3 Rain drainage

In combination with the urban landscape, the main way of drainage in Wulijie is to use the grass ditches along the both sides of the roads to conduct water to relative watersheds. The township should also make full use of the existing canals and river courses to discharge rainfall. The principle is to discharge water to the nearest watershed, so the majority of water can be guided to Xingfu Harbor and Dongba Harbor.

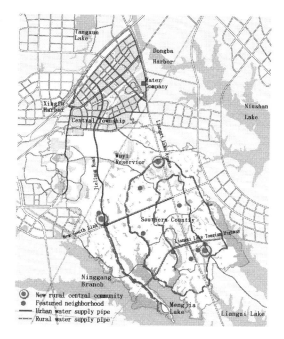

Figure 3. Water Supply System in Wulijie.

The drainage in the rural residential areas mainly takes the path along Jieliang Road, Liangzi Lake Avenue, New South Link, Liangzi Lake Tourism Highway, and other travel lanes, to flow into the Liangzi Lake.

The irrigation and drainage project in agricultural areas uses the storage capacity of ponds, lakes, reservoirs, ditches and soil, and selects the ways of integrating horizontal or vertical drainage, with gravity or pumping drainage, in accordance with local conditions in Wulijie. The project will construct open conduits, blind conduits, and shafts individually or assembly. Farmland drainage buildings should be planned along with the drainage system, and be located in the field. Pumping stations should be built with the functions of both irrigation and drainage.

4.3.1 Sewage drainage

The sewage drainage system is divided into three parts, which are urban sewage drainage system, rural sewage drainage system, and wetland purification system, as shown in Figure 4.

All domestic and production sewage in the township is sent via underground pipes to the sewage treatment plant. After the treatment, it is sent to artificial wetlands for further treatment, so as to reach the standard of landscape water. Then the water will be used for roadside greening, road cleaning, and the supplement of landscape water.

The sewage from new rural central communities, featured rural communities, and agricultural production in the country will all be sent to the sewage treatment plant in the township. Some communities

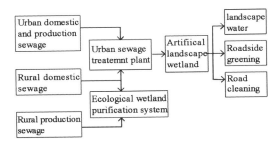

Figure 4. Sewage Treatment System in Wulijie.

far from roads can use the wetland for purification, which is mainly used for the treatment of agricultural wastewater in the southern farmland. It is suggested to build a purification system on the spot. The farmland drainage will be natural and assisted by the guide of a few ditches.

4.3.2 *Flood control*

According to the long-term planning, the total population will be 85,000 in 2030. According to the "Flood Control Standards (GB50201-94)" and "Wuhan City Flood Control Management Regulations", the flood control standard will be 20-to 50-year return period in Wuhan. Since Wulijie has abundant water system and is on the verge of Liangzi Lake, the flood control standard should be 50-year return period, and the drainage standard should be 20-year return period. In the urban communities, there are many ways to manage the flood. Firstly, large areas of grass and infiltration pavements will be maintained for water to go underground. Secondly, rainfall will be collected and reused in communities. The rainfall in communities has been little polluted, for it only gets in touch with roofs, streets and squares. Rainfall from the roof can be collected by the roof garden and be used for household flushing, cleaning and washing after simple treatment. Rainfall on the street or square can be collected by grass ditches and open conduits, and be used for green spray, flushing roads, or flow downward. Thirdly, green spaces and landscape ponds will be the main facilities for rainfall storage. The overflow port can be connected with municipal drainage system to meet the discharge requirements. The design of storage and infiltration facilities should ensure the safety of the young and the old.

In the countryside, the scattered natural ponds are the main support of watershed management. Based on the different distribution characteristics, the ponds can be divided into three types: residential ones, farmland ones and tributary ones. The ponds will be connected organically with surrounding residential area, woodland, farmland, and water flows, to achieve the function of rain and flood storage, water purification, soil and water conservation, pollution interception, so as to transfer this "green sponge" into an ecological water network.

4.4 *Water resources reservation*

According to the evaluation results of ecological sensitivity in Wulijie, the core ecotope is the Liangzi Lake wetland zone. The environmental conservation of Liangzi Lake should be strengthened by decreasing the disturbance as far as possible in the zone. At the same time, according to relevant requirements, an area of 1153.24 hectares within the protection boundary, which extends the water line of Liangzi Lake by 30 meters, is defined as the core protection zone. To achieve integrity of the water cycle system, environmental restoration projects will be implemented in Dongba Harbor, Xingfu Harbor, Ninggang Branch, Mengjia Lake and reservoirs.

5 SUGGESTIONS

5.1 *Take active measures to save water*

Firstly, Wulijie should increase the recycling rate of industrial water, and enhance the treatment and reuse of wastewater. At the same time, factories could collect surface water nearby for industrial production, to reduce the usage of tap water. High water consumption enterprises should be limited. Secondly, it should increase the installation rate of water-saving appliances, such as water saving faucet and water-saving toilet in both households and public places. Thirdly, it should build reclaimed water utilization system in the township, which is constituted by collection, storage, treatment and supply system, to achieve the complete use of water resources. Last but not the least, residents should be encouraged to develop a water-saving habit in daily life.

5.2 *Strictly prohibit any pollution to Liangzi Lake*

To maintain the sustainability of water resources in Wulijie, any destructive or impolite behaviors are strictly prohibited in the Liangzi Lake protection zone by relative rules, regulations and even laws. Firstly, any domestic and production pollution is forbidden in the zone. Polluting enterprises and projects are not allowed to enter this zone. Domestic and production sewage, industrial waste, and other poisonous waste shouldn't be poured into the lake without strict treatment. Pesticides and other toxins to kill aquatic organisms shouldn't be used for farming in this area. Secondly, any destruction to the environment and species are banned, such as illegal quarry, deforestation, and activities damaging vegetation, wetland and aquatic organisms. Moreover, both fishing and fish farming are forbidden here. Building cross-bridges or dikes in the water or around water lines are not permitted. Thirdly, polluting commercial development is forbidden in the zone. For instance, water recreational project with pollution, new waterfront real estate development, water catering business, and building golf course are all banned here.

5.3 Appropriately develop recreational functions

The waters are a significant extension of the public space of land and part of urban landscapes in terms of boundary, landmark, and an open and publicly usable territory (Lynch, 1960). However, the accessibility of water is very poor in the whole area due to the shortage of roads. And the recreational and educational potentials have not been dug up yet. When developing a new city, residents' needs of appreciating water and getting in touch with water should be paid special attention to. Recreation refers to tourism strategies, it can be concluded that recreation near or on water is one of the most popular leisure activities, even with internally conflicting groups (Pēteris, 2010). So more open space, theme parks, river banks, and travel lanes should be built in both the township and country for people to get close to nature.

6 CONCLUSIONS

As an old saying, a city is prosperous by water. With abundant water resources, Wulijie has a great advantage to become an ecological and livable city. Thus, to develop a sustainable water system is very important during its urbanization process. The construction of ecological water system in Wulijie is aimed to ameliorate the severe water shortage in this area, and establish a water system that meets the requirements of sustainable and environment-friendly development. The core of the system is water saving, and the main approach is to store, reuse and control water, being assisted by the retention disposal of wetlands and the infiltration of green space. In the future plan, Wulijie is going to build a circulating utilization system of rain and sewage, and optimize the allocation of water supply, in order to maintain the continuity and integrity of water system, improve the service functions of water resources, as well as promote the harmony between water and environment, water and human beings.

REFERENCES

Gui, P. & Kong, Y.H. 2009. Urban Water System Planning Based on Ecological Security Pattern. *Urban Planning* (4): 61–64.

Li, S.P. & Liu, S.Q. 2007. Requirements for Sustainable Management of Urban Water Systems. *China Water & Wastewater* (11): 159–163.

Lynch, K. 1960, *The Image of the City*. Cambridge, Mass: MIT Press: 202.

Pēteris, S. 2010. Integrated Water Planning System. *Scientific Journal of Riga Technical University* (4): 106–111.

Song, H.L. 2005. The Planning of City Water System. *Urban Planning Newsreport* (12).

Sun, J.H. 2010. The Present Conditions and Strategies of Water Resources in China. *Public Communication of Science & Technology* (13):31.

Research on vibration characteristics of building construction while tunnel crossing underneath

Zheng Li
Key Laboratory of Transportation Tunnel Engineering, Ministry of Education, Southwest Jiaotong University, Chengdu, Sichuan, China

Ziqiang Li
Southwest Jiaotong University, Chengdu, Sichuan, China

ABSTRACT: In order to reduce the influence on surface construction, which is caused by large quantity construction of urban shallow tunnel, cautious consideration should be put into the choice of construction method. Compared with mechanical excavation, blasting construction can not only decrease engineering cost efficiently but also improve construction progress; however, it did augment the disturbance of the surrounding environment. Due to the complexity of ground constructions and the characteristics of poor surrounding rock condition, choosing an appropriate blasting construction method has great significance for urban shallow tunnel. Combining with Guanhui intercity transportation project that shallow tunnel pass through underneath the dense residential areas, the author simulate the impact of blasting construction to surface construction when adopting CD method and reserved core soil method via finite element method. Based on the vibration velocity and settlement of surface construction, the paper compared and analyzed the suitability of two different blasting construction methods. Finally, the author concluded that CD method blasting excavation is more suitable for urban shallow tunnel; the paper can also provide references and suggestions for similar urban undercrossing tunnel construction.

Keywords: urban tunnel; vibration characteristics; masonry structure; frame structure

1 INTRODUCTION

In the engineering of tunnel passing through underneath the building, how to guarantee the safety of existing building is not only the starting point but also the ultimate purpose of engineering construction, especially for urban shallow tunnel with numerous underground pipelines and complicate existing buildings which can be easily damaged by tunnel construction (ZHANG Dingli. 2009). It is so important for us to acknowledge the vibration pattern of the adjacent structure, which is also the premise of safely passing through, that we can take corresponding reinforcement measures (XU Chuanhua. 2004).

The impact to surrounding environment during the tunnel excavation process is also a significant research subject (DIMMOCK Paul Simon & MAIR Robert James. 2008). Vibration caused by the excavation of underground structure and its effects on surface construction had drawn attention of engineering field, thus domestic and foreign researchers had done a lot of research. Glenn concluded the relationship between blasting vibration frequency and structural damage degree via research (LIU Shibo. 2006); Stephen. D. Butt believed that the damage of blasting vibration on the tunnel is not only related with the intensity of vibration, but also with frequency closely (FU Shigen. 2006). Therefore safety criterion combining with vibration frequency has been the main part of safety evaluation system at present. Hongjun Teng researched the influence of tunnel excavation on the upper structure, established the security risky analysis process of the structure and explored the risk control measures during the construction phase (HOU Yanjuan. 2007). From the existing researches, we can infer that the influence of undercrossing tunnel on the upper structure had achieved some results, but there are no certain conclusions for the vibration characteristics of the upper structure during the construction phase of the undercrossing tunnel.

Based on Guanhui intercity transportation project that shallow tunnel pass through underneath the dense residential areas, the author simulated the influence of undercrossing tunnel on surface construction via numerical simulation; and analyzed the vibration frequency of the upper existing structure combining with actual engineering environment, and also the vibration characteristics of masonry structure as well as frame structure. The paper can provide references and suggestions for other similar engineering.

2 THEORETICAL ANALYSIS

2.1 Project profile

The route length of Guanhui urban railway transport projects is approximately 97 kilometers, starting in Dongguan Hongmei station and end of Huizhou City, the structures of interval form is complex. The Guanhui urban railway transport projects in close proximity to above many important ground buildings and underground pipelines clouds. Most foundation of the ground building above tunnel are natural foundation and man-power hole pile foundation, Most building are the frame structure and masonry structure. It is strict with the structure vibration frequency of when the section tunnel passes through large area of residential areas and factory. Tunnel sits mostly on completely-weathered migmatitic gneiss, it is belongs to the urban shallow tunnel that the minimum thickness of covering soil on the vault is only 8m.

2.2 Construction method introduction

Depending on the nature of each segment surrounding rock of Guanhui urban railway, mainly choose CD method for the V~VI class surrounding rock. CD method is mainly used in the rock mass which the geological conditions is poor and the steady ability is low, it also applies to underground engineering construction when surface subsidence must be strictly controlled. The excavation method is as: excavate one side of the tunnel → construct the temporary middle-wall → after a certain distance in advance excavation side, partially excavate the other side of the tunnel. To support each part after the excavation so making each part to be an independent closed structure.

2.3 The determination of blasting load

In this paper assume that blasting load on the excavation boundary surface and face of tunnel by the triangular load method (G. I. Taylor. 1941). Taking rise time as 0.012 s and time of descent segment as 0.100 s on the basis of numerous investigations. Total duration of the simulation takes 2.0 s in order to understand the situation of surrounding rock-supporting system after the blasting load.

The peak stress P_{max} of blasting load USES the following empirical formula to solve:

$$P_{max} = \frac{139.97}{Z} + \frac{844.81}{Z^2} + \frac{2154}{Z^3} - 0.8034 \text{ (kPa)} \quad (1)$$

$$Z = R^* / Q^{1/3} \quad (2)$$

where Z = reduced distance; R^* = the distance from explosion center to the load acting surface (m); and Q = explosive charge (kg), take the total explosive charge when simultaneous blasting, take the maximum period of explosive charge when segmented blasting.

2.4 Structural natural frequencies

Based on structural mechanics, we can infer that the response of single degree of freedom system to the wave ($x_{0k}(t) = A_k \sin(\omega_k t + \phi_k)$) is:

$$x_k(t) = e^{-\xi \omega t} \frac{\omega^2 A_k \sin(\alpha - \varphi_k)}{\sqrt{(\omega^2 - \omega_k^2)^2 + 4\xi^2 \omega^2 \omega_k^2}} \cos \omega' t$$

$$- e^{-\xi \omega t} \frac{\omega^2 A_k \left[-\xi \frac{\omega}{\omega'} \sin(\alpha - \varphi_k) + \frac{\omega_k}{\omega'} \cos(\alpha - \varphi_k)\right]}{\sqrt{(\omega^2 - \omega_k^2)^2 + 4\xi^2 \omega^2 \omega_k^2}} \sin \omega' t$$

$$+ \frac{\omega^2 A_k \sin(\omega_k t + \varphi_k - \alpha)}{\sqrt{(\omega^2 - \omega_k^2)^2 + 4\xi^2 \omega^2 \omega_k^2}} \quad (3)$$

$$\alpha = \tan^{-1} \frac{2\xi \omega \omega_k}{\omega^2 - \omega_k^2} \quad (4)$$

$$x_k(t) = -e^{-\xi \omega t} \frac{A_k}{2\xi} \sqrt{\sin^2(\alpha - \varphi_k) + (\xi^2 + 1)\sin^2(\varphi_k - \alpha + \beta)} \sin(\omega t - \theta)$$

$$+ \frac{A_k}{2\xi} \sin(\omega_k t + \varphi_k - \alpha) \quad (5)$$

where ω = seismic frequency (Hz); ω_k = structure frequency (Hz); and ξ = damping ratio.

There are three situations as follows:

1. $\omega_k > \omega$: For certain structures, when blasting seismic harmonic circular frequency is bigger than structural inherent circular frequency, thus the bigger the difference between seismic harmonic circular frequency and structural inherent circular frequency is, the less intense the vibration of structure in response to the harmonic wave will be.
2. $\omega_k < \sqrt{1 - 2\xi^2}\omega$: The amplitude of the structure unsteady free vibration and Steady forced vibration will decrease as frequency decreases.
3. $\sqrt{1 - 2\xi^2}\omega \leq \omega_k \leq \omega$: The unsteady free vibration will resonate with blasting seism, and the resonant response should be Eq. (5).

The three situations mentioned above indicate that resonance effect will happen to the building and the amplitude will increase substantially when natural frequency of structure and blasting seismic frequency is relatively approaching, therefore, the research on the structural seismic behavior involves their own behavior (natural frequency).

For the natural frequency of ordinary building, its relationship with height should be:

$$T = 0.0168 (H_0 + 1.2) \quad (6)$$

where T = period (s); H_0 = building height (m). The above formulas show that the increase of building height will lead to the decrease of natural vibration frequency.

Table 1. The parameter of numerical simulation.

Designation	Elasticity modulus (Pa)	Poisson's ratio	Yield strength (Pa)	Failure strain	Density (kg/m^3)
Weak-weathered migmatitic gneiss	1e9	0.28	1e6	0.02	2500
Strong-weathered migmatitic gneiss	0.4e9	0.32	0.8e6	0.02	2300
Completely-weathered migmatitic gneiss	0.1e9	0.33	0.5e6	0.02	2200
Silty clay	0.26e9	0.35	0.2e6	0.02	1950
Rough sand	0.20e9	0.35	0.5e6	0.02	1900
Clayed silty sand in allurial-diluvial deposit	0.14e9	0.36	0.1e6	0.02	1850
Plain fill	0.05e9	0.37	5e4	0.02	1800
Sprayed concrete	2.8e10	0.2	–	–	2300
Middle wall	2.8e10	0.2	–	–	2500

3 THE ESTABLISHMENT OF THE FINITE ELEMENT MODEL

The numerical simulation calculation adopts Druck-Prager yield criterion. In the calculation, surrounding rock adopts 3-D solid element (soild 45), anchors adopt 3-D rod element (link 8), initial projected concrete, temporary supporting (middle-wall etc.) and building floor adopt spatial shell element (shell 63), arcs and pillars adopt beam element (beam 4), masonry buildings adopt elastic structure + rigid foundation vibration system. Based on tunnel mechanics and existing research materials, the width of model is 80 m (extending 40 m from the position of tunnel central line on both sides), the height is 55 m (from the ground surface which is also called free surface downward), and the thickness is 56 m (in the tunnel axial direction).

Apply vertical normal constraints on the bottom boundary of the calculation model and horizontal normal constraints on both left and right boundary. Considering the influence scope of tunnel excavation and minimizing the "boundary effect" (LUO Jianjun. 2007), respective 12m which adopting whole section excavation is arranged in the pre-and-post position of the tunnel axial direction to eliminate the boundary effect. The calculation model is shown in figure 1.

Based on Guanhui intercity geological materials and the field material test and combined with relevant codes, the selection of calculation parameters is shown in table 1.

The existing building is shown in figure 1: the four-floor frame structure is reinforced concrete structure with the height of 12 m, the width of 8 m and the length of 16 m; the two-floor frame structure is also reinforced concrete structure with the height of 6 m, the width of 8 m and the length of 16 m. The parameters of each structure is shown in table 2.

4 THE ANALYSES OF THE RESULTS OF MODEL CALCULATION

Numerical simulating mainly chooses modal analysis to solve the frequency characteristic of the structure.

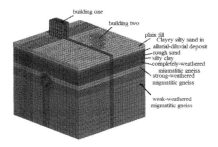

a. Full finite element model

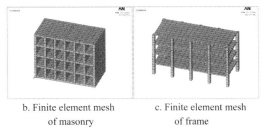

b. Finite element mesh of masonry

c. Finite element mesh of frame

Figure 1. Schematic diagram of the finite element model.

Using the subspace iteration method to solve the fifth-order modal by ANSYS. The finite element calculation results of natural vibration frequency of the masonry and frame structure houses in the fifth-order listed in the table 3.

From table 3 mentioned above, we can infer that the natural vibration frequency of each structure is less than 10 Hz. Based on the rules, the vibration frequency of structures increases while the number of iteration increases, the natural vibration frequency of the third to the fourth step increases substantially shows that the frame structure has the greatest impact when the tunnel under crossing. The natural vibration frequency of the two-floor frame structure is always higher than four-floor frame structure, it shows that the increase of building height will lead to the decrease of natural vibration frequency, the results coincide with theoretical analysis as 2.4 section.

Table 2. Building mechanical parameters.

Name	Density (kg/m³)	Elasticity modulus (GPa)	Poisson's ratio	Compressive strength (MPa)	Tensile strength (MPa)	Shear strength (MPa)
Masonry structures	1.8	10	0.2	8.5	1.5	1.7
Frame structure	2.30	28.5	0.25	–	–	–

Table 3. The natural frequency of the first five order vibration type.

Name	Mode of vibration	1	2	3	4	5
Four-floor masonry structure	Frequency (Hz)	5.8	12.4	12.9	18.2	31.9
Four-floor frame structure		2.1	2.5	2.5	7.9	10.1
Two-floor frame structure		4.8	6.2	6.2	10.9	11.4

The natural vibration frequency of four-floor frame structure is lower than that of four-floor masonry structure, it indicates that the natural vibration frequency of masonry structure is higher than that of frame structure when tunnel under wear the structure. Analysis of blasting vibration frequency, found its value generally located in the 10~100 Hz, thus to reach the lower masonry structure are more likely to produce resonance phenomenon and damage. So we must strengthen the monitoring of stratum brick structure during construction.

5 CONCLUSION

From the initial analysis of the structure vibration characteristics, we can infer that the structure vibration characteristics involved its own behavior and external load etc. Based on theoretical analysis and numerical simulation, we can draw some conclusions as follows:

1. The increase of building height will lead to the decrease of natural vibration frequency.
2. Compared with blasting vibration frequency (10~100 Hz), natural frequency is less than 10 Hz in the simulation process, which indicate that resonance phenomena will not happen in general.
3. In the same conditions, the natural vibration frequency of masonry structures is bigger than that of frame structures. It shows that tunnel-crossing will be even greater impact on masonry structure.
4. Finally, based on the above analysis, Low-rise masonry structure happens resonance occur most easily when shallow buried tunnel under passing densely populated areas. So we must strengthen the monitoring of stratum brick structure in order to prevent damage of the structure during construction.

REFERENCES

DIMMOCK Paul Simon & MAIR Robert James. 2008. Effect of building stiffness on tunneling-induced ground movement.
FU Shi gen. 2006. Analysis of the effect of blasting vibration and vibration prediction.
G. I. Taylor. 1941. Analysis of explosion of a long cylindrical bomb detonated at one end.
HOU Yanjuan. 2007. Safety risk analysis and assessment of complex building in shallow large-span tunnel construction.
LUO Jianjun. 2007. Direct back analysis algorithm for deflection distribution control during the construction of subway.
LIU Shibo. 2006. Test and evaluation of tunnel and the blasting vibration.
XU Chuanhua. 2004. Fuzzy-synthetic evaluation on stability of surrounding rock masses of underground engineering.
ZHANG Dingli. 2009. Experimental study on safety control of buildings during construction of shallow-buried soft rock tunnel with large-section.

Review and assessment of urban Car Free Day program in China

Yu Zhang
China Academy of Urban Planning & Design

ABSTRACT: In order to address the double challenges of urbanization and the increasing use of private motorized transportation, Car Free Day activity has now been in operation in Chinese cities for seven years since 2007, accompanied by the popularization of Green Transportation concept and the formulation of sustainable transportation development policies and measures. This study first provides a concise review of the development process of Car Free Day movement and its core content, followed by the quantitative assessment, which is the focus of this study, of Car Free Day movement. The assessment includes the cities participating in the program, vehicle free areas, the long-term measures adopted, public opinion polls, and the monitoring of traffic environmental standards. The author believes that Car Free Day movement promotes the growth of public transportation passenger flow and raises the service quality of public transportation, improves the efficiency of traffic circulation, benefits the environment, and helps with the promotion of the sustainable development and transformation of urban transportation policies. Finally, the following recommendations are made: continuous expansion of Car Free Day movement participation scale and depth, reasonable designation of car-free zones, strengthening of information provision for car users, and long-term promotion of formulation of long-term sustainable measures by cities.

Keywords: urban transportation, Care Free Day activity, Green Transportation, long-term measures, sustainable development

1 INTRODUCTION

Urbanization and motorization are both growing rapidly, and have been defining characteristics of the urban development in China over the past 20 years, showing the tendencies of accelerated growth in demand for urban transportation and a shift in urban transportation towards private motorized transportation. As of 2011, the urbanization rate in China was 51.27%, with an urban population of 690 million. In 2012, there were 240 million cars on the roads in China, with the numbers exceeding 1 million in 23 medium and large cities, and the number in Beijing, 5.2 million, ranked the first. The numbers of cars exceeded 2 million in Shanghai, Guangzhou, Shenzhen, Hangzhou and Tianjin (Lu Huapu, 2012). During the years before the launch of Car Free Day, the rate of public transportation use in large cities declined by 6 percent, and the rate of bicycle use between 2 and 5 percent (Ministry of Housing and Urban-Rural Development et al. 2012). The public transportation system has many problems including slow vehicle speed, poor station coverage, poor punctuality, inconvenient transfer connections, and overcrowding. The rate of the use of bicycles for transportation has been continuously decreasing in the past few years. Shenzhen is a typical example; the rate of bicycle usage in Shenzhen declined from 30% in 1995 to 4% in 2007. The growth of motorized transportation and enlargement of parking spaces have further deteriorated the environment for both pedestrian and bicycle mobility. Unbalanced transportation system, intensification of traffic jams, increasing natural resource consumption, and the deterioration of the environment are all common problems currently facing the development of urban traffic in China.

In order to promote the sustainable development of urban transportation, the Ministry of Housing and Rural-Urban Development, having studied the successful experience and methods used in European Mobility Week, sent out a nationwide recommendation to launch the Car Free Day movement to the governments of all officially established cities, especially those with population over 500,000. In the run-up to 2012, the China City Car Free Day has been held six consecutive times, with the total of 162 cities committed to the movement, including 4 directly governed municipalities, 21 provinces (excluding Qinghai province), and 5 autonomous regions, and over 200 million urban residents involved.

2 CAR FREE DAY MOVEMENT AT PRESENT

2.1 *The themes of Car Free Day activities*

The main theme of the movement in 2007 was Green Transportation and Health. The emphasis was made on prioritizing the development of urban transportation strategy, which is vital for sustainable urban development. The responsibility and duty of the government, transportation enterprises and private individuals were made clear, forming the green mobility concept in public consciousness.

The main theme of the movement in 2008 was Humanizing the Streets. The emphasis was made on the reallocation of road spaces, pointing out that the reduction of road space allocated to cars is an efficient solution to traffic problems and conducive to sustainable development. Priority was given to the continuous enlargement of pedestrian areas and exclusive bicycle and public transportation lanes.

The main theme of the movement in 2009 was Walking and Cycling – Healthy and Good for the Environment. The emphasis was made on the establishment of safe, convenient, comfortable and well-organized pedestrian and cycling environment. The public had to be made aware that walking and cycling play an important role in the overall urban transportation system, and that short- and medium-distance trips are a means of interchange with other transportation methods, and also have a recreational and health function.

The main theme in 2010 was Green Transportation, Low Carbon Living. The emphasis was made on urban transportation as a priority in national energy saving and emission reduction strategy. It was made clear that green mobility is a healthy and active method of transportation. Travel habits, once formed, are very difficult to be changed, and this crucial period has to be seized in order to shape people's travel behavior. Result-centered policy measures must be implemented.

The main theme in 2011 was Green Transportation – Urban Future. The emphasis was that the development of green transportation optimizes urban transportation system: transportation and land use planning are integrated, public transportation priority development strategy implemented, fair allocation of road space carried out with priority given to pedestrians and bicycles, and public awareness and the spirit of cooperation are cultivated.

The main theme in 2012 was Love Your City – Travel Green. The emphasis was made on the implementation of transportation policies that promote sustainable urban development, including reasonable street parking fees, encouragement of park-and-ride services, exploring congestion charge policies, and formulation of travel demand management policies leading to the change from commuter transportation to green transportation.

2.2 Participation in the activity

An increasing number of cities has been identifying with the green transportation concepts proposed by Car Free Day movements. The ratio of the number of cities that have actually launched the activity to the number of cities that have pledged to do so has been growing year by year (Table 1).

Table 1. Year by year China Cities Car Free Day activity statistics.

Year	Number of pledged cities	Number of cities that have carried out the activity①	Participation ratio②
2012	152	134 (12)	80%
2011	149	104 (8)	64%
2010	132	78 (9)	52%
2009	114	82 (3)	69%
2008	112	88 (6)	73%
2007	110	82	75%

① The data in parenthesis refers to cities that are non-signatories to the activity pledge, but have independently carried out Car Free Day activity. Those cities are referred to as "supporting cities". This number is included in the corresponding number of cities that have carried out the activity for that year.
② The number of supporting cities is not included into the calculation of participation ratio.

Table 2. The number of cities with designated Car Free Areas and the percentage of the total number of cities.

Year	The number of cities with designated Car Free Areas	The number of cities participating in Car Free Day	Percentage
2012	19	134	14%
2011	38	104	37%
2010	50	78	64%
2009	45	82	55%
2008	22	88	25%
2007	65	70	93%

2.3 Car free areas

During the Car Free Day activities held successively between 2007 and 2012, 134 cities planned and designated Car Free Areas (roads) on the day of the 22nd of September (Table 2). In other words, 1/5th of all cities in the country made an attempt to set up Car Free Areas, actively responding to the recommendations put forward by Car Free Day. Undoubtedly, the designation of Car Free Areas is an excellent experiment for setting up pedestrian areas in a city.

2.4 Long-term measures

During the Car Free Day activities carried out between 2007 and 2012, the 330 long-term measures adopted by the cities fall into five major categories: public transportation, non-motorized transportation, transportation centers, road traffic infrastructure facilities, or traffic management. The public transportation and non-motorized measures account for nearly 80% of all the measures (Figure 1), reflecting the goals improving green transportation services and safety by Car Free Day activities. Every city adopted approximately three different long-term measures, and this average exceeds what was agreed in Car Free Day Pledge. This clearly

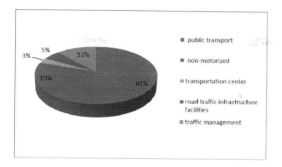

Figure 1. Proportions of all categories of long-term measures adopted between 2007 and 2012.

Table 3. Ten long-term measures adopted as part of China City Car Free Day movements between 2007 and 2012.

Category	Long-term measure	Number
1	Raising the level of public transportation services	37
2	Opening of new public transportation routes	19
3	Increasing the number of public transportation vehicles	16
4	Establishment of specialized public transportation lanes	16
5	Increase in vehicles using new energy sources	15
6	Optimization of public transportation network	14
7	Construction of the sites for public transportation parking and maintenance	13
8	Remediation of traffic orders Traffic regulation adjustments	13
9	Special remediation of road-occupying parking	11
10	Additional facilities of separation of motorized and non-motorized	10

demonstrates the importance attached by the cities to the implementation of long-term measures, with their ability to carry out the measures exceeding the original expectations. Other cities possess large potential that is still to be tapped.

Ten different long-term measures implemented between 2007 and 2012 are listed in Table 3, illustrating very well the orientation of urban traffic policy: seven of them are in the area of public transportation, with the remaining three related to traffic management, including those designed to safeguard the conditions and safety for non-motorized.

2.5 *Public opinion poll*

In the 2012 public opinion poll, 2410 valid questionnaires were collected across 10 cities (Ministry of Housing and Urban-Rural Development et al. 2012). The overall awareness rate of Car Free Day amongst city residents stood at 78.3%, with a 3.8% increase from the 74.5% in 2011. Over 75% of the city residents did not think that Car Free Day could affect their shopping trips, with 95% expressing support for the governments' implementation of traffic improvement measures.

2.6 *Monitoring of traffic environmental indicators*

Traffic environmental indicators include traffic volume, vehicle speed, accidents, air pollutants and noise pollution. In 2012, there were 20 cities submitting traffic environmental indicators monitoring reports. The comparison of the environmental indicators before and after the implementation of Car Free Day has revealed the following:

1) In the cities that set up Car Free Areas, the volumes of traffic in both the Car Free Areas and the cities proper showed a marked reduction, with drop rates of 30–80%. Moreover, the volume deduction was accompanied by an increase of 5%–40% in walking, cycling and using public transportation as the means of mobility. In the cities that did not set up Car Free Areas, Car Free Day had essentially no effects on the volume of traffic in the city area.
2) The citywide accident rates on Car Free Day showed a decrease, with vehicle speed increased in Car Free Areas by 21% to 28%. The accident rates in the city proper also showed a decline when compared with normal rates, with an average decrease of 20%. This demonstrates that the reduction in private car volume is conducive to increasing city traffic safety levels, reducing damage to life and property of the residents, and bringing down the traffic management costs.
3) The concentrations of airborne pollutants and noise pollution levels were relatively low in Car Free Areas. In Car Free Areas, the levels of CO, CO_2, SO_2, nitrogen oxides, and particulate matter showed different degrees of reduction, ranging from 5% to 50%. The noise pollution levels reduction was usually equivalent to 2–3 decibels. Internationally, the noise and air pollutions are currently the most important indicators for assessing the levels of urban sustainable development. Therefore, the reduction in private car volume plays a very active role in raising the levels of sustainable urban development and improving the living environment of the residents.

3 THE ACHIEVEMENTS OF CAR FREE DAY MOVEMENTS

3.1 *Public transportation passenger flow increase and improvement of the quality of services*

The achievements of Car Free Day activity are directly reflected in the use of public transportation. During the Car Free Day in the city of Kunming in 2010, the passenger traffic volume reached 2.45 million, with the public transportation contribution to the total mobility modes exceeding 36%, setting a new record. Moreover, the cities had various initiatives, and the quality of public transportation service showed marked improvement. In 2010, during the Car Free Day in the city of Changsha, the entire city's fleet of 3,500 buses and 6,280 taxis underwent a comprehensive internal and external inspection as well as a service facilities inspection in order to guarantee good vehicle conditions and the aesthetic appearance of the vehicles (Ministry of Housing and Urban-Rural Development et al. 2009). The increased passenger flow further demonstrated a raise in the public awareness of green mobility, environmental protection and health, while also demonstrating the importance and urgency of the need to raise both the efficiency and service quality of public transportation.

3.2 Improvements in urban transportation operation effectiveness and environmental protection

During the 2007 Car Free Day, the volume of traffic decreased by 41% in Car Free Areas in the city of Kunming. In the city of Hefei, the volume of motorized transportation entering the area within the city ring road decreased by 79.6%. In the city of Changzhou, the vehicle volume on Nan Da Street was reduced by approximately 90% with bus travel time on that road section reduced by an average of 50%. Moreover, according to the environmental monitoring reports, in 2009, in the city of Suzhou, the SO_2 levels were reduced by 12.2%, NO_2 by 26.67%, CO by 30.96%, ozone by 21.05% and suspended particles by 35.6% (Ministry of Housing and Urban-Rural Development et al. 2011). The levels of noise pollution decreased by 2–3%. Comparative analysis of environmental indicator monitoring data collected in each city before and after Car Free Day clearly shows that holding Car Free Days can efficiently reduce ineffective traffic demands, improve traffic congestion, reduce the number of accidents, bring down the levels of air pollution and alleviate noise pollution. These are some of the most direct and obvious achievements of Car Free Day activities.

3.3 Preliminary results of publicizing and popularizing green traffic concept

Looking at the influence of green traffic concept on the public, it can be said that the extent of popularization and the promotion of Car Free Days have been increasing year by year. In 2010, the Development Bank in the city of Shenzhen collected a million pledges from car owners nationwide, appealing to everybody not to drive their cars on Car Free Day on the 22nd of September, and received an enthusiastic response from nearly 1.2 million people pledging to do so. In 2011, during the Car Free Day in Shanghai, approximately 200,000 car owners put up Car Free Day themed car stickers and changed to public transportation mobility modes (Ministry of Housing and Urban-Rural Development et al. 2010). Moreover, the intense reaction towards Car Free Day by the public sparked discussions on major websites, forums and Weibo social networking site, with overall support expressed for holding the activity, and people actively offering their opinions and suggestions. This clearly demonstrates the awakening of green traffic culture awareness of the public and once again proves the effectiveness of green transportation.

3.4 Urban traffic policy change and the promotion of sustainable development

The continuation of Car Free Day movement not only raises the quality of public transportation services and its contribution to people's mobility, improves the conditions of non-motorized and promotes energy saving and emission reduction, but at the same time it also plays a very active role in promoting the urban traffic policy change and the formation of sustainable model of urban traffic development. Strengthening green traffic system construction has already become a key value in organization of Car free Days by municipal governments. For example, in 2009, the city of Hanzhong banned the retailing business which occupies the roads, and prohibited parking of cars on pavements and on lanes reserved for non-motorized. In 2011 the city of Dalian set up a number of long – term measures aimed at the promotion of public transportation, including setting up exclusive use of public transportation lanes, construction of integration public transportation hubs and stops, extension of rail transit network, restriction of motorized vehicle volume, and optimization of the integrated bus network serving the newly constructed parts of the city and out-of city areas.

4 RECOMMENDATIONS FOR ORGANIZATION OF CAR FREE DAY

4.1 Extension of participation scope and depth

Cooperation between different organizations is very important to ensure the success of Car Free Day, all the more so for a multi-faceted Car Free Day activity. The flexibility and creativity of the public is something that government departments often lack. Therefore, during the planning, preparation, publicity, execution, traffic maintenance, environmental monitoring and public questionnaire survey stages, the organization by the public and individual participation can be utilized. A business mechanism can even be conceived, creating a Car Free Day public and private cooperation model, with the three parties: the public, the government and business all benefitting. The future direction in which the efforts must be applied lies in the enlargement of the number of participating cities, organizations and groups, increasing media access to Car Free Day activities to provide more in-depth coverage, forming an interactive relationship between the government, the public and the media.

4.2 Reasonable designation of Car Free Areas

The formulation of the plan for a Car Free Day activity must be in accordance with the practical realities of the city. The designation of Car Free Areas must be reasonable, and the scale of Car Free Areas must be very carefully considered, together with the effect they might have on the mobility of citizens living on the periphery of the city. It must be ensured that the Car Free Day is held at a time convenient for the city residents. Moreover, in case of a time clash between Car Free Day and another large-scale activity in the city, the organic integration of the two can be considered. A good example is the integration of the 2010 World Expo held in Shanghai with that year's Car Free Day. The roads within the traffic area under the management of the Expo became Car Free Area (lanes) with the total area of 7 km^2. A series of activities were also organized, such as bicycle parade, optimization of pedestrian and bicycle traffic and public transportation facilities displays, all attracting vigorous and extensive participation by the city residents, with the interaction of Car Free Day and the Expo activities creating an excellent atmosphere.

4.3 Publicity and information provision aimed at private car users

Judging by the publicizing experience of previous Car Free Days, greater emphasis must be made on the creativity of publicity aimed at private car users. One of the objectives of holding Car Free Day is to allow the urban residents to personally experience an entirely new mobility environment, especially those residents who overly rely on their private cars for transportation. Therefore, specific promotional strategies on how to make car users abstain from using their cars on Car Free Day must be researched. For example, influencing the mobility habits of parents by influencing their children at school, recognizing and rewarding their staff who continue not to use their vehicles after Car Free Day by enterprisers, and organizing social car pooling programs.

4.4 Promoting the formulation of long-term sustainable development measures

Promotion of long-term measures for sustainable urban traffic development during Car Free Day periods is of special importance and is a very important part of Car Free Day. Those long-term measures can subtly deepen the new awareness gained by the public on the subjects of public transportation and urban traffic sustainable development, forming completely new concepts of mobility. The city must, therefore, continuously promote one or several long-term measures during the duration of Car Free Day activity, as well as carefully address the relationship between the long-term nature of such measures and the short time availability during the Car Free Day activity. Therefore, the crucial time granted by Car Free Day activity for the formulation of long-term sustainable measures must be used to vigorously push ahead the construction of green transportation system, promote energy saving and emission reduction in urban traffic zones, and increase the capacity of sustainable urban development.

5 CONCLUSIONS

Car Free Day originates in Europe and is today the world's largest sustainable traffic activity. The aim of the activity is to promote the adoption of sustainable traffic development policy measures by local governments, as well as to invite the public to try the models of transportation that provide an alternative to cars. China City Car Free Day has been held since 2007, and its national and global influences have been growing continuously. Car Free Day is not just effective over a single day, but extends long-term effects on the quality of urban life. Car Free Day encourages municipal governments to rethink urban traffic policies and to adopt sustainable traffic improvement policies, thus transforming urban traffic development model, and producing livable cities with green transportation methods such as a suitable public transportation system and walking and cycling facilities. Car Free Day activity also advocates and encourages urban residents to more actively choose high-efficiency, low-cost, low-consumption and low-pollution green mobility methods, and makes people aware that the transportation choices made by an individual can influence public health and community welfare.

REFERENCES

Lu Huapu 2012. Urban Transportation in China: A Retrospective Look and Future Prospects. Urban Transport of China, 10(1): 5–8.

Ministry of Housing and Urban-Rural Development, China City Car Free Day Steering Committee 2009. 2007–2009 China City Car Free Day Data Compilation. Beijing: China Academy of Urban Planning & Design.

Ministry of Housing and Urban-Rural Development, China City Car Free Day Steering Committee 2010. 2010 China City Car Free Day Data Compilation. Beijing: China Academy of Urban Planning & Design.

Ministry of Housing and Urban-Rural Development, China City Car Free Day Steering Committee 2011. 2011 China City Car Free Day Data Compilation. Beijing: China Academy of Urban Planning & Design.

Ministry of Housing and Urban-Rural Development, China City Car Free Day Steering Committee 2012. 2012 China City Car Free Day Data Compilation. Beijing: China Academy of Urban Planning & Design.

Review of prestressed concrete technology in flexural members

Yixiang Yin, Yuanbing Cheng, Zilong Guo & Guanghua Qiao
North China University of Water Resources and Electric Power, China

ABSTRACT: This paper reviews the definitions and structures of prestressed Reinforcement technology, and discusses the characteristics and calculations relevant to this field. The structures studied include the composition and classification in prestress concrete technology. The difference between prestressed tie bar reinforcement and high strength-low relaxation strand wire reinforcement is discussed. Finally, this paper analyses two calculation methods and puts forth topics that call for notice.

Keywords: prestressed reinforcement; definition; characteristics; structure; calculation of prestressed reinforcement

1 DEFINITION

The prestressed reinforcement method is also called the external prestressed strengthening method. It is an indirect reinforcement method that adjusts the stress of the original structure and supports it by applying external pre-stressing. This method works by imposing an opposite balanced force on the original members, offsetting the original force and improving the capacity of the components. (Chunsheng Zhang & Baoshan Yan, 2013).

2 CHARACTERISTICS

2.1 *Advantages*

The prestressed reinforcement method has the following advantages: (1) It requires simple construction; (2) It produces obvious reinforcement effects; (3) It is economically sound due to the availability of product and the low cost; (4) It does not reduce the net structure space; (5) It has short construction time and little influence on production and living.

2.2 *Disadvantages*

The method also has the following disadvantages: (1) It requires Relatively difficult protection; (2) It easily produces stress concentration and possesses high requirements for anchoring construction; (3) It cannot bring high strength external cables into full play; (4) It easily causes the loss of prestress.

3 STRUCTURE

3.1 *Composition*

3.1.1 *Anchorage*
The external prestress system only transmits force at the anchor end, thus the security and reliability of the externally prestressed anchor system are much higher than those of other general prestressed anchor systems.

3.1.2 *External cable*
Unbonded strands are frequently applied to external cables. PE epoxy resin coated steel strands have good corrosion resistance and applicability, so it is widely used in bridge reinforcement.

3.1.3 *Anchorage and steering block*
The external prestress system only transmits force by anchors and steering blocks, so they must have the ability to effectively connect the original structure and transfer stress. Anchors and steering blocks are commonly used in reinforcement concrete and steel structures.

3.1.4 *Vibration damper*
An external prestress system damping device is applied to the external cable when the distance between the anchor and steering block is too long. The main purpose of this device is to reduce the vibration influence on external cable under dynamic loads.

3.1.5 *Construction equipment*
According to the different tension requirements, construction equipment usually can be classified into single jacks and overall jacks. The single jack is used when construction space is narrow or the external cable must have single tensioning, and the overall jack is applied to the external cables which must all be tensioned.

3.2 *Classifications*

3.2.1 *Prestressed tie bar reinforcement*
The prestressed bar reinforcement methods are classified in three types: horizontal tie rods, stay tie rod, and

combined tie rod. (Chaofei Wei, 2009) The horizontal tie bar method is suitable when the flexural bearing capacity is insufficient. The stay tie rod is appropriate for flexural members lacking the bearing capacity both in shear strength and normal flexure. A combined tie rod is suitable for serious shortages of normal bending capacity while the shear bearing capacity is also insufficient. It includes two tie beams and four down-stayed tie rods. All the above methods can reduce the deflection and crack width of original members.

3.2.2 *High strength and low relaxation strand wire reinforcement*

The cross bundling method reinforces the member with a prestressed steel strand under the bottom of the beam to exert external prestress to improve the bearing capacity. (Jinfeng Xiang, 1998)

Setting the steering block at the bottom of the beam with prestressed wire line tension improves the bearing capacity of the beam. (Zhanji Zhao, Chuang Li & Junbo Tian, 2008)

3.2.3 *Differences*

The methods previously mentioned are explained as follows: (1) The first way is to utilize cross bundling to exert external prestress, where the position of the steering device is set depending on the section size and the span of the beam. The steering device can also be randomly placed to offset the beam load itself. (2) Another way is to add steering device at the mid-span position, as generally the destruction of the beam is caused by cracks and deflection at the mid-span position, which leads to the concrete crushing at the upper part. Adding steering blocks has no effect, yet it can unify external restrained deformation and structure deformation. It also reduces the midspan deflection partly. (3) Special attention is needed to anchor a zone at the highly prestressed area and apply a steering block, since stress is higher in these positions. If these positions are handled improperly, it can lead to damage of the members. It also leads to stress relaxation and impacts reinforcement. (4) Loss of externally prestressed reinforcement beams is much lower than those of common prestressed concrete beams. Reinforced steel strands at the bottom of the beam can stretch and bend constantly, which increases the strength of prestressed reinforcements. (5) When prestressed steel is exposed, regular checks for corrosion should be performed. (Li Li, 2003)

4 BEARING CAPACITY CALCULATION METHODS

4.1 *Calculating eccentric compression members*

The equivalent load method was employed in this experiment, taking tension as the external force on the original beam. However, this method does not take into account the stress increments of prestressed tendons in the process of loading. Stress of prestressed tendon uses the proof stress σ_k when prestressed tendon is damaged. According to the calculation of eccentric compression members, we can check the bearing capacity of the original beam. The deflection strength is equal to the pre-tensioning deflection subtracting the invert arch caused by prestress and adding the deflection under load after reinforcement. There are errors between this calculation method and the simplified mathematical model because the original beam has some bending capability. The actual deflection should be calculated on the original beam. This calculation is complex and needs to combine internal force, which is determined after reinforcement.

4.2 *Calculating the unbonded partial prestressed concrete structure*

This method is based on the prestress ratio λ to ensure that the prestressed tendon area and the non-prestressed steel bar can yield, and the prestressed tendon does not reach ultimate strength when the section hits flex failure. Therefore, it is sensible to perform calculations based on an unbonded partial prestressed concrete structure. (Yanhe Li & Gui Chen, 1998)

The prestress ration is $\lambda = M_0/M$, where M is the service load, not including the bending moment of control section under pre-tension; M_0 is the decompression moment which offsets stress on the tensile edge of control section to zero when service load is working. $M_0 = \sigma_k W_0$, σ_k is effective compressive prestress on tensile edge of flexural member, and W_0 is the section modulus in bending. (Jiancong Han & Wei Zhao, 2007)

$$\sigma_k = \frac{N_{\rho_e}}{A}\left(1 + \frac{e_\rho y_0}{r^2}\right) \quad (1)$$

In the above expression, $r^2 = I/A$; e_ρ is the length between the prestressed tendon application point of resulting force and the section gravity axis; y_0 is the length between the edge of the beam bottom and the section gravity axis. We can calculate N_{ρ_e} in light of the formula, thus the prestressed tendon area $A_\rho = N_{\rho_e}/\sigma_{\rho_e}$.

It is important to know whether the beam is a reinforced beam because of the steel in the original beam. Then, the bearing capacity of the reinforced structure, the stress, and the structural deformation in the usage phase are obtained. In addition, it should be ensured that the effective height decrease of external prestressed tendons and referencing calculation formula for unbonded prestressed steel do not contain errors.

The stress increment and the position of external prestressed steels change as prestressing when buildings put to use external prestressing technology following the load and the member deformation. It is necessary to perform nonlinear analysis when precise calculations are needed. (Yumei Zhang, 2013).

4.3 *Design steps*

(1) Draw the internal force chart under remnant loads and all other loads respectively;
(2) Based on the total bending moment M, calculate the bending moment ΔM, and then calculate the reinforcement area needed in the midspan section of the beam;
(3) Determine the tension control stress and calculate the loss of prestress;
(4) Based on the tension control stress, calculate prestressed internal force;
(5) Check the section bearing capacity;
(6) Calculate the prestressed effect and tension.

5 CONCLUSION

External prestressed structures have the advantages of flexible arrangement of reinforcement, convenient construction, safety and reliability. They can effectively improve the bearing capacity of bridge structures; and cable adjustment or replacement can even be done under normal working conditions. The characteristics of prestressed reinforcement technology are superior to those of other strengthening methods. In the coming years, with the development of civilization, there will be more and more old buildings and old bridges that could possibly fail and cause accidents, and they need identification, maintenance and reinforcement. The cost of maintenance, reconstruction and reinforcement of the old buildings and bridges will increase gradually. Prestressed reinforcement technology is simple and produces good reinforcement effects. It will be imperative that the application and development is used to its potential.

REFERENCES

Chunsheng Zhang & Baoshan Yan. The review of engineering structure reinforcement method. Fiber reinforced polymer/Composite material, 2013(02).

Chaofei Wei. The study of externally prestressed reinforcement technology of concrete bridge. Shanxi Transportation Technology, 2009(05):26~27.

Guoming Shu & Tingchen Wang. The reinforcement design of external prestressed reinforcement bridge based on prestressed ration. Road Transportation Technology, 2006, 6(03):62~65.

Gui Chen, Yanhe Li & Mingbao Min. The application of unbonded prestressed steel strand reinforcement technology [J]. Architecture Technology, 1998, (23):388~399.

Hongguang Jiang, Tingchen Wang & Hui Xu. The experimental study of external prestressed reinforced concrete simple beam. Road Transportation Technology, 2006, 03, 23(03):107~111.

Jiancong Han & Wei Zhao. The discussion on calculation of reinforced concrete flexural member. Architecture Structure, 2007, 07 (the supplement of 37 volume).

Jinfeng Xiang. Prestressing reinforcement metahod of High strength steel strand in reinforced concrete girder. The eighth annual meeting proceedings of China civil engineering society, 1998, 3.

Li Li. The study of external prestressed reinforced concrete flexural behavior and the experimental research of external prestressed strengthening compares with the method of sticking steel strengthening [D]. Beijing: Beijing University of Technology, 2003.

Wenbin Lu. The technical analysis of external prestressed reinforced concrete beam. Journal of Xiangtan Normal University (Natural Science), 2009, 12, 31(4).

Yanhe Li & Gui Chen. The research of prestressed reinforcement design and calculation method [J]. Engineering Quality Supervision, 1998(2):21~29.

Yi Wan. The application of reinforcement and reconstruction method in the existing buildings, Shanxi Architecture, 2013, 05(13):54~55.

Yumei Zhang. The analysis of structural reinforcement method. Journal of Taiyuan Urban Vocational College, 2013(03).

Zhanji Zhao & Shaowu Zhang. Externally prestressed reinforcement technology of reinforced concrete beam. Shanxi Architecture, 2007, 33(31):64~65.

Zhanji Zhao, Chuang Li & Junbo Tian. External prestressed reinforced concrete beam. Shanxi Architecture, 2008, 34(03):106~107.

Roads in urban areas: Limits to regulations and design criteria

A. Annunziata & F. Annunziata
Department of Civil and Environmental Engineering and Architecture, Cagliari, Italy

ABSTRACT: Design and/or adjustment of urban roads need to be based on a preliminary functional classification of roads within the general transportation infrastructure system. It can only be referred to the metropolitan area, considering the infrastructure system aimed at a territorial rearrangement, with an objective of balancing the overall system of settlements and services. This paper suggests assigning functions to different elements of an infrastructure network, not only when its features are adequate, but also when the other elements of the network help ensure the conditions for fulfillment of traffic needs of the different roads types in urban areas.

1 INTRODUCTION

Design and/or adjustment of the urban roads are to be seen as an important part of transportation infrastructure systems in urban areas, aimed at sustaining a more equitable territorial rearrangement, but also, at reinforcing various levels of scale relations among people, among peoples and spaces, and at avoiding segregation of disadvantaged users. The main aim of a rigorous concept of road infrastructures must focus on people, and must propose the promotion of an intense transfer of ideas and resources. Design or adjustment of urban roads must focus on edges as a neglected resource useful to preserve forces that permeate a region of the urban fabric, and to enable new modes to interpret a site. This effort aims at producing new meanings, i.e. at modulating a fecund transformation of a landscape, in order to forge a logo, an identity that supports a cohesive community. The future of cities, of European cities in particular, expands to metropolitan areas: the function of urban roads cannot neglect a strong connection among different urban systems in order to build robust relations within and among different areas of the same urban district.

1.1 General considerations

The Decree of the Ministry of Infrastructures and Transportation dated November 5, 2001 "Functional and geometric guidelines for road construction" (D.M. n. 6792/2001) contains conceptual details and design guidelines for construction of all types of roads included in new Highway Code located on morphologically complex areas, with the exception of mountain roads, for which compliance with the design criteria is generally impossible. Moreover, regulations overlook particular categories of urban roads, such as those with peculiar functions or those located in residential areas requiring, for example, speed calming devices. The elements related to the geometric design features of roads are governed in relation to their role in the urban planning as well as in a more specific transportation plan. Actions on existing roads should prioritize adaptation of the geometric features to the regulations in order to better meet traffic needs.

The abovementioned decree references and the limits of the new regulations need some more clarifications in order to develop into a more comprehensive regulatory framework. It should take account of functional adjustment of the existing road systems: not only refurbishing of road elements, rebuilding of superstructures or implementing actions aimed at higher operating speed, but rather reconstruction of network systems, in which territorial functions are defined and performed to achieve homogeneity of supply and of levels of service for the various infrastructures in the area. According to this approach, the first step of functional adjustment should be assigning specific functions to single roads elements. The definition of a methodology for classification of existing roads, therefore, requires comprehensive and specific planning. A standardization of the features of different types of road infrastructure throughout the national territory is the mail goal: classification will therefore be a verification of the validity of road functional hierarchy defined by these planning tools. If these are absent, classification process requires prior identification of provisionally expected functions of infrastructure based on criteria for geometric and functional standards aimed at ensuring that users allowed on each type of road can travel safely.

2 POSSIBLE FRAMEWORK FOR NEW REGULATIONS

2.1 General considerations

The above mentioned Ministerial Decree (D.M. n.6792/2001) defines the criteria for design and functional aspects of road geometric features, in relation to their classification in accordance with the Highway Code. Classification is also based on the types of users and activities allowed on the roads, as well as the environmental context to which the roads belong. This suggests that road design choices are strongly influenced by physical and human-related features, and not only by the quality of the environment where infrastructures are located.

The urban road is defined as part of a road network to be planned within a comprehensive framework of functional relations, in order to simultaneously satisfy the different movement needs, both with collective and private means. It is due and proper non to conceive roads as a mere space for motion, or as a poor and arid car realm. Specially in urban areas roads must be seen as a part, a consistent portion of the urban space. This statement drives us to consider roads as a stage aimed to promote the rise of a *civitas*, of a social tissue sutured by a common baggage of memories, ideas, rites, values. Roads are to be designed for people, and primarily, or as a consequence, they must be designed as a space that encloses scenarios aimed to promote relations among people, and among people and features, meanings, concealed in the urban landscape.

As a result, the first required step, in order to recover a more profound and sensitive idea of road, is that the concept of movement of people, vehicles, or goods should be replaced by a more fecund idea of exchange, as a wider phenomenon nourishing the life of an urban area. A road is the set of a wide osmosis of notes, of content, at various levels, among nodes of a site, among people, and between people and site. These are a significant part of the real essence of a road, evoking a range of activities varying in tone and breath, for which that road should be designed. The importance of the site in sanctioning the role of a road also is, thus, fundamental. The result of the design process, both planning new roads and recovering existent ones, is that the infrastructure is a sign aimed to reinforce and foster relations that permeate and structure a landscape. Those relations must be considered as a precious and unique resource to be maintained, since they are the driving force and the condition of those vast processes of osmosis that support life in an area; the meaning and the function of a road, both in urban and rural areas, cannot disregard this fecund and fragile connective tissue that pervades the site. This becomes particularly meaningful when a new route allows planners to rethink the function of existing roads and branches which, due to deficiencies in the road network, have been overloaded with unsuitable uses making circulation unsafe and causing or exacerbating the degradation of the site. It follows that the different design phases are to be developed in the light of an environmental analysis and, in turn, adjusted to specific purposes, considering as fundamental maintenance of existing territorial relations and exploration of latent ones. This is achieved by an iterative process in which technical choices depend on the verification of their possible effects, in order to attain the best compromise of functional requirements and of economic and environmental issues, which ensure the compatibility of any action.

2.2 *Road networks and territorial organization*

Roads should be seen as part of an overall infrastructure network at the service of mobility which also includes public transportation on dedicates lanes, other public transportation systems, and parking facilities. The classification of different types of road within the abovementioned system mainly refers to conurbation or metropolitan areas.

The design of transportation systems in these areas needs necessarily to be based on the assumption that they cannot normally be studied in restricted territorial areas: their aim is integrating the city with its highly populated suburbs. Planning should consider environmental requirements and structural guidelines which call centrality of the city into question focusing on a new understanding of urban and territorial relationships. The static form of the city, which has been identified with the urban culture for centuries, as well as the operational models related to a formal static conception of actions, have therefore been overcome by a dynamic view of city-region or city-territory as a set of dynamic relations within a general process of population growth and social advancement. This centripetal development pattern focused on chief towns is therefore outdated, however a pattern of settlements highly dispersed in the territory should be avoided, since a number of urban entities, despite territorial contiguity and common interests, lack a specific connotation in their disorder. A polycentric settlement organization and network system, open outwards, should thus be built: "border" nodes are to be given a function of spatial and relational interface with the external territories of the surrounding area (the conurbation which is reaching the connotation of metropolitan area), and they should also be partially redundant, in order to replicate some functions and thus counteract the attracting and hierarchizing effects of major towns (Annunziata et al. 2004).

The organization and planning of services across the country should be based on the interconnection of services and housing, mobility and environment, and on a strategic organization of services deployed across the area according to demand. A framework planning of infrastructure works is therefore needed to solve mobility problems of the metropolitan area. Planning should consider:

0* the concentration of new settlements in the vicinity of the routes of public transportation systems, with the enhancement of public transportation on dedicated lanes;
1* the creation of a logistics platform based on the networking of transportation poles in the area;
2* integration of the different modes of public transportation on dedicates lanes, using the existing tracks in connection with surface

public transportation services and with parking areas/facilities, in order to create an extended logistics platform aimed at a reduction of vehicular traffic on existing roads.

In summary, over time the framework plan should achieve the following results:

3* reorganizing the modal distribution of transportation demand, creating a system of public transportation on dedicates lanes as the backbone of the infrastructure network serving mobility;
4* reducing commuter flows to the central urban area, avoiding location of new traffic attractors or implementation of the existing ones;
5* reducing commuter flows to the central urban area, placing metropolitan area services outside central areas with poor accessibility, and locating them near public transportation on dedicates lanes, closer to the source of transportation demand and therefore more accessible.

In order to solve mobility problems in an urban area, it is essential to carry out actions within a far-reaching framework, which is in turn the result of a pondered strategic conception, giving transportation policy a pragmatic, yet intelligent, honest and far-sighted vision of social facts. Planning a transportation infrastructure system in an urban area requires preliminary knowledge and understanding of the city and its current problems, that is a reflection on the history of the urban area. Planning the transportation infrastructure system of an urban area requires consideration of environmental sustainability as preparatory work for preliminary design, and identifying the areas that are not to be crossed by new infrastructures or impacted by adjustment of existing ones.

Mobility is influenced by the acceleration of the enlargement of the European towns in the last hundred years, particularly the thirty years following the II World War, and by the increase of private car traffic, which is causing a paralysis of circulation in many cities. Historical centers, developed for pedestrian traffic, are being destroyed by mere adaptation to cars. Motorization rates have undoubtedly grown in pathological terms, but especially pathological is that people are forced to choose individual private transportation means in the absence of alternatives. The radiality of an urban area is not a mere issue of roads systems and networks, but it implies that most activities and interests orbit around the center and create movements towards it. Towns are such a complex organism that a single action, even the best one, may not be suitable to match its needs. For example, a subway network needs bus service for the suburb stations to be reached, but suburb stations also need parking areas for those who cannot reach them by bus. Approaching the city center, it is important that stations give passengers a chance to interchange between different subway lines or other surface transportation means. Parking of private vehicles around the center is to be seen in relation to their distance from it, and pedestrian paths should be proportional to the town size. Uses of public space and transportation means should be suitable to the structure of the different urban areas. In this sense, historical centers are most critical because of their historical and architectural significance and due to the density of both residents and cultural and economic activities. Identifying citizens with this part of town requires great care for values, such as the right to use the common areas. Common areas are to be primarily considered as meeting places meant for city-life activities, well served by public transportation and where moderate and controlled private traffic is allowed. Parking areas for residents, increased pedestrian safety, and a drastic reduction of pollution are ways of encouraging residence in historical centers, which needs to be preserved as an element of balance of city life.

2.3 *Regulations and urban roads*

Regulations, however, expressly omit dealing with urban roads serving residential areas. No specific regulations concern issues related to the needs of vulnerable users: design methods that are specifically intended for the people involved and take into account their needs should be defined. If quality is defined as "the degree of performance compliance with the requirements of the purpose the product was made for" and all users have a right to travel safely but also pleasantly, it is obvious that regulations are not intended for attaining design of quality roads. Although quality control and measurement should be regulated by legislation, current regulations set no reference for the design of visual sequences, road furniture, and for what allows traveling to be not only safe but also pleasant and fully satisfying the needs of all road users. Urban roads need in the first place to be conceived as scenarios where people gather, talk, take a break: a scenario aimed at giving new breath, new sense, new reach to the living theme promoting uses and episodes based on it. Roads should no less serve the urban fabric they belong to, and mass transport should be implemented without neglecting controlled private traffic (Annunziata & Annunziata 2013). Some roads open on squares, others narrow into alleys, some achieve fame and notoriety, but most are known only to the residents and those who use them regularly. Residential roads are an important part of the urban fabric and their proper operation can influence the quality of the city itself, its safety and comfort, and its citizens' well-being. The primary function of residential roads is to serve the area they belong to, providing access for those who live or work there. Design of residential roads should therefore be aimed at these objectives, considering their peculiar significance in town administration, attentive to its inhabitants' needs and well-being. The design of residential roads and traffic control should therefore safeguard the neighborhoods and their quality of life, avoiding the three types of traffic that are unwanted and penalizing:

6* Traffic on residential roads as shortcuts and deviations from congested arterial roads;
7* Fast traffic;

8* Parasite traffic, looking for and coming in and out from parking lots, outsiders using sidewalk space for parking.

3 THE MANAGEMENT OF INFRASTRUCTURAL SYSTEMS

3.1 *Road networks*

Typically, road systems consist of roads, i.e. the infrastructures necessary for movement of vehicles, as well as all the functional appurtenances under the responsibility of a road owner. The above definition mainly concerns technical and administrative competences and restricts an overall evaluation of the road network. This mostly consists of several roads, managed by various organizations, interconnecting with each other and with different networks, and therefore requires a territorial management, at a higher level than the owner's prerogatives. This concept and the subsequent setting of work therefore are worth rethinking so that the different managing organizations identify the needs of their own infrastructures as part of a more complex system serving a given territory seen in its unity, although subject to the decisions of different organizations. An authority to be given the financial resources to coordinate management should be identified, subject to prior verification of needs and consequent identification of measures suitable to the different elements of the network infrastructure and of services aimed at meeting mobility needs.

The prevailing purpose of the management of infrastructure systems is keeping the system appropriate to the functions assigned to each of its components within a given life: when an infrastructure no longer meets all or part of its requirements, adjustment is needed in order to amend functionally abnormal situations. Functional adjustment, in particular, will be required due to an unforeseen change in the functions required to a component of the network infrastructure, resulting in a reduction in service life and/or in the inadequacy to meet the new needs generated by the changed conditions of use. Planning functional adjustment actions is often more difficult than designing new infrastructures as it needs to consider the existing situation and the marked heterogeneity of the elements composing a system designed and constructed in times when current technical design standards were not available. In this perspective, when designing functional adjustment, criteria used in the planning of new infrastructure cannot be used because they are based on a set of prescriptive guidelines meant for its individual elements.

These considerations imply infrastructure system management as including design, maintenance and adjustment actions, aimed at maintaining/improving the service provided by an infrastructure network, which also consists of elements added in time, referred to different modes transportation, as required by needs for increased functionality resulting from changes in the local context. The different management activities usually refer to a time horizon called useful life. The concept of road useful life is generally bound by the conditions of superstructure and artworks, but it should more properly refer to the infrastructure in its complexity. As a result, if infrastructure management aims at ensuring service with reference to the project flow for a number of years, actions may be needed in order to guarantee continuance of life in view of the changed environmental conditions.

Taking the road infrastructure into account, it may be considered decayed when it is affected by deterioration, subsidence, functional and/or structural deficiencies, with particular reference to its physical elements. If, instead, the infrastructure is seen as part of a network and/or a local context, or if social, urban and ecological issues are considered, it is clear that in evaluating degradation, according to different needs, the following should be considered:

9* local context;
10* road network and routes.

The deterioration of a road network thus implies that it is no longer able to perform the assigned functions at a given level of service, because, for example:

11* The demand for transportation has changed due to a different territorial organization or transportation system rearrangement;
12* The cultural objectives and development trends in the local context have changed.
13* Geometric and project features assigned to the different segments of the infrastructure network thus prove inadequate to meet the expected new features of the traffic flow.

This comprehensive approach to network management can result in the need for functional adjustment, assigning the existing road network elements functions involving the adaptation and/or partial design of some of its sections.

4 FUNCTIONAL ADJUSTMENT OF ROAD NETWORKS

4.1 *Analysis of urban roads*

Taking urban roads into consideration, they can be distinguished from the extra-urban roads by two features:

14* the components of traffic;
15* the areas adjacent to the road.

There are four traffic components:

16* movement of private cars;
17* movement of public transportation vehicles;
18* parking of private vehicles;
19* pedestrian flow.

Designers should take these components into account with their specific needs and assigned space, in order to avoid the current deleterious situations of promiscuity of functions performed by the urban roadways and travel lanes.

The areas adjacent to the road, placed at the two sides of the roadway, are part of the complete road space. It includes all appurtenances aimed at meeting the needs of the additional components besides automobile traffic, which are essential for urban roads to comprehensively fulfill the functions they are expected to perform. They are intended to maintain vehicular traffic volumes required for each type of road. Surprisingly this general statement seems to identify private vehicles as the most important users of urban roads.

Basic classification of urban road types focuses primarily on the context they belong to, and their definition is further specified according to the functions provided by territorial planning and, consequently, in relation to the features of the traffic they gather. It follows that classification, as in relation to the different functions in a specific area, is the result of the Transportation Plan: the road is to be classified and therefore adjusted to the new function, as part of an integrated and intermodal system.

It is also necessary for the classification of a single element of an existing road network to consider the availability of areas for design/construction on the edges of the road: if these do not exist, the road network should be considered to find out if, with the other supporting routes, that element can perform the function requiring side areas with given features. For example, it is important to examine whether there are areas or structures for parking facilities to ensure intermodality and fluidity of the traffic flow. In classifying a road network in relation to the interconnected network, location and type of intersections is equally important in order to ensure operation of the network as a whole and of all its elements.

Regulations (such as B.U. C.N.R. n. 60/ 1978) provide for the design of new urban roads, including ring roads, under the responsibility of any authority, as well as town private roads open to public use (industrial areas, ports and airports, parceling, etc.). They, however, also provide elements for the adjustment of existing urban road networks, which can be summed up in:

20* suitable assignment of specific functions to individual road elements;
21* consequent arrangement of intersections, with possible limitation in number and access;
22* regulation of driving directions for the various types of vehicular traffic (public and private);
23* organization of appropriate rules for parking and pedestrian traffic.

The urban area was the context considered for the types of road. It was defined by the limit of the urban perimeter: the built-up (or urban) center bounded, for each inhabited cluster, by the uninterrupted perimeter that includes all the continuous built-up areas and in-between lots (L. n. 865/1971, Article 18). This statement is to be considered outdated: the urban area to be considered, as said, is the metropolitan area.

4.2 *Types of urban roads*

The regulatory framework considers the following road types:

24* Arterial roads: referred to as urban highways in the abovementioned 2001 Decree;
25* Collector roads;
26* Neighborhood roads;
27* Local roads.

Arterial roads are the end sections of highways or passing extra-urban roads. They mainly gather and channel traffic flows between the urban and suburban areas. They relieve urban roads of transit traffic (ring roads) and they only connect with the collector roads.

Collector roads are completely integrated in the urban area. They ensure fluidity of vehicular traffic towards or within the urban environment: they distribute the traffic from arterial roads and gather the neighborhood road flows. In medium to large cities, this type of roads may also serve individual urban areas and provide for fast connections between distant points of the larger surrounding neighborhoods. Collector roads, by means of service lanes (as appropriate points for turning and merging needs and for access and exit to parking areas and driveways) may serve, although never in a direct way, settlements and road space of particular interest. Service roadways have then the features of the lower class roads (neighborhood roads).

Neighborhood roads are those fully belonging to one sector of the urban area, and serve the connection between neighborhoods in the same urban area. They gather traffic from collector roads and channel it towards local roads. This type also includes those roads serving the main settlements of one district (services, facilities, etc.) through complementary or subsidiary road elements, as well as ensuring easy connections between the outer points of the same district.

Local roads spread entirely within a district serving that settlement exclusively. They gather traffic directed to neighborhood roads. For these roads only, reference is to the urban area, as set by the L. n. 865/1971 and subsequent amendments and additions. As a result, it is arguable that a classification based on size and scope of the connection provided by each element of the road network, according to its measures, form, and geometry, is fundamental in order to deduce the limit values of scale and scope of the paths and movements that a road can sustain. Nevertheless this classification must be completed by a more specific definition based on contents and essence of the piece of urban fabric or landscape, that encloses a road. It is this content that defines the spirit of a space. From this spirit it is to be deduced the vast and various range of modes of use, of episodes, that the road is supposed to promote. As a result, the first step of a sensitive recover is to recognize regions of space, pervaded of a specific spirit, and defined by data emanating from boundaries or via boundaries from nodes. "It is proper to discern regions defined by nodes of net cultural value, residential zones, regions connoted by boutiques or stores, or by recreational functions, regions defined by areas of great natural value, spaces connoted by presence of educational nodes, zones defined by spaces for major events (opera houses, sport or concert arenas), spaces defined by presence of poles hosting services of large scale (premises of banks, government offices, universities, hospitals), and roads located in areas signified by production plants" (Annunziata 2013, p. 280).

4.3 *Proper fruition of urban roads*

Having defined the basic types of urban roads with reference to the territorial context, their functional ranking is then detailed in relation to types of traffic they manage:

28* Movement of private vehicles (cars, trucks, articulated trucks, tour buses, etc.);
29* Movement of public transportation vehicles with stops (both urban and suburban buses and trolleybuses, trams, etc.);
30* Parking of private cars;
31* Movement and stop of pedestrians.

However, it is due and proper to propose a broader image of the functions of urban roads. A careful consideration about episodes to be pondered and sustained is required in order to define the function and structure of an urban road. Car traffic is to be taken into account considering access, i.e. small-size motion, penetration, medium-size motion entering a specific urban portion, and distribution and transit, to be considered as a large-size relation among distant points of an urban area. Brief car stops can be seen as not onerous uses, while longer car stops, i.e. parking, are medium-size uses that are onerous for the urban fabric in terms of soil consumption. Transfer from pedestrian realm to cars and vice versa are a consequence to be pondered of cars stops. Furthermore, transit and stop of public transport means is to be considered in order to enhance pedestrian routes. An obvious consequence is the need for people to wait at bus stops, and to get on and off the means of transport. An urban road should also provide spaces to allow people to move by bicycle, which can be considered as a medium scale episode. Walking for basic tasks is the most frequent episode, its scale and size is modest but its emotive outcome is significant, like paths traveled for leisure whose impact is small in scale and size but based on the prospect of a large emotive gain. An urban road is also to be conceived to persuade people to stop, to linger there, for a rest, for leisure, or to observe a detail and to meet: above all, users must be able to sit, so as to recover from fatigue, or to evade in a brief pause, to perform leisure activities, or to be part of broader events, spending time among people. The scale of pedestrian activities varies according to essence and relevance of contents that permeate the piece of urban fabric that encloses, structures and signifies a given road or a certain space. Moreover, it is proper to point out that pedestrian activities, even if of modest scope and size, are, if frequent, source of an intense emotive gain and of clues about tone and essence of a site and about opinions, ideas, values, rites and trends that permeate the social fabric.

Those episodes and activities are fundamental to recover and reinforce relations among people and among users and sites or pieces of an urban landscape, i.e. in order to build a *civitas*, a community. As a result, an urban space is presumed to invite people to listen, to sense scents and sounds, to observe, to meditate, to read, to converse, to perform sport activities and games, to rest to have a snack. It is also important to consider uses aimed at connecting private and enclosed ambits and services of a broader scale, i.e. loading and unloading, storage and waste collection. Moreover, a main concern is to allow access, penetration or transit of emergency vehicles (Annunziata 2013).

From what provided for by regulatory framework, in particular, it follows that on collector roads:

32* public transportation network has to have dedicated lanes and/or defined stops;
33* parking is allowed on separate spaces with concentrated access and exit paths.

On neighborhood roads:

34* public transportation has dedicated stop areas and may run on dedicated lanes;
35* parking is permitted with open driveways;
36* public transportation may be admitted on local roads, but the size and geometry of the lanes has to be suitable to the needs of the vehicles used.

It follows that a line of public transportation on dedicated lanes, as a rule, is only permitted on collector roads and neighborhood roads, and thus they should be designed with particular attention to public transportation, if the government of urban mobility is meant to be based on it.

Moreover, Contents and essence of a region of the urban fabric can produce a vast process of osmosis that involves and signifies the urban space. As a consequence, it is due and proper to deduce from nature and relevance of services and contents, aligned along a road, the nature and scale of modes of use to be enabled and their distribution inside a space, giving priority to a rigorous definition of nature and size of pedestrian activities.

The outcome is a solid basis, a cogent reference that entails the plot of a road, its spaces, their content, scale and structure: a residential area is a space to be conceived as a place for modest and frequent social episodes. Spaces are to be provided to allow people to meet, converse, and perform games or sports. Spaces connoted by acute cultural or natural value are primarily to be conceived to reinforce medium scale episodes: it is supposed to allow a large number of people to sense the site, observe, and notice its content of notes, scents, sounds, and forms. Spots conceived to invite people to stop and linger so as to observe are needed in stores and boutiques areas, as spaces designed for people to sit or to stand, so as to rest, are to be provided in areas where major services, educational poles, or event venues are located. The definition of measure, structure, and scenarios conceived for people is supposed the first step. The second step consists in defining cars space, ascertaining that these spaces, unsafe and unable to foster and sustain pedestrian routes, do not cut or ruin the core of the space or separate its regions. Carload, size and purpose of car motion are to be deduced primarily from measures, form, and geometry of a road, i.e. from size and scope of the tie it provides inside an urban area. Moreover, carload, and size and nature of car motions define the grade of separation among car space and promenades or spots conceived for people. Local roads do not require or, even more, claim a tenuous distinction of car spaces and pedestrian realms. A tenuous boarder could be useful so as to protect spots conceived to invite people to meet. For roads aimed to sustain a more intense cars presence, i.e. motion of modest scale, a more net border is needed. In case of an arterial road overloaded by copious car traffic, definite borders are required. Separate spaces or side roads are required in case it serves access to contiguous nodes or services. It is also useful to recover residual voids generated along the road, as safe scenarios, as containers of urban spaces, of services and contents, or as focal points of a dense fabric of unforeseen relations. A continuous border is needed to separate spaces. Car space should be placed at a different level, in order to create a network of promenades and spaces designed for people aimed at suturing the urban realm and providing a spontaneous, safe, and frequent interaction among its parts (Annunziata 2013).

The regulatory framework does not address the novelty represented by suburban corridors. They differ from other roads systems because they usually have a limited number of exchanges with urban and extra-urban roads, generally by means of equipped junctions, and it is generally complex. Suburban corridors can include several roads of different types and functions, with different configurations (sometimes overlapping), and non-homogenous (motor roads, railways, subways, monorails, electricity pipelines, etc.): this definition suggests that they can be considered a variant of the collector roads. In the past, suburban corridors were generally considered as complex road systems with the organization functions for the entire urban transportation network and serving as drainage for traffic crossing and penetrating the city. Given the expected concentration of directional facilities, as well as residential and office buildings, they were seen as the backbone of a new functional and figurative city open to the territory.

Regulations in force do not give explicit guidance for design/adjustment of roads in urban areas as for they do for extra-urban roads. Therefore, regulations, intended exclusively for extra urban roads and mainly focused on design of plano-altimetric alignment, can be used for planning of arterial and collector roads, as defined above, since they are mainly involved in one type of traffic: the movement of motor vehicles.

4.4 The design of urban roads

As a consequence, a more sensitive design of roads must focus on edges, as a neglected resource, useful in order to structure vivid, precious scenarios aimed to promote relational episodes, so as to fill roads of a more fecund and intense content, and to transform infrastructure spaces in pieces of public space. This implies that nodes and scenarios of a road are to be designed to promote a vast range of uses and activities. The main aim is to create safe and attractive spaces.

Soil, nodes or spots, and boundaries are basic concerns. Soil must be regarded in its latent visual potential, and also as a medium and as a support. As a medium it must be studied in its materials, colors and patterns, in order to reveal the structure and subdivision of a space, in zones designed for a specific group of users. As a support, it must be conceived in order to ensure a safe and comfortable fruition of a space. Boundaries are to be considered as a sign aimed to separate and protect realms diverse for content, purposes and scale, to set and moderate transfers of people, data, resources, to mould and structure spaces. Boundaries are also to be designed as the source of the content that permeates a scenario and forges its essence. A sumptuous facade creates a vivid public space, and a porous front can put in relation episodes, activities stored in a building and outside spaces, attracting pedestrians and promoting cluster of activities (Salingaros 2005). It is also due and proper to create porous protections between the areas on which vehicles travel and the pedestrians areas: Line of bollards, or arboreal curtains, are useful so as to define a net boarder as well as to punctuate the space, to mark its nodes, to evoke varied landscapes full of notes and possibilities (Annunziata 2013). Nodes must be considered as pauses that punctuate pedestrian routes, and must be conceived to persuade users to stop, to sit, to sense and observe episodes, people, and pieces of the urban scape. Nodes must be located in points of the urban realm permeated of a more vivid and intense content, i.e. where boundaries enclose and preserve a more fecund range of activities. It is important to observe that a structure of the pedestrian realm consistent with the trends of contents emanating from boundaries produces a mutual reinforcement among enclosed spaces and public scenarios.

Finally, each type of road requires a focus on the side hardscaped areas and medians, in order to achieve a more correct environmental integration aimed at recovery and enhancement of the contexts.

4.5 Aims of a rigorous classification

The clear goal of classification is to standardize the characteristics of the different types of infrastructure. It implies a validity check of the functional hierarchy of the various segments of the network identified by the planning tools. Therefore, plans and planning framework, if any, should be studied, while functions performed by or assigned to the existing infrastructure network (arterial, major, minor, local functions) should be identified as a medium or long-term objective. The performance goals are therefore specified in the definition of the functional class expected after adjustment actions. The following can be considered as goals of the functional adjustment of a road network:

37* contributing to the territorial rearrangement aimed at harmonizing the overall system of settlements and services;
38* giving the network new connectivity features within the transportation system to which it belongs;
39* improving the accessibility of the area;
40* improving conditions of traffic safety and service levels;.
41* fostering relations among people;,
42* reinforcing relations among users and urban realms, or landscapes;,
43* avoiding segregation of disadvantaged users.

It is important to observe that the last proposed purposes, are to be considered as an effort aimed to mould a *civitas*, or a community, to give her a specific and, consistent logos or identity, reflected in, and sustained and nurtured by, a relation with the landscape, modulated by an intense process of transformation, and by a dense combination of events and uses, that produces new meanings, contents and values.

Environmental sustainability of the action can also be added as a further target-condition.

Considering the abovementioned goals, it is essential that the infrastructure system be used to govern the location of activities in the area: enhanced connectivity of transportation networks is the essential condition for supporting a more reticular territorial structure and a better diffusion of development.

Among all the different design alternatives, infrastructural transportation system aimed not only at speeding up connections between suburb and center but also at supporting a different organization of a given region can be considered even more a convenient choice in order to achieve territorial harmony. However, the above considerations and the regulations settings prove a severe limitation: the service offered by the single road element is central; pedestrian traffic is only seen as one of the ways to go from one place to another, and pedestrian routes are seen as synergistic to other types of routes. Yet people do not just move to go somewhere, they often just enjoy walking, lingering and get pleasure from the environment around them and choose routes designed in this respect, pleasant to walk on and to linger.

5 CONCLUSIONS

Concluding remarks concern road infrastructures in urban areas, considered as exchange infrastructures, since rather than serving movement of people and goods they provide for connection of the different parts of an urban area, and their inadequacy may lead to degradation of the area and of its inhabitants' quality of life.

What presented above shows that when designing, for instance, public transportation on dedicated lanes, designers' goals are usually related to the creation of an integrated intermodal transportation system, to be realized by also pursuing the goal of optimizing the available infrastructure resources, with the greatest attention to using and reusing of the existing ones in the area. This results in improved road safety and levels of service and improved territorial accessibility, by reducing travel time and travel costs in respect of the collective use of specific services located in larger urban centers. Urban areas surely benefit from the creation of infrastructures, as it reduces car pressure on roads and the need to build parking areas and facilities, undoubtedly increasing the quality of life in the urban area. And yet road infrastructure systems should be getting increasing attention, and public transportation on dedicated lanes in particular should be used to govern the location of activities in the area.

Among the different design options, it could be convenient to opt for a transportation infrastructure system aimed not only at speeding up connections between suburbs and city center, giving an alternative to private cars, but also at laying the foundations for a different organization of the urban area, promoting localization of additional services and new traffic attractors in a perspective of territorial balance. Another goal not to be underrated in the reorganization of an urban area and its relational infrastructure system is the creation of a community bond symbolizing and materializing a common future within the urban area, a sense of belonging to that very area.

The significance of this objective aimed at territorial reorganization should be evaluated with respect to the abovementioned traditional principles. Quality of life in an urban area may be ensured, and its deteriorating conditions retrieved, by means of a supply of transportation supporting a different and more balanced distribution of services, corrective of traditional relations of dependence between center and suburbs. It may be the time for urban areas no longer intended as consisting of a center of quality surrounded by marginal suburbs, but as interacting centers, though with different weight: an infrastructure network no longer designed just for transportation, but also relational, could encourage this kind of development.

Scholars and designers have been speculating for quite a long time whether formal quality can be considered one of the goals of road design, and if roads are to be considered architectural objects, or whether performance goals (listed above) should be central, and thus environmental sustainability and formal quality should be conditions to be met during planning phases. A possible answer to these issues may be based on how the road is perceived by users, however, distinguishing new roads from the existing ones, and suburban from urban roads.

In the relationship between road and users, it is important to consider drivers' behavior with respect to the geometric and design features of the road: during the last decade studies on the changes of driving behaviors in relation to road environment have increased. In this regard, national and international literature can supply many studies and models that consider the relationship above as a major issue, and provide excellent insights on the significance of the role of road environment during the design phase. This includes everything adjacent to the road perceived by drivers that affects not only their driving attitude, but also the choice of a route over another, depending on their traveling reasons.

When various options connecting the same nodes are available, drivers prefer a route to another depending on traveling time, driving comfort, pleasantness of route, etc. In case of high environmental quality sites, on which attention needs to be drawn as well as investments related to cultural tourism, particular care should be put in the recovery/adjustment of existing infrastructures that were built with no consideration of environmental sustainability. This choice of a route affects territory more or less significantly, both in suburban and urban areas. With regard to the latter, interesting cause for reflection can be developed with reference to different types of users, the complexity of the affected environment, the urban and architecture peculiarities and, equally importantly, to socio-economic and cultural issues.

In an urban environment, drivers perceive the road differently than pedestrians, whose perception varies in relation to age, cultural level, or motivation. In order to investigate the perception of the road, designers should, in turn, identify themselves in the role of drivers, of pedestrians, and observers "at height z".

Drivers will probably evaluate the road in relation to the materials used for paving and to its state of decay, levels of congestion, intrinsic safety, illumination, overall driving comfort, and thus the mental workload required by driving (Annunziata et al. 2007). Mental workload is felt even when the surroundings are pleasant and linearly and harmonically varied, because drivers' attention tends to move outward, and this can be done safely only if traffic conditions, characteristics of the road section, and maintenance status of the pavement allow it. Level of perceived risk should not be underestimated: drivers not always allow themselves to be enthralled by the surrounding environment and they therefore could take driving attitudes and speeds not suitable to the context, due to a perceived high level of security.

Vulnerable users, such as cyclists and pedestrians, have a different perception of the road according to their different needs, lower travel speed and higher levels of perceived risk. Cyclists are less likely to get distracted by the external environment, since they need to keep their balance and this requires attention. Pavement decay, shadow cones, non-functional or unsuitable intersections, etc., are sharply perceived by cyclists. Pedestrians, on the other hand, walk along the side edges or central areas of the roads and are more sensitive to details but deeply disturbed by decay conditions, which they experience harder and longer. Damaged pavement on reserved pathways, for example, affects pedestrians more or less seriously, often in inverse proportion to their motor skills. They also have a different perception of the road furnishings, which often determine preference of a walk path to another, because it is more pleasant, more relaxing, or safer (for example, consider the presence and the type of lighting, green areas, etc.).

Finally, static observers at "height z", in a privileged position with respect to the layout of an urban route, perceive the harmony between the road and built-up environment in different ways, often with a more or less detached and less involved attitude. Certainly, they will not worry about rutting, but they will perceive road furniture, possibly just the most perceptible items, such as greenery, lighting, etc (Annunziata et al. 2007).

As observable in the urban environment, when it comes to quality of roads, the relation between space and time is important: speed, but not only. Therefore, works in urban areas should involve analysis of materials, furniture, and improvement of traffic safety, such as study of roundabouts, traffic calming systems, lighting, etc. The quality of roads in urban areas is closely related to the performed functions, to traffic flow composition, as well as to the presence of urban, historical or archaeological constraints.

Squares, for example, should be seen as meeting places, not as parking areas, road widening, or "non-places" in which traffic flow on intersections are somehow regulated. Utilities, furniture, and traffic calming systems should be designed in harmony with the environment in which they are located, not simply comply with regulations. Regulations are largely deficient on the issues of urban road design and it is desirable that design of new roads as well as planning of recovery and adjustment works consider that roads may also be the only public place in some urban areas, the only space shared by the whole community.

In residential neighborhoods and on local roads or streets, but not only, pedestrians cannot be put at the same level as cars, as if they were just as another transportation mode. The spaces reserved to pedestrians cannot be considered residual and separated from the other traffic components. The infrastructure space, both in the case of arterial roads and in case of more modest streets, must be recovered proposing original models and forms of urban space, pondering possibilities to create a fabric of safe and pleasant scenarios, conceived to foster relations and interactions among people and the urban realm. It is necessary to rediscover the street's real meaning and significance: public places reserved for people, designed and manufactured to allow pleasant walk and lingering, conceived to be places of sociality.

A valid case to cite, at this regard, is the Woonerf. Even though this archetype must be reputed proper and peculiar of residential areas, it points out how the urban road can be a rich scenario of a broad range of social events, a common land aimed at protecting and reinforcing relations and processes vital for the formation of a robust and cohesive social fabric. A strict adhesion to standards of aesthetic quality, comfort, and road safety, can promote and create a pleasant place where pedestrian is prior and cars are considered guests. Vice versa, the Gran Via de les Corts Catalanes, in Barcelona, product of the creative estrus of Carmen Fiol and Andreu Arriola, is evidence and echo of a Poetics tended to avoid the sense of net fracture of the urban area produced by large boulevards and to enounce a severe and suggestive use of green and materials to convert edges and areas contiguous to the route in space suitable for a vast and various placing at a lower level the lanes aimed at sustaining the most intense and copious traffic, suggests and fosters the perception of a strong cohesion and of a continuity of the urban fabric. It is useful to note how the 3.5 meters projection over the central carriageway, formed by the service roads, reduces noise and air pollution, and resolves the absence of sea-mountains relations. Furthermore, sound screens posed alongside the route protect pedestrian areas and urban fabric from the onerous noise produced by traffic, and can be considered a strong sculptural sign, able to give a more poetic value to the place. Moreover, the walkways that extend at the end of all the perpendicular streets and cross the Gran Via reinforce a strong and secure relation between the two opposite sides of the urban fabric. It is not arduous to presume that this work could be reputed apex of an appropriate and consistent method tended to restore the role, proper in the past of urban roads, as pleasant scenario of a fertile urban life, and as crucial part of urban fabric, able to evoke a secure and refined "ratio", source of a strong and peculiar *genius loci* (Annunziata 2011).

Moreover, it is important to point out that respecting the environment does not mean considering it untouchable, but living in harmony with it, although this often may imply significant costs for its protection, especially when carrying out certain works. A road, either urban or suburban, is unique and needs be studied as such in relation to the environment to which it belongs, according to criteria related to its function, to the traffic type of the network it is part of, to the choice of materials, to the quality of the natural environment, to the presence or absence of urban, archaeological and historical constraints, as well as to social and cultural needs of the involved area.

Design is an iterative process: the solution is the result of several efforts as well as of deep analyses, and the outcome cannot be tested only at the end of process leading to the definition of the work and to the estimate of its costs. Environmental and formal assessments are part of the planning process and should therefore be considered as halfway checks, increasing in depth as the planning process evolves.

The goals of design should not be distinctly environmental and/or of formal significance. It should rather be a sum of performance objectives, which may also be part of a reorganization planning of the area, reached by a road layout inserted in the surrounding environment in a sustainable way and formally consistent with the context.

In conclusion, these reflections lead to inquiry of whether the Decree of 2001 (D.M. n.6792/2001) (and subsequent amendments and supplements) can be strictly applied to design urban roads. It may be used for design of arterial roads or collector roads, but it seems hardly applicable to design of neighborhood or local roads.

It is not just an issue of environmental integration or formal values, nor is the design of its geometric elements, since they developed, in time, depending on the availability of voids. The adjustment and design of neighborhood or local roads mainly require attention to the abovementioned additional components, and the areas adjacent to the roads should be reassessed as sustaining the satisfaction of the needs connected to the role these roads play in the urban network.

REFERENCES

Annunziata A. 2011. *Sustainable urban roads*. Proceedings of the XXIV road world congress, Mexico City, September 26–30, 2011.

Annunziata A. 2013. Social sustainability. A key concern for the recovery of urban roads. In: Bartolo H.M. et al. Eds. *Green Design, Materials and Manufacturing process*. Proceedings of the 2nd International Conference on Sustainable Intelligent Manufacturing – SIM 2013. Lisbon, Portugal, June 26–29, 2013. London: CRC Press/Balkema, 277–282.

Annunziata F., Coni M., Maltinti F., Pinna F. & Portas S. 2004. *Progettazione stradale integrata*. Bologna: Zanichelli.

Annunziata F., Coni M., Maltinti F., Pinna F. & Portas S. 2007. *Progettazione Stradale. Dalla ricerca al disegno delle strade*. Palermo: Dario Flaccovio Editore.

Annunziata F. & Annunziata A. 2013. Recovery of roads in urban areas. From an indistinct feature to a specific function. In: Bartolo H.M. et al. Eds. *Green Design, Materials and Manufacturing process*. Proceedings of the 2nd International Conference on Sustainable Intelligent Manufacturing – SIM 2013. Lisbon, Portugal, June 26–29, 2013. London: CRC Press/Balkema, 283–288.

Bollettino Ufficiale del Consiglio Nazionale delle Ricerche n. 60 of April 26, 1978. *Norme sulle caratteristiche geometriche e di traffico delle strade urbane*.

Bollettino Ufficiale del Consiglio Nazionale delle Ricerche n. 150 of December 15, 1992. *Norme sull'arredo funzionale delle strade urbane*.

Decreto Ministeriale n. 6792 of November 5, 2001. *Norme funzionali e geometriche per la costruzione delle strade*. Supplemento ordinario alla Gazzetta Ufficiale (2002) n. 3.

Legge n. 865 of October 22, 1971. *Programmi e coordinamento dell'edilizia residenziale pubblica; norme sulla espropriazione per pubblica utilita'; modifiche ed integrazioni alle Leggi 17 agosto 1942, n. 1150; 18 aprile 1962, n. 167; 29 settembre 1964, n. 847; ed autorizzazione di spesa per interventi straordinari nel settore dell'edilizia residenziale, agevolata e convenzionata*.

Salingaros N. 2005. *Principles of urban structure*. Amsterdam: Techne.

Route choice behavior model under time pressure

Feng Gao
School of Electric and Electronic Engineering, Wuhan Polytechnic University, Wuhan, China

ABSTRACT: The impact of time constraints is quite evident in en route decisions. The en route deliberation process is restricted to the available time frame prior to the divergence point. Understanding route choice behavior under time pressure and predicting route choice decisions in such an environment are important components in the overall goal of building a more reliable and efficient Advanced Traffic Information System (ATIS). As previous route choice models do not mention time strain, the purpose of this paper is to describe and model driver route choice behaviors under time pressure at the individual component level and from a psychological decision-making perspective. The route choice behavior model under time pressure is introduced and describes the random and dynamic nature of route choice behavior. This model can account for the observed changes in choice probabilities and preference reversal phenomena under time pressure.

1 INTRODUCTION

The effectiveness of ATIS relies heavily on the drivers' response to the real-time information it offers. Therefore, understanding drivers' reactions to the information is critical to improve the accuracy of the existing ATIS. The impact of time pressure constraints is quite evident in drivers' en route choices. The en route deliberation process is restricted to the available time frame prior to the divergence point. Traffic information can be distributed to drivers through different technologies. Some of which, such as radio reports, can reach a driver just before the diversion node, which results in much-reduced deliberation times. Therefore, the potential of various traffic communication technologies in broadcasting useful and usable traffic information can vary significantly. As such, there is a need for enhancing our understanding of the impact of time pressure constraints on drivers' compliance behavior. It is evident that the Microeconomic Theory approach to route choice modeling is dominant in literature (CG Chorus, et al., 2006). Although several modeling attempts have departed from the formal utility-maximization paradigm and adopted more behaviorally realistic frameworks (Eran Ben-Elia, Yoram Shiftan, 2010, Bekhor, Albert, 2014), there remains a lack of an explanatory mechanism of the decision process itself. Modeling of drivers' choices is mainly perceived from a structure-oriented perspective wherein a relationship is formulated between a set of inputs and outputs without a realistic understanding of the underlying psychological process. The time deliberation dimension seems to have been completely ignored.

There are two main factors that affect route choices. One is the drivers' route decision is the outcome of complex deliberation processes involving uncertainty. The other is that the decision is not instantaneous, but rather time-consuming. The direct influence of the length of a deliberation process on decisions cannot be ignored. Drivers are commonly faced with divergence decisions while driving. The length of the deliberation process is restricted to a time frame prior to tentative bifurcation or divergence points. Available time frames might vary according to many factors, such as driver familiarity with the network geometry, daily traffic conditions, and the timing and location of information dissemination. Limited deliberation times may pressure drivers into making choices before they are completely certain.

Absent in previous route choice models are any mention about time pressure, so this paper proposes the framework of the routing choice and the route choice behavior model under time pressure on the Decision Field Theory (DFT) (Busemeyer, Townsend, 1993). The DFT is a dynamic behavioral theory that is able to capture the psychological processes involved in general decisions under uncertain conditions. It is one of very few process-oriented behavioral decision theories that explicitly accounts for varying degrees of uncertainty, as well as time pressure, in a unified, scientifically-sound framework. The DFT attempts to abstract the deliberation process based on a realistic representation of the underlying motivational and cognitive mechanisms. The developed DFT-based model is used to study the impact of time pressure constraints on drivers' route choice behavior.

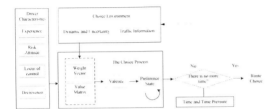

Figure 1. Route choice behavior under time pressure abstraction.

2 ROUTE CHOICE BEHAVIOR MODEL UNDER TIME PRESSURE

2.1 Route choice model frameworks

Drivers' route choices are the outcome of complex interactions of several psychological processes. On one hand, drivers make their choices through a mental deliberation process that includes a trade-off between the perceived attributes of available alternatives. Drivers form perceptions about these attributes based on previous experiences, day-specific aspects, and, in many cases, traffic information sources. Driver characteristics influence the operation of the underlying psychological processes and the resulting choices. On the other hand, drivers' route choices are performed within a unique choice environment. The complexity of the choices stems from several contributing factors, the most prominent of which are uncertainty and time pressure. In addition, situational conditions (such as road work), and environmental conditions (such as weather-related obstructions) further impact drivers' perceptions of the options. The highly intertwined aspects of the overall route choice behavior mandate the abstraction of the main underlying processes and/or factors for understanding and modeling purposes. Figure 1 illustrates an abstract representation of the main contributing processes and/or factors and their interrelations.

2.2 Route choice behavior model under time pressure

Drivers' route choices are performed within a challenging environment characterized by uncertainties and time constraints. The simultaneous interaction of these processes within the available time constraints influences and ultimately gives rise to drivers' decisions. When choosing among routes, drivers compare and trade off the perceived attributes of available routes. Considering the attributes of routes may differ from one driver to another, so standardizing the measurable set of attributes is necessary. For the proposed route choice behavior model, two main attributes are represented; Travel Time (T) and Distance (D). These two aspects of the specification are based on extensive reviews of the relevant literature, and informal discussions with a number of drivers (GAO and WANG, 2011, GAO, 2011).

In this work, DFT is adopted to investigate drivers' route choice decision processes under time pressure. Throughout the deliberation process, drivers' preference level toward each route, from time $t-h$ to time t, can be described as linear difference equation (1).

$$P(t) = S \cdot P(t-h) + C \cdot M \cdot W(t) \qquad (1)$$

where column vector (n × 1) $P(t)$ (n is the number of routes) represents the preference state at time t, where $p_i(t)$ represents a preference state value for option i, and h is a time step. The feedback matrix S provides memory of the previous preference state for a given route (the memory effect) in the diagonal elements as well as the effect of the interactions among the routes in the non-diagonal elements. The Eigen values λ_i of the feedback matrix are restricted to less than 1 in magnitude to ensure system stability (Busemeyer, et al., 2006). The value matrix M (n × m matrix, where n is the number of routes, and m is the number of attributes) represents the personal evaluations of a driver for each route on each attribute (e.g. travel time and distance). The weight vector $W(t)$ (m × 1) containing weights corresponding to each column of M, represent the joint effect of the importance of an attribute. The matrix product $M \cdot W(t)$, is a vector of weighted average values of each route, based on momentary evaluations at time t. M and $W(t)$ can be inferred with the Bayesian belief network (BBN) (GAO, 2011). The matrix C (n × n) is the contrast matrix comparing the weighted evaluations of each route $M \cdot W(t)$. The contrast matrix has elements $c_{ii} = 1$ and $c_{ij} = -1/(n-1)$ for $i \neq j$ where n is the number of routes.

2.3 Route choice rules

There are two DFT stopping rules (Busemeyer, et al., 2006):

(1) The externally controlled stopping decision process. The stopping of the route choice process is controlled by stopping time: Set T_D as the stopping time. If $p_m(T_D) = \max(p_i(T_D), i = 1, \ldots, n)$, $p_m(T_D)$, with the maximum preference state at T_D, is chosen.

(2) The internally controlled stopping decision process. The stopping of the route choice process is controlled by threshold θ, and the route choice process continues until one of the preference states exceeds this threshold. If $p_m(t) \geq \theta$, $p_j(t) < \theta, j \neq m$, $p_m(t)$, the first one to reach the threshold is chosen.

Within DFT abstraction of the route choice deliberation process, drivers make their choices when their preference strength exceeds a threshold bound for a specific alternative. If the available deliberation time is less than what they need, the choice process is cut short with a premature decision prior to reaching that bound. The imposed interruption of the deliberation process could, therefore, result in different choice decisions.

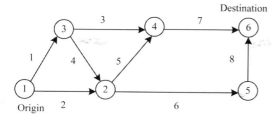

Figure 2. The example road network.

Table 1. Link characteristics.

Link	Link Travel Time (min)			Distance (km)
	short	medium	long	
1	30	42	70	35
2	32	47	84	37
3	54	90	135	45
4	35	53	84	33
5	28	38	60	35
6	74	107	192	80
7	60	93	168	70
8	20	32	51	28

Accordingly, Rule (1) is more suitable when modeling en route choice decisions under time constraints.

3 SIMULATION AND ANALYSIS OF ROUTE CHOICE BEHAVIORS

3.1 Numerical examples

To illustrate route choice behavior under time pressure, an exemplary road network used for this study, shown in Figure 2, consists of six nodes, eight links and one Origin-Destination (OD) pair. The OD pair is served by a total of five routes: Links 1, 4, 5, and 7 for Route A; Links 1, 4, 6 and 8 for Route B; links 1, 3, and 7 for Route C; Links 2, 5, and 7 for Route D; Links 2, 6, and 8 for Route E. The travel times and the distance for each link on the road network are shown in Table 1.

3.2 Simulation scope

Gao (2011) explored a real-time planning algorithm for en route choice processes. Using the algorithm, a driver makes use of his route until he achieves his destination. Matlab is used to simulate the route choice behavior model under time strain. The evolution of a driver preference level with times during the route choice process is presented in Figure 3. The abscissa represents deliberation time, and the ordinate represents preference state. The three trajectories (labeled A, B, and C) represent the preferences of each route as they evolve randomly over time. The three vertical lines represent the stopping times, respectively, for the route choice deliberation process according to

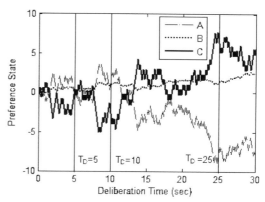

Figure 3. The route choice process for a choice among three routes.

the stopping rule (1). Before the driver's preference matures in the direction favoring Route C, it oscillates back and forth among the Routes A, B, and C in the route choice deliberation process. Terminating the route choice deliberation process is performed by stopping time T_D. In this example, Route B would be chosen at the stopping time $T_D = 5$ and Route A would be chosen at the designated time $T_D = 10$.

As can be seen in Figure 3, time pressure constraints directly affect the decision process outcome. The imposed interruption of the deliberation process can, therefore, result in different decisions. A short deliberation time frame may result in an immature and possibly wrong route choice where relaxation will lead to a better, informed choice. In real life route choice situations, drivers often make wrong or 'regrettable' route choices if the available deliberation time is less than what they need. This situation primarily arises if the decision time is tightly constrained before an impending bifurcation. If the driver is allowed more time to think and deliberate, more mature route choice decisions may evolve. The route choice behavior model under time pressure is able to account for the observed changes in choice decisions, including the preference reversal phenomenon under time pressure.

3.3 Time pressure impact on switching behavior

Generation of route choice data under time constraints is based on the simulation of the deliberation processes, in accordance with the developed DFT-based route choice model. Simulation of the deliberation process is a second-by-second estimation of the evolution of drivers' preference levels toward each of the alternative routes. In an un-constrained deliberation process, route choice decisions are made based on the first preference strength exceeding the upper-threshold bound. Under a time pressure constraint, the deliberation process is limited to a specific deliberation frame. If a decision is not reached within the specified frame, the route with the highest preference strength, at the deliberation time limit, is chosen. Observed mean deliberation time frames ranged from 5 to 30 seconds.

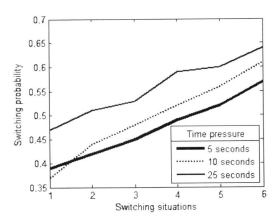

Figure 4. Switching probability of each route situation under varying conditions of time pressure.

A time pressure of 5, 10, and 25 seconds are externally imposed to restrict the deliberation process. Choice percentages are based on 100 repeated simulations of the respective deliberation processes. Choice percentages of the base case (with no time pressure constraint) are preserved for comparison purposes.

In order to investigate how time constraints affect route choice behaviors, consider the situation where a driver chooses among three routes A, B and C. Real-time traffic information could be provided to drivers in the form of advice to take a certain route. Suppose the driver's initial preference is Route C. Choice percentages are estimated for two information scenarios: C → A (take Route A) and C → B (take Route B). Each of the two scenarios represents a challenging situation, where disseminated information opposes drivers' intuitive biases to take another route (A or B), i.e., the real-time traffic information advises that the travel time of Route B or C is shorter than that of Route A. Then, there are six possible scenarios: C → A, C → B, B → A, B → C, A → B, A → C, namely switching situations: 1, 2, 3, 4, 5 and 6. Different choice percentages (switching probabilities) are estimated for different sets of information. Switching is to act according to the real-time traffic information.

In Figure 4, the switching probability is estimated for each scenario, under each condition of time pressure. As the adopted scenarios are challenging, the switching probability is estimated to be the choice percentages, for real-time traffic scenarios.

A completely different behavioral trend is revealed under different scenarios. Real-time traffic information provides drivers with network conditions (travel time) and it is up to the driver to make a decision. The incorporation of traffic information into the route choice process is, thus, time-consuming. Under tight time constraints, the degree of influence of disseminated information in route choice behavior is less pronounced. Longer deliberation times allow for more in-depth deliberation processes. Under a higher perceived information reliability level, the switching probability increases with the relaxation of the time constraint. As the information reliability level decreases, drivers' perception of the significance of provided traffic information decreases. The reduced confidence in provided information discourages drivers from undertaking in-depth deliberation processes, regardless of the available time frame (GAO and WANG, 2010). The switching probability is, therefore, lower and less influenced by time pressure levels under the reduced reliability level. The practical implication of this finding is that if travelers are not provided with sufficiently accurate information and sufficient time to deliberate it, they are less likely to comply.

4 CONCLUSION

ATIS applications require a thorough understanding of drivers' route choice behavior in a complex network using real-time traffic information. In the past decade, understanding and modeling of route choice behavior have been the focus of numerous research efforts. Most of the models lack a choice process mechanism that is capable of explaining the dynamics and high variability of drivers' preferences and choices, especially under time pressure (GAO, 2011). The DFT provides such a mechanism. This paper describes a study investigating the role of time pressure in understanding drivers' reactions to real-time traffic information under time pressure. The route choice behavior model under time pressure was used for analyzing the dynamics of deliberation by studying the effect of time pressure constraints on route choices. The imposed interruption of the deliberation process could, therefore, result in different decisions. A short deliberation time frame might result in an immature and possibly wrong route choice, where relaxation could lead to a more informed choice. A completely different behavioral trend was revealed under different information scenarios. Under tight time pressure constraints, the degree of influence of disseminated information in route choice behavior was less pronounced. And longer deliberation times allowed for more in-depth deliberation processes.

REFERENCES

Busemeyer, J. R. & Townsend, J. T. 1993. Decision field theory: A dynamic-cognitive approach to decision making in an uncertain environment. *Psychological Review* 100:432–459.

Busemeyer, J. R. et al. 2009. The dynamic interactions between situations and decisions. In P. Robbins, M. Aydede (Eds.), *The Cambridge Handbook of Situation Cognition*: 307–321. Cambridge University Press, New York.

CG Chorus et al. 2006. Travel information as an instrument to change car-drivers' travel choices: A literature review. *European Journal of Transport and Infrastructure Research* 6(4):335–364.

Eran Ben-Elia & Yoram Shiftan. 2010. Which road do I take? A learning-based model of route-choice behavior with real-time information. *Transportation Research Part A* 44(4): 249–264.

GAO Feng. 2011. A Real-Time Planning Algorithm for Route Choice Behavior and Processes. The 1st International Conference on Transportation Information and Safety, ICTIS 2011, ASCE, USA: 537–544.

GAO Feng & WANG Mingzhe. 2010. Route Choice Behavior Model under Guidance Information. *Journal of Transportation Systems Engineering and Information Technology 2010 Bound Edition*: 310–315. Elsevier Ltd.

GAO Feng & WANG Mingzhe. 2011. Modeling En-route Driver Route Choice Behavior under Real-Time Traffic Information. *Journal of Information and Computational Science* 8 (7): 1053–1062.

Johnson, J.G. & Busemeyer, J.R. 2010. Decision making under risk and uncertainty. *WIREs Cognitive Science* 1:736–749.

S. Bekhor & G. Albert. 2014. Accounting for sensation seeking in route choice behavior with travel time information. *Transportation Research Part F: Traffic Psychology and Behaviour* 22:39–49.

Self-vibration characteristic test and finite element analysis of spring vibration isolated turbine-generator foundation

Tie Jun Qu, Hong Zhen Yang & Gang Zhao
College of Architecture and Civil Engineering, North China University of Technology, Beijing, China

ABSTRACT: In order to get each mode and the corresponding frequency of a spring vibration isolated turbine-generator foundation by three-dimensional three-point excitation method, a test model was designed and fabricated in this study. Afterward, finite element method was used in modal analysis of the foundation, which proved that the computed result was in good agreement with the test result. For further study of the influence on foundation self-vibration frequency caused by vibration isolated springs, finite element method was further applied to analyze the self-vibration characteristic of the same foundation but without vibration isolated springs. According to the analysis result, responses of the foundation in vertical vibration as well as horizontal earthquake action were both reduced by using vibration isolated springs.

1 INTRODUCTION

The research of turbine-generator foundation has proved that for rigid foundation without vibration isolated springs, natural vibration frequency of the foundation and working frequency of the equipment are close, which can easily cause sympathetic vibration of both. To avoid destruction caused by sympathetic vibration, vibration isolated springs were applied to change the natural vibration frequency of foundation, thereby to keep it away from the working frequency of the equipment. This measure will reduce operating vibration of the equipment.

Modal test research of spring vibration isolated turbine-generator foundation is firstly discussed in this paper. On this basis, the models with and without springs were created successively by ANSYS finite element analysis software. The result proved that vibration isolated springs can reduce the responses caused by vertical vibration as well as horizontal earthquake action.

2 TEST MODEL

The test model of spring vibration isolated turbine-generator foundation was a frame structure. It consisted of baseboard, columns, beams and top plate. Material characteristics of the model were consistent with those of the practical engineering foundation. Concrete strength class was C35, and the class of longitudinal carrying bars was HRB400. Similarity constant of the spring stiffness of the springs was 1:8. The spring stiffness converted according to similarity relation are shown in Table 1.

Table 1. Stiffness of springs on model columns.

Column Number	Vertical Direction kN/mm	Horizontal Direction kN/mm
1	13.58	2.06
2	13.58	2.06
3	30.15	4.53
4	30.15	4.53
5	45.66	6.75
6	45.66	6.75
7	65.46	9.67
8	65.46	9.67
9	65.46	9.67
10	65.46	9.67
11	46.9	6.93
12	46.9	6.93
13	15.95	2.37
14	15.95	2.37

Dimensional similarity relation between the model foundation and the practical one was 1:8. According to this relation, the model's length was 8.59 m and the width was 2 m. Taking no account of the thickness of baseboard, the model's height was 3.3 m. The built-up test model is shown in Fig. 1.

3 MODE RESULTS OF TURBINE-GENERATOR FOUNDATION WITH VIBRATION ISOLATED SPRINGS

3.1 *Test results*

Three-point random excitation method was applied in the test. In X, Y and Z directions, exciting force was

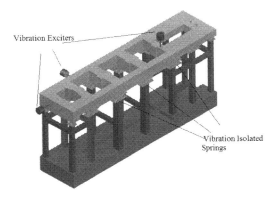

Figure 1. Positions of vibration exciters and springs.

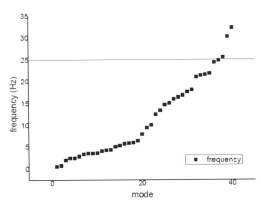

Figure 2. Relation between the test mode and frequencies.

exerted on the foundation model by vibration exciter. In this process, in order to obtain the frequency-response function of the test model, exciting force was measured by force sensor and model accelerated speed was measured by accelerated speed sensor (Zhang Li 2011). The positions of the vibration exciters and springs are shown in Fig. 1.

The analyzed mode frequencies of the test model ranged from 0 Hz to 300 Hz. In this range, 40 modes were recognized. According to the frequency similarity ratio, the mode frequencies were converted to those of the practical foundation. With mode order as horizontal ordinate and frequencies as vertical ordinate, the relation between them can be described as in Fig. 2. The first three modes are shown in Figs. from 3(a) to 3(c).

As seen in Figs. 3(a) to 3(c), the first three modes were global vibration of the foundation. Working frequency of the equipment on the practical foundation was 25 ± 1.25 Hz. There were only three modes in this range and all of them were local vibration of the foundation. However, in such situation, the low-frequency modes cannot be motivated.

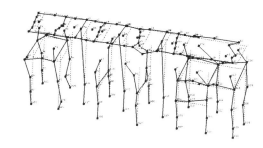

Figure 3a. First-order mode.

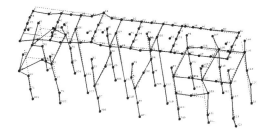

Figure 3b. Second-order mode.

3.2 Finite element mode results

In order to do comparisons of the test results, the finite element analysis software ANSYS 14.0 (Hu Renxi 2011) was applied in model mode analysis. In the modeling process, the adopted top plate element was solid95 and, the element beam188 was applied in columns and beams. Elasticity modulus value adopted was 31500 MPa and poisson's ratio was 0.2. The density of materials was 2500 kg/m³.

For model foundation without springs, CERIG command was used to create rigid zones between column tops and the top plate. For foundation with springs, the rigid zones were created between springs and the top plate, ensuring the rigid connection between the top plate and its nether structure. Matrix element MATRIX27 (Wang Xinmin 2007) was used to simulate three-dimensional springs. The element

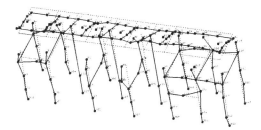

Figure 3c. Third-order mode.

has only two nodes but can simultaneously simulate stiffness in X, Y and Z directions. (Wang Xinmin 2007)

The first three modes of the model with springs are shown in Figs. 4(a) to 4(c). Comparison between finite

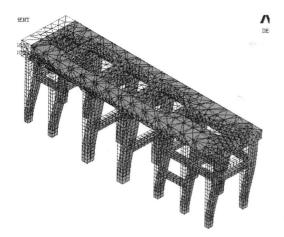

Figure 4a. First-order mode.

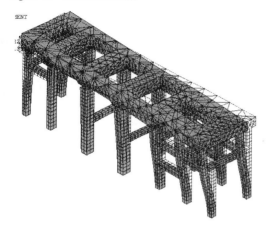

Figure 4b. Second-order mode.

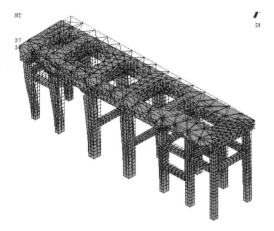

Figure 4c. Third-order mode.

element modal frequencies and the test frequencies are shown in Table 2.

By inspection of the first two modes, the finite element modes are in good agreement with the test modes.

Table 2. Comparison between finite element mode frequencies and the test frequencies (with springs)

No.	Finite element frequencies (Hz)	Test frequencies (Hz)
1	0.483	0.650
2	0.499	0.871
3	0.528	2.106
4	2.606	2.533
5	3.376	2.538
6	3.474	2.939
7	3.595	3.434
8	3.833	3.628
9	4.001	3.629
10	4.018	3.681

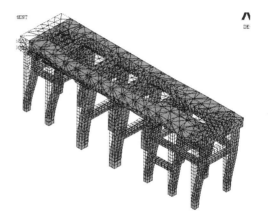

Figure 5a. First-order mode.

The first mode was longitudinal whole-vibration horizontally. The second mode was torsional whole – vibrations horizontally. There was a little discrepancy after the second mode between finite element modes and test modes, but consistent variation trend was obvious, proving that the finite element model selection and numerical calculation result were reasonable.

4 COMPARISON OF MODE RESULT BETWEEN SPRING MODEL AND SPRINGLESS MODEL

With the vibration isolated springs removed from the described finite element model, the top plate was connected with the columns directly. After calculation several modes and their corresponding frequencies were obtained. The first three modes are shown in Figs. 5(a) to 5(c).

Fig. 6 and Table 3 show the frequencies of the first 20 modes of the models with and without springs. Through inspection, frequencies of the model without springs are obviously higher by comparison with

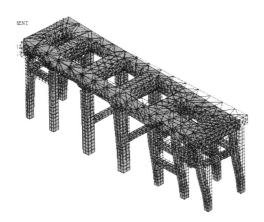

Figure 5b. Second-order mode.

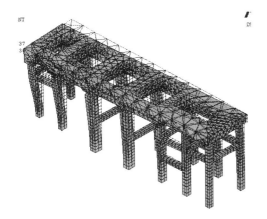

Figure 5c. Third-order mode.

Table 3. Frequencies of finite element models

No.	Without springs (Hz)	With springs (Hz)
1	0.543	0.483
2	0.548	0.499
3	0.620	0.528
4	2.656	2.606
5	3.705	3.376
6	4.098	3.474
7	4.218	3.595
8	4.558	3.833
9	4.619	4.001
10	4.789	4.018

the model without springs. Ignoring the fourth-order mode, the mode frequencies of the model without springs are 1.1 higher than those of the model with springs. It is demonstrated that vibration isolated springs can reduce the natural vibration frequency of the foundation effectively.

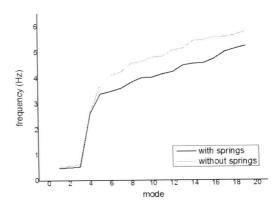

Figure 6. Comparison of mode frequencies between two finite element models.

5 RESPONSE OF FOUNDATION IN HORIZONTAL EARTHQUAKE ACTION

By the assumption that the damping ratio of the foundation is $\xi = 0.05$, site classification is I, in the third seismic design group, the earthquake fortification intensity is 6. It can be calculated that the horizontal earthquake influence coefficient $\alpha_{max} = 0.04$ g, and the design characteristic period of ground motion $T_g = 0.35$ s. The periods of the first-order modes of the model with springs and without springs are respectively $T_1 = 1.84$ s and $T_1' = 2.07$ s. According to seismic influence coefficient curve, the corresponding seismic influence coefficients can be obtained:

$$\alpha_1 = [1.0 \times 0.2^{0.9} - 0.02 \times (1.84 - 5 \times 0.35)] \times 0.04\,g$$
$$= 0.00932\,g$$
$$\alpha_1' = [1.0 \times 0.2^{0.9} - 0.02 \times (2.07 - 5 \times 0.35)] \times 0.04\,g$$
$$= 0.00914\,g$$

(1)

Obviously, α_1' is 1.9% lower than α_1. Therefore, vibration isolated springs can reduce the response of the foundation in horizontal earthquake action (GB50011-2010).

6 CONCLUSIONS

(1) In the case where vibration isolated springs were installed, each mode frequency of the foundation was lowered effectively. Frequencies of the first three modes were reduced by over 10%, keeping the first several vibration modes far away from the working frequencies of the equipment, thereby avoiding the risk of sympathetic vibration between equipment and foundation.

(2) Through observation of the first 20 modes, the global vibration modes all have lower frequencies. In higher mode frequencies, only one or several beams or columns vibrate locally. In other words, working frequency cannot motivate global vibration of the foundation. So the vibration isolated

springs can lower the risk of whole overturning destruction of the foudation.

(3) Vibration isolated springs reduce the horizontal earthquake action by 2% or so, effectively alleviating the destruction caused by horizontal earthquake action on the foundation.

ACKNOWLEDGEMENT

This study was financially supported by the Funding Project for Academic Human Resources Development in Institutions of Higher Learning under the Jurisdiction of Beijing Municipality (PHR201107106) and the National Undergraduates' Innovation and Entrepreneurship Training Program (201210009021). The authors deeply express sincere appreciation to them.

REFERENCES

Ministry of Housing and Urban-Rural Development of the people's Republic of China. GB50011-2010 Code for seismic design of building. Beijing: China Building Book Shop.

Hu Renxi. ANSYS 13.0 Finite Element Analysis in civil engineering. Beijing: China Machine Press, 2011.

Wang Xinmin. ANSYS engineering structures numerical analysis. Beijing: China Communications Press, 2007.

Wang Xinmin. ANSYS structural analysis element and its application. Beijing: China Communications Press, 2007.

Zhang Li. Modal Analysis and Modal Test. Beijing: Tsinghua University Press. 2011.

Shaking table test study on bottom-business multi-storey structure

Yongduo Liang
Earthquake Administration of Liaoning Province, Liaoning Shenyang, China

Xun Guo
Institute of Disaster Prevention, Hebei Sanhe, China

Fu He, Yang Zhou & Weisong Yang
Institute of Engineering Mechanics, China Earthquake Administration, Harbin, China

ABSTRACT: Bottom-business multi-storey masonry buildings were destroyed severely in the earthquake. The rate of collapse was very high except some specials, for instance, the Apartment of Beichuan Telecommunication Bureau, which behaved well in earthquake resistant capacity and only suffered a moderate damage during the earthquake. The difference between this building and others lies in the winged columns set in the front longitudinal wall of the first floor. To study the effect of this special element in the earthquake, two 1/5 reduced scale models were designed and shaking table tests were carried out. From the data of tests, such as acceleration, relative displacement and strain analysis and so on, it is found that the earthquake resistant capability of the building with winged columns is better than that without winged columns. The method of balancing stiffness and increasing ductility may provide reference for retrofitting design of the existing houses.

1 INTRODUCTION

The intensity of Beichuan County was XI in 2008 Wenchuan large earthquake, where vertical PGV over 1.0 g and was encountered almost destructive damage (Li Xiaojun et al. 2008). Among the heavily damaged or collapsed buildings, the most serious is the bottom-business multi-storey masonry structures. But the Apartment of Beichuan Telecommunication Bureau (as shown as Fig. 1), a multi-storey masonry structure, experienced only medium damage. A careful investigation was made and it was found that this building had winged columns set in the front longitudinal wall of the first floor while others don't. By carrying out shaking table tests, we might be able to attain some important results and put forward some valuable advices to retrofit the existing buildings.

Figure 1. The Back View of Apartment of Beichuan Telecommunication Bureau.

2 SHAKING TABLE TESTS

2.1 Test preparation

The prototype was built in 2000, including one bottom storey, 4.0 m high, for business, and five storeys, 2.8 m high, for residential purpose. Two 1/5 reduced scale models that take 1.3–1.9 axes as object of study and some necessary simplification were designed. Model 1 is a two storey test model with winged columns in the front longitudinal wall of the first floor, and Model 2 is a similar model without winged columns. The plan views of the two models are as shown as Fig. 2.

Before test, 5 displacement sensors, 6 acceleration transducers and 26 strain gauges were set at proper positions. During the tests, Wolong wave was used and the tests were carried out in 5 steps.

2.2 Test phenomenon

In Model 1, the damage degree of the three longitudinal walls in descending order is: mid longitudinal wall, back longitudinal wall and front longitudinal wall. The failure details after PGA = 1.05 g are as shown in Fig. 3.

In Model 2, the cracks appeared earlier than that in Model 1, one is at PGA = 0.45 g, and the other is at PGA = 0.65 g; the damage level is serious than Model 1, too, for example, the damage of Model 2 after PGA = 0.82 g is more severe than Model 1 after

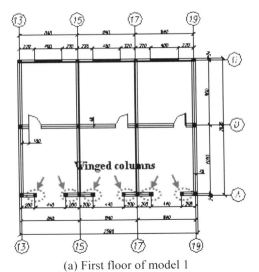

(a) First floor of model 1

(b) First floor of model 2

Figure 2. First Floor Plan of Models.

(a) The front longitudinal wall

① Left bay

② Mid bay

③ Right bay

(b) The mid longitudinal wall of the first floor

(c) The back longitudinal wall

Figure 3. Cracks Distribution of Model 1 after PGA = 1.05 g.

PGA = 1.05 g. The details are shown in Fig. 4 and Fig. 3.

3 EFFECT ANALYSIS OF WINGED COLUMNS

3.1 Comparative analysis on acceleration response

The dynamic amplification effect can be attained through the acceleration earthquake response of the models during earthquake simulation shaking table test. The data curves for comparison of acceleration amplification of the two models are shown in Fig. 5.

From the curves, the acceleration amplification of the two models of the first floor is very low, all not exceeding 1.2. Among all the values related with second floor, except the value of Model 2 at PGA = 0.60 g is 1.92, others are all between 1.2 and 1.5. Because the frequency of Model 2 after it suffered 0.1 g, 0.35 g and 0.45 g is close to the frequency of earthquake motion PGA = 0.60 g, the acceleration amplification value reach to the maximal data, that is 1.92, and this consist to the serious damage state.

(a) The front longitudinal wall

① Left bay

② Mid bay

③ Right bay

(b) The mid longitudinal wall of the first floor

(c) The back longitudinal wall

Figure 4. Cracks Distribution of Model 2 after PGA = 0.82 g.

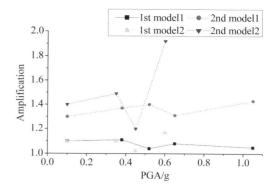

Figure 5. Acceleration Amplification.

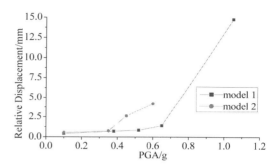

Figure 6. Relative Displacement.

3.2 Comparative analysis on relative displacement

The curves of relative displacement are shown in Fig. 6. From the curves, it can be seen that the maximum value maybe 14.8 mm when Model 1 suffered PGA = 1.05 g, and at this time, Model 1 is perhaps in the plastic deformation, the corresponding storey displacement angle is 1/73, but it didn't collapse. As we know, the elastic-plastic storey displacement angle limit of a concrete frame structure is 1/50 (GB 50011-2010, China Architecture Press, 2010), that means the bottom-business multi-storey masonry with winged column has some ductility.

3.3 Comparative analysis on strain

The strain values of the three longitudinal walls of the two models at their corresponding representative positions are shown in Fig. 7.

In Fig. 7, all values are compressive strain values, which will increase as the acceleration of ground motion increases. In general, the values of the two models show a consistent rule, i.e., the value of mid longitudinal wall is the largest, followed by the back longitudinal wall and the mid constructional column of transverse wall, while the smallest is the front longitudinal wall (frame column and constructional column); in addition, most of the values in Model 2 are larger than that in Model 1.

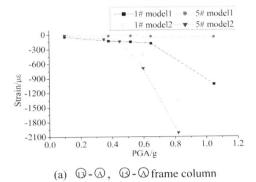

(a) ①-Ⓐ, ⑲-Ⓐ frame column

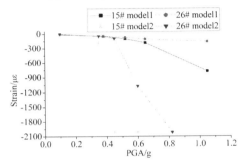

(b) ①-Ⓓ, ⑲-Ⓓ constructional column

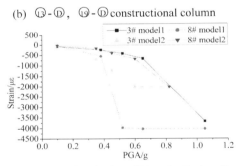

(c) The top of door in the mid longitudinal wall

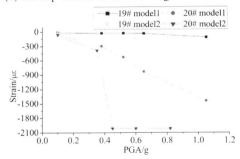

(d) The edge of window in the back longitudinal wall

Figure 7. Strain Comparison in Corresponding Positions of the Two Models.

4 SUMMARY

The damage degree of the model with winged columns is obviously smaller than that without winged columns. The stiffness of the front longitudinal wall maybe strengthened because of winged columns and this can balance distribution of earthquake inertial forces among the walls. The method to add winged columns in the front longitudinal wall maybe a effective measure in retrofitting existing normal bottom-business multi-storey masonry buildings.

ACKNOWLEDGEMENTS

This work was financially supported by the Doctoral Fund of Liaoning Earthquake Administration (LNDZBSJJ002) and Ministry of Science and Technology (2012A04).

REFERENCES

Li Xiaojun, Yu Aiqin, Gan Pengxi, etc. Survey and Analysis of the Disaster and Engineering Damage of Beichuan County Seat in Ms8.0 Wenchuan Earthquake [J], *Technology for Earthquake Disaster Prevention*, 2008, 3(4): 352–362

Seismic Code for Buildings (GB 50011-2010), China Architecture Press, Beijing, 2010.

Simulation of wireless sensor for monitoring corrosion of reinforcement steel in concrete

Gaowa Xu, Jin Wu, Zhe Wang & Zhong Wang
Department of Civil Engineering, Nanjing University of Aeronautics and Astronautics, Nanjing, China

ABSTRACT: Reinforcement corrosion has a great impact on the durability of the concrete structures. The assessment, remaining life prediction and maintenance schedules of concrete may depend on the data of steel corrosion-monitoring to a great extent. With the aid of the circuit simulation program NI Multisim software, in the paper, the simulation of the sensors with both single switch and three switches equivalent circuit models has been carried out, by which the relationship between the resonant frequency and different thresholds of corrosion. The good agreement between the simulation results and the theoretical values supports the feasibility of the wireless sensors.

Keywords: reinforcement corrosion, wireless sensor, equivalent circuit, resonant frequency, simulation analysis

1 INTRODUCTION

Corrosion of reinforcing steels in concrete is a thorny problem throughout the world, as it will not only degrade the integrity of the steel but also cause extensive damage to the concrete. Structures near the coast are vulnerable to corrosion because of salt in the air, and bridges that are exposed to deicing salts are also susceptible to corrosion. The development of comprehensive health monitoring systems for civil infrastructure is thus indispensable. Many methods have been devised to detect corrosion in concrete structures. While health monitoring systems have been installed on several key bridges in China, the costs associated with installation, maintenance, and interpretation of the data are prohibitively expensive for the majority of civil infrastructures.

A commonly used electrical chemistry method for monitoring corrosion in steel reinforced concrete is the half-cell potential method (B. Assouli et al. 2008). Another electrical chemistry technique is the use of linear polarization resistance, which provides an indication of the corrosion current in the rebar. Due to the complexity of steel corrosion, however, the data of these methods are often unreliable and unstable.

Beyond doubt, the sensors, which can be widely used, should possess three features: low cost, long life, and reliable output. The wireless sensors may be an ideal choice to monitor structural health and detect the early signs of corrosion which may appear far before visible damage occurs. The mechanism of the wireless sensors for the monitoring of reinforcement corrosion based on LC circuit is investigated in this paper. And the simulation of the wireless sensors with single switch and three switches equivalent circuit models is carried out.

2 MECHANISM OF WIRELESS CORROSION SENSOR

As shown in fig 1, the whole sensor system includes a reader and a resonant sensing circuit, the sensing circuit on the right in fig 1 composed of the inductance component, and the capacitance component and an external switch (Simonen J T et al. 2004). Therefore, the sensing circuit can be idealized as a LC circuit. The resonant frequency of the circuit will shift due to a change in capacitance, which can be realized by connecting a second capacitor to the circuit via an external switch. When the external switch is closed, the circuit resonates at its initial frequency:

$$f_{\text{Initial}} = \frac{1}{2\pi\sqrt{L_2(C_1+C_2)}} \quad (1)$$

When the external switch is opened, the circuit resonates at its final frequency:

$$f_{\text{Final}} = \frac{1}{2\pi\sqrt{L_2 C_1}} \quad (2)$$

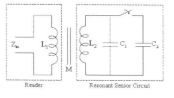

Figure 1. Simplified Schematic of the Sensor.

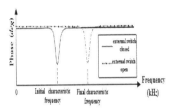

Figure 2. Principle Diagram of the Impedance Response.

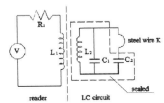

Figure 3. Refined Schematic of the Sensor for Monitoring Corrosion of Reinforcement.

From fig 2, it can be seen that the measured initial resonant frequency of the sensor is smaller than its final resonant frequency. When the external switch is opened, the final resonant frequency shifts to an appreciably higher value.

In the case of the corrosion sensor, in the paper, the external switch is replaced by an annealed steel wire K which is put out of the sealed portion of the sensor, as shown in fig 3. The intact state of the steel wire corresponds to the case of closing the external switch mentioned above, and the broken state of it corresponds to the case of opening the external switch.

The sensor is embedded in the concrete and the wire is exposed to the same environment conditions, i.e. the same corrosive environment, as the steel reinforcement. As the corrosion process begins in the steel wire and in the reinforcement simultaneously, the cross-sectional areas of both are decreased. Because the diameter of the wire is much smaller, the wire will fracture due to corrosion before appreciable corrosion damage occurs in the steel reinforcement. Therefore, the corrosion detection to the reinforcing steel can be performed by reading the resonant frequency data of the sensor that reflects the state of the steel wire.

3 SIMULATION FOR WIRELESS CORROSION SENSOR

The circuit simulation program NI Multisim (version 10.0) software is used to analyze the results about working process of corrosion-monitoring wireless sensor.

3.1 Resonant frequency of circuit modeling in the simulation method

As a function of fig 2, the resonant frequency of the LC circuit corresponds to a minimum in the phase of the impedance which is called the phase dip. The depth of the phase dip depends on the coupling efficiency between the reader coil and the coil in the resonant circuit as well as the resistance of the LC circuit. As the resistance increases, the amplitude of the phase dip decreases.

3.2 Identifying the variable states of corrosion

In the paper, a resistor is applied to model external switch (i.e. steel wire K) in the simulation. As a result, a large resistor represents an open switch while no ressistor represents a closed switch.

4 ANALYSIS ON THE RESULTS OF SIMULATION OF THE SENSOR WITH SINGLE SWITCH EQUIVALENT CIRCUIT MODELS

As mentioned above, the phase dip depends on the coupling efficiency between the reader coil and the coil in the resonant circuit as well as the resistance of the LC circuit. The coupling efficiency k (dimensionless) is defined as follows:

$$k = \frac{M}{\sqrt{L_1 L_2}} \quad (3)$$

where M is the mutual inductance between inductances L_1 and L_2. As k value approaches 1, the coupling between the two inductors becomes stronger and stronger.

The coupling between two different coils, see fig 3, is described by mutual inductance (Ulaby, 1999). As the coupling between two coils increases, the mutual inductance of them increases as well. By Neumann's Form, the mutual inductance can be computed in the following manner:

$$M = N_1 \times N_2 \times \mu_0 \times \sqrt{A_1 \times A_2} \times [(\frac{2}{t} - t) \times K - \frac{2}{t} \times E] \quad (4)$$

where

$$t = \sqrt{\frac{4 \times A_1 \times A_2}{d^2 + (A_1^2 + A_2^2)}} \quad (5)$$

where A_1 and A_2 are the radius of each coil, d is the distance between them, N_1 and N_2 are the number of turns of each coil, and K and E are complete elliptic integrals of the first and second kinds, respectively (Ramo S et al. 1984).

Table 1. Wireless Sensor Design Parameters.

	Outer diameter	Inner diameter	Thickness	Maximum recognizable distance	Diameter of copper magnet wire	Turns	Inductance	Coupling efficiency k
Reader loop	50 mm	30 mm	20 mm	40 mm	0.4 mm	12	$L_1 = 28.2\,\mu H$	0.1633
Sensor coil	50 mm	30 mm	20 mm	40 mm	0.4 mm	26	$L_2 = 102.0\,\mu H$	0.1633

4.1 Production and optimal physical parameters of sensors

A low-cost, passive and wireless sensor has been produced at the Nanjing University of Aeronautics and Astronautics to monitor corrosion of reinforcing steel in concrete (Jin Wu et al. 2010). A ferrite bead is connected to the coil in order to enhance the coupling efficiency between the reader coil and the sensor coil. The outer and inner diameters of the ferrite bead are 50 mm and 30 mm while the thickness and maximum recognizable distance are 20 mm and 40 mm, respectively. Inductors are produced by winding enamel-insulated copper magnet wire with 0.4 mm diameter around the ferrite bead. The reader and the inductor with 12 turns and 26 turns of the wire were selected for the sensor coils. With the help of electric inductance measuring-testing instrument, the inductance of the reader and the inductor are measured to be $L_1 = 28.2\,\mu H$, $L_2 = 102.0\,\mu H$, respectively. In accordance with the Equations (3) and (4), the coupling efficiency k calculated between the reader and the inductor is taken as 0.1633. The capacitors of 500 pF and 10 pF are separately selected for the sensors with single switch and three switches, respectively, in the simulation using NI Multisim. The values of sensor design parameters are listed in tab 1 (He Lei et al. 2011). The coil and capacitance are assembled inside a plastic box, which is filled with encapsulating compound later.

4.2 Analysis on the results of simulation program NI Multisim

The idealized circuit model using normalized component values as discussed in the previous section is useful for examining basic circuit behavior and formulating general design considerations. In order to further investigate how an actual circuit may perform in the real environment, several simulations are carried out using NI Multisim software in advance.

The values of sensor design parameters listed in tab 1 are used in the following simulations. Fig 4 shows a simulation of the wireless sensor as the resistance of the wire increases. As the wire begins to corrode, thereby increasing in resistance, the phase dip of the impedance becomes smaller and begins to shift in frequency. The AC Analysis (see fig 5) as a main approach including several key parameters, such as the start frequency sweep, end frequency, sweep type,

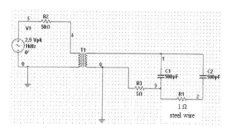

Figure 4. Simulating of Multisim of the Sensor with Single Switch Equivalent Circuit Model.

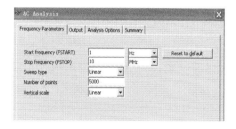

Figure 5. AC Analysis Parameters.

number of points, vertical scale, are set before the conduct of the simulation. In addition, it is ideal to make the number of points in appropriate range, which can get the maxium phase depth. The resonant frequencies in the simulation are displayed in tab 3, and the resistance of the steel wire varies from $1\,\Omega$ to $10\,k\Omega$.

Tab 3 shows that capacitor and the inductor of the sensor are 500 pF and $102.0\,\mu H$ before the simulation, which results in initial resonant frequency with an intact steel wire of approximately 506.62 kHz (fig 6 and fig 7) and the final resonant frequency with a fractured steel wire of approximately 714.56 kHz (fig 8 and fig 9). In comparison with the theoretical values calculated by Equations (1) and (2) of 498.33 kHz and 704.75 kHz, respectively, it can be seen that they are similar to each other.

In addition, the resonant frequency of the sensor is unstable due to the existence of the "transition zone" that indicate the steel reinforcement experiences a slight amount of corrosion near the wireless sensor in the concrete structure. If resonant frequency fluctuates around initial theoretical value when a sensor is interrogated, the possibility of corrosion at the

Table 2. Results of Simulation of the Sensor with Single Switch.

Steel wire resistance R1 (Ω)	1	500	1000	1800	2000	5000	8000	10000
Resonant frequency f1 (kHz)	506.62	506.62	506.62	–	–	–	714.56	714.56

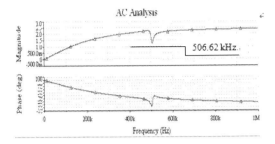

Figure 6. Phases and Magnitude vs. Frequency with an Intact Steel Wire.

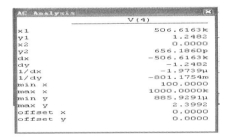

Figure 7. Results from AC Analysis with an Intact Steel Wire.

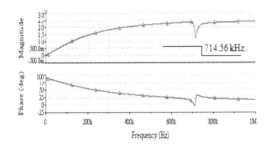

Figure 8. Phase and Magnitude vs. Frequency with a Fractured Steel Wire.

location of the sensor is low. In contrast, if resonant frequency fluctuates around final theoretical value with an fractured steel wire, the possibility of corrosion is high.

It can be observed that it is feasible to determine whether the steel wire is broken or not by the data of the resonant frequency in the circuit.

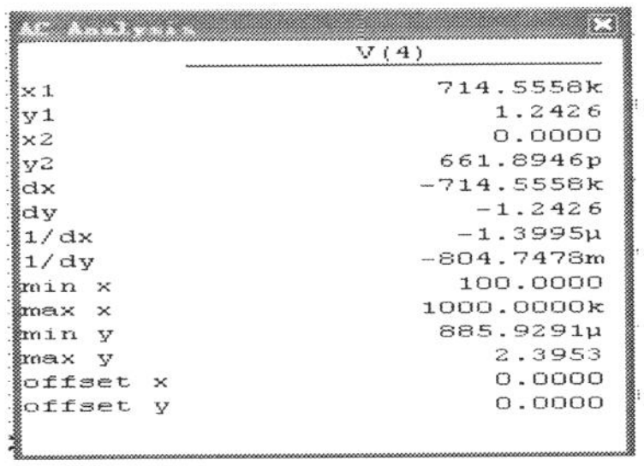

Figure 9. Results from AC Analysis with a Fractured Steel Wire.

5 ANALYSIS ON THE RESULTS OF SIMULATION OF THE SENSOR WITH THREE SWITCHES EQUIVALENT CIRCUIT MODEL

5.1 Optimal circuit of wireless sensor for identifying different corrosion state

As previously stated, once the sensing wire fractures, it can indicate that the steel reinforcement is corroded to some extent but unable to tell how corroded it is. There is a strong correlation between the fracture of the sensing wire and the corrosion of the steel reinforcement in the area surrounding the sensor. To detect different corrosion thresholds, three kinds of steel wire gauges are used to fabricate the prototype sensors. Fig 10 illustrates the optimal circuit of the wireless sensor for identifying different corrosion states. The resonance frequency will change when each wire fractures due to corrosion. Therefore, it can be concluded that three steel wire gauges can indicate four types of corrosion state, such as non-corrosion, slight corrosion, obvious corrosion and severe corrosion.

5.2 Analysis on the results of simulation of the sensor with three switches equivalent circuit models

The equivalent circuit with three steel wire gauges in simulation was shown in fig 11, in this case the AC analysis (see fig 5) design parameters are the same as those in the previous simulation with single steel wire. The analysis of the data returned from this simulation revealed several results which are all depicted in tab 4. Monitoring the corrosion sensors over time for changes in the resonant frequency may allow estimation of the corrosion state. When the three

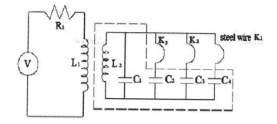

Figure 10. Circuit Diagram for the Sensor.

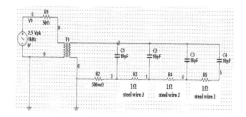

Figure 11. Simulating of Multisim of the Sensor with Three Switches Equivalent Circuit Model.

Table 3. Results of Simulation of the Sensor with Three Switches.

Corrosion state of reinforcement steel	Resistance			Resonant frequency f1 (MHz)
	R3 (Ω)	R4 (Ω)	R5 (Ω)	
Non-corrosion	≤ 90	≤ 90	≤ 90	2.5209
Slight corrosion	≤ 90	≤ 90	$\geq 1 \times 10^6$	2.9146
Obvious corrosion	≤ 90	$\geq 1 \times 10^6$	$\geq 1 \times 106$	3.5678
Severe corrosion	$\geq 1 \times 10^6$	$\geq 1 \times 10^6$	$\geq 1 \times 106$	5.0503

steel wires fracture one by one due to corrosion, the switches in the sensing circuit open and the changes in resonant frequency of the sensing circuit can be determined. In addition, the resonant frequency of the sensor can clearly indicate the visible corrosion state (above 3.5678 MHz in this simulation) when the sensor is interrogated, which demands some realistic measures to be taken immediately.

6 SUMMARY

The sensors discussed in this paper were fabricated to detect the onset of corrosion in reinforced concrete structures. The purpose of this paper was to explain the functionality of two different corrosion sensor models and to simulate them. The results from the simulation demonstrate that the wireless corrosion sensors provide an economic, nondestructive means to evaluate the corrosion state within a reinforced concrete structure. The condition of the steel wire can be easily determined through the resonant frequency of the wireless sensor. The simulation of optimal circuit of wireless sensor with different wire gauges indicates that different corrosion thresholds can be identified according to the changes in resonant frequency. In addition, diligent consideration of the couping factor will produce designs that will work successfully under the most rigorous conditions. All in all, the future of this technology is bright and will soon make its way into daily use.

ACKNOWLEDGMENTS

The work carried out in this paper was sponsored by National Natural Science Foundation of China through grant number 51279074 to the Nanjing University of Aeronautics and Astronautics. The opinions expressed in this paper are those of the researchers and do not necessarily represent those of the sponsors.

REFERENCES

Assouli, B. 2008. Influence of Environmental Parameters on Application of Standard ASTM C876-91: Half Cell Potential Measurements. *Corrosion Engineering* 3(1): 93–96.

He, L. 2011. Simulation of Wireless Sensor for Monitoring Corrosion of Reinforcement in Concrete. *Journal of Yangzhou University* (Natural Science Edition) 14(4): 41–43.

Jin, W. 2010. Study on Wireless Sensing for Monitoring the Corrosion of Reinforcement in Concrete Structures. *Measurement* 43(3):375–380.

Ulaby, F.T. 1999. Fundamentals of Applied Electromagnetics. *New Jersey: Prentice Hall.*

Ramo, S. 1984. Fields and Waves in Communication Electronics, Second ed. New York: John Wiley & sons.

Simonen, J.T. 2004. Wireless Sensors for Monitoring Corrosion in Reinforced Concrete Members. *Proceedings of SPIE* (5391):587–596.

Slope's automatic monitoring and alarm system based on TDR technology

Canyang Lin
Northwest Research Institute Co., Ltd of C.R.E.C., Lanzhou

ABSTRACT: Based on time domain reflectometry (TDR) technology, this paper describes the research centering on slope's automatic monitoring and warning alarm. Upon TDR's fundamental theory and carrying out coaxial-cable deformation and destruction simulating experiments, the features of coaxial-cable deformation and destruction and their respective reflection rules were understood. Moreover, both coaxial-cable arrangement and grout, data automatic collection and remote transmission, were studied. In addition, deformation threshold values ere investigated. Finally, a complete set of software, including data processing and monitoring and warning alarm, was developed. Slope's real-time monitoring and warning alarm were realized at last.

Keywords: coaxial-cable deformation experiment; automatic collection; remote transmission; real-time monitoring; security early warning

1 INTRODUCTION

Due to complicated geologic conditions, multi-influenced factors and maintenance difficulty, once deformation or landslide occurs, mountainous expressway's slope will cause great trouble to expressway operation and impose high security threat to passing drivers and passengers. In the past, slope monitoring was achieved by administration and maintenance staff through on-site examination. However, on-site analysis not only requires professional equipment, but also is expensive and time-consuming. Moreover, potential security threat of slope body and preventive measures might not be able to find out on time. Under the situation that maintenance funds are limited, studying on some advanced monitoring technology so that slope administration and maintenance can be achieved in a scientific way is necessary.

Time Domain Reflectometry (TDR) is an electric measuring technology, and its research began at the beginning of the 20th century. Because of its multi-function feature, it was first used to determine the fault and breakdown of cable and electric wire in the power and communication industry. Its utilization in national defense and telecommunication to localize the position of aerocraft, telephone cable fault and other material's property, started many years ago, too. After 1970s, TDR technology began to be applied in geologic investigation, and in the middle of 1990s, it began to be utilized in the monitoring of geologic hazard[1~5].

2 BASIC THEORY OF TDR TECHNOLOGY

Time domain reflectometry (TDR) technology takes coaxial-cable as a transmission medium to transport instantaneous pulse; electric pulse signal transports along the coaxial-cable and when the pulse signal finds out that cable characteristic impedance is changed, it will produce a reflected signal, which is called TDR signal. Through measure and comparison computation of transmitted signal and reflected signal in the cable, the properties of cable, both fault location and fault characteristics, can be verified. During TDR test, the transmitted signal is defined as Vt and the reflected signal is defined as Vr, and the ratio (ρ) of Vr to Vt is called reflectance, written as

$$\rho = Vr/Vt \quad (1)$$

Because it can state the variation of reflected signal, the condition of the cable can be described by reflectance. When the cable termination is open, the reflected signal is similar to the transmitted signal and their phases are alike, so the reflectance is +1. In contrast, when the cable termination is short, the reflected signal is equivalent to the transmitted signal, but their phases are inverse, so the reflectance is −1. When cable characteristic impedance is varied, its reflectance changes between −1 and +1. If the characteristic impedance is minified, then the magnitude of the reflected signal decreases and the reflectance is negative, otherwise the reflectance is positive[6~8].

3 FIELD EXPERIMENT OF COAXIAL-CABLE DEFORMATION FEATURES

3.1 *Aims and significance*

As stated above, there is some relationship between the deformation of coaxial-cable and its TDR reflected signal. By field experiment, the relationship can be used to analyze and estimate the slope displacement trend. The main purposes of the field experiment are as follows:

(1) Deciding coaxial-cable buried process based on TDR technology;
(2) Deciding mortar mix ratio that can sensitively seize coaxial-cable deformation;
(3) Understanding deformation rules of coaxial-cable;
(4) Confirming alarm threshold of coaxial-cable deformation.

Necessary supports for the utilization and promotion of TDR technology in slope automobile monitoring and safety alerting is achieved by means of numerous simulation experiment of field destruction, overcoming various influencing factors of monitoring results, and analyzing and processing experiment data.

3.2 Deformation experiment

Because of the long period of boring and that the actual deformation condition in the hole cannot be seen directly, in the field experiment, horizontal wiring was adopted so that the coaxial-cable deformation can be observed directly and easily. First of all, two ditches were dug, and then two coaxial-cables were set in the ditches, and mortar was paved at last. After 7–10 days of maintenance, the mortar obtained its 75 to 80% of strength, and meanwhile, the destruction simulation experiment was started. After substantially reduplicative experiment, the coaxial-cable buried method and mortar mix ratio were confirmed. The experiment procedures are as follows:

(1) Choosing an appropriate area in the experiment field to dig ditches in the section size of 10 cm × 20 cm and length of 20 m, and the ditches must be direct.
(2) Waterproofing must be finished before coaxial-cable was placed so that underground water seepage will not influence the experiment.
(3) Mortar (mortar mix ratio is 1–2) in thickness of 5–7 cm must be grouted at the bottom of the ditches and leveled. Then, set the coaxial-cables, and place another 5–7 cm of mortar afterward. At the same time, some small trenches, every five meters, along the ditches to place jack, need to be dug as antiforce structures, as shown in Figure 1.
(4) Increasing jack loads gradually. As the force imposed on the mortar increased, both the mortar and the coaxial-cable deformed imperceptibly; the coaxial-cable deformation curve is as shown in Figure 2. When the load reached a certain degree, the mortar-body began to crack and the coaxial-cable deformed too, as shown in Figure 3 (triggering alarm threshold was set as 60 $m\rho$; red dots on the curve indicate that the coaxial-cable deformation exceeded 60 $m\rho$). The experiment also proved that when relative reflectance reached certain values (500~550 $m\rho$), the coaxial-cable reflected curve did not transform synchronously

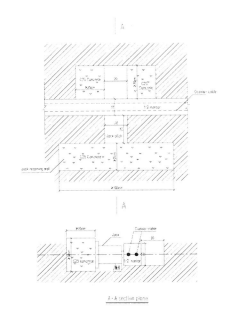

Figure 1. Coaxial-cable place sketch map in filed.

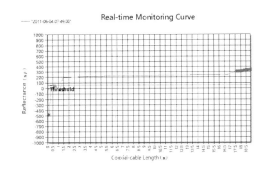

Figure 2. Coaxial-cable deformation curve corresponding to initial deformation of mortar-body.

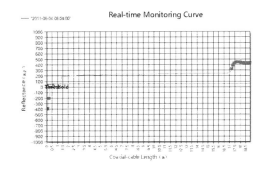

Figure 3. Coaxial-cable deformation curve increased as mortar-body deformed gradually.

with the mortar body; it stayed in a relative stable status, as shown in Figure 4. Nonetheless, when the mortar-body's deformation was big enough, the coaxial-cable in the mortar-body was broken,

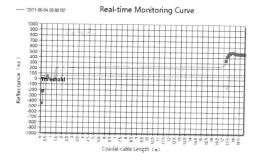

Figure 4. Coaxial-cable deformation curve did not change synchronously with mortar body's deformation.

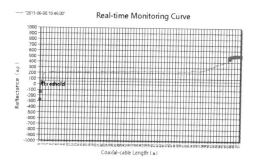

Figure 6. Coaxial-cable initial deformation curve at 68 m when alarm threshold is 50 $m\rho$.

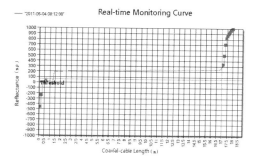

Figure 5. Coaxial-cable broken.

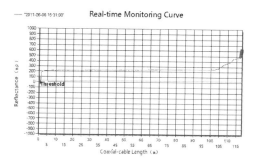

Figure 7. Coaxial-cable initial deformation curve at 118 m when alarm threshold is 30 $m\rho$.

then the deformation curve changed dramatically (Figure 5).

(5) Confirming triggering alarm threshold when the coaxial-cable lengthens. The basic principle was to lengthen different lengths of coaxial-cables and measure their triggering alarm thresholds when their deformation quantity is equivalent as they are not lengthened.

Figure 6 shows the alarm threshold of a 18 m coaxial-cable, corresponding to triggering deformation, was 60 $m\rho$; after lengthened extra 50 m cable, alarm threshold is 50 $m\rho$, and alarm threshold is only 30 $m\rho$ when extra lengthened cable is 100 m as shown in Figure 7.

3.3 Results analysis

The deformation stimulation experiment indicated that the deformation rule of coaxial-cable (bonded in mortar) could be understood. The appropriate mortar mix ratio is 1–2, and under this ratio, the coaxial-cable initial triggering alarm threshold can reflect practical deformation.

The coaxial-able triggering alarm threshold is 60 $m\rho$, but the triggering deformation is minute and cannot be seen by naked eyes. Therefore, it is suggested to set the alarm threshold at 120 $m\rho$ in practice, so that the mortar-body and surrounding soil deformation can be observed and false alarms can be avoided greatly.

Coaxial-able attenuation features can be comprehended by lengthening different cable lengths. When the lengthened length is less than 30 m, the coaxial-cable attenuation is negligible; when the lengthened length is 50 m, the signal attenuation is about 50%; and when the lengthened length is 100 m, the signal attenuation is about 75%.

Moreover, the experiment showed that the deformation rules of coaxial-cable and lengthened coaxial-cables are alike. That is, the deformation curve changes as the mortar-body deforms, and when the reflectance reaches certain values (500~550 $m\rho$), both cable and mortar deformations increase, but the deformation curve remains steady. The deformation curve changes dramatically till coaxial-cable is broken when the mortar-body's deformation is significant enough.

4 SLOPE AUTOMATIC MONITORING

4.1 Monitoring data collection

The TDR monitoring system produced by Campbell scientific Co. Ltd, U.S. was adopted in the experiment; the system consists of power supply, CR1000

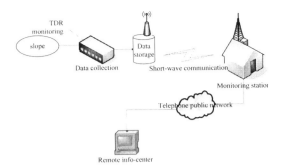

Figure 8. Sketch map of TDR remote monitoring and data transmission.

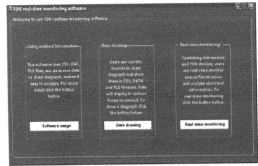

Figure 9. Host interface of TDR real-time monitoring software.

data collector, TDR100 reflector, SDMX 8-channel router, coaxial-cable (sensor), software PCTDR and LOGGERNET, etc.

The TDR100 reflector could collect one-channel monitoring data; via connecting to SDMX 8-channel router and CR1000 connector, it can collect eight channels' data, and the data content varies, it could be hundreds of species. This study focuses on slope deformation monitoring, thus the major data collected included collecting time, battery level, device temperature, channel number and some other data relative to slope monitoring. All of the collected data is stored in CR1000 collector and withdrawn via LOGGERNET. When editing the acquisition program, the monitoring length can be edited as required; for those important slopes, 0.1 m is supposed, and 0.2 m or 0.5 m for usual slopes. The collecting interval can to be 12 h or 24 h or others according to the monitoring requirement.

4.2 Remote data transmission

After receiving reflected signals, TDR reflector stores them in data collecting station, and transmits to the PC in remote info-control center through wireless public network by using wireless communication device. A typical TDR slope remote monitoring and data transmission system is shown in Figure 8.

The connection between the PC and public network includes two aspects, one is the PC-wireless communication device by network, and the other is the LOGGERNET-network. When these two connections are established, monitoring software (LOGGERNET) could collect on-site monitoring data via network and transmit data to info-center.

5 TDR SAFETY ALARM

Slope auto-monitoring is to reduce staff workload and alarm timely when hazard happens, so that slope potential risk could be avoided and passing vehicles and passengers security can be granted.

Slope safety alarmings bases on TDR technology. First of all, data collection needs to be dealt, by analyzing monitoring data and finding abnormal data, and alarming depending on the magnitude of deformation. By monitoring the deformation of coaxial-cable (bonded in mortar) to reveal slope's deformation, it was found in this study that the ideal mortar mix ratio is 1:2, which can sensitively reflect coaxial-cable deformation, and the corresponding alarm threshold is 120 $m\rho$. When the reflectance reaches 120 $m\rho$, it means the slope deforms, and administration and maintenance staff should go to the slope locality and put forth fundamental handling procedure according to the alarm level. If the deformation is small, monitoring could be continued with increased alarm threshold; nevertheless, if the deformation is large, not only temporary treatment is needed, but also the situation needs to be reported to supervisory department to get treatment counter-measures. According to this principle, a functional software application was developed; it can monitor multiple cables simultaneously. This software can draw monitoring curves according to historical monitoring data and analyze deformation trend, meanwhile, real-time monitoring of every cable deformation is also available. By setting the alarm threshold, slope auto-alarm can be realized when deformation reaches or exceeds the alarm threshold. Figure 9 is the host interface of the TDR real-time monitoring software.

5.1 Monitoring curve drawing

The curve drawing module of this software could present historical monitoring curves. There are eight channels in the TDR monitoring system, and the software will discern each channel's data based on collected data. In addition, both coordinate scale and range could be adjusted. Regularizing curve display scale, specific curve's information (date, distance and reflectance) could be obtained by marking curve's event. Figure 10 is an example of the curve drawn with display scale of 20%. Monitoring curves could be printed for archive purpose.

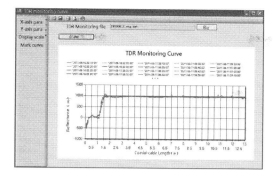

Figure 10. Regularizng display scale and marking curve.

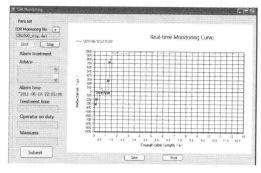

Figure 12. System alarm interface.

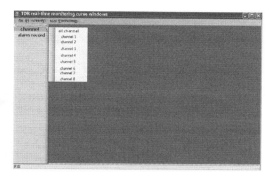

Figure 11. TDR real-time monitoring interface.

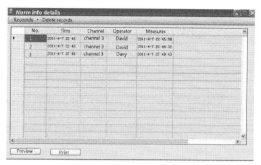

Figure 13. Alarm treatment detail list.

5.2 *Real-time monitoring and alarm*

A real-time monitoring module was developed for slope real-time and dynamic monitoring. It can display eight channels' real-time deformation conditions simultaneously. Fig. 11 is the interface of real-time monitoring module.

In the real-time monitoring interface, coordinate scale, display scale and range could be adjusted, and curve's form, such as polyline, pie chart, histogram could be chosen, too.

In addition to real-time monitoring, the real-time monitoring module can also alarm, with alarm threshold and alarm mode set in advance. Figure 12 is the alarm interface; once an alarm occurs, "alarm deal time", "operator on duty", "alarm treatment", and other information must be filled, or the alarm interface will continue.

Because the reflectance is collected as distributed, that is, collecting monitoring data alongside coaxial-cable at equidistance, anywhere in coaxial-cable is deformed, reflectance collected at this place will vary. So alarm threshold means the difference of subsequent reflectance and primary reflectance.

Each alarm treatment will auto-saved, when we need to look for historical alarm records just utilizes "alarm treatment review", Figure 13 shows alarm treatment detail list, beside, print is available too.

6 SUMMARY AND SUGGESTION

(1) The advantages of TDR monitoring system are its low cost, time-saving and high security.
(2) There are relations between TDR reflectance and coaxial-cable situation. The magnitude of coaxial-cable deformation could be known by analyzing reflectance signal, and TDR monitoring technology could be utilized in slope deformation monitoring in different stages.
(3) Mortar mix ratio of 1:1.5–1:2 is optimal for grouting coaxial-cable, but a ratio exceeds 1:2 is inappropriate. In addition, the water for stirring must be clear. The deformation reflected signal increases as the mortar and coaxial-cable deformation increase; when the signal reaches a certain degree, it will remain stable and change sharply when the coaxial-cable breaks.
(4) During grouting coaxial cable, mortar's work ability should be taken into account, so that the gap between coaxial-cable and surrounding soil could be filled and that deformations of cable and soil are ensured simultaneous.
(5) TDR monitoring system could automatically collect and save monitoring data, and store 8-channel data separately, which is very convenient for analysis. Monitoring data could be transmitted remotely by applying remote data transmission technology.

(6) A functional TDR monitoring software was developed so that monitoring curves could be drawn and slope real-time monitoring and alarm once a hazard happens could be realized.

Studies have shown that the utilization of TDR technology in the geotechnical area is promising. In slope monitoring, it can accurately and quickly localize deformation location. However, the application of TDR technology in slope monitoring is still in initial stage. It cannot confirm displacement direction and displacement quantity, thus a lot of experiments and practical applications are needed to promote the development of TDR monitoring technology.

REFERENCES

Chen Yun, Chen Renpeng, Chen Yunmin. 2003. Calculation model of TDR slope monitoring system and preliminary experiment. Industrial construction. 33(8):37–41.

Chen Yun, Liang Zhigang, Chen Yunmin. 2004. Application of TDR to rock and soil deformation measurements. Central South Highway Engineering. 12(29), 1–5.

Chih-Ping Lin, Shr-Hong Tang, Ph.D, Wen-Chin Lin, Chih-Chung Chung. 2009. Quantification of cable deformation with time domain reflectometry-implications to landslide monitoring. Journal of Geotechnical and Geoenvironmental Engineering, 143–152.

Dowding, C. H. and Huang, F. C. 1994. Early detection of rock movement with time domain reflectometry. Journal of Geotechnical Engineering, American Society of Civil Engineers, 120(8), 1413–1427.

Dowding C.H, Su M.B and K.M. O'Connor. 1988. Principles of time domain reflectometry applied to measurements of rock mass deformation. International Journal of Rock Mechanics, Mining Science, & Geomechanics Abstract, 26:287–297.

O'Connor, K.M., Dowding, C.H., and Su, M.B. 1987. Quantification of Rock Caving Within Sinkholes by Time Domain Reflectometry [C]//Proceedings, Second Multidisciplinary Conference on Sinkholes and the Environmental Impacts of Karst, Orlando, FL, 157–160.

Wu Xiaola, Tu Yaqing. 2002. New approach of landslide activity monitoring-probing into TDR technology. Chinese Journal of Rock Mechanics and Engineering. 21(5):740–744.

Zhang Qing, Shi Yanxin, Zhu Rulie. 2001. The study on landslide monitoring with TDR technology. The Chinese Journal of Geological Hazard and Control. 12(2):64–66.

Author info: Lin Canyang, male, engineer, engaging in research of slope and landslide engineering. Tel: 139 59199148.

Speed variations of left-turning motor vehicles from minor road approach while merging with mainline motor vehicles at non-signalized at-grade intersection

Guoqiang Zhang
Jiangsu Key Laboratory of Urban ITS, Southeast University, Nanjing, China
Jiangsu Province Collaborative Innovation Center of Modern Urban Traffic Technologies, Nanjing, China

Qingyuan Zhang & Yuli Qi
Jiangsu Key Laboratory of Urban ITS, Southeast University, Nanjing, China

ABSTRACT: At non-signalized at-grade intersections, left-turning motor vehicles from the minor road approach have to merge with the mainline traffic. During the process, their speeds undergo great variations. This paper, by using statistical methods, has studied the various factors influencing speed variations of vehicles making such movements. The results show that vehicles usually accelerate while merging with mainline traffic. Different types of vehicles have different accelerations in which small sized passenger cars have the highest accelerations and the highest degree of dispersion of accelerations. In addition, accelerations of the vehicles left-turning from the minor road during the merging process are also affected by speeds of anterior vehicles and posterior vehicles on the major road traffic gaps, into which the vehicles from the minor road are trying to enter. Based upon the data, two nonlinear models have been set up to predict impacts produced by speeds of anterior vehicles and posterior vehicles.

1 INTRODUCTION

In China, with the ever-increasing traffic demand of motor vehicles, traffic problems such as safety and traffic congestion are becoming more and more serious. It is very urgent to find efficient solutions to solve or alleviate these problems through law enforcement, education and engineering.

At-grade intersections are important components of road network and their performance has great influence on the whole transportation system. Compared with signalized intersections, non-signalized at-grade intersections are widely used in rural areas, where traffic volumes are low. Motor vehicle drivers at non-signalized at-grade intersections are self-organized and experience little or no control delay. However, traffic accidents at those locations are usually very high and severity of traffic accidents is higher than other parts of the roadway system. According to an investigation of all kinds of traffic accidents happened in China, 40% of them were within areas of intersection and 28% of them were within areas of non-signalized at-grade intersections. Therefore, it is very urgent to investigate the causes behind those traffic accidents and find out effective solutions to improve traffic safety at non-signalized at-grade intersections.

Usually, traffic accident data are used to study traffic safety and it has been used for many years. Numerous researches of traffic safety have been carried out by this kind of approach (Lundy RA 1967, Cirillo J A 1970 & Chen H et al. 2011). However, because the collection of accident data demands years of data accumulation, many impacting factors may have changed during this period. In order to shorten time spent at data collection and set up a more efficient study approach, traffic accident analysis was put forward and has been used (Allen B L et al. 1978, Zhang G Q et al. 2008 & Lu GQ et al. 2011). Besides, speed variations have been proved to be one of the major reasons for most of the traffic accidents and therefore, some researchers have begun to analyze speed variations for the study of traffic safety (Aljanahi A A M et al. 1999 & Tsui MA & Garcia A 2013). This approach has been widely used in places where vehicles have to change speeds frequently. At non-signalized at-grade intersections, drivers tend to change speeds more frequently in order to avoid all kinds of conflicts. Therefore, speed variations are the main cause underlying many traffic accidents occurring in these places. This paper aims to study speed variations of left-turning motor vehicles from minor road approach while merging with mainline motor vehicles. Results of the study will shed more light on the mechanisms of traffic safety at non-signalized at-grade intersections.

2 DATA COLLECTION AND METHODS

2.1 Data collection

A large number of at-grade intersections were previewed carefully and finally, a typical non-signalized at-grade intersection was chosen as the object of research. The major road has two lanes on each side. The minor road approach has two lanes, one for right turn movement and another for left turn movement. Traffic markings and signs at the intersection are maintained in good condition. Traffic volumes on the major road are moderate and sometimes interfered by traffic from the minor road approach.

To facilitate data collection, two video recorders were used to record the movements of vehicles at the intersection. Marks were made upon the pavement in the research area to help to recognize the distances traveled by vehicles. Later on, the videos were played on computers for several times to recognize and record the time when vehicles reached these marks on the pavement. The running time of vehicles over the distances between these marks were then calculated. Average vehicle running speeds between these marks were calculated by dividing distance by running time. At last, average accelerations of these speeds were calculated.

Apart from the data concerning speeds and accelerations of vehicles from the minor road approach, other data concerning impact factors were also collected. In order to understand how various factors influence speeds of motor vehicles left-turning from the minor road approach while merging with traffic of the major road, data such as motor vehicle types, the space distance and time distance of the gaps in traffic streams of the major road, into which vehicles from the minor road approach intended to merge, and speeds of anterior vehicles and posterior vehicles of the gaps were also gathered.

2.2 Statistical methods

Scatter diagrams are most useful for examining the relationship between two continuous variables. In some cases, when one variable depends (to some degree) on the value of the other variable, then the first variable, the dependent, is plotted on the vertical axis. The pattern of the scatter diagram provides information about the relationship between two variables. In this research, the statistical method has been used to study the relationship between acceleration of left-turning motor vehicles from the minor road approach while merging with mainstream traffic, and various impact factors.

A scatter plot can show a positive correlation, no correlation, and a negative correlation between two variables. Nonlinear relationships between two variables can also be seen in a scatter diagram and typically will be revealed as curvilinear. Scatter diagrams are typically used to uncover underlying relationships between variables, which can then be explored in greater depth with more quantitative statistical methods such as linear regression.

Linear regression is one of the most widely studied and applied statistical and econometric techniques. There are numerous reasons for this. First, linear regression is suitable for modeling a wide variety of relationships between variables. Second, the assumptions of linear regression models can be satisfied properly in many practical applications. Third, regression model outputs are relatively easy to interpret and communicate to others, numerical estimation of regression models is relatively easy, and software for estimating models is readily available. The simple linear regression model with only one independent variable is given by the follows:

$$Y_i = \alpha_0 + \alpha_1 X_i + \varepsilon_i \ (i = 1,2,3,\cdots,n) \qquad (1)$$

where Y_i = value of dependent variable for observation i; X_i = value of independent variable for observation i; ε_i = value of disturbance term for observation i; α_i = coefficient of independent variable; and α_0 = intercept (constant) term.

It is neither always expected nor defensible, however, that physical, engineering, or social phenomena can be best represented by linear relationships; nonlinear relationships, instead, are often found more appropriate. Fortunately, nonlinear relationships can be accommodated within the linear regression frameworkthrough variable transformations, which provides a great deal of flexibility for finding defensible and suitable relationships between two or more variables. In this paper, by using variable transformations, the nonlinear relationships between accelerations of left-turning motor vehicle from the minor road approach and impact factors were studied within the linear regression framework. Based upon it, predict models were then set up.

3 DATA ANALYSIS RESULTS

3.1 Types of left-turning motor vehicles from the minor road approach

According to the characteristics of motor vehicles in highway systems, vehicles in the study has been classified into four types: small sized passenger car, medium sized passenger car and small sized freight car, large sized passenger car and medium sized freight car, and large sized freight car. The types are represented by 1, 2, 3 and 4 respectively. Different types of left-turning motor vehicles from minor road approach have undergone different speed changes while merging with mainline motor vehicles. An analysis has been carried out to study the accelerations of these vehicles. The results are shown in Figure 1.

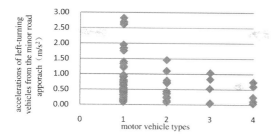

Figure 1. Scatter of relationship between accelerations and vehicle types for left-turning motor vehicles from the minor road approach while merging with major road vehicles.

From Figure 1, it is obvious that accelerations of all types of motor vehicles from minor road approach during the merging process were within the range of 0 and 3 m/s². The fact that accelerations were more than zero shows that motor vehicles from minor road approach speeded up while they were merging with vehicles on the major road. This phenomenon was brought about by the fact that left-turning motor vehicles usually ran at a rather slow speed before they merged with major road traffic. Therefore, they must increase their speed in order to be in agreement with the mainstream traffic. Otherwise, significant speed difference between motor vehicles would happen and the slowly moving left-turning vehicles would be in danger of being hit by the fast moving vehicles closely behind them.

Among the different types of motor vehicles from minor road approach, small sized passenger cars represented by 1 had the highest accelerations and for them the degree of dispersion of accelerations was also the highest. Such performance characteristics of small sized passenger cars can be explained by the following reasons: First of all, compared with other types of motor vehicles, small sized passenger cars have better power performance and braking ability. Therefore drivers of small sized passenger cars have more choice during the process of merging. Besides, most of the drivers of small passenger cars are not full-time drivers and their driving skills are not as good as those of full-time drivers, which could cause unsteady and diverse driving behaviors. Some of these non full time staff might have riskier performances, such as the sudden change of lanes and running speeds. Others might have too conservative operations, such as changing speeds too cautiously. The variety of driving behaviors of small sized passenger cars partly explained the extremely high degree of dispersion of accelerations.

The performance characteristics of small sized vehicles will bring some risks to them when they are turning left from minor road approach and finishing merging operation. Many studies have shown that higher levels of acceleration will bring about more turbulence to traffic stream and greatly increase the possibility and severity of automobile rear end collisions. Higher levels of acceleration also mean higher running speeds, which can contribute to more serious traffic crashes. Moreover, higher degree of dispersion of accelerations shows that movements of small passenger cars are not steady when they turn left from minor road approach and this means additional safety risks. Therefore, compared with other motor vehicle types, small passenger cars are more likely to encounter traffic accidents while making left-turn movements from minor road approach and the traffic crashes might be more serious if they did take place.

3.2 Speeds of anterior vehicles on the major road

When drivers of motor vehicles from the minor road are entering the major road by turning left, they need to choose an adequate gap in the traffic stream of the major road, with which they are going to merge. During the merging process, their driving behaviors are greatly affected and constrained by the anterior vehicles in the gap. If the speed of the anterior vehicle is lower, the driver of the left-turning vehicle from the minor road will be constrained and cannot choose to reach a fast running speed by quick acceleration. However, if the speed of the anterior vehicle is higher, under the influence it, the driver of the left-turning vehicle will tend to accelerate more quickly when merging so as to follow the anterior vehicle in front of him more closely.

In order to study the assumption discussed above, data concerning accelerations of left-turning motor vehicles from the minor road approach and speeds of anterior vehicles on the major road during traffic merging process have been analyzed and the results are shown in Figure 2. It is obvious that when speeds of anterior vehicles are higher, accelerations of left-turning motor vehicles from the minor road approach tend to be higher, too. This tendency is in agreement with the assumption. In order to study the relationship more accurately, a non-linear statistical model as follows has been set up based on the data:

$$Y = 0.0059X^2 - 0.2014X + 2.1484 \quad (2)$$

where Y = acceleration of left-turning motor vehicle from the minor road approach (m/s²); and X = speed of anterior vehicle on the major road (km/h).

The statistic R^2 is 0.7182, which shows that the goodness of fitting is acceptable and the model can be used to predict the impact of speeds of anterior vehicles on the major road. In order to understand the meaning of the equation better, first order derivative of the variable Y with respect to independent variable X is derived as follows:

$$\dot{Y} = 0.0118X - 0.2014 \quad (3)$$

Equation 3 shows that when the independent variable is larger than 17 km/h, first order derivative of the dependent variable is positive and will increase with the independent variable. Therefore, when the speed of anterior vehicle on the major road is larger

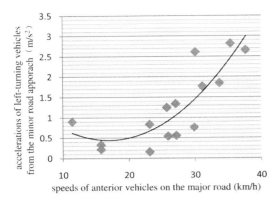

Figure 2. Scatter of relationship between accelerations of left-turning motor vehicles from the minor road approach and speeds of anterior vehicles on the major road during traffic merging process.

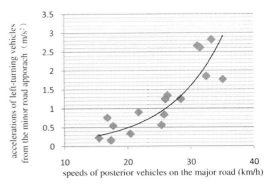

Figure 3. Scatter of relationship between accelerations of left-turning motor vehicles from the minor road approach and speeds of posterior vehicles on the major road during traffic merging process.

than 17 km/h, acceleration of left-turning motor vehicle from the minor road approach will increase with the speed of anterior vehicle on the major road. And this tendency will be strengthened continually with the increasing of the speed of anterior vehicle on the major road. The analysis shows that the model is also in agreement with the assumption.

3.3 Speeds of posterior vehicles on the major road

Similarly, when motor vehicles from minor road approach turn left and enter the major road, drivers' behaviors are also constrained and influenced by the posterior vehicles in the gap into which they are trying to merge. If the speeds of posterior vehicles are very high, vehicles left-turning from minor road approach will be threatened potentially and are forced to accelerate urgently. If the speeds of posterior vehicles are low, drivers of vehicles left-turning from minor road approach will feel much safer and therefore tend to perform acceleration more gently.

In order to study the assumption discussed above, data concerning accelerations of left-turning motor vehicles from the minor road approach and speeds of posterior vehicles on the major road during traffic merging process have been analyzed and the results are shown in Figure 3. It is obvious that when speeds of posterior vehicles are higher, accelerations of left-turning motor vehicles from the minor road approach tend to be higher, too. This tendency is in agreement with the assumption. In order to study the relationship more accurately, a non-linear statistical model as follows has been set up based on the data:

$$Y = 0.046 e^{0.117x} \qquad (4)$$

where Y = acceleration of left-turning motor vehicle from the minor road approach (m/s²); and X = speed of posterior vehicle on the major road (km/h).

The statistic R^2 is 0.7515, which shows that the goodness of fitting is acceptable and the model can be used to predict the impact of speeds of posterior vehicles on the major road.

Equation 4 shows that the dependent variable increases with the independent variable of the exponential function. Therefore, acceleration of left-turning motor vehicle from the minor road approach will increase with the speed of posterior vehicle on the major road. This tendency will be strengthened continually with the increasing of the speed of posterior vehicle on the major road. The analysis shows that the model is also in agreement with the assumption.

In addition to the factors discussed above, this paper, by using statistical approach, has also analyzed the influence of the space distance and time distance of the gap into which motor vehicles from the minor road approach are trying to merge. The statistic R^2 is 0.0236 for the time distance of the gap and 0.1611 for the space distance of the gap. The analyses show that the impacts of the two factors are trivial and can be ignored.

4 CONCLUSIONS

At non-signalized at-grade intersections, left-turning motor vehicles from the minor road approach have substantial impact on the performance of traffic streams on the major road, especially when they are merging with mainline traffic. This paper has studied operation of left-turning motor vehicles from the minor road approach and has analyzed impact factors influencing their speed variation while merging with major road vehicles, using statistical approaches.

The analysis shows that motor vehicles from minor road approach will speed up while they are merging with vehicles on the major road and different types of motor vehicles have different accelerations. Small sized passenger cars have the highest accelerations and the highest degree of dispersion of accelerations.

Moreover, accelerations of the vehicles left-turning from the minor road during the merging process are also affected by speeds of anterior vehicles and posterior vehicles on the major road traffic gaps, into which the vehicles from the minor road are trying to enter. Data analysis shows that accelerations of the vehicles from the minor road will increase when speeds of anterior vehicles and posterior vehicles in the traffic gaps on the major road increase. Based upon the data, two nonlinear models have been set up to predict impacts produced by speeds of anterior vehicles and posterior vehicles.

Results of the paper can be used to analyze traffic performance at non-signalized at-grade intersections. They are also helpful for the analysis of traffic safety at those locations. In the future, similar researches should be carried out to further explore the complicated and potentially dangerous interactions between traffic from major road and traffic from minor road.

ACKNOWLEDGEMENTS

This research was jointly supported by Humanity and Social Science Youth Foundation of Ministry of Education of China (Project Number: 10YJCZH214) and National Natural Science Foundation of China (Project Number: 51278103).

REFERENCES

Aljanahi A A M, Rhodes A H & Metcalfe A V 1999. Speed, Speed Limits and Road Traffic Accidents under Free Flow Conditions. *Accident Analysis and Prevention* 31(2): 161–168.

Allen B L, Shin B T & Cooper D J 1978. Analysis of Traffic Conflicts and Collision. *Transportation Research Record* 667: 67–74.

Chen H, Zhou H, Zhao J & Hsu P 2011. Safety Performance Evaluation of Left-Side Off-Ramps at Freeway Diverge Areas. *Journal of Accident Analysis and Prevention* 43:605–612.

Cirillo JA 1970. The Relationship of Accidents to Length of Speed-Change Lanes and Weaving Areas on Interstate Highways. *Highway Research Record, Report HRR* 312.

Lu GQ, Cheng B, Kuzumaki S & Mei BS 2011. Relationship between Road Traffic Accidents and Conflicts Recorded by Drive Recorders. *Traffic Injury Prevention* 12(4): 320–326.

Lundy RA 1967. The Effect of Ramp Type and Geometry on Accidents. *Highway Research Record* 163: 80–119.

Tsui MA & Garcia A 2013. Use of Speed Profile as Surrogate Measure: Effect of Traffic Calming Devices on Crosstown Road Safety Performance. *Accident Analysis and Prevention* 61: 23–32.

Zhang G Q, Wang W, & Lu J 2008. Safety Performance Evaluation of Highway Intersection based upon Traffic Flow Theories. *2008 IEEE International Conference on Service Operations and Logistics, and Informatics*: 1459–1463.

Stability impact analysis about the layout of tie beam for thin-walled pier

Xiaomei Dong
Luoyang Institute of Science and Technology, Luoyang, China

ABSTRACT: Degraded solid element was used to simulate double thin-walled piers. Two working constructions including the maximal cantilever stage and the completed bridge stage were taken into account. The effect of the different location and number of tie-beams on the stability of high thin-walled piers was analyzed. It turns out to be that setting one tie-beam on the middle of pier shaft can assure stability safety and avoid intense stress.

1 INTRODUCTION

Double thin-wall pier is widely used in long-span bridges with high piers. Its overlarge slenderness ratio poses great stability problem to the bridges that could not be neglected (Baolin, Ma. 2001). Degree of indeterminacy of the structure can be increased by some tie beams between the two limbs of pier. Owing to the restricted deformation, the stability of double thin-wall pier would be improved. Degraded solid elements and beam elements are adopted to establish the finite element models. The impact of the number and position of tie beams on stability was analyzed by utilizing current FEM software.

Figure 1. Finite Element Model at Stage of the Maximal Cantilever.

2 FINITE ELEMENT MODEL

In this paper, an engineering case – Yupogou Bridge, a continuous rigid frame bridge with double thin-wall piers, is analyzed. The pier of the bridge is 72 meters high; its wall thickness is 0.5 meters along the bridge and 0.8 meters across the bridge. The characteristics of the high pier and thin wall meet thin sheet deformation hypothesis. So thin shell element was adopted to establish the model (see Fig. 1 and Fig. 2). Stress and deformation were calculated. Shell element (SHELL143) was made to simulate nonlinear buckling. This kind of plastic element has four nodes which has six degrees of freedom including translation and rotation in x, y, z direction. It allows for large deformation, small strain, creep and stress intensification (Lianyuan, Wu. 1996, Bathe, K. J, 1996). This shell element is actually a degenerate solid element (Shouyi, Xue. 2005). Coefficient of stability was calculated considering material nonlinearity and geometrical nonlinearity (Mender, J. B. & Priestley, M. J. N. & Park, R. 1988).

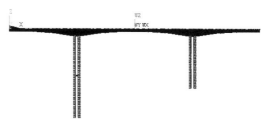

Figure 2. Finite Element Model at Stage of the Completed Bridge.

3 QUANTITY FACTOR OF TIE BEAMS

The pie of bridge belongs to the upper intermediate lever (72 meters high). Stability coefficient of the double thin-walled piers was calculated respectively according to the three schemes including zero tie beam scheme, one tie beam scheme and two tie beams scheme. The tie beams were equally spaced. The maximal cantilever stage and the finished phase were reckoned as the main working conditions (The Ministry of Communication in China. 2004). The calculation results are shown in Table 1 and Table 2. The curves of stability coefficient under different numbers of tie beams are shown in Figure 3.

The calculation shows that the stability coefficient at the stage of the maximal cantilever constructing is greater than that of the finished phase. The more beams of the pier, the greater the stability coefficient is. Considered geometric nonlinearity and material nonlinearity, the stability coefficient descends consumedly. Stability coefficient with one tie beam is significantly greater than that with zero tie beam during the finished phase. The difference between stability coefficient with one tie beam and that with two tie beams is smaller. Owing to the tie beam, the pier stiffness becomes greater and the ability to resist deformation is enhanced. Accordingly, critical collapsing load increases greatly.

Table 1. Stability Analysis at the Stage of Maximal Cantilever.

Number of tie beams	Eigenvalue	Stability coefficient
0	8.1	3.28
1	13.23	3.94
2	16.14	4.21

Table 2. Stability Analysis at the Finished Phase.

Number of tie beams	Eigenvalue	Stability coefficient
0	26.65	4.51
1	33.55	7.54
2	42.5	8.97

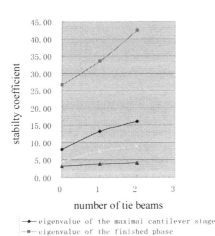

Figure 3. The Curve of Stability Coefficient Corresponding to Different Number of Tie Beam.

4 LOCATION FACTOR OF TIE BEAMS

4.1 Stability analysis at the age of maximal cantilever

Stability coefficient was calculated under the several conditions where the tie beams were set at the different location of pier shaft during stage of the maximal cantilever. There are two sets of schemes to be designed. In the first set of schemes, one tie beam was placed in different location, the corresponding stability coefficients were calculated (see Table 3). The curve of stability coefficients are shown in Figure 4. In the second set of schemes, two tie beams were placed symmetrically to the midpoint of the pier shaft. The calculation results are shown in Table 4 and Figure 5. The distance in Figure 4 is from the tie beam to the bottom of the pier shaft. The distance in Figure 5 is from the tie beam to the midpoint of pier.

The charts above show that the coefficient of stability reaches maximum when the single tie beam is placed at the midpoint of pier shaft. For single tie beam schemes the coefficient of stability goes down with the increase of offset distance from the midpoint of pier shaft. For the two tie beams scheme the coefficient of stability reaches maximum when the pier shaft is divided averagely by the two tie beams. The stability weakens if the two tie beams deviate to the ends of pier from the trisection point.

Table 3. Coefficients of Stability with One Tie Beam.

Distance from tie beam to the bottom of pier m	Coefficient of stability
10	3.43
19	3.64
30	3.9
36	3.94
40	3.86
50	3.66
59	3.47

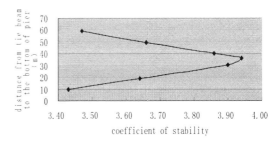

Figure 4. The Curve of Stability Coefficient with One Tie Beam.

4.2 Stability analysis at the finished phase

Critical collapsing load at the age of finished bridge is much larger than that at the stage of maximal cantilever. This is a safe working condition relatively. Likewise, two sets of schemes were designed to analyze its stability, including the single tie beam and the double ones. Firstly, coefficients of stability with single tie beams were calculated at different locations. The results are listed in Table 5. The curve of stability coefficients was shown in Figure 6. Secondly, in the second set of schemes, two tie beams were placed symmetrically to the midpoint of the pier. The calculation results are shown in Table 6 and Figure 7.

For the single tie beam scheme, analysis has come to similar conclusions. Table 5 shows that the coefficient of stability reaches maximum when the single tie beam was placed at the midpoint of pier shaft. The coefficient of stability decreases when the tie beam is away from the midpoint of pier shaft. For the two tie beams scheme different conclusions are reached. Coefficients of stability with the two tie beams near to the midpoint of pier are larger than that near to the end of pier shaft. If the two tie beams are at the trisection point of pier the stability coefficient would be close to the maximum (see Table 6). The relation curves between the location of the two tie beams and stability coefficient are plotted in Figure 7.

Table 4. Coefficients of Stability with Two Tie Beams.

Distance from the tie beam to the midpoint of pier	
m	Coefficient of stability
36	3.28
26.4	3.59
16.8	4.04
12	4.21
5.6	4.06
0.8	3.97

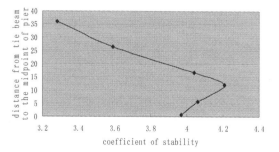

Figure 5. The Curve of Stability Coefficient with Two Tie Beams.

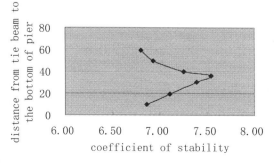

Figure 6. The Curve of Stability Coefficient with One Tie Beam.

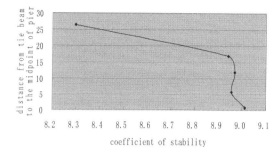

Figure 7. The Curve of Stability Coefficient with Two Tie Beams.

Table 5. Coefficients of Stability with One Tie Beam.

Distance from tie beam to the bottom of pier	
m	Coefficient of stability
10	6.86
19	7.11
30	7.39
36	7.54
40	7.25
50	6.93
59	6.8

Table 6. Coefficients of Stability with Two Tie Beams.

Distance from tie beam to the midpoint of pier	
m	Coefficient of stability
26.4	8.30
16.8	8.94
12	8.97
5.6	8.96
0.8	9.01

5 CONCLUSIONS

Through stability analysis it can be concluded that the midpoint of pier shaft is the optimal position for the single tie beam scheme; and placing the two tie beams at the trisection point of pier shaft will be the optimal layout; the coefficient of stability is lowest during cantilever construction. Besides, both economy and stability need to be taken into account synthetically to set the number and position of tie beams. In fact, the internal forces are so complex that the principal stress would exceed the limit of rupture (Hognestad, E. & hlanson, N W. & McHenry, D. 1955). It is not necessary to set more tie beams on the premise of stabilization. Temporary cross bracing can be used to ensure stabilization at the age of maximal cantilever so that too much stress of the tie beam can be avoided at the age of finished phase of bridge.

REFERENCES

Baolin, Ma. 2001. *Continuous Rigid-frame Bridge with High Piers and Long Span*. China Communications Press. China.

Bathe, K. J, 1996. *Finite Element Procedures*. Prentice-Hall. Englewood Cliffs, NJ.

Hognestad, E. & hlanson, N W. & McHenry, D. 1955. Concrete Stress Distribution in Ultimate Strength Design. *ACI Journal Proceeding*: 455–479.

Lianyuan, Wu. 1996. Plate and Shell Stability Theory. Huazhong University of Science and Technology Press. China.

Mender, J. B. & Priestley, M. J. N. & Park, R. 1988. Theoretical Stress-strain Model for Confined Concrete. *Structural Eng.* ASCE Vol. 114.

Shouyi, Xue. 2005. *Finite Element Method*. Beijing. China Building Materials Industry Press.

The Ministry of Communication in China. 2004. *Code for Design of Highway Reinforced Concrete & Prestressed Concrete Bridges and Culverts*. China Communications Press. China.

Studies concerning the guidance in curves of bogies with elastic driven wheelsets

I. Sebeşan & M.A. Spiroiu
Rolling Stock Department, Transports Faculty, University Politehnica of Bucharest, Bucharest, Romania

ABSTRACT: Providing guidance in curves, in complete safety and with minimum wear of wheels and rails is a basic requirement of railway vehicles. In this paper is made an analysis of the circulation conditions in curve of a bogie with elastic driven wheelsets, type Y 32, used by railways in Romania. The system of wheelsets elastic driving allows its quasi radial position in curves, leading to the reduction of friction between wheels and rails and to lower wear. The presented mathematical model is original, taking into account the wheel loads transfer and the creep coefficients evaluated according to Kalker theory. It is found that high elasticity causes a reduction of the hunting critical speed. Therefore the paper presents also an original study model of the hunting movement of a high speed bogie.

1 INTRODUCTION

The study of the motion in curve of a railway vehicle aims to establish conditions which ensure safe vehicle guidance.

From the point of view of the guidance, the vehicle is an assembly consisting of a number of wheelsets connected rigidly or elastic on a frame.

In the present paper is presented the case of bogies with steerable axles that besides transverse displacement have the possibility to rotate the wheelset from the chassis due to its longitudinal and transversal elastic connection to the chassis.

This rotational motion allows the wheelsets to adopt a quasi-radial position in curves, which has the effect of reducing the wheel-rail contact forces and also the wear of wheels and rails tread and guiding surfaces.

As the elasticity of the longitudinal axle guiding system is higher, the leading wheelset of the bogie will be closer to the radial position thus being created conditions for a "pure" rolling – in the case of wheels with wear profile.

A higher elasticity in the longitudinal direction can lead to an unstable hunting movement of the wheelset at a smaller travelling speed of the vehicle than the stipulated one.

Taking into account the contradictory requests regarding the elasticity for high speed stability on straight track and for good curving, the design of the vehicle must be made by selecting the best compromise that would be acceptable from both points of view previously presented.

2 THE INFLUENCE OF ELASTICITY IN THE DRIVING SYSTEM OF THE WHEELSET ON THE BOGIE GEOMETRIC POSITION IN CURVES

In this section it will be analyzed the motion in curve of a bogie with elastic driven wheelsets, assuming a stationary, quasi-static travelling regime. A reference paper that deals with this problem has been developed in (Newland 1969).

It is assumed that under the action of external force and of contact forces between wheels and rails the bogie is adopting a position in curve as shown in Figure 1, the angles of wheelsets with the normal to the curve being α_1 and α_2 (the attack angles).

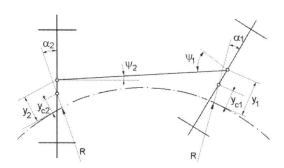

Figure 1. Elastic driven wheelset bogie motion in curve.

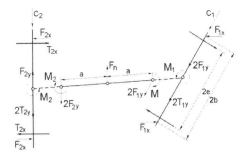

Figure 2. Forces and moments acting on the axles and bogie frame.

It is also assumed that there are no large sliding of the wheels (with wear profile), but only creep proportional to contact forces and that the centering forces can ensure the running of wheelsets without contact on flanges.

To the axis of the track, the wheelsets centers are shifted towards the outside with y_{c1} respectively y_{c2} and the bogie chassis, reduced to its longitudinal axis, is shifted with y_1, respectively y_2.

During the motion, the friction forces T_x, T_y, the centering forces C and the forces from the guidance system on each wheelset must be balanced.

Moreover, the forces acting on the steering system of the chassis of the bogie must be in equilibrium with the external force applied to the bolster F_n (Fig. 2).

Relatively to the chassis, the axles are rotated with the angles:

$$\psi_1 = \alpha_1 + \frac{a}{R} + \frac{y_1 - y_2}{2a}$$
$$\psi_2 = \alpha_2 - \frac{a}{R} + \frac{y_1 - y_2}{2a} \quad (1)$$

The longitudinal forces of the suspension springs will be:

$$F_{1x} = k_x b \psi_1 \; ; \quad F_{1x} = k_x b \psi_2$$

where k_x = the longitudinal stiffness and $2b$ = the distance between the axle boxes, which can be reduced to the moments:

$$M_1 = 2bF_{1x} = 2k_x b^2 \psi_1 = 2k_x b^2 \left(\alpha_1 + \frac{a}{R} + \frac{y_1 - y_2}{2a} \right)$$
$$M_2 = 2bF_{2x} = 2k_x b^2 \psi_2 = 2k_x b^2 \left(\alpha_2 - \frac{a}{R} + \frac{y_1 - y_2}{2a} \right) \quad (2)$$

The transverse forces of the suspension springs are:

$$F_{1y} = k_y (y_1 - y_{c1}) \quad F_{2y} = k_y (y_2 - y_{c2}) \quad (3)$$

If it is denoted in this case $f_x = \chi_x Q$ and $f_y = \chi_y Q$ (units of force), χ_x and χ_y the creep coefficients (dimensionless), Q being the load on the wheel, then the creep forces will be:

$$T_{1x} = f_x v_{1x} = f_x \left(\frac{\gamma}{r} y_{c1} - \frac{e}{R} \right)$$
$$T_{1y} = f_y v_{1y} = f_y \alpha_1 \quad (4)$$
$$T_{2x} = f_x v_{2x} = f_x \left(\frac{\gamma}{r} y_{c2} - \frac{e}{R} \right)$$
$$T_{2y} = f_y v_{2y} = f_y \alpha_2$$

where v_x & v_y represents the wheel – rail creep, γ – the effective conicity of the wheel profile; and r – the radius of the wheel.

The centering forces of the two wheelsets are:

$$C_1 = k_g y_{c1}; \quad C_2 = k_g y_{c2}$$

where it was denoted with k_g the elastic centering constant, which is given by:

$$k_g = \frac{2Q\gamma}{\gamma_0 \rho_r} = \frac{2Q}{\rho_r - \rho_s}$$

where Q is the wheel load and for the central (median) position of the wheelset, in the wheel-rail contact point – γ_0 is the flank angle of the wheel tread (conicity); ρ_r and ρ_s are the radius of curvature of the profile of the wheel and, respectively, of the rail (Sebeşan 2011).

With forces established thereby, equilibrium equations can be written for the axle and for the bogie frame, which takes into account the small values of the angles involved:

$$C_1 - 2F_{1y} + 2T_{1y} = 0$$
$$2eT_{1x} + M_1 = 0$$
$$C_2 - 2F_{2y} + 2T_{2y} = 0 \quad (5)$$
$$2eT_{2x} - M_2 = 0$$
$$2F_{1y} - 2F_{2y} - F_n = 0$$
$$M_1 + M_2 + 2F_{1y}a - 2F_{2y}a = 0$$

which, after substitutions, becomes of the form:

$$k_g y_{c1} - 2k_y (y_1 - y_{c1}) + 2f_y \alpha_1 = 0$$
$$2e \cdot f_x \left(\frac{\gamma}{r} y_{c1} - \frac{e}{R} \right) - 2k_x b^2 \left(\alpha_1 + \frac{a}{R} + \frac{y_1 - y_2}{2a} \right) = 0$$
$$k_g y_{c2} - 2k_y (y_2 - y_{c2}) + 2f_y \alpha_2 = 0$$
$$2e \cdot f_x \left(\frac{\gamma}{r} y_{c2} - \frac{e}{R} \right) - 2k_x b^2 \left(\alpha_2 - \frac{a}{R} + \frac{y_1 - y_2}{2a} \right) = 0$$
$$2k_y (y_1 - y_{c1}) + 2k_y (y_2 - y_{c2}) - F_n = 0$$
$$2k_x b^2 \left(\alpha_1 + \frac{a}{R} + \frac{y_1 - y_2}{2a} \right) + 2k_x b^2 \left(\alpha_2 - \frac{a}{R} + \frac{y_1 - y_2}{2a} \right) +$$
$$+ 2k_y (y_1 - y_{c1})a - 2k_y (y_2 - y_{c2})a = 0 \quad (6)$$

where $2e$ is the nominal distance between the nominal rolling circles of the wheels.

Equations 6 are enabling the determination of the position in curve of a vehicle with 2 axles, in the general case when the wheelsets are connected to the chassis by longitudinal and lateral elastic elements.

Considering in this case that cg = 0, from the system of equations 6, the following are obtained:

$$A = 1 + \frac{f_x \cdot a}{k_x \cdot b^2} \cdot \frac{e \cdot \gamma}{r}\left(1 + \frac{k_x \cdot b^2}{k_y \cdot a^2}\right)$$

$$B = \frac{f_y \cdot a}{(k_x \cdot b^2) - 1}$$ (7)

$$C = 1 + \frac{f_x \cdot f_y \cdot a^2}{(k_x \cdot b^2)^2} \cdot \frac{e \cdot \gamma}{r}\left(1 + \frac{k_x \cdot b^2}{k_y \cdot a^2}\right)$$

$$y_{c1} = y_{c0}\left(1 + \frac{f_y}{f_x} \cdot \frac{a^2}{e^2} \cdot \frac{A}{C}\right) + a \cdot \frac{F_n}{4f_y} \cdot \frac{B}{C}$$

where $y_{c0} = e\gamma/\gamma R$ is the transversal shift value for which the wheelset has no longitudinal slide (Sebeşan 2011).

The first term in equations 7 highlights the deviation from the track axis and the second is the radial displacement of the axle due to lateral force Fn.

If kx b2 > fx a, the suspension elasticity does not improve the position in curve of the vehicle. If kx b2 < fx a, the deviation from the track axis and will be reduced and, for a suspension elastic enough, the deviation will be close to the minimum (possible for free wheelsets) yc1 = yc0.

The deviation from the track axis when kxb2 < fx a, is reduced also by increasing the value of the term:

$$\frac{e\gamma}{r} \cdot \frac{1 + k_x \cdot b^2}{k_y \cdot a^2}$$

meaning a high effective conicity γ and a decreased transversal stiffness k_y

The displacement of the rear axle from the track axis yc2 is given by the relation:

$$y_{c2} = y_{c0}\left(1 + \frac{f_y}{f_x} \cdot \frac{a^2}{e^2} \cdot \frac{2 - A}{C}\right) + a \cdot \frac{F_n}{4f_y} \cdot \frac{B + 2}{C}$$ (8)

For Fn = 0, the angles of wheelsets with the normals to the curve are:

$$\alpha_1 = -\alpha_2 = -\left(\frac{1}{C}\right) \cdot \frac{a}{R}$$ (9)

The maximum force of creep will occur on the two front wheels. For these, if Fn = 0, the creep are given by the equations:

$$v_{1x} = \frac{\gamma}{r}\left(1 + \frac{f_y}{f_x} \cdot \frac{a^2}{e^2} \cdot \frac{A}{C}\right)y_{c0} - \frac{e}{R} = \frac{f_y}{f_x} \cdot \frac{a^2}{e \cdot R} \cdot \frac{A}{C}$$

$$v_{1y} = \frac{(1/C) \cdot a}{R}$$ (10)

and the creep force T considering that $f_x = f_y = f$:

$$T = \sqrt{(f_x \cdot v_x)^2 + (f_y \cdot v_y)^2} = \frac{f \cdot a}{RC} \cdot \sqrt{1 + \frac{a^2}{e^2} \cdot A^2} \leq \mu Q$$ (11)

where μ is the coefficient of the rail wheel adhesion (the maximum value of the coefficient of friction).

It follows therefore that a bogie with elastic driven wheelsets will travel without slides on any curve whose radius is R:

$$R \triangleright \frac{f \cdot a}{C \cdot \mu \cdot Q}\sqrt{1 + \frac{a^2}{e^2} \cdot A^2}$$ (12)

As the suspension is more elastic, the radius of the curve R is lower, the wheelsets tending towards a radial position.

Comparing the displacements produces by lateral force Fn, it is noticed that the rear axle is shifted more than the front axle, this displacement being independent of the radius of the curve.

The displacement under the effect of force Fn is also independent of the deviation from the track axis, which occurs even if on the bogie is not exerting any lateral force and which actually indicates the inherent ability of the bogie to be self-guided in track by creep forces between wheels and rails.

3 THE INFLUENCE OF ELASTICITY FROM THE DRIVING SYSTEM OF THE WHEELSET ON THE BOGIE HUNTING STABILITY

The study of hunting motion for stability of an elastic driving bogie axle is based on relationships obtained after linearization of hunting phenomenon. Linearization of the hunting phenomenon is achieved by:

– considering that the contact forces are varying linearly with lateral displacement of the wheelset;
– neglecting friction and clearances between various elements of the bearing structure of the vehicle;
– neglecting the irregularities and discontinuities of the track;
– considering the wheel profile equivalent conicity as constant and proportional to the tangential creep force in the wheel-rail contact point.

This study aims to determine the travelling speed at which the hunting stable motion of a vehicle equipped with bogies with elastic driven wheelsets will turn into an unstable motion, respectively the determination of the critical speed, which once exceeded will result in a rapid deterioration of ride quality. In other words, we aimed to determine the maximum speed that can be reached safely by vehicle.

Considering the general case of the hunting motion of a bogie in which the axle suspension consists of springs of stiffness kx, ky, and of linear characteristic dampers (viscous type) with damping constants cx and cy (Fig. 3). The centre of mass of the bogie is considered located at the level of axle's axis. Equations of motion are:

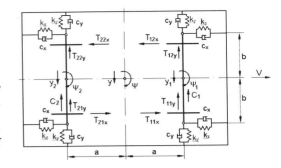

Figure 3. Forces acting on a bogie with elastic driven axles.

$$m_0 \ddot{y}_1 + \frac{2\chi Q}{v} \dot{y}_1 + k_y^* y_1 - k_y^* y_2 - (k_y^* a + 2\chi Q)\Psi_1 -$$
$$- k_y^* a \Psi_2 = 0$$

$$m_0 \ddot{y}_2 + \frac{2\chi Q}{v} \dot{y}_2 + k_y^* y_1 - k_y^* y_2 - k_y^* a \Psi_1 -$$
$$- (k_y^* a - 2\chi Q)\Psi_2 = 0$$

$$I_{0z} \ddot{\Psi}_1 + \frac{2\chi Q e^2}{v} \dot{\Psi}_1 + (k_x b^2 + k_y^* a^2)\Psi_1 - (k_x b^2 - k_y^* a^2)\Psi_2 -$$
$$- \left(k_y^* a - \frac{2\chi Q e \gamma}{r}\right) y_1 + k_y^* a y_2 = 0;$$

$$I_{0z} \ddot{\Psi}_2 + \frac{2\chi Q e^2}{v} \dot{\Psi}_2 + (k_x b^2 + k_y^* a^2)\Psi_2 + k_y^* a y_1 +$$
$$+ \left(k_y^* a - \frac{2\chi Q e \gamma}{r}\right) y_2 - (k_x b^2 - k_y^* a^2)\Psi_1 = 0,$$

(13)

where it was denoted: k_x and k_y – longitudinal and transversal elastic constants, m_o – the wheelset mass, Q – the wheel load, a – the bogie wheelbase, b – half of the transverse distance between the suspension springs, e – half of the distance between the nominal rolling circles, r – the wheel radius, v – the speed, γ – equivalent conicity, I_{oz} – the moment of inertia of the sprung mass about the vertical axis which passes through the center of mass of the bogie and

$$k_y^* = \frac{k_y k_x b^2}{k_y a^2 + k_x b^2}$$

which signifies an equivalent transverse elastic constant.

Determination of critical speed and critical frequency in the case of neglecting the sprung mass of the bogie and the damping forces can be made based on the equations of motion of the bogie frame, and of the wheelsets, by neglecting also the spin and gyroscopic effects. Changing the variables by introducing:

$$2y_1^* = y_1 + y_2 \qquad 2y_2^* = y_1 - y_2 \qquad (14)$$

the equations of motion are of the form:

$$2m_0 \ddot{y}_1^* + \frac{4\chi Q}{v} \dot{y}_1^* - 4\chi Q \Psi_1^* = 0;$$

$$2m_0 \ddot{y}_2^* + \frac{4\chi Q}{v} \dot{y}_2^* + 4k_y^* y_2^* - 4k_y^* a \Psi_1^* - 4\chi Q \Psi_2^* = 0;$$

$$2I_{0z} \ddot{\Psi}_1^* + \frac{4\chi Q e^2}{v} \dot{\Psi}_1^* + 4k_y^* a^2 \Psi_1^* + 4\chi Q \frac{e\gamma}{r} y_1^* -$$
$$- 4k_y^* a y_2^* = 0$$

$$2I_{0z} \ddot{\Psi}_2^* + \frac{4\chi Q e^2}{v} \dot{\Psi}_2^* + 4k_x b^2 \Psi_2^* + 4\chi Q \frac{e\gamma}{r} y_2^* = 0.$$

(15)

Neglecting the mass of the axle, i.e. considering $I_{oz} = m_0 e^2$ and using the notations:

$$A_1 = 4k_y^* \left(1 - \frac{a^2}{e^2}\right) + 4k_x \frac{b^2}{e^2}$$

$$B_1 = 4k_y^* \cdot 4k_x \frac{b^2}{e^2} \left(1 - \frac{a^2}{e^2}\right) + 2(4\chi Q)^2 \frac{\gamma}{e \cdot r}$$

$$C_1 = (4\chi Q)^2 \frac{\gamma}{e \cdot r} \left[4k_y^* \left(1 + \frac{a^2}{e^2}\right) + 4k_x \frac{b^2}{e^2}\right]$$

$$D_1 = (4\chi Q)^2 \frac{\gamma}{e \cdot r} \cdot 4k_y^* \cdot 4k_x \frac{b^2}{e^2} + (4\chi Q)^4 \left(\frac{\gamma}{e \cdot r}\right)^2$$

(16)

results

$$A_1 = K_x + K_y$$
$$B_1 = K_x \cdot K_y + 2\Gamma$$
$$C_1 = \Gamma \cdot (K_x + K_y)$$
$$D_1 = \frac{K_x \cdot K_y}{1 + a^2/e^2} \cdot \Gamma + \Gamma^2$$

yields – using equations 15, the equation of natural pulsations:

$$16m_0^4 p^8 + 32m_0^3 \cdot \frac{4f}{v} p^7 + 8m_0^2 \left[3\left(\frac{4f}{v}\right)^2 + m_0 A_1\right] p^6 +$$

$$4m_0 \frac{4f}{v} \cdot \left[2\left(\frac{4f}{v}\right)^2 + 3m_0 A_1\right] p^5 +$$

$$+ \left[\left(\frac{4f}{v}\right)^4 + 6m_0 \left(\frac{4f}{v}\right)^2 \cdot A_1 + 4m_0^2 B_1\right] \cdot p^4 +$$

$$+ \frac{4f}{v} \cdot \left[\left(\frac{4f}{v}\right)^2 A_1 + 4m_0 B_1\right] p^3 +$$

$$+ \left[\left(\frac{4f}{v}\right)^2 B_1 + 2m_0^2 C_1\right] p^2 + \frac{4f}{v} C_1 p + D_1 = 0 \quad (17)$$

Considering that the stability limit was reached when v = vc, and making the substitution p = jωc in equation 17, it results the final form of the equation of natural pulsations:

$$16m_0^4 \omega_c^8 - 8m_0^2 \cdot \left[3\left(\frac{4f}{v_c}\right)^2 + m_0 A_1\right] \omega_c^6 +$$

$$+ \left[\left(\frac{4f}{v_c}\right)^4 + 6m_0 \left(\frac{4f}{v_c}\right)^2 \cdot A_1 + 4m_0^2 B_1\right] \cdot \omega_c^4 - \quad (18)$$

$$- \left[\left(\frac{4f}{v_c}\right)^2 B_1 - 2m_0 C_1\right] \omega_c^2 + D_1 = 0$$

The following functions are defined:

$$g(\omega_c) = \left(\frac{4f}{v_c}\right)^2 = \frac{32m_0^2 - 12m_0^2 A_1 \omega_c^4 + 4m_0 B_1 \omega_c^2 - C_1}{(8m_0 \omega_c^2 - A_1)\omega_c^2} \quad (19)$$

$$f(\omega_c) = 16m_0^4 \omega_c^8 - 8m_0^2 \cdot [3 \cdot g(\omega_c) + m_0 A_1] \omega_c^6 +$$

$$+ [g^2(\omega_c) + 6m_0 \cdot g(\omega_c) \cdot A_1 + 4m_0^2 B_1] \cdot \omega_c^4 - \quad (20)$$

$$- [g(\omega_c) B_1 - 2m_0 C_1] \omega_c^2 + D_1$$

allowing the calculation of critical speed v_c and critical pulsation ω_c. Thus, the critical pulsation value results as a root of the equation $f(\omega_c) = 0$ and critical velocity is resulting from equation 19.

For the wheel–rail friction coefficients it can be considered the researches in (van Bommel 1968) where there are recommended approximate values of the creep coefficients. Thus based on the results of Kalker (Kalker 1990) it is found that (for Q expressed in tons)

$$\chi_x = \chi_y = \chi = \frac{300}{\sqrt[3]{Q}} \ldots \frac{400}{\sqrt[3]{Q}} \quad (21)$$

4 NUMERICAL APPLICATION

For example, it is considered a passenger car equipped with bogies Y 32 R type at which $m_0 = 2000$ kg, $Q = 59.65$ kN and with the following geometrical characteristics: $a = 1.28$ m, $b = 1$ m, $e = 0.75$ m, $r = 0.46$ m. The creep coefficients calculated with equation 21 have the values $\chi_x = \chi_y = \chi = 175$ and $f = 10.43 \cdot 10^6$ N. For adhesion coefficient $\mu = 0.36$

A special importance for the transversal stability of the bogie has the elastic characteristics of the wheelset driving system. R. Joly (Joly 1971) indicates for high speed bogies with elastic driving axles values $k_x = 10^7$ N/m and $k_y = 5 \cdot 10^7$ N/m. The equivalent transversal stiffness is set ky = 1.481 kN/mm.

The wheel profile has also an important influence on the stability of the vehicle, by its effective conicity γ. A reduced conicity generally contributes to an increased critical speed and it is noted that the influence of the effective conicity on the critical speed is dependent on the values of rigidities k_x and k_y. For values of k_x and k_y larger than 10^7 N/m, the optimal effective conicity is between 0.10 and 0.15. It is adopted the effective conicity $\gamma = 0.15$.

From relation 12 results that the vehicle is travelling without sliding in curves R > 1552 m.

Taking as an example R = 1600 m we obtain yc0 = $1.437 \cdot 10^{-3}$ and A = 1.367, B = 0 and C = 1.018.

For Fn = 0, according to the relation 7, results yc1 = $1.438 \cdot 10^{-3}$ m, yc2 = $1.438 \cdot 10^{-3}$ m, angles of the axles according to 9 are: $\alpha 1 = -\alpha 2 = -0.786 \cdot 10^{-3}$ rad = $-45.057 \cdot 10^{-3}$ deg.

The creep of the front wheel, calculated with equation 10 are v1x = $1.833 \cdot 10^{-3}$, v1y = $0.786 \cdot 10^{-3}$.

For the comparison there were studied also the motion in curve of a bogie with fixed wheelsets. Taking in consideration that $k_x = k_y = \infty$ and $k_g = 0$, we obtain y1 = yc1, y2 = yc2 and for Fn = 0 results: $\alpha 1 = -\alpha 2 = -a/R = -0.8 \cdot 10^{-3}$ rad = $-45.86 \cdot 10^{-3}$ deg.

yc1 = yc2 = yc0 = (fy/fx) · (a2/e · R) = $1.365 \cdot 10^{-3}$ 1y = 2y = y = a/R = $0.8 \cdot 10^{-3}$.

There is a decrease of the angle of attack to the elastic driven wheelsets compared to the case of fixed wheelsets. The difference is small because the comparison was made for a relatively large curve radius. The wheelset elastic driving system is then an advantage while travelling in small radius curves but it also increases the risk of unacceptable sliding.

Critical speed and critical pulsation calculation was based on equations 19 and 20. The positive real roots of the equation 20 (the critical pulsations) are: $\omega c = 20.79836$ rad/s and $\omega c = 139.40631$ rad/s.

Since the first value marks the transition from stability to instability of the wheelset hunting motion, it will be taken into account in calculating the critical speed. Therefore, according to this pulsation the critical speed is

$$v_c = 72.5 \, m/s = 259.4 \, km/h \quad (22)$$

5 CONSTRUCTIVE CONDITIONS TO IMPROVE STABILITY ON BOGIES WITH ELASTIC DRIVEN WHEELSETS

Relationships established before are allowing to analyze the influence of various constructive parameters on the hunting motion of the bogie. In this way, it is possible to establish the conditions which would lead to the extension to higher travelling speeds of the hunting stability domain.

A particular importance for the transversal stability of the bogie has the elastic characteristics of the driving system of wheelsets. In (Sebeşan 2011) is presented the influence of kx and ky of the bogie Y32 on the hunting stability. In (Joly & Laurent 1974) are indicated for speed bogies with elastic driven wheel sets the values $kx = 10^7$ N/m and $ky = 5.10^7$ N/m.

By reducing the wheel set mass m0 and its inertia radius i0z the value of critical speed increases. An important role has also the bogie unsprung mass m, which has to be as small as possible. This aspect must be taken into account especially for motor bogies designed to run at high speeds. Fixing the traction motor on vehicle body chassis is a solution to reduce the bogie mass and hence improves its stability.

Increasing the wheelbase of the bogie leads to an increased critical speed. On the other hand it must be noted that by increasing the wheelbase, it is possible to obtain inadmissible driving forces when the vehicle is running in curves.

The wheel profile influences the stability of the vehicle by its effective conicity γ. A reduced effective conicity generally contributes to increasing the critical speed, but the effective conicity influence on the critical velocity is dependent on the values of rigidities kx and ky. For values of kx and ky larger then 10^7 N/m, optimal effective conicity is between 0.10 and 0.15 (Joly & Laurent 1974).

As shown before, transverse elasticity of central suspension has to be adopted so as to minimize the influence of the bogie hunting on the vehicle body. It is also necessary to achieve a hunting damping torque between the bogie and the body in order to increase the stability of the bogie at high speed. The hunting torque is limited taking into account that it contributes to the increase of the forces exerted on the track in curves.

The report ORE B10, RP 15/F indicates for coaches whose critical speed is below 200 km/h, the adoption of a hunting damping torque of 18 … 25 kNm.

6 CONCLUSIONS

From the study of the motion in curve of the vehicle results the minimum value of the radius for which there is no sliding at the wheel-rail contact. In the present case it can be observed a tendency of a radial positioning of the first axle. The presented relationships are indicating the advantages of a wheelset radial driving for small radius curves but with appropriate adjustment for the elastic characteristics of the axle vehicle guidance system.

The results obtained from the study concluded that, when overcomes the speed of about 260 km/h, the vehicle motion become unstable, phenomena which would lead to unacceptable transversal loads of the track and even to endanger traffic safety.

The maximum safe speed of a vehicle will be lower by 10–15% than the critical speed, taking into account a possible change in time in the elastic characteristics of the wheelset guiding system.

The presented equations allow to analyze the influence of various parameters on the bogie hunting motion and the establishment of constructive conditions for the extension at higher speed of hunting stability regime. Although the applied method is based on a series of simplifying assumptions, it can be used for faster engineering evaluation of bogies performance.

There are also presented the constructive conditions in order to improve stability in general of bogies with elastic driven wheelsets.

The calculation model belongs to the authors of the paper, validated experimentally on a number of a high speed bogie analyzed in the department of Railway Rolling Stock of Polytechnic University of Bucharest.

ACKNOWLEDGEMENT

The present paper is part of the researches included in the PN II PCCA Contract 192/2012 – "Solutions to improve the dynamic performance and the crashworthiness of traction railway vehicles to align with the requirements of European regulations", contract financed by UEFISCDI.

REFERENCES

Gilchrist, A. O. 1997. The long road to Solution of the railway hunting and curving problem, Proceedings of the Conference *From Rocket to Eurostar and beyond*

Joly, R. 1971. Etude de la stabilite transversale d'un vehicule ferroviaire circulant a grande vitesse, *Rail Interntional*

Joly, R., Laurent, M. 1974. Etude de la dynamique transversale d'un vehicule ferroviaire. Banc experimental de Vitry, Rapport SNCF, division des essays de materiel

Kalker, J.J. 1990.Three-dimensional elastic bodies in rolling contact., *Kluwer Academic Publishers*, Dortrech E.A.

Lee, S.Y., Cheng, Y.C. 2006. Influences of the vertical and the roll motion of frames on the hunting Stability of trucks moving on curved tracks, *Journal of Sound and Vibration*

Muller, S. 1998. Linearized Wheel-Rail Dynamics-Stability a Corrugation, *Fortschitt – Berrichte VDI*, Dusseldorf

Newland, D.E. 1969. Steering a Flexible Railway Track in Curved Track, *Transaction of ASME*.

Sebesan, I. 2011. Dynamics of Railway Vehicles. *MATRIX ROM*, Bucharest (in Romanian)

Van Bommel, P. 1968, Considerations lineaires concernant le mouvement de lacet d'un vehicule ferroviaire, UIC/ORE, C9

Study on bicycle flow characteristics at intersections

Xiaohan Deng
Jinan Municipal Engineering Design Institute (Group) Co. Ltd, Jinan, China

Jian Xu
Jinan Urban Planning and Design Research Institute, Jinan, China

ABSTRACT: The distribution characteristics of density, speed, and volume of bicycle flow at intersections are studied, and then a quantitative study for bicycle expansion characteristics is conducted. The conclusion will provide a basis and guidance for traffic and phase design for bicycle traffic at intersections.

1 INTRODUCTION

The characteristic parameters of bicycle traffic flow, such as density, velocity, and volume have been investigated in previous research. Table 1 lists these investigation results.

There has been little research done on the elements of bicycle traffic at intersections, or the spatial distribution characteristics of density, speed, and volume. As for bicycle expansion methods, qualitative descriptions are much more than quantitative study. The conclusion of this paper will provide basis and guidance for traffic and phase design for bicycles at intersections.

2 DISTRIBUTION CHARACTERISTICS OF BICYCLE DENSITY AT INTERSECTIONS

There are three research purposes for density distribution of bicycle flow at intersections: the first one is that it reflects the actual state of bicycle flow at different locations in the intersections; the second one is that instant density at one location will indicate the requirement of bicycle for dynamic distance; the third one is to provide relative data for bicycle lane expansions at intersections.

Unlike the definition of vehicle flow density, instant density of bicycle flow is defined as the number of bicycles per area at a certain moment.

$$k_i(t_0) = \frac{N(t=t_0)}{A_i} \quad (1)$$

where
$k_i(t_0)$ = Bicycle flow density when $t=t_0$ for area i, bicycles/m^2
$N(t=t_0)$ = Number of bicycles when $t=t_0$ for area i, bicycles;
A_i = Area of area i, m^2.

Using the video of an intersection in Chengdu, see figure 1. The following will introduce the method of survey and calculation for bicycle density at intersections.

A grid network is overprinted on the original video which can be seen in figure 2. Instead of the south approach, the east approach was chosen because of acceptable distortion. These girds are numbered as A1, B1, etc.

The dimension of each gird is determined based on the following consideration: the length of a bicycle is 1.9 m, and the average long-direction distance is 0.5 m. Therefore, the minimum long-direction length of each zone must be 2.4, which is the sum of the bicycle length and the long-direction distance, to ensure that there is at least one bicycle in a zone. The width of a bicycle is 0.5 m, and the average lateral distance is 0.5 when moving. And therefore the minimum lateral distance of a zone must be 1.5 m to avoid a bicycle that moves on the line between the grids. Finally, with satisfaction for minimum long-direction and lateral length, the dimension of the zones is also determined by the precision of video. In summary, the video is grided into 5 ∗ 6 zones, which are numbered from west to east beginning with 1, and from north to south beginning with A.

According to the computing formula of instant density, the number of bicycles in each zone for each second should be counted. Counting starts when the bicycles cross the stop line in saturation, and ends when the bicycle flow becomes unsaturated. Counting is done every second during rush hour. Table 2 shows the instant density distribution of the surveyed intersection in Chengdu.

And Figure 3 illustrates the tendency as this data.:
After quantitative analysis, some important conclusions are as follows:

(1) Density varies among different zones at the intersection.

Table 1. Summary of research on bicycle flow characteristics (ZHOU 1988, National Research Council, 2000; XU 2002; LIU 2002; SUN 2003; JING, 2005; JING, 2007).

Publication Year	Researcher	Parameter investigated	Investigation sites	Conclusion
1988	ZHOU Rongzhan	Single lane width	Intersections in many cities in China	Single lane width equals 1.5 m
		Capacity of single lane	Intersections at Nanjing	Capacity of single lane equals 750 bicycles/h
2000	TRB	Single lane width	Intersection in Many countries	Single lane width of approach equals 1.2–1.8 m
		Capacity of single lane	Intersection in Many countries	Capacity of single lane equals 2000 bicycles/h
		Bicycle speed	Roads in Many countries	Bicycle speed equals 18 km/h with variance 3 km/h
2002	XU Jiqian	saturation flow rate	Intersections at Beijing	Saturation flow rate equals 2200 bicycles/h
2004	LIU Dong	speed	Roads at Beijing	Bicycle speeds at intersections equal 4.06 km/h.
		Headway	Roads at Beijing	Jamming density equals 0.56 bicycle/m^2
		Volume	Roads at Beijing	The max volume equals 2,395 bicycles/h*m
2003	SUN Mingzheng	Length of queue and number of parking bicycles	Intersection at Shanghai	The length of bicycle queue is linear to number of parking bicycle.
		Length of queue and lane width	Intersection at Shanghai	The density of queue decreases linearly as lane width of approach increases.
		Bicycle speed	Intersection at Shanghai	Bicycle speed at intersection equals 2.11 m/s.
		Dynamic lateral space between bicycles	Intersection at Shanghai	Dynamic lateral space equals 1.012 m, increasing as proportion of E-bikes rises.
		Saturation flow rate	Intersection at Shanghai	Saturation flow rate during beginning period of green light is apparently linear to lane width.
2005	JING Chunguang	Distribution of arrival	Roads at Tianjin and Shijiazhuang	/
		Distribution of speed	Roads at Shijiazhuang	Speed at arterial with separation of car and bicycle is between 14 and 16 km/h and speed at arterial without separation is between 12 and 14 km/h.
		Traffic flow model	Roads at Shijiazhuang	Free flow speed is 4.02 m/s. Jamming density is 0.63 bike/m^2.
2007	LIANG Chunyan	Saturation flow rate	Intersections at Tianjin and Shijiazhuang	Saturation flow rate equals 1.0 bike/s/m.
		Jamming density Speed	Intersection at Shijiazhuang Road and intersections at Tianjin and Shijiazhuang	Density equals 0.65 bikes/m^2. Speed at intersection is between 3 and 5 m/s.
		Space between bicycle flocks	Road and intersections at Shijiazhuang	Headway at intersection equals 0.6 s. Headway at road equals 6 s.
		Length of bicycle flocks	Roads at Shijiazhuang	/

Figure 1. Original video of an intersection in Chengdu.

Figure 2. Processed video of the above mentioned intersection.

Table 2. instant density of each zone for surveyed intersection (bicycles/m²).

Zone	A	B	C	D	E
1	0.11	0.15	0.16	0.14	0.12
2	0.11	0.16	0.18	0.15	0.12
3	0.12	0.18	0.19	0.18	0.12
4	0.12	0.20	0.22	0.22	0.12
5	0.20	0.22	0.24	0.23	0.21
6	0.14	0.15	0.16	0.16	0.15

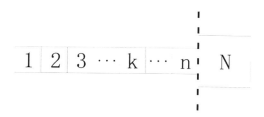

Figure 4. Assumption of one column of bicycle flow.

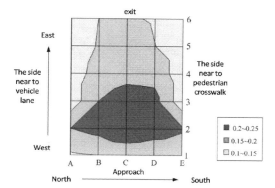

Figure 3. Average instant density during rush hour for each zone at the intersection in Chengdu.

(2) There presents a tendency that the central area has a higher density than the marginal area. That is because the bicycle flow expands without physical constraints on both sides.
(3) The zones near the pedestrian crosswalks have a higher density than those near the vehicle lanes. That is because pedestrians and bicycles interact with each other when they cross the intersection together and the vehicles are far from the bicycles, which means less constraint. This also indicates the extent to which bicycles expand is affected by the intension of constraint.
(4) Bicycle density near approaches is greater than that near exits. And the reason is that bicycles start from the approaches with approximate jamming density, and bicycles move at some speed near exits, which leads to a relatively low density.
(5) The maximum bicycle density does not occur at stop lines. Instead, it takes a while to reach the maximum value after bicycles start. And then the bicycle flow expands to a wider extent, and the dark green area in Figure 3 disappears. The reason maximum value is reached near approaches is that the back row of bicycles move faster than the front row, and the density reaches maximum when they move side by side. This also explains why after reaching maximum density bicycle expansion becomes obvious.

3 THE DISTRIBUTION CHARACTERISTICS OF BICYCLE SPEEDS AT INTERSECTIONS

There are three study purposes for speed distribution of bicycle flow at intersections. The first one is that it reflects the actual operating state of bicycle flow at different locations in the intersections. The second one is that instant speed distribution provides the basis for intersection safety design. The third point is to provide relative data for bicycle expansions at intersections.

The same video gridding method was adopted with some modification. As for one column of bicycle flow, the following assumption can be made, as Figure 4 shows:

Assumption 1: All bicycles moves at long-direction, and there are no lateral moves.
Assumption 2: All bicycles move at most one grid during one period.
Assumption 3: During one period, bicycles cross grids at the same speed.
Assumption 4: The speed of bicycles that do not cross grids during one period are assumed to be 0 m/s.
Assumption 5: The ending grid of one column is named N, which entering bicycles will not exit.

Some variables are defined as follows:

$a_i(j)$ = number of bicycles of grid i at moment j (bicycles);
$N(j)$ = number of bicycles of grid N at moment j (bicycles);
$\Delta a_i(j)$ = variation of bicycle number of grid i from moment j − 1 to j (bicycles);
$\Delta N(j)$ = variation of bicycle number of grid N from moment j − 1 to j (bicycles);
$b_i(j)$ = from moment j-1 to j, the number of bicycles move from grid i − 1 to i (bicycles);
$v_k(j)$ = instant speed in grid K at moment j (m/s);
L = length of grid (m);
T = investigation time (s);
T_n = number of periods during investigation;

According to the assumptions:

$$\Delta a_i(j) = a_i(j) - a_i(j-1) \quad (2)$$

$$\Delta N(j) = N(j) - N(j-1) \quad (3)$$

For grid N, there is:

$$\Delta N(j) = b_N(j) \quad (4)$$

For grid n, there is:

$$\Delta a_n(j) = b_n(j) - b_N(j) \quad (5)$$

And then

$$\Delta a_n(j) = b_n(j) - \Delta N(j) \quad (6)$$

$$b_n(j) = \Delta a_n(j) + \Delta N(j) \quad (7)$$

In a similar way, there is:

$$b_{n-1}(j) = \Delta a_{n-1}(j) + \Delta a_n(j) + \Delta N(j) \quad (8)$$

For a universal grid k ($1 \leq k \leq n$), there is:

$$b_k(j) = \sum_{l=k}^{n} \Delta a_l(j) + \Delta N(j) \quad (9)$$

From assumption 3 and 4, the instant speed of grid k($1 \leq k \leq n$) at moment j is:

$$v_k(j) = \frac{b_k(j) * L/(t_j - t_{j-1})}{a_k(j)} \quad (10)$$

And for the universal grid k ($1 \leq k \leq n$), the average instant speed during investigation is:

$$\overline{v_k} = \frac{\sum_{j=1}^{T_n} v_k(j)}{T_n} \quad (11)$$

The investigation method is similar to that for density distribution. However, different calculation methods lead to different zone gridding standards. For the lateral length, it should be enlarged when the bicycle expansion is obvious based on the assumption that bicycles only move on the long-direction and no bicycles cross zones laterally. And a method that tests whether lateral length is reasonable. If $b_k(j)$ is less than zero, it shows that bicycles cross zones laterally, which is the result of bicycle expansion. And when it comes to long-direction length, according to the previous study (LIU, 2002), the lower speed when bicycles cross the intersection is 1.13 m/s. Therefore, the long-direction length of each zone should not less than 1.13 m for Assumption 2.

The zone gridding is the same as that for density distribution because the bicycle expansion is not obvious in this example. The Table 3 shows the calculation results for average instant speed distribution of the surveyed intersection in Chengdu.

And the Figure 5 illustrates the tendency as this data.

After quantitative analysis, some important conclusions are as follows:

(1) Speed for each zone at the intersection are not well-distributed. The speed at the central location (Column C in this example) is the lowest. For lateral direction, the average instant speed increases as the distance from the central line extends. The volume increases when the zone is farther from the central location, while for long direction, the zone near the approach has a lower speed than that at the exit. And this distribution tendency is inverse to that of density, which can be explained by: $q = k \cdot v$.
(2) Speeds of zones near pedestrian crosswalks are lower than that near the vehicle lanes. That is because pedestrian flow has an obvious effect on bicycle flow.
(3) Combined with the density distribution, it can be found that speeds reach the minimum a short distance from approach.

Table 3. Average instant speed during peak hour for each zone at the intersection in Chengdu (m/s).

Zone	A	B	C	D	E
6	2.37	1.41	1.32	1.36	2.27
5	3.03	2.21	0.90	2.01	2.71
4	2.85	2.81	1.78	2.57	2.77
3	3.90	2.35	2.07	2.13	3.10
2	3.20	2.72	1.88	2.52	3.00
1	3.73	3.53	2.33	3.33	3.60

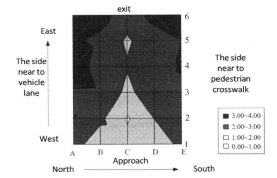

Figure 5. Average instant speed during peak hour for each zone at the intersection in Chengdu.

4 DISTRIBUTION CHARACTERISTICS OF BICYCLE VOLUME AT INTERSECTIONS

There are two study purposes for volume distribution of bicycle flow at intersections: one is that it reflects the actual operating state of bicycle flow at different locations in the intersection, and the other is to provide relative data for bicycle expansions at intersections.

The instant volume of Zone k at moment j is:

$$q_k(j) = b_k(j) \quad (12)$$

Table 4. Average instant volume during peak hour for each zone at the intersection in Chengdu (bicycles/s).

Zone	A	B	C	D	E
6	1.05	1.00	0.65	0.95	1.00
5	1.10	1.15	0.80	1.15	1.10
4	1.25	1.30	0.95	1.10	1.00
3	1.35	1.50	1.10	1.45	1.25
2	1.40	1.75	1.30	1.65	1.35
1	1.55	1.90	1.45	1.80	1.50

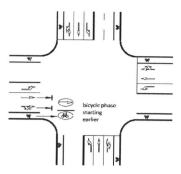

Figure 7. Bicycle flow starting earlier.

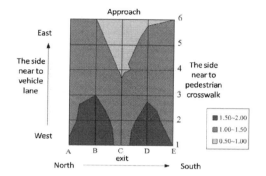

Figure 6. Average instant volume during peak hour for each zone at the intersection in Chengdu.

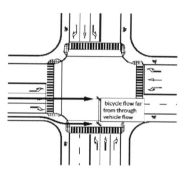

Figure 8. Bicycle flow far from through vehicle flow.

Then the average instant volume during the investigation period T_n of Zone k $\overline{q_k}$ is:

$$\overline{q_k} = \frac{\sum_{j=1}^{T_n} q_k(j)}{T_n} \quad (13)$$

The Table 4 lists the calculation results of the average instant volumes during peak hour for each zone.

The Figure 6 illustrates the tendency as these data:

After quantitative analysis, some important conclusions are as follows:

(1) Volumes for each zone at the intersection are not well-distributed. The volume at the central location is the lowest. The volume increases when the zone is farther from the central location.
(2) The zones near the pedestrian crosswalk are lower than that near the vehicle lane. That is because the effect of vehicles on bicycle flow becomes negligible with enough waiting-area for right turning vehicles, and saturated pedestrian flow has a large effect on bicycle flow at the start of green time.
(3) Volume decreases at the length direction, and the average instant volume at approach is larger than that at exit of the intersection. The reason is that bicycle flow cross the approach in saturated levels at the beginning of green time.
(4) The central zones of the intersection have a more balanced volume distribution than that at approach and exit. That is because after starting at approach, the bicycles expand to both sides without limitation and the bicycles shrink when they come to exit, which means space constraints.

5 EXPANSION CHARACTERISTICS OF BICYCLE FLOW AT INTERSECTIONS

The study of bicycle expansion characteristics can be used to guide traffic design and bicycle phase design at intersections.

Without physical limitations on both sides, the saturated bicycle flow will expand. When it comes to an intersection, if the bicycle phase starts earlier (such as Figure 7), or the through vehicle flow is far from the bicycle flow (such as Figure 8), the expansion of bicycle flow is easily occurs. Sun Mingzheng studies the expansion by the dynamic lateral distance of bicycles. However, the expansion does not only depend on the lateral distance, but also on the column numbers of bicycle flow. This paper will use expansion width to study the expansion of bicycle flow at intersections of its affected factors.

Table 5. Expansion width of bicycle expansion at the intersection in Changzhou.

Name of intersection	Approach	Approach width (m)	Crosswalk length (m)	Maximum expansion width (m)
Heping Road–Boai Road	North	7.2	60.5	11.6
Heping Road–Yanlingxi Road	South	3.8	59.7	12.4
Huaide Road–Yanlingxi Road	West	10.8	67.9	12.6
Jinlingzhong Road–Boai Road	East	4.3	46.4	13.2
Laodongxi Road–Guanghua Street	West	11.4	90.2	18.7
Laodongxi Road–Guanghua Street	South	12.7	50.8	14.7
Yanlingxi Road–Beida Street	East	2.7	34.7	6.0

Table 6. Bicycle expansion rate at intersection of Changzhou.

Name of intersection	Approach	Width of Approach/ Length of Crosswalk	Expansion Rate
Heping Road–Boai Road	North	0.12	62.57%
Heping Road–Yanlingxi Road	South	0.06	225.13%
Huaide Road–Yanlingxi Road	West	0.16	16.94%
Jinlingzhong Road–Boai Road	East	0.09	205.32%
Laodongxi Road–Guanghua Street	West	0.13	64.07%
Laodongxi Road–Guanghua Street	South	0.25	16.30%
Yanlingxi Road–Beida Street	East	0.08	120.74%

Firstly, the expansion rate is defined as the ration of maximum expansion width to the width of bicycle approach.

$$\rho = \frac{D_{max} - d}{d} = \frac{D_{max}}{d} - 1 \qquad (14)$$

Where:
ρ = Expansion rate (%);
D_{max} = Maximum expansion width (m);
d = Width of bicycle approach (m).

Secondly, the investigation site where bicycle expansion is obvious with less effect of right turning vehicles was chosen. In this paper, video at intersections in Changzhou is used to extract data to study bicycle expansion. And the data is listed in Table 5. And expansion rates are calculated which can be seen in Table 6.

With the ratio of width and length of the bicycle approach as independent variable x and expansion rate as induced variable y, there is:

$$y = 6.682 e^{-17.820 x} \quad R-Square = 0.8229 \qquad (15)$$

Indicating that the explanatory capability of x to y is acceptable. The fit curve is drawn in Figure 9. The expansion rate of bicycle traffic flow decreases as the ratio of width and length of the bicycle approach increases. There are two reasons that explain this: Firstly, narrower bicycle approaches lead to narrower bicycle crosswalks, and there is less chance for the faster bicycles to take over others that have to search for taking-over opportunities from both sides, resulting in bicycle flow expansion. And this result matches

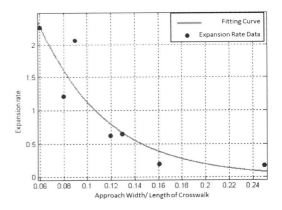

Figure 9. Fit curve of expansion rate.

up to the finding that bicycles far from the central line of approach move faster. Secondly, the longer the bicycle crosswalk is, the larger the difference for crossing time of faster and slower bicycles, which increases the take-over manner of the faster bicycles and expansion becomes obvious.

6 CONCLUSION

This paper put forward an investigational method for bicycle traffic flow characteristics at intersections, such as density, speed, and volume, and demonstrated the study results. The bicycle flow characteristic distributions at intersections were revealed by taking one intersection in Chengdu city as a working example.

Then, the expansion rate was defined to describe bicycle expansion, and the data collected in Changzhou was used to do regressive analysis of the findings that there was an exponential relationship between the expansion rate and the ratio of approach width and crosswalk length. These conclusions will provide basis and guidance for traffic and phase design for bicycles at intersections.

REFERENCES

National Research Council. *Highway Capacity Manual 2000*, Washington D.C.

C.Y. LIANG. 2007. *Research on the theory and application of bicycle flow characteristic*. Jilin: Jilin university

C.G. JING. 2005. *Research on the theory and application of vehicle-bicycle conflict at intersections*. Jilin: Jilin university

D. LIU & F.C. HAN & Y.S. CHEN. 2002. Study on bicycle flow characteristics at signalized intersection. *Journal of Chinese People's Public Security University*: 89–92

M.Z. SUN. 2003. *Study on theory of bicycle traffic design of signalized intersection*. Shanghai: Tongji University.

J.Q. XU. 2002. *Traffic Engineering*, Beijing: China Communications Press.

R.Z. ZHOU. 1988. *Urban Road Design*. Beijing: China Communications Press.

Study on characteristics of typical vernacular dwellings in Mount Emei

Kezhen Chen, Yunzhang Li & Yating Jiang
Architectural and Environment Department, Sichuan University, Chengdu, Sichuan, China

ABSTRACT: In order to rescue the culture of mountainous vernacular dwellings located in Mount Emei, we made researches and drew maps about them, analyzed their features and got conclusion about the feature of the residential houses. That is: use of three-section courtyard layout, emphasis on building of outer space, local material used on the elevation, clear structure and variability of the space, good adaption of the local production, lifestyle and feature of mountainous climate which is often moist and rainy. We did these to offer technical support for protective repair in the future.

Keywords: Mount Emei, mountainous vernacular dwellings, west-Sichuan, three-compound section

Mount Emei is a famous travel destination, which enjoys the name: 'beauty spot under heaven.' She is one of the four famous Buddhist mountains in China, sharing the name with other three mountains which are Wutai Mountain in Shanxi, Putuo Mountain in Zhejiang, Jiuhua Mountain in Anhui. Mount Emei is located at the southwest edge of Sichuan basin. She faces Chengdu plain and sits west to east, from north to south. The mountainous area is cloudy, foggy, and receives an abundant amount of rainwater, and gets moist atmosphere all over the year. Mount Emei has a high reputation for the Buddhist architecture there. The ancient architectural complex has got a wide range of attention and has been protected properly as the national priority cultural relic protection sites [2].

However, the mountainous vernacular dwellings which have been built according to the topography of Mount Emei are not that lucky. With typical features of Sichuan mountainous vernacular dwellings, the mountainous residential houses in Emei keep their own features at the same time. However, in recent years, the traditional dwellings have gradually decayed because they don't suit the modern lifestyle, and their number is decreasing year by year. If the trend continues, the Emei mountainous vernacular dwellings would disappear in a few years. In order to keep the existing data, the writer made studies and drew pictures about the typical mountainous vernacular dwellings and analyzed their characteristics, hoping the study results can be of technical support for protective repair in the future.

Figure 1. Outer-space.

1 GENERAL LAYOUT

Emei mountainous vernacular dwellings are mostly three-section compound. For local people living in a scattered community, the cohesiveness and centrality of the three-section compound form make them feel more safety. As for the reason why they are not four-section compound, it's that the three-section compound makes it easier to engage in production activities, and at the same time it reflects the reception of nature and the neighborhood of local people more. And it is the reveal of the natural and simple custom. Besides, orientations of the houses are different according to the topography of the mountain. They are generally located at the southerly exposure of the mountain, lower in the south and higher in the north. Therefore, they can resist the north wind, embrace the south wind, and receive the sunshine in a maximum amount. That follows the basic principle to select sites in the geomancy [6][7] (figure 1).

The dwellings and courtyards are built on a platform artificially made [8]. Local people choose the way as the easiest to manage the sites because there are plenty of stones in the mountain, so that the platform can be made by stones carried here in a small distance.

Figure 2. The connection between the house and road.

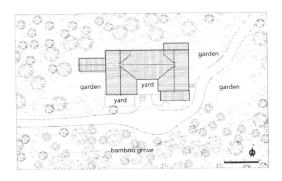

Figure 3. Site-plan.

Compared to the method of using wood as support, the choice of stones as the foundation may be more stable and grave. Therefore, with this construction, a living area is higher than the path before the front door and can be recognized easily from the surrounding nature. As the south part is a little lower than the north part, the inhabitants can get a good scenery in the courtyard and obtain the warm and natural sense of being and belonging from the living "field" made by themselves (figure 2).

Mount Emei enjoys a beautiful natural environment. Rich in plants, she is green all over the year. The dwelling at which the writer made his study is surrounded by lots of plants and enjoys a quiet environment. There is usually a gentle slope right in the front of the road before the dwelling and bamboos are planted all over the slope, and they will be sold to make money to support the family. It reflects local people's intelligence to accord with local condition and make good use of the natural resources in the mountain, while people down the hill make their living on agriculture. A few deciduous trees with beautiful branches are planted in front of the courtyard, so that they can provide shade in summer and be appreciated in winter, shaping a simple and unsophisticated lifestyle of local people after work. Beside the dwelling, there's a vegetable garden, so that the vegetable is enough for them and can be used at any time. Behind the dwelling there's Emei's natural landscape, a steep slope, providing the inhabitants with good background scenery.

The population is thin in Mount Emei, and the dwellings are distributed along the way, leaving people the impression of leisure far away from the realistic world. Moreover, the dwellings are completely different from common settlement and vernacular dwellings. That is because levels of the roofs are different and dwellings are randomly scattered.[2][8] And the courtyards are different from the big houses with deep courtyards. Besides, the scale of the mountainous vernacular dwellings is mostly small but it fits people, and the three-section compound on which the writer make his study is about 401.4 m^2 (figure 3).

2 PLAN LAYOUT

The design is a typical three-section compound and symmetry of the axis. The width is about 21.7 m and the depth is about 11 m. To adapt to the topography, room in the west is shorter than the one in the east. Depths of the principle room and the rooms on the east and west are similar with the same column layout. The principle room is connected with the two rooms through veranda and the function design of the plan is to provide the residents with convenience. The courtyard is the center of outdoor production and activities, serving several basic functions as a place to do housework, pile up things and as a transport center. Being spacious, open and enough of sunlight, the principle room is the real center of daily life. The wide terrace and roof provide enough space for residents in rainy mountain areas. The elders live in the principle room and in the west and east house each lives a younger family. Because the living style is of small family system based on the distribution of dwellings in Sichuan countryside, the three-section layout usually lives a few small families. When the younger generation grew up they start to cook themselves but still live with the elders. Instead of keeping the younger ones from having their own lives, this way does good to keep family connection. The subsidiary rooms are added around the three-section design for efficient production, such as rooms to keep the hens, cows, pigs and to keep firewood and so on. The open space in the front of the two houses, in the west and east, can be used to keep things. The design is under principle and flexible as well, representing people's autonomy of construction [3] (figure 4).

Rooms on the two sides are not open spaces and serve as transport functions. The inhabitants can go

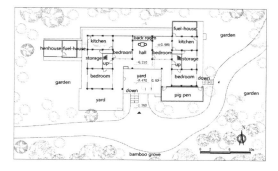

Figure 4. First-floor plan.

Figure 6. Section.

Figure 7. Elevation.

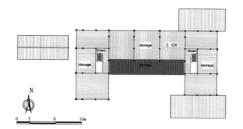

Figure 5. The attic plan.

upstairs through the wooden stair in the two side rooms. The attic space of the vernacular dwellings can be used as open store space for agricultural production. It's rainy and moist in mountain areas, so corps are kept there after harvest due to the good ventilation of attic. The vertical height of eaves is small and bamboo poles are paved closely there by the local people. On the bamboo, some corps can be set to dry and under the bamboo corn or other food can be hung. It is not sultry in the attic for there's not much sunlight in Mount Emei; therefore, some inhabitants take the attic as bedroom (figure 5). Diameters of columns are not the same, with a size of about 200 mm each, for the material of the columns are obtained locally—cedar wood which can be seen anywhere in the mountain.

3 SECTION AND ANALYSIS

Structure of the principle room and that of the two side rooms are nearly the same, all has six beams and five bujias, four columns, two extended eaves and the distance between beams is 1 m. The columns are connected by floor beam. The height of hips is 5.1 m and the height of eaves is 2.64 m, the same as the eaves of the side rooms. The roof grade of the principle room is about 60%, comparatively steeper than the others, while that of the side rooms are 47.5%, comparatively smoother. So the roof grade is ready to handle with frequent and abundant rainwater.

The side close to the three-section courtyard is two-cross cantilever out the door and reaches a distance of 2.5 m, providing enough space for activities. However, the structure is not good, as the bearing element stands too much weight in sacrifice to emphasize the outer space. Therefore, the inhabitants have to add brick masonry column under the roof as a support to avoid house collapse.

There's a row of beams on the height of eaves in the section, realizing flexible separation between the two floors. This construction is often seen in vernacular dwellings located in the southern part of Sichuan and residents can build or remove the floor according to their need, thus realizing flexibility in the use of spaces [5].

The clear height of the bedroom is 2.3 m. Rooms that do not need dampproof like central room and side rooms are a little bit higher, with about 2.48 m in height. The difference is caused by the elevation of floor of the bedroom made by the stone beam (figure 6).

The elevation scale of vernacular dwellings is very graceful. The roof made of Chinese-style tile occupies a larger proportion. The walls are mostly made of wood. Because the eaves extend out a large distance and the house has some open spaces like central room, the elevation is in the shadow in most time of the year. The dwelling is particularly in need of good ventilation because it is very moist in mountain areas and there's no enclosure on the side face of the upper attic, thus the air can circulate freely. Bamboo wall is not commonly seen in Sichuan vernacular dwellings. Walls are mostly made of wood plank, which is simple and natural [1] (figure 7).

4 DETAILS

The vividness of the roof art depends greatly on the decoration of ridge. In the study about the vernacular

Figure 8. The style of laying tiles.

Figure 11. Suspension.

Figure 9. Concave.

Figure 12. Window style.

Figure 10. Overhanging jack.

dwelling, laying tiles in the middle of the ridge belongs to the typical middle-flower decoration of Sichuan vernacular dwellings [4] (figure 8). At the end of the ridge the tiles are made in concave, which is very vivid (figure 9). With no extra for overhanging eaves, the supporting object overhangs right on the overhanging jack. The dwelling overhangs on both sides with no left-hanging and is maintained by fence where different architrave and can be seen. The suspension is fixed right on the overhanging jack (figure 10). This kind of decoration of the dwelling is very common in Sichuan areas. It not only ensures the basic need of people, but also beautifies the dwelling. The suspension down the eaves is very simple, imitating the shape of lantern and representing a happy atmosphere. The work represents the love of happy life of toilers (figure 11) Windows and doors are made in the pattern of turtle back (figure 12). On the meeting between back ridge and cornice, pumpkin-like decoration are made, which is a picture of simplicity of local people (figure 13). Wood planks are paved in the rooms which call for dampproof and under the wood plank scattered stones are put in as supports. Holes can be seen from behind the house, which is good for ventilation and dampproof (figure 14). To beautify the front elevation, strip-type stone girders are paved closely and only several small holes can be seen. Plinths are made by earthly square stone piers, with a little decoration added to the plinths under the eaves of central room.

Figure 13. The detail of the connection.

Figure 14. The detail of the floor.

5 CONCLUSION

Compared to town areas in Sichuan, the mountainous areas are most used for agricultural production. Under the background of a small agricultural prosperity, farmers choose to live separately; therefore, the vernacular dwellings in mountainous areas are usually located here and there. The design of the vernacular dwellings is well adapted to natural environment. It is suitable for the agricultural production, as well as adapting to the land tense of the mountainous area. The function plan is most of three-section design, including the colorful outer space which is built by courtyard, wide eave gallery and central room. The vernacular dwellings in Mount Emei are more focused on the building of outer space than other common dwellings [8].

Chuandou structure is simple and appropriate, using small girder to build the variability of the space. Wood is used on the elevation. The attic elevation is open, so as to adapt to the moist and rainy mountain climate. The vernacular structure and decoration is not heavy and complicated, reflecting the taste of natural, simple, and close to life. The Emei's mountainous vernacular dwellings are simple and decent, as well as unique and mature, reflecting strong emotion bond between pragmatism and naturalism.

Now researches on Sichuan vernacular dwellings are more focused on important buildings and towns. These buildings are big, delicate and of high value. However, little attention is paid on small buildings scattered in the mountain like Emei's mountainous vernacular dwellings. As primitive vernacular dwelling in Mount Emei, the dwellings can reflect local lifestyle and environment feature most. Today, the society is rapidly changing, and a great number of buildings are being built. The dwelling scattered in mountain are disappearing as time passes by, facing the fact of being torn down or abandoned. Therefore, the importance of research on the dwellings cannot be underestimated.

REFERENCES

[1] Menggang LI. 2009.00. A Brief Discussion About The Features of Sichuan Vernacular Dwellings. [J]
[2] Mingjia TANG. 2007.02. "The Holy Land Emei"—The Treasure of Ancient Architectural Complex. [J]
[3] Qiang LI, Keisuke KITAGAWA. 2005.05. The Classification and Distribution of The Arch Member of Construction System in Eurasia. [J]
[4] Survey and Design Institute of Sichuan Province. 1996. Sichuan Vernacular Dwellings [M]
[5] Shigeru WAKAYAMA, Kenichi KATAOKA. 2005.07. The Composition and Decoration of The Intermediate Parts Between Vertical and Horizontal Members—Classification and Distribution of Building Construction in The Euriasian Region. [J]
[6] Xiao HUANG. 2008.06. The Geomantic Study of Bayu's Traditional Vernacular Dwellings. [J]
[7] Xiankui LI. 2009. Sichuan Vernacular Dwellings. [M]
[8] Zhaoxia WANG. 2004.S1. The Regional Features and Formation of Traditional Vernacular Dwellings in Sichuan Basin. [J]

Study on space structure change of land used for carbon source and carbon sink in urban fringe areas

Menglin Qin, Hang Li, Jing Ya, Jing Zhao & Weichu Chen
Department of Architecture and Urban Planning, Civil Engineering and Architectural School of Guangxi University, Nanning, China

ABSTRACT: Urban fringe areas play an essential role in balancing carbon emission and carbon sink of urban regions. Between 1990 and 2013, the space structure of the land used for carbon sink in urban fringe areas of Nanning City has changed greatly by reducing by 534 km^2, among which, gardening plots, woodlands, cultivated lands, unused lands and water areas have been reduced by −87,484, 109,198, −11,599, 21,871 and 214,44 hectares, respectively. During the three periods of time, i.e. from 1990 to 2000, from 2001 to 2006 and from 2007 to 2013, the increment/decrement, the annual average occupation ratio and the increasing/decreasing rate of the land for carbon sink are −88.75 ha^2/year, −2,592.57 ha^2/year and −5,271.11 ha^2/year; 89.99%, 88.70%, 84.62%; and −0.15%, −2.42%, -5.74%, respectively. The occupation ratios by land for carbon sink in 1990, 2000, 2006 and 2013 are respectively 90.07%, 89.91%, 87.49%, and 81.74%, dropped by 18.33% in total from 1990 to 2013.

Keywords: fringe area, land used for carbon source and carbon sink, space structure, low-carbonizatio

1 INTRODUCTION

Development of low carbon cities is one of the significant countermeasures in dealing with global climate change, while increasing the capacity for carbon sink is a core pathway for development of low carbon cities. Urban fringe areas play an essential role in balancing carbon emission and carbon sink of urban regions, at where shift between carbon emission and carbon sink is realized in a concentrated manner. Therefore, studying the space structure of the land used for carbon source and carbon sink in urban fringe areas and investigating its changing characteristics will be helpful to provide theoretical basis for low-carbonized planning and design for urban fringe areas and realize the development of low-carbon cities. At present, the studies concerning space structure change of land used for carbon source and carbon sink in urban fringe areas are proceeding from the aspects of land use and land cover. Especially since the launch of the "International Geosphere and Biosphere Plan" (IGBP)[1] and the "International Human Dimensions Programme on Global Environmental Change" (IHDP) in 1995, as well as the "Land-utilization and Land-cover Change" (LUCC) proposed by UN, a lot of scientific achievements[2–3] have been obtained. With the help of GIS and RS technologies, by making use of multi-period remote sensing images and detailed surveyed data of land utilization, from the dimensions of such urban fringe areas, municipal administrative regions and regions etc., the changing characteristics, modes and momentum of the space structure of land-use and land-cover have been studied, and systematic study results[4–7] have been acquired on such respects as urbanization level, development of cultivated acreage, conversion efficiency of cultivated land and driving factors for cultivated land conversion etc. On the aspect of the influence on the ecological environment, associated studies indicate that under the comprehensive action of multiple factors, urban land-use and land-cover would bring about associated environmental changes. It is believed that urbanization will bring out different levels of increase in surface runoff[8] in the expanded areas, the expansion of different administrative units during the process of urbanization will produce varied influences on soil diversity, and the expansion of urban areas will deteriorate the overall environmental quality of soil, especially the pollution of heavy metals[9–12]. In the meantime, studies also show that the change of land-utilization brought out by urbanization has negative impacts on the ecological functions of plantation and regional flood carrying capacity[13–14].

The above-mentioned studies investigated the changing characteristics, modes and momentum of land-use and land-cover, neglecting the fact that keeping the balance between the lands for carbon source and carbon sink is also important in guiding the low-carbon development of cities. This paper studied the space structure change of lands used for carbon source and carbon sink in urban fringe areas, hoping to acquire some essential changing characteristics of it.

Table 1. Categorization and Constitution of the Land Used for Carbon Source and Carbon Sink[1].

	Category	Sub-category	Implications
Land used for carbon source and carbon sink	Land used for carbon source	Urban construction land	Including lands used for construction of urban and rural residential areas, regional public facilities, land for special purposes, land for mining activities and other construction land
		Land for regional	Including lands for such regional communication and transport facilities and their subsidiary facilities as railways, highways, ports, airports and pipeline transportation, etc.
	Land used for carbon sink	Water areas	Including rivers, lakes, reservoirs, ponds, irrigation channels, intertidal zones, glaciers and permanent snow covers.
		Plowland	Including paddy fields, irrigated lands and dry lands
		Gardening plot	Including gardening plots, tea gardens and other gardening land
		Woodland	Including woodlands, shrublands and other forestlands
		Grassplot	Including natural pastures, artificial pastures, and other grasslands
		Greenbelt	Including greenbelts in parks and green buffers
		Other non-construction lands	Including such lands as idle lands, saline and alkaline lands, sandy lands, bare lands and lands not used for husbandry purpose etc., and lands for agricultural facilities such as rural roads and raised paths through farming fields etc.

[1] 1. Table 1 "Categorization and Constitution of the Land Used for Carbon Emission and Carbon Sink" is prepared by studying and integrating two standards – *Categorization of Urban Land and Land Standard for Planning and Construction* and *Categorization of Current Status of Land-utilization*. The land used for carbon emission is mainly based on the former while the land used for carbon sink on the latter.
2. The "Lands for regional public facilities", "Lands for special purpose" and "other lands for construction" in the *Categorization of Urban Land and Land Standard for Planning and Construction* are classified into the "construction land for urban and rural residential areas"; and the "Park green space and protective green belts" in the "Lands for green belts and squares" of urban construction lands is moved into the "Green belts" item under the land used for carbon sink, i.e. one sub-category of "green belts" is created;
3. The detailed category of agriculture and forestry in the *Categorization of Current Status of Land-utilization* (GB/T21010-2007) is raised as a sub-category, i.e. the cultivated lands, gardening plots and grassplots are taken as a sub-category; the cultivated lands include "Paddy Fields", "Irrigable Lands" and "Dry Lands", in which "Such Cultivated Lands for Facilities, Raised Paths Through Fields and Rural Roads etc." are reckoned into Other Non-construction Lands.

2 DEFINITION OF ASSOCIATED CONCEPTS

2.1 *Urban fringe areas*

Urban fringe areas are such zones at where the features such as land utilization, social and population etc. are undergoing changes. They are located in the transition areas between continuous built-up areas, suburbs and pure agricultural land where there are almost no non-farming houses, non-agricultural land-occupation and land-utilization[15]. It is also considered that urban fringe areas are an important constituent part of the regional structure of a city, the transitional zones from the urban environment to a rural environment, as well as the most complicated and impermanent zones during the urban and rural construction process[16]. According to the definition of urban fringe area, the spatial range of it may be divided into two levels: one is the urban-rural ecotone, i.e. the urban-rural junction part, and the other one is the backland of urban development, i.e. those regions that can be driven or radiated on by the urban development, including the inner fringe areas, outer fringe areas and rural backlands. Based on the research requirements and data availability, the range of this study is confined to the fringe spatial zones of built-up areas of city development, i.e. those zones that may be controlled and utilized by the city, or those fringe zones at the city outskirts that may be driven or radiated on by the city, including inner fringe areas, outer fringe areas, urban shadowed zones and rural backlands.

2.2 *Land used for carbon source and carbon sink*

The land used for carbon source and carbon sink covers all the urban-rural land, comprising the land used for carbon source and the land used for carbon sink. The land used for carbon source refers to the land providing activity places to human beings and releasing carbon dioxide to the atmosphere through human activities, namely those lands that are not playing the role of carbon sink, but may be supplied to human beings for their carbon source activities. It includes two sub-categories land used for urban construction and regional communication facilities. The land used for carbon sink refers to those lands where the carbon dioxide in the atmosphere is converted into carbohydrate through photosynthesis and fixed inside plants or soil in the form of organic carbon. It includes seven sub-categories: lands used for water areas, cultivated lands, gardening plots, woodlands, grassplots, greenbelts and non-construction lands and so on. See Table 1 for the detailed categorization and constitution of land used for carbon source and carbon sink.

2.3 Related research indexes

(1) Increment/decrement of carbon sink land (S, in the unit of hectare/year): refers to the acreage increased or decreased within one year, and the equation is: $S = (CS_{end} - CS_{begin})/Year$. If $(CS_{end} - CS_{begin})$ is positive, S denotes an increment in carbon sink land and if negative, a decrement. The increment or decrement of carbon sink land expresses the annual increasing or decreasing velocity of carbon sink land, the bigger the absolute value of which is, the faster the absolute increasing or decreasing speed is and the larger its influence on the lands used for carbon source and sink is.

(2) Occupation ratio of carbon sink land (R, in the unit of %): refers to the percentage of carbon sink land (CS) against the total land acreage used for carbon source and carbon sink (CES), and the equation is: $R = CS/CES * 100$. This equation expresses the overall occupation volume of carbon sink land. The bigger this number is, the more favorable to the balance of urban carbon emission and sink.

Increasing/decreasing rate of carbon sink land (I, in the unit of %): refers to the ratio of increment/decrement of carbon sink land against the total land acreage used for carbon source and carbon sink, and the equation is: $I = (CS_{end} - CS_{begin})/CES4 * 100$. If the value of $(CS_{end} - CS_{begin})$ is positive, I denotes increase of carbon sink land and if negative, decrease of carbon sink land. I expresses the relationship between the increased or decreased amount of carbon sink land and the total amount of lands used for carbon source and sink. The bigger its absolute value is, the heavier it's impact on the balance of urban carbon source and sink is.

3 ANALYSIS ON CHANGING CHARACTERISTICS OF SPACE STRUCTURE OF LAND USED FOR CARBON SOURCE AND CARBON SINK

3.1 Study area and method

3.1.1 Study area
This paper selects Nanning as a region for study. Nanning, the capital of Guangxi, is a second-tier city in China. The population in its central districts in 2013 is about 2.74 million, the GDP per capita is CNY 35,138 and the urbanization rate is 55%. It has large potential in economic development with fast speed of urbanization. Therefore, we choose Nanning as the representative for study on small cities. Based on the research requirements and data availability, the areas to be studied is confined to the urban area of Nanning with an acreage of 6,422.3 km^2, whose length in the east-west and south-north direction is 96 km^2 and 85 km^2, respectively.

3.1.2 Study approach
This study adopts TM images as information source to obtain information about carbon source and carbon

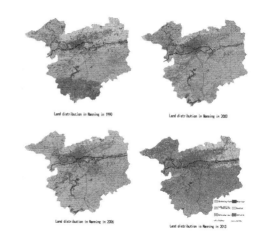

Figure 1. Spatial Distribution of Lands Used for Carbon Emission and Carbon Sink in the Four Periods of Time from 1990 to 2013 *Study approach*.

sink land. The data of space used by the researched area comes from the same seasonal aspects (from September to December) of remote sensing images of Nanning City in 1990, 2000, 2006 and 2013, with an image resolution of 30 m. Erdas software is used to extract the distribution data of carbon source and carbon sink land in different periods of time for supervision and categorization purpose. After interpretation, spatial interpretation figure and structural data of carbon source and carbon sink land of the four periods of time are shown in Figure 1 and Table 2. Herein, because Nanning's grasslands are less due to its location in the South and so are the greenbelts in the city, which are also difficult to distinguish, this study omits grasslands and greenbelts and will not take them into statistics.

3.2 Analysis on changing characteristics of space structure

Incessant decline in carbon sink land is one of the most prominent changing characteristics of space structure of lands used for carbon source and carbon sink, an analysis is conducted from three study indexes, i.e. the occupation ratio of carbon sink land (S), the increment/decrement of carbon sink land (R) and the increasing/decreasing rate of carbon sink land (I). See Table 3 for the results of the analysis.

3.2.1 Analysis on increment/decrement of carbon sink land (S)
During the three periods of time of 1990~2000, 2001~2006 and 2007~2013, the increment/decrement of carbon sink land is −88.75 ha^2/year, −2,592.57 ha^2/year and −5,271.11 ha^2/year, respectively. That is to say, the change of increment/decrement of carbon sink land is very little from 1990~2000. However, after 2001, the carbon sink land began to reduce drastically by about 26 square kilometers per year, and it had reached its

Table 2. Change in Carbon Emission and Carbon Sink Land during the Four Periods of Time (unit: hectare).

Land used for carbon sink and emission (unit: ha^2)		Year			
		1990	2000	2006	2013
Carbon sink	Woodlands	169621.47	116137.16	93880.95	60423.41
	Gardening fields	112139.44	80494.00	245067.08	199623.05
	Cultivated lands	191783.54	321764.53	200333.22	203382.99
	Land unfit for construction	74727.05	39432.06	73905.87	52856.55
	Water areas	30158.52	19625.97	15108.70	8714.54
	Total	578430.02	577453.72	561898.28	525000.54
Carbon emission	Construction	59972.69	59918.75	73905.87	105507.19
	Communication	3828.04	4858.28	6426.60	11723.02
Total		63800.73	64777.03	80332.47	117230.21
Total		642230.75			

Table 3. Analysis on the Changes of Carbon Sink Land.

Analytical Indexes	Year		
	1990–2000	2001–2006	2006–2013
Increment/decrement (S, ha^2/year)	−88.75	−2592.57	−5271.11
Annual average occupation ratio (R, %)	89.99	88.70	84.62
Increasing/decreasing rate (I, %)	−0.15	−2.42	−5.74

climax in 2006~2013, with an annual reduction of 53 kilometers. On the whole, the increment/decrement of carbon sink land in Nanning is not balanced, which are reduced by about 156 square kilometers in 2001~2006 and 369 square kilometers in 2007~2013.

3.2.2 Analysis on occupation ratio of carbon sink lands (R)

Viewing from the periods of time, during the three periods of time of 1990~2000, 2001~2006 and 2007~2013, the occupation ratio of carbon sink land is 89.99%, 88.70%, and 84.62% respectively. Viewing from the point of time scale, the occupation ratio of carbon sink land in 1990, 2000, 2006 and 2013 is 90.07%, 89.91%, 87.49% and 81.74%, respectively. From the end of the last century to the beginning of this century, the carbon sink land didn't change much. The main causes are that in this time, the city did not begin its large-scale construction and the tide of construction of new rural areas did not begin yet. That is why the carbon sink land had not been reduced by a significant amount. Yet to the end of the first ten years of this century, the urban construction began to meet its climax while the construction of new rural areas was also carried out with a large scale. That is why the average occupation ratio of carbon sink land was reduced to 88.70% from 89.99%, dropped by 2.22 percentage points in total. During 1990~2006, the change of carbon sink land was still quite little but in the years of 2006~2013, the carbon sink land was reduced drastically. And during this period of time, large-scale construction of urban and new rural areas was conducted in Nanning City. The occupation ratio of carbon sink land declined to 81.74% in 2013 from 87.49% in 2006, totally decreased by 5.75 percentage points.

3.2.3 Analysis on increasing/decreasing rate of carbon sink land (I)

During the three periods of time of 1990~2000, 2001~2006 and 2007~2013, the increasing/decreasing rate of carbon sink land is −0.15%, −2.42% and −5.74%, respectively, indicating little change during 1990~2000, but large ones during 2001~2006 and 2007 ~ 2013. As a whole, compared to the total amount of lands used for carbon source and carbon sink, its increasing/decreasing rate is not very large. But viewing from the point of balance of carbon source and sink of the city as a whole, the carbon source has enormously exceeded the capacity of carbon sink. This is because that on one hand, the lands used for carbon sink is declining; on the other hand, a large amount of carbon has been emitted due to the industrial development of the city. So the key cause of the carbon imbalance in the urban areas still lies in the industrial development of the city instead of the decline in the capacity of carbon sink. We should note that the carbon sink land always possesses the capacity of carbon sink and is very hard to be recovered once disappearing. Therefore, they are a huge treasure to be cherished during the low-carbon development of the city.

4 CONCLUSIONS

The following preliminary conclusions are obtained by this study: (1) By studying the change in space structure of urban fringe areas based on the lands used for carbon source and carbon sink, a new sight of research for design and planning of urban low-carbonized spaces has been found. (2) The study indexes – Occupation ratio of carbon sink land, Increment/decrement of carbon sink land and Increasing/decreasing rate of carbon sink land have been created, providing a new pathway for the study on the changing characteristics of space structure of carbon source and sink at urban fringe areas. In the meantime, certain problems still exist in the study, mainly including: 1. In the process of supervised classification, the used satellite remote sensing images are of low resolution, and limitation of other supporting materials has caused certain errors in the accuracy of categorization; 2. In categorization, for simplifying the categorization and due to the actual conditions of Nanning, the acreage of grassplots and greenbelts were not taken into statistics; 3. To be more accurate, the datum of more years should be taken into statistics. In the future, in-depth study on the mode and momentum mechanism of space structure change will be done in order to acquire associated theoretical basis for the plan and design of low-carbonized spaces.

ACKNOWLEDGEMENTS

This work was financially supported by the National Natural Science Foundation of China (51208119), Natural Science Foundation of Guangxi (2013GXNSFBA019240), Foundation of Guangxi Educational Committee (2013YB010), and Natural Science Foundation of Guangxi University (XBZ120394).

REFERENCES

[1] Turner I. BL, Skole D., Sanderson S. et al. Land-use and Land Cover Change Science/Research Plan, IGBPR Report No.35 and HDP Report No.7. Stochkholm: IGBP, 1995.
[2] Li Xiubin. The Kernel Field of the Research on Global Environmental Change: International Research Tendency of Land Use/Cover [J]. 51(6): 553–558, 1996, *The Geographical Journal*.
[3] Yan Xiaopei, Mao Jiangxing, Pu Jun. Analysis on Humanity Factors in the Change of Land-utilization at Mega-city Regions: Pearl River Delta Taken as an Example. 61(6):613–623, 2006, *The Geographical Journal*.
[4] Yu Bohua, Lue Changhe. Spatial-temporal Features of the Change of Cultivated Land Acreage at Urban Fringe Areas and the Driving Mechanism: Taking Shunyi District of Beijing City as an Example. 28(3): 348–353, 2008, *Geoscience*.
[5] Luan Weixin, Wang Maojun. Study on the Relationship between Elevation of Urbanization Level and Change of Acregae of Cultivated Land: Empirical Study on Dalian City. 22(2):208–212, 2002, *Geoscience*.
[6] Bian Xuefang, Wu Qun, Liu Weina. Associated Analysis on Urbanization and Land-use Structure of Cities in China. 27(3):73–78, 2005, *Resource Science*.
[7] Wei Suqiong, Chen Jianfei. Comparative Study on Cultivated Land Conversion and Its Correlation Factors for Fujian and Taiwan [J]. 19(5): 568–576, 2004, *Journal of Natural Resources*.
[8] Bian Xuefang, Wu Qun, Liu Weina. Associated Analysis on Urbanization and Land-use Structure of Cities in China. 27(3):73–78, 2005, *Resource Science*.
[9] Zheng Jing, Yuan Yi, Feng Wenli et al. Simulation Study on the Influences of Change of Land-utilization on the Depths of Surface Runoffs: Taking Shenzhen Region as an Example. 14(6): 77–82, 2005, *Journal of Natural Disaster*.
[10] Sun Yanci, Zhang Xuelei, Cheng Xunqiang et al. Grey Correlation Analysis on the Influences of Urbanization on Soil Diversity in Nanjing Region. 61(3): 311–31, 2006, *The Journal of Geographical Science*.
[11] Li Guilin, Chen Jie, Tan Manzhi. Spatial-temporal Features of Influences of Non-agricultural Land Expansion of Suzhou City on Soil Resource. 23(4): 674–684, 2008, *Journal of Natural Resources*.
[12] Li Ling, Lu Jie, Feng Xinwei et al. Change of Cultivated Land Resource and Soil Environment during the Urbanization Process of Zhengzhou City. 30(6): 949–954, 2008, *Resource Science*.
[13] Chen Yujuan, Guan Dongsheng, PEATM R. Study on the Influences of Fast Urbanization in Pearl River Delta on the Carbon Sink and Oxygen Release Capacity of Regional Vegetation. 45(1): 98–102, 2006, *Journal of Sun Yat-Sen University: Natural Science Edition*.
[14] Wu Yunjin, Zhang Ganlin, Zhao Yuguo et al. Study on the Influences of Land-utilization Change in the Urbanization Process on Regional Storage Capacity of Flood Detention: Taking Hexi Region of Nanjing City as Example. 28(1) 30–34, 2008, *Geoscience*.
[15] PRYOR R T. Defining the Rural-urban Fringe [J]. *Social Forces*, 1968(47).
[16] Cui Gonghao, Wu Jin. Features of Space structure and Its Development of Fringe Areas of Cities in China: Taking Cities as Nanjing etc. as Examples [J]. 45(4):400–411, 1990, *Journal of Geographical Science*.

Study on the mechanical performance of bridges affected by the creep of concrete in bridge widening

Qikun Zhang
Tianjin Railway Technical and Vocational College, Tianjin, China

Guangyin Zhang & Maoqi Li
Tianjin Municipal Engineering Research Institute, Tianjin, China

ABSTRACT: In order to study the bridge widening, the force performance of the creep of concrete between the old bridge and new bridges, back of the project of a highway bridge widening, analysis through modeling model by the finite element analysis software ANSYS, the bridge structure have basically the same deformation trend in the conditions of the bridges had built and creep of concrete, due to the creep of concrete action in the connection between the new bridge and old, the deformation of the new bridge section in vertical, lateral and longitudinal direction will occur, but the amount of deformation within the allowable range.

1 INTRODUCTION

With modern economic development, traffic volume has been increasing rapidly. Some old bridges can no longer meet the requirements of modern transport due to their low design capacities and over-use (Hao Wang & Ruoxi Zhu. 2013, Qidi Guo. 2006, Yi Hao. 2007), which requires immediate improvement. To widen the original bridges is a fast and low-cost way to promote traffic capacity (Xiaoni Gao. 2009, Xinyun Xiang. 2008, Zhiqiang Xu. 2005). Many bridge researchers have made a lot of design-related researches, and gained a lot of achievements. But research on concrete creep force between new bridges and old ones rarely appeared. In this paper, we have made a study on concrete creep between new bridges and old ones and drawn some conclusions that may serve as references for other researchers.

2 OVERVIEW OF THE BRIDGE

The bridge intersects with the road and the angle is 73°, the span arrangement of the bridge is 5 × 20 + 16 + 25 + 10 × 20 m. The bridge width is 12.5 m. Upper structure is pre-stressed T beam; the height of the beam is 1.3 m. The old bridge has 5 beams. The substructure of the bridge is double-column piers and drilling piles.

Remain the middle beams and the inner edge beams; Replace the outer edge beam of old bridge for the new beam, while the new bridge was built outside edge beam. There is a rigid connection between the old and new T beam's cantilevered, but the substructure with no connection.

Table 1. Load calculate table.

No	Condition	Load condition
1	Working condition	Structural weight + prestressed (New bridge prestress discount by 10%, Old bridge discount by 20%) + Deck and baluster weight + Car Loads (Highway load − I).
2	Creep condition	Calculate the creep of the No 5, No 6, No 7 beams of the new bridge and the connection between the new and the old bridge in the condition of all the loads.

3 MODELING AND ANALYSIS

We chose the 6th Cross as the research object, using the finite element solid model to analysis the creep. Finite element model of the concrete unit uses SOLID45 element simulation, steel unit with LINK8 element simulation. In order to facilitate the study of the role of creep on the structure, set two calculation condition, the first condition is the bridge have built; the second condition is creep after three years after built. Load calculation in Table 1.

4 DATA ANALYSIS

4.1 *Structural deformation analysis*

Structural deformation is the major force affecting the bridge forms; we carried out the vertical, longitudinal and lateral deformation analysis.

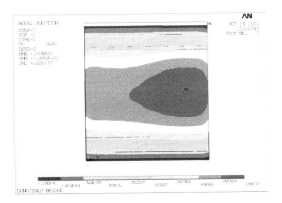

Figure 1. Vertical deformation in working condition.

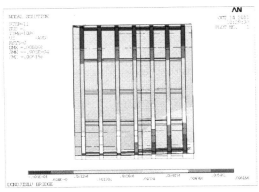

Figure 3. Longitudinal deformation in creep conditions.

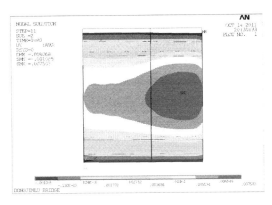

Figure 2. Vertical deformation in creep condition.

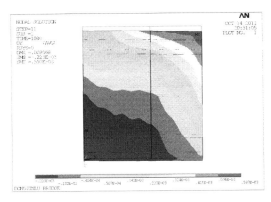

Figure 4. Lateral deformation in creep condition.

4.1.1 Vertical deformation analysis

In the have working conditions, the vertical deformation of the bridge shown in Figure 1; in creep conditions, the vertical deformation of the bridge shown in Figure 2. The figure on the right side is the new bridge.

From the Figure 1 and Figure 2 we can see that the red part of the deformation value is positive, the structure up bending under the two conditions. The up bending of the new bridge is larger because of the new T beam pre-stressing decrease very less, but the pre-stressed of the old T-beams have lost a lot after long-term used. Comparisons between the two, the up bending of the new bridge become larger and larger as the role of concrete creep of new bridge which continues up bending. The largest up bending of the new bridge in creep condition is 7.5 mm, 1.1 mm larger than in working conditions, but increase by only 18%, this is because the new bridge bound by the old bridge, unable to up bending freely. 1.1 mm deformation poor little impact on traffic.

4.1.2 Analysis of longitudinal deformation

The longitudinal deformation of the bridge in creep conditions as shown in Figure 3.

From Figure 3 we can see that the longitudinal length of structure becomes shorter because of up bending, the longitudinal deformation of the new bridge is larger than the old bridge which has the same trend to the vertical deformation. The maximum longitudinal displacement of the new bridge's free end is 6.16 mm in creep conditions, by 24% larger than in working conditions. The deformation of the edge beam which near the old bridge is significantly greater than not near the old bridge, the binding effect of old bridge as far as farer as smaller. But the longitudinal deformation value of the bridge still in the level of acceptable range.

4.1.3 Analysis of lateral deformation

The lateral deformation of the bridge in creep conditions as shown in Figure 4.

When a new main beam arch bridge bound by the old bridge, its free end occurs a slight lateral displacement is about 0.6 mm, little impact on the structure. On working conditions is 0.3 mm.

4.2 Analysis of longitudinal stress

In accordance with the structure theory, it only produces deformation but not produces secondary forces

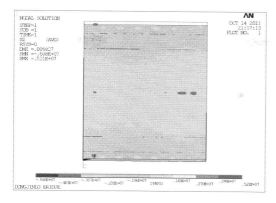

Figure 5. The longitudinal stress of bridge deck in working conditions.

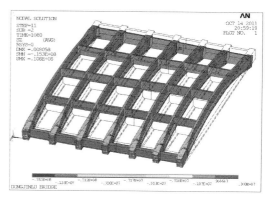

Figure 7. Dong-Jin Bridge creep conditions lattice girder longitudinal stress diagram.

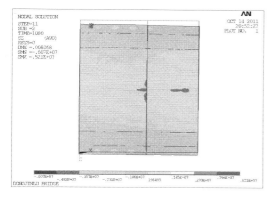

Figure 6. The longitudinal stress of bridge deck in concrete creep condition.

4.2.2 T beam web and diaphragm

In creep conditions, grillage (T beam web and diaphragm) longitudinal stress situation shown in Figure 7.

Form the Figure 7 we can see that the force of each piece beam of the bridge is uniform, the maximum compressive stress at the bottom of T-beam less than 11 MPa, the compressive stress of the web less than the bottom. The diaphragm basic does not participate in the longitudinal force, there is does not appear longitudinal tension area except the anchorage zone of the beam end. The longitudinal stress figures of grillage same as to the working condition. The longitudinal stress value of the grillage changes are not obvious in creep condition same as the roof.

5 CONCLUSION

By modeling calculations and data analysis we can get the following conclusions:

- First, from the results we can see that the deformation the bridge in working condition same as to in creep condition, the results same as to the theory of creep;
- Second, due to the effect of connection concrete creep, in the vertical, lateral and longitudinal of the new bridge will take place deformation, but the value of deformation in the allowable range.
- Third, as the role of the connection of the upper part of the old and new parts, the old bridge will play the role of restraints to the binding deformation of the new bridge, the will become less and less with the distance from the old bridge to the new bridge increase.

for simple structure as the role of concrete creep. But only partially present effect to the new bridge as the role of concrete creep. Taking into account the impact of this uneven effect, it is necessary to examine creep in the longitudinal stress.

4.2.1 Bridge deck

The longitudinal stress of the bridge decks in working condition is shown in Figure 5, the longitudinal stress of the bridge deck in concrete creep conditions is shown in Figure 6.

From Figure 5 and Figure 6 we can see that, the whole deck in stress compression condition and the stress value mostly less than 3.5 MPa, the maximum compressive stress appears in the middle section of the bridge, the slight tensile stress appears at the end of the bridge. There is no significant displacement of the bridge deck as the reason of the new bridge concrete creep. Overall, it has little effect to longitudinal stress of the bridge deck as the role of concrete creep, force distribute uniform.

REFERENCES

Hao Wang & Ruoxi Zhu. 2013. The study of connection stiffness effect to lateral load distribution for widening bridge. *Highway*: 66–71.

Qidi Guo. 2006. The widening theory and application of concrete bridge [D]. *Huazhong teleology University master thesis*: 10–14. Wuhan: China.

Xiaoni Gao. 2009. The study of interaction of bridge in widening. *Chang'an University master thesis*: 24–42. Chang'an: China.

Xinyun Xiang. 2008. The study of creep and shrinkage effect to the new and old bridge in high way widens. *Beijing Jiaotong University master thesis*: 25–33. Beijing: China.

Yi Hao. 2007. The study of widen technology in reinforced concrete arch bridge. *Chang'an University Master Thesis*: 30–40. Chang'an: China.

Zhiqiang Xu. 2005. The study of widening technology in T beam highway, *Southeast University Master Thesis*: 15–19. Nanjing: China.

Study on the traffic prediction of Dezhou-yu'e province boundaries highway

Minglei Li & Hongbin Liu
Liaoning University of Science and Technology, Liaoning, China

ABSTRACT: In view of the problems existing in the traffic volume prediction, this paper used Foxpro software package and combined with "series and parallel rules" to check the original traffic data of the case, optimized predictions of economic and traffic occurrence, calculated the induced and transfer traffic volume prediction, simulated and analyzed the traffic volume flow direction of interchange bridges in order to optimize the existing prediction methods. This optimization method in the analysis of traffic volume prediction for the future has high precision and strong applicability.

Keywords: highway, traffic volume prediction, optimization methods

1 ENGINEERING BACKGROUND

Dezhou-yu'e Province Boundaries highway is located in Dengzhou, Henan province, between the Sanxi highway and the Er'guang highway. It is the route that will become a new channel between Henan and Hubei Province. The highway starts from the east of Zhaolou village, Pengqiao town, Dengzhou (K0+000). It is connected with the K51+043 point of the Neixiang-Dengzhou highway by Zhaolou interchange bridge. The route, to the west, passes by Xiaoli hill, Qigan temple, crossing Yindan main canal at the north of Pengqiao town, then, to the southwest, crosses Liushan, Sunlin hill, ending in the west of Zhai village, Pengqiao town, Yu'e provincial boundaries (K13+147), and connecting the Laohekou to Gucheng section of Laoyi highway, Hubei province. The total length of the route is 13.147 kilometers.

2 ROAD NETWORK DATA PROCESSING OPTIMIZATION

Based on Origin and Destination (OD) survey and historical traffic data, the paper analyzes and processes the original data. After encoding, data the Foxpro software package was employed to analyze and check the OD original data repeatedly (Dongmei Peng. 2013). The main work process is shown in Fig. 1.

OD investigation is a gap survey, which is a sampling survey on specific date. However, traffic volume prediction is based on average annual daily traffic volume. Hence, it is necessary to revise and check the OD data.

The summary in the process uses "series and parallel rules" (Jingfeng Chen. 2008) namely:

Series rule: the same path in traffic survey, when appear more data, take larger values;

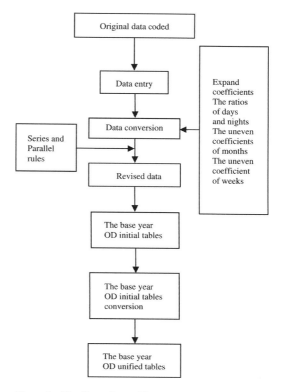

Figure 1. The flow of compiling OD tables.

Table 1. The results of OD check.

No.	The section position	The observations in 2012	Distribution of value of OD data	Errors
1	S231 Xiaodian in Dengzhou	3372	3547	5.19%
2	S249 Miaogou in Dengzhou	7043	7391	4.95%
3	S335 Kafang in Xichuan	2543	2416	−5.00%
4	G207 Xinzhuang in Dengzhou	5362	5571	3.90%
5	G209 Xiping in Xixia	4864	4641	−4.58%

Parallel rule: for many roads' traffic investigation, take the sum.

To ensure the accuracy of survey data, this paper makes use of the traffic volume of historical data, amplifies the OD data of high grade highways, which according to the volume, is uneven coefficients of amplification of the passages, and then removes the traffic volume unrelated to the project area, repeatedly contrasting and adjusting the data between the OD data of affected area in the provinces, then distributes the base year's OD data of affected areas which are adjusted to the present highway network, calculating the passage traffic volume of each sections, checking with the actual investigation of traffic volume.

Due to the investigation in the process of human factors involved and other objective factors, it is difficult to guarantee that all the data are true and reliable. Therefore, it is necessary to check with the initial OD tables in order to control the error within a certain range (Yaohua Fu. 2008). Re-examination and analysis of the initial data can be used the following formula:

$$M'_{ij} = M_{ij} \cdot S_j / \left(\sum_i M_{ij} + \sum_i M_{ji} \right) = M_{ij} \cdot S_j / \left(2 \cdot \sum_i M_{ij} \right) \quad (1)$$

The formula: M_{ij}—the distribution traffic volume traffic of the communities i to j, according to the principle of equilibrium;

S_j The traffic survey all day of the community j.

M_{ij} Should satisfy the constraint $\sum_i X_{ij} = S_j / 2$.

Part of the data which is used in the formula (1) do not satisfy the constraints, and can adopt the formula (2) to revise again.

$$M'_{ij} = M_{ij} \cdot S_j / (2 \cdot \sum_i M_{ij}) \quad (2)$$

Through (1) and (2) modified repeatedly, the error narrows about 5% to conform the requirements. Then, the paper gets the modified OD tables, compares with the observations of the present. The results as shown in Tab. 1.

As seen in the table above, the errors between the observations and distributions of the relevant sections about within ±5%. Hence, the base year's OD data meet with the traffic volume of current network section, which is required the traffic volume prediction of this project.

3 THE FUTURE TRAFFIC VOLUME PREDICTION OPTIMIZATION

In term of economic forecasting, the paper adopts quadratic polynomial with three smooth index combination methods and combines with qualitative analysis to get the project's future economic growth rates of the areas. Quadratic polynomial method by using Excel software is simple and efficient (Wei Li. 2010). In view of the economic data in line with the characteristics of curve distribution, there are the high accuracy by using three smooth index (Yankun Zhang. 2011). So combining with both can accurately predict the future economic growths.

In term of traffic attract prediction, the paper studies the relationship of the economy and transportation, by combining with local conditions, modifying the elastic coefficients in the future, determining the passenger and freight elastic coefficients of the affected areas in order to obtain the occurrence attract traffic volume and strengthen the induced traffic volume prediction. Based on calculating the generate traffic volume, according to the distribution of OD survey for traffic volume and the present situation of time distance between the OD communities, the paper calculates separately for the two kinds of circumstances: one the regional traffic volume is zero, the other is not. Through multiple regression analysis, obtain the parameters of gravity model value (Hang Zhang et al. 2006), then use TransCAD software to calculate the induced traffic volume.

In view of the proportions of other modes of transportation, they are too small in affected areas, so the transfer traffic volume internal transfer is given priority. Base on the characteristic years' OD tables, the paper uses the logit model and TransCAD software to calculate the transfer traffic volume (Ai'hua Wang. 2011), then gets the unified characteristics years' OD tables.

In the traffic distribution, the paper uses equilibrium assignment model, according to the principle of Warddrop which is commonly used Beckmann model and adopts the objective function of minimum value method to solute. As the interchange bridges are important parts of the highway, TransCAD software is used for simulating the traffic volume flow direction of interchange bridges (Shan Yan et al. 2009). The traffic volume flow direction of Pengqiao and Zhaolou interchange bridges in 2016 are shown in Fig. 2 and Fig. 3.

Figure 2. The traffic volume flow direction of Zhaolou interchange bridges in 2016.

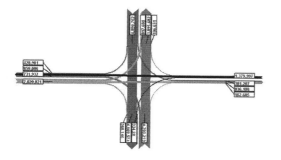

Figure 3. The traffic volume flow direction of Pengqiao interchange bridges in 2016.

Table 2. The error analysis of traffic growth rates.

The project	2016	2020	2025
Traditional prediction method for the four stages (%)	8.29%	7.14%	5.90%
Optimization method (%)	8.45%	7.34%	6.19%
Errors	0.16%	0.20%	0.29%

Table 3. The error analysis of traffic growth rates.

The project	2030	2035	2045
Traditional prediction method for the four stages (%)	4.71%	3.83%	2.94%
Optimization method (%)	5.04%	4.19%	3.48%
Errors	0.33%	0.36%	0.54%

4 THE ERRORS ANALYSIS OF THE TRAFFIC IN THE FUTURE

Through using optimization method for instance traffic prediction compared with the traditional prediction method for the four stages, the prediction errors are as shown in the following Tab. 2 and Tab. 3.

The tables show that the traffic volume of the project, which is applied to the prediction of the traditional four stage method and optimization method overall deviation, is not big. However, as time goes by, the various uncontrollable factors prediction will increase, the deviation will gradually increase (Guanglei Guo. 2011). On the other hand, optimization method to predict the future traffic growth rates is slightly larger than the traditional method of four stages, mainly because the optimization method strengthens the induced and transfer traffic volume prediction, adopts the qualitative analysis of local actual, and conform the laws of traffic volume development.

5 CONCLUSIONS

Through the project example of highway traffic volume predicting, the conclusions are as follow.

Checking and processing original data are very important. Traffic volume predicting involves a large amount of data, the prediction with reliable parameters can only be provided by accurately modifying and checking original data.

Selecting the model in the course of prediction is the key point. Everything has two sides, so it is significant to choose the right models and parameters according to the actual situations.

The qualitative and quantitative analysis must be unified. Traffic volume prediction involves many factors, only combining with the qualitative and quantitative analysis can get more accurate results.

REFERENCES

Ai'hua Wang, 2011. The transfer traffic volume prediction methods based on abstract model way. *Entrepreneurs*, 3(1): 61–62.

Dongmei Peng, 2013. The method of traffic volume flow direction interchange bridges. *Heilongjiang Traffic Science And Technology*, 7(2): 32–33.

Guanglei Guo, 2011. The discussed and analysed of the traffic volume forecast deviation. Traffic Engineering, 9(1): 60–63.

Hang Zhang & Ling Zhang, 2006. Based on the gravity model induced traffic volume prediction methods research. Highway Traffic of Science and Technology, 2(1): 111–113.

Jingfeng Chen, 2008. OD matrix estimation in the analysis of regional transportation applications. The Road of traffic, 7(4): 19–24.

Shan Yan & Ai'qiang Nan. 2009. TransCAD software application in the highway construction project traffic analysis and predicting. The Chinese and Foreign Road, 31(6): 273–276.

Wei Li, 2010. Computer statistical technology application in highway traffic volume elasticity coefficient of regression analysis. Qinghai Traffic of Science and technology, 5(2): 7–8.

Yankun Zhang, 2011. Regional economic analysis and local highway traffic volume forecasting analysis. The North of Traffic, 12(3): 81–83.

Yaohua Fu, 2008. The statistical analysis software development of traffic volume and OD survey. Standardization of Traffic, 9(1): 81–83.

Study on traffic safety risk assessment of expressway in cold areas

Jian Wang, Kun Zhou & Xiaowei Hu
School of Transportation Science and Engineering, Harbin Institute of Technology, Harbin, Heilongjiang, China

ABSTRACT: Statistical analysis of the expressway traffic accidents data of Heilongjiang Province from 2007 to 2011 was performed. With the help of SPSS software, a safety risk assessment model with methods of factor analysis and multiple regression was obtained. Finally, the accidents at one section of G1 expressway was taken as example to verify the feasibility of the established model.

1 INTRODUCTION

In recent years, expressway traffic mileage in China has been rising rapidly, from 10,000 km in 2000 to 95,000 km in 2012. In 2010, expressway mileage only accounted for 1.85% of total highway mileage, yet the number and death toll of expressway accidents accounted for 7.76% and 13.54% of total numbers (*Traffic Management Bureau of the Ministry of Public Security. 2011*). Along with the formation of the expressway network, expressway has brought serious and large number of traffic accidents, which caused great casualties and property losses. Therefore, the way to evaluate the expressway traffic safety situation scientifically and effectively, the analysis of highway traffic safety problems and the improvement of the expressway traffic safety situation had become the current problems to be solved.

Scholars at home and abroad have studied and established many traffic safety risk assessment models, but seldom applied models in cold areas. Hence, this study built an expressway traffic safety risk assessment model based on 10 factors, and the model could be used in cold areas.

2 DATA COLLECTION AND CLASSIFICATION

The Heilongjiang Province expressway network mainly includes 10 expressways and the total mileage was about 4378 kilometers in 2013, ranking the ninth place in China. The expressway traffic accident data of Heilongjiang Province from 2007 to 2011, from the database of Communications Department and the traffic administrative department of Heilongjiang Province, was used in this study. The accident data was classified into following six categories through comparison and analysis.

(1) Road data: traffic signal way, road physical isolation, road type, pavement behavior, crossing section type, cross-sectional location expressway classification, road alignment;
(2) Environment data: weather, terrain, visibility lighting condition, road surface condition, pavement structure, visibility;
(3) Person data: age, gender, career, drinking, driving years;
(4) Vehicle data: number of vehicles, vehicle trajectory, vehicle type, usage;
(5) Accident data: accident type accident form accident cause;
(6) Result data: death toll, economic losses, injury

3 FACTOR ANALYSIS

In view of transportation system, traffic accidents are caused by mismatch among people, vehicles, road and environment. From the perspective of time, accident rate not only has U-shaped relationship with volume, but also has relationship with traffic and bad weather (*Zhong L D, Sun X R, Chen Y S. 2007; Zhou E B, Liu L. 2008*). From the perspective of space, road longitudinal linear and nonlinear combinations affect traffic accident. Therefore 14 candidates were firstly selected from the factors listed in section 2 as independent variables, and then the influencing factors that could affect traffic accident more greatly were determined through the Principal Component Analysis (PCA) in SPSS software. The 14 candidates for independent variables are shown in Table 1.

The function of factor analysis was used in this study to get the result of factor analysis through importing accident data in the format of Excel spreadsheets into SPSS software.

3.1 *The variance contribution rate test*

Through the Principal Component Analysis (PCA), the explanation to the total variance table of the candidates for independent variables was obtained, as shown in Table 2.

The following two conclusions were made based on Table 2.

(1) In the "total" column below "Initial Eigenvalues" seven numbers were greater than 1, which means seven common factors can be extracted from the results of factor analysis and their amounts of variance are 13.619%, 10.983%, 9.752%, 8.321%, 7.616%, 7.480% and 6.897%.

(2) In the "accumulation" column below "Rotate Sum of Squared Loadings", the last data was 64.669%, which means that the data in the table could well explain the independent variables.

3.2 Rotating component matrix

Through orthogonal rotation method, the "rotated component matrix" was obtained as shown in Table 3. Then, independent variables whose numbers were greater than 0.5 were merged into one common component. For example, in column 2, the "weather" and "road surface condition" were merged into environmental component, as shown in Table 4.

Table 1. The List of Candidates for Independent Variables.

Number	Independent variables	Number	Independent variables
1	Traffic signal way	8	Road alignment
2	Road physical isolation	9	Lighting condition
3	Road type	10	Weather
4	Pavement behavior	11	Road surface condition
5	Crossing section type	12	Visibility
6	Cross-sectional location	13	Terrain
7	Expressway classification	14	Accident form

Table 2. Total Variance Explained.

Component	Initial Eigenvalues			Extraction Sum of Squared Loadings			Rotate Sum of Squared Loadings		
	Total	% of variance	Cumulative %	Total	% of variance	Cumulative %	Total	% of variance	Cumulative %
1	2.043	13.619	13.619	2.043	13.619	13.619	2.021	13.470	13.470
2	1.647	10.983	24.602	1.647	10.983	24.602	1.495	9.966	23.436
3	1.463	9.752	34.354	1.463	9.752	34.354	1.447	9.646	33.082
4	1.248	8.321	42.675	1.248	8.321	42.675	1.254	8.362	41.444
5	1.142	7.616	50.291	1.142	7.616	50.291	1.186	7.903	49.348
6	1.122	7.480	57.772	1.122	7.480	57.772	1.160	7.734	57.082
7	1.035	6.897	64.669	1.035	6.897	64.669	1.138	7.587	64.669
8	.984	6.559	71.227						
9	.928	6.184	77.411						
10	.869	5.796	83.208						
11	.678	4.521	92.669						
12	.628	4.186	96.855						
13	.472	3.145	100.000						
14	−1.059E−1	−7.062E−16	100.000						

Table 3. Rotated Component Matrix.

	Component						
	1	2	3	4	5	6	7
Road physical isolation	.995	−.011	.004	.008	.043	.008	−.005
Road type	−.995	.011	−.004	−.008	−.043	−.008	.005
Weather	.010	.853	−.100	−.012	.050	.060	−.150
Road surface condition	−.045	.689	.162	.128	−.056	−.028	.320
Lighting condition	−.023	−.154	−.760	−.088	.241	.156	.043
Visibility	−.023	−.374	.706	−.021	.151	.094	.013
Accident form	.021	.278	.503	−.102	.338	.241	−.022
Terrain	.013	.048	−.025	.749	−.044	−.176	.129
Traffic signal way	.000	.032	.036	.740	.107	.228	−.094
Expressway classification	.061	.014	.101	−.169	−.656	.224	.175
Crossing section type	.157	−.057	.020	−.017	.559	.162	.100
Cross-sectional location	−.077	.167	.245	−.128	.438	−.035	.362
Road alignment	−.017	.092	−.003	−.124	.106	.589	−.003
Pavement behavior	.012	.007	−.087	.057	.008	−.002	.902

Table 4. The Results of Comprehensive Analysis.

Component	Independent variable factor contains	Component	Independent variable factor contains
1	Road physical isolation	5	Crossing section type
2	Weather (weather, road surface condition)	6	Road alignment
3	Visibility (visibility, accident form)	7	Pavement behavior
4.	Signal (traffic signal way, terrain)		

Table 5. Component Score Coefficient Matrix.

	Component		
	3	4	5
Traffic signal way	.016	.590	.099
Road surface condition	.097	.070	−.091
Weather	−.092	−.043	.030
Visibility	.488	.007	.086
Accident form	.308	−.088	.237
Terrain	.002	.600	−.026

Table 6. Model Summary.

Model	R	R Square	Adjusted R Square	Std. Error of the Estimate
1	.138a	.019	.012	1.01306270

Table 7. ANOVA.

Model	Sum of Squres	df	Mean Square	F	Sig.
1 Regression	18.949	7	2.707	2.638	.011a
Residual	981.139	956	1.026		
Total	1000.088	963			

3.3 Component score coefficient matrix

After determining the number and content of common factors, the coefficients of common factors that contain more than one independent variables were determined from the "component score coefficient matrix" shown in Table 5.

The coefficients of components 3, 4 and 5 could be concluded from Table 5, as shown in Equations (1), (2), and (3) below.

$$X_3 = 0.436 \times \text{road surface condition} + 0.595 \times \text{weather}, \quad (1)$$

$$X_4 = 0.488 \times \text{visibility} + 0.308 \times \text{accident form}, \quad (2)$$

$$X_5 = 0.590 \times \text{triffic signal way} + 0.60 \times \text{terrain}, \quad (3)$$

Table 8. Regression Coefficient.

	Unstandardized Coefficients		Standardized Coefficient		
Model	B	Std. Error	Beta	t	Sig.
1 (Constant)	.006	.033		.194	.846
Component 1	.030	.033	.030	.925	.355
Component 2	.011	.033	.011	.346	.730
Component 3	.035	.033	.034	1.068	.286
Component 4	.014	.033	.014	.435	.664
Component 5	.030	.033	.029	.920	.358
Component 6	.124	.033	.122	3.800	.000

where X_3, X_4 and X_5 stand for the components 3, 4 and 5, respectively.

4 MODELING

After the factor analysis, the common factors were analyzed, and the results of linear regression are shown in Tables 6, 7 and 8, respectively.

According to Table 8, the regression equation can be obtained as Equation (4).

$$Y = 0.006 + 0.030X_1 + 0.011XX_2 + 0.035X_3 + 0.014XX_4 + 0.030X_5 + 0.124X_6 + 0.031X_7, \quad (4)$$

where Y stands for the danger coefficient, X_1, X_2, X_3, X_4, X_5, X_6 and X_7 stand for components 1, 2, 3, 4, 5, 6 and 7, respectively.

5 APPLICATION EXAMPLE

The G1 expressway section from the stake mark of K1143 + 500 to K1152 + 500 was selected with totally eight accidents, as an example to verify the regression equation. The results are shown in Table 9.

Based Table 9, it was concluded that the section between stake marks of K1143 + 500 and K1147 + 55 had high danger coefficient, and corresponding improvement measures must be immediately taken to prevent the occurrence of major accidents.

Table 9. The Result of Accident Danger Coefficient.

Stake Mark	Danger Coefficient	Stake Mark	Danger Coefficient	Stake Mark	Danger Coefficient	Stake Mark	Danger Coefficient
K1143+500	5.08	K1143+600	1.53	K1144+900	2.43	K1145+450	1.52
K1146+200	1.55	K1147+55	4.85	K1148+919	1.53	K1152+500	1.57

6 CONCLUSION

In this study, the influence components of expressway traffic accident in cold areas were studied, and an expressway traffic safety risk assessment model based on multiple regression was established. The model was verified with data of the G1 expressway section from the stake mark of K1143 + 500 to K1152 + 500 The analyses showed that the danger coefficient has relationship with traffic signal way, road physical isolation, crossing section type, road alignment, weather, road surface condition, visibility, terrain and accident form. Due to the deficiency in traffic speed and volume data, a comprehensive analysis of expressway traffic safety risk assessment model is yet to be further discussed.

REFERENCES

Traffic Management Bureau of the Ministry of Public Security. 2011. Road traffic accident satistic annual report of the People's Republic of China. Wuxi: Traffic Management Research Institute of the Ministry of Public Security.

Zhong L D, Sun X R, Chen Y S. 2007. The highway, the relationship between accident rate and car speed difference in expressway. Journal of Beijing Polytechnic University. 33(2):185–188.

Zhou E B, Liu L. 2008. The influence of severe weather on expressway traffic safety and countermeasures. Transportation Technology. (4):94–96.

Study on urban spatial development and traffic flow characteristics in Tianjin

Ying Wang
School of Highway, Chang'an University
School of Civil Engineering, Tianjin Chengjian University

Kuan-Min Chen
School of Highway, Chang'an University

ABSTRACT: Tianjin has a unique urban spatial development mode called Twin City. By using the main component analytical method, a case study on the relationship between urban spatial structure and urban transportation in Tianjin was carried out. Effects mechanism between location selecting action and traffic behavior was analyzed based on multiple spatiotemporal perspectives. Characteristic of transport structure, traffic volume traffic distribution in different areas were investigated, and in-depth analysis were carried out based on the survey data. The results show the urban form of Twin City directs the distribution of traffic demand, the evolution of urban spatial structure is the main factor to the traffic characteristics change.

1 INTRODUCTION

The number of people and the amount of goods moving from one place to another is due to the production and consumption of a city, and the movements are embodied in urban traffic demands. Many production and consumption activities must travel across the urban land. Fundamentally, business activities are the driving force of traffic demands. So, the city's economic activities will promote the coordinative development of the urban spatial form and urban traffic. Studying the interactive relationship between the urban spatial form and urban traffic is very helpful to improve the rational development of the urban economic layout, optimize the urban space structure arrangement, and promote traffic efficiency (Liu Lu, 2010, 2008).

Tianjin has a unique urban spatial development model called Twin City. According to Strategic Planning of Tianjin Urban Spatial Development (2008–2020) (Figure 1), the overall spatial development strategy of Tianjin is "Twin City-Twin Port, Relative Growth, One Axis-Two Belts, Eco-development in the South and North." With the implementation of the spatial development strategy and relative growth between the Central Tianjin and Binhai New Area, the features of the urban traffic demands have been transformed. In this paper, features of traffic demand under the influence of the Twin City Spatial Development Strategy in Central Tianjin are analyzed, and this research will provide a basis for adjusting transportation management as appropriate and construct transportation development strategy adapting to strategy of the city spatial development (Strategic planning of Tianjin Urban Spatial Development (2008–2020)).

Figure 1. Schematic diagram of Twin City-Twin Port, Relative Growth.

2 EVOLUTION OF SPATIAL STRUCTURES IN CENTRAL TIANJIN

Strategic Planning of Tianjin Urban Spatial Development (2008–2020) is based on the Orientation of Urban Spatial Development in Tianjin general planning (2005–2020), and it makes a further promotion to the general planning in the part of urban spatial planning. As two development points, central Tianjin and Binhai New Area have been growing steadily, and the development of city interspaces will focus on the Wuqing-Central Tianjin-Tanggu development axis (General Urban Planning of Tianjin (2005–2020)).

According to Figures 2 and 3, Central Tianjin will develop along the Wuqing-Central Tianjin- Tanggu main development axis, with relative expansion to Binhai New Area, and a new corridor of transport and space growing will be formed between Central Tianjin and Binhai New Area.

Figure 2. Land use of Tianjin in 2004.

Figure 3. Planning of Tianjin land use for 2005–2020.

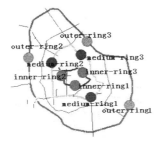

Figure 4. Distribution of survey road section.

At the same time, according to the plan, industrial base will be relocated. Binhai New Area is the main point in new industrial bases, and the function of Tianjin central city will be transformed to provide intellectual support and service, Tianjin central city is going to form a commercial and educational center. Land use type is residential land mainly in the Central Tianjin which differs from industrial land in Binhai New Area, and it would lead to a separation between living and work, which must develop coordinately with efficient transport corridor.

3 FEATURES OF TRAFFIC DEMAND IN TIANJIN CENTRAL CITY

A great change in the features of traffic demand occurs due to the gradual implementation of the urban development strategy and the transformation of the urban spatial form. Framework of the road system in Tianjin central city is three rings-forty rays. Therefore, in this paper, the traffic survey sites are distributed over all rings. To guarantee the reliability and comparability of data, there are three traffic survey sites in each ring and distribution of survey road section is shown as Figure 4. Survey time: April 13 (Wednesday, Sunny).

Table 1. Modal splits of survey sites (full-traffic mode).

Traffic mode location	Non-motorized transport	mini-bus	large bus	light-van	HGV
Inner-ring 1	0.38	0.57	0.05	0.00	0.00
Inner-ring 2	0.42	0.55	0.02	0.00	0.00
Inner-ring 3	0.31	0.64	0.05	0.00	0.00
Medium-ring 1	0.31	0.65	0.04	0.00	0.00
Medium-ring 2	0.39	0.52	0.08	0.00	0.00
Medium-ring 3	0.50	0.38	0.11	0.01	0.00
Outer-ring 1	0.09	0.63	0.07	0.09	0.12
Outer-ring 2	0.13	0.52	0.01	0.15	0.16
Outer-ring 3	0.12	0.57	0.03	0.11	0.17

Table 2. Modal splits of survey sites (only motor vehicle).

Traffic mode location	minibus	large bus	light-van	HGV
Inner-ring 1	0.92	0.07	0.00	0.00
Inner-ring 2	0.95	0.04	0.01	0.00
Inner-ring 3	0.93	0.07	0.00	0.00
Medium-ring 1	0.94	0.06	0.00	0.00
Medium-ring 2	0.85	0.13	0.01	0.01
Medium-ring 3	0.76	0.23	0.01	0.00
Outer-ring 1	0.69	0.07	0.10	0.14
Outer-ring 2	0.54	0.01	0.19	0.21
Outer-ring 3	0.65	0.03	0.12	0.19

3.1 Features of traffic mode

There will be different traffic mode for different land-use. From table 1, we can include that Non-motorized transport had a higher proportion in inner-ring and medium-ring; at some survey sites the proportion is raised to above 40%. The reason lies in the fact that high and new technology industries and modern services become the central development industry on the outer-ring periphery. And there are rarely residential areas; most of the commuter traffics have a long travel distance. Non-motor vehicles are ideal traffic mode only for a short distance, so the proportion of non-motor vehicle is more likely to be lower at the outer-ring survey sites. Recreation and culture are the industrial development direction in the inner-ring area, the mainly land-use is residence in the medium-ring area. In table 2, it shows that passenger transportation is predominant, and the characteristic feature is particularly evident in the east area in comparison with the west area. Most workers in Binhai New Area as industrial centre mainly come from Tianjin central city. With the further industrial development and the economic growth of Binhai New Area, commuters flow between Tianjin central city and Binhai New Area will increase tremendously in South Eastern regions.

The traffic function of outer-ring goes to devote to cross-boundary traffic, lessen traffic pressure for the central city, and mainly serve for freight transport. However, the feature of traffic demand has changed enormously in the outer-ring area. Table 3

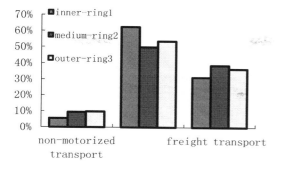

Figure 5. Modal splits of survey points in outer-ring.

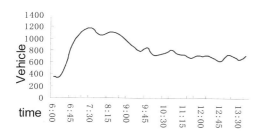

Figure 6. Map of traffic time-varying in inner-ring (6:00–14:00).

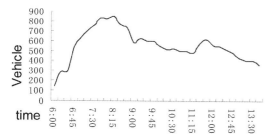

Figure 7. Map of traffic time-varying in medium-ring (6:00–14:00).

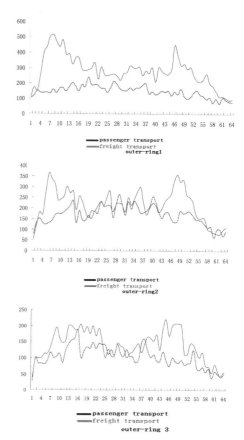

Figure 8. Map of traffic time-varying in outer-ring (6:00–22:00) P.S: investigation data statistics in 15-minute unit, so every section is 15 minutes in horizontal abscissa.

indicates that passenger transport and freight transport all account for a rather high percentage. The proportion of passenger transport is about 60%, which is only 24% in 2003. So the proportion of traffic modes has changed significantly in outer-ring, passenger transport becomes dominant, especially in the outer-ring 1 section (Figure 5), the proportion of passenger transport to freight transport is 2:1. The function of the outer-ring has transferred from cross-boundary traffic to urban traffic. An often startling array of bus and truck moving at different speeds bring hidden trouble to driving. So the outer-ring traffic condition has been further worse.

3.2 Features of time and space distribution

Figure 6 and Figure 7 show the time-varying of traffic. There are two kinds of passenger traffic demands, one is rigid traffic demand and the other is flexible traffic demand. Rigid traffic demand consists mostly of commuting. Travel time concentrates on short time in the morning and evening, just like the tide in a day. On the other side, non-commute trip is flexible traffic demand, travel time of which is flexible. Freight transport is greatly influenced by traffic management, its travel time is flexible for a certain degree, and travel time feature of which is unique.

Figure 8 shows that passenger transport is the main components of traffic in inner-ring and medium-ring, travel time of which have features of two peak periods, tide phenomenon is evident. There is a unique sign that a travel peak appears at noon, so there are quantities of short distances commute trips at medium-ring.

In Each survey point, Traffic time-varying characteristics of outer-ring have two peak periods similar with inner-ring and medium-ring (Figure 9), but more concentrated, especially at south and east sections. In Non-peak periods, traffic volumes of passenger transport are significantly decreased, so passenger transport of outer-ring consists mainly of long-distance commute travel between Tianjin central city and Binhai New Area. Another aspect is that there are many flexible trips excepting commuter traffic at the outer-ring

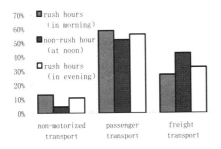

Figure 9. Modal splits at different times in outer-ring.

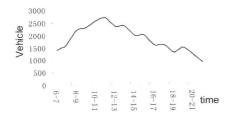

Figure 10. Map of Traffic time-varying in outer-ring-ring (6:00–22:00, 13/8, 2003).

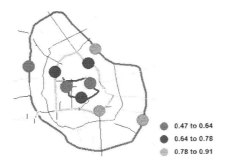

Figure 11. Saturation of peak hours at survey points.

1 sections, which lies between Tianjin central city and Binhai New Area. It shows that there are some residential areas. And with the land development between Tianjin central city and Binhai New Area, economic and living links between them will become closer and closer.

Travel time of freight transport is relatively well-distributed. Traffic volume of freight transport rises gradually after 7:00, and decreases by degrees after 20:00, the rise and fall is slowly, steadily, one peak in a day. Traffic time-varying is very similar with that in 2003. Traffic consisted mostly of freight transport in 2003, makeup just over 70%. Comparing Figure 8 with Figure 10, features of Traffic time-varying have no significant change (Li Yong, 2006).

3.3 Features of level of service

Level of service is service quality which travelers get from the aspects such as state of the roads, condition of traffic, road environment, etc. Level of service is an important factor to evaluate traffic operational situation and the feeling that driver and passenger get from the road. Commonly it is evaluated by saturation that is the ratios of traffic volume to highway capacity (Highway Capacity Manual 2000).

From Figure 11, it can be concluded that at each ring, traffic at southeastern sections is heavier than others (seen here superiorly). South of the inner-ring and medium-ring, southeastern of the outer-ring are in the state of congestion, unlike that, the traffic is smooth in most west and north area. The implementation of the urban spatial development strategy has begun to function in traffic.

4 CONCLUSION

In the process of urban development, urban spatial form and urban traffic will affect each other and promote cooperative development. Ribbon group development form is inevitable trends for the urban spatial forms in Tianjin, which will lead to the increase of traffic demand and developing along the urban spatial corridor. Regional transportation corridor is the direction of spatial layout of urban road system. With the evolution of urban road system, features of traffic demand on many roads have changed, some transportation infrastructure facilities and method of traffic management is no longer applicable. The new function of road has to be re-orientated. With the relative growing between the Central Tianjin and Binhai New Area, traffic demand is still exploding, rigid and long-distance commute trip will be the main traffic demand along the urban development axis. So traffic facilities, traffic mode, and transportation management needs to re-planning and re-designing. In the future, the transport planning will not only meet the requirement of urban spatial development, but also guide the optimization of urban spatial arrangement to promote harmonious development between people and city.

REFERENCES

General Urban Planning of Tianjin (2005–2020) [R].
Highway Capacity Manual 2000. 2000. Transportation Research Board.
Li Yong, 2006. Study on the construction of inboard way of Tianjin outside route. Tianjin University.
Liu Lu, 2010. The Relationship between the Urban Spatial Structure and the Transportation Development—taking Tianjin as an Example. Forum on Science and Technology in China, (11):11–115.
Lu Liu, 2008. Research on the Correlation between the Urban Spatial Structure and the Transportation Development in Tianjin. East China Normal University.
Strategic planning of Tianjin Urban Spatial Development (2008–2020) [R].

Survey and analysis of energy consumption of shopping mall buildings in Chengdu

Zhong Liu & Yong Xiang
School of Architecture and Civil Engineering, Xihua University, Chengdu, Sichuan, China

ABSTRACT: By investigating shopping mall buildings in Chengdu, the results show that shopping mall buildings have an average annual electricity consumption per unit area of 92.82 kWh. The consumption of lighting electricity accounted for 49.34% of the total power. After the implementation of new standards, mall buildings designed and constructed in Chengdu have lowered their power consumption by 31.93% on average annually per unit area. Although the new standards are successful in creating energy-efficient new construction, there are some options available for upgrading older, less-efficient buildings. This paper outlines several steps to lower the energy consumption of older shopping mall buildings in Chengdu.

1 PREFACE

China is currently undergoing a construction boom, and the growth of housing construction has accelerated markedly. From 2000 to 2010, the national floor space increased from 27.7 billion to 45.3 billion square meters, with an average annual increase of 1.33 billion m^2 and a growth rate of 11.3%. If calculated in accordance with the present building energy consumption, in the year 2020 the energy consumption of urban structures is expected to reach more than 35% of the total consumption [1]. The total area of large public buildings account for a small portion of the total area of urban construction, but with an annual power consumption of 70–300 kwh/m^2, the proportion of energy consumption in urban construction is more than 20% of the total energy consumption. is the general public buildings 4–6 times [2]. According to *"The State Council issued a comprehensive work program on energy saving"* and *"Implementation opinions on strengthening state office buildings and large public building energy management work"*, Chengdu has been included in the scope of the state office buildings and the large public buildings energy monitoring system construction demonstration provinces. Investigation showed, in Chengdu City, an area of large department stores accounted for 28.83% of the total area of large public buildings. so, do a good job of building energy consumption survey large shopping malls of Chengdu city, which can provide data for all levels of government developing policies and standards.

2 SURVEY OVERVIEW

2.1 Climatic characteristics

Chengdu belongs to a subtropical humid climate zone. By the influence of its geographical location,

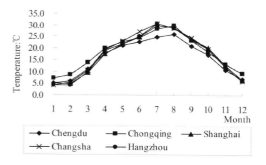

Figure 1. Temperatures each month in the cities in 2012.

underlying surface, and other topographic geographical conditions, it suffers more vertical climate and complex local climate phenomenon. It has a longer frost free period, less sunshine, a mild climate, abundant rainfall, and four seasons. In order to understand the climate difference between Chengdu and other cities, it was compared to the cities of Fuzhou, Hangzhou, Chongqing, Shanghai, and Changsha by monthly temperatures and hours of sunshine in 2012, as shown in Figure 1.

Figure 1 shows that during the hottest months in Chengdu (July and August), the temperatures were lower than several other cities (3.7–6.1°C, 2.6–3.8°C), and during the coldest months (December, January and February), the temperatures were not the lowest, nor the highest. Thus, the climatic conditions in Chengdu are dissimilar to other cities.

2.2 Scope

Limited to the Chengdu City administrative districts, and December 31, 2010 before the completion of the shopping mall and the monomer construction area of greater than or equal to 20,000 m^2.

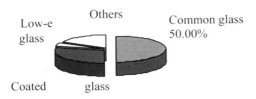

Figure 2. Classification and the proportion of outer windowpanes.

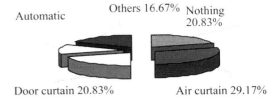

Figure 3. Classification and the proportion of door barriers.

2.3 Investigation process

2.3.1 Survey data

Investigation methods include comprehensive surveys, on-site investigations with organizational personnel, and onsite investigations of the administrative departments (such as the area of the department in charge of construction, the State Grid and the gas company).

The content of this investigation is divided into two categories: The first category is the basic construction information, including the building name, completion time, the construction area, shading, walls, windows, lighting systems. The second category includes the building energy consumption information, including (from January 1, 2006 to December 31, 2011) the power consumption, gas consumption, and water consumption.

2.3.2 Data review and summary

Survey data review refers to the self, peer review, on-site review steps of registration or copying of data verification, to correct the errors found in order to ensure accurate data.

3 THE RESULTS OF THE SURVEY ANALYSIS

Through the survey, the basic information of the 54 buildings and monthly energy consumption information covering a total area of about 2,821,700 m² was obtained.

3.1 Basic information analysis

3.1.1 Completed building time

Large shopping malls in Chengdu City completed before 1990 account for 2.22%, those completed in the years 1991–1995 account for 7.78%, those completed in the years 1996–2000 accounted for 27.78%, 2001–2005 account for 38.89%, and those completed after 2006 equal 23.33%.

3.1.2 Retaining structure

Exterior wall heat preservation measures of building is only 33.33%, while the use of clay brick wall materials up to 66.67%, aerated concrete block is only 4.17%. 50% of the building roof without thermal insulation measures.

Outside the window glass building 50% use of ordinary glass (see Figure 2). The outer window using aluminum alloy materials accounted for 62.50%, the

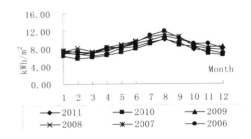

Figure 4. Monthly power consumption per unit area of shopping mall buildings.

single glass accounted for 66.67% and 75% of the building did not take the measures.

A door block classification and proportion as shown in figure 3.

3.1.3 Air conditioning system

Heating and cooling air conditioning system in the building monomer concentration for hot and cold mode, accounting for all the way for cold and hot 71.29%, 4.62% use of CCHP for hot and cold way.

3.1.4 Lighting system

Between the various shopping malls, the average daily business hours ranged between 10–14 hours, the running time of the indoor lights was 12.5–16 hours, and outdoor lights' running time was 6–15.5 hours. The lighting systems were divided into energy-saving lamps and ordinary lights, where the indoor lighting was energy-saving lamps and accounted for 87.50%, and ordinary lights used outdoors accounted for 12.50%.

3.2 Information analysis of energy consumption

3.2.1 Energy consumption of every month

The large shopping malls' monthly electricity consumption per unit area is shown in Figure 4, where the average monthly electricity consumption was 5.72–11.97 kWh/m². Power consumption was more obvious seasonally, annual 7, September 8, and the largest power consumption was the air conditioning used in summer months.

Large shopping malls' average monthly gas consumption per unit area was 0.07–0.26 m³/m², and the monthly water consumption was 0.07–0.13 t/m².

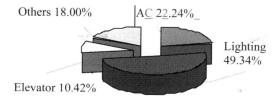

Figure 5. Composition and the proportion of the electricity consumption.

Table 1. Construction completion and age of the average annual energy consumption per unit area.

Completed building time	The average energy consumption per unit area		
	Electric power kWh·a^{-1}·m^{-2}	Gas m^3·a^{-1}·m^{-2}	Water t·a^{-1}·m^{-2}
Before 1990	16.19	2.39	0.37
1991–1995	170.10	5.52	1.72
1996–2000	83.00	2.69	1.08
2001–2005	101.34	3.22	1.32
After 2006	68.98	1.69	1.14

3.2.2 Each single energy consumption distribution

The main energy consumption of large shopping centers is the electricity, survey data shows. Total effective electricity sub 14 building, electrical components and proportion in Figure 5. As can be seen in the buildings, the electricity is mainly composed of three parts: air conditioning, lighting, and elevators. These accounted for 82% of the total power, in which the lighting and air-conditioning accounted for the largest portions at 49.34% and 22.24%, respectively.

3.2.3 The relationship between the construction completion time and energy consumption

The construction completion time and average annual energy consumption per unit area is shown in Table 1. From this, it can be seen that the average annual energy consumption per unit area of the building completed before 1990 is the lowest. There are only two wholesale stores in the survey, whose lighting and air conditioning energy consumption are not large. The energy consumptions of the buildings completed from 1991 to 1995 are the highest and the average annual power consumption per unit area is as high as 170.10 kWh/(a·m^2). The energy consumptions of the building completed from 1996 to 2000 are not very high, most of which are low energy stores such as furniture, medical supplies, wholesale markets, etc. The energy consumptions of the buildings completed after 2006 are relatively low, and the average annual power consumption per unit area is 68.98 kWh/(a·m^2). These buildings were built after the "*Public building energy efficiency design standards*" (GB50189-2005) had been enacted. After the promulgation of the new standards, the average annual power consumption per unit area of the buildings designed and constructed in

Table 2. The city shopping mall annual electricity consumption of per unit area comparison.

City	Statistical Years	The average power consumption per unit area kWh·a^{-1}·m^{-2}
Chengdu	2006~2011	92.82
Chongqing [3]	2007~2008	279.36
Fuzhou [4]	2007	208.95
Hangzhou [5]	2007~2008	324.46
Changsha [6]	2008~2009	178.00
Shanghai [7]	2007~2009	228.80

Chengdu is 31.93% lower, with the average annual gas consumption 47.52% lower, and the average annual water consumption 13.64% lower.

4 EACH CITY SHOPPING MALL BUILDINGS ENERGY CONSUMPTION

The Chengdu City, Fuzhou City, Hangzhou City, Chongqing City, Shanghai City and Changsha City average power consumption of large shopping malls are shown in Table 2. This shows the annual electricity consumption per unit area, and Chengdu is equivalent to other city of 2/7~1/2, is the latitude of Shanghai's 2/5, is the main cause of the special climatic conditions in Chengdu area.

5 ENERGY SAVING MEASURES ANALYSIS

According to the plans of Chengdu Tianfu New Area as well as the "North" area, large shopping malls will greatly increase in the next few years, and the demand for energy will also increase dramatically. The existing building plus, Chengdu City shopping mall at least 10 million m^2. This area is estimated, the power consumption is about 928 million kWh/a. Energy saving operation management, reasonable optimization, achieving 20% energy saving target, a year can save electricity 186 million kWh.

5.1 Lighting system

Buildings use energy-saving lamps, but lighting electricity still accounts for about half of the consumed electricity, is designed according to the maximum demand, and energy-saving lamps to replace the ordinary lamp, without considering the new light source illumination, cause indoor illumination is too high. To reduce electricity consumption for lighting, must take the sub metering, reduce the operation time, reduce indoor lighting and outdoor lighting reasonable management measures.

5.2 Air conditioning system

Air conditioning systems have the following two problems: the total energy consumed is very high due to the outdated equipment not being energy-efficient; the other problem is also due to the age of the equipment where the level of automation is low, resulting in inefficient use of temperature settings. The Chengdu climate is unique, and the benefits of new technology should be applied in ways of low-energy machinery with programmable temperature settings.

5.3 Both the building enclosure structure energy saving reconstruction measures

An important part of the existing energy saving reform is finding ways to refine building characteristics in order to reduce the building load [8–10]. According to the Chengdu area climate characteristics and the results of survey and analysis, suggestions for energy saving include the windows and doors.

5.3.1 Outside the window transformation

Reconstruction of external windows needs to start with the window frame materials, glass type and sun shading measures. For the frames, thermal insulating aluminum alloy or UPVC plastic profiles are good energy saving materials. Glass types that have coatings or low-e double or hollow glass, and shading measures are all great ways to improve the exterior insulation [11].

5.3.2 Door transformation

Heat loss door will be the gate material, structure and the influence of flow etc. The whole building energy saving reconstruction of the door, turning door, air curtain, automatic induction door and door curtain electronic.

6 CONCLUSION

Through the investigation and analysis of the energy consumption of 54 large department stores in the years 2006–2011 in Chengdu, the following conclusions can be drawn:

(1) In large shopping malls, lighting accounts for 49.34% of the total electricity use.
(2) The average annual electricity consumption per unit area of large shopping malls is 92.82 kWh/(a·m^2), which accounts for 28.5–50% of other cities with hot summers and cold winters, and 40% of Shanghai, which shares a similar latitude.
(3) If the electric power is converted by 0.3300 kgce/kWh and natural gas is converted by 1.2143 kgce/m^3 as the standard coal, the average annual energy consumption per unit area of large shopping malls is 33.24 kgce/(a·m^2).
(4) In order to produce energy-saving results of the existing large shopping malls in Chengdu, the authors suggest the lighting systems and windows be updated in the ways outlined in this work, which are capable are generating very noticeable changes in energy consumption.

ACKNOWLEDGEMENTS

The research work was supported by "11th five-year plan" building energy saving special major science and technology development plan of Chengdu under Grant No. 07YTZD976SF-020 and Xihua university graduate student innovation fund under Grant No. YCJJ201372.

REFERENCES

[1] The People's Republic of China Ministry of Housing and Urban-Rural Development. Energy-saving renovation of existing residential buildings Hundred Questions [M]. China building industry press: Beijing, pp. 1, 2013
[2] DAI Xuezhi, WU Yong. Government regulation of the market economy and its innovation in a large public building energy monitoring system. HVAC, 8, pp. 2–7, 2007
[3] TAN Yin, LI Baizhan, DI Yong, etc, Status and problems class shopping malls in Chongqing public buildings energy use and management analysis, Doors and Windows, 1, pp. 48–51, 2009
[4] LU Wenying, Building energy consumpti on investigation and suggestions in Fujian, Fujian Construction Science and Technology, 3, pp. 26–28, 2009
[5] DUAN Xiaoping, LANG Ying, Energy consumtion investigationand energy-saving analysis on pulic buildings in Hangzhou, Building Scence, 26(8), pp. 36–39, 2010
[6] YANG Song, ZHU Qingsong, Analysis government office buildings and large public building energy consumption survey, Chinese and foreign construction, 7, pp. 193–195, 2010
[7] Xu Qiang, Zhuang Zhi, Zhu Weifeng, etc. Large public building energy consumption statistics and analysis of Shanghai [C]//Urban development research. Beijing, 2011: 322–326
[8] Hsu Che-Chiang. Study Oll the construction rand applicatinn of egional peak load forecasting annlysis system [J]. Monthly JoBmai of Taipower's Engineering, 2005, 686: 120–33
[9] Hang-Sik Kitu, Seung-Pil Moon, Jae-Scok Choi. A Study on Construction of the CMELDC at Load Points [J]. Transactions of the Korean Institute of Electrical Engineers, 2000, 49(4): 195–198
[10] Mason, M.D. Extension of Available Critical Design Conditions for AC Load Estimaoon Australian Refrigeration [J]. Air Conditioning and Heating, 1993, 47: 33–34
[11] Wroblaski, Kylie. Embrace External Shading [J]. Buildings, 2011, Vol. 10, 5(9), pp. 28

Sustainable high-rise buildings

Linxue Li, Zhendong Wang & Ren Qian
CAUP Tongji University, Shanghai, China

ABSTRACT: Climate change and the rapid growth of energy consumption have caused some urgency on the subject of sustainable high-rise buildings. This paper's aim is to explore the main issues of sustainable high-rise design based on research, utilize local climate data to develop ecological design strategies, and provide the path of research for sustainable high-rise buildings from four aspects: wind, radiation, recycling, and vertical landscape.

1 EMERGING SUSTAINABLE DESIGN FOR HIGH-RISE BUILDINGS

High-rise buildings are designed by copying a plane in a vertical direction, or designed as isolated sculptures. Particularly in developing countries, the external, vertical and air-conditioned box has become a standard model (Wood, 2010). Huge energy expenditure is always a problem for tall buildings, in both the construction and function phases. Methods for reducing energy use are of great interest for the development of urban growth.

A high-rise building is not strictly an ecological building, and in fact, it is one of the most non-ecological building types. Due to its height, the building implies many additional requirements. For example, it needs to use more materials to resist the high wind loads to its structure system; it needs to consume extra resources for water provision or other supplies to the upper floors against gravity (Yeang 2007). In addition, huge energy will be consumed for the vertical transportation systems.

However, the high-rise building is inevitable. It can meet the urgent need of more space as it covers less land than a sprawling building. High-rise buildings built in the transport hub contribute to the reduction of transportation energy consumption. The key point is that skyscrapers meet the needs of urban growth and a growing population.

Based on the inevitability of high-rise buildings, architects should actively seek a sustainable development method of high-rise rather than ignoring them. The production "Towards Zero Carbon—Hi-Rise Design with the Chicago Decarbonization Plan" is based on this opinion. The studio sought to establish a method to consolidate and advance the understanding of sustainable high-rise design, and put it into practice.

2 SUSTAINABLE ISSUES

To create a sustainable high-rise building means to design its sustainability from the onset of the project. This generates great differences compared to the traditional high-rise design. Beginning with the first steps, the most important energy-saving strategies must guide the morphological and functional design later. This paper started by studying the research of Adrian Smith + Gordon Gill Architecture, which contains nine aspects: architecture, urban matrix, mobility, smart infrastructure, water, waste, community engagement, energy, and funding (Smith 2011). From this, it was found that the energy consumption of architecture is heavily impacted by the natural environment, equipment, and operation systems. The climate parameters, space strategy, and a variety of natural resources such as wind, thermal radiation, and water were studied, and tried to establish the most vital factor to further our design.

Having an accurate understanding of meteorological data is a prerequisite for sustainable design. Chicago is known as the "Windy City", and is rich in wind resources. The annual and monthly wind roses of Chicago were analyzed, and it was found that the main wind direction from January to March is northwest, from April to August it's northeast, and from September to December it's southwest (Fig. 1). Consequently, the main wind directions in Chicago are northwest, southeast, and southwest (Fig. 2–3), with an average wind speed of 10 m/s. Chicago is in a gale period throughout the year. As a result, the wind has to be considered an important factor on high-rise design.

From the annual solar radiation chart (Fig. 4), it can be seen that Chicago Reaches hot temperatures from June to August with high amounts of solar radiation, and the coldest period is from December to February, with low solar radiations level. As such, it's essential

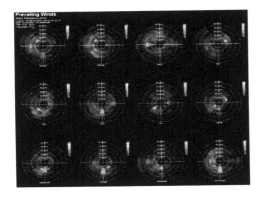

Figure 1. Monthly wind roses in Chicago.

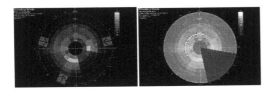

Figure 2–3. Main wind direction.

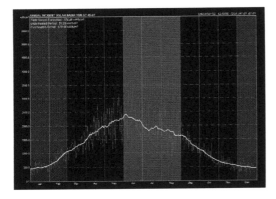

Figure 4. Annual solar radiation chart.

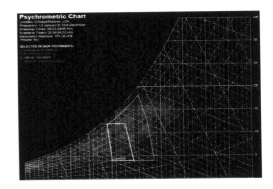

Figure 5. Psychrometric chart.

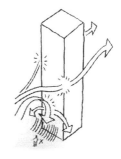

Figure 6. Diagram of wind shadow area.

to consider solutions for both summer heat insulation and the winter heat preservation.

While searching for building strategies to deal with the environment, the psychrometric chart was discovered to help make a preliminary judgment (Fig. 5). The psychrometric chart provides four initial energy-saving measures: passive solar heating, thermal mass effects, natural ventilation, and direct evaporative cooling. The yellow frame in the chart represents the comfort period without energy-saving measures. The natural ventilation and thermal mass effects are the most effective measures in the hotter months, and passive solar heating can be rather effective in winter. These key points will provide a guide and criteria for further ecological study.

With a deep understanding of climatic characteristics, it's possible to proceed with the sustainable high-rise design. That is how to transform the climate language to architecture language. Further study will be made in the following aspects: wind, thermal radiation, and water.

3 WIND STRATEGY

Natural ventilation is an important strategy of passive design, which can reduce the energy consumption of air conditioning. There are many evidences to indicate that people can tolerate higher indoor temperature in naturally ventilated buildings than in mechanically ventilated buildings. The tolerance for higher temperatures further strengthens the ecological benefits of natural ventilation. The following case shows the strategy of natural ventilation used in high-rise design.

The higher sections of high-rise buildings are subject to more wind loads. When the wind blows on a vertical surface, it tends to be deflected upward and downward (Fig. 6), which will create a wind shadow behind the high-rise building. If the building plane is the shape of a spindle, the wind can easily deflect away from both sides. which will effectively reduce the wind shadow behind the building (Fig. 7). According to the annual wind roses of Chicago, the wind comes mainly from the northwest, northeast, and southwest, so the strategy will be to set three buildings in the shape of a spindle facing the three main directions.

Due to the difference of uses, the treatments of upper and lower parts in high-rise buildings should be

Figure 7. Diagram of transforming shapes.

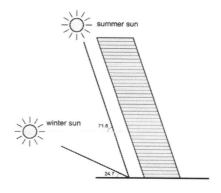

Figure 11. Inclined facade.

Figure 8. The shape of the bottom part of the three buildings.

Figure 9. The shape of the upper part of the three buildings.

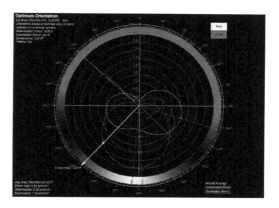

Figure 12. Best orientation chart.

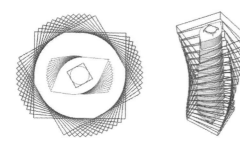

Figure 10. Twisting strategy.

4 SOLAR RADIATION STRATEGY

For solar radiation, the most common strategy is to use a self-shading system. Architecture achieves self-shading by adjusting the orientation and inclining the facades. An inclined facade is a very effective strategy, if well designed, because it can block summer sun but allow winter sun so that it can simultaneously reduce energy consumption of heating and cooling (Abalos and Ibanez 2012) (Fig. 11).

Building orientation is also an important reference index. ECOTECT and other energy-saving software can easily obtain the site data of sunshine. The chart shows the best orientation for architecture in Chicago (Fig. 12). The red line represents the direction of maximum solar radiation during the hot period, and the blue line represents the direction of maximum solar radiation during the cool period. At last, the green line represents the direction of maximum solar radiation during the whole year. The goal is to minimize solar radiation in the hot periods and maximum it during the cool periods. The software gives its conclusion after calculating all the information above, using the yellow line to represent the best orientation for architecture in Chicago. However, it is not an absolute indicator.

different. There can be outdoor activities in the bottom areas of the high-rise buildings, so the three buildings should be able to block the wind with their edges facing the directions of the wind (Fig. 8). The influence made by severe wind pressure for the upper part was considered, so the angles of the buildings should be facing the wind directions (Fig. 9). Combining the two aspects together, the twisting strategy should be chosen as our basic form of architecture (Fig. 10).

These are the preliminary studies of the architectural form guided by natural factors. However, further studies have to be made during the design process because the wind direction and speed will change unpredictably with time and date.

There are more complicated factors that need to be considered in the design. However, using energy-saving software for analysis provides a good reference in the beginning.

5 RECYCLING STUDIES

High-rise buildings will consume large amounts of energy per day, including energy loss. Therefore, studying the recycling of energy should also be an important part of the method in sustainable design. There are two main energy sources related to this: thermal energy and water. Thermal stability reduces energy loss through heat exchange between different spaces. We know that different functions have different requirements for temperature. So, the room with the highest demand can be located in a position where it can receive more solar radiation, and the auxiliary rooms can be placed on the other side so that they can take advantage of the heat loss from the main room and can also be regarded as buffer zones for the main ones (Song, 2003).

Water recycling is also a very important sustainable strategy for high-rise buildings. Large amounts of water carried from the bottom to the top consume large amounts of energy. A complete water recycle system can help save energy. First of all, the rain is a clean source of water which can be stored using artificial technology and after filtration, it can be used as non-drinking water (Fig. 13). A good strategy could be based on the collection and storage of rainwater on the roof level, and later transported to the purification system by pipes. A reclaimed water system also assures water savings. After deep treatment and purification of sewage, the water whose quality reaches the standard can be reused as non-potable water. In order to apply the system, a double pipe structure must be installed so as to separate the recycled water with the waste water (produced by toilets) (Chen et al., 2013).

6 VERTICAL LANDSCAPE STUDIES

Plants are aesthetic and ecological, and can affect the climate. Vegetation can reduce the sun's radiation when placed in interior spaces and on exterior walls, and can also reduce glare directed into a room. Transpiration of vegetation becomes an effective cooling device on the facade and can regulate the microenvironment of a building surface.

The external surfaces of the high-rise building are 4–5 times its covered area. So, if the entire external surfaces are covered with greenery, the ecological effect is much more than in low-rise buildings. At the same time the vertical landscape of high-rise buildings can play an important role in slowing down the urban heat island effect (Zhou, 2012). Traditional tall buildings' air conditioning systems just transfer the heat from indoors to outdoors and consume much energy. In fact, there is no change in the value of heat. However, plant transpiration can reduce heat in the whole city on a macro level, which can effectively slow down the heat island effect. Plants can reduce the temperatures around the city about 10°C, and the temperatures under the shade can be further reduced by 2°C. A tree can evaporate 450 L of water every day. Using a mechanical system to achieve the same evaporation effect would require the use of 5 air conditioners at 2500 kcal per hour, working continuously for 19 hours (Yeang, 1999).

A continuous green facade is strategically important for the vertical landscape system of high-rise buildings. The plants in the system should be selected so to assure variety and ecological balance, and as a continuation of the ground vegetation to integrate them with the entire ecosystem (Fig. 14).

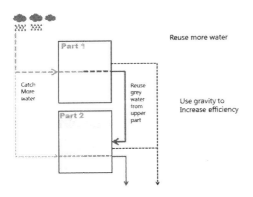

Figure 13. Diagram of water recycling system.

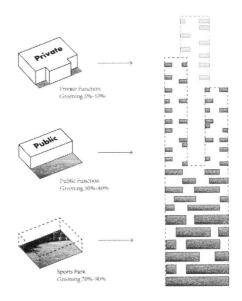

Figure 14. Diagram of vertical landscape.

7 CONCLUSIONS

Based on this study, multiple technologies were searched for ways to reduce carbon emissions and to find a course of sustainable high-rise building design. The study began with an analysis of climate data and focused on the possible responsive performances of high-rise buildings in unique climate conditions. All these studies may become part of a new architectural design methodology for the development of sustainable high-rise buildings in the future.

8 GRAPHICS

All graphics are from the studio directed by Professors Linxue Li, Zhendong Wang, and Antony Wood at Tongji University.

ACKNOWLEDGEMENTS

We are thankful for the support of the National Natural Science Foundation of China (Project No. 51278340). This paper is part of the research project.

REFERENCES

Chen W.P. et al. 2013. Reclaimed water: A safe irrigation water source. *Environmental Development* 8: 74–83.

Smith, A. & Gill, G. 2011. Toward zero carbon: the Chicago central area decarbonization plan. Australia: Images Publishing.

Song, D.X. 2003. Ecological architecture design. In Song, D.X. (ed), *Design method and technology for ecological architecture*: 25–34. Shanghai: Tongji University.

Wang, B. 2012. Design strategy using principles of thermodynamics. In Abalos I & Ibanez, D (eds), *Thermodynamics applied to high-rise and mixed use prototypes*: 133–145. Boston: Harvard GSD.

Wood, A. 2011. Tall buildings: search for a new typology. PhD Thesis, University of Nottingham: 1–3.

Yeang, K. 1999. Designing and planning the green skyscraper. *World Architecture* 2: 21–29.

Yeang, K. 2007. Green design. In Yeang, K (ed), *The green skyscraper*: 20–27. Australia: Images Publishing.

Zhou, M. 2012. Talking about the theory and design of ecological architecture *Construction Materials & Decoration* 7: 14–15.

Temperature deformations of the ultralong frame structure of MIXC-Qingdao

Nana Han, Shan Ke & Yujie Ge
Qingdao Hotel Management College, Qingdao, China

Qiang Li
Qingdao Teng Yuan Design Institute Co., Qingdao, China

ABSTRACT: At present, Qingdao MIXC is the largest commercial building complex structure. The structure is L-shaped, and the main vertical length is over 300 meters. The temperature expansion joints are not set in the main structure, and only part of the main structure is connected to the short side of the L-shaped split separated by a seismic joint. In this article, the finite element model of the MIXC structure is created to analyze the temperature stress deformation and deformation of the whole structure, and compare these with actual site measurements of the structure. This paper proposes to measure to strengthen control of the temperature stress and deformation. The design provides a valuable reference for similar structures.

Keywords: Ultralong frame structure; temperature deformation; temperature stress; expansion joints; finite element analysis

1 INTRODUCTION

JGJ-2010 (Gu 2004, Sun et al. 2006) stipulates that the maximum distance between expansion joints on frame structures is 55 meters; otherwise it should adopt effective measures to control temperature deformations. To a certain extent, the setting of the temperature expansion joints can influence the function of the building, so ultralong frame construction in industrial complexes can be used in many applications.

In this paper, the finite element model of Mixc ultralong frame structure of Huarun is built based on the software PMSAP. Then the temperature stress and deformation of various regions for the whole structure are analyzed. Some measures are proposed to strengthen the structure when designing Mix according to the change of stress. This provides related suggestions for the whole structure of building decoration and equipment installation

2 PROJECT OVERVIEW

The construction of Qingdao Mixc is invested by Huarun (Shandong) Business Co. Ltd.. The building covers an area of 58625 m², and the total construction area of is about 489533 m². The overall planning design of the Mixc of Qingdao Huarun is shown in Figure 1.

The Mixc project has 3 floors below ground, and it also has L-shaped shopping center above the ground

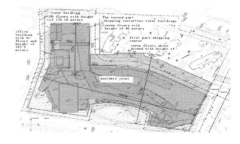

Figure 1. Qingdao Huarun the Mixc overallplanning layout.

plane. The two towers locate in the plane of the short side of the two L-shaped. The roof height of the shopping center with seven floors above ground is 46 m. The office has 41 floors above ground, and its roof height is 183.2 m. The maximum height is 195.8 m. Apartment building has 36 floors above ground, and its roof height is 135.19 m. The maximum height is 143.19 m. Through seismic joints, the above ground structure is divided into two parts: The first part is a long side L-shaped shopping and it is also a ultralong frame structure. And in the middle it does not set any expansion joints. The second part is the short side of the L-shaped shopping center and the two towers consisting of large multi-tower chassis construction, the podium of the outsourcing of size is 78 m × 127 m, and the outsourcing of size is 49 m × 36 m. Figure 2 shows the plane design of the Mix seismic joints.

Figure 2. The plane design of the Mix seismic joints.

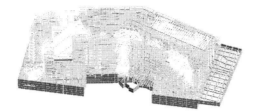

Figure 3a. The Mix L-shaped plane long edge model.

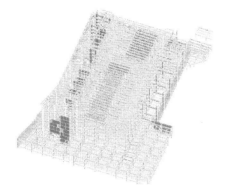

Figure 3b. The Mix L-shaped short edge plane model.

3 THE CALCULATION RESULTS OF TEMPERATURE DEFORMATION AND TEMPERATURE STRESS

The Mixc project of Huron in the L-junction set up seismic joints, and based on the seismic joints boundaries are established. The Mixc L-shaped design of the long side of the plane and the short side of the L-shaped planar model are shown in Figure 3. Because the length of the long side of the L-shaped structure is more than three hundred meters and the intermediate expansion joints are not set, the temperature stress and deformation temperature in this section are the main focus in this paper.

In the Load code for the design of building structures (GB50009-2012), the minimum temperature of Qingdao is −9 degrees, and the maximum temperature of 33 degrees. According to the code, this paper applies the temperature load on the finite element model of the

Figure 4a. Absolute value of the cloud structure deformation temperature is positive.

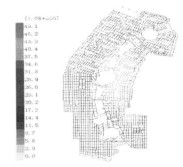

Figure 4b. Absolute value of the cloud structure deformation temperature is negative.

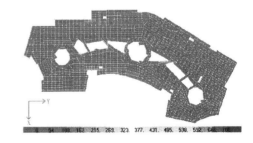

Figure 4c. Typical plane temperature tensile stress.

Mix. The structure of the L3 layer is a regular structure plane. The regular plane temperature deformation and the structure under the action of temperature are shown in Figure 4, and the temperature stress value is shown in Figure 4c.

As shown from the deformation cloud and stress cloud, when the floor appears the overall negative temperature, the slab member produce shrinkage deformation. However, the vertical part of structure formation level is fixed on the beam and plate horizontal part, and the beam slab horizontal component produces much tensions, and the vertical members subjected to horizontal is shearing at the same time. When there is a difference between the whole floors, beam plate horizontal component produces compressive stress. Because the concrete slab structure is subjeted to compressive stress from the concrete slab auto consumption, and is subjected to tensile stress of reinforcement. Therefore, it should be increased to offset the value of tensile stress, so as to avoid crack in construction.

From the deformation program of the Mix, the deformation, local stiffness and the stiffness of smaller parts for local large deformation are known. However, the deformation of the entire structure approximates to the centerline axis under temperature effect on the complete symmetric contraction and stretching. The maximum deformation is about 50 mm. The deformation of the corresponding local large tensile stress is larger, and the Mixc regular surface temperature maximum tensile stress value is about 0.4 MPa. Thus, we should use 8@200 double two-way reinforcement which can generate enough tensile stress so that it can resist negative difference. Due to the influence of temperature stress, reinforcement of the structure should be a double two-way reinforcement, with lack of local additional reinforcement way instead of just using separate reinforcement process.

4 STRENGTHEN MEASURES TO CONTROL THE TEMPERATURE OF DEFORMATION AND TEMPERATURE STRESS

The Mixc structure is an ultralong structure, and the deformation of it may lead to cracks in the slab under temperature shrinkage. Therefore, measures should be taken to reduce the shrinkage of concrete and reduce the rate of cooling structure (Gong et al. 1988, Fu et al. 2010). The concrete measures are as follows:

(1) The expansion of support cast strip can release early shrinkage stress of concrete. As the project size is large, the complex plane can support pouring belt to control the spacing distance in 70 meters, as shown in Figure 5. Post pouring belt should be on both sides of the concrete with pouring two months after placement. After pouring concrete with the same grade micro expansion concrete, the micro expansive shrinkage of concrete can be properly compensated.
(2) The support pouring belt is divided into grids in the same of the warehouse construction. The concrete performance is not stable with no early thoroughly before solidification on the stress release, and the release of further shrinkage stress is shown in Figure 6.
(3) By reducing the hydration heat, there are some measures including selection of low hydration heat 425 ordinary Portland cement and incorporation of fly ash in concrete. Furthermore, in order to ensure the design strength of concrete, there is some need to reduce the amount of cement, and to reduce the water cement ratio, aggregate by 5 mm~40 mm continuous gradation that containing strict control of sand mud and stones within 1.5%.

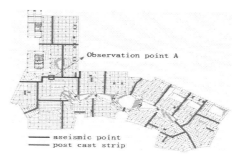

Figure 5. Late poured band layout.

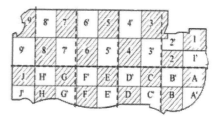

Figure 6. Partition method construction diagram.

(4) Strengthen maintenance work is need after the pouring of concrete, and the heat preservation measures (such as film insulation or insulation covering bag) to extend the hydration heat radiating time. The relaxation properties of concrete is to use the time to relieve the stress caused by temperature changes. At the same time, it should also be paid attention to regular watering maintenance when keeping the initial pouring concrete in wet.

5 ANALYSIS OF THE MEASURED RESULTS OF DEFORMATION TEMPERATURE

To verify the correctness of the model, take the model shockproof joints A point as observation points (Figure 5). They are L long edge endpoints with large deformation and a double column and L shaped structure of plane short edge disconnection. By measuring the relative displacement of double column, the calculating results are compared with the measuring results to validate the model reliability.

Field measurements of the actual deformation observation points at A: A in the construction of Qingdao in summer concrete molding temperature of the reference temperature is 28 degrees. Field measurements in time for the Qingdao winter, environmental temperature is −1 degrees. Temperature is 29 degrees. The measured relative displacement of the observation point and two pillars is 46 mm (Figure 7). The temperature difference applied to the Mix L-shape plane is calculated. And the relative displacement A points is 41 mm, and the calculated results and

Figure 7. The two column graph of displacement at observation points at A.

the measured results are almost the same which can verified the model.

6 CONCLUSION

This paper establishes the Mixc overall design and has carried on the analysis. The reliability of the design are verified by comparing the calculation results and the field measurements. The the deformation and temperature stress are analyzed. Some measures are taken to control the temperature stress and deformation, which include setting the double two-way reinforcement and the temperature expansion joints.

Although the result of deformation has little effect on the local temperature, for ultralong frame structure, the horizontal deformation accumulation value will affect some high precision size component installation, such as elevator track installation, the vertical pipe installation and so on. In addition, the horizontal deformation size decorates in the building process should also be considered. If the renovation process without considering the temperature effect, the temperature effect in the structure after the temperature deformation may destroy the indoor and outdoor decoration effect.

REFERENCES

Fu Xueyi, Yu Weijiang, Huang Yongjun. Ping An finance center vertical deformation analysis [J]. Twentieth National Academic Conference of high-rise building structure of 2010, 876–883

Gu Xianglin. Principle of reinforced concrete structures [M]. Shanghai: Tongji University press, 2004

Gong Luoshu, Hui Yang Bei. Full marks, shrinkage and creep of concrete and practical mathematical expressions of concrete [J]. Journal of building structures, 1988, 5:37–41

Sun Xunfang, Fang Xiaoshu, Guan Thailand. Mechanics of materials [M]. Beijing: Higher Education Press, 2006

Terminal departure passenger traffic forecast based on association rule

S.W. Cheng
Department of Traffic Information and Control Engineering, Harbin Institute of Technology, Harbin, China

H.B. Zhang, J. Xu & Y.P. Zhang
Department of Traffic Engineering, Harbin Institute of Technology, Harbin, China

ABSTRACT: Terminal departure passenger traffic is the basis of dynamic allocation of passenger service resources. This article applies association rule to study the relationship between the departure passenger traffics of various airlines and combine it with fuzzy theory to construct a table of airlines' daily departure passenger traffics. This table can be used to deduce terminal departure passenger traffics of other airlines. Compared with the multiple regression model, this method has incomparable advantages in data acquisition, and the process is simple and easy to operate.

1 INTRODUCTION

The forecast of terminal departure passenger traffic is important to an airport. It is the basis of the dynamic allocation and scheduling of passenger service resources. The forecast is based on the historical and current situations of terminal departure passenger traffic. Scientific methods are used to analyze the trend and provide evidence for configuring passenger service resources. Therefore, the accuracy and operability of the forecast is related to the level of service. There are a great many methods for the farecast. Grosche et al. (2007) proposed two Gravity models for the estimation of air passenger volume. Those two methods could be applied to airports that are currently established or being established. Tsui et al. (2014) employed the Boxe-Jenkins Seasonal ARIMA (SARIMA) model and the ARIMAX model to forecast passenger traffic for Hong Kong; both the models were highly accurate with small forecasting errors. Xie et al. (2014) proposed hybrid seasonal decomposition and least squares support vector regression model to forecast short-term air passenger traffic. These algorithms have problems like overly long computing time, difficult for understanding and lack of simpleness or accuracy, which seriously affect the application of these methods in large hub airports.

This paper, based on association rule data mining method, explores the relationship between daily departure passenger traffics of different airlines. Then, combining with fuzzy theory, a derivation table of airlines was constructed. According to this table, daily departure passenger traffics of other airlines can be inferred.

2 ASSOCIATION RULE

Association rule (Li at el. 2014) reflects regularity between two or more sets of data. It is a kind of important law that can be implied and found.

Association rule mining can be expressed as: $I=\{i_1, i_2,...,i_m\}$ is a set of m numbers of different items $T=\{t_1, t_2,...,t_n\}$ is a database of transactions in which each t_i is an item set and $t_i \subseteq I (1 \le i \le n)$. Association rule is a relationship like $X \to Y$, in which $X \subset I, Y \subset I$ and $X \cap Y = \phi$.

People only pay attention to the association rules that meet certain degree of relevance. Therefore, to find meaningful association rules, appropriate thresholds must be determined. Only when these thresholds are met, it is meaningful to study the association rules.

Common indexes to measure the strength of association rules are *support* and *confidence*.

Support (s): It represents the percentage of the transactions including $X \cup Y$ in T. Support determines how often the rule is used in given data set.

Confidence (c): It represents the percentage of transactions that include X and Y in transactions that only contain X. Confidence determines how often Y appears in transactions that include X.

3 TRAFFIC PREDICTION METHOD OF ASSOCIATION RULES

In all association rule mining algorithms, the most classic is the Apriori algorithm proposed by Agrawal (Agrawal & Strikant 1994) in 1994. Apriori algorithm uses an iterative searching method. According to apriori principle, all subsets of a frequent item set must also be frequent. Therefore, the $k + 1$-item set comes from the k-item set. The specific procedure of the algorithm is described (Anwar & Ahmed 2014) as follows.

(1) Calculate and get all C_1 (Candidate 1-item set).
(2) Scan the database and delete the non-frequent subsets. Collect all items whose support is no less than *minsup* and constitute L_1 (1-frequent item sets).
(3) Form candidate 2-item set by all the elements of L_1. Delete the non-frequent subsets and produce L_2 (2-frequent item sets).
(4) In this analogy, form C_{k+1} (Candidate $k + 1$-item set) through L_k (k-frequent item set), until no frequent item sets generate so far.

This paper fuzzifies the daily departure passenger traffics and passenger load factors of various airlines. Taking each day's passenger traffics and load factor of the two airlines as a transaction, a transaction set was formed and passenger traffic was linked with the concept of association rule (Gong & Chen 2003). Apriori algorithm was used to mine all the association rules in the transaction set that meet the conditions and the derivation table of airlines' daily departure passenger traffics was constructed.

4 EXAMPLE

4.1 Data source

The data of this paper comes from the terminal departure passenger of an international airport of a provincial capital of China in May 2012. The data of the first 20 days in May was used as a training set to construct the derivation table of two airlines' departure passenger traffic based on association rule, and data of the last 11 days as a test set to verify the accuracy of forecast.

4.2 Association rule mining of passenger traffic

Two airlines, China Eastern Airlines and China Sichuan Airlines, were taken as example. The daily departure passenger traffics and load factors of two airlines were regarded as fuzzy variables (Cui & Kim 2014) q and o. The domain of q is $Q = \{200, 400, 600, 700, 800, 1000, 1200, 1400, 1600, 1800\}$.

Letters A to E were used to describe the ratings of departure passenger traffics of the two airlines, with $A > B > C > D > E$. These levels were assigned as Table 1 shows.

The domain of o is $O = \{10, 20, 30, 40, 50, 60, 70, 80, 90, 95\}$.

Also, the 5 levels of A to E were used to represent the ratings of passenger load factor and assigned as Table 2 shows.

The departure passenger traffic q' of another airline was assigned with this approach. To distinguish the two airlines, they were indicated by codes. For example, the level of China Eastern Airlines' departure passenger traffic on one day is A and the passenger load factor is b, while the level of China Sichuan Airlines' departure passenger traffic is B, then the item set can be expressed like {A (MU), b, B (3U)}.

The goal of this paper is to identify the state's rate of the two airlines' departure passenger traffics that is greater than *minsup* and *mincon*. According to the particularity of departure passenger traffic, the *minsup* was set between 75%–85% and the *mincon* greater than 50%. Then, the next task is to find the rate's law of the state of the two airlines' traffics and load factor.

Apriori algorithm was used to mine the departure passenger traffics and load factors of the two airlines in the first 20 days in May. The two-dimensional fuzzy rules of the two airlines' departure passenger traffics and load factor were obtained as shown in Table 3.

Table 1. The ratings of departure passenger traffics of airlines.

Levels		A	B	C	D	E
Departure passenger traffic	200					
	400					
	600					0.2
	700			0.2	0.1	0.7
	800			0.8	0.6	1
	1000		0.2	1	1	0.7
	1200	0.2	0.7	0.8	0.6	0.2
	1400	0.6	1	0.2	0.1	
	1600	0.9	0.7			
	1800	1.0	0.2			

Table 2. The ratings of passenger load factor of airlines.

Levels		a	b	c	d	e
The ratings of passenger load factor	10					
	20					
	30					0.2
	40			0.2	0.1	0.7
	50			0.8	0.6	1
	60		0.2	1	1	0.7
	70	0.2	0.7	0.8	0.6	0.2
	80	0.6	1	0.2	0.1	
	90	0.9	0.7			
	95	1.0	0.2			

Table 3. The two-dimensional fuzzy rules.

	The passenger load factor of MU				
Levels	a	b	c	d	e
A (MU)	–	–	A (3U)	–	–
B (MU)	–	–	B (3U)	B (3U)	–
C (MU)	–	D (3U)	C (3U)	C (3U)	–
D (MU)	E (3U)	E (3U)	–	D (3U)	D (3U)
E (MU)	E (3U)	–	–	–	E (3U)

Table 4. The derivation table of MU and 3U's departure passenger traffics.

q \ o	95	90	80	70	60	50	40	30	20	10
1800	—	—	—	—	—	1800	—	—	—	—
1600	—	—	—	—	—	1600	1600	—	—	—
1400	—	—	—	—	1200	1400	1400	—	—	—
1200	—	—	—	—	1000	1200	1200	1200	—	—
1000	—	—	—	—	800	1000	1000	1000	—	—
800	—	—	—	—	700	—	800	800	800	—
700	—	—	—	600	600	—	—	700	700	—
600	—	—	400	400	400	—	—	600	600	600
400	—	—	200	200	200	—	—	—	400	400
200	200	200	200	200	200	—	—	—	200	200

In this table, "—" represents the situation that did not appear in this mining.

For combinations of all the elements in domain Q and O, the following formulae were used (Liao & Kao 2014).

$$R = \bigcup_{i=1}^{n} R_i, \quad (1)$$

$$q'_i = [(q_i) \times (\Delta o_j)]^{T_2} \circ R, \quad (2)$$

where R_i is the i-th fuzzy relation and R is the total fuzzy relations.

According to the data, the fuzzy sets of Chinese Sichuan Airlines' departure passenger traffic were obtained. The maximum membership degree method was applied to calculate the fuzzy sets clearly. This way, the derivation lookup table of MU and 3U's departure passenger traffics can be constructed, as shown in Table 4.

As long as there is daily departure passenger traffic of MU, according to this table, the data of 3U can be deduced. This way, other airlines' departure passenger traffics can be predicted. Then, all the data can be summed to obtain the terminal departure passenger traffic.

In this paper, MU's departure passenger traffic of the last 11 days in May was regarded as actual data. According to this table, 3U's departure passenger traffic was obtained and compared with the actual data. The results are shown in Table 5.

In the same way, the rest airlines' departure passenger traffics of the last 11 days were forecasted. These data were added to obtain the departure passenger traffic of terminal. The obtained data were compared with actual data, and the curves in Figure 1 were drawn.

As can be seen from the curves, the forecast based on association rules has deviations compared with the actual data. However, considering the instability of the terminal departure passenger traffic, the errors are within an acceptable range. The average error is 16.5%. With the multiple regression model (Huang et al. 2013), the average error is 14.7%. Thus, the accuracies of association rule and multiple regression are almost the same. However, in terms of data availability, association rule has incomparable advantages. The data extracted from the departure control system can be used without parameter estimation and statistical tests, which is quite cockamamie in using multiple regression. It is obvious that the forecast of terminal departure passenger traffic based on association rule ensures the accuracy with easier process.

Table 5. The forecast of 3U's departure passenger traffic.

Date	Actual data	Forecast data	Relative error
21st May, 2012	1169	1287	0.100940975
22nd May, 2012	851	1023	0.202115159
23rd May, 2012	1262	1082	0.142630745
24th May, 2012	1109	924	0.166816952
25th May, 2012	1004	901	0.102589641
26th May, 2012	972	1085	0.116255144
27th May, 2012	1086	1224	0.127071823
28th May, 2012	883	1008	0.141562854
29th May, 2012	1113	1232	0.106918239
30th May, 2012	1130	992	0.122123894
31st May, 2012	904	1005	0.111725664

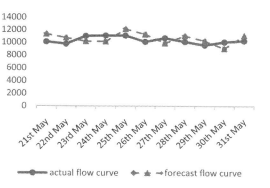

Figure 1. Caption of the actual and forecast flow curves in the last 11 days.

5 CONCLUSIONS

This paper studies the departure passenger traffic of various airlines by mining association rules and constructing a derivation table of airlines' daily departure passenger traffics so that from the known or predicted departure passenger traffic, the data of other airlines can be forecasted. Then, these data can be summed to obtain the departure passenger traffic of the terminal. By researching the data of an international airport of a provincial capital of China in May 2012, it was proved that the forecast based on association rule is feasible. Meanwhile, it can improve the efficiency of prediction algorithm. Due to the ability and time constraints, the accuracy of this method still needs to be improved. This method can be considered in the future to combine with neural networks, trend extrapolation and so on, to improve the forecast accuracy.

ACKNOWLEDGEMENTS

This study was sponsored by the National Natural Science Foundation of China (Project No. 61179069, U1233124) and the Natural Science Foundation of Heilongjiang Province, China (Project No. E201114).

REFERENCES

Agrawal, R., & Srikant, R. 1994. Fast algorithms for mining association rules. *Proc. 20th int. conf. very large data bases, VLDB*. Vol. 1215: 487–499.

Anwar, M. A., & Ahmed, N. (2014). Analyzing Lifestyle and Environmental Factors on Semen Fertility using Association Rule Mining. *Information and Knowledge Management*. Vol. 4: No. 2: 15–21.

Gong X. Y. & Chen W. X. 2003. Study of relationship among adjacent road segments traffic based on association rules mining. *Highway and Transportation Research and Development*. 20(4): 63–66.

Cui, R., & Kim, & H. J. 2014. A new approach to preserve privacy data mining based on fuzzy theory in numerical database. *Fifth International Conference on Graphic and Image Processing International Society for Optics and Photonics*. 90691A–90691A.

Grosche, T., Rothlauf, F. & Heinzl, A. 2007. Gravity models for airline passenger volume estimation. *Journal of Air Transport Management*. 13(4): 175–183.

Huang B. J., Lin J. S., Zhen X. Y. & Fang X. D. 2013. The prediction for civil airport passenger throughput based on multiple linear regression analysis. *Mathematics in Practice and Theory*: 172–178.

Li, Y., Thomas, M., & Osei-Bryson, K. M. 2014. Using Association Rules Mining to Facilitate Qualitative Data Analysis in Theory Building. *Advances in Research Methods for Information Systems Research Springer USA*: 79–91.

Liao, C. N., & Kao, H. P. 2014. An evaluation approach to logistics service using fuzzy theory, quality function development and goal programming. *Computers & Industrial Engineering*. 68: 54–64.

Tsui, W. H. K., Ozer Balli, H., Gilbey, A., & Gow, H. 2014. Forecasting of Hong Kong airport's passenger throughput. *Tourism Management*. 42: 62–76.

Wei, Y., & Chen, M. C. 2012. Forecasting the short-term metro passenger flow with empirical mode decomposition and neural networks. *Transportation Research Part C: Emerging Technologies*. 21(1): 148–162.

Test and analysis of the vacuum degree transport within plastic drainage board

Tianyun Liu
Tianjin Port Engineering Institute Ltd. of CCCC, Key Laboratory of Port Geochnical Engineering of Ministry of Communications, Key Laboratory of Port Geochnical Engineering of Tianjin, Tianjin

Huizhen Kang
Military Transportation University, Tianjin

Aimin Liu
Tianjin Port Engineering Institute Ltd. of CCCC, Key Laboratory of Port Geochnical Engineering of Ministry of Communications, Key Laboratory of Port Geochnical Engineering of Tianjin, Tianjin

ABSTRACT: In this paper, the vacuum degree transport patterns along depth within the B-type and C-type plastic drainage boards were tested and analyzed. The results showed that due to good longitudinal water passing capacity in both the plastic drainage boards, the vacuum degree transport losses along depth are few, and higher vacuum degree is still left in deep level. Moreover, the vacuum degree transport needs certain time to process, thus the vacuum preloading pumping time should be increased in order to facilitate the reinforcement of the deep soil.

Keywords: vacuum preloading; plastic drainage board; vacuum degree

1 INTRODUCTION

The basic principle of the vacuum preloading method is to use vacuum equipment in the foundation laid by the sealing film to withdraw the gas and water in the horizontal drainage blanket level under the sealing film gradually, and then form the vacuum degree in horizontal drainage blanket. Then, the vacuum degree gradually extends downward through the vertical drainage channel set in the foundation, promotes the drainage channel and reduces boundary pore water pressure. Furthermore, pressure difference and hydraulic gradient can form between the pressure and pore water pressure in the soil. With the difference of pressure and the hydraulic gradient, gas and water in the soil pore flow from the soil to the vertical drainage channel, then to the horizontal drainage blanket, and finally are brought together to the vacuum filter tube to be pumped. Thus the foundation soil is settled, consolidated and reinforced (Chen Huan. 1991, Gao Zhiyi. 1989). Because of its good water passing capacity, plastic drainage board has been widely used as the vertical drainage channel vacuum when the soft ground is processed by vacuum preloading. The field practice has proved that there will be the vacuum degree losses during the transfer process in the plastic drainage board, and such losses are caused by the resistance within the plastic drainage board. The vacuum degree losses have important impacts on the speed of the consolidation of the soil and the reinforcement effect, especially for deep soft soil reinforcement (Lou Yan, 2003).

Currently, there are few measured results about the vacuum degree transport within plastic drainage board. Zhang Zepeng et al. (2002) have compared the vacuum degree transport capacities of plastic drainage plate and packed sand within 10 m-thick reinforced soil. The vacuum transfer loss rate is 2.2 kPa/m within the plastic drainage board (domestic SPB-l-type), and is 6.3 kPa/m within the packed sand. The vacuum degree of the plastic drainage board at the depth of 10 m accounted for 68% of the vacuum degree under the film. The vacuum degree of the packed sand wells is 20% of the degree of vacuum (Zhang Zepeng et al. 2002). Zhu Qunfeng et al. (2010) have tested the vacuum degree within plastic drainage plate at a maximum test depth of 7.4 m, and the attenuation average value of the vacuum degree along the depth was 10 kPa/m (Zhu Qunfeng et al. 2010).

As seen from the existing research, the test depth of the vacuum degree in plastic drainage board is relatively shallow, and the measurement results are quite different and need further study. Moreover, in recent years, the engineering cases requiring using the vacuum preloading method to reinforce deep soft ground emerging are emerging continually, and the driven depths of the plastic drainage board are also increasing continually. Therefore, it is of great significance to study the transport and loss of vacuum degree within plastic drainage board, in order to promote the further development of vacuum preloading.

2 TEST

To understand the vacuum degree transport pattern along the set depth of plastic drainage board, experimental study was conducted in the vacuum preloading at the construction site to measure the vacuum degree values along different depths within the different types of plastic drainage board. The method of probe observations of the pore water pressure was adopted to measure the vacuum degree. Firstly, embed the pore water pressure probes in the plastic drainage board. Then observe changes of pore water pressure at different depths within the plastic drainage board during the vacuum preloading the process. At last, indirectly anti-calculate transport law of the vacuum degree along the depth of the plastic drainage board.

2.1 Project overview and experimental program

The engineering test area lies at the Dongjiang Port in the Tianjin Port Logistics Processing Zone. The vacuum preloading method was used for foundation treatment. The area belongs to the reclamation areas, and soil of the test areas consists of reclamation silt, silty clay, silty clay, and silt from top to bottom.

B-type and C-type plastic drainage plates were selected as vertical drainage channel of the test area. The performance indicators of the plastic drainage boards are shown in Table 1. Different types of drainage board were set up to different depths; the depths were 25 m and 35 m for the B-type and C-type plates respectively. Along the plastic drainage board, the pore water pressure measuring probes were buried in the plastic drainage boards at different depths. The number of probes and the spacing between adjacent probes for both types of plastic drainage boards are both shown in Fig. 1. Probes with small size and high accuracy were used in the pore water pressure measurements.

The plastic drainage boards were clipped with corresponding buried length and a small mouth was cut on the membrane, where the pore water pressure probe was set at a distance of 30~40 cm to the plastic drainage board. The wire probe entered into the mouth of the membrane and was placed at the designed depth, and the wires stay out of the plate. Next, seal the cut mouth with dense lines and waterproof adhesive plaster, and fix the wires at cut mouth bilateral and probes inner plates with a fine lead wire. Then cover another layer of filter membrane of the same model with the drainage board to prevent the soil enters into the plastic drainage board from the cut to the blocking the water during the test. Straightening the wires from different positions and lashing together with adhesive tape, and then fixing them on the drainage board. The plastic drainage plates which have been set up with the pore water pressure measuring probes are shown in Fig. 2.

2.2 Test results and analysis

2.2.1 Analyses about the variation pattern of vacuum degree at different depths with time

The vacuum degree in the plastic drainage boards are measured after the gas is exhausted formally. The measuring frequency is one time per two to four days, and the measurement lasted for 76 days. The test results

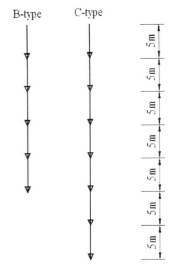

Figure 1. Pore water pressure measuring probes distribution along the depth of the plastic drainage boards.

Table 1. Main performance indicators of the plastic drainage boards.

Main performance indicator		Unit	Types B	Types C	Remarks
The applicative driven depth		m	≤25	≤35	
Sectional dimension	Thickness	mm	4	4.5	
	width	mm	100	100	
Longitudinal flow capacity		cm^3/s	35	45	Lateral pressure 350 kPa
Permeability coefficient K_{20}		cm/s	4.6×10^{-3}		The specimen was soaked in water for 24 h

were plotted as a curve of the vacuum degree at different depths in the plastic drainage plate versus time, as shown in Figs. 3 and 4.

As Figs. 3 and 4 show, the vacuum degree under the membrane increased rapidly with the vacuum pumping time. It reached more than 80 kPa after 15 days of pumping. However, there is a greater variation in the early values of the vacuum degree under the membrane due to construction reasons, and the values at later period maintained above 80 kPa.

Moreover, the vacuum degree in the plastic drainage board quickly increased once the jet pump started to work. At the beginning, the increments of vacuum degree at various depths were almost the same. The differences among the values of vacuum degree at various depths increased with time; the values at shallow depths increased sharply and those at deep depths increased slowly. In the later period, the increase of the vacuum degree values at the shallow depths became slow, while there is still upward trend in that at the deep depth. It means that the transport of the vacuum degree needs time. Therefore, increasing the pumping time of vacuum preloading is propitious to the reinforcement of the deep soil.

The vacuum degree at the shallow depths is prone to be affected by changes of the values of the vacuum degree under the membrane. The shape of the curve of the vacuum degree at a depth of 5 m is almost identical with that under the membrane. The effect applied by the vacuum degree under the membrane to the vacuum degree at the shallow depths gradually decreases along with the increasing depth, and the curve shows a gradual upward trend. It is due to the existence of the well resistance, the fluctuations of the vacuum degree at the shallow depths are gradual decreased during the process transferring to the deep depths.

2.2.2 Analyses on the curves of the vacuum degrees vary along the depth in plastic drainage board

Taking the measurement results of pumping vacuum for 76 days for example, curves of the vacuum degree vary along the depth for two types of plastic drainage plate are drawn as Fig. 5 shown.

It shows that the decline of the vacuum degree is the biggest at the depth 0~5 m in the plastic drainage board, which is because the vacuum degree will have a greater local loss when it is varied from the sand cushion into vertical plastic drainage board. Besides the mutation in the top of the plastic drainage board due to partial loss, the curves of the vacuum degree along the depth are approximate to the straight-lines under 5 m. The transmission loss rate of vacuum degree along the depth in B-type plate is about 1.39 kPa/m, and that

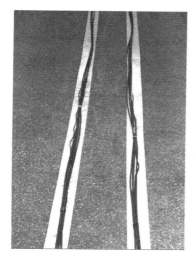

Figure 2. The plastic drainage board set with the probes.

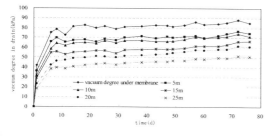

Figure 3. Variations of vacuum degree at different depths in the B-type plate with time.

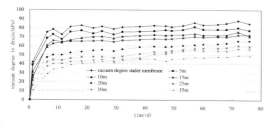

Figure 4. Variations of vacuum degree at different depths in the C-type plate with time.

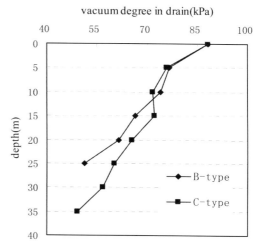

Figure 5. The vacuum degree varies along with the depth.

in C-type plate is about 1.02 kPa/m. Moreover, it shows that the friction loss of the vacuum degrees along the plastic drainage board is less. Thus the vacuum degree is still 49.2 kPa in the C-type plastic drainage board at underground 35 m depth with the loss of 44.1%. It indicates that the well resistance is much smaller using the plastic drainage board as vertical drainage channel.

The test results show that when the vacuum degree under membrane is between 80~90 kPa, the vacuum degree in the plastic drainage plate is between 50~80 kPa substantially. However, the current design is that the vacuum degrees along the depth within the plastic drainage board are all counted as 80 kPa generally, which will lead to larger soil residual settlement after vacuum preloading.

Besides, the vacuum degree at the same depth in the B-type plate is less than that in the C-type plate, and friction loss are larger than that in the C-type plate, which is due to the thickness of the C-type plate is greater than the B-type plate, the longitudinal drainage ability is preferable, and the well resistance is relatively small.

3 CONCLUSIONS

To conclude, the test results of the vacuum degree in the B-type and C-type plastic drainage boards were analyzed. It was shown that the longitudinal drainage ability of the plastic drainage plate was preferable, and the well resistance was relatively small. The vacuum degree decreased along the depth, yet the decline was relatively small. Therefore, there was still relatively high vacuum degree in deep level. The results indicate that when using the vacuum preloading method to reinforce the deep soil and ensure the deep reinforcement effect, the C-type plastic drainage board which has preferable longitudinal drainage ability should be used as the vertical drainage channel. At the same time, because the vacuum degree needs a pass time process, the vacuum preloading pumping time should be increased in order to facilitate the reinforcement of the deep soil.

REFERENCES

Chen Huan. 1991. Study on The Mechanism of Vacuum Preloading Pressure Method for Ten Years, *Port Engineering*, 4: 17–26.

Gao Zhiyi. 1989. Analysis on Mechanics of Vacuum Preloading Method, *Chinese Journal of Geotechnical Engineering*, 4: 45–56.

Lou Yan. 2003. Influence of Vertical Drain on Effect of Vacuum Pre-loading, *Highway*, 6: 90–93.

Zhang Zepeng, Li Yuejun. 2002, The Effect of Plastic Drain Board in Atmospheric Pressure Softbase Reinforcement, *Journal of Guangzhou University (Natural Science Edition)*, 1:68–71.

Zhu Qunfeng, Gao Changsheng. 2010, Transfer Properties of Vacuum Degree in Treatment of Super-Soft Muck Foundation, *Chinese Journal of Geotechnical Engineering*. 9:1429–1433.

The analysis of anti-seismic safety behavior of split columns with a core of reinforced concrete

Xiujuan Chu & Yang Liang
Faculty of Civil Engineering and Mechanics, Kunming University of Science and Technology, Kunming, China

ABSTRACT: Combine the big shear-span ratio of the split reinforced concrete column with high axial-compression ratio of the core column, and the short column is changed into "long" columns directly, at the same time we can reduce the cross-sectional area of the column. Thus the concept of split core-column is put forward. It recommends the peculiarities of split core-column, covering bearing capacity experimentation and application investigation. It is shown that the compressive bearing capacity of the column has increased, and the capacity of shearing unchanged, but the deformation capacity and ductility of the column have also increased remarkably. Besides, the failure pattern of the column is changed from the shearing pattern to the flexural pattern, In addition, improving the anti-seismic behavior of the reinforced concrete short columns and the anti-seismic safety of the reinforced concrete tall building structures significantly.

1 INTRODUCTION

Anti-Seismic behavior-based design demands reinforced concrete structures to meet the requirements that they should have strong column and weak beam, strong cut and weak bending, strong nodes, strong anchoring, and strong column root. The purpose of this design is to meet three designed criteria, which are, behaviors of the structure – no collapse in case of large earthquakes, being repairable in the middle, and no damage in small earthquakes. This way, the purposes of safety can be achieved by preventing accidents, eliminating hidden dangers, saving costs, and obtaining the best results. However, because of the characteristics of high-rise buildings, such as tallness, being multi-floored, big volumes, the lower floors' column axial force and shear are very strong, resulting in the problems of short column or sometimes even super short columns (easily occurring in the equipment layers, because of the small distance between floors), and increasing possibility of shearing brittleness in the case of earthquakes. To meet the limit requirements of the axial compression ratio and shear span ratio, researchers (Jinglong et al. 2005, Jinglong & Yuguang 2001, Junqing & Tingting 2005, Zhong et al. 2001, Cuilian et al. 2003) solve the problem of the ration limitation of shear span using the technology of splitting columns, and solves the problem of the axial compressive ratio (Zhongxian 2005) using the reinforced concrete core column technology.

This research was conducted based on the above two achievements; the reinforced concrete split core-column technique was used to completely solve the problem of low ductility and large sections of short columns, by making full use of the ductility of the reinforced concrete split-column and the strong bearing capacity of the reinforced concrete column with a core.

2 WEAKNESS OF THE SHORT COLUMN

2.1 Definition of the short column

According to the current architectural earthquake-resistant design code (GB50011-2010 2010) and concrete structure technique code for tall buildings (JGJ3-2010 2010), the shear-span ratio was used to define the reinforced concrete short columns. For the lower columns of high-rise buildings, the criteria of the shear-span ratio is $\lambda = M^c/(V^c h_0)$, where Mc is the bending moment calculation of the combination of the end-column cross-section, which can be the bigger value for the upper and lower sections of the column; Vc is the value of the combined shear force corresponding to M^c; h_0 is the calculated height of the column section. Inflection-point was in the middle of the frame column and the shear-span ratio was defined as $\lambda = H_{C0}/(2\,h_0)$, where H_{C0} is the net height of the frame column.

Therefore, (1) a short column has the shear-span ratio $\lambda \leq 2$; (2) a super short column has the shear-span ratio $\lambda \leq 1.5$ (Aifeng 2003, Zhongxian 2003).

2.2 Dangers of short column

A reinforced concrete short column has the properties of low rigidity, low deformation capacity, and low ductility (Martirossyan & Xiao 2001, Yashiro et al. 1990, Jianzhuang et al. 2002). The investigation of earthquake damage (Watanabe 1997, Anon 1994) and the analysis of the simulative experimental results have shown that the damage conformations of a short column are significantly different from that of a long column (Jinglong et al. 2005). Brittle fractures, such as slipping along obliquely fissured sections, severe concrete peeling and so on, easily happen and endanger the whole structural safety (Yongchang & Zhongxian 2002, Qingchang & Yunfei 1989). Typical brittleness damages include the slip along oblique fissure sections and severe concrete peeling. The main damage conformations are oblique dragging damage, oblique pressing damage, shear dragging damage, shear pressing damage, cementing damage, and high axial shearing damage. For all the damages, cracks spread along nearly entire height of the column. After the oblique cross cracks went through the column, the strength decreases dramatically and the damage occurs suddenly, causing collapse and disastrous accidents.

3 SPLITTING COLUMNS AND CORE COLUMNS

3.1 *The characteristics of splitting columns*

Shearing damage is liable to happen if the shear-span ration is too big. Therefore, a splitting column technique has been put forward to improve the shear-span ration (Jinglong et al. 2005, Jinglong & Yuguang 2001, Junqing & Tingting 2005, Zhong et al. 2001, Cuilian et al. 2003). Of this technique, clapboards are used to split the short columns of the whole section into two, four or more even-sectional unit columns, and every unit column is reinforced separately (Jinglong et al.2005). A number of experiments and researches (Martirossyan & Xiao 2001, Yashiro et al. 1990, Jianzhuang et al. 2002, Watanabe 1997, Anon 1994, Jinglong et al. 2005, Jinglong & Yuguang 2001, Junqing & Tingting 2005, Zhong et al. 2001, Cuilian et al. 2003, Shuting et al.1997) have shown that although splitting column technique decreases the bending capacity of a column a little bit, the shearing capacity remains unchanged. However, the deformation capacity and ductility of the column increase a lot and its damage conformation changes from shearing to bending, which is the implementation of the concept of changing short columns to "long" columns. This change can effectively improve the earthquake-resistance of short columns, especially super short columns. On the other hand, since a splitting column itself is an ordinary column, the axial compression ratio limits its section, which remains big.

3.2 *The characteristics of core columns*

Researches and engineering practices have shown (Zhongxian 2005) that a concrete frame column with a core column added in the middle can effectively increase the axial bearing and deformation capacity, and decrease the compression of the column (Shuting et al. 1997, Yongchang & Zhongxian 2002, Aifeng 2003). Moreover, the frame can still be able to resist the vertical load and prevent structural collapse in case of a great earthquake, when cracks, peeling or even destructive concrete appear surrounding the column. The reason is that the vertical steel bar of the column does not bend when the pressure is imposed on it, and can keep the good sticky condition. Architectural earthquake-resistant design code (GB50011-2010 2010) says that the limit of the capacity of the section can be increased but only up to 0.05, if the sizes of the cross sections of the core column adhere to the code and the total area of the longitudinal steel bar of the core column is no more than 0.8% of the area of the column's cross section. The fifth footnote of Figure 6.4.2 of the concrete structure technique code for tall building (JGJ3-2010 2010) says "when this step is taken with the forth footnote (compound pinch, screw pinch, continuous compound screw pinch), the limit of axial-compression ratio can be 0.15 more than the given value in the table. But the value of λ_v is determined as the value of the axial-compression ratio is increased 0.10". The setting of core column is conducted on the condition that the axis-compression ratio is satisfied. By doing so, the column section area is decreased, and the construction is more convenient than those of steel-tube concrete column and steel core concrete column.

4 SEISMIC RESISTANT PROPERTY AND LOADING TEST OF THE REINFORCED CONCRETE SPLITTING CORE-COLUMN

4.1 *Model test*

As was expressed in figure 1 and 2, four splitting core columns were designed for test (Jinglong et al. 2005).The net height of the samples was 800 mm and the cross sections were squares with the sides of 400 mm. A 10 mm-paper plasterboard was used to divide each column into 4 dependent reinforced unit core-columns. Each unit column had a standard longitudinal bar (SRB335) on each side. The core column had a standard longitudinal bar of 4Φ10, and a stirrup (SPB235) of $\phi 6@40(30)$. The thickness of the masonry protective layer was 12mm. A whole cross sectional core domain was set on both the top and the bottom edge of the sample columns. This core domain was equipped with $4\phi 8@30\#$ compound pinches.

A load-repeat equipment made in Japan was used in test. First, axial-force was added to make the axis-compression ratio 10% more than the limit value of the criterion, and the axial-force was kept invariable during the experiment. Then, low-circle flat repeating load was added. The low-circle flat repeating load was controlled by the system of load-distortion. It means that before the submit load, load was added by load, repeating once every time, whereas after the submit load, load was added by distortion, and by multiple submit

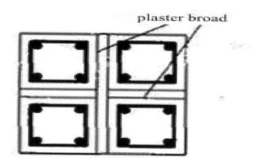

Figure 1. Dimension and reinforcement of split core-column.

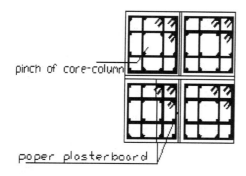

Figure 2. Sections of split core-column.

displacement, repeating three times every adding-load, till every first adding-load of that class load decreased to 85% of the top load. By then the split column was destroyed.

In the experiment, resistance emergency patches were stuck to the portrait steel bar and stirrup to test their change. At the top and bottom of this kind of places, the change of the concrete surface was tested by millenary watch. The relative horizontal shift was tested by shift sensing machine at the top and bottom of the column. And the crack distribution of load-horizon-shift (P-Δ shift) are noted by function machine.

4.2 Data analysis result

According to the experiential result, conclusions were as follows: the designed axis-compression ratio was 0.75 for specimens Z-1 and Z-2, and 0.80 for Z-3 and Z-4. The rates of stirrup were different; when it was increased from 0.91% to 1.21%, specimens forms of destruction changed. Z-1 was destroyed by the big partiality compression on one side and high compression shearing on the other. Z-3 was destroyed by the big partiality compression on one side and boundary partiality compression on the other. Compared with Z-1, Z-3 had a larger area of sluggish loopback, good capacity of wasting energy, and decreased bearing capacity of the attenuation rate, at the same time, a larger shift angle between the limit layers. Namely, it could be prevented from splitting core-column destroyed from slanting section by the rate of stirrup increase (Xiu-juan & Gui 2006a, b). Z-4 was destroyed by the high compression sheer on one side, then Z-2 was destroyed by the big partiality compression on one side and small partiality compression on the other. Compared with Z-1, the area of the sluggish loopback of Z-2 enlarged, the capacity of wasting energy increased, the bearing attenuation rate decreased, and the limit shift angle between layers enlarged. The designed axis-compression ratio was 0.8 for specimens Z-3 and Z-4, the rates of stirrup were different, and both specimens were both crushing binding destruction. Compared with Z-2, with the rate of stirrup of Z-4 increased, the ductility modulus increased from 5.44 to 4.75. However, the bearing capacity was not influenced greatly, the bearing attenuation decreased, and the limit shift angle between layers enlarged. That is to say, the seismic capability of splitting core-column was improved by the increase of the rate of stirrup. The designed axis-compression ratio of specimens Z-1 and Z-2, Z-3 and Z-4 were different, yet with little contrast among the shift ductility moduli, and the limit shift angle between layers decreased slightly. It was shown that the bearing capacity and the seismic capability of splitting core-column were not influenced distinctly by the increase of axis-compression ratio.

There were full transition sections on the top and bottom of the four pieces, which decreased the shear-span ratio. The limit horizon shift values were not large, the bearing attenuation kept steadily, and the limit shift angle between layers kept small. At the same time, seen from the change of four stirrups in the full transition section, the nearer to the clapboard slot, the more change of the stirrup, which shows that if there weren't full transition section, parting crack would have entered the node section (GB50011-2010 2010). This is to say that they have the same bearing capacity and transmutation.

5 THE ANALYSIS OF INTEGER DOMINO EFFECT OF THE SPLITTING CORE-COLUMN AND THE CALCULATION OF AXIS-COMPRESSION RATIO

5.1 The overall domino effect modulus of the splitting core-column

According to the reference (Jinglong et al. 2005), to calculate the side shift under horizon load, the fully domino effect of clapboard should be taken into account. The bearing of splitting core-column should be predigested according to the inner bent characteristic of the column section. Under the influence of horizon load P, the bending moment of the end column with the single split column with a core is (Jinglong et al. 2005):

$$M_c = \frac{PH}{8} - M_f = \frac{PH}{32}(4-3\beta) \quad (1)$$

where P is the horizon load, H is the net height of the column, β is the integer domino effect modulus, and

$$\beta = \frac{\frac{\Delta}{H}}{1.2(\frac{\Delta}{H}-0.3)^2 + \frac{\Delta}{H}} - 0.07 \quad (2)$$

5.2 The account of axis-compression ratio of the split core-column

The formula to account the axis-compression ratio of the splitting core-column is as follows:

$$\mu = \frac{N}{f_c A} = 1.69\frac{C_c}{f_{ck}A} + 1.69\frac{N_{sx1}+N_{sx2}}{f_{ck}A} = n_1 + n_x \quad (3)$$

where n_x is the increase of the axis-compression ratio to the splitting core-column, and it approximately adopts $h_0/h = 0.93$, then

$$n_x = 0.845\rho_x \frac{E_s}{f_{ck}}(0.93\varepsilon_{cu} - 1.07\varepsilon_y) \quad (4)$$

where ρ_x is the reinforced rate of vertical bar, $\rho_x = A_{scor}/A$ (currently it's more than 1.5%). When SRB335 and SRB400 steel are used, $Es = 2.0*10^5$ N/mm², then ε_y is 0.0017 or 0.002 separately. According to literature (JGJ3-2010 2010), when $f_{cuk} \leq 50$ MPa, substituting $\varepsilon_{cu} = 0.0033$ in Eq. (4), and when the vertical bar is Class II SRB335,

$$n_x = 211.25\frac{\rho_x}{f_{ck}} \quad (5)$$

when vertical bar is class III SRB400,

$$n_x = 157.13\frac{\rho_x}{f_{ck}} \quad (6)$$

With Eqs. (5) and (6), reinforcement ratio n_x and core column linear correlation ρ_x are known. However, when the concrete strength grade of split column with core improves, the effect of column core reinforcement ratio ρ_x on n_x relatively decreases (Zhongxian Li 2005).

$$n_x = 0.845\rho_x \frac{E_s}{f_{ck}}(0.93\varepsilon_{cu} - 1.07\varepsilon_y). \quad (7)$$

It's informed by Eqs. (5) and (6) that n_x is correlated in linearity with the reinforced rate ρ_x of core column. However, when the concrete intensity of the splitting core-column increases, the infection degree of the reinforced rate ρ_x of core column to n_x decreases accordingly (Anon 1994).

6 CONCLUSIONS AND EXPECTATION

The reinforced concrete split core-column has the common virtues of the reinforced concrete splitting column and the reinforced concrete core column.

(1) The short column is changed into "long" columns directly, improving the anti-seismic behavior of the reinforced concrete short columns and the seismic resistance of the reinforced concrete tall building structures significantly. Their failure pattern is changed to ductility from crack failure. There is no collapse in the case of large earthquakes, and the aim of the accident loss minimum is achieved.

(2) Using the increase of core column to axis-compression ratio, combining high concrete, the section dimension of column can be decreased properly, then good useful room and economy benefit are achieved. (The increased value of axis-compression ratio of column is connected with the reinforced rate of core column, the class of concrete intension and the class of vertical bar and so on.)

(3) The core column reinforcement ratio (x) is the main influencing factor for improving the axial compressive ratio under the condition where the boundaries are destroyed.

(4) It is proved that the whole section transition zone will be set on the top and low end of the split column, and can effectively prevent the split crack of reinforced concrete column with a core stretching and expanding towards the core area of the top and low node sections, thus ensuring the anti-seismic performance of the frame beam-column joints.

(5) To ensure safety of the mining engineering, producers leave a considerable amount of Tiberium spikes, ore wall, so the waste of resources is serious. Using reinforced concrete column with a core fission can realize the Tiberium spike, ore wall underpinning, and meet the demand for bearing capacity and the anti-seismic behavior. Currently with coke reserves steelmaking limited, using this method, the underpinning will get satisfactory results.

REFERENCES

Aifeng Wei, 2003. Treatment of Reinforced Concrete Short Column, Special Structures, Vol. 20(4): 35–36.

Anon, 1994. Lessons learned from the Northridge Earthquake. *Modern Steel Construction* 34(4): 24–26.

Cuilian Wan, Zhongchang Wang & Shiwei Xie. 2003. Improving Technical Measures for the Strength of Reinforcing Bar Concrete Column. *Construction & Design for Project* 8: 37–38.

GB50011-2010.2010: National Standard of the People's Republic of China. *Code for Seismic Design of Buildings* [S]. Beijing: China Building Industry Press.

JGJ3-2010.2010: The industry standard of the people's Republic of China. *Technical specification for concrete structures of tall building* [S]. Beijing: China Building Industry Press.

Jianzhuang Xiao, Jun Yao & Lei Xia. 2002. Analysis of Shear Resistance Capacity of Extremely Short R.C. Stub Post. *Special Structures* 19(1): 5–7.

Jinglong Pan & Yuguang Wang. 2001. Effect of Sectional Shape of Concrete Column on Bearing Capacity of Short Columns Wrapped with FRP. *Industrial Construction* 31(2): 17–19.

Jinglong Pan, Wei Wang, Xinan Jin & Chenyuan Wang. 2005. A Study on the Properties of FRP-Confined Short Reinforced Concrete Column Subjected to Eccentric Loading. *China Civil Engineering Journal* 38: 46–50.

Junqing Guo & Tingting Xu. 2005. The Analysis Core Column Influences on the Limit Values of Axial Compressive Ratio for the Concrete Columns. *Industrial Construction* (35) Sup. 175–178.

Martirossyan A. & Xiao Y. 2001. Flexural-shear behavior of high-strength concrete short columns. *Earthquake Spectra* 17(4): 679–695.

Qingchang Hu & Yunfei Xu. 1989. A New Way to Improve the Seismic Performance of Rectangular Section Reinforced Concrete Short Column. *Building science* 3: 1–8.

Shuting Liang, Yongsheng Jiang, Minjie Zhu, Zongque Lu & Xiaofeng Yang. 1997. Experimental Research of Earthquake Resistant Behavior for High Strength Concrete Short Column with X Shaped Reinforcement. *Journal of Southeast University* 27 Sup. 33–38.

Watanabe F. 1997. Behavior of reinforced concrete buildings during the Hyogoken~Nanbu earthquake. *Cement & Concrete Composites* 19(3): 203~211.

Xiu-juan Chu & Gui Fu. 2006a. The Seismic Safety Behavior of Reinforced Concrete Split Core-columns. *Journal of Lanzhou University (Natural Sciences)* Vol. 42: 640–643.

Xiu-juan Chu & Gui Fu. 2006b. The Seismic Safety Behavior of Reinforced Concrete Split Core-columns. *Twenty-third Annual Meeting of The Society for Organic Petrology (TSOP)* (Abstracts and Program, Volume 23). 61, September 15–22, Beijing, China

Yashiro H., Tanaka Y. & Nagano M. 1990. Study on shear failure mechanisms of reinforced concrete short columns. *Engineering Fracture Mechanics* 35(1–3): 277–289.

Yongchang Hao & Zhongxian Li. 2002. The study of the method to improve seismic performance of reinforced concrete short columns. *Building structures* Vol. 32(10): 8–10.

Zhong Fan, Jiaru Qian & Xuemin Wu. 2001. Experimental Study on Seismic Behavior of Concrete Columns with Central Reinforcement. *Journal of Building Structure* Vol. 22, 22(1).

Zhongxian Li, Wenzhang Yuan & Yongchang Hao. 2003. Problem of Short Columns and Technology of Split Columns in Industrial and Civil Reinforced Concrete Buildings. *Industrial Construction* Vol. 33(11): 63–66.

Zhongxian Li. 2005. Theory and Technology of Split Reinforced Concrete Columns. *Engineering Mechanics* Vol. 22 Sup. Jun. 127–141.

The analysis of urban-rural transit trip characteristics using a Structural Equation Model

Pengfei Li & Xiaohong Chen
School of Transportation Engineering, Tongji University, Shanghai, China

Yingfei Tu
School of Economics and Management, Tongji University, Shanghai, China
Institute of Railway and Urban Rail Transit, Tongji University, Shanghai, China

ABSTRACT: With the accelerating pace of urbanization in China, passenger travel demand between urban and rural areas has been rapidly increasing. This has led to significant changes in the traditional city transit and highway passenger transport systems along with the diversification of trip purposes. Based on survey data from Rui'an city on urban-rural public transit, qualitative and quantitative analysis was conducted on the trip purpose and individual characteristics. A Structural Equation Model (SEM) was then implemented between the trip purpose and individual characteristics, using AMOS (Analysis of Moment Structure) software. Results show that there is a correlation between individual characteristics and trip purpose. Type of traveler and traveler's age are directly related to the trip purpose while latent personal and work attributes are indirectly related to trip purpose. This study improves on past literature by focusing on the relationship between urban-rural transit trip purpose and individual characteristics for the urban-rural transit planning, construction and services.

Keywords: Structural Equation Model, urban-rural transit, trip characteristics

1 INTRODUCTION

With the acceleration of urbanization in China, passenger travel demand between urban and rural areas has increased along with an increased diversity in trip purposes. It has been demonstrated that there are correlations between individual trip behavior and individual characteristics[1]. Research on the relationship between trip behaviors, individual characteristics and trip purposes of urban-rural travelers is important to improve the planning, construction and provision of public transit services for urban-rural passenger transportation.

Ding wei al. (2008) have reviewed the research on individual trip behavior theory[2]. Traditional analysis methods on trip behavior often ignore that the trip behavior is a derived need of an individual's activity[3–5]. Therefore, when investigating this for different groups, (occupation, income, etc.), the traditional method may not be an accurate reflection of the impacts on the individual's choice behavior.

Lu Xuedong & Eric (1999) discussed the relationships among socio-demographics, activity participation and travel behavior by using the structural equation modeling (SEM) methodology[6]. Using trip survey data of residents in Portland in 1994, Golob analyzed the relationship between trip chain quantities, activity duration and travel time by using SEM[7]. With the development of software for SEM, Golob reviewed the use of SEM for travel behavior research[8]. Li et al. investigated the relationship between residents' activities and trips during holidays using SEM and the impacts of residents' socioeconomic attributes on their activities and trip generations[9]. Zhou et al. proposed SEM for travel demand analysis and investigated the relationship among travelers' characteristics, activity participation and travel behaviors[10].

In all, most research about individual's travel behavior concern the trip chain, mode choice and impacts while focusing on automobiles in cities. Research on urban-rural transit has largely been limited to network layout, infrastructure planning, and operation management and policies. The lack of research on the relationship between individual characteristics and travel demand characteristics which impact infrastructure planning and bus line operations will be addressed in this paper.

Urban-rural transit is a new kind of passenger transportation mode arising from the need for urban-rural integration; between the traditional urban transit systems and highway passenger transportation systems. This new system serves the passengers traveling between urbanized areas and rural areas within the same administrative area. Its management system is different from traditional urban transit and the traditional highway passenger transportation for several reasons. Since the passenger flow of the urban-rural transit is a unique situation, the cost cannot be covered through the fare income alone. Also, these systems do not have enough government support because there is no urban transit subsidy. However, urban-rural transit plays an important role in the urbanization process and urban-rural integration. Therefore, research on the characteristics of urban-rural trips is useful for the development of urban-rural transit.

The city of Rui'an is located in the southern part of Zhejiang province. It ranked 36 in the list of top 100 Chinese counties in 2011 for population with typical industry-promoted urbanization. During the urbanization process of the city, a large number of external long-term migrant workers were attracted to Rui'an for economic opportunities. Many local rural residents from the same administrative area have also moved to the city to work. These rural residents generally focus their trip characteristics on commuting, business trips and relaxation. This paper establishes a structural equation model for trip purpose and individual characteristics based on survey data of urban-rural transit in Rui'an. The software AMOS (Analysis of Moment Structure) is applied to solve the model.

2 ANALYSES ON TRIP PURPOSE AND INDIVIDUAL CHARACTERISTICS

Data used in this study were obtained from the Rui'an urban-rural bus passenger trip characteristics survey. The survey was conducted at the urban-rural passenger transportation lines and hubs and includes the traveler's residence and their personal status. The survey sent out 1,985 questionnaires. After correcting for errors and missing data, 1,719 valid samples have been obtained. In the questionnaire, four attributes about the individual were collected: type, age, monthly income and occupation along with seven types of trip purpose: work, school, business, shopping, entertainment, visit and other.

2.1 Relation between traveler's type and trip purpose

There are four types of traveler in the sample: local urban resident, local rural resident, external resident (more than 6 months), and external floating population (less than 6 months).

The results from the survey show that 47.2% are local rural residents and 31.5% are local urban residents. This indicates that urban-rural transit serves mainly local residents and especially local rural residents.

Table 1 shows that for the four traveler types, rigid travel demand comprises the majority (63%–80%) of travel demand. With the development of economic and urban-rural integration, more rural residents are living in rural areas and working or studying in urban areas. This leads to traveling for work, school and business accounting for 74% of the total travel demand.

2.2 Relation between the traveler's age and trip purpose

Table 2 shows the ratio of trip purposes across different age groups. For travelers who are between 20 years old and 45 years old, trips for work and business account for 70.1%. For elder travelers, the elastic trips are more than others due to many in this age group being retired.

2.3 Relation between monthly income and trip purpose

The monthly income have been classified into 6 classes: (1) <1000¥; (2) 1000~2500¥; (3) 2500~3000¥; (4) 3000~4000¥; (5) 4000~5000¥;

Table 2. Ratio of Trip Purposes for each traveler age.

		<20	20~45	>45
Rigid demand	Work	6.8%	49.1%	29.1%
	School	84.3%	2.3%	1.4%
	Business	1.3%	21.0%	20.0%
	Subtotal	92.4%	72.4%	50.5%
Elastic demand	Shopping	1.8%	7.5%	18.6%
	Entertainment	2.2%	5.9%	8.6%
	Visit	2.8%	6.2%	8.6%
	Others	0.9%	8.1%	13.6%
	Subtotal	7.6%	27.6%	49.5%
Total		100%	100%	100%

Table 1. Ratio of Trip Purposes for each traveler type.

		Local urban residents	Local rural residents	External residents	External floating population
Rigid demand	Work	41%	28%	46%	38%
	School	24%	33%	6%	4%
	Business	15%	13%	22%	21%
	Subtotal	80%	74%	74%	63%
Elastic demand	Shopping	6%	9%	6%	5%
	Entertainment	3%	6%	6%	11%
	Visit	5%	5%	6%	11%
	Others	6%	6%	8%	11%
	Subtotal	20%	26%	26%	37%
	Total	100%	100%	100%	100%

Table 3. Ratio of Trip Purposes for each class of income.

		(1)	(2)	(3)	(4)	(5)	(6)
Rigid demand	Work	7.1%	49.7%	46.5%	48.3%	46.9%	36.8%
	School	76.1%	2.2%	0.8%	0.6%	0.0%	0.0%
	Business	3.1%	16.4%	21.2%	25.3%	29.6%	35.3%
	Subtotal	86.3%	68.3%	68.5%	74.2%	76.5%	72.1%
Elastic demand	Shopping	5.2%	8.2%	10.3%	8.4%	3.7%	4.4%
	Entertainment	2.1%	6.8%	5.7%	7.3%	6.2%	8.8%
	Visit	3.1%	7.4%	6.8%	3.4%	8.6%	7.4%
	Others	3.4%	9.4%	8.7%	6.7%	4.9%	7.4%
	Subtotal	13.7%	31.7%	31.5%	25.8%	23.5%	27.9%
Total		100%	100%	100%	100%	100%	100%

Table 4. Ratio of Trip Purposes for each kind of occupation

		(1)	(2)	(3)	(4)	(5)	(6)
Rigid demand	Work	83.1%	26.0%	63.2%	17.0%	0.7%	44.4%
	School	0.0%	0.0%	0.7%	3.4%	95.1%	0.0%
	Business	6.0%	34.9%	13.1%	18.4%	0.9%	0.0%
	Subtotal	89.1%	60.9%	77%	38.8%	96.7%	44.4%
Elastic demand	Shopping	1.2%	11.5%	4.8%	29.3%	0.5%	11.1%
	Entertainment	1.2%	7.8%	6.6%	6.1%	1.2%	0.0%
	Visit	3.6%	10.1%	4.2%	10.2%	1.4%	22.2%
	Others	4.8%	9.7%	7.4%	15.7%	0.2%	22.2%
	Subtotal	10.9%	39.1%	23%	61.2%	3.3%	55.6%
Total		100%	100%	100%	100%	100%	100%

(6) >5000¥. Table 3 shows that for travelers whose income is less than 1000¥/month, the main trip purpose is to study. Those with income between 1000 and 3000¥/month account for 50.6% of the sample. Their trip purposes are mainly work and business with elastic trips comprising about 31.5%. For travelers with income higher than 3000¥/month, work and business trips are 75%. In the sample, 73.8% of travelers' incomes are lower than 3000¥/month and only 8.7% are higher than 4000¥/month. This indicates that for lower income people, transit is their main travel mode, while for higher income people the proportion taking bus is very low.

2.4 Relation between occupation and trip purpose

In the survey, six kinds of occupations have been provided: (1) government employee; (2) private owner; (3) company employee; (4) peasant; (5) student; (6) other.

The results are shown in Table 4. For employees, work is the primary trip purpose. For private owners, business is their main trip purpose. For peasants, urban-rural transit is mainly utilized for their elastic demands, which account for 61.2% of their trips.

3 MODELS FOR TRIP PURPOSE AND INDIVIDUAL CHARACTERISTICS AND SOLUTION

From the above analysis concerning trip purpose and individual characteristics, we now consider if there is correlation between traveler's type, age, income, occupation and trip purpose. In this section, a Structural Equation Model is applied to discover this correlation.

3.1 Structural Equation Model

Structural Equation Modeling (SEM) is a type of statistical method often incorporated to analyze the relationship between variables based on the covariance matrix, also referred to as covariance structure analysis. The purpose of this technique is to investigate the causal relationship between things and express this relationship through a causal model, path graph, etc. SEM has the following advantages compared with traditional statistical methods: 1) hidden variables can be measured through one or more manifest variables; 2) the relationship between unobservable variables can be calculated according to observed manifest variables; 3) measurement errors of variables are allowed in the model; 4) the relationship between hidden variables and manifest variables can be estimated simultaneously by the model; 5) the model can be expressed both by a mathematical model and a path graph. Therefore, SEM is often applied in travel behavior analysis research. The model equation is as follows[12–15]:

1. Measurement equation:

$$\chi = \Lambda_\chi \xi + \delta \qquad Y = \Lambda_y \eta + \varepsilon \qquad (1)$$

where χ is a q × 1 vector with q exogenous variables; Y is a p × 1 vector with p endogenous variables; Λ_χ is q × n factor loading matrix that χ on exogenous latent variable; δ is q × 1 vector composed of q measurement errors of χ; Λ_y is p × m factor loading matrix that Y on endogenous latent variable; ε is p × 1 vector composed of q measurement errors of Y.

2. Structural equation:

$$\eta = B\eta + \Gamma\xi + \zeta \qquad (2)$$

where η is an m × 1 vector with m endogenous latent variables; ξ is n × 1 vector with n exogenous latent variables; B is m × n coefficient matrix, describing the impacts among endogenous latent variables; Γ is m × n coefficient matrix, describing the impacts between exogenous and endogenous latent variables; ζ is m × 1 residual vector.

The evaluation of the SEM is conducted through the comparison of the following indices with anticipated values in order to determine model fit.

1) P-value, a probability based on diversity factor and degree of freedom;
2) CMIN/DF;
3) IFI, Incremental Fit Index;
4) TLI, Tucker-Lewis Index;
5) CFI, Comparative Fit Index;
6) RMSEA, Root Mean Square Error of Approximation.

3.2 *Model construction*

For the survey data, individual information is considered a latent variable, which includes traveler type, age, monthly income and occupation. Trip purpose is considered as the manifest variable. The initial model is hypothesized as shown in Figure 1.

After solving the model with AMOS, the goodness of fit test results is shown in Table 5. It shows that the fitted values of IFI, TLI and CFI are all close to 1 as desired with a P-value less than 0.05; indicating a rejection of the null hypothesis. Also, the value of CMIN/DF is much greater than 2 while the value of RMSEA is also not fitted well. Therefore, it appears that the initial model needs to be modified.

3.3 *Model corrections*

There are two methods to correct the model: 1) add new hidden variables or manifest variables in order to modify the relationship between the variables and

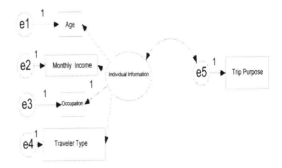

Figure 1. Hypothesized Structural Equation Model.

Table 5. Hypothesized Structural Equation Model Test of Goodness of Fit.

Fit Index	Fitted Value	Desired Value	Explanation
P	0.000	>0.05	Not Good
CMIN/DF	12.922	<2	Not Good
IFI	0.964	Close to 1	Good
TLI	0.928	Close to 1	Good
CFI	0.964	Close to 1	Good
RMSEA	0.083	<0.08	Not Good

2) change the initial fixed parameters to free parameters and reevaluate the model. In the AMOS software, modification indices (MI) are provided to reflect the consequence on the estimated covariance after adding parameters to the model in order to indicate the modification direction. Larger MI means better modification. The threshold value of MI is generally 3.84 or 6.63, corresponding to level of significance of 0.05 or 0.01 under degree of freedom of 1.

In the initial model shown above, for e4, the value of MI is 13.940; for e1, the value of MI is 42.047. This indicates that traveler type and age are correlated with trip purpose. In general, it is accepted practice to modify one variable in one model correction in order to better judge the precision of the correction.

Here, two corrected models were proposed as shown in Figure 2.

In model (a), a new latent variable 'job attribute' was proposed, which is directly related to monthly income and occupation. This new variable along with traveler type and trip purpose is impacting each other. In model (b), the direct correlations between traveler type and trip purpose and between age and trip purpose were considered. The difference between the two corrected models concerns whether the variables traveler type and age belong to one latent variable.

Table 6 shows the corrected models' goodness of fit tests. The table shows that both of the models are accepted with high goodness of fit. Therefore, the corrected models are suitable approaches that can be applied to analyze travel behavior.

Based on the corrected models, the structural relationship between trip purpose and individual

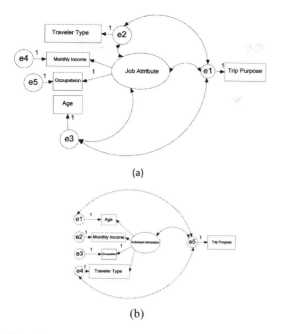

(a)

(b)

Figure 2. Two Corrected Structural Equation Models.

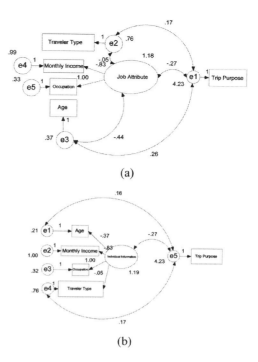

(a)

(b)

Figure 3. Corrected Structural Equation Models Paths Chart.

Table 6. Corrected Structural Equation Models Test of Goodness of Fit

Fit Index	Fitted Value Model (a)	Fitted Value Model (b)	Desired Value	Explanation Model (a)	Explanation Model (b)
P	0.227	0.253	>0.05	Good	Good
CMIN/DF	1.446	1.358	<2	Good	Good
IFI	0.999	0.999	Close to 1	Good	Good
TLI	0.997	0.998	Close to 1	Good	Good
CFI	0.999	0.999	Close to 1	Good	Good
RMSEA	0.016	0.014	<0.08	Good	Good

characteristics is illustrated in Figure 3. In the figure, the direction of arrows indicates causality while two-way arrows indicate correlation. The figures at the arrows are the correlation coefficients and those besides the circles are the residual variances of the variables.

Corrected Model (a) considers the correlation between job attribute and the other three manifest variables, which is relatively complex. Corrected Model (b) considers the traveler type and age in the individual characteristics, which is easier to understand. According to the goodness of fit tests, Model (b) is slightly better than Model (a).

4 CONCLUSIONS

Using survey data of urban-rural transit in Rui'an, this study analyzed the correlation between trip purpose and individual characteristics. AMOS was then used to apply SEM to further understand this relationship. The findings are:

(1) There is a significant correlation between trip purpose and traveler's individual characteristics. Traveler type and age is directly correlated with trip purpose. Other variables were found to be indirectly correlated with trip purpose through the latent variable "individual information". The direct correlation between traveler type, age and trip purpose can be useful for predicting urban-rural transit system's passenger flow, network planning, operation and management.

(2) Urban-rural transit plays an important role in urban-rural area integration. The main purpose is to serve residents' rigid travel demand, which accounts for 74.8% of all trips. Analysis has shown that urban-rural trips have gradually turned from elastic demands (such as shopping and visit), to rigid demands (such as commuting). In the industry-promoted urbanized areas, it is thought that this proportion will increase in future years. Thus, the development of urban-rural transit should adjust according to new passenger flow characteristics in order to satisfy the travel demand of different types of travelers.

REFERENCES

Ben-Akiva M, Bowman J L. Activity-based Disaggregate Travel Demand Model System with Activity Schedules [J]. Transportation Research A, 2001, 35(1):1–28

Bhat C R, Koppelman F S. Activity-based Travel Demand Analysis: History Result and Future Directions [C]. Paper Presented at the 79 Annual Meeting of the Transportation Research Board, Washington, DC, January

Chen J. Research on Travel Characteristics of Urban Residents in Public Transport [D] 2008. Beijing: Beijing Jiaotong University

Ding W., Yang X. and Wu S. A review of activity-based travel behavior research [J]. Human Geography, 2008, 23(3): 85–91

Golob T F. A Simultaneous Model of Household Activity Participation and Trip Chain Generation [J]. Transportation Research B 2000, 34:355–376

Golob T F. Structural Equation Modeling for Travel Behavior Research [J]. Transportation Research B. 2003, 37:1–25

Hagerstrand T. What about people in regional science [J]. Regional science, 1970, 24(2): 7–21

Jin Y. From Regression to Structural Equation Model: Linear causal modeling methodology [J]. Shandong Economy, 2008, 3(2): 19–24

Li X., Shao C., Sun Z. et al. Modeling Correlation of Holiday Trips and Activities Based on Structural Equation Model [J]. Journal of Transportation Systems Engineering and Information Technology, 2008, 8(6): 91–95

Lou Z. Basic Steps of Structural Equation Modeling [J]. Productivity Research, 2005, 6: 201–202

Lu Xuedong & Pas E T. Socio-demographics, Activity Participation and Travel Behavior [J]. Transportation Research A, 1999, 33A:1–18

Ren Y. Research on Going-on-work Trip Behavior [D]. Tianjin: Tianjin University, 2009

Tian F. The Construction of Social Indicator System with Structural Equation Modeling Technology [J]. Journal of Anhui University (Philosophy and Social Sciences), 2007, 31(6): 92–95

Zhang J. Comparison of Structural Equation Model Building Methods [J]. Statistics and Decision, 2007, 9:137–139

Zhou Q, Li Y., Meng C. et al. Analysis of travel demand based on a structural equation model [J]. Journal of Tsinghua University (Science and Technology), 2008, 48(5): 879–883

The application of creeper and overhang plants on exterior facades of Guangfu dwellings

Shiyang Zhang
Beijing Normal University, Zhuhai, China

ABSTRACT: There has been vertical greenery in China since ancient times. Because of the warm climate and type of building materials used, Guangfu dwellings have created an abundant plant facade landscape. This paper took village dwellings in parts of the Pearl River Delta for the research location. The distribution and forms of creeper and overhang plants on exterior façades were summarized, their functions analyzed, and their advantages and disadvantages summed up. The application of creeper and overhang plants on exterior facades of ancient Guangfu dwellings lends inspiration to renovations of ancient buildings and design of modern architecture.

Keywords: Guangfu dwellings; creeper and overhang plants; vertical greening

1 INTRODUCTION

Guangfu dwellings are located in the Pearl River Delta, which is the widest distribution of traditional houses in the Lingnan area. Historically, a combination of wood, soil, and rock was used for building materials, while present day dwellings are made of brick and stone construction.

The tropical and subtropical monsoon maritime climate of the Pearl River Delta is characterized by hot summers and relatively warm winters, while the annual rainfall is abundant and the air humidity is high. Affected by climate, the local vegetation has rapid growth and high species diversity. The plants expanded from the ground and water surfaces to the surface layers of buildings. The large gaps in the building structures became the cradles of creeper and overhang plants. The rainy and moist air was an essential nutrient for the plants' growth, contributing to the fusion of the dwellings with the greenery that includes mosses, herbs, and small ligneous plants.

However, there are several hazards that arise from plants growing on the surfaces of ancient buildings. One is that the roots of small ligneous plants burrowing deep inside the building walls may keep growing and expanding, forming large cracks and even resulting in partial rupture or collapse. Another problem is the plants provide breeding grounds for mosquitoes that may spread disease to the residents.

Taking into account of all the above aspects, it is considered that the least disruptive to the ancient buildings and the minimal interference to local residents are the creeper and overhang plants.

1.1 *Creeper and overhang plants*

Creepers, also known as climbing vines, are commonly characterized by slender stems that generally need to cling to something in order to grow upward. Overhanging or epiphytic plants hang over or downward while they grow on structures or the branches of other trees. Plants that grow downwards due to cantilevered planting beds are also classified as overhang plants (Zang de kui, 2000).

2 DISTRIBUTION OF CREEPER AND OVERHANG PLANTS

2.1 *Eaves*

Vertical plants of the Guangfu dwellings originated from the eaves of the buildings. The soil in the gaps of the tiles mixed with rain and moisture vapor in the air created the perfect environment for growth, so the eaves became the concentrated growing areas of climbing and hanging plants. When the residents gradually realized the aesthetic rewards of the climbing and hanging plants, they began to cultivate them to create the affects of green doorways. Commonly used plants for this purpose are *Scindapsus aureum* (Figure 1), and *Clerodendrum thomsonae* and *Nephrolepis auriculata (L.) Trimen* (Figure 2).

2.2 *Surface layers of exterior walls*

To get the effect of a large green facade, there is a method which uses the exterior layer of the walls as the

Figure 1.

Figure 2.

Figure 3.

Figure 4.

Figure 5.

Figure 6.

starting point of the climbing plants. *Parthenocissus tricuspidata* (Figure 3) and *Hedera nepalensis K,Koch var.sin ensis (Tobl.) Rehd* are commonly used in this way. Another way to achieve this look is by planting vines near the inside walls, and leading the vine over the top to hang down the exterior wall. Commonly used plants for this purpose are *Pyrostegia venusta (Ker-Gawl.) Miers* (Figure 4) and *Campsis grandiflora*.

2.3 Planting beds in the wall corners

Parts of side streets have no separate courtyards or prominent eaves above the doors and are relatively simple. In order to break up the monotonous gray tones, residents often create planting beds in the corners between the exterior walls, or even insert small planting beds in the walls themselves. These planting beds provide soil and water to allow the creeper to grow up the wall or overhang to the ground. All of these can create a beautiful lush green landscape on the exterior facades (Figure 5). Plants commonly used for this purpose are *Lonicera japonica Thunb.*, *Parthenocissus tricuspidata* and *Bougainvillea spectabilis Willd* (Figure 6).

Figure 7.

Figure 8.

3 FUNCTION

3.1 Heat insulation

The existing Guangfu dwellings were mostly built with bricks and stone. The shortcomings of these materials can be the high conductivity of heat. After the absorption of heat in summer, the exterior wall will store heat and relay it to the interior of the wall, which leads to the temperature of the entire dwelling being increased. It was discovered that, between similar scale dwellings, the ones that had creeper and overhanging plants on the exterior facade would be cooler than the houses without plants by approximately two degrees. Large areas of western-facing facades covered by creeper and overhang plants may be particularly shadier and cooler.

3.2 Beautification of dwellings

The facades of Guangfu dwellings are mostly shades of gray, which is monotonous and may be slightly oppressive. Planting creeper and overhanging plants can increase the beauty of a building significantly and can result in changes of color and texture that add beauty to the entire landscape. Another thing it can do is make a building more noticeable and recognizable where there is no house number or name. The facade of dwellings can be distinguished with a different variety of creeper and overhang plants or different colors of plants (Figure 7). Meanwhile, many commonly used creeper and overhang plants are flowering shrubs in southern China, such as *Bougainvillea spectabilis Willd*. These flowering shrubs can ensure a varied scenery in all seasons.

4 DISADVANTAGES

Through surveys of Guangzhou Xiaozhou, Foshan Gaozan, and Zhuhai Huitong villages, there was a common legacy of large old Guangfu residential facades which are mostly of a single color and artificial vertical greening is relatively uncommon. The newly renovated houses are usually covered with creeper and overhang plants, while the ancient dwellings put more emphasis on showing the status of the owner by using heavy masonry and meticulous carving details. Unlike large scale buildings, the small scale dwellings will consider using vertical greening to decrease the inside temperatures and increase the building's identity. On the other hand, with the rapid development of tourism and creative industries, more and more old villages have started to renovate ancient buildings for carrying out commercial activities. People now are more focused on the modern aesthetic sense of color and pay more attention to green plants, ecology and other factors. Many businesses use innovative plants in the building facades to attract customers, which spawn a growing number of creeper and overhang plants on ancient buildings (Figure 8).

5 TECHNOLOGY AND APPLICATION

To ensure the safety of exterior walls and to avoid the problems of plant roots, creeper and overhang plants mainly have two forms: inside the wall and outside the wall.

If a courtyard is connected to an exterior wall, the internal corner can be used to set up a large planting bed for large vines to grow over the top of the wall and hang over the exterior wall. This scenario would need at least 2.2 meters of soil for commonly used *Pyrostegia venusta (Ker-Gawl) Miers*, *Campsis grandiflora*, Campsis and *Lonicera japonica Thunb* (honeysuckle). For *Pyrostegia venusta (Ker-Gawl) Miers*, for example, arrange the planting bed with approximately 30–60 cm of soil thickness. After the vine grows to 70 cm, set up a trellis for the vine to climb. After about three years of growth, this provides beautiful scenery.

Planting outside of the wall can also be broken down into two types, corner planting and planting in the exterior layers of walls. The second type is primarily used for small hanging plants which often utilize ordinary flower pots hanging on the walls or inlayed in the exterior walls.

Corner planting by exterior walls is different than planting inside the walls. Usually creepers climb walls or use pergolas to climb, and therefore, the plants usually do not grow above the tops of walls. Plants commonly used in this situation include *Parthenocissus tricuspidata, Hedera nepalensis K,Koch var.sinensis (Tobl.) Rehd* and *Bougainvillea spectabilis Willd*. *Parthenocissus tricuspidata*, for example, is a shallow-rooted plant and its planting soil thickness can be reduced to just 20–30 cm. There can be a dense green wall after two years made up of plants arranged at intervals of approximately 50 cm.

These two methods do not require special anti – seepage treatments or water protection. Compared with roof greening and vertical geometric greening, these plantings of creeper and overhang plants are easier for people to maintain and cultivate to their own landscaping method. The only maintenance is on the ground and more convenient than on the tops of the dwellings.

6 SUMMARY

The exterior facades of dwellings have been used for planting vegetation for a long time, where wise ancestors learned that eaves can grow plants quite easily. By cultivating greenery, they not only improved the microclimate of the dwellings, but also beautified the landscape and enriched the visual effects. Due to limited ground space, today's plant life leans more toward roof gardens than vertical greening, but the concept uses both vertical greening research and what can be learned from ancient buildings. Representatives of vertical greening, Nanhaiyiku and Shenzhen form beautiful scenery from buildings, using different varieties of creeper and overhang plants such as *Parthenocissus tricuspidata* and *Pyrostegia venusta (Ker-Gawl.) Miers*. Another benefit aside from the beauty is the temperature control. Visitors can feel at least two degrees cooler indoors than in surrounding buildings without exterior plant life (Wu Yuqiong, 2012).

In the future, the development of vertical greening is optimistic. The use of creeper and overhang plants on ancient Guangfu dwellings can be used in conjunction with scientific research and cultivation technology advancements to benefit the public.

REFERENCES

Wu Yuqiong, The application of new vertical greening technology in architecture, South China University of Technology, 10–16, 2012.

Zang dekui, Zhou shujun, creepers and vertical greening, Chinese garden, Vol. 16, 2000 (5).

Civil Engineering and Urban Planning III – Mohammadian, Goulias, Cicek, Wang & Maraveas (Eds)
© 2014 Taylor & Francis Group, London, ISBN 978-1-138-00125-1

The comparison of multi-ribbed frame structure filled with phosphogypsum with conventional frame structure

Fei Liu
Guizhou Building Science Research & Design Institute limited company of CSCEC, Guizhou Guiyang, China

Fujie Wang
Guizhou Electric Power Design Institute, Guizhou Guiyang, China

ABSTRACT: This paper takes multi-ribbed frame structure made of phosphogypsum, which is a kind of energysaveing material, as the researching object., via the mathod of ANSYS model analysis, combining with the scale model experiment of lateral force test and repeated horizontal load test, in order to get the he failure pattern, the hysteresis curve and the force distribution rule of this new structure system, provid experimemtal basis and technical parameter to the establishment of practical analysis method.

Keywords: gesso, energy-saving, multi-ribbed frame structure, new type

1 INTRODUCTION

Phosphogypsum was favorable for the formation of the wall materials of energy saving and the integration of

The multi-ribbed grid frame structure system is a new type of vertical structure, taking phosphogypsum as the wall material, with the characte of both energy-saving and integration with the structure. It is different from the conventional reinforced concrete frame structure, by casting in spot, the three main energy-saving materials of reinforced concrete building (Yunshou Huang. 2007), the reinforcement, concrete and phosphogypsum, formed a multi-ribbed net format frame structure system with higher lateral stiffness against general frame structure, in the vertical direction, along the floor height each wall is divided by $n \geq 4$ or more lines of grid frame beams, the size of the beam cross section is far less than the horizontal continuous beam of the convention frame structure. (Jianke Ma etc. 2006) While in the horizontal direction, within the original large column grid distance, the wall is divided by $m \geq 6$ lines of vertical grid columns, the column section size is much smaller than conventional frame column (Jianke Ma etc. 2006). The original structure of 1×1 grid in each floor, is now divided into $n \times m$ grids which formed the grid frame structure, under the horizontal loads, the lateral force resistant structure has a much larger shear stiffness than the conventional frame structure, its lateral stiffness is also increasing rapidly. Because the concrete and gypsum is cast in spot and segment by segment, in this kind of structure, gypsum is not only the filling wall inside the grid, but also the cover of concrete columns and beam, the cover is usually as thick as 50mm, which can stop the effect of cold bridge, heat preservation and heat insulation effect of this kind of structure is better than the conventhion method (Jianke Ma etc. 2007).

2 RIBBED GRID FRAME COMPARED WITH CONVENTIONAL FRAMEWORK ANSYS8.1 EXAMPLE ANALYSIS

On the base of the conventional framework, the concrete frame wall in each floor is divided into $n_x \times n_y$ grids concrete wall with small cross section columns and beams. Because of the differences of the calculation diagram, their mechanical properties vary greatly. To illustrate their respective effect of the mechanical properties, based on the condition of the same lateral force, the same span and floor height,an analysis of the finite element method (fem) is adopted to tell the different mechanical properties.

2.1 Calculation conditions

The element type of Beam4 is adopted of both beams and columns in the calculation and analysis process. In order to consider the mechanical characteristics in the high-rise building components, make sure that the wall height to width ratio is more than 3, the cross section size of 1/5 of the original size of the structure is

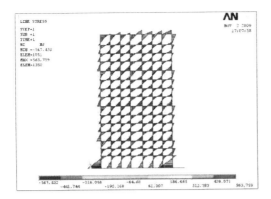

Figure 1. Bending moment of multi-ribbed grid wall.

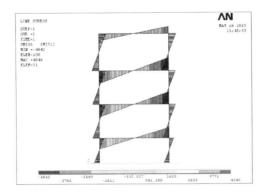

Figure 2. Bending moment of conventional frame wall.

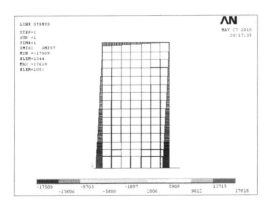

Figure 3. Axial force of multi-ribbed grid wall.

adopted in experiment to analyze two kinds of horizontal displacement and internal force distribution rule of the wall.

2.2 The calculation results

A finite element model is set up via the software of ANSYS8.1, the force distribution of multi-ribbed net

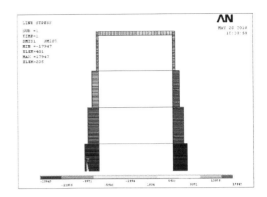

Figure 4. Axial force of Conventional frame wall.

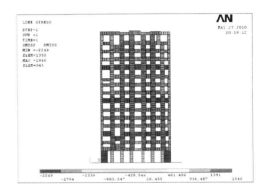

Figure 5. Shear diagram of ribbed grid wall.

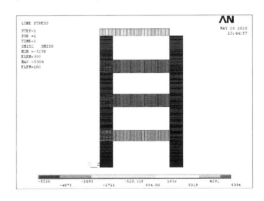

Figure 6. Shear diagram conventional frame wall.

format frame wall and conventional frame wall under the same action of lateral force is respectively shown in the figure below Fig. 1~Fig. 6.

3 ANALYZING THE RESULTS OF TWO WALL CALCULATION

3.1 Vertex displacement value comparison

ANSYS calculation example above, the multi-rib grid frame structure's vertex displacement is 3.49 mm, while conventional frame structure is 25.26 mm, obviously the biggest displacement of the former is far less than the latter.

3.2 Internal force value comparison

In ANSYS numerical example above, to the multi-ribbed reticulated format frame structure, the maximum bending moment is 0.60 kN · m, the maximum axial force is 17.62 kN, the maximum shear force is 2.25 kN, while to the conventional framework, the maximum bending moment is 10.53 kN · m, the maximum axial force is 17.95 kN, the maximum shear force is 5.38 kN, obviously the internal force effect of conventional frame structures is much more significant than multi-ribbed grid frame structure.

4 SCALE MODEL TEST

4.1 The test content and purpose

This article mainly research in quasi static test of multi-ribbed net format frame structure made of phosphorus and convertion frame structure under the same conditions, the experimental analysis and theoretical research of both kind of structures are conducted. Internal force, hysteretic curve, ductility, energy dissipation capacity are analyzed in comparison. the mechanical properties and deformation characteristics of the multi-ribbed net format frame structure is verified through the experiment. Due to space limitation, in the guarantee of good aspect ratio (>3) of multi-ribbed grid plate frame structure, only small component (1/5 scale model) is experimented.

4.2 The experimental process

Take a piece of the multi-ribbed net format frame made of phosphorus gypsum as a specimen, Numbers for ML – 1; Column section is the rectangular, side column section is large, while the middle column section is small, beam section is also rectangular. Take a piece of conventional frame structure wall as another specimen for experiment. Numbers for CK – 1; Column and beam section are all rectangle. Limited by the height of the reaction wall, the height of each specimen is 3600 mm, divided into four floors, the floor height is 900 mm. before casting concrete, the two specimens are fixed to respective support beam, with the fixed connection as a whole, making it the cantilever support conditions, to imitate the effect of a fixed support.

4.3 Test apparatus and loading system

This experiment conducted in the lab with 1000 kN reaction wall and 500 kN hydraulic servo control system. Horizontal loading used laboratory MTS loading system for low frequency horizontal load. Specimen loading process is divided into two stages: the first stage is completely elastic stage, the stage is divided into five loading levels. when the stress in the wall at the bottom of the column reinforced reaches to yield, we take the displacement control the loading, which is also the second phase, the displacement of the control values in horizontal displacement is the integer times of Δ_y, according $1\Delta_y$, $2\Delta_y$, $3\Delta_y$, $4\Delta_y$, cyclic loading until the specimen is damaged. Due to the limitation of equipment, no vertical force is considered in the experiment.

4.4 Test point selection and layout

The position of foil gage is decided according to ansys analysis of the internal force distribution, displacement meters are fixed at the top and middle of each spcimen. resistance strain gauge was laid at the end of each coupling beam, and some key points.

5 THE MAIN ANALYSIS OF EXPERIMENTAL RESULTS

5.1 The whole process of loading

From the beginning of loading to the destruction of the whole specimen, both of the two specimens have experienced three stages, namely uncrack stage, crack stage and failure stage.

Coverd with phosphorus gypsum, the wall of grid frame structure can't see the crack in the concrete column and beam. When the horizontal force p = 10 kN, at the bottom of the specimen ML-1 on the site, small vertical and oblique micro cracks appeared in phosphogypsum. With the continuous increasing of horizontal loading, new diagonal cracks apear in a higher floor of the multi-ribbed structure, and then developed into horizontal cracks, and the original cracks become larger (Jianke Ma etc. 2008), lower floor of transverse cracks even well versed in, and in each floor, there are two inclined cracks by the central of the joint, gradually changed into horizontal crack, these are all big cracks, just like two "V" shapes. After cracking, the specimen goes into the stage of fracture, the lateral stiffness is reduced, the vertex horizontal displacement increases, more and more horizontal cracks appears at the bottom of the wall, crack width increases, susequently, the longitudinal reinforcement yields (Jianke Ma etc. 2008). When Yielding, specimen vertex horizontal load P was 20 kN, vertex horizontal displacement for Δy was 2.5 mm.

To the conventional frame structure wall. when the horizontal force P = 8.5 kN, vertical cracks appeared at the end of the upper and lower of the first floor joints in the specimens of CK-1. As the load increasing, the beam-column joint of the 2nd, 3rd, 4th floor also appeared angle cracks in the vertical. After cracking, specimen get into the stage of fracture, the lateral stiffness is reduced, the vertex horizontal displacement increases, at the bottom of the concrete column more and more cracks appeared on the horizontal direction, crack width increased, then the wall at the bottom of

Table 1. Ribbed grid frame compared with the conventional frame lateral stiffness.

Methods	Specimens	Lateral stiffness K(10^3 N/mm)
Test	ML-1	(2.1/0.12 + 4.2/0.33 + 6.3/0.59 + 8.4/0.69 + 10.5/0.96)/5 = 12.8033
	CK-1	(2.1/2.38 + 4.2/4.32 + 6.3/11.37 + 8.4/15.06 + 10.5/24.65)/5 = 0.6785
The finite element of ANSYS analysis	ML-1	(2.1/0.695 + 4.2/1.394 + 6.3/2.093 + 8.4/2.792 + 10.5/3.491)/5 = 3.0122
	CK-1	(2.1/2.575 + 4.2/5.095 + 6.3/15.171 + 8.4/17.692 + 10.5/25.257)/5 = 0.5891

Table 2. Two kinds of specimens ductility ratio.

Specimens	ML-1	CK-1
The yield displacement ΔU_y (mm)	2.5	24.65
Destroy the displacement ΔU_p (mm)	20.0	150
Ductility ratio U	8	6.1

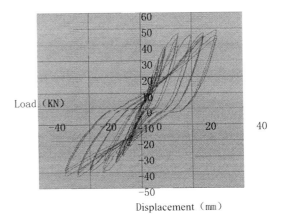

Figure 7. ML-1 hysteresis curve.

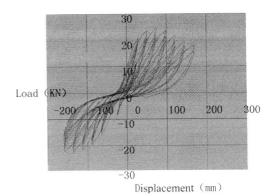

Figure 8. CK-1 hysteresis curve.

the column of longitudinal reinforcement tensile yield. When yielding, the specimen of CK 1 vertex horizontal load P was 10.5 kN, vertex horizontal displacement for Δy was 24.65 mm.

5.2 Specimen lateral stiffness

Comapred with the value of force and displactment under different level of latteral force which we get from experiment and the ansys mode analysis. We can get the lateral stiffness of the two kind of specimens. (Jianke Ma etc. 2008) Because of ignoring the axial deformation, according to the vertex displacement of the elastic cantilever component, peak horizontal force P, the r stiffness K is approximate concluded. Lateral stiffness formula $K = P/\Delta y$.

From table 1, the comparison shows that under the same level of load, apparently, the lateral stiffness of multi-ribbed grid framework is much larger than that of conventional frame structure.

5.3 Ductility of the specimens

Ductility, refers to capatility of deformation under the same force when the structure or component exceds elastic stage, This article uses the ratio U of vertex horizontal displacement to measure the ductility of the structure. While using the vertex displacement when the structure is damaged ΔUp (i.e., the load down to 85% of the maximum load when the corresponding displacement) divided by the structure yield vertex displacement ΔUy (that is, the maximum load of the displacement). (Jianke Ma etc. 2008)

From table 2, it can be concluded that the ductility ratio of greater than 5, both have good ductility, but $U_M/U_C = 1.3$, grid frame structure wall ductility ratio is 1.3 times that of the conventional framework, the ductility of the former obviously better than the latter.

5.4 Hysteresis characteristics

Picture below Fig. 7~Fig. 8 shows the ribbed net format framing wall specimen and conventional frame structure wall specimen of vertex horizontal force $P-\Delta$ horizontal displacement hysteresis loop back curve.

Can be seen from the graph, though, hysteresis loops is an s-shaped, but grid frame structure with phosphorus gypsum as the wall has a fuller hysteresis loops curve than conventional frame structure wall, more horizontal forces, smaller displacement, the relative area of energy dissipation is larger, which have better energy dissipation capacity.

6 CONCLUSIONS

Via the experiment of a multi-ribbed frame structure and a conventional frame structure both of whose ratio is larger than 3, compared with the corresponding analysis of the two kind of structure with the software of ansys. The following conclusions are got: muti-ribbed frame structure with phosphorus gypsum as wall shows the stress characteristics of frame structure, inflection point appeared in each floor of the wall, under the condition of the same span and height, the lateral stiffness, bearing capacity, energy dissipation, ductility and seismic performance of multi-ribbed frame structure is superior to the conventional frame structure. Because of this experiment is conducted under the action of horizontal force, the vertical axial force effect is not considered.

REFERENCES

Jianke Ma etc. Ribbed reinforced concrete shear wall and making method. Invention patent of China 200610200093.8[P]. 2006.7.26

Jianke Ma etc. Phosphorus gypsum template as a ceiling for reinforced concrete waffle slab floor. Invention patent of China 200620200099.0[P]. 2007.2.7

Jianke Ma etc. The wall material of reinforced concrete structural system cast-in-situ construction method using phosphogypsum as [Z]. Invention patent of China (Application number 2008103054544). Beijing: The state intellectual property office, 2008

Jianke Ma etc. With phosphorus gypsum as the wall materials type firm on the soft curved reinforced concrete shear wall structure and its producing method [Z] Invention patent of China (Application number 2008103054563). Beijing: The state intellectual property office. 2008

Jianke Ma etc. With phosphorus gypsum as the wall materials type firm on the soft curved reinforced concrete shear wall structure [Z]. Invention patent of China (Application number 2008203027097). Beijing: The state intellectual property office. 2008

Yunshou Huang. With phosphorus gypsum as a template and building and auxiliary materials of reinforced concrete ribbed tic-tac-toe board and ribbed grid plate shear wall structure system of the research and application [J]. Guizhou university master's degree thesis. 2007

The correlation of P-wave velocity and strength of solidified dredged marine soil

Chee-Ming Chan, Kok-Hoe Pun & Lek-Sing Hoo
Universiti Tun Hussein Onn Malaysia, Batu Pahat, Johor, Malaysia

ABSTRACT: To examine the reuse potential of dredged marine soil via solidification, a series of tests were conducted on a dredged sample treated with cement-flyash. The addition of flyash to the commonly used cement was to incorporate a "green" and sustainable value to the solidification process, as the flyash is a byproduct of coal combustion in power plants. The dredged marine soil in the present study was a high plasticity clay with natural moisture content of approximately 166%. From standard compaction tests, the maximum dry unit weight and the optimum moisture content were found to be 14.62 kN/m^3 and 24% respectively. Solidification was achieved by adding 10% binder (by dry mass of the soil) with various ratios of cement: flyash, i.e. 7:3, 5:5, 3:7 and 0:10. The lightly compacted cylindrical specimens (38 mm in diameter and 76 mm in height) were then cured for 3, 7, and 28 days before being subjected to the bender element and unconfined compressive strength test. The bender element test gave measurements of P-wave velocity in a non-destructive manner, hence the same specimen was allowed to undergo the latter test to obtain the strength. Overall, it was shown that the P-wave velocity and strength increased with time, though with variations depending on the cement-flyash ratio. Charts of the parameters were established as a quick guide to the strength of the solidified soil by the P-wave velocity measurement alone. It could enable trial specimens with various binder dosages to be quickly examined in the laboratory before actual solidification on site. This could encourage reuse of the otherwise waste geomaterial with low-cost solidification using cement-flyash in various civil engineering applications, especially as a backfill material.

Keywords: dredged marine soil, solidification, cement, flyash, strength, P-wave velocity

1 INTRODUCTION

Dredging is defined as the removal of material from the bottom of lakes, rivers, harbours and other water bodies. The process involves loosening or dislodging the materials, disposing them to the open water, and transporting them to the site where they are to be relocated (Riddel, 2003). Indeed, dredging is an ancient engineering practice dated as far back as the Roman times (Mountford, 2000). Most dredging is carried out to maintain or deepen water depths for safe and efficient navigation of vessels. The traditional handling of dredged materials, considered a geo waste, is either discharge into a confined disposal facility inland or designated open waters. Inland disposal requires immense costs at risks of leaching and contamination, while offshore disposal has been shown to bring negative physical, chemical and biological impacts on the marine environment. At the same time, the advancements in scientific knowledge and heightened public awareness in nature conservation have led to the recognition of dredged materials as a potentially valuable resource for reuse, which could replace the traditional "dredged and disposed" approach. Some areas of applications include habitat creation and restoration, Landscaping, road construction as well as land reclamation (Pebbles and Thorp, 2001).

Malaysia is essentially a peninsular (Peninsular Malaysia) and island (Borneo) teeming with marine activities. It is unsurprising that an enormous amount of dredged soils was generated from the rivers, lakes and seas. These dredged soils have poor geotechnical properties and are generally classified as "useless materials" (Goldbeck, 2008). For instance, over 3.5 million m^3 of dredged material were recently generated from the rehabilitation of the Lumut waters of Perak state in Malaysia to maintain and enlarge the navigation channels for commercial, fishing and national defense purposes. The dredged soil was disposed offshore at a designated sea disposal site. While located at an adequate distance from disrupting the local fishermen's livelihood, such a disposal method could still create disturbance to the aquatic ecosystem (Snyder, 1976). In addition, Bogers and Gardner (2004) reported that light attenuation by suspended

Table 1. Physical properties of the dredged marine soil.

Soil Classification (USCS)	High plasticity clay, CH
Natural moisture content (w), %	166
Specific gravity (Gs)	2.60
Liquid limit (LL), %	95.8
Plastic limit (PL), %	34.4
Plasticity Index (PI), %	61.4

sediments can affect the amount of light available to seagrass plants, coral reefs and other marine organisms. Also, Cruz-Motta and Collins (2004) found that soft bottom macrobenthic assemblages may respond quickly to the disturbance associated with the dumping of dredged materials and affect the overall marine ecosystem.

The environmental concerns due to poor handling of the dredged marine soils have made it imperative to formulate a more sustainable solution, such as reuse of the geomaterial with certain pre-treatment, like solidification with hydraulic binders. The treatment is necessary to improve the strength and stiffness to acceptable levels befitting good geomaterials for civil engineering applications. If the treated soil is able to withstand traffic loading under all weather conditions without deformation, it is considered as stable (Flaherty, 2002). The present study set out to examine the improved strength and P-wave velocity of the dredged marine soil collected from Lumut using the conventional unconfined compression test and the non-destructive bender element test, with the aim of establishing a plausible relationship of the parameters to aid in the implementation of solidification on site, particularly at the design and quality control stages.

2 MATERIAL AND METHOD

2.1 Dredged marine soil

The dredged marine soil was collected from 8–12 m below the seawater level by using a trailing suction hopper dredger at the Lumut waters mentioned earlier. Grab samples were retrieved from the storage tank of the dredger manually and were placed in double-layer of plastic sampling bags. They were then immediately transported to the laboratory and stored in covered containers at average room temperature of 30°C. To prepare the mixture of soil and cement-flyash, the soil was first scooped from the storage container and mixed uniformly in a kitchen mixer. It was then left overnight to ensure uniform redistribution of the pore water, before predetermined amounts of cement-flyash were added for mixing and solidification. Physical properties of the dredged marine soil are given in Table 1.

2.2 Cement

Cement is arguably the most popular binder's choice. In this study, the cement quantities were proportioned

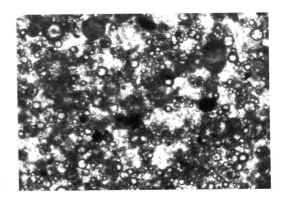

Figure 1. 20× magnification of the fly ash; note the distinctly spherical particles (using the CANON Eclipse T100 inverted microscope).

on a weight basis and quoted as a percentage of the oven-dried soil (i.e. weight of solids). Dallas and Nair (2009) pointed out that many soils can be successfully treated and improved with considerably low cement contents. Similar effective solidification of soft clay soils with small dosages of cement were illustrated in the findings of Chan (2012) and Mokhtar and Chan (2012). A similar approach with binder content no more than 10 % was adopted in the present study.

2.3 Fly ash

Flyash is a byproduct of coal combustion for power generation, collected by electrostatic precipitators in the plant system to avoid contamination of the atmosphere. This results in the ash particles being spherical and non-uniform in size (Figure 1). Class F ash (as used in the present study) is produced from burning anthracite or bituminous coal, which is usually non-cementitious when mixed with water alone (Halstead, 1986). It is unlike Class C flyash, which can be used on its own to solidify moderately plastic soils due to the presence of 20–35% calcium compounds (CaO), especially when mixed with activators like lime and Portland cement (Ferguson, 1993).

2.4 Test specimens

The binders were introduced to the soil at its natural water content for 10 minutes of mechanical mixing. The soil-binder mixture was then lightly compacted in 3 layers in a cylindrical split mould to form standard test specimens of 75 mm in height and 38 mm in diameter. The consecutive layer compaction technique was adopted to avoid bottom-heavy specimens commonly encountered using the single-layer static compaction method. Each layer was first tamped 50 times with a steel cylindrical rod, followed by the same number of tamping with a bent fork-like miniature compaction tool, which simulated kneading and pressing simultaneously. Table 2 gives a summary of the test specimens examined in this study. The prepared specimens were

Table 2. List of test specimens.

Specimen	Cement (%)	Fly Ash (%)	Total binder by dry weight of soil (%)
10FA	0	10	10
5CFFA	5	5	10
7C3FA	7	3	10
3C7FA	3	7	10

carefully wrapped in cling film and stored in sealed containers (on raised platforms) to prevent surface drying and moisture loss. The sealed containers were partially filled with a diluted bleach solution to prevent fungal growth on the specimens. The specimens were then subjected to the tests on day 3, 7 and 28 of curing.

2.5 Measurements and tests

The bender element (BE) test was conducted according to the procedure in the Bender Element Test System Manual by GDS Instruments Ltd. (2005) to obtain the P-wave velocity of the solidified specimens. It consists of a transmitter and receiver, attached to the top and bottom surface of the specimen for measurement. For time-saving benefits, the automatic stacking method with manual trigger was preferred (Mokhtar, 2011). A single sinusoidal wave (5 kHz frequency, ±10 V amplitude) was used to trigger the transmitter of bender element. The rate at which readings were taken per channel was 100,000 samples per second with 10 ms sampling time. The P-wave velocity (v_p) is simply calculated by dividing the travel distance between the transmitter and receiver bender elements by the arrival time (t). The unconfined compression test (UCT) was conducted according to procedures prescribed in BS 1377-7:1990. The top and bottom surface of the specimen were kept smooth and square to avoid bedding error during compression. The load was applied at a constant strain rate of 1.5 mm per minute.

3 RESULTS AND DISCUSSIONS

3.1 Unconfined compressive strength (q_u)

Figure 2 shows the vertical stress – curing period plots for the solidified specimens. The strength attained may not be very dramatic, but they are within the acceptable range for reuse as sound geomaterials, such as stipulated in the requirements by Japan's Ministry of Land, Infrastructure and Transport (2005). Apart from the 10FA specimen showing negligible strength improvement with time, the percentage of strength increment appears to increase with higher flyash content (Figure 2). This is interesting as large fly ash content has been reported to negatively affect the strength due to presence of fine particles and unburned carbon in the

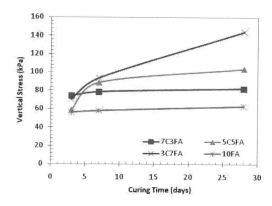

Figure 2. Vertical stress–curing time plots.

Figure 3. FESEM images of specimen 5C5FA at 3 and 28 days (5000× magnification factor).

fly ash (Wang et al., 2011). The results obtained in the present study seem to suggest otherwise.

The strength increment is attributed to the formation of gelatinous cementing compounds, which occupies the voids within the soil spaces and binds the soil particles together (Zentar et al., 2012). Cement treatment typically leads to flocculation of the fractions in soils, consequently increasing the particle size and modifying the plasticity of the original soil (Rekik and Boutouil, 2009). This expedient effect of solidification can be clearly observed in Figure 3, where the images of specimen 5C5FA were captured using field

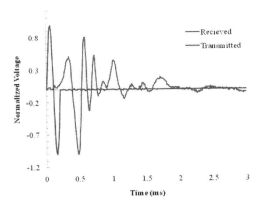

Figure 4. Determination of P-wave arrival time.

emission scanning electron microscopy or FESEM. At 5000× magnification factor, the large voids at early stage of solidification (3-day) were apparently filled up by the cementing compounds derived from cement-flyash (28-day). The large voids are distinguished in the 3-day micrograph of Figure 3 (marked with thick-line boxes), which are no longer visible in the same specimen a little over 3 weeks later. Also, it can be seen that the soil's microstructure changed significantly with prolonged curing time, with the 28-day specimen showing larger adjoined lumps of solids. While it remains unclear if the flyash contributed to the gelatinous filler, it is almost certain that they at least helped to form the solid mass to strengthen and stiffen the originally weak soil structure.

3.2 P-wave velocity (v_p)

Three methods were explored to determine the P-wave arrival time from the waves captured with the BE measurement system, i.e. peak-to-peak (p-p), trough-to-trough (t-t) and cross-correlation (cr), as illustrated in Figure 4. The purpose of carrying out all 3 methods was to cross-check the arrival time and to minimize errors. The P-wave velocity calculated is denoted as v_{p-p}, v_{t-t} and $v_{p(cr)}$ respectively. However the differences were found to be small (i.e. not exceeding 3.5%), as shown in the comparison plots in Figure 5. As such, for convenience, v_{p-p} was used for subsequent analysis and discussions.

Looking at the v_{p-p} – curing period plots in Figure 6, it can be observed that the pattern is generally not dissimilar to those of the strength's (q_u) evolvement with time (Figure 2). Specimen 10FA registered the lowest velocities, followed by 5C5FA, 7C3FA and 3C7FA. Note that specimen 5C5FA started out having a lower v_{p-p} than 10FA, but eventually overtook it at around 18 days. As the 7C3FA specimen lies above that of 5C5FA, it does not correspond with the relationship between strength increase and flyash content in the specimens. Nonetheless the specimen with the least cement content (3C7FA) attained the highest

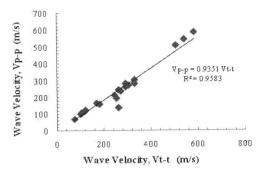

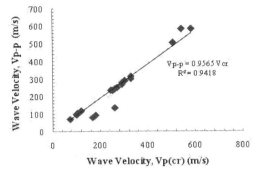

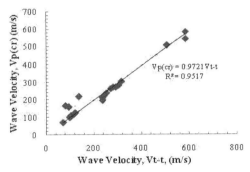

Figure 5. Comparisons of P-wave velocity using tPP, tTT and tCC.

v_{p-p}. Considering that v_{p-p} is an indicator of stiffness, albeit at small strain levels (i.e. strain not exceeding 0.001%), the strength and stiffness values do match up to a certain extent. The discrepancies mentioned earlier may be implausible at the moment, but they are very likely due to masking of the actual arrival time commonly encountered in less than satisfactory waveforms received. This could be caused by loose contact between the bender element and specimen, uneven end surfaces of the specimen leading to poor interface, and interference of the received signals by external factors. The mismatch notwithstanding, it can be noted that prolonged curing did not result in marked increase in v_{p-p}, as observed in the qu plots, especially in specimen 3C7FA.

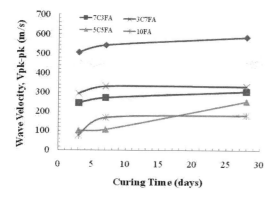

Figure 6. P-wave velocity–curing time plots.

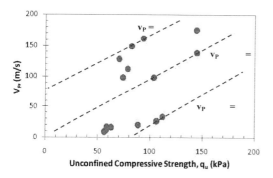

Figure 7. Relationship between P-wave velocity and q_u.

3.3 Correlation between P-wave velocity (v_p) and unconfined compressive strength (q_u)

Figure 7 shows the correlation between v_{p-p} and q_u, with a linear correlation derived for the data set, i.e. $vP = 0.92q_u$. Taking into account the rather dispersed nature of the data points, an upper and lower limit lines were included in the same figure, i.e. $v_{p-p} = 0.86q_u$ and $v_{p-p} = 0.89q_u$ respectively. These two lines were simply drawn parallel to the original regression line. Taking a closer look, the strength and stiffness (represented by v_P) of the specimens are relatively low, suggesting possible experimental errors creeping into the measurements. This is not impossible, especially for the unconfined compression test, which makes an indirect measurement of the strength and displacement easily marred by small imperfections of the specimen. These include uneven ends and inherent weak zones, generally undetectable visually or overlooked prior to tests. Irrespective of the possible errors (which can be validated with further tests), the relationship, however tenuous, serves as a promising tool for quick estimation of the improved strength of solidified dredged marine soil intended for reuse. For example, a P-wave velocity measurement of 100 m/s would correspond with an estimated strength of $25\,kPa \leq q_u \leq 185\,kPa$, with the average at approximately 105 kPa. The margins may seem large, but considering the irregularity and non-uniformity often reported of in situ soil mixing, the estimated range may yet prove acceptable.

4 CONCLUSION

The following conclusions are drawn from the study to relate the strength and P-wave velocity of a solidified dredged marine soil.

Fly ash alone is ineffective to solidify the dredge marine soil.

Prolonged curing did not result in significant increase in strength and P-wave velocity (stiffness), most probably due to the small binder dosage added to the soil.

The three methods adopted for P-wave arrival time identification (i.e. peak-to-peak, trough-to-trough and cross-correlation) showed almost similar results, suggesting their applicability as a cross-check measure.

In spite of some mismatch between the strength and P-wave velocity evolution with time, the qu-vP correlation established is useful for performance monitoring of the solidified soils, particularly at the trial mixing stage.

Further work is clearly needed to shed light on the qu-vp-p relationship over a wider range of binder dosages, if not to explain the measurement discrepancies made in the present study. An extended test series is in the pipeline for that purpose and future data are expected to complement the current exploratory findings.

ACKNOWLEDGEMENT

The study was funded by the Malaysian Ministry of Education's research grant, RACE Vot 1115.

REFERENCES

Bogers. P. and Gardner. J. (2004). Dredging near livecoral. Proceedings of the 17th World Dredging Congress (WODCON XVII 2004), Hamburg, Germany. Paper A31.

British Standard 1377. (1990). Methods of test for soils for civil engineering purposes. Part 2: Classification tests. United Kingdom: British Standard Institution.

Chan, C-M. (2012). Variation of shear wave arrival timein unconfined soil specimens measured with bender elements. Journal of Geotechnical and Geological Engineering, 30(2):419–430

Cruz-Motta, J.J. and Collins, J. (2004). Impacts of dredged material disposal on a tropical soft-bottom benthic assemblage. Marine Pollution Bulletin, 48(3–4): 270–280.

Dallas, N.L. and Nair, S. (2009). Recommended practice for stabilization of subgrade soils and base materials. National Cooperative Highway Research Program.

Ferguson, G. (1993). Use of self cementing flyashes as a soil stabilization agent. American Society of Civil Engineer.

Flaherty, C. A. (2002). The location, design, construction & maintenance of pavements. Edward Arnold Ltd.

GDS Instruments Ltd. (2005). The GDS bender elements system handbook for vertical and horizontal elements.

Goldbeck, S. (2008). State of the Estuary Report, A Greener Shade of Blue. San Francisco Estuary Report and CALFED State of the Estuary Conference Proceedings.

Halstead, W.J. (1986). Use of flyash in concrete. Synthe-sis of Highway Practice 127, National Cooperative Highway Research Program (NCHRP), Transportation Research Board, Washington DC.

Ministry of Land, Infrastructure and Transport, Japan. (2005). Current state of construction by-products.

Mokhtar, M. (2011). Mechanical properties of soft clay stabilized with cement-rice husk (RH). Universiti Tun Hussein Onn Malaysia: Master's thesis.

Mokhtar, M. and Chan, C-M. (2012). Settlement controlof soft ground with cement-ricehusk stabilisation. Civil Engineering Dimension, Research Centre of Community Outreach, Universitas Kristen Petra, Indonesia, 14 (2): 69–76.

Mountford, K. (2000). History of dredging reveals deeper need to understand Bay's bottom line. Bay Journal, 10(5): 8–10.

Pebbles, V. and Thorp, S. (2001). Waste to resource: Beneficial use of Great Lakes dredged material, Great Lakes Commission.

Snyder, G.R. (1976). Effect of dredging on aquatic organisms with special application to areas adjacent to the Northeastern Pacific Ocean: A marine fisheries review.

Rekik, B. and Boutouil, M. (2009). Geotechnical Properties of dredged marine sediments treated at high water/cement ratio. Geo Marine Letters, 29: 171–179.

Riddell, J.F. (2003). Dredging for development. International Association of Dredging Companies (IADC).

Wang, D.X., Abriak, N.E. and Zentar, R. (2011). Durability analysis of fly ash/cement-solidified dredged materials. Coastal and Maritime Mediterranean Conference.

Zentar, R., Wang, D.X., Abriak, N.E., Benzerzour, M. and Chen, W.Z. (2012). Utilization of siliceous–aluminous fly ash and cement for solidification of marine sediments. Construction and Building Materials 35: 856–863.

The discussion of multi-channel emergency management pattern in the north bank of Wenzhou Ou River

Yunhao Yao & Yuanfu Li
School of Civil Engineering, Southwest Jiaotong University, Chengdu, China

ABSTRACT: Wenzhou Ou River bridge is grand bridge which is shared by railway, highway and municipal road. This makes the bridge more complex than others. As the bridge is shared by three parties, the emergency management pattern is a key problem which makes the bridge in a good condition. This paper summarizes domestic and abroad related materials and analyses the framework of the emergency management mode.

1 PROJECT OVERVIEW

Wenzhou Oujiang bridge is a controlling project, which comprises expressway from Yueqing to Ruian, municipal road of Wenzhou and S2 urban railway. The bridge connects Qitou mountain and Lingkun island.

2 OVERALL FRAMEWORK OF THE MULTI-CHANNEL'S EMERGENCY SYSTEM

According to the expected objectives and requirements of the multi-channel emergency system, the system should achieve function of each subsystem (i.e. warning, command and disposal, information and security subsystem) through a series of technical methods to decrease accident loss, prevent and handle emergency. In order to improve the risk warning capability of emergency system, we take safety and reliability as principal and take emergency response and disposal capability as objectives.

Finally, the system is divided to two parts shown in figure 1.

The main function of the risk warning subsystem is to monitor, identify and evaluate the external and internal environment. Internal environment mainly refers to system factors including management, equipment and technique. The change of these factors should be monitored by warning system. External environment mainly refers to natural environment and social environment such as weather and urban planning.

The emergency response subsystem is mainly for the management needs of occurred emergencies. The subsystem acquires optimal handling model and emergency scheme through information collection, real-time data analysis. Each emergency security department should coordinate under united command.

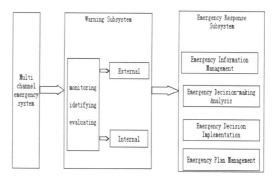

Figure 1. Overall framework of the emergency system.

3 MANAGEMENT ORGANIZATION

The organization structure of emergency management is shown in figure 2. This paper will analyze the six parts in the following part.

3.1 *Organization structure*

Command organization is a central department in emergency response process. The effectiveness of command and the rationality of decision are closely linked to the effectiveness of emergency management. Command organization plays a key role in resources allocation, mobilizing social forces and international cooperation. When emergency incident occur, department such as police, firefighters and civil affairs can be integrated in a emergency response mechanism. The command organization can be set up temporarily. It can also be set up based on regions. Members of command organization include transportation department, local

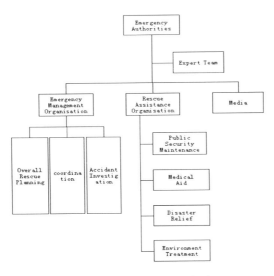

Figure 2. Emergency management organization.

railway bureau, public security department, rescue department, meteorological department and environmental protection department[1].

Management organization is a special organization in the emergency management. It undertakes core responsibility of emergency management. Its main function is to announce various transportation information coordinate the relationship between surface and air. It is also responsible for emergency prevention, information transmission, rescue, evacuation, accident investigation and rehabilitation.

The assistance rescue department refers to various departments, for example: public security bureau is responsible for public order maintenance, health bureau is responsible for medical care and environment monitoring department is responsible for environmental treatment[2].

In order to make reasonable decision, consulting relevant expert is essential. Senior professionals and senior managers comprise the expert group and its main duty is to provide decision consultation, suggestions and participation in emergency command.

After emergencies occur, emergency organization should utilize media such as broadcast, television to announce the rescue condition, evacuate passengers and guidance of public opinion.

3.2 Material and equipment guarantee

When emergency occurs, according to material requirement planning proposed by emergency command organization, production, storage, transfer scheme of daily necessity should be formulated to ensure supply of necessity and rescue equipment. Emergency organization should grasp relevant emergency supply production condition and ability of manufacturing enterprise, establish archives and be responsible for rapid production. Emergency management department strengthen stock management, establish information management system, clarifying various material storage capability.

3.3 Information and communication security

Emergency management system should achieve information interaction. Owing to abundant relevant departments, information systems and database platforms are various too. Information of different system should be integrated seamlessly to form a powerful and covered transportation emergency information support network. The key of modern rescue is the rapid acquisition of rescue resources and data, so an all-around modern information network should be built. In order to achieve breakthrough in dynamic information collection and monitoring, traffic information service, traffic emergency command and traffic operation, we should set up perfect traffic information security system in following aspect.

Firstly, traffic information network platform should be built. Combing with public social network, we should a comprehensive platform including municipal road, railway and highway. It links government and transportation department with other related department.

Secondly, traffic information resources Centre should be improved. Several databases should be established including traffic geographic information management database, transportation asset management database and comprehensive law enforcement database. Meanwhile, a set of transportation geographic information platform should be established, attribute information should be managed hierarchically and provide visual technique platform support for traffic emergency command, traffic security supervision and traffic information service.

Thirdly, we should optimize transportation information system and travel information website and other information inquiry through broadcast, text message[5].

4 PRE-ARRANGED PLAN

According to different emergency situation, pre-arranged plans are essential. It can be used in daily training, guaranteeing various materials in a good condition, leading emergency action in order and preventing work delay caused by chaos.

4.1 The main content of emergency plan

Emergency plan should include the identification and evaluation of emergency events, confirmation of rescue human and material resources, confirmation of action scheme and formulating restoration measures after accident rescue. In addition, other requirements should be taken into account, including clarifying referred laws and regulations, confirming obligation of each department, designing training program and establishing special plan.

4.2 *Emergency plan level*

The emergency plan level can be dived by accident frequency, degree of danger and economic loss. Then, we can divide 3 levels: general accidents, major accidents and serious accidents, which can be colored by red, orange and yellow. On the basis of this method, equipment and personnel required by size of accidents and warning level can be determined. Moreover, emergency organization can be decided.

4.3 *Design of emergency plan system*

As a complete file system, emergency plan should be an effective tool for emergency action. A complete emergency plan includes overall plan, disposal procedure, acting instruction and action record.

Overall plan is a programmatic document. It includes management policy of transportation emergency, emergency plan aim, emergency organization and responsibility.

Disposal procedure presents scope and purpose of specific action. The content of procedure is specific, for example action time, action position, action personnel. It provides guidance for emergency action, but the program and format of disposal procedure is concise, which avoids implementation procedure being misunderstanding[3].

Action instruction illustrates specific task and action details for emergency staff, for example emergency monitoring equipment instructions, emergency staff responsibility instruction.

Emergency action record refers to communication record and record of each step of emergency action. It can restrain rescue action and provide reference for later period.

5 TECHNICAL SUPPORTS FOR EMERGENCY SYSTEM

Technical supports include detection technology, communication technology, computing and intelligent technology and control techniques.

Detection technology can be used to collect real-time information such as road condition information, weather information, accident prediction and confirmation and rescue task. It includes video monitor system, meteorological monitor, emergency telephone and electronic toll collection system.

Communication system can be used to ensure smooth information transmission, avoid rescue delay caused by information congestion. It includes wire or wireless communication system, networking toll collection system, wireless network equipment and microwave communication devices.

Computing and intelligent technology can provide basis for instant and accurate rescue measures. It includes the use of database technology, reasonable allocation of emergency resources, network analysis techniques based on GIS, reasonable guidance to on-the-spot traffic dredge and reasonable formulation of the optimal route for rescuers[4].

Control techniques include traffic inducer equipment which avoids former accidents.

REFERENCES

[1] Imoussaten, Abdelhak Montmain Jacky, Mauris, Gilles. A multicriteria decision support system using a possibility representation for managing inconsistent assessments of experts involved in emergency situations, International Journal of Intelligent Systems 2014
[2] Peng Yi, Yu Lean. Multiple criteria decision making in emergency management. Computers and Operations Research 2014
[3] Wang Bin, Li Haijiang, Rezgui, Yacine. Intelligent building emergency management using building information modeling and game engine. ICIC Express Letters 2013
[4] Fogli, Daniela Guida, Giovanni. Knowledge-centered design of decision support systems for emergency management. Decision Support Systems 2013
[5] Kremers Enrique, González de Durana, José María, Barambones Oscar. Emergent synchronisation properties of a refrigerator demand side management system. Applied Energy 2013

The Florence charter and the conservation of Chinese classical gardens

Yue Zhong
Guangzhou Nanyang College, Guangzhou, China
South China University of Technology, Guangzhou, China

Jianjun Cheng
South China university of technology, Guangzhou, China
South China University of Technology architectural design and Research Institute of cultural heritage protection, Guangzhou, China

ABSTRACT: As the Florence charter released, it has been correctly guiding the overall direction of the world's garden heritage conservation, particularly for western developed countries. What mental and which methods should China take to cope with the conservation challenges for classical gardens, and how to apply the Florence charter correctly to classical gardens conservation in China are important problems require attention.

Keywords: Florence charter; authenticity; classical gardens; conservation

1 INTRODUCTION

It has been more than 30 years since the Florence charter was released. As one of the important charters in a series of convention charters of historical and cultural heritage protection, it guides the historical sites protection of garden, which is a specialized field. So how to better apply this important international charter to the classical gardens conservation practice in China? This is the actual focus of this paper.

2 THE BACKGROUND AND MAIN CONTENT AND CHARACTERISTICS OF THE FLORENCE CHARTER

2.1 Background of the Florence charter

In 1964, the Venice charter expanded the concept of cultural relics on the basis of the Athens charter, making conceptual beddings for classical gardens. In 1972, the Convention on the protection of world cultural and natural heritage put forward the concept of natural heritage, and the natural heritage was also included in the scope of protection for the first time. In 1976, the Nairobi Suggestions was in a view of the world. It aimed at expanding and modernizing as an excuse, causing a large number of demolition and unreasonable, improper reconstruction, which had brought serious damage to the historic heritage. And it clearly pointed out the common value of the historical block protection in the social, historical and practical aspects. In 1977, the Machu picchu charter raised what governments could do and should also be related to the adopted policies and measures of improvement in human settlements quality of the world, at the same time for countries with the traditional of classical garden reparation, reconstruction and transformation. (Jing Wang, 2012) Therefore, the world heritage protection in constant attention and also the scope of the heritage protection are constantly expanding. However, in the field of classical garden, the special protection has not formulated the corresponding protection measures, and the reconstruction of the world is merciless torment of classical gardens, which requires that there must be a special specification to protect classical gardens. Under the urgent situation, the international council on monuments and sites of the eighth plenary meeting was held on May 21, 1981, and the Florence charter was passed in Florence, and it was registered as the Venice charter in the attachment on December 15, 1982. (The International Council on Monuments and sites and the historic garden Committee 1982)

2.2 Characteristics of the Florence charter

The Florence charter was made in the field of classical garden; it is a programmatic document is applicable to any classical gardens.

2.3 The main content of the Florence charter

The charter is made up of 25 sections and comments. Articles 1 to 9, as the definition and goal, clearly define that the classical gardens should be protected as monument. On the one hand, in the process of protection, it should be on basis of the authenticity spirit of the Venice charter; but one the other hand, classical gardens are "live" sites which are known to everyone, so it should be also based on the different specific rules for other heritage protection. These articles also define the intricate problems of gardens compared to other heritage. There are four elements to make up gardens, namely, water, plants, rocks, buildings. Each element has its own characteristics, especially for plants with vivid vitality. It is obvious that the protection of classical garden is quite complex, thus a certain law cannot be applied, and specific case needs be treated particularly. Gardens are also products of particular social environment related to economic level, social phenomena, and cultural and aesthetic aspects. As a result, classical gardens also have strong time efficiency. However, history is changing all the time; in different periods, gardens have their different characteristics and unique creativity, embodying the artists' ability and skill level at that time. The concept of classical gardens is not for a garden in a specific period. As long as it meets the historical conditions, it is applicable to all gardens in different size levels or in different nature. Classical gardens also carry the in-depth insight of history and they are merged with surrounding environment.

Articles 10 to 17 describe the maintenance, conservation, restoration and reconstruction of classical gardens from different aspects, emphasizing to pay attention to the integrity to maintain their authenticity in the process of maintenance, protection, restoration and reconstruction. Living plants are important factors of gardens, so the maintenance work should be done continuously. A library of the seedling should be established and suitable plants should be selected. As for longevity and alternate plants, they need different treatments, and should be kept base on the principle of dynamic programming and the ecological balance of the garden surrounding. The elements that make up gardens should not be moved freely and alternated. If the gardens' elements need be moved or altered, they should also be done in accordance with the laws of the Venice charter with description of any alternative dates. Before the restoration and reconstruction work, investigation and study should be carried on in order to obtain real comprehensive information. If the physical elements disappear completely, the reconstruction is not to be called classical gardens.

Articles 18 to 22 specify the usage of classical gardens. In order to ensure the sustainable utilization of a classical garden, its use should be carried out under the premise of protection; the patronage must be limited within the range of the garden's available capacity and corresponding measures must be taken to protect it even on holidays or anniversaries. The use of crowd need be considered in the planning and design time.

The planning work should meet different levels of crowd in the premise of the protection of classical gardens. Reasonable partitions of functions need be carried on in order not to deviate the original intention of this paper, namely the authenticity protection of classical gardens.

Articles 23 to 24 propose legal and administrative protections of classical gardens. Besides that the corresponding legal protection should be formulated, the professional quality of practitioners is also very important.

In addition, the charter also describe corresponding instructions in detail about subsequent maintenance and management problems of plants, stimulating public enthusiasm through effective ways such as the declaration of the world cultural heritage, and scientific propaganda issues. The charter also suggests the protection of all world classical gardens be in accordance with the Charter of the spirit. (The International Council on Monuments and sites and the historic garden Committee 1982)

Therefore, the Florence charter comprehensively expounds the general problems that should be paid attention to in the protection of classical gardens, and the characteristic that is very insightful. The charter is indeed a worthy reference as international programmatic document, which is why it still has a guiding significance for decades.

3 THE GUIDING SIGNIFICANCE OF THE FLORENCE CHARTER FOR CHINESE CLASSICAL GARDEN PROTECTION

Florence is the world of art, European cultural center, the birthplace of Renaissance, and also a cultural tourism resort. The Florence charter was born in the land to evoke old memories for classical gardens, to better call on everyone's enthusiasm to protect them.

The protection of the Chinese classical gardens started late and experienced a rough and bumpy road. The research of landscape protection and development is mainly combined with the theory of the foreign classical garden protection and practice experience in landscape design, protection, restoration and development and utilization. Because foreign landscape protection is mainly based on the spirit of the Florence charter, the protection of the Chinese gardens is also inseparable from the Florence charter's spiritual blessing. In China, there are several significant Landscape Protection Conferences, namely, West Lake Declaration, From Traditional Gardens to City Declaration, Suzhou Gardens Conservation and Management Regulations, etc., and the subjects are in line with the spirit of the Florence charter in order to carry on the elaboration of the actual situation of China's classical gardens. (The Scenic Professional Committee Chinese society of Landscape Architecture 1992) (The Standing Committee of the eleventh Suzhou Municipal People's Congress 1996)

Hence, the Florence charter has profound significance for the guidance of classical garden protection in China.

4 CHINA NEEDS TO MAKE CLASSICAL GARDEN PROTECTION THEORY TO CONFORM TO OUR COUNTRY'S NATIONAL CONDITION IN THE FOUNDATION WITHOUT VIOLATING THE FLORENCE CHARTER SPIRIT

4.1 The development and the present situation of Chinese classical gardens

Chinese classical garden has a long history, with abundant culture content, bright personality and significant artistic charm, and it is the largest of the world's three largest garden systems. In the history of the development of the Chinese garden, it experienced Shang, Zhou Eras to bounded, nursery, pool, and marsh as the representative of the embryonic stage, forming stage in the Spring and Autumn and the Warring States, Qin, Han. The Wei Jin, and Southern and Northern Dynasties were the China garden system's completion stage. The Sui, Tang and Five Dynasties, and Song were the Chinese garden's freehand brushwork stage. Yuan, Ming and Qing Dynasyties were the mature stage of the Chinese garden. (Jianhua Zhu 2012)

Three different regional landscape patterns of the most obvious style characteristics have formed, namely the North Garden, South Garden and Lingnan Garden. The excellent landscape works are legion, for example, Shanglin Garden, Gusu Platform, JianZhang Palace, Wei Yang Gong, Epang Palace, Xiyuan, Shougen yue, the Ming tombs, Wutai mountain, Yuanmingyuan, the Summer Palace, Chengde Summer Resort, Qing dongling, Suzhou Garden, Lingnan Four Big Gardens, etc. However, preserved intact gardens are not many; a large number of them have disappeared in the history. Many of gardens saved were damaged because of the war, disrepair or other reasons. Although some have been repaired, many of them are beyond recognition and the original historical value they bear have faded. This means that the protection of the classical gardens must be continued, or the original intention will deviate.

The Chinese nation, with thousands of years of history of gardening, has accumulated rich classical gardens. These classical gardens, carrying great historical and cultural value and social value, as well as historic buildings, are one of the most intuitive carrier of history and culture. They are non-renewable, thus protection of the classical gardens, in order to let their value and function play a role in the sustainable development in today's modern world, is a necessary trend and tide. The protection of classical gardens in recent years gained more and more attention from experts and scholars; but because of the late start, the protection theory and regulations are not mature enough.

4.2 Classical garden protection in developed countries

In view of the present situation of Chinese classical garden, drawing lessons from foreign classical garden protection is important. In the process of landscape protection abroad, some good methods and theories are worthy as reference. For example, France's classical garden protection system is based on the historical heritage protection theory system, after the 1930s through carrying on series of legal measures to strengthen the protection of classical gardens. In addition, through the "garden open day", "garden", "calling all private contribution" and other activities, public participation is achieved in the protection of classical gardens. In academia, by hosting a number of seminars, a variety of constructive suggestions are put forward for the theory of classical garden protection, including classical garden repair, restoration and reconstruction. Britain, by making monuments, logging in architecture, reserving, logging in classical garden and other ways, strengthens the protection of classical gardens. Their protection objects include private garden, public gardens, cemetery, water pumping station, hospital affiliated green space and square connection between urban green space and residential district. Britain also carries on the fusion of historical garden and city green space system, and through the connection of green gallery, the classical gardens and modern green belt park, garden, tree-lined trails, street green space, park green space together, a natural, diverse, high-efficiency and self-maintaining network of city green space system capacity is formed. Western countries also actively promote the implementation of measures, and carry on international academic cooperation and academic research. Many of those methods are not existed in China or are not perfect, but their attention to the classical garden protection, the combination of protection and utilization and management systems are relatively perfect, and the maximum activation of cultural and historical value of the classical gardens attitude and methods are worth for our reference.

4.3 Understand the differences between Chinese and foreign gardens and garden protection theories

There are great differences between Chinese gardens and western gardens in origin and cultural background, thus foreign experience should not be simply copied, the differences need to be recognized to lay a foundation for future reasonable absorption.

4.4 The differences between Chinese and foreign gardens

4.4.1 The differences of the origins

The functions of the Chinese gardens are mainly for hunting, god worshiping, and production at first. Gradually, they were transformed into recreation and ornamental purposes. Gardens are based on simple awareness to conform and return to nature. "Imitate the nature", "Although people do, Wan since opening day" and so on are as basic ideas of succession of Chinese gardens. Hence, Chinese gardens started in respect of nature on the basis of the imitation of nature and natural reproduction, and they used natural sustainability in the service to themselves and "create" a natural style garden to form a garden style of harmony between man and nature with harmonious natural beauty.

Western gardens originated three thousand years ago. The Nile River alluvial fertile soiled and was suitable for farming. But because of its annual flooding, refunding after water to measure the cultivated land developed geometry. The ancient Egyptians designed their gardens with flexibility according to their own needs; hence their gardens became the world's first regular gardens. It is easy to see that from the beginning, the west gardens struggled against nature, and attempted to conquer the nature to create the harmonious beauty they considered, and the ideas of "man can conquer nature" and the pursuit of rational were both embodied in western gardens.

Differences of the origins between Chinese and western gardens have produced two distinct garden styles of "natural" and "rules".

4.4.2 *Differences in cultural background*

Chinese attach great importance to the overall harmony, show a kind of emotion and reason between human and the nature, and emphasize the whole and pay attention to unity, while the westerns pay attention to the analysis of the differences, and attach great importance to the exploration and innovation of nature. Chinese traditional mainstream philosophy is the harmony between human and nature, while the westerns pay attention to personality, promote human dignity, emphasize people's value and idea, certain individuals, certain real life and advocate the survival competition. Those different cultural backgrounds make the east and west home gardening and garden designers show great differences in the gardening rules, and direct response in the structure and form of the gardens. (Jianhua Zhu 2012)

4.4.3 *Making classical garden protection theory conform to China's national condition*

Summaries of western developed countries' long theoretical and practical experiences can save detours in the classical garden protection in China, and a lot of essences are worth learning. However, the characteristics of Chinese and western gardens should be considered. This kind of learning and reference should also be on the premise of maintaining the authenticity of Chinese classical garden. As the Florence charter in the spirit of classical garden original genuine stressed, we could not blindly copy western things. For instance, we could not bring their "conquer" concept to the garden protection in our country, and the view of "more than 30 years of the botanical garden has historical value" of the British is not applicable in our country. We should advocate the spirit of innovation, but can never base on the past with completely negative or completely positive attitudes; we should base on critical innovations according to scientific and reasonable basis.

5 SUMMARY

Chinese classical garden protection must be based on the "Florence charter" to protect the classical gardens' important spirit in the premise of maintaining the authenticity of the gardens, at the same time, draw lessons from the advanced technology and theoretical knowledge, but should base on the reality of China. The authors think that the following several measures should be applied: (1) The legal protection of garden cultural heritage should be strengthened, tend to make cultural heritage protection work; (2) The scientific and cultural studies of classical garden protection should be paid attention, in order to establishing more reasonable and perfect protection mechanism. (3) Parallel development and protection should be carried out to strengthen the planning of garden and cultural heritage protection. (4) In addition to large scenic spots at the provincial level, the classical gardens protection in local small areas should also be focused. (5) Personnel training should be intensified to protect classical gardens and to call for the public to participate in the protection. Classical garden protection in China has a long way to go, and constantly efforts and continuous exploration are needed in order to gain maximum benefits of the protection enterprise, and the classical gardens of historical, social, and practical value for inheritance.

REFERENCES

Jing Wang, 2012.4, The city historical and cultural landscape protection – Xiamen modern Zhongshan Park research [D]. Xiamen: Huaqiao University.

Jianhua Zhu, 2012.3, History of Chinese and foreign landscape [M]. Chongqing: major press.

The International Council on Monuments and sites and the historic garden Committee. 1982.12.15. "Florence charter".

The Scenic Professional Committee Chinese society of Landscape Architecture, 1992.11.14, "West Lake declaration".

The Standing Committee of the eleventh Suzhou Municipal People's Congress, 1996.12.31, "Suzhou landscape protection and management regulations".

The impact of the rocking wall layout on the structural seismic performance

S.B. Yang, Y.B. Zhao, Z.T. Wei & J.H. Jia
School of Civil Engineering, Hebei University of Engineering, Handan, Hebei, China

ABSTRACT: Using finite element method a static nonlinear analysis of a frame structure and embedded rocking wall reinforcement structure consisting of four schemes has been carried out in this paper. The analysis results show that the different layout schemes of the wall in frame-rocking wall structure have significant effect on stiffness of the structure, seismic bearing capacity, injury characteristics, deformation and deformation between layers as well as other aspects of the degree of concentration.

1 INTRODUCTION

Because of distinct advantages of flexible layout, construction convenience and so on, frame structure is widely adopted in various high-rise residential buildings such as offices, classrooms, shopping malls and residences. However, it also has many drawbacks such as low integral rigidity, weak seismic collapse resistance and difficulty of post-quake maintenance that is usually expensive. It is just these drawbacks that put people's lives and properties at risk in case of disaster. When being subjected to earthquake, the phenomenon of frame structure failure can be observed at a variety of members, mainly including destructed components such as frame columns, nodes, infill walls, stairs, as well as damaged frame beam, etc. In particular, the destruction of the frame column is the most common failure of frame structure. The consequence of the frame layer failure due to the destruction caused by the column is the most serious. Considering the phenomena of the conventional reinforced concrete frame structure layer yielding and failure, a new type of structural system, namely concrete frame-rocking wall system [1], is developed and studied by many researchers. Rocking wall is a wall with a special structure hinged at the bottom of the wall, so that the wall has a certain ability to rotate [2]. Moreover, the rocking wall can be combined with frame structure to form frame-rocking wall architecture. The rocking wall substructure can also be attached to the frame-rocking wall structure system, which can effectively control the structure layer yielding mechanism, give full play to energy dissipation capacity of the entire frame structure, and enhance the seismic capacity of the structure [3].

The frame-rocking wall structure and the general frame shear wall structure differs in their joint at the bottom of the wall, the former is a hinged joint while the latter is a solid one. It makes the basic period of the structure significantly longer, and wall bearing capacity lower. Meanwhile, bending moment at the bottom of the wall can be released and therefore reinforcement measure is not necessary to be taken. It also reduces foundation requirements as well.

Frame-rocking wall structure can be used in new buildings and also as reinforcement for the existing framework [4]. According to the connection method of rocking wall and the frame, frame-rocking wall structure is divided into two types, namely the embedded and the pluggable. With respect to the former type, the rocking wall is directly embedded inject between frames, connecting the rocking wall and the frame consolidation. The latter can hinge the rocking wall with the frame structure.

For better understanding the seismic performance of the embedded rocking wall under different arrangements of the rocking wall in terms of its location, a static nonlinear comparative analysis of a frame structure model and the embedded rocking wall consisting of four reinforcement schemes has been conducted by the finite element software, *SAP2000*.

2 DESCRIPTION OF THE STRUCTURAL MODEL

A hotel is assumed to be a 6-storey frame structure with a total height of 21.9 meters. The story height of the first floor is 3.9 meters and 3.6 meter for remaining ones. The sizes of columns from the first storey to the third storey are 650 mm × 650 mm and 600 mm × 600 mm; the sizes of columns from the fourth storey to the sixth storey are 550 mm × 550 mm and 500 mm × 500 mm. The sizes of main beam are 250 mm × 700 mm and 250 mm × 700 mm; The size of secondary beam is 250 mm × 500 mm [5]; the structural plane is shown in Figure 1. Slab thickness is 110 mm. In terms of design, the seismic fortification intensity is 7 degree and the hotel is classified as building of category C, design earthquake grouped into the

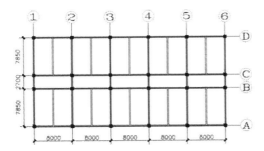

Figure 1. Layout plan of frame structure.

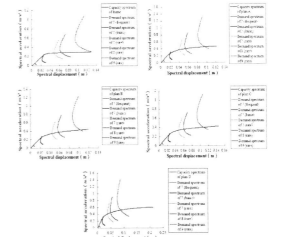

Figure 2. Performance point of each model.

first group, and the basic design earthquake acceleration is 0.15 g. The anti-seismic level for frame is three. The live load is designed to be 2.30 kN/m². According to relevant Chinese design codes, construction site of the hotel falls into category of class III. Referring to the codes, concrete strength grade C30 (Axial compressive strength standard values of concrete standard test block is 30 MPa) and rebar is HPB300 (Yield point is 300 MPa) and HRB400 (Yield point is 400 MPa).

There are four reinforcement schemes of embedded rocking wall introduced as follows.

Plan A: the walls are arranged on the middle secondary beam between the axes 3, 4 and C, D. The intermediate position of the main beam is between axes A, B and 2, 5;

Plan B: the walls are arranged at the same location as the plan A, but set both ends of the main beam to be hinged;

Plan C: the walls are arranged in the secondary beam between the axes 3, 4 and C, D. While the secondary beams are set to be located between axes A, B and axes 2, 3 and axes 4, 5, respectively;

Plan D: the walls are arranged in the same position as the plan C, but the original piece of wall becomes two walls connected by coupling beams that form the double leg rocking wall having the same size as the secondary beams.

The size of rocking wall is identified by the wall and the frame stiffness ratio of 7%. Section of the wall in the first three schemes is 3.35 m high and 0.25 m thick, but 2.64 m high and 0.35 m thick for plan D. The walls are modeled by layered shell element, and the rocking wall is connected through the wall split points with each frame beam to ensure the integrity of the wall and the beam. The bottom of the wall is set to be hinged.

3 COMPARATIVE ANALYSES ON THE RESULTS OF DIFFERENT MODELS

3.1 Comparison of structural stiffness

Each model is pushed over along the Y-axis (the short side), and the model analysis is made for each plan. The first-order vibration period of original frame structure for the plans from A to D, is 0.7244 s, 0.5746 s, 0.6261 s, 0.681 s and 0.5468 s, respectively. If the walls have no coupling beam in the plan D, the first-order vibration period is 0.6352 s. it indicates that the wall's stiffness becomes larger with the coupling beam in the plan D, which makes the greatest impact on the stiffness of the original frame structure. Since the two pieces of the wall are arranged in the main beam in the plan A, the impact on the stiffness of the structure is also significant. While in plan C, the walls are arranged underneath the secondary beam and the effect on the stiffness of the structure is minimal. Given the above analysis, the arrangement position of the rocking wall and the layout form of the wall may have a significant effect on the stiffness and vibration period of the original structure.

3.2 Comparison of structural performance point

With respect to the frame structure and embedded rocking wall which contains 4 plans, performance points associated with a single demand spectrum of variable damping ratios under the condition of different structure seismic intensity have been obtained (see Figure 2). It shows that original frame structure can be withstand severe earthquake with seismic intensity of 8 degree, however it will suffer structural damage under a rare earthquake with seismic intensity of 9 degree. Compared with the frame structure, bearing capacity of embedded rocking wall structure in those plans has greatly improved and it can meet requirement of a rare earthquake with seismic intensity of 9 degree.

The comparison of the equivalent damping ratio corresponding to performance point for each structure model under different seismic intensity has made, as shown in Table 1. It shows that equivalent damping ratio increases of the original frame under the basic seismic intensity of 7 degree. It means the frame has entered the plastic deformation subjected to the earthquake with seismic intensity of 7 degree. Under the

Table 1. Structural plasticity parameters of each performance point.

Seismic intensity	Equivalent damping ratio ξ (%)				
	Frame	Plan A	Plan B	Plan C	Plan D
7 (frequent)	5.0	5.0	5.0	5.0	5.0
7 (basic)	6.0	7.5	7.1	6.1	6.4
7 (rare)	20.7	16.9	17.0	16.0	16.2
8 (rare)	24.8	20.2	20.2	19.9	19.3
9 (rare)	—	25.1	24.9	25.5	23.8

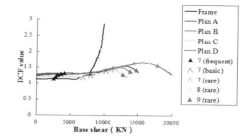

Figure 4. DCF and Base shear.

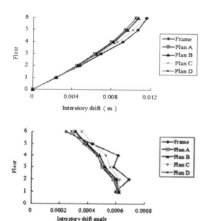

Figure 3. Interstory drift of each model under earthquake with intensity of 7 degree (frequent).

same earthquake action, the equivalent damping from plan A to D is greater than that of the frame. It implies the overall plastic deformation of structure reached a higher extent than that of the frame.

Under a rare earthquake of 7 degree, damping ratio of rocking walls for each plan is lesser than that of the frame. However, with the continuous push-over, the equivalent damping ratio of rocking wall for each plan is always lesser than that of the original framework. Amongst plans A, B, C and D under different seismic intensity, the equivalent damping ratio in plan D has gradually become the minimum one. It means that with the increase of the seismic action, undergoing plastic deformation of the structure in plan D is moderately lesser than that in other three plans, so it also means a lesser extent of structural damage. In other words, it can withstand a higher seismic action.

3.3 *Analysis of deformation between the structural layers*

3.3.1 *Comparison of interstory drift and drift angle*
Compared with the original frame in Figure 3, interstory drift of rocking wall in those four plans is prone to distribute uniformly under the same earthquake action [6]. The minimum interstory drift is observed in plan D, which presents the most uneven deformation between consecutive floors. By contrast, the maximum interstory drift is shown in plan C where the deformation between consecutive floors tends to be more evenly. As for plan A and plan B, they are in-between cases. Therefore, it has revealed that the deformation of the structure is closely related with the extent of the wall's impact on the structural stiffness. For the wall with an identical stiffness, the lesser the impact on the structural stiffness due to the arrangement position, the more uniform interstory deformation of the structure becomes. On the contrary, the greater the impact on the structural stiffness due to the arrangement position, the smaller the overall displacement of the structure is.

3.3.2 *Comparison of interstory drift concentration*
In order to reflect lateral deformation of each model, interstory drift concentration factor value, denoted by DCF, is introduced as follows.

$$DCF = \frac{\theta_{\max}}{u/H} \quad (1)$$

where $\theta_{\max}$ is the maximum interstory drift angle in all structural floors, u is the vertex displacement of structure, and H is the total height of the structure.

The closer the DCF value is to 1, the more consistent the interstory drift angles of the all structural floors become. This positive correlation is also applicable to the control of lateral deformation modes of the structure and the overall seismic performance.

As shown in Figure 4, each model's DCF value remains unchanged at the elastic stage. It indicates that the change of base shear coefficient at the elastic stage has no effect on the degree of the interstory drift concentration. DCF value of frame structure changes described that with the plastic hinge emerging its stiffness increasingly uneven distribution, part of plastic hinge into the plastic are very serious, and distortion centralized serious too, and so will easily cause interstory fracture mechanism.

For embedded rocking wall in those four plans, DCF versus base shear force curves show the same trend that a slow increase first and decline at last with nearly the same shape. The peak values of DCF sorted in descending order just correspond with the influence degrees of the wall layout on the structural stiffness from high to low. It can be inferred that the peak values of structure DCF are associated with the influence

Table 2. Ductility coefficient of all structures

Structural model	Ductility coefficient
Frame	5.33
Plan A	13.85
Plan B	8.67
Plan C	7.30
Plan D	14.02

degrees of rocking wall layout plan on structural stiffiness. Graphically, the DCF value corresponds to the end of the curve. Compared with corresponding DCF versus base shear force curves in the first three plans, there a long decline part of the curve in the plan D, which can be interpreted that structure in Plan D has better earthquake resistance.

3.4 Comparison of structure ductility

Ductility coefficient is the ratio of the corresponding vertex displacement and yield displacement when the largest interstory drift angle of structure is equal to 1/50.

Compared with all models listed in Table 2, it found that ductility coefficient of the frame, 5.33, is the minimum, which indicates the frame can reach the ultimate bearing capacity although the plastic deformation is small. However, structure ductility of the embedded rocking wall increased more or less. This demonstrates that structure ductility factor has apparently increased due to contribution of rocking wall; the structure can undergo large plastic deformation under a strong earthquake. Similar to DCF value, Ductility coefficient sorted in descending order is consistent with the stiffness of the wall, which fully shows that the ductility coefficient of the structure is related with layout of rocking wall. The greater influence of the wall layout plan on the structure stiffness, the larger energy dissipation capacity of structure at plastic stage, the greater the relative plastic deformation can undergo.

3.5 Comparison of shear force among frame floors

As shown in Figure 5, the shear force among frame floors is compared under earthquake with seismic intensity of 7 degree and 8 degree. It indicates that the shear force at bottom and top of frame structure in the four plans of embedded rocking wall is greater than that of the original frame structure under frequent earthquake with seismic intensity of 7 degree and shear force at other floors is smaller than that of the original structure. By Contrast, under rare earthquake with seismic intensity of 8 degree, the shear force at bottom and top of structure frame in the four plans of rocking wall is smaller than that of the original frame structure but shear force at other floors is greater than that of the original frame structure. Therefore, it can be inferred that shear force of frame structure is changed due to attachment of rocking wall.

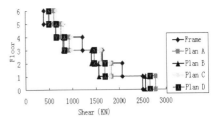

(a) 7 (frequent) degree

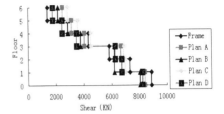

(b) 8 (rare) degree

Figure 5. Layer shear of each model.

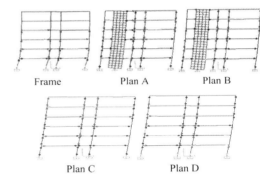

Figure 6. Final plasticity distribution of structural 5-axis.

3.6 Comparison of the structural plasticity

From Figure 6 and Figure 7, it concludes that frame structure belongs to layered yield structure, and plastic hinges mainly appear in the first four floors. Among the plans, more uniform plastic area distribution of the frame structure and plastic zone expansion status in both plan C and plan D is observed. Particularly in plan D, plastic hinges which shows uniform distribution and expansion status, are formed not only in the frame column but also in the main beams connected with the rocking walls (see Figure 7). The main cause is that a rotation identical to rotation of wall base will be attached to each structural floor due to the rigid displacement of the wall in the process of pushover. As a result, vertical displacement is produced in the frame beams, which are connected to the side of the wall and which also suffer a large shear force and bending moment, finally leading to the generation of plastic hinges in the frame beams.

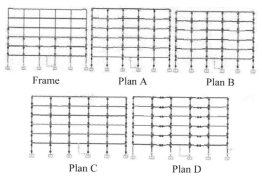

Figure 7. Final plasticity distribution of structural B-axis.

Since distribution uniformity of the structure plastic hinge and the development degree directly affect the capability of the structure absorption and dissipation of seismic action. Then, the embedded wall in plan C and plan D can make a better use of overall capacity of energy dissipation.

4 CONCLUSIONS

Through the above comparative study on the seismic performance of the original frame and the embedded rocking wall in four schemes, there are some conclusions can be drawn as follows.

(1) Under the same conditions, compared with the rocking wall arranged in the main beams, the one arranged in the secondary beams of frame will have a smaller effect on the structural stiffness and the period.
(2) If the embedded walls are doubled rocking wall and arranged in the secondary beams of frame, the seismic capacity of the structure can be enhanced obviously.
(3) Deformability of structure controlled by the rocking wall is related to the influence degree of the wall on the structural stiffness. For the wall with the same stiffness, the smaller impact of layout plan on the structural stiffness, the interstory drift of the structure is prone to be more uniform. The greater impact of layout plan on the structural stiffness, the smaller the overall displacement of the structure becomes.
(4) Ductility coefficient of embedded rocking wall structure is related to the influence degree of the wall on the structural stiffness. The greater impact of layout plan of rocking wall on the stiffness, the larger relative plastic deformation the structure can withstand.
(5) Under the same conditions, if the rocking walls are arranged in the secondary beams and affect the structural stiffness less, the shear force between all floors of the frame part by the rocking wall becomes more apparently.
(6) Under the same conditions, compared with walls arranged in the main beam, rocking walls arranged in the secondary beam favor the uniform distribution of the structure plasticity hinge and expansion status. It can play the energy dissipation capacity of the whole structure better.

REFERENCES

[1] Ajrab J J, Gokhan P, Mander J B. Rocking wall-frame structures with supplemental tendon systems [J]. Journal of Structural Engineering, 2004, 130(6): 895–903.
[2] Housner GW. The behaviour of inverted pendulum structures during earthquakes. Bulletin of the Seismological Society of America 1963; 53(2):404–417.
[3] Midorikawa M, Tatsuya A, Tadashi T, et al. Earthquake response reduction of buildings by rocking structural systems [C] // Smart Structures and Materials 2002: Smart Systems for Bridges, Structures, and Highways. San Diego, CA: SPIE, 2002: 265–272.
[4] Wada A, Qu Z, Ito H, et al. Seismic retrofit using rocking walls and steel dampers [A]. ATC&SEI Conference on Improving The Seismic Performance of Existing Buildings and Other Structures [C], San Francisco, US: ASCE, 2009: 1010–1021.
[5] Building code requirements for structural concrete and commentary (ACI 318M-05) [S]. Farmington Hills, Michigan, US: American Concrete Institute. 2005
[6] Ji X, Kato M, Wang T, et al. Effect of gravity columns on mitigation of drift concentration for braced frames [J]. Journal of Constructional Steel Research, 2009, 65(12): 2148–2156.

The loads variations on the locomotives axles

Ioan Sebeşan
Rolling Stock Department, Transports Faculty, University Politehnica of Bucharest, Bucharest, Romania

Marian Mihail Călin
University Politehnica of Bucharest, Bucharest, Romania
The National Railway Freight, Transport Company, Romania

ABSTRACT: The current trend of increasing power diesel locomotives require more efficient use of their weight especially during the startup, when does the hazard happen slip axles to be unloaded.

Keywords: wheel slip, starting, adherence, gallop, non pitching, stick–slip

1 INTRODUCTION

The individual locomotive drive force of the drive motor for each axle can be greater than the force of adhesion to the axle at more than discharged as excess adherence to one of axles would cause slippage, and the traction power needed would be broken down other axles that will slip and they.

2 THE LOADS VARIATION STATIC NATURE

The loads variation static nature occurs due to rotation box locomotive bogie rotation and traction motors action. In equation (1) is shown the actual amount Q of load on the axle, where components are static relationship Q_0 per axle, of the axle load variation ΔQ_s due to factors such as static of the axle load variation ΔQ_d due to factors such as dynamically. This depends on various factors such as mechanical, mainly on the type of connection between the box and the bogie of the locomotive slurry, suspension mode of the traction motor as well as any devices "non pitching".

The influence of these factors are considered a type of locomotive Co – Co, considering the bearing and alignment line, coupled bogie vertical tractive force equal to all axles.

Considering released locomotive box links with external forces and moments acting on it (figure 1), the conditions of equilibrium of moments, in relation to A and B support box bogies, bogie vertical reactions are obtained, presented in relation (2) where $2l$ is the wheelbase of locomotive, H is the height above the rail the draw hook, h is the height of point the traction force transmission from the locomotives box and the bogie M' and M'' are moments reaction forces on the box due to the device "non pitching".

Considering their longitudinal axes represented bogies with low torque forces and moments at the points A' and B' (that is, centers of rotation of the bogie, as schematically shown in figure 2), the conditions of static equilibrium and deformation will result reactions ΔP_i ($i = 1, ..., 6$) of the suspension of the axles, as shown in fact in the relationship (3) in which $c_{a_1}, c_{a_2}, c_{a_3}$ are the suspension of the axles are rigidities (1, 6), (2, 5) and (3, 4) while (F', M') and (F'', M'') are the torsion points reduction A' and B'. Likewise, c it represents the distance between the middle axle of the bogie and the center of rotation of the bogie.

Given the forces and moments acting on the bogie [5] will results the relations (4) where h_1 is the height of the point of transmission of the drive force from the bogie to the box while e is the distance of point of application of the vertical reaction. Also, $\lambda_0 F_0$ is the reaction engine to the bogie (λ_0 being a coefficient which depends on the traction motor suspension [5]) while M'_1, M''_1 represent the reaction times on the bogies of the "non pitching" device.

Between the loads variations on the springs ΔP_i ($i = 1, ..., 6$) given by the system of equations (3) and the axle load variations ΔQ_{is} ($i = 1, ..., 6$) is the relationship (5) where the positive sign (+) corresponding positioning electric engine the traction before the front axle of the running and the negative sign (−) is used when the engine is positioned after the axle. To note is the fact that variations in axle loads given by the relation (5) after solving the system of equations (6) leading to the system of equations (6) wherein the variables N' and N'' are defined as mathematical expressions by the form (7).

The individual Training axle reduces the possibilities to use the full weight of the adherence, so the use of appropriate means to minimize download axles (by the "non pitching" phenomenon), because of high traction effects have now become a necessity in modern locomotives construction type [5], [6].

3 THE LOADS VARIATION DYNAMICALLY NATURE

Variations dynamic axle loads occur due to fluctuations locomotive during the startup. Out of these the most influential have oscillations "gallop" of the locomotives box due to longitudinal forces [5]. Considering negligible oscillations of electric engines bogies and traction differential equation of oscillations will be of the form (8) while Ψ is the angle of rotation of the vertical of the locomotives box, I_c is the moment of inertia of the locomotives box to the center of gravity, $\Delta V'_d$ and $\Delta V''_d$ are the vertical reactions of the bogies to the locomotives box, F_d is the force on the coupling hook locomotive, F_{bd} is horizontal reaction locomotive bogie over the box while M'_d and M''_d are the moments given by the "non pitching" phenomenon devices.

To note is the fact that, in the equation (8) were only considered dynamically nature forces and moments, their expressions are given in relations (9), (10) and (11), where dV/dt is the acceleration of railway vehicle, γ is a coefficient that takes into account the mass inertia in rotation, m_L is the mass of the locomotive, m_b is the mass of the bogie, R_L locomotive is the resistance to progress while c_c is the stiffness of the suspension locomotives box (on bogies). Also, taking into account the relationship (9), then the equation (8) can be written in the form (12) whose solution can be explicit as (13) expression of the factor can be deduced that Ψ_0 defining in the form (14), The equilibrium position about which the oscillation takes place "gallop" whose own pulsation ω is given in equation form (15).

The variations in the maximum dynamic axle load are obtained by replacing the factor Ψ_{max} from the relationship (13) in the equation no. (9) where the negative sign (−) take the first three axles of the locomotive while the positive sign (+) for the next three. Because in general the locomotive drive systems with "shaft torsion" [5], [6] we have $\omega \langle\langle p$, where p is the angular frequency of the oscillations of stick–slip, we can neglecting the influence of the "gallop" oscillations of the locomotives box over the stick–slip oscillations.

4 ESTABLISHING THE VARIATION OF TASKS TO START AND DRIVE AXLE LOCOMOTIVES CLASS O6O EA

As noted above, slip axle locomotive tasks depends not and does not remain constant during walking. Knowing the variation of static and dynamic tasks axle is absolutely necessary because they depend only on the mechanical construction of the locomotive. For this case study was taken as an example such as electric locomotive type 060 EA, which have been carried out some experiments with the train and the power of the diesel type 060 *Carpatia*, electric traction with engines into alternating current – alternating, who performed a test sample train in October 2010 to the distance between Berlin Est and Postdam. The parameters of these kind of diesel electrical type locomotive, are: $l = 5.15$ m; $a = 2.25$ m; $b = 2.1$ m; $c = 0.05$ m; $e = 0.438$ m; $H = 1.05$ m; $h = 0.59$ m; $h_l = 0.484$ m; $h_c = 2.3$ m; $r_0 = 0.625$ m; $c_{a1} = c_{a3} = 228.10^4$ N/m; $c_{a2} = 134.10^4$ N/m; $c_c = 320.10^4$ N/m; $I_c = 1.4 \cdot 10^6$ kg·m²; $m_L = 126 \cdot 10^3$ kg; $m_b = 24.5 \cdot 10^3$ kg; $\gamma = 0.135$; $\lambda_0 = 1.427$. Because locomotive traction is low moments due to the mode of transmission of the thrust will be to form the system of equations (17), where $d = 3.23$ m represents the points of articulation of drawbar on the locomotives box, $d_1 = 2$ m is the distance between of hinge points on the bogie drawbars while $\alpha = 10^0$ is the angle of inclination from the horizontal drawbars.

The values of static variations axle loads calculated with relations (6) and (17) depending on tractive force F_0 are summarized in Table 1 wherein positive sign (+) corresponding axle load while the negative sign (−) corresponds to its unloading. From this table it can be seen easily that the download of the locomotive axle is axle 1, it having so therefore the first tendency to skate.

The pulsation own oscillations "gallop" of the box locomotive, calculated with equation (15) will have the value $\omega = 11.011$ rad·s^{-1}, this value is much lower pulsation due to the phenomenon of stick–slip which is generally the value $p = 180, \ldots, 375$ rad·s^{-1}. It thus follows therefore that the axle load at the time of slip can be considered constant.

The unloading one axle maximum dynamic nature because of tasks will be given by (16) where Ψ_0 depends on the starting of Vehicle acceleration dv/dt who are made explained in the relations (10), (11) and (14) which is determined by the equation of the train motions (18), where $\varphi = g/(1 + \gamma)$; R is the total resistance to the train progress; G_L, G_V are respectively the weight of the locomotive and the weight wagons. It also will consider and locomotive towing a train consisting of freight cars in alignment and landing tier. In this case, $R = r_L \cdot G_L + r_V \cdot G_V$, where r_L; r_V represents the specific resistance of the locomotive forward wagons respectively, are determined by the following relations respectively $r_V = 1.6 + V^2/2700$[daN/10^3daN] and $r_L \cdot G_L = R_L = 296 + 7.068(V/10)^2$[daN], where in V is expressed in km/h.

The force of the axle 1 such are limited by the adherence, most unloaded will be given by the relation no. (19), where $\mu_a = \mu_a(V)$ represents for varying the adherence coefficient depending on speed V [5]. It may reveal the influence of constructive parameters of the locomotive, and resistance to progress R_L; R_V of the locomotive and wagon respectively adhesion coefficient $\mu_a(V)$ over dynamic load Q_l the axle downloaded if you take into account the dynamically nature of the loads variation of. For this system to be solved formed by equations (16),

(18) and (19), thereby achieving the equation (20) in the canonical form.

Because the phenomenon of stick–slip axle occurs with the most discharged after passing this axle and as the adhesion force of the drive motor for each axle can be greater than the adherence force to the axle at more than unload, the load variations length was calculated for $F_0 = F_a$, where $F_a(V)$ is given by the relation (19) for $\mu_a(V)$ determined by the *Curtius–Kniffler* relations and $G_V = 1500 \cdot 10^3$ daN. Comparing calculated and limited adherence strength, taking into account only the variations of static axle loads ($\Delta Q_{ld} = 0$) and the results were shown in the table no. 2.

The locomotive force will be limited by adherence and became $F_{aL} = 6F_a$ and the weight adherent will be $G_a = 6 \cdot Q_l$, both of which are functions of the speed V by running of the train. To be able to observe the influence of train velocity on the emergence of the phenomenon of stick–slip, were represented (in figure no. 2 and in figure no. 3), the variation curves of forces $F_{aL}; F'_{aL}; R$ and respectively $Q_L, -\Delta Q_s$ and $-\Delta Q_d$ to overcoming the adhesion for speeds between 0 and 50.4 km/h.

Considering also that the locomotive speed control is constant tensile force during the startup and since the adherence force decreases as walking speed, axle slippage will occur at the speed corresponding to the thrust intersection with adherence forces. The locomotive traction force value during the startup, determine the size of which will depend on vehicle acceleration directly proportional to the speed of movement of the train speed. This can be seen from the diagram shown in Figure no. 4, in which, have been presented the curves C_i ($i = 2, \ldots, 8$) the variation of acceleration with walking speed V of the train and that have been calculated using the relationship (18) for different values of constant traction force. The boundary points $A_1; A_2; \ldots; A_8$ of acceleration correspond to the velocity at which the train the locomotive slip curve C which linking them actually representing acceleration variation for adjustment after the limited adherence force whose experimentally determined values were summarized in table no. 3.

In order to highlight the influence of the coefficient of adhesion μ_a (to $V = 0$) over the variation of the loads on the axles are calculated the forces F_a and the loads, for values of the coefficient of adhesion between 0,340 and 0,486 such us the apparent fact of table no. 4. With these values, in Figure no. 4 were represented the variation curves of functions $F_a(\mu_a)$ and $Q_l(\mu_a)$.

5 THE EMERGENCE OF THE STICK–SLIP PHENOMENON

Movement axle motors can be attached to intermittent or stuttering, a phenomenon known in the literature as the stick–slip. Because of stick–slip-'s occurring so important dynamic axle overload and its drive system and variations of tensile forces on the periphery of the wheels, producing gait disturbance and therefore reduction of vehicle traction performance. Aspects of the phenomenon of stick–slip will be analyzed if the axle drive system with electric motor drive torque is transmitted to the gearbox of the drive spindle shaft via a "torsion elastic".

The vibrations of stick–slip, whose root cause lies in the allure characteristic friction wheel – rail and traction force occur at low sliding speeds generally-start the vehicle, it is possible to slip axle are beyond the limited by the force the traction force adherence. The variation in time of the sliding speed of the wheel in the manufacture of stick–slip-'s, and Schönenberger Schröter determined experimentally, is seen in Fig. 6. In general, the vibrations that occur under the influence of dry friction at the wheel – rail presents different manifestations depending on sliding speed. The occurrence and timing of stick–slip – it's dependent on the variation of the coefficient of friction wheel – rail by sliding speed. When producing slip axle intense mechanical action occurs between the contact surfaces particles wheel – rail, with significant heat generation. Significant changes in the contact surfaces makes the friction coefficient varies with sliding speed.

The measurements were carried out by Frederick revealed that in the slide small coefficient of friction increases with speed of sliding up to a maximum value, and the larger slide, it is not maintained constant, but decreases with increasing the speed of slip.

6 CONCLUSIONS

By analyzing the characteristics shown schematically in the figures above it can be concluded that the limited force adherence F_{aL} is less than the adherence force F'_{aL} calculation which was not taken into account the variation of the dynamic axle loads (shown schematically in Figure 3).

Likewise, given the dependence of dynamic load train acceleration will result in a worsening of the locomotive the traction feature with increasing the train acceleration.

Finally it should be mentioned the fact that slip axle and consequently, the emergence and manifestation of the phenomenon of stick–slip will occur especially in the case of "strength avulsion/pull-out" in place of locomotive train that occurs when starting with a jolt powerful practice that can be amplified by linking locomotive first railway vehicle car of the train without proper tightening torque (of the hook) the traction, allowing a wider broad coupling between the buffers of the locomotive and the first car (vehicle) of the train. This effect can be enhanced also in the case in which load is the minimum up to the axle of the locomotive. In this context, it is important to note that the load Q_l the axle locomotive downloaded directly proportional to speed at which the train, as shown in the diagram shown in Figure 2. This is due in particular to lower acceleration with decreasing walking speed according to the graph in Figure 4, and so consequently, and because of the variation of the dynamic nature of the locomotive loads.

7 EQUATIONS

$$Q = Q_0 + \Delta Q_s + \Delta Q_d \quad (1)$$

$$\begin{cases} \Delta V' = -\left[\dfrac{1}{2J}.6.F_0(H-h) + M' + M''\right] \\ \Delta V'' = -\Delta V' \end{cases} \quad (2)$$

$$\begin{cases} \Delta P_1 + \Delta P_2 + \Delta P_3 = F' \\ (a-c)\Delta P_1 - c.\Delta P_2 - (b-c)\Delta P_3 = M' \\ \dfrac{b}{c_{a_1}}\Delta P_1 + \dfrac{a+b}{c_{a_2}}\Delta P_2 - \dfrac{a}{c_{a_3}} = 0 \end{cases}$$

$$\begin{cases} \Delta P_4 + \Delta P_5 + \Delta P_6 = F'' \\ (b+c)\Delta P_4 + c.\Delta P_5 - (a-c)\Delta P_6 = M' \\ \dfrac{a}{c_{a_2}}\Delta P_4 - \dfrac{a+b}{c_{a_2}}\Delta P_5 - \dfrac{b}{c_{a_1}} \end{cases} \quad (3)$$

$$\begin{cases} \Delta Q_{1s} = \dfrac{1}{N'}\left\{F'\left[\dfrac{(a+b)(b+c)}{c_{a_2}} + \dfrac{a.c}{c_{a_1}}\right] + M'\left(\dfrac{a+b}{c_{a_2}} + \dfrac{a}{c_{a_1}}\right)\right\} - \lambda_0.F_0 \\ \Delta Q_{2s} = \dfrac{1}{N'}\left\{F'\left[\dfrac{(a+b)(b+c)}{c_{a_2}} + \dfrac{a.(a-c)}{c_{a_1}}\right] + M'\left(-\dfrac{b}{c_{a_2}} - \dfrac{a}{c_{a_1}}\right)\right\} + \lambda_0.F_0 \\ \Delta Q_{3s} = \dfrac{1}{N'}\left\{F'\left[\dfrac{b.c}{c_{a_1}} + \dfrac{(a+b)(a-c)}{c_{a_2}}\right] + M'\left(\dfrac{b}{c_{a_1}} - \dfrac{a+b}{c_{a_2}}\right)\right\} + \lambda_0.F_0 \\ \Delta Q_{4s} = \dfrac{1}{N''}\left\{F''\left[-\dfrac{b.c}{c_{a_1}} - \dfrac{(a+b)(a-c)}{c_{a_2}}\right] + M''\left(\dfrac{b}{c_{a_1}} - \dfrac{a+b}{c_{a_2}}\right)\right\} - \lambda_0.F_0 \\ \Delta Q_{5s} = \dfrac{1}{N''}\left\{F''\left[\dfrac{b.(b+c)}{c_{a_2}} + \dfrac{a.(a-c)}{c_{a_1}}\right] + M''\left(-\dfrac{b}{c_{a_2}} - \dfrac{a}{c_{a_1}}\right)\right\} - \lambda_0.F_0 \\ \Delta Q_{6s} = \dfrac{1}{N''}\left\{F''\left[-\dfrac{(a+b)(b+c)}{c_{a_2}} - \dfrac{a.c}{c_{a_1}}\right] + M''\left(\dfrac{a+b}{c_{a_2}} + \dfrac{a}{c_{a_1}}\right)\right\} + \lambda_0.F_0 \end{cases} \quad (4)$$

$$\Delta Q_{is} = \Delta P_i \pm \lambda.F_0; (i=1,\ldots,6) \quad (5)$$

$$\begin{cases} F' = \lambda_0.F_0 + \Delta V' \\ M' = -3.F_0.(h_1 - r_2) - \\ \quad -3.\lambda_0.e.F_0 + \lambda_0.F_0.(a+b+c) - c.\Delta V' + M'_1 \\ F'' = \lambda_0.F_0 + \Delta V'' \\ M'' = -3.F_0.(h_1 - r_2) - 3.\lambda_0.e.F_0 + \\ \quad + \lambda_0.F_0.(a+b+c) - c.\Delta V'' + M''_1 \end{cases} \quad (6)$$

$$\begin{cases} N' = -\dfrac{b^2}{c_{a_1}} + \dfrac{(a+b)^2}{c_{a_2}} + \dfrac{a^2}{c_{a_3}} \\ N'' = -N' \end{cases} \quad (7)$$

$$I_c \ddot{\Psi} = (\Delta V'_d - \Delta V''_d).l + F_d.(h_c - H) - 2.F_{bd}.(h_c - h) + M'_d + M''_d \quad (8)$$

$$\Delta V'_d = -\Delta V''_d = -c_c.l.\Psi \quad (9)$$

$$F_d = (l+\gamma).m_L.\dfrac{dV}{dt} + R_L \quad (10)$$

$$F_{bd} = \left(m_b + m_L.\dfrac{\gamma}{2}\right).\dfrac{dv}{dt} + \dfrac{R_L}{2} \quad (11)$$

$$I_c.\ddot{\Psi} + 2.c_c.l^2.\Psi = F_d.(h_c - H) - \\ - 2.F_{bd}.(h_c - h) + M'_d + M''_d \quad (12)$$

$$\Psi = \Psi_0.(1 - \cos\omega t) \quad (13)$$

$$\Psi_0 = \dfrac{1}{2.c_c.l^2} \\ .[F_d(h_c - H) - 2.F_{bd}.(h_c - h) + M'_d + M''_d] \quad (14)$$

$$\omega = \sqrt{\dfrac{2.c_c.l^2}{I_c}} \quad (15)$$

$$\Delta Q_{id} = \mp\dfrac{2.c_c.l.\Psi_0}{3}; (i=1,\ldots 6) \quad (16)$$

$$Q_1 = \left\{Q_0 + \dfrac{1}{3J}\left[(1+\gamma)m_L.(h_c - H) - \right.\right. \\ \left. - 2\left(m_b + m_L.\dfrac{\gamma}{2}\right)\!\!\left(h_c - h - \right.\right. \\ \left.\left. - d.\dfrac{\tan\alpha}{2}\right)\right].\dfrac{R.\varphi}{G_L + G_V} - \dfrac{R}{3J} \\ \left(h - H + d.\dfrac{\tan\alpha}{2}\right)\right\}/\{l + \mu_a\{0.707 + \\ + \dfrac{1}{3J}\left[(1+\lambda)m_L.(h_c - H) - \right. \\ \left. - 2\left(m_b + m_L.\dfrac{\gamma}{2}\right)\!\!\left(h_c - h - d.\dfrac{\tan\alpha}{2}\right)\right]. \\ \dfrac{6.\varphi}{G_L + G_V}\right\}; \quad (17)$$

$$F_a = \mu_a.(Q_0 + \Delta Q_{is} + \Delta Q_{id}) \quad (18)$$

$$\dfrac{dV}{dt} = \varphi.\dfrac{6.F_0 - R}{G_L + G_V} \quad (19)$$

$$\begin{cases} M' = M'' = (3/2).F_0.\tan\alpha.d \\ M'_1 = M''_2 = (3/2).F_0.\tan\alpha.d_1 \\ M'_d = M''_d = (1/2).F_{bd}.\tan\alpha.d \end{cases} \quad (20)$$

8 TABLES

Table 1.

Osia	1	2	3	4	5	6
Q_{is}	$-0.707.F_0$	$+0.447.F_0$	$-0.311.F_0$	$+0.311.F_0$	$-0.447.F_0$	$+0.707.F_0$

Table 2.

V [km/h]	R [daN]	F_a [daN]	F'_a [daN]	$-\Delta Q_{ls}$ [daN]	$-\Delta Q_{ld}$ [daN]	Q_l [daN]	$\frac{dV}{dt}$ [m·s^{-2}]
0	2696	5405	5639	3821	872	16307	0.158
7.2	2728	5098	5304	3604	816	16580	0.148
14.4	2826	4860	5045	3436	772	16792	0.140
21.6	2988	4671	4840	3302	733	16965	0.133
28.8	3215	4516	4672	3193	702	17105	0.127
36.0	3507	4388	4533	3102	673	17225	0.121
43.2	3864	4281	4416	3027	642	17331	0.116
50.4	4285	4183	4315	2962	616	17422	0.111

Table 3.

V [km/h]	0	7.2	14.4	21.6	28.8	36.0	43.2	50.4
$\frac{dV}{dt}$ [m·s^{-2}]	0.1581	0.1481	0.1400	0.1331	0.1269	0.1213	0.1160	0.1106

Table 4.

μ_a	0,340	0,360	0,380	0,400	0,420	0,440	0,460	0,480	0,486
F_a [daN]	5513	5759	5998	6232	6459	6681	6897	7108	7170
Q_l [daN]	16215	15996	15785	15580	15379	15184	14993	14807	14752

9 FIGURES

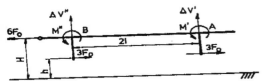

Figure 1. The forces and the moments acting on the box locomotive.

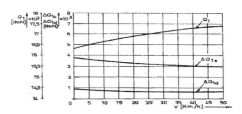

Figure 4. The curves of variation of axle loads to overcome adhesion.

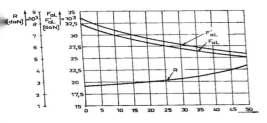

Figure 2. The adhesion force characteristics of the locomotive and train drag.

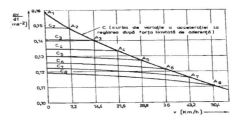

Figure 5. Acceleration variation curves for adjusting the $F_0 = $ const.

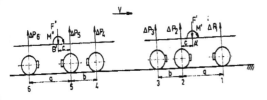

Figure 3. The forces and the moments acting on the bogie.

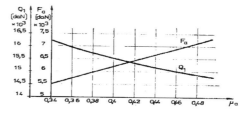

Figure 6. Variation of adhesion strength F_a and axle load Q_l according to the coefficient of adhesion μ_a at $V = 0$.

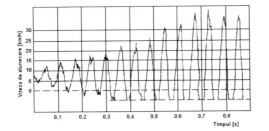

Figure 7. The variation in time of the sliding speed of the wheel.

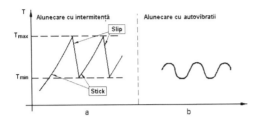

Figure 8. Vibrating under the influence of friction: a – the stick–slip. b – vibrating at high sliding speeds.

REFERENCES

[1] Kalker, J.J., *A strip theory for rolling with slip and spin*, Proceedings Kom. Akad. Wet., Amsterdam, Section B. 70, 1967.
[2] Nadal, M.J., Locomotives a Vapeur, Collection Encyclopédie Scientifique, Bibliothèque de Mécanique Applique' et Génie, Paris, 1908.
[3] Prud'homme, M. A., *La Voie*, Revue Générale des Chemins de Fer, ian., 1970.
[4] Scheffel, H., *Conceptions Nouvelles relatives aux dispositifs de suspension des* véhicules ferroviare, Rail International, dec., 1974.
[5] Sebeşan, I., *Dinamica vehiculelor de cale ferată* (The Dynamics Of Railway Vehicles), Editura Matrix Rom, Bucureşti, 2010.
[6] Sebeşan, I., Mazilu, T., *Vibraţiile vehiculelor feroviare* (Vibrations of the railway vehicles), Editura Matrix Rom, Bucureşti, 2010.
[7] Van bommel, P., *Consideraions lineaires concernant lemouvement de lacet d'un vehicule ferroviare*, UIC / ORE C9, nov., 1968.

• UIC Code 513: Guidelines for Evaluating Passenger Comfort in Relation to Vibration in Railway Vehicles, 1st ed., 1.7. 1994, International Union of Railways, Paris 1995.

• UIC Code 518: Testing and Approval of Railway Vehicles from the Point of View of their Dynamic Behaviour – Safety – Track Fatigue – Ride Quality, Paris, October 2005.

Lucrarea face parte din cercetările efectuate în cadrul contractului PCCA tip II, nr. 192/UEFISCDI, cu titlul "Soluţii pentru îmbunătăţirea performanţelor dinamice şi a securităţii la impact a vehiculelor de tracţiune feroviară pentru alinierea la cerinţele impuse de normativele europene".

The work is part of research conducted under the contract PCCA type II, no. 192/UEFISCDI, entitled "Solutions for improving dynamic performance and security impact of rail vehicles to align with the requirements of European standards".

The performance of ground resource heat pump unit in a long time operation in summer hot and winter warm district

Yingning Hu, Fan Ruan & Shanshan Hu
College of Mechanical Engineering, Guangxi University, Nanning, China

ABSTRACT: In summer hot and winter warm district, when ground-source heat pump (GSHP) unit operates for a long time, the buried pump absorbs heat from the ground mat lead to that the ground temperature around will reduce the stability, which will make the performance of GSHP unsteady. This paper collects the data of GSHP operating for a long time, and compares the recovery of ground temperature under different stop-operate ratio and different length of buried pump, which can be used as reference fro for the research of GSHP.

Keywords: GSHP, continuous operation, COP ground temperature recovery, stability

1 BRIEF INTRODUCTION

Summer hot and winter warm district needs hot water supplied all year. The cop of boiler is 0.7~0.9, but the GSHP is as high as 3~4.5. Besides, the GSHP makes use of the shallow geothermal energy which is renewable and non-polluting, so the use of ground source heat pump technology in this kind of district has a great significance for energy conservation and environmental protection.

The operation theory of GSHP shows that when GSHP operates intermittently, the soiled temperature recovers well, and the unit can keep efficient and stable (Xianying Liu et al. 1999). Researches on the stability of GSHP are focus on numerical simulation (Lei T K. 1993). Jiaxiang Wei has done some researches in a stop-operate ratio in GSHP. The result shows that if the ratio of unit stop-operate is 2:1 in a day, the unit would have the highest efficiency (Jiaxiang Wei et al. 2008). Haiwen Shu Fan in Dalian university of technology has done research on the recovery of ground temperature with numerical simulation. The result shows that the ground temperature will back to the start level when unit operates in a stop-operate ration (Haiwen Shu et al. 2006). Fenghao Wang of Xi'an Jiaotong University made a research in the heat transfer performance in summer (Fenghao Wang et al. 2009). And Huajun Wang in Tianjing University has finished an experiment with a GSHP operating in a long time (Huajun Wang et al. 2007). But little of the researches have involved in the situation which unit operates in a long time in summer hot and winter warm district, and how the performance and the recovery of ground temperature are. And the research in the effect in length of buried pipe on the temperature recovery of ground temperature. So this paper aims to these questions, and to make some tests to consummate the research.

2 EXPERIMENTAL PLATFORM AND EXPERIMENTAL METHOD

2.1 Experimental platform

The research target is a GSHP system in Guangxi University which works for hot water's supplying in student apartment. This system has one GSHP unit and two pumps working for water's supplying. Each of them has a rated power of 3 kW. There are two soil heat exchanger circulating pumps which have rate power of 3 kW. The average length of buried pipe is 50 m and the total depth of buried pipe is 4180 m.

2.2 System structure diagram

Figure 1 is a direct heat type of GSHP system. The tap water directly flows into condenser unit and the unit will heat the water to a setting temperature and make them into the heat preservation water tank. There is a hot water cycle system in the unit. When the temperature in tank falling under the setting temperature, the cycle pump will inhale the water in tank into the unit and heat them again. All of these are controlled by

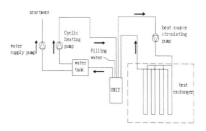

Figure 1. Direct heat type of GSHP system.

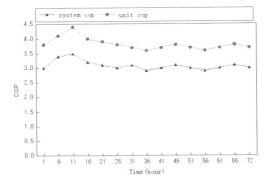

Figure 2. The temperature date of parameter when unit operates in a long time in summer.

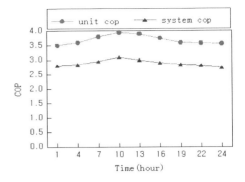

Figure 3. The cop curve of unit and system.

electronic systems, and tap water flows into the unit directly, so the system can supply water to consumer when it operates, and the design standard is according to the largest water consumption by 70%, which could reduce the investment of unit.

2.3 Experimental apparatus

In this experiment there are high precision thermometer, a stopwatch and clamp type multimeter.

2.4 Experiment scheme

In order to study the unit's performance and temperature of ground when GSHP operation in a long time, the experiment scheme is to keep the unit operates 72 hours and collect the needed data, then start the unit again after it stops 24 hours and 48 hours and collect the needed data. At the same time, compare the unit with a different depth length in the same experiment option, observe the unit's performance.

3 THE PERFORMANCE AND RECOVERY OF GROUND TEMPERATURE WHEN GSHP HAVE A LONG TIME OPERATION

3.1 Unit operate in 72 hours test

The experiment begins on 7 am June 3 in 2013. The research objects are the temperature of environment, hot water outlet and tap water. The data was recorded every 5 minutes and the result is shown in Figure 2.

Figure 2 shows that, when the unit starts, the inlet temperature of heat source side decreases a little until keeps stable at 16, and the outlet temperature is 13. The outlet and inlet temperature difference is 3. And the outlet temperature of hot water is stable at 46 to 47, with tap water at 26. So the outlet and inlet temperature's difference is 3.

In order to survey the variation of unit's heating performance, power consumption and the volume of hot water produce in every hour would be collected. Compute the COP, and it is shown in the next figure.

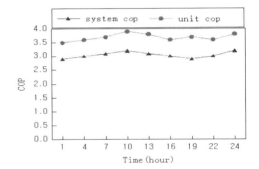

Figure 4. The unit performance under the 2 options.

Figure 3 show that, the cop of system rises in the first 10 hours and reaches the maximum at the 11th hour. Then in the following 35 hours it decreases slowly to 3.8. When the unit becomes stable, we observe the system performance have a little fluctuate and the environment temperature also had obvious fluctuate. Thus the environment temperature has some affect on the unit performance. When unit becomes stable, the COP always stays at 3.5 to 4. The result shows that in summer hot and winter warm district, the unit performance can keep stable and high efficient when unit operates in a long time.

3.2 The recovery of heat source

After it operates 72 hours, stop the unit to recover the heat source. Restart it after 24 hours to observe its performance. In the same way, after the unit operates for 72 hours, stop it and restart it after 48 hours to observe its performance. The results are shown in Figure 4.

Figure 4 shows when start the unit after it stops 24 hour and keep it operates for another 24 hours, its cop curve is similar with the curve of which the units runs in the first 24 hours in its continuously operating 72 hours. And it reaches stable at 3.5 to 4. The performance is stable.

Figure 4 also shows that when unit stops after continuously operates for 72 hours and restarts after 48 hours, its COP is stable at 3.5 to 4. It shows that in

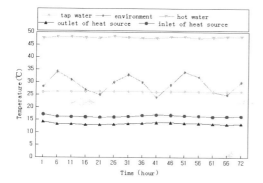

Figure 5. Unit COP under different length of buried pipe.

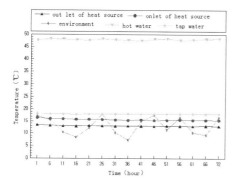

Figure 6. The date of temperature in winter operation model.

summer hot and winter warm district, the unit has only a small effect on heat source when it operates 72 hours.

3.3 The effect of different average length of buried pipe in unit performance

In the same district, there is a total depth of buried pipe same unit. But the average buried pipe is 17 m, and this unit is for hot water supply too. So we choose those two units as comparison.

Figure 5 shows that in the same district, during 72 hours' operation of unit with 50 m buried pipe length, the COP rises first and then reduces and keeps stable at last. But when it comes to the unit with 17 m buried pipe, the COP rises first but keeps declining, though the declined speed tends to be slow. Comparing to the 50 m buried pipe, its stability is worse. The result shows that the average length of buried pipe has some effect on unit COP when unit operation in a long time.

3.4 Winter operation model

Figure 6 shows that on 12 am February 26th 2014, when unit starts, the inlet temperature of heat source side reduces a little. At last, the temperature keeps stable at 15 and the outlet temperature at 13. The outlet

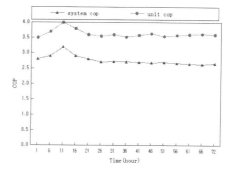

Figure 7. The cop curve of unit and system.

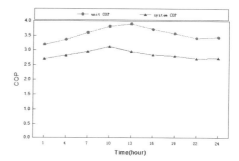

Figure 8. The unit performance when restart after 24 and 48 hours.

and inlet temperature difference is around 2.5. The outlet temperature of hot water is at 46 to 47 and the tap water is at 28. So the unit can keep high efficiency when it operates in winter.

We also survey the unit COP, and through which we can make use of the COP to judge the stability of unit. The result is shown in Figure 7.

Figure 7 shows that the COP of system rises in the first 10 hours and reaches the maximum at the 11th hours. Then in the after 35 hours it reduces slowly to 3.5 when the unit keeps stable. The result shows that in summer hot and winter warm district, the unit performance can keep stable and efficient even when unit operates in a long time in winter operation model.

Figure 8 shows that start the unit after stop the unit 24 hour, and keep it operates 24 hours, its COP curve is similar with the curve of which the first 24 hour when unit operate 72 hour. And it reach stable at 3.5 to 4.

Figure 8 also shows that start the unit after stop for 48 hours, it reach stable at 3.5 to 4. It shows that in summer hot and winter warm district, there is little effect when unit operates 72 hour in winter model.

4 CONCLUSION

This paper is a test on GSHP unit operating in a long time. In this test, we survey the recovery of heat source; compare in different buried pipe length, the outlet and inlet temperature parameters COP; draw the curve and so on. In theory the unit will release heat to ground which may lead to the soil cooling that make the unit performance worse. But in our experiment in summer hot and winter warm district, when the average length of buried pipe is over 50M, even unit operate 72 hour. The unit will keep highly efficient. After start the unit after 24 hours or 48 hours when unit operates 72 hour. The unit still keeps highly efficient. We can summarize that the depth of buried pipe have some effect to the unit performance in long time operation.

The conclusion is that if we have a long enough average length of buried pipe, the heat absorb of buried pipe have little influence on ground temperature, so the unit operate in a long time can keep highly efficient.

5 TOPIC SOURCE

(1) The center of Guangxi laboratory in Manufacturing system and advanced manufacturing: The application and technical standards ground source heat pump.
(2) Guangxi housing and urban rural construction project: The application system efficiency assessment technical guideline of Guangxi shallow low-energy building.

REFERENCES

Fenghao Wang, Liang Yan, Yu Bin. In different stop-drive model, the performance of GSHP. [J] China association of refrigeration. 2009, In Chinese

Haiwen Shu, Mulin Duan, Pingping Fan. The impact of ground heat recovery to GSHP operation [J]. Cooling and air condition. 2006 6(1), In Chinese

Huajun Wang, Liang Shen, Liang Zhao, Li. The research on GSHP performance when it operate in a long time [J]. Journal of north China electric power university: natural science edition. 2007 34(2), In Chinese

Jiaxiang Wei, Wenxue Liu, Zhiwei Tang. The research about GSHP operate in interval model. [J]. Renewable energy. 2008 26(1), In Chinese

Lei T K. Development of a computational model for a ground-coupled heat exchanger [J]. ASHRA E Translations, 1993, 99(1): 149–159

Xianying Liu, Yong Wang. The research about GSHP heat exchanger [J]. Journal of Chongqing Jianzhu University, 1999, 10(5): 21–25, In Chinese

The planning bases on landscape-oriented proposal stems

Limin Bai
Jilin Jianzhu University, Changchun, Jilin, China

Jianmin Hou
Design and Research Institute, Changchun Institute of Technology, Changchun, Jilin, China

ABSTRACT: Through the analysis of the international bidding scheme about Qingdao International Horticultural Exposition, designed by the Valley Crest Design Group, the characteristics of urban planning based on landscape-oriented proposal stems is introduced in this paper. In addition, the relationship between urban development and nature has been put forward and suggestion that more attention should be paid to the nature for urban planners is made in order to create more harmonious environment between human and nature.

1 INTRODUCTION

In April 2014, the International Horticultural Exposition had been held in Qingdao. Review the International bidding planning scheme, designed by Valley Crest Design Group & Jianhua, Jiang's Team was made. The Phase-II planning, as a further development of the Expo, has combined with the exhibition area. Michael Braden, the landscape planner, put forward "landscape-oriented" proposal stems. The planning scheme reflects the perspective of landscape planners, namely, concern about plants and animals and primitive natural environment. Therefore, the long-term planning presents the characteristics of urban construction in respect of the natural environment, which is worth considering for the urban planners.

2 INTERNATIONAL HORTICULTURAL EXPO

The Garden Festival or International Horticultural Exposition began in 1962 and was organized by the International Association of Horticultural producers (AIPH) in Rotterdam Floriade; this event is now also known as the Floriade Garden Festival.

The popularity of the Garden Festival has spread and the focus of the Festival has created the needs in life. Particularly in the urban environment, it has expanded. Concerns over pollution and global warming indicates environmental crisis we are facing in reality, indirectly leading to the increasing need for the Horticultural Exposition that aims to address not only the beauty of nature, but necessity of environmental sustainability.

The World Horticultural Expositions has been successfully held four times in China. Such events also earned the honor and respect for the country.

The first Expo was held in 1999 in Kunming followed by the 2006 Expo in Shenyang and the third Expo in Xi'an in 2011. Something different from previous Expos, the Qingdao International Horticultural Exposition 2014 (hereinafter referred to as "Qingdao Expo") is the Garden Expo held in the Mountains at first time. The environment of Qingdao belongs to subtropical marine climate. In cold winter, the city is currently undergoing a downturn in tourist numbers; the Horticultural Expo can offer visitors an opportunity to develop year-round experiences and acts as a catalyst for sustainable development.

3 OVERALL PLANNING OF THE BIDDING SCHEME

The Qingdao Expo covers a total area of 5.3 square kilometers, including the site area of 1.6 square kilometers and the remaining belonging to the plan of second phase. Also, the Qingdao Expo offers a unique opportunity to examine the "natural systems" that support our urban lives and provide inspiration for future development.

Nature and human settlement is a theme that will present in the master plan. Our approach stems from an interdisciplinary understanding of the complex forces and living systems required to sustain an ecologically sensitive system. Our influency in ecology, landscape architecture, architecture, interpretive education, engineering and design provides a strong framework for resolving the complexities and dual nature of the Expo.

Our landscape-oriented proposal stems not only focus on from the need of express horticultural advances and floral beauty, but from a passionate desire to create sustainable and holistic living environments for works, lives and plays. The Expo and post-Expo development can be mutually strengthen and provide an inspirational model for the future by embracing the need for environmentally balanced development that establishes a new frontier for sustainably scaled communities.

4 PHASE-II DEVELOPMENT PLANNING

4.1 Objectives in planning

Creating an ecologically sustainable garden, which is inspiring and which can also be actively integrated into the surrounding beautiful mountains and residential areas.

Qingdao is being shaped as one of the most popular tourist attractions. There are something should not be ignored. Creating a high-quality public space with excellent and humanized structure, outstanding space orientation, gestalt space and pedestrian-friendly parks, where communication and social life is encouraged and which should be built as the driving force for Qingdao's economic development in order to make Qingdao an ideal living and tourist destination.

4.2 Concept development diagram

Our approach to the Phase-II (Fig. 1) Development followed the same process as the Phase-I Expo Site and occurred concurrently. Significant consideration was taken to the sustainability and viability of utilizing the Expo buildings as key architecture for public space within the late-stage development. While a dynamic program of architecture and landscape design has been proposed that will transform the existing Baiguoshan Forest Park and Scenic Area in Laoshan Mountain into a sustainable botanic park, wilderness experience area, and learning center in the Qingdao Expo, the ambitions of a growing city and the inherent successes of the Phase-I development also deserve attention.

Specifically, a series of broader initiatives and sustainable systems is being proposed that will promote fiscal viability and creation of an ecological footprint in a mountainous landscape, which is sensitive to overdevelopment; As a reference, the vision and successes of the Expo development might be applicable in the wider region. One of the goals during the Phase-II development is to form a larger network of open space by connection of smaller sustainable development zones, and ecological habitat for preservation of the mountain landscape and the enjoyment of the end users.

Our proposal for the Phase-II development capitalizes on the environmental features of the Laoshan Mountain while integrating the needs of a growing city. The wetland treatment facilities initiated in the Expo development are integrated and expanded into the Phase-II development that will increase habitat for the local and migratory birds of Qingdao. The existing and proposed roads have been realigned in order to create a naturally landscaped parkway in harmony

Figure 1. Master plan.

with the natural terrain and reduction of the impact on the arterial roads.

4.3 Study on industrial composition

Based on the industrial strategy of Qingdao and the survey of industrial status of Licang, one urban district of Qingdao, the industries of the Expo area and surrounding areas consist of real state, tourism and modern service industry. Detailed types of land use include villa, multi-story apartment buildings, public housing, holiday resorts and hotels, residential and office buildings, commercial buildings, mixture of residential and office, night entertainment centers, theaters, ski resorts, outdoor experience park, flower markets, maintenance center, sustainability research center and environmental protection schools.

A mixture of various functions is required in the proposed industrial plan. The land development is also diverse. A 24-hour shopping center and a commercial street will be built based on infrastructure including commercial and residential building and retail stores. Combined with the existing commercial zone in Licang, this area will become a business center in the eastern part of Qingdao. The development of the Expo site and its surrounding areas will include a film studio, thematic tourism such as mountain adventure tourism, mountain holiday resort and hotels and high-end tourism service facilities. This will be the new growth point that promotes tourist industry Qingdao to higher level. The Qingdao tour will be not only limited to the city itself but also the surrounding areas. The advantage of its geographical location and natural resources will fuel its booming flower industry and it will be beneficial for forming a flower industrial chain covering production, exhibition, research and trade. Education institutions such as junior school and senior school for environmental protection will be established in order to improve its cultural service function. The development of European-style towns and high-end villas is considered, featuring a combination of residence and natural design idea. The height of the city will be determined by high-rise buildings in certain parts. Villas are usually low-rise buildings. The combination of land use and mountain topography requires that the development of town should take into account the local feature of mountain nature park.

4.4 Analysis on existing land-use

The existing land-use in the planning area mainly consists of village lands, collectivity-owned lands, orchard, flood discharge ditch and vacant lot. The efficiency of the land-use in the planning area is very low, as the countryside houses here are mainly single storey buildings but few multi-storey buildings.

Existing land-use in the north part of mountain is covered with forest of a higher crown density, but only a certain level of developments in the foot of mountain, including Chinese Candock World, Moth Orchid Growing Greenhouse and an education training base in the west side. Service facilities are built in the surrounding slopes and their land-use area is small.

4.5 The layout planning

Once the Expo site is incorporated into the development plan of Phase-II, the overall site will be connected by a green space that links each land to the Expo site and the Laoshan Mountain. This space will be used for both pedestrian access and recreational activities.

4.6 Analysis of the roadway and pedestrian movement systems

The main site can be accessed from Tianshui Road to each of land use and parking garage. For the pedestrian system, it consists of two parts: one is street accessing all public open space, the other is the pedestrian path through the landscape network connecting all land uses at the phase-I and phase-II together. The access systems shall satisfy the requirement of walking, sightseeing and relaxation.

Grade separation is designed for the intersection of pedestrian paths and vehicular roadways, such as pedestrian underpass or pedestrian bridge, etc.

Underground tunnel is proposed and utilized at Phase-II when different land uses are equipped with underground parking garages.

4.7 Analysis on green landscape system

Green land and water system distribute like a web, integrated with mountain area and water system that forms a natural ecological system. By reserving the natural mountains and valleys, an open parking system is formed; the main corridor runs through south to north. The water system includes two reservoirs and wetland system at phase-I. At development phase-II, it stores the water in the original land and wetland system is formed gradually as a result.

Sustainable development, particularly in sustainable energy supply is no doubt in promotion of the health and happiness of people. For example, the use of "green" materials and search of alternative energy does should be encouraged instead of those materials harmful to environment. In addition, financial success is of great importance to sustainable development. Successful urban developments have scattered pedestrian zones linked to environmental corridors that enrich daily life of the people who live and work in these areas; critical habitats for plants and animals must be preserved. Considering the geomorphology of the mountain and surrounding foothills, the approach has been proposed, aiming to create an anamorphic plan that preserves the physical attributes of the site and give landmark to those landforms.

4.8 Analysis on open space

Open space in layout is divided in to two types, including natural open space formed by water body and the central open space system. Nature and city is linked organically in the open space system and it forms a rich special experience. The scale of open space is based on the European village that provides a warm and attractive pedestrian connection. There are two night entertainment districts, which are connected by walkways that form the core of the open space.

5 SPACE AND ARCHITECTURAL DESIGN

5.1 Open space for night entertainment and businesses

Bars, restaurants, fashion retails and clubs are set up at the first floor of buildings and multi-functional buildings, as the commercial and entertainment center for the community. The buildings located in the night time entertainment center will be built at an appropriate scale. The European-style buildings, together with the streets and the street view, will create an extremely attractive commercial space and will become the commercial center of the area.

5.2 Ecological wetland

The ecological corridor consisting of wetland and mountain green areas will be a vital part for the large regional park system. Other land use and the ecological corridor will create an uninterrupted open space. The ecological corridor will integrate complete with the walkway system, which will allow pedestrians and cyclists to move around the area. The combination of the wetland waterfront and the functional land use of the city will offer people a comfortable tourist space.

5.3 Villas in the mountains

The villas that are well designed and built in harmony with the mountains (Fig. 2) and the trees will become part of the mountain. This emphasizes the harmonious relation between buildings and landscape. The layout of the buildings pattern will disappear, which will result in a beautiful natural environment.

Figure 2. The Bird's Eye View of Phase-II.

5.4 Towns

Multi-story residential buildings adopt the European-style town and mainly have 3~6 stories. The town center is nursery or district center, shaping a centripetal pattern. The town streets will create a warm and comfortable environment. The buildings and street view are well integrated. Its building pattern is correspondence to that of multi-national building such as historical street of Qingdao.

5.5 Film studio

Unlike the traditional Chinese theaters, the film studio will refer to the style of film studio in California of America. Indoor scenes are shot in an array of big studios and outdoor scenes will be shot among the buildings with various styles. Meanwhile, the studio links with the regional big park, night entertainment center, small town and arts district, which will become a larger shoot location.

5.6 Outdoors experience park

The combination of outdoors experience park and regional big park weaken its effect in thematic orientation. For instance, combination with wetland supports the activity like water kayak. Sports like rock-climbing, indoor ski facilities and cableway highlights the feature of outdoor experience and adventure.

5.7 Holiday resorts and hotels

Land is developed for high-end holiday resorts, hotels, conference centers and amenities. Holiday villas include two types of villas, that is, small-area mountain villa and luxurious holiday villa. These villas are built in modern style and are diversified.

6 CONCLUSION

Nature and our place is an eternal theme the modern planners should focus on. Being fiscally responsible to the realities of the city, the development of Phase-II serves as a model of urban design & sustainable development that meets the environmental demands of a fragile habitat. Landscape-oriented proposal stems and focus on the nature as well as peoples' desire to return to nature should be a new perspective of planner planning in the new period.

REFERENCES

Appleyard. 1979. *The Conservation of European Cities*. USA: The MIT Press.
Kevin, Lynch. 1981. *Good City Form*. USA: The MIT Press.
Valley Crest Design Group & Jianhua, Jiang's Team. 2011. *The International Bidding of 2014 Qingdao International Exposition Planning*.

The research and application of reinforced truss slab in steel framework

Wenxin Xia & Haiying Wan
School of Civil Engineering, Hefei University of Technology, Hefei, China

ABSTRACT: This paper describes the analysis of the bending bearing capacity of reinforced truss slab floor under different conditions, and a realistic finite element analysis of the overall effects of reinforced truss slabs in a steel frame structure. This paper also discusses the great effects of reinforced truss slab on improving the bending stiffness of a floor, which was considered, from the angle of the truss, as a single reinforced truss slab and of that as a part of an entire steel framework. The research results provide reasonable basis for design of the corresponding frame structural floor slabs.

Keywords: reinforced truss slab; framework; bending bearing capacity; ANSYS finite element analysis

1 INTRODUCTION

As is well known, in the development of building floors, entire casting floor slab and assembled precast floor slab have been playing an important role. Structure of entire casting type has many advantages of its fine integrity and stiffness, while its defects are also obvious, especially that it would seriously affect the construction period in steel structure construction. The prefabricated reinforced concrete structure can realize the industrialization of structure members, it is not affected by season or weather, thus speeding up the construction progress with high quality, using precast floor slab can reduce use of formwork support and thereby save the materials. However, there are still some disadvantages in the prefabricated reinforced structure: the integrity and permeability of the structure are not good, neither is its seismic performance, especially in high-storied buildings the precast concrete slab can hardly meet the seismic requirements (Fang Lixin 1999).

It has been the key of the slab design to utilize resources rationally and strengthen the seismic performance under the steady development of building structure. In recent years, many new slab forms appeared one after another in the field of engineering, such as cold-formed steel deck and concrete composite slab and pre-stressed slab.

This paper introduces a new slab form called Reinforced truss slab, then analyzes and compares the stress and deformation of the structure under different conditions with ANSYS software to study the advantages of the reinforced truss slab.

2 REINFORCED TRUSS SLAB

Reinforced truss floor slabs (Gong Shuguang et al. 2004) were made in the following way: firstly the upper and lower reinforced were made into space truss form, then they were connected together with tilted belly poles and the galvanized steel sheet is welded together into steel truss, finally it is concreted on the steel plate (Fig. 1).

In this floor system, steel truss plays the decisive role; it reduces the construction stage support and increases the stiffness of the floor, meanwhile it changes the original way of binding reinforcement in the slab to welding, which makes the connection performance more reliable. In the reinforced truss slab, stable space truss takes the place of parallel stressed steel, thus improving the overall stiffness of the components, at the same time the diameter of the steel bar and the height of the truss are allowed to change to meet the needs of different practical projects. In general, reinforced trusses promote the standardization productions of components, effectively guarantee the precision and quality of components, and increase the safety of the structure system.

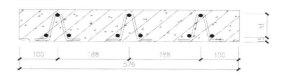

Figure 1. A cross-sectional view of the reinforced truss slab.

In order to get more theoretical understanding of the function of reinforced truss, the ANSYS software was used to study the performance of the steel truss in a single floor and the overall framework.

3 NUMERICAL ANALYSIS OF THE BASIC THEORY

3.1 Basic assumptions

Before the nonlinear analysis of the reinforced truss slab under loading, the following assumptions were made:

(1) The material is elastic-perfectly plastic;
(2) In order to facilitate convergence of the second model, the model is simplified by omitting the dislocation in the short cross direction between the steel plate and the concrete;
(3) The sections of the model are made by straight lines, ignoring the impact of steel knuckle at the arc, and without considering the geometric imperfections and residual stresses.

3.2 Material nonlinearity

In this study, the yield criterion of von Mises was used to determine the yield load of various materials.

3.3 The establishment of the connection elements

The ways of connections in the finite element mainly include three kinds, contact, coupling and constraint. The connection used in this study consists of three parts: (1) The nodes of steel trusses and the concrete were coupled; (2) The contact between the analog interfaces of the concrete and the steel plate was simulated by the spring element of combin 39 (Liu Tao et al. 2002), which describes the approximate bonding properties between the two materials; (3) The approximate rigid connection was set between different elements among the entire slabs and steel frames.

In this article, the setting of the parameters of spring elements and the coupling of nodes is very important; a close connection exist between the setting and the results of the degree of convergence.

4 MODELING

In this study, the model mainly includes two aspects: a single steel truss floor (Fig. 2) and the overall framework of reinforced girder slab structure system (Fig. 3).

The establishment of the models was divided into the following steps: geometric modeling, element meshing, the definition of material properties, the definition of the load and boundary conditions, the definition of the contact conditions, the definition and analysis of the parameters for solving analytical results and so on (Li Weibin et al. 2006).

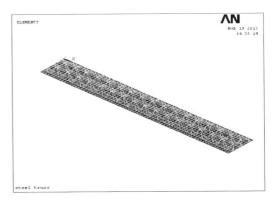

Figure 2. Meshed model of steel trusses.

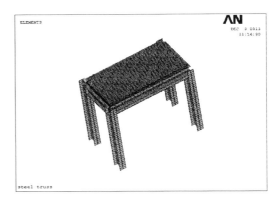

Figure 3. Meshed model of the overall framework of reinforced girder slab.

In a single slab of steel trusses, steel truss and the concrete were simulated by different elements. Steel trusses were simulated by link 8 element, and the concrete was simulated by solid 65 element. The material properties were determined by using multi-linear isotropic hardening model (miso), and the bottom of the plate was simulated by shell 63 element. The models were meshed by the approach of body mapped meshing. In order to make the shape more regular and to carry out a more detailed analysis of the mechanical behavior of the reinforced truss slabs, the models should adopt the "constraint equation method", which establishes constraint equations between the nodes of the concrete elements and the nodes of the truss elements by the command of "ceintf" (Shen Wei et al. 2011). Through multiple sets of constraint equations, the steel truss elements and the concrete elements were connected into a whole. This method is in line with the actual situation and the result is more accurate.

Throughout the steel frame structure system, steel beams and steel columns were simulated by beam 188 element, and the floors were simulated by shell 63 element. The key of establishing models is to deal with the connection (Sun Bin. 2007) between the shell elements and beam elements. In this study, in order to simulate the rigid connection between the beam and

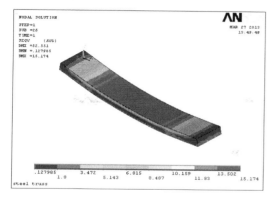

Figure 4. Stress contours of slab A.

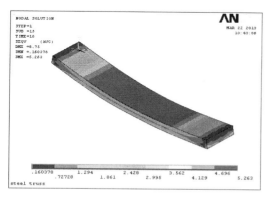

Figure 5. Stress contours of slab B.

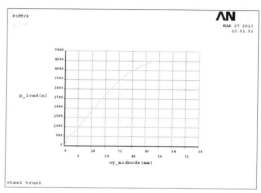

Figure 6. Load–displacement curve of slab A.

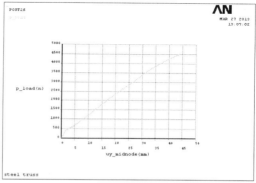

Figure 7. Load–displacement curve of slab B.

the slab, shared nodes can be used without making constraint equations and appropriate beam offset can be set.

5 NUMERICAL RESULTS

In this study, in order to study the stress and the deformation of reinforced truss slabs, a mid-span load was set in the single slab and the overall framework. The comparison included the following two parts: (1) the comparison between slab A (no steel truss) and slab B (with steel truss); (2) the comparison between of frame system A (no steel truss slab) and frame B (with steel truss slab).

5.1 The comparison between the presence and absence of a single slab of steel truss

In the comparison groups, the dimensions of slab A and slab B are both 3400 mm × 576 mm × 160 mm. The difference is that slab A had no steel truss, while slab B had steel truss. From the comparison groups of Figs. 4 and 5, it can be seen that reinforced concrete floor trusses had no greater impact on the stress distribution. Then it can be found that the presence of steel truss had a certain impact on the flexural

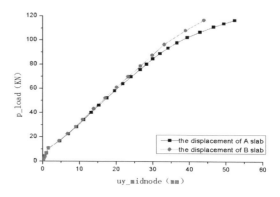

Figure 8. The comparison of the deflection curves of slabs A and B.

rigidity of steel truss floor in Figs. 6 and 7. From the curve in Figs. 8, we can see that when the load was less than 60% of the ultimate load (Wang Xinmin. 2007), steel truss was not obvious, but as the load increased, the deflection of the slab A span was gradually larger than that of floor B, which shows an increasing trend. This indicates that in the loading stage, the steel truss played a certain influence on not only increasing the

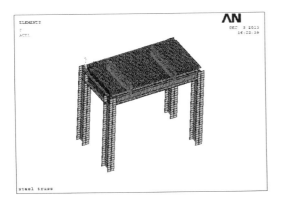

Figure 9. The frame load chart.

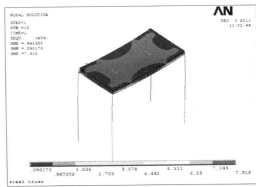

Figure 11. Stress contours of framework B.

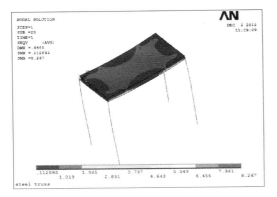

Figure 10. Stress contours of framework A.

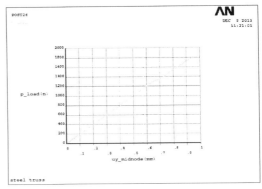

Figure 12. Load–displacement curve of frame A.

stiffness of the floor but also preventing the development and extension of the fracture, which made some contributions on the crack resistance of the floor (Yuan Fashun et al. 1996).

At the same time, according to the process and the final ANSYS numerical analysis, the limit loads of slab B were significantly greater than those of slab A; to some extent, steel trusses improved the bearing capacity of the slabs.

5.2 *The analysis of the entire system of the slab frame structure with steel trusses*

In the comparison groups, the dimensions of frame A and frame B were both 3400 mm × 1728 mm × 160 mm, which are horizontal spread by three single slabs. The height of steel columns was 3300 mm and the four column feet were all fixed. The difference was that frame A had no steel truss, while frame B had steel truss. Figs. 10 and 11 are stress contours, respectively of frame A and frame B, applied by the line load of about 1800 N/mm, which can be seen in Figs. 9. From the following stress contour data of the steel frame model in Figs. 12 and 13, it can be seen that the mid-span displacement of frame B was smaller than the one of frame A, which also can be checked by the slopes of the curves in the initial stage (Zhao

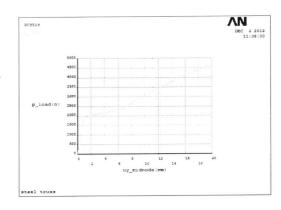

Figure 13. Load–displacement curve of frame B.

Lei. 2007). Meanwhile, the stresses of frames A and B were distributed evenly, but the maximum stress value of frame A was slightly larger than the one of frame B, indicating that the maximum stress of the frame with steel trusses is smaller, the stress difference is small and the corresponding structures are more secure and reliable.

According to Fig. 14 and the analysis of ANSYS finite element for frame B, the ultimate load could

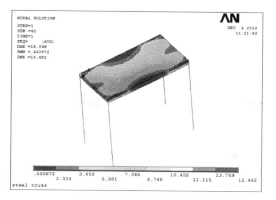

Figure 14. Stress contours of framework B under the ultimate load.

reach about 4600 N/mm, but once the load is greater than 2000 N/mm for frame B, the nodal solution appeared offset, which cannot effectively come in line with the real situation of the static solution; this situation reflects that the bearing capacity of frame B has been greatly improved due to the steel trusses, and the improvement undoubtedly proves that the steel trusses play a very important role in the bearing capacity of the system (Zhao Peng et al. 2011).

6 CONCLUSIONS

Through finite element analysis, it was validated that the use of the reinforced truss slab form is feasible and rational.

From the contrast analysis of a single slab of steel bar truss, it can be seen that for a single floor, bending rigidity of steel truss has been slightly improved. On the other hand, it can be seen that in the steel truss framework for the entire system of stress distribution, stress and deformation trend also plays a better role in the whole frame structure system and the bending bearing capacity of the slabs has been substantially increased.

In a word, steel bar truss floor improves the rigidity and bending bearing capacity of the whole structure, in the mean time the stress of the steel bar truss floor model is more reasonable and it can be designed for bidirectional board after additional transverse reinforcement. Then, the advantages of overall bending bearing capacity performance of the framework are more apparent, including that the height and the rebar diameter can be changed to adapt to different spans, different heights and different stress distributions of slabs. On the whole, the study on the floor of steel truss structure has a constructive meaning.

REFERENCES

Fang Lixin. 1999. Nonlinear FEM analysis of steel-concrete composite structure. *Journal of Southeast University* 29 (2):76–80.

Gong Shuguang et al. 2004. *ANSYS command and parameter programming*. Beijing: Machinery Industry Press.

Liu Tao et al. 2002. *Proficient ANSYS*. Beijing: Tsinghua University Press.

Li Weibin et al. 2006. Application of steel bar truss deck in steel structure. *Construction Technology* 35(12):105–107.

Shen Wei et al. 2011. Steel truss floor construction technology. *Tianjin Construction Science and Technology* (4):8–17.

Sun Bin. 2007. *Reinforced concrete composite slab nonlinear finite element analysis and experimental research*. Hefei: Hefei University of Technology.

Wang Xinmin. 2007. *Numerical analysis of ANSYS structures*. Beijing: People's Communications Press.

Yuan Fashun et al. 1996. Nonlinear analysis of composite panels. *Industrial Building* 26(10): 28–33.

Zhao Lei. 2007. *Self-supporting composite slab of reinforced concrete truss design calculation method*. Changsha: Central South University.

Zhao Peng et al. 2011. An accurate model for the stiffness evaluation of spring interface elements used for the interfacial layers of laminated plates. *Chinese Journal of Computational Mechanics* 28:131–135.

The research of construction engineering life

Wei-shu Zhao
School of Management, Anhui Jianzhu University, Hefei, China

ABSTRACT: The short service life of construction buildings directly affects people's lives and social stability. Extending the building service life is necessary for building a resource-saving society. Thus, it has been widely concerned by the society. To fundamentally solve the problem of short service life of buildings, the factors affecting construction life must be found out. The current phenomenon of short service lives of architectures in China was first introduced in this paper. The factors affecting building life were concluded by engineering case induction. Finally, the methods and measures to improve building life were put forward according to actual problems. Those methods and measures provide theoretical basis for improving the service life of buildings.

Keywords: construction engineering; life; affecting factor; method

1 INTRODUCTION

The building is one of the most important products, it must meet the needs of the community, such as service life, security, reliability, etc. For example, according to the provisions of China's Code for Design of Civil Buildings (GB50352-2005), the design service life of civil architecture structures should conform to the following standards in Table 1.

As shown in Table1, the durability of memorial-buildings and particularly important structures is 100 years, and for ordinary houses and buildings it is 50 years. However, the average service life of Chinese residential buildings is only 30 years, the service life of some important buildings and memorial buildings is even shorter. There is a big disparity between the real service life of Chinese building and requirement of the national standard. It also has obvious gap with that of developed countries.

The statistics show that the average life expectancy of British buildings reaches 132 years, and that of American building is 74 years. Ephemeral buildings refer to those demolished before reach the design service life or durability period (Zhang Xu et al. 2010).

China's Construction is in a stage of rapid development, the national energy is consumed in a high speed and the "conservation-oriented society" is initiated by the government. However, buildings with short service lives waste massive social resources, and generate construction garbage. According to statistics, the number of construction garbage accounts for 30%–40% of total city garbage, which is not consistent with the social development and also cause many social contradictions.

The building life in China should be considered from a comprehensive. The factors influencing construction life are not only the construction quality problems; there are many other factors that need to be studied and explored. Scientific methods must be used to manage strictly the building demolition behavior and regulate the construction market, in order to gradually reduce and eliminate ephemeral buildings.

The reasons causing ephemeral building were summed in this study according to author's many years of engineering practice experience and theoretical study. The methods and measures to improve the construction life were put forward correspondingly.

2 CASE STUDY AND REASON ANALYSIS OF EPHEMERAL BUILDING

In recent years, not only concerned landmark buildings, but also a large number of ordinary residential buildings were removed soon after the completion due to various reasons. Recent reports show a lot of typical cases of representative buildings with short service lives, as described in Table 2.

As shown in Table 2, the common feature of those buildings is their short service lives, which are a great

Table 1. Design service life.

Class	Design life (years)	Example
1	5	Temporary structure
2	25	Easy to replace structure
3	50	Ordinary houses and buildings
4	100	Memorial buildings and particularly important structures

Table 2. Representative buildings with short service lives.

Building name	Service life	City	Reason for short service life
The Five Mile River stadium	18 years	Shenyang	In pursuit of economic benefits
Qing dao Le Grand Large Hotel	20 years	Qingdao	Defects in architectural design
Wenzhou Bank of China Tower	6 years	Wenzhou	The unqualified construction
Qingdao railway building	16 years	Qingdao	Image project
Nantong train station	2 years	Nantong	Image project
Zheda the Lakeside Campus Building No. 3	13 years	Hangzhou	In pursuit of economic benefits
Shenyang Summer Palace	15 years	Shenyang	Unscientific planning
Wuhan the Bund Garden District	4 years	Wuhan	Unscientific planning
Hubei Shou Yi sports training center training hall	10 years	Wuhan	Image project
Hefei Vienna Forest Garden District	0 years	Hefei	Unscientific planning

Figure 1. Reason analysis of short-life buildings.

Fifty representative buildings with short lives were selected for the case study. The reasons for short life were analyzed according to the author's many years of engineering practice and theoretical study, as shown in Fig. 1.

Several results can be seen in Fig. 1.

2.1 The pursuit of economic benefits

It can be concluded from Fig. 1 that the main cause of short-service-life buildings is the pursuit of economic benefits. Real estate industry can drive a lot of industries and support economic growth. The building demolition and reconstruction can double the GDP. In order to pursue economic benefit, the phenomenon demolishing a four-star hoteland building a five-star hotel always appear. With the sound of blasting, a large number of "young" architectures died abnormally with smoke and debris.

However, the new GDP growth created by building removal is superficial and does not bring substantial increase of social wealth and economic value. On the contrary, abnormal construction and demolition are great waste of wealth and resources, and bring nothing to the social and economic benefits except political achievements.

2.2 The image projects

Many government and people lack the awareness of the protection of historical and cultural city buildings, as well as the sense of old building reuse. Especially driven by interests, many governments use the ideology of "big demolition, big construction" to reform the city image. Because of the short period of local government leaders, many city leaders do not consider the economic capacity and the objective need of the city, or the long-term development. Some local leaders would take their preference into consideration when making city planning and construction decisions. The feasibility and rationality of city planning and construction are inevitable poor.

2.3 The planning problems

Many buildings are removed because of not quality problems, but the irrational, unscientific, unsustainable city planning. City planning is the start of city construction, and the city construction and development must be guided by scientific planning. After the formulation of a reasonable planning, big housing and city construction make a lot of changes to the city

waste of social resources and energy, and cause environment pollution problem. There are many reasons for short service lives of buildings, such as unqualifiedconstruction, image projects, planning issues, unperfected legal system, etc. Many measures, such as establishing legislation and accountability, strengthening the construction of people's livelihood strategy, encouraging scientific and forward-looking city planning, and strengthening the construction market supervision, can be applied to improve the construction quality and reduce the occurrence of ephemeral building. Vigorously developing the removable building component and the building maintenance industry, and reducing construction waste can help achieve the purposes of energy saving and sustainable development. Life extension of construction building is the largest saving, which is much more than a few percent of energy saving (Shen Jin-zhen 2008).

In a word, the ends of those buildings' service lives are not normal. Even worse, the Hefei Vienna Forest Garden District was demolished before completion. According to the official statement, this district influenced the city landscape axis between. Huangshan road and Shushan mountain. Therefore, there are many reasons related to the short life of Chinese building. It is necessary to study and resolve this problem, which is important to people's livelihood.

planning, which is also a main reason of short-life buildings. There are other reasons such as political projects, image projects and improper political view on achievement, for example the change of industrial district to commercial or residential areas. The main reason for such situation is that the city planning lacks of forward-looking, resulting in the inconsistence between overall planning of the city and the rapid development of economy. City planning is a predictive science, which can set out the development route based on the need of cities. However, the city planning policies of many cities lack of foresight and understanding of buildings of historical value.

2.4 *The investment problems*

Some short-life buildings appear due to investment problems, for example, there was an "official shareholders" in "fragile building" in 2009.

2.5 *Engineering quality problems*

The service lives of buildings depend mainly on the architectural design quality, construction quality, usage and maintenance and other factors. Engineering quality is one of the most important factors affecting the housing life, safety, applicability and anti disaster ability. Therefore, the construction quality must be improved by all means to extend the service life. However, there are a lot of short-life buildings due to the poor engineering quality. For instance, many affordable houses in Beijing were demolished because of serious quality problems.

The frequent construction quality problems should give the government a wake-up call. Driven by huge profits, some construction enterprises do not provide good quality, and construct the buildings with scamp work and stint material, leaving a lot of hidden troubles to the building life. Some architectural design has problems that are doomed to lead to short service life. On the other hand, the construction regulators do not have strict criteria upon approval of construction quality. This also affects the engineering quality.

3 PROBLEMS CAUSED BY SHORT-LIFE BUILDINGS

Short-life buildings will bring a series of problems to the society.

3.1 *Waste of resources*

The building construction consumes a lot of resources. Short-service-life buildings result in repeated construction and increase of energy consumption, which cause a great waste of resources, especially the cement, water, steel and other non-renewable natural resources. The repeated construction also brings serious damage to the riverbed, and destruction of vegetation, soil, water loss, as well as natural disasters.

3.2 *Environment pollution*

Large-scale demolition of old buildings results in a huge amount of building garbage. Those garbage not only takes up a lot of land and increase environmental load, but also contains harmful chemicals that can seriously pollute water and environment. This means that dealing with construction waste is also an important task.

3.3 *Destruction of culture*

City culture is the essence of the city history. Old buildings, especially particular historical buildings and monuments, are the context of a city reflecting the local characteristics. Blindly removing old buildings and building new ones will split the city's historical context, weaken the city's cultural tradition and damage the urban History and culture.

3.4 *Contradiction*

In recent years, many families were destroyed during the demolition incidents, which caused a very bad influence on the society. Therefore, the prevention and reduction of ephemeral buildings is very important. Coupled with the improper treatment from government, this phenomenon is easy to cause social contradiction and destroy social stability.

4 METHOD AND MEASURES TO IMPROVE CONSTRUCTION ENGINEERING LIFE

Many buildings are demolished in the early and middle phases of their life cycles because of various reasons. The urban building lives greatly reduce and affect the healthy development of cities. According to previous analysis and existing problems, the problem of ephemeral building should be resolved from four aspects.

4.1 *To strengthen the Urban and Rural Planning Law*

China has already formulated the *Urban and Rural Planning Law*, thus the critical step now is to strictly implement the law and regulations to ensure scientific and forward-looking city planning. Legal regulations should become the conduct norm of government officials for executing plans. Important city construction projects, new construction or removal, must convene public hearings and solicit public opinions on the basis of strict expert argumentation. Once made, city planning cannot be changed without any legal procedures. All the procedures including demolition proposal, application, and approval must have a specific responsible person. The planning and major projects must implement hearing and accountability, and accept social supervision.

4.2 To put people's livelihood strategy at first

To emphasize the livelihood of people, we must build hundred years of building and extend building life. Therefore, the government should give priority to the people's property rights and the residence right. To ensure people's life quality, the quality of the environment must be improved, the carbon emissions reduced and resources saved. At the same time, people's livelihood strategy should be considered in official performance evaluation systems. The government should use taxpayer's money for people's livelihood or promoting people's life level. Public funds from the people should serve the people, receive public supervision and realize the public investment construction goal.

4.3 To improve the scientific cadre examination management system

GDP is one of the most important indexes for cadres' assessment, which stimulates officials to blindly pursue GDP growth. Thus, the city update is submitted to the need of economic interests while the real demand of city residents for livable environment and happy life is ignored. This causes great negative impact on the society. Therefore, it is imperative to abolish GDP standard index in cadre examination management system and perfect the scientific cadre examination management system.

4.4 To improve the construction quality

Construction engineering quality directly affects the service life of buildings. The main factors affecting the construction quality are survey design, construction method, project personnel, building material, system management, working environment, project funds, project progress and market regulation (Zhang Liang-cheng, 1999). Therefore, a life cycle responsibility system should be established for architectural design personnel, construction personnel, supervision personnel and government department's responsible person in building service life. Beforehand controls, in-process controls, afterwards controls should be accomplished at the same time.

Beforehand controls mainly include physical factors control and human behavior control. The physical factors include procurement, acceptance and approval of engineering material, semi-finished products, component, permanent equipment, and quality control of design drawings.

Process controls mainly include the control of construction organization design, construction scheme, construction methods and process and technical measures. Quality supervision institutions should implement engineering quality supervision.

Afterwards controls mainly include completion of acceptance, approval, approved for sale. Housing presale institution must be cancelled to prevent the selling of residential housing without legal approval or with unqualified acceptance.

5 CONCLUSIONS

There are lots of problems of construction service life in recent years in China. The factors affecting the construction building life were analyzed to find out the critical factors. Finally the methods and countermeasures to solve the problem of short building life were given, providing theoretical basis to resolve the problem in the future. Only by controlling those influencing factors of construction engineering life, can the problem of building short service life be reduced and eliminated. Not far in the future, building lives can be extended, and century building standard and sustainable development strategy can be realized with the exploration of the government and the public. Therefore, this paper has important theoretical and practical significance for solving the problem of short-life building.

REFERENCES

Shen Jin-zhen. To solve the problem can not be solved for building short-lived ephemeral building on the construction of buildings, Research on the development of city [J]. 2008 (15): 117–122.

The related national codes, regulations, standard, «code for design of civil buildings» GB50352-2005.

Zhang Xu, Shi Ling-Lin, Zhang ben-niu. The causes and preventive measures of ephemeral building Chongqing architecture [J], 2010, 1 (10): 18–21.

Zhang Liang-cheng, The quality control of construction project, China water conservancy and Hydropower Press [M], 1999.

Foundation item: key research base for Humanities and social science of Anhui Province Department of education project (project number: SK2013A043).

The research on structural design to resist progressive collapse

Zilong Guo, Yuanbing Cheng, Yixiang Yin & Guanghua Qiao
North China University of Water Resources and Electric Power, China

ABSTRACT: The definition of progressive collapse of structures is given in this paper. The research on structural design to resist progressive collapse at home and abroad is recounted. This paper recommends the design method of concept design, tie force design and demolition widget design and provides the principle and procedure of these designs.

Keywords: progressive collapse; concept design; tie force design; demolition widget design

1 INTRODUCTION

Structural progressive collapse is defined as the local damage caused by accidental load, and it leads to the development of the structural non-stable damage. Progressive collapse of a structure will result in serious loss of life and property and produce significant social impacts. The measures to ensure the integrity of structures and prevent progressive collapse have been widely investigated by engineers and researchers.

2 CURRENT RESEARCH STATUS OF PROGRESSIVE COLLAPSE

2.1 Research by foreign scholars

Since the collapse of Ronan Point apartment in 1968, researches on progressive collapse have been carried out for more than 30 years abroad. Some major foreign specifications and provisions on the improvement of structures' ability to resist progressive collapse have been presented.

Ellingwood and Leyendecker discussed the advantages and disadvantages of three kinds of methods on account of a structural progressive collapse in 1978 (G. Kaewkulchai and E.B. Williamson 2006); the three methods are damage control method, indirect design method and direct design method, and direct design method and damage control method were recommended. In 1998, Corleyetal discussed the progressive collapse of the Murrah Federal Building caused by burst in Oklahoma, and some structure measures were presented to improve structural resistance to progressive collapse under immense load based on their research. Corleyetal argues that strengthening the ductility of structures can improve their resistance to progressive collapse. Collapse of reinforced concrete frame structure was analyzed by Wenjun Guo and Ramon Gilsanz (Wenjun Guo and Ramon Gilsanz 2003). Serkan Sagiroglu analyzed the large displacement collapsed of plane frame structure by using beam model. The measure put forward to resistance progressive collapse by Abolhassan Astaneh-Asl is to set cables on the position of the beam on the floor, so that the loads can be delivered by the cables when the frame columns became invalid (Abolhassan Astaneh-Asl and Erik A 2005).

2.2 Research by domestic scholars

Structural progressive collapse is a common occurrence in CHINA. Typical of them include the collapse of the major structure caused by gas explosion in Panjin, Liaoning province in 1990, the houses destroyed by earthquake in Sicuan province in 2008. Until now, researches on structural design to resist progressive collapse in CHINA are rare, and the relevant research is mainly on the load of earthquake. Only a few scholars do the research of progressive collapse caused by accidental loads. Hongquan Sun, professor of Hebei Poly-technic University, using variable stiffness stem cell model of numerical simulation on collapse process of reinforced concrete frame under earthquake, defined the principle of collapse by using damage model (Xiaobin Hu and Jia Ru 2006). Xilin Lv, Xianglin Gu and Gang Xuan, researchers of Tongji University, analyzed the collapse of reinforced concrete frame structure by using discrete element method on seismic action; they used an improved model of story cell with the method of dynamic relaxation (Sufen Zhang 2008). Xinzheng Lu from Tsinghua University provided a design process referred to United States Department of Defense compiled by the resistance of progressive collapse design. He analyzed the ability to resist progressive collapse of a three-story reinforced concrete frame based on the current code for design of concrete structure simulation of progressive collapse, and the fiber model analysis of

Table 1. The requirements of structure design method to resist progressive collapse.

Importance of structure	Concept design	Tie force design	Demolition widget design
General structures	Y	—	—
Important structures	Y	Y	—
Very important structures	Y	Y	Y

frame structures was based on the program THU-FIBER developed by Tsinghua University (Xinzheng Lu 2008). The knot strength method and dismantled components method was used to design the ability to resist progressive collapse of the frame. He established a design example of frame structure that is based on the progressive collapse resistance design method in foreign norms.

3 PRESENTATION OF DESIGN METHODS TO RESIST PROGRESSIVE COLLAPSE OF BUILDINGS

3.1 General provisions

Based on the researches on design methods of progressive collapse of reinforced concrete structure, and combining with the structures designed under the code in China, some scholars presented the design method to resist progressive collapse of the framework structure (GB50010 2010; GSA2005; JG-J3 2010). The method includes concept design, tie force design and demolition widget design; the selection of the methods mainly depends on the importance of the building. The tie force design method and demolition widget design method can be adopted with existing instruments of structure analysis (Table 1). For particularly complex or special request structure, nonlinear dynamic analysis method should be carried out.

The combined design load can be determined according to the following formula in process of tie force design and demolition widget design.

$$S = A(S_{Gk} + \sum \psi_{qi} S_{qik}) + \psi_{cw} S_{qwk} \quad (1)$$

In the equation, S_{Gk} is the standard values of permanent load; S_{qik} is the standard values of the vertical variable load (including floor, roof live load and snow load); ψ_{qi} is the quasi-permanent coefficient of variable load; ψ_{cw} is the combination coefficient of wind load; S_{qwk} is the characteristic value of wind load; and A is the dynamic magnification factor.

It is a dynamic process from the occurrence of initial partial damage to the new state of equilibrium reached by remaining structure. It should be considered to multiply the dynamic magnification coefficient of dynamic effect after dismantle components, if the internal force of the remaining structure is calculated according to the static method. This article suggests that the dynamic amplification coefficient of A should take the following values: When components are directly connected to the demolition component or demolition component is located in the upper part, the value of A is 2; otherwise, the value is 1.

3.2 Concept design

The conceptual design mainly depends on the alternate paths of structural system, wholeness, ductility, connection of structural members and distinguishes of key elements. And the purpose of design is to avoid the weakness that could easily lead to progressive collapse of structures. Details are as follows.

(1) Increase structural redundancy to enable the system to have a sufficient number of alternative load transfer path

Possessing with enough spare load transfer path is a basic requirement of structure to resist continuous collapse. Using reasonable structure scheme and structure arrangement could avoid provoking continuity weak positions, increasing the redundancy in structure to form multiple and multidirectional load transfer path structure system. Usually, it can be determined whether the structure has a spare load transfer path by dismantling method. When the structure scheme is determined, a particular component should be removed, then it should be checked whether the remaining structure have alternate load transfer path, and estimated whether the alternate load transfer path has the bearing capacity of the corresponding according to the experience.

(2) Set the integrated strengthening component or set structure seam

The goal of prevention continuous collapse after local components damage is to control the scope of resulting damage. Therefore, the integrated strengthening component or structure seam can be set to partition in the entire structure. Once a certain component damaged, the damage can be controlled in a single partition to prevent the collapse of continuity from spreading. The overall strength member is a key component; its safety should be higher than the general component.

(3) Strengthen the connections to ensure the integrity of the structure

Effective connection of the structural members can enhance the integrity of the structure, and increasing the redundancy is very important to improve the structure ability to resist continuous collapse. For frame structure, when a column is damaged and loses its bearing capacity, beams on it should be able to cross two spans and not fall. This requires the beams to have enough tensile strength, so that the load will transfer to the adjacent column by linking steel catenary transmission mechanism. Draw knot structure composition is the easiest measure to ensure structural integrity.

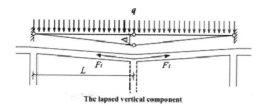

Figure 1. The spanning capacity of girder after column failure.

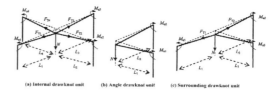

Figure 2. The free-body diagram of frame beam's spanning ability in different positions.

For concrete frame structure, the general requirements in the peripheral component of longitudinal force reinforced drawing shall be submitted to decorate (vertical, horizontal and vertical), the structure of the internal Rachel should be along the two mutually perpendicular direction distributions in each floor, with effective connection with external draw knot component.

(4) Strengthen the structure to ensure the ductility of the remaining structure

Some members of the remaining structure will come to the inelastic phase after the structure partially failed. Therefore, progressive collapse can be avoided by choosing good ductile material, using the ductile structural measures, improving the plastic deformation ability of the structure, and strengthening the redistribution ability of the remaining structures. According to the concept of structural failure modes of demolition of component after discrimination, the ductility of the parts that need to be improved can be confirmed.

The conceptual design of the defect is hard to quantify, because it depends on the level and experience of the designer. However, for general structure, the integrity of the structure can be effectively enhanced with the above guidance, and the structure can be improved to resist the progressive collapse.

3.3 Tie force design

Tie force design is done by calculation of the connection strength between the members. The members should meet the requirement to ensure the integrity of the structure and the bearing capacity of the alternate load path. The basic principle is that the across vertical components of frame beam have enough capacity to avoid continuous damage after the failure of a vertical member, as shown in Fig. 1.

After the vertical component failed, the ultimate bearing capacity of the across vertical component of the frame beam is determined by two mechanisms: In small deformation stage, the ultimate bearing capacity of the frame beam is provided by the flexural bearing capacity of the beam-end plastic hinge, it is referred to as "Beam mechanism"; In large deformation stage, the ultimate bearing capacity of beam-end plastic hinge forfeited, the ultimate bearing capacity of the frame beam is provided by the vertical component of the ultimate tensile of the continuous longitudinal reinforcement in beam, it is referred to as "Catenary system". To ensure that the continuous longitudinal bars in the beam can reach their ultimate tension in a reliable way, longitudinal reinforcement need to have enough anchorage at the beam-end bearing, that is called upon to longitudinal reinforcement is continuous at the beam-end bearing in the beam which play "Catenary system". Under normal circumstances, the mechanisms of "beam" and "catenary system" do not appear at the same time; therefore the spanning ability of frame beam prefers the larger values of "beam mechanism" and "catenary system".

According to the position of the frame beam, there are three main substructure units to resist continuous collapse in frame structure, as shown in Fig. 2. For the internal structure unit as shown in Fig. 2(a), the frame beams crossing the demolished vertical component have reliable anchorage at both ends. At this time, the bearing capacity provided by the "catenary system" is higher than that of the "beam mechanism" generally, so it is available to calculate the across ability of internal structural units as "catenary system" only. For the angle structure unit as shown in Fig. 2(b), the spanning ability of frame beam can only rely on the "beam mechanism", thus it is calculated as "beam mechanism" only. For the surrounding structure unit as shown in Fig. 2(c), the crossing ability can be calculated through the larger values between the "beam mechanism" of three frame and the "catenary system" side frame beam.

Tie force design is simple because it is not required to analyze the whole structure; but the computational model is too simplified, and it does not consider wind load, in addition, it is based on calculation experiences. For a complex structure, its reliability and economic problems also exist. Based on the research of Xinzheng Lu, a simple introduction of tie force design method of the concrete frame structure is as follows.

The basic principles and underlying assumptions of draw knot concrete frame structure strength design method: 1) After the column failed, the column-supported beam maintains its ultimate bearing capacity in the condition that it able to pass the load on the beam directly; 2) Beam's spanning capacity can be calculated through the plastic hinge mechanism (the plastic hinge formed at the beam-end and mid-span) and continuous penetration of steel catenary mechanism (the tensile strength of continuous transfixion steel bar), respectively, as shown in Fig. 3(a), (b); 3) For

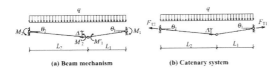

Figure 3. The calculation diagram of draw knot strength.

beam mechanism, for the purpose of calculating simply, only the flexural capacity of negative moments of plastic hinge at the beam-end is considered; 4) If the longitudinal bar in the beam is cut-through along one direction, the catenary system is needed; 5) For the internal structure unit as shown in Fig. 2(a) and the angle structure unit as shown in Fig. 2(b), the bearing capacities of "beam mechanism" and "catenary system" are checked and the larger value is taken, while for the surrounding structure unit as shown in Fig. 2(c), only the bearing capacity of the beam mechanism needs to be checked; 6) The plastic hinge of beam-end should have enough deformation ability, and the beam shall have a sufficient shear capacity.

The calculation method of bearing capacity of two kinds of mechanism is as follows.

(1) According to the catenary system calculation, the draw knot strength F_T provided by the beam is required to meet the following requirement:

$$F_T > \beta q L_i L_j / \Delta. \quad (2)$$

In the equation, β is the internal force reduction factor considering the effects of nonlinearity, with value of 0.67; q is uniformly distributed load on the beam according to Equation (1), L_i and L_j are the spans of the beam which was removed the column in crossed directions; Δ is the permissible limit displacement of node, valued as 1/5 of the short beam span.

According to the mechanism of beam calculation, only the flexural capacity of negative moments of plastic hinge at the beam-end is considered, the flexural capacity M_i provided by the plastic hinge at the beam-end is required to meet the following requirement:

$$M_i > \beta q L_i^2 / 2. \quad (3)$$

In the equation, L_i is the L_1 or L_2 in Fig. 3(a); the significances and values of q and β are the same as in Equation (2).

In addition, each column (wall) shall be continuous of the vertical draw knot from the foundation to the top of the structure; the draw knot force must be greater than the largest floor load standard values of the column (wall) subordinate floor area.

3.4 Demolition widget design

Demolition widget designs of Structure design to resist progressive collapse is removed one by one vertical component of the structure based on certain rules, then the spanning ability of the remaining structure is calculated and ensured. Demolition widget design method proceeds as follows.

(1) Remove the side column, angle column and the bottom inner column of the structure from the top layer to the bottom layer one by one, analyze the inner force of the remaining structure after removing columns, and check whether the remaining structure member fails or not.

(2) Use elastic static analysis in the analysis of internal forces of the remaining structures, and consider the dynamic effect of dismantle the vertical component, then multiply by the dynamic amplification factor A.

(3) The remaining structural component shall meet the following requirement:

$$R \geq \beta S. \quad (4)$$

In the equation, S is the internal force of the remaining structure component which is analyzed by the combination of load in Equation (1); R is the resistance of the remaining structural elements, β is the internal force reduction factor considering the plastic energy dissipation of horizontal member, β is valued as 0.67 when horizontal member considering the plastic hinge occurs on both ends, and valued as 1.0 to angle and cantilever horizontal component.

After the demolition of artifacts, if adjacent elements do not meet the requirement of Equation (4), then the adjacent widget is considered invalid. It should be removed before continue analyzing the rest of the structure by the same method. According to this analysis, if all lapsed area is no larger than 70 m^2 or 15% of the gross floor area, the structure is considered to meet the requirements of progressive collapse resistance, otherwise the first removed elements should be regarded as the key elements, and their safety reserves should be enhanced.

4 CONCLUSIONS

At present, studies on progressive collapse of structures under accidental loads have made remarkable achievement. Those researches have played a very important role in the awareness of structural damage and collapse behavior, prevention and reduction of earthquake disaster. Based on relevant domestic and foreign researches, this paper summarized several design methods to resist progressive collapse; tie force design and demolition widget design are operational methods for engineering design. Due to the complexity of progressive collapse, further researches on some important issues are still needed.

REFERENCES

Abolhassan Astaneh-Asl, Erik A. Madsen, Charles Noble. Use of catenary cables to prevent progressive collapse of buildings.

G. Kaewkulchai and E.B. Williamson. Modeling the Impact of Failed Members for Progressive Collapse Analysis of Frame Structures. Journal of Performance of Construction Facilities, ASCE, 2006, 20(4):375–383.

GSA2005, Progressive collapse analysis and design guidelines for new federal office buildings and major modernization Project.

Specification for concrete structures of tall building design JGJ3-2010. Beijing: China architecture and building press, 2010.

Sufen Zhang. Progressive collapse resistance analysis of reinforced concrete frame structure. Hunan University, a master's degree. 2008.5.

The concrete structural design standard GB50010-2010. Beijing: China architecture and building press. 2010.

Wenjun Guo, Ramon Gilsanz. Nonlinear Static Analysis Procedure-Progressive Collapse Evaluation. Design Engineers of Gilsanz Murray Steficek, LLP. 2003.

Xiaobin Hu, Jia Ru. Overview of progressive collapse analysis and design method of structure [j], building structures, 2006, 36(S1):79,283.

Xinzheng Lu. Study on design method to resist progressive collapse for reinforced concrete frames engineering. Mechanics. 2008–25.

The study of closed surface-water-source heat pump system in hot summer warm winter zone

Weibin Lian
Guangxi JunFuHuang ground-source heat pump Co. Ltd

Yang Jiang & Yingning Hu
College of Mechanical Engineering, Guangxi University, Nanning, China

ABSTRACT: This paper introduces the application in hot summer and warm winter zone closed surface water source heat pump air conditioning installations-with hot water system based on a prison in Guangxi as an example which was designed by the author's team independently. The system operation characteristic was analyzed through experiment, and it was proved that the system of refrigeration and air conditioning in summer is highly stable, the year-round hot effect is good, with high temperature heating capacity, and the energy saving effect is remarkable.

Keywords: closed surface-water-source heat pump, performance characteristic, COP

1 INTRODUCTION

Energy and environment problem is a major social problem that the world is facing today. In China, building energy consumption accounts for over 28% of the total energy consumption in (Shenggong Feng et al. 2010), 65% of the consumption was air conditioning system, in commercial buildings in developed countries, the central air conditioning energy consumption accounts for more than 50% of the total energy consumption, in some areas reached 70% (Dexing Sun et al. 2007). Thus, to develop energy conservation, the environmental protection and sustainable development of HAVC technology are imminent. Large reserves with no pollution of surface water are one of the suitable for the application of new energy.

Starting in the 1950's, developed countries in Europe and the United States began water source heat pump research and engineering practice. Kavanaugh S P carried out a study for the southern region using lake water as heat transfer system (Kavanaugh S P. 1989), and tested the performance of air source-lake water heat pump designed for the south (Kavanaugh S P. 1990). The results show that in the warm south, the water source heat pump system' cop can achieve 2.8 above using lake water as heat source, and it has a broad prospect and good running effect. Antero A M from Finland carried out a study on using lake water source heat pump as heat source in cold areas, and the results show that in the cold region where water is not suitable as heat source, freezing heat collection system should be used when using this application (ANTERO A M. 2003). In China, Qin Hong (Hong Qin et al. 1998) analyzed surface water energy utilization in Wuhan area, and made the conclusion that in the typical hot summer and cold winter area in China, water source heat pump can be used as low heat source. Chen Xiao (Guoqiang Zhang 2004), performed a simulation analysis on the use of lake water source heat pump in southern region in China, and the results show that the application in filled water of hot summer and warm winter area the water source heat pump system' cop can reach 2.6 above, energy saving is about 60% more than the electric heating boiler, and have promotion prospects.

The current research of water source heat pump at home or abroad is active, and the engineering applications are also becoming more complete. However, from the view of the present research, basing on the fact that research of closed water heat recovery system in hot summer and warm winter area of research is rare, relatively few related engineering applications have been carried out.

This paper combined with a prison in Guangxi is applied in office and residential district of closed surface water source heat pump, analysis the applicability applying the heat recovery system of closed water source heat pump in our country, in the hot summer and warm winter area, make a good demonstration effect that the technology of large area for the future promotion.

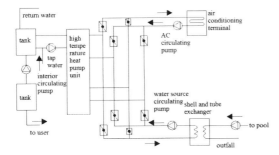

Figure 1. System principle diagram.

2 PROJECT PROFILE

The project is located in Liucheng county of DaPuZhen in the east of the old Liucheng cannery. Figure 1 shows the schematic system diagram for the project designed by the author's team using high temperature plain surface water source heat pump-closed shell and tube heat exchanger heat pump.

2.1 System operation strategy

The system used in the project is shell and tube heat exchanger with screw type high temperature heat pump series of closed water source heat pump. In winter conditions, the evaporator system against the side of the river water is used to absorb heat from the river; the condenser terminates at the hot water tank, making hot water. The whole system uses the method of cyclic heating to heat the water for living. When the hot water tank temperature sensors detect that the hot water temperature is low, the system boots and exchanges heat with the river, making hot water; when the temperature reaches set value, the system automatically stops in order to achieve the purpose of energy saving, and at this time the water circulation pump and the heating circulation pump work. In the summer installations mode, the system provides air conditioning for complex buildings, and recovers condensing heat for making hot water at the same time. The air conditioning circulating pump and the heating circulation pump work, and the heat pump units work in full heat recovery mode. For water cooling during the summer, the water circulation pump and the heating circulation pump start, and the heat pump unit work in cooling mode.

2.2 Principle of the system

In Fig. 1, the high temperature water source heat pump system consists of three Climaveneta 1801 type unit; the unit refrigerating capacity is 585.1 kW, the heating capacity is 611.8 kW, total heat recovery is 576.1 kW.

2.3 Design for experiment

In order to explore closed surface water source heat pump running effect in hot summer and warm winter area, the experimental design was as follows.

2.4 Hot water in winter condition

(1) High temperature heating system intermittent operation: According to the standard of start-stop set before, start from the hot water temperature 50°C running 2 hours after the temperature rise to 54°C then automatically shut off the unit and the water circulating pump, heating circulation pump. Turn it off after 4 hours and open units and water pump, and test and record the unit and the power of the pump parameters, the inlet and outlet of evaporator and condenser temperature differences, calculate the unit and the system coefficient of performance.

(2) Continuous operation: Let the system run before the peak water time in the evening, then start from the hot water temperature 45°C and stably operating for 6 hours until the hot water temperature reaches set temperature then stop automatically, test and record the unit and the power of the pump parameters, the inlet and outlet of evaporator and condenser temperature differences, calculate the unit and the system coefficient of performance.

2.5 Hot water and cold air conditioning in summer condition

(1) The heat recovery mode: Supply air conditioning for the office building making living hot water at the same time, test and record the unit and the power of the pump parameters, the inlet and outlet of evaporator condenser temperature differences, calculate the unit and the system coefficient of performance after the system run stably.

(2) Water cooling mode: Separate the water side heat transfer to provide air conditioning for the office building, test and record the unit and the power of the pump parameters, the inlet and outlet of evaporator condenser temperature differences, calculate the unit and the system coefficient of performance after the system run stably.

The instrument used in the test: Type TR118 timer; XMTJ1602K temperature inspection instrument; Accuracy of ±0.1°C TP3001 thermometer; Testo175~T1 temperature and humidity recording instrument; DTFX1020PX1 series ultrasonic flow meter; rating accuracy for level hot water meter; CA8335 power quality analyzer; Victor DM6266 type clamp multimeter.

3 REACH ON CLOSED SURFACE WATER SOURCE HEAT PUMP OPERATING CHARACTERISTIC IN WINTER CONDITION

3.1 Intermittent operation characteristics

According to the scheme design, here is intermittent operation of the system running characteristics.

There are the inlet and outlet temperature and evaporator temperature changes under high temperature

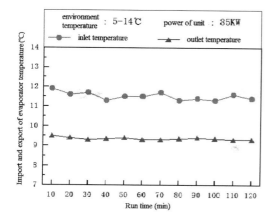

Figure 2. Relationship between inlet-outlet temperature of evaporator and system run time (stable for 2 hours).

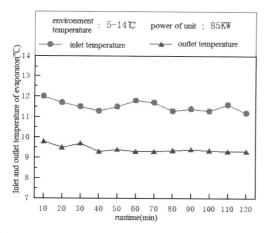

Figure 3. Relationship between inlet-outlet temperature of evaporator and system run time (run 2 hrs. after shut down for 4 hrs.).

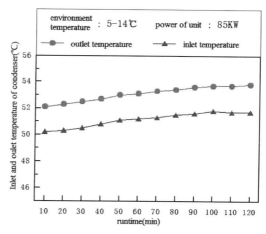

Figure 4. Relationship between inlet-outlet temperature of condenser and system run time (stable for 2 hours).

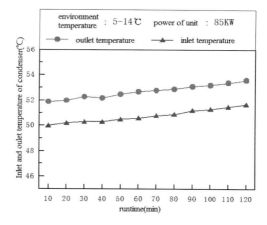

Figure 5. Relationship between inlet-outlet temperature of condenser and system run time (run 2 hrs. after shut down for 4 hrs.).

heating when intermittent operation as shown in Fig. 2 and Fig. 3. It can be seen from the curve that evaporator inlet temperature is 11.9°C outlet temperature is 9.5°C in the beginning. As the system running, the inlet and outlet of evaporator temperature are descend slowly. After an intermittent operation cycle, the outlet and inlet temperature of the evaporator are 11.4°C and 9.3°C, the reduction of the inlet and outlet of evaporator temperature is only about 0.5°C in the whole process, the difference between inlet and outlet temperature is 2°C, the circulating medium flow rate of evaporator is 140 m³/h. This diagram illustrates the application of shell and tube heat exchanger coupling with closed surface water source heat pump hot water system in winter conditions, the performance of the entire water side heat transfer is stable, and the operation of system is stable.

The curve in Fig. 4 and Fig. 5 explained the variation of the inlet and outlet temperature of condenser side in the winter heating intermittent operation mode. Because of the cyclic heating system, the condenser heats the water in the tank inside-out continuously, so the condenser side inlet and outlet temperature is on the rise. Outlet and inlet temperature of the hot water are 50.2°C and 52.1°C in the beginning. After the end of a cycle, outlet and inlet temperature of the hot water are 51.7°C and 53.8°C. In the whole process, the difference between inlet and outlet temperature is basically kept in 2°C. The hot water flow rate of the condenser side is 164 m³/h. After the end of two operation cycles, the inlet and outlet temperature of condenser and the difference between them remained constant. Intermittent operation has no effect on the system basically, condenser side heat transfer effect is good, and operation of system is stable.

The curve of intermittent operation is shown in Fig. 6. System reach steady state after running for an hour from the start, the coefficient performance of unit is basically stable in 4.75, while that of system is 3.5.

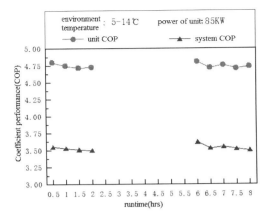

Figure 6. System COP in winter intermittent operation.

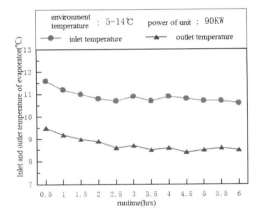

Figure 7. Relationship between inlet-outlet temperature of evaporator and system continuous running time.

COP's downward trend is not obvious, so using this system for winter hot water preparation is effective and stable in hot summer and warm winter area.

3.2 Continuous operating characteristics

Curve in the Fig. 7 shows the change process of inlet and outlet temperature of evaporator and the difference between them during continuous run time. In the beginning, the outlet and inlet temperature of the evaporator are 11.6°C and 9.5°C, as the system running, the inlet and outlet evaporator temperature decline slowly. System reached steady state in about 2 hours. After running for 6 hours, outlet and inlet temperature of evaporator are 10.6°C and 8.5°C. As you can see, the inlet and outlet temperature of evaporator is only about 1°C lower after 6 hours. In the whole process, the difference between the inlet and outlet temperature is basically kept in 2°C. The circulation medium flow rate of evaporator side is 141 m³/h. Illustrates the application of this system runs stably, and the heat transfer of evaporator effect is very efficient and stable in the process of continuous operation.

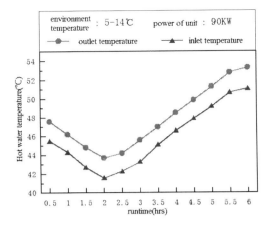

Figure 8. Relationship between inlet-outlet temperature of hot water tank and system continuous running time.

Curve in the Fig. 8 shows the change of the inlet and outlet temperature of hot water in condenser side. It is not hard to see from the picture that in the beginning outlet and inlet temperature of the hot water are 45.1°C and 47.9°C. After running for 2 hours, living quarter and prisoners' hot water consumption reach peak in the project, and the inlet and outlet water temperature of heating water tank showed a trend of decline at this time. After the rush hour, inlet and outlet temperature of hot water began to gradually pick up. When running after 6 hours, temperature reach the set value in the hot water tank, the inlet temperature of hot water is 51°C, outlet temperature is 53°C. the inlet and outlet temperature, The water tank water temperature is always 52°C or above so that can fully meet the requirements. When heating the water tank temperature meet the requirement, new heating water will inflow the heat preservation water tank by the circulation pump to ensure the living hot water supply. The flow rate of hot water side is 164 m³/h. This could explain that heat transfer effect in condenser side is stable; the system runs stably in the continuous runtime.

The curve in Fig. 9 shows the change of coefficient performance of the closed surface water source heat pump system in the continuous runtime. It shows that, system reach steady state after operating for 2 hours. In this state, COP of unit continuous operation is 4.7, while COP of system is 3.5. There is not much difference between intermittent operation mode. This shows that the system in the condition of hot water in winter can maintain an efficient and stable state whether it is in continuous or intermittent operation mode.

Fig. 10, Fig. 11 and Fig. 12 show that the inlet and outlet water temperature of evaporator do not dramatically change along with the change of the lake water temperature by using closed water source heat pump in hot summer and warm winter area. The unit and system COP also not dramatically change along with the temperature. Heat transfer in water side is stable because of the shell and tube heat exchanger in the

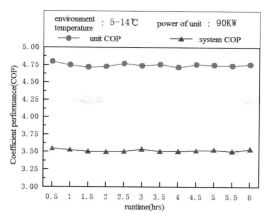

Figure 9. COP of hot water system continuous operation in winter condition.

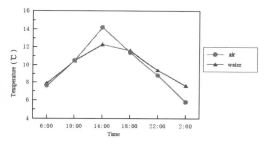

Figure 10. The relationship between water temperature and time.

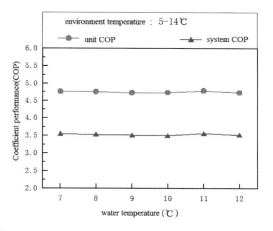

Figure 11. Relationship between COP and water temperature.

system. Inlet and outlet water temperature of evaporator do not fluctuate substantially along with the lake temperature what make the heat transfer stability and efficiency. The result has high reference value in order to carry on more closed surface water source heat pump projects in hot summer and warm winter area in the future.

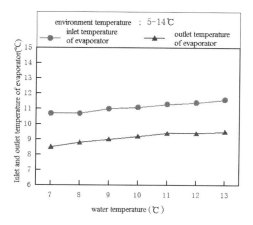

Figure 12. Inlet and outlet temperature of evaporator's relationship with the water temperature.

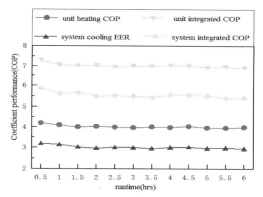

Figure 13. Relationship between runtime and COP in the heat recovery mod.

4 RESEARCH ON CLOSED SURFACE WATER SOURCE HEAT PUMP IN SUMMER CHARACTERISTIC

4.1 The heat recovery mode

Fig. 13 shows the relationship between the system runtime and COP in the heat recovery mode. Because of the living water is made by condensing heat generated by air conditioning refrigeration, so the curve of conditioning COP corresponding to refrigeration COP curve show the same trend. When the indoor temperatures meet set value, cooling and heating COP also tend to be stable, the system will be balance. You can see from the figure that refrigeration EER and heating COP is respectively stable in about 3.0 and 4.0. While the unit and system COP is respectively stable in about 7.0 and 5.5.

4.2 Water cooling mode

Fig. 14 shows the EER of unit and system separated refrigeration only use lake water. At the beginning

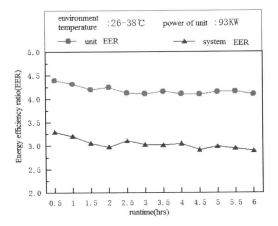

Figure 14. Relationship between runtime and EER in water cooling mode.

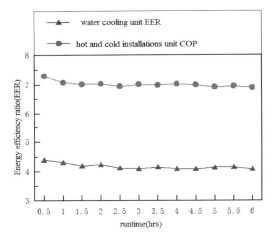

Figure 15. Unit COP of two kinds of conditions.

of operation, cooling capacity is larger, system heat transfer is well, and EER is higher. As the office building's indoor temperature is reduced, cooling load is falling. When indoor temperature drop to a stable range, the system is becoming more stable. When stable, EER of unit is about 4.1 and the system EER is about 3.0.

4.3 Comparison the COP of unit and system between two kinds of conditions of refrigeration

It is not hard to see in Fig. 15 and Fig. 16, EER of unit and system is far higher than separate water refrigeration when using heat recovery mode. Living hot water consume no heating energy consumption, it not only meet the requirements of office building cooling capacity but also satisfy the living area and the prisoners of hot water demand, kill two birds with one stone. So it can greatly improve the utilization rate of

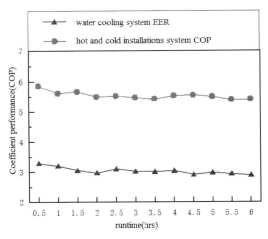

Figure 16. System COP of two kinds of conditions.

energy and achieve efficient environmental protection by using all heat recovery mode.

5 CONCLUSION

(1) Adjust measures to local conditions to adopt Closed Water Source Heat Pump system in view of a prison in hot summer and warm winter area. It can extremely reduce the water temperature affect COP of system by using the secondary heat transfer and guarantee the unit clean and safe.
(2) In this paper, we study the closed water source heat pump system in hot summer and warm winter area. It can be integrated to realize summer cooling, summer all heat recovery, and perennial supply of living hot water applications, practical effect is obvious.
(3) In summer, unit and the system's COP is respectively about 6.9 and 5.5 in all heat recovery mode; Unit and system EER are respectively about 4.1 and 3.0 in separate water cooling mode. In winter, unit and the system COP are respectively about 4.7 and 3.5 when making living hot water. Experimental results show that it is stable, high efficiency and energy saving in preparing refrigeration, heating and living hot water by using Closed Water Source Heat Pump system.
(4) Successful implementation of this project provides a good reference for the future in hot summer and warm winter area gradually to promote
Closed Water Source Heat Pump system.

REFERENCES

Antero AM. Lakes as a heat source in cold climate [C]. International Congress of Refrigeration, Washington DC, 2003:1–8.

Dexing Sun, Jianfeng Qian. A new type of surface water source heat pump and its related technical analysis [J]. Journal of Harbin university of commerce, 2007.

Guoqiang Zhang, Xiao Chen. Use of lake water source heat pump system analysis [C]. The national HAVC refrigeration 2004 academic essays, Beijing: China building industry press, 2004.

Hong Qin, Wenhua Zhang, Yuangao Wen. Surface water use of air-conditioning systems and energy conservation and environmental impact analysis [C]. The national HAVC refrigeration 1998 academic essays, Beijing: China building industry press, 1998.

Kavanaugh SP. Design considerations for ground and water source heat pumps in southern climates [J]. ASHRAE Transactions, 1989, 95(1):1139–1149.

Kavanaugh SP, Pezent MC. Lake water applications of water-to-air heat pumps [J]. ASHRAE Transactions, 1990, 96(1):813–820.

Shenggong Feng, Wei Liu. The analysis of application of river water source heat pump in air conditioning engineering [J]. HAVC, 2010, 12(40):51–54.

Urban study on physical environment and social migrants networking focused on Russian-speaking Town in Seoul, Korea

Ekaterina Shafray
Urban Planning and Design Lab, Korea University, Seoul, Korea

Kim Seiyong
Urban Planning and Design Lab, Architecture Department, Korea University, Seoul, Korea

ABSTRACT: Due to international migration many "foreign villages" emerged in Seoul over the last two–three decades. Compared to old and relatively well-established Chinese communities, studies concerning these new international settlements are few. Particularly, paper focused on Russian-speaking, Central Asian and Mongolian town in Gwanghuidong located near Dongdaemun markets. Together with the Korean residents, these migrants from Post-Soviet countries and Mongolia live work, or visit this area. The problem of managing the area and value for Russian migrants' community becomes important after the short-term visa-waiver agreements have been adopted. This paper introduces the results of our social study conducted at this area in June–November 2013. In the survey the possible ways for improvement were investigated by collecting responses on buildings, facilities and existing problems. Overall, 77 respondents participated in the survey, with 15 selected for interviews. This paper discusses problems of urban typology for immigrants social networking closely connected with the physical environment, and a potential of "foreign villages" in general.

1 INTRODUCTION

1.1 Situation of "foreign villages"

International migration in Asian countries and worldwide becomes a social challenge affecting the physical environment of large cities. In Korea, new foreign minorities, without considering the old, relatively numerous and well-established Chinese and Japanese communities (half of all of the foreigners are Chinese, 71% of them are ethnic Koreans), have appeared and become more noticeable. According to Doyoung (2012), these areas with high number of international migrants were called "diaspora foreigners' space" (Kim, 2005) or "ethnic villages" (Kim and Kang, 2007) in previous research. The foreign communities started to emerge in Seoul around 1980s. These ethnic groups and communities act as new cultural players in the city and transform physical spaces for their social needs.

At the same time, management of "foreign villages", often insufficient strategies towards these areas and other social variables determine problem of study. Aiming to outline the situation of foreign "villages" in Korea, this paper focused on those appeared in Seoul over the last two-three decades in particular. It selects one Russian-speaking, Central Asian and Mongolian town in Gwanghuidong, located near Dongdaemun markets as its particular research object.

In Post-Soviet period over the past two decades, influx of migrants from Russia, Mongolia, Uzbekistan, Kazakhstan, and Kyrgyzstan, and ethnic Russian Koreans has led to the establishment of distinct ethnic enclaves within Seoul. Over four thousand Russians are currently living in South Korea. The urban conditions of their settlement near Dongaemun markets, however, were not specifically studied in previous research.

1.2 Location and characteristics of "foreign villages" in Seoul

The international areas in Seoul start to become an issue after about 1980s. Kong, Yoon, Yu (2010) noted that the labour market, the embracement of returning ethnic Koreans and increase in international marriages after immigration in Korea are features of recent international migration. Lim (2006) argued that composition of international migration has changed since the influx of transnational migrant workers came to the country. Industrial trainee system for foreign workers (FITP), established in November 1991, and other further policies were crucial for international migration process.

As for spatial condition of "foreign villages", Kim and Kang (2007) reported that the foreign communities could be divided into "1) foreign residential communities formed around schools since the 1970s, which are represented by the Japanese and Chinese communities, respectively; and 2) "foreign cultural communities", nonpermanent, but regularly forming one." As the authors suggest, both communities are

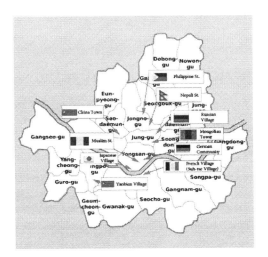

Figure 1. Foreign Communities in Seoul (Source: Korea in the World, World in Korea, 2011).

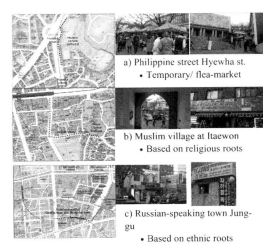

Figure 2. a) Philline street, temporary market, Hyewha station b) Muslim Village at Iraewon c) Russian-speaking, Cental Asian and Mongolian town, Gwanghui-dong, Jung gu.

considered as important spaces for long-term immigrants. Location of "foreign villages" in Seoul, for instance, is described (Korea in a world, World in Korea, 2011). The most known areas in Seoul are listed (Global villages in Seoul, 2006), and their locations are shown in Figure 1.

The urban character of these "foreign villages" is different. Mostly these areas have a visual presentation of foreign companies on facades or billboards as their signs. Frequently these settlements are grouped around a community center, church, school or migrants' employment place. They have public spaces for migrant's gathering, traditional food stores or markets. At the same time, these areas are located in residential districts and Korean host society. Transnational commerce exist along with Korean businesses and residences, so visual representation of these areas is an important issue to outline their clear evidence in urban prospective.

2 VISUAL VIEW OF SELECTED "FOREIGN VILLAGES" IN SEOUL

We selected Philippine street near Daehak-ro, Hyewha-dong rotary, Islamic Village in Itaewon-dong, Hannam-dong (Usadan-ro, Usadan-ro 10 gil) and Russian-speaking, Central Asian and Mongolian town in Gwanghui-dong, Jung-gu in Seoul in particular for consideration.

These areas have different urban characters. Philippine street is a temporary flea-market, where migrants are gathering on Sundays after Hyewha Catholic cathedral religious service. It has over eleven years history (Korea in the World, World in Korea, 2011). Lee (2006) described problems of Philippine migrants' community. Islamic Village in Itaewon is much more established and based more on religious roots. It was developed around Seoul Central Mosque.

Nowadays the area consists of various halal restaurants, commercial stores, Islamic bookstores, electronic stores and agencies along the street; and near the central Mosque there are many traditional food markets. Seoul Central Mosque (Masjid), the only mosque in Seoul, was opened in 1976 in Itaewon and has stored lectures in English, Arabic, and Korean. Friday prayers attract up to 800 worshipers at 1pm regularly, the majority of them are Arab, Indian, Pakistani or Turkish (Seoul Central Masjid). Russian-speaking, Central Asian and Mongolian town appeared near Dongdaemun market in Post-Soviet period after 1990s. Russian Koreans and migrants opened traditional restaurants and businesses at this area. Frequent trips of Russian wholesale buyers to Dongdaemun Market generated a demand for Russian and Central Asian companies (Jun, 2011). Social gathering of Uzbek migrants provides business and networking opportunities. Cyrillic font can be seen on many facades at the area. Recently, with the establishment of short-term visa-waiver agreement between Russia and South Korea from 2014, the number of Russians visiting Korea tends to increase.

Figure 2 shows a visual view of these areas, the photographs in which were made in Jan–Feb 2014.

3 PHYSICAL CHARACTERISTICS OF RUSSIAN-SPEAKING, CENTRAL-ASIAN AND MONGOLIAN TOWN IN JUNG-GU, SEOUL

The area contains many companies and important facilities for migrants (such as traditional Russian or Uzbekistan restaurants and the post office) which are frequently marked with Cyrillic font in their facades. Figure 3 shows a fragment of Mareunaero Street with foreign (Russian and Central Asian) businesses.

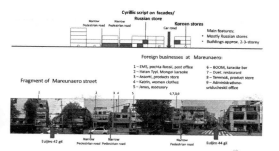

Figure 3. Fragment of the Mareunaero Street with Foreign (Russian and Central Asian) Business.

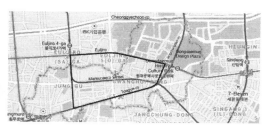

Figure 4. Russian-speaking, Cental Asian and Mongolian Town Area, Gwanghuidong, Jung-gu.

4 SOCIAL STUDY AT RUSSIAN-SPEAKING, CENTRAL-ASIAN AND MONGOLIAN TOWN IN JUNG-GU, SEOUL

4.1 *Social survey aim*

The purpose of the social survey is to analyze the conditions of this area and the satisfaction of the residents, employees, visitors and customers in this area towards their life. Possible ways for improving the area are investigated by questioning respondents about necessary buildings and facilities to improve life quality in the area. The area's boundary is shown in Figure 4 (black color. Gwanghui-dong area is of 0.74 km^2 with an overall population of 5019 people in 2008 (Gwanghui-dong).

4.2 *Method and limitation of the social survey*

The social survey was held in Gwanghuidong area in June–November 2013 and 77 respondents were interviewed. The survey includes writing questionnaire responses (including personal survey at the site and online survey of the target respondents), and personal interviews with selected respondents at the site (15 persons).

According to the official statistics, in Jung-gu District there are 391 Uzbekistan and 399 Mongol; and in Dongadaemun-gu District there are 101 Uzbekistan and 536 Mongol (KOSIS, 2012). The Korean statistical information service shows that these districts have more Russian-speaking migrants from Post-Soviet Countries than other districts in Seoul.

Working full time	Working part-time	Having own business	Registered unemployed	Not registered, but seeking work	At home/ not seeking work	Long-term seek/ disabled	Full-time student	Fully retired	Other variant (please, specify)
35	8	11	0	0	3	0	19	1	0
45,45	10,39	14,29	0	0	3,9	0	24,67	1,3	0

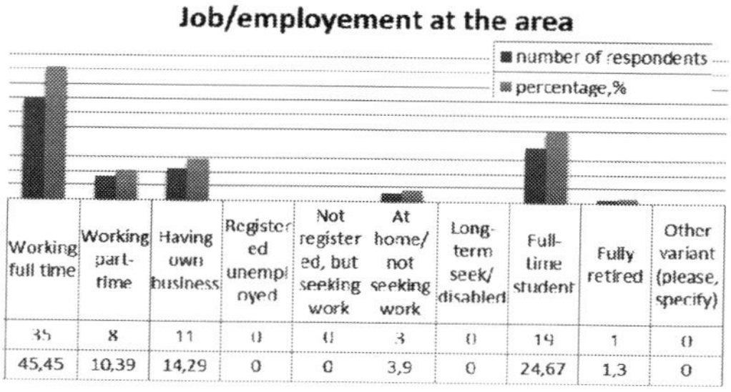

Figure 5. Russian-speaking, Cental Asian and Mongolian Town Area, Gwanghuidong, Jung-gu.

Since the percentage of migrants from Russia is relatively small, the information of their area distribution for Seoul is not available there. The limitation of the study is related with consideration of Seoul-resident Russian-speaking foreigners.

The following reasons contribute to the limitation of the social study: people are reluctant to participate in a survey made in the street; the area is short of open public spaces where people can seat and talk; and people in the area don't have sufficient time to answer the questionnaire because they are always busy with the work or do not want to lose customers.

4.3 *Analysis on the social study and results thereof*

Several questions including age, occupation of respondents, their nationality and ability in languages (Korean, English, Russian language), transportation at the area, frequency of use of different facilities and others were analyzed in the survey. In this paper we will focus specifically on employment in the area and way of spending time in the area.

The survey questioned residents of the site, employees, visitors, and Russian students. According to the survey, 43 of 77 respondents are employed (full-time or work part-time). This does not necessarily mean that they are employed at the area, because some area users or visitors also took part in the survey. 19 of 43 respondents work at the area, and 11 respondents have their own businesses. The employment of respondents (and employment in the area) is shown in Figure 5.

A high percentage of students in the jobs/employment chart (19 people, 24.67% of respondents) could be explained by the general trends in international migration. They come to the area for general interest or as its visitors and usually visit traditional area restaurants and product stores.

Figure 6 presents ways of spending time in the area. The high percentage of respondents who claimed to mainly visit traditional restaurants indicates that the area lacks other entertainment options. Besides, Russian restaurants are visited not only by locals, but also by Russian students or tourists.

Other questions about necessary area facilities and buildings showed that language classes and centers, children playground or kindergarten, sports fitness center, international health center and

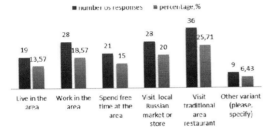

Figure 6. Ways of spending time at the site.

consultation center are the facilities needed most by the respondents.

5 RESULTS AND DISCUSSION

A few decades ago many new "foreign villages" emerged in Seoul due to trends of international migration. The physical environment of these settlements is different; they are often formed around a community center, religious buildings, school or migrant's employment place and serve as a space of their social networking. The literature review on "foreign villages" shows that some social policies such as FITP were important for this process.

This paper studied the international settlements in Seoul–Philippine street, Muslim Village in Itaewon, and the Russian-speaking, Central Asian and Mongolian Town near Dongdaemun markets, trying to define their distinct features. The Philippine street works as a temporary market for Philippine workers, while
Muslim and Russian-speaking settlements put together foreigners based on their religious and ethnic roots respectively. These areas are characterized by relatively undeveloped spatial conditions, although some visual evidence of foreigners staying at these areas could be seen through facade fonts and signs.

The Russian-speaking, Central Asian and Mongolian town that emerged around 1990s near Dongdaemun markets in Gwanghui-dong is analyzed in terms of its visual characteristics. Facades of the Mareunaero Street with foreign (Russian and Central Asian) business signs are shown. The social survey conducted in the area also analyzed the conditions of the area and the satisfaction of the residents, employees, visitors and customers in this area towards their life. As the area is an important place for networking and socialization of Russian-speaking and Central Asian migrants, existing and necessary facilities were considered. However, due to the area's poor conditions and shortage in social infrastructure for migrants' socialization, necessary buildings and facilities need a detailed consideration.

Considering many social variables and host society, problems of urban typological objects for immigrants' social networking closely connected with the physical environment, and a potential of "foreign villages" to shift from needs-based to cultural-diversity based area require further research.

ACKNOWLEDGEMENT

This work was supported by the BK21 Plus Program (Education Program for Urban Regeneration toward Sustainable Communities, Korea University).

REFERENCES

Amuse-bouches in Seoul's French Quarter (2011). Korea Joongang Daily Retrieved March, 5 2014, http://korea-joongangdaily.joins.com/news/article/Article.aspx?aid=2935365.

Doyoung, Song. (2012). Spatial Process and Cultural Territory of Islamic Food Restaurants In Itaewon, Seoul. Migration and Diversity in Asian Contexts, 233.

Global Villages in Seoul, Retrieved Feb, 7 2014 from http://www.investkorea.org/ikwork/iko/eng/cont/contents.jsp?code=10205050207.

Gwanghui-dong, Doopedia, Retrieved March 25, 2014 http://www.doopedia.co.kr/photobox/comm/community.do?_method=view&GAL_IDX=101012000719527.

Kim, H.M. *Global sidaeui munhwa beonyeok* [Cultural interpretations of global age]. Seoul: Ttohanauimunhwa.books, 2005.

Kim, E.M. and Kang, J.S. (2007). Seoul As a Global City with Ethnic Villages, Korea Journal, v.47 n.4, 64–99.

KOSIS (2012) Retrieved January 24, 2014 http://kosis.kr/statisticsList/statisticsList_01List.jsp?vwcd=MT_ZTITLE&parmTabId=M_01_01#SubCont.

Korea Muslim Federation, Retrieved March, 5 2014 http://www.koreaislam.org/index.jsp.

Kong, D., Yoon, K., & Yu, S. (2010). The Social Dimensions of Immigration in Korea. Journal of Contemporary Asia, 40(2), 252–274.

Korea in World, World in Korea. (2011). Retrieved Feb, 7 2014 from http://mizycenter.tistory.com/entry/Korea-in-World-World-in-Korea.

Lee, M. (2006). Filipino Village in South Korea. Community, Work and Family, 9(4), 429–440.

Lee, M. (2006). Invisibility and Temporary Residence Status of Filipino Workers in South Korea. Journal for Cultural Research, 10(02), 159–172.

Lim, T. C. (2006). NGOs, Transnational Migrants, and the Promotion of Rights in South Korea. Local Citizenship in Recent Countries of Immigration: Japan in Comparative Perspective, 235–269.

Little France on the Han River (2006). The ChosunIlbo Retrieved March, 5 2014 http://english.chosun.com/site/data/html_dir/2006/04/11/2006041161007.html.

Seoul Central Masjid, Retrieved March, 5 2014 http://english.visitkorea.or.kr/enu/SI/SI_EN_3_1_1_1.jsp?cid=1702328.

Seoul Global Center (2013), Retrieved March, 5 2014 http://www.visitseoul.net/en/article/article.do?_method=view&m=0004007002010&p=07&art_id=42942&lang=en.

Statistics Korea (2011). Retrieved September 30, 2013 from http://kostat.go.kr/portal/english/index.action.

Yun, J. (2011). Becoming Like the World: Korean Articulations of Globalization in the Global Zones, 1987-present.

Waste water treatment through public-private partnerships: The experience of the regional government of Aragon (Spain)

S. Carpintero
Polytechnic University of Madrid, Madrid, Spain

O.H. Petersen
Roskilde University, Roskilde, Denmark

ABSTRACT: This paper analyses the experience of the regional government of Aragon (Spain) that has extensively used public-private partnerships for the construction and operation of waste water treatment plants. The paper argues that although overall the implementation of this PPP program might be considered successful, most projects have experienced significant difficulties and long delays until they have entered in operation. The analysis shows that the main reasons for the problems in the implementation of the program were the rush in implementing it, the inadequate transfer of some risks, and the wrong allocation of some tasks. The paper also illustrates two features of this PPP program that arguably have strongly influenced its successful implementation: the mitigation of demand risk and the rigorous estimations of demand carried out by the regional government.

Keywords: instruction; table; figure; headings; deadline; Kyoto

1 INTRODUCTION

In early 2004, the regional government of Aragon (Spain) faced the daunting challenge of building 131 waste water treatment plants. The reason behind this initiative was to comply with the European Union regulations that made it compulsory by December 2005 to treat waste water in all municipalities that had over 2,000 equivalent inhabitants. The regional government decided to use the public-private partnership (PPP) formula because of the lack of financial resources to build the plants.

In the context of building and maintaining infrastructure assets, PPPs can be defined as 'an agreement between the government and one or more private partners (which may include the operators and the financers) according to which the private partners deliver the service in such a manner that the service delivery objectives of the government are aligned with the profit objectives of the private partners and where the effectiveness of the alignment depends on a sufficient transfer of risk to the private partners' (OECD, 2008). The very notion of PPP thus builds on the principle of risk transfer and integration of key project characteristics including design, finance, construction, operation and maintenance into a single (and long-term) contract between a public sector organization and a so-called Special Purpose Vehicle (SPV) (Yescombe, 2011).

The motivations of governments for embarking on PPPs for the delivery of public infrastructure are manifold, and include on-time and on-budget delivery and access to private project management experience (Grimsey and Lewis, 2002; Kwak et al., 2009). By integrating key project elements into a single contract structure, optimizing long-term incentives and letting each of the partners do what they do best, PPPs are often seen as a panacea to avoiding time- and budget-overruns in large-scale infrastructure projects. Indeed, one of the key arguments underlying PPP, compared to traditional procurement methods, is that an efficient risk transfer and task integration provides the private partner with a clear incentive to develop innovative solutions that can deliver more infrastructure with fewer resources in the long-run: for instance, by designing infrastructure assets that are cheaper or more cost-effective to operate and maintain (Yescombe, 2007; Tang et al., 2010).

There is an abundant literature about PPP projects. A large number of studies have examined risk transfer and risk-sharing in PPPs (Grimsey and Lewis, 2002; Shen et al., 2006; Ng and Loosemore, 2007; Marques and Berg, 2011; Demirag et al., 2012). But PPP projects have also been examined in the literature under many other perspectives. For example, some studies have analyzed PPPs as a way of effective planning and timely implementation of large-scale infrastructure projects (Flyvbjerg, 2009; Siemiatycki and Friedman, 2012). Some other authors have focused on the organizational integration of public and private actors under the public-private partnership (PPP) label (Klijn and Teisman, 2003; Kwak et al., 2009; Tang et al., 2010).

However, there are very few studies that have analyzed the use of PPPs to build and operate waste water treatment plants. Ali et al. (2012) focus on the valuation of minimum revenue guarantees in this kind of PPP projects. Memon (2002) discussed the use of PPPs for water supply and waste water treatment in Japan. The only case study we have found in this respect is by Zheng and Tiong (2010), who examined the first PPP application for waste water treatment in Taiwan. This paper adds to the literature a case study that analyzes the most relevant application worldwide of PPPs to build and operate waste water treatment plants.

The article's empirical base consists of data provided by the regional government officials in charge of those projects and information collected through fifteen face-to-face interviews with public and private representatives of the stakeholders participating in some of the PPP schemes. Interviewees include government officials of the regional government of Aragon, private project managers of the concessionaires and representatives of banks involved as financiers in the projects. Unstructured in-depth interview was adopted as a means of investigation for this study because of its powers to achieve honest and robust responses (Whitehead, 2002) and to ensure realism in the collection of an overall impression of stakeholders' perspectives. The unstructured approach encouraged participants to openly express their viewpoints based on their experience in managing PPP projects for waste water treatment plants.

The paper argues that overall the implementation of this PPP program might be considered successful, yet most of the projects have experienced significant difficulties and long delays until they have entered in operation. The analysis shows that the main reasons for the problems in the implementation of the program were the rush in implementing it, the inadequate transfer of some risks, and the wrong allocation of some tasks. Probably, all those were influenced by the lack of experience of the regional government—and everywhere else—in using the PPP formula to build and operate waste water treatment plants. The paper also illustrates two features of this PPP program that arguably have strongly influenced its successful implementation: the mitigation of demand risk and the rigorous estimations of demand carried out by the regional government.

2 IMPLEMENTATION OF THIS PPP PROGRAM

During 2004, the regional government prepared the preliminary designs of the plants and elaborated the terms of reference of the tenders. The 131 plants to be built were distributed in 13 contracts of around €20–30 million—each of them covered a specific area of

Table 1. Phases of the concession program.

Phase	# contracts	# plants	Investment (€M)
1	7	77	194
2	3	17	43
3	3	37	93
Total	13	131	330

Table 2. Phases of the concession program.

Phase	Bidded out	Awarded	Contracts signed
1	July 2005	Dec. 2005	Feb. 2006
2	June 2006	March 2007	May–Jun 2007
3	Jan–Feb 2008	Nov 2008	Dec. 2008

the region. The concession tenders were distributed in three phases, as shown in Table 1 and Table 2. The total investment was €330 million. In all cases, the concession period was 21.5 years (1.5 years for the design and construction, and 20 years for the operation).

There was high competition for the projects. The number of bidders for each contract was between 13 and 18 in the first phase, between 16 and 19 in the second phase, and between 10 and 13 in the third phase. Most consortiums included companies with extensive experience in waste water treatment, many of them big companies operating at national level. Some consortiums also included small local companies.

All projects were financed through project finance although the banks asked for recourse to the sponsors until the plants were in operation and had all authorizations. Most of the projects were financed before the global financial crisis. Therefore, most of the concessionaires got the financing in a period of time usual in project finance (10–12 months). Leverage was around 75% of the initial investment in most cases and the spread over Euribor was around 100 basis points. However, in some cases, this spread increased sharply up to 300 basis points because of delays in payments to the banks.

As of early 2014, most of the plants have been built and are in operation. However, there have been long delays in all cases, most notably in the projects of the first phase. The terms of reference established a period of 18 months between the sign of the contract and the entering in operation. But the projects have experienced delays that are in the range of 20%–50% of this period. Arguably, the main reasons of the delays were: 1) Problems in getting the lands available; 2) Problems in getting all permits and authorizations related to construction and to entering in operation; 3) Problems in getting the financing for the projects.

As of early 2014, the outcome of each phase of the program is as follows:

Phase 1. All projects have been built, but one of the contracts has been renegotiated because the municipality did not build some main sewers it was supposed to build.

Phase 2. One of the contracts, which involved only one big plant, has not been built and was terminated by mutual agreement. The main reasons were: 1) Some problems with the foundations; it was not clear who should assume that risk; 2) The plant had to be built for 133,000 equivalent inhabitants but the government realized that the projections of population's evolution were too optimistic and that it was enough to design the plant for 70,000 equivalent inhabitants.

Phase 3. As of early 2014, all projects of one of the contracts have already been built. The projects of another contract are under construction. The third contract of this phase was terminated by mutual agreement in January 2012. The main reason was that the concessionaire did not get the financing for the projects. This contract has been divided into three smaller contracts. One of them has already been put out for bidding and the others will be shortly.

3 RISK TRANSFER IN THESE PPP SCHEMES

The examination of risk transfer in this paper was carried out with reference to three key PPP risk categories: construction risk, revenue risk and availability risk. The terms of reference established that the concessionaires assumed all these risks.

3.1 Construction risk

The terms of reference established that the concessionaires had 18 months for the elaboration of the construction designs and for building the plants. However, there were long delays in all projects, mostly because of problems with the availability of the lands, as well as with the authorizations for the construction and the entering in operation. In a few cases there were also problems related to the geotechnical risk.

In some cases, there were also delays that were related to responsibility of the concessionaires. According to the interviews conducted for this research, the main reasons were: 1) Problems between the companies involved in the consortiums; 2) Lack of enough human resources managing the projects in most consortiums (each contract involved 8–10 plants); 3) The companies had to provide financial resources for the construction in the cases where getting the loans approved by the banks took more time than anticipated.

The regional government was responsible for providing the lands to build both the plants and the main sewers. The terms of reference established that the lands should be available within four months from the contract signature—otherwise the concessionaire would be entitled to an extension of the concession period.

The municipalities were in charge of making the arrangements to make the lands available for the concessionaires. The initial approach of the municipalities was not to expropriate but to negotiate with the owners of the lands (in order to reduce the political cost of getting the lands). However, this procedure proved very lengthy and, in many cases, fruitless. In the end, the officers of the regional government in charge of the program realized that it was necessary to expropriate and that it had to be done by the regional government (not the municipalities). But the expropriation procedure needed a long period to be implemented.

The officers of the program learnt from the experience of the early projects that they should start the expropriation procedure from the very beginning. And this is what they did in the second and third phases of the program.

Most of the projects experienced significant delays because of problems in having the lands available. Those delays were in the range of 20–50% of the period established from contract signature until starting operation (18 months). The delays in having the lands available were compensated by the regional government with extensions of the concession period.

Another source of delays in the projects was the difficulty in getting the permits and authorizations for the construction and the entering in operation. Most of the projects had problems with getting the permits and authorizations, but the delays produced because of these problems were in all cases shorter than the delays produced because of the problems in getting the lands available.

The concessionaires were in charge of getting all permits and authorizations. It was necessary to ask for them to many different public bodies at local, regional and national level. And it took very long to get them. The most problematic authorizations were the ones related to the electricity connection.

Another authorization that took more time than anticipated was the one related to the supervision of the construction design. The regional government had established a period of one month to approve the construction design. But in most cases it took much longer. One of the reasons is that during the supervision of the design, the regional government requested some changes because of technical reasons. Another reason is that the officers were overwhelmed with workload, especially in the first phase where seven contracts (totaling 77 plants) were launched at the same time. The public body in charge of supervising both the design and the construction process was reinforced with more staff but even though it was impossible to avoid all delays.

The concessionaire was supposed to assume geotechnical risk. However, in practice it was not so clear. The bidders had a short period of time to prepare the bids (around two months). In addition, they had no access to the lands at that time because they were not available yet. Therefore, they did not have the opportunity of carrying out tests to check the geotechnical conditions of the land.

In one of the projects of the first phase, there were severe problems with the foundations and it was not clear who had to assume this risk. Since there were problems also with the demand projections, this contract was terminated by mutual agreement and it was expected to be put out for bidding in April 2014.

3.2 Revenue risk

The retribution of the concessionaire was calculated according to the following formula:

$$\text{Revenue} = Q_A P_A + (Q_{measured} - Q_A) P_B, \quad (1)$$

where $Q_{measured}$ is the real flow in each plant. The variables Q_A, P_A, P_B had to be offered by the bidders for each plant. All these variables were capped in the terms of reference with specific values for each plant.

Demand risk was mitigated through a smart way: the maximum amounts allowed for these variables were established in such a way that the concessionaire had to get most of the revenue (roughly 95%) from the component $Q_A P_A$. The maximum amount allowed for Q_A was low and the maximum amount allowed for P_A was high. At the same time, the maximum amount allowed for P_B was low. In addition, the maximum amount of flow ($Q_{measured}$) that generates revenue was capped at $1.1 Q_A$ (which means an increase of 10% over Q_A). This way, the concessionaire was quite sure that they were going to get 95% of the forecast revenue even with low flows.

The size of each plant was dependent on the design flow (Q_D) that was estimated for it. The regional government carried out thorough assessments of the demands estimated for each plant. They took into account the current population (both the usual one and the one in vacation periods), the existing industries, and the estimated growth of both population and industries. Those rigorous studies made possible to get good estimations of the future flows in each plant, particularly Q_A and Q_D.

Regarding Q_D, in some cases the assumptions were proved too optimistic. This has led to building some plants bigger than needed. At the time of carrying out the demand studies (in the period 2005–2007), the construction of new houses was booming in Spain and the perspectives of population growth were very high. A few years later the perspectives were much gloomier because of the burst of the housing bubble and the global financial crisis.

The revenue risk has several components: 1) The flow Q_A and the tariff P_A; 2) The formula to update the tariff to be paid to the concessionaire during the concession period.

The variables Q_A, P_A, P_B had to be offered by the bidder for each plant. All these variables were capped in the terms of reference with specific values for each plant. These variables had a great weight in the awarding criteria, as shown in Table 3.

As already explained, the concessionaire obtains roughly 95% of their revenue through the component $Q_A P_A$. It was a smart way of mitigating the revenue risk because Q_A is quite low compared to Q_D (Q_A is around 30%–50% of Q_D in most cases), which means

Table 3. Awarding criteria.

Criteria	Points
Economic criteria	
– Q_A, P_A, P_B	30
– The lowest investment cost	5
– Certificate of a bank securing financing of the project	5
Technical criteria	
– Related to construction	30
– Related to operation	30
Total	100

that the real flow is going to be above Q_A most of the time in all plants.

As of early 2014, the real flow in 95% of the plants is somewhere between Q_A and Q_D, which shows that the estimations of the regional government were good.

In the concessions of the first phase, the formula for the yearly update of the tariff P_A to be paid to the concessionaire was:

$$I = 0.75 + 0.25 \, CPI_n/CPI_0. \quad (2)$$

Therefore, only 25% of the revenues are indexed to inflation. This raised a lot of complaints by the concessionaires. They claim that the percentage of variables costs is much higher than 25%.

In the second and third phases of the program, this formula was changed to:

$$I = 0.58 + 0.42 \, CPI_n/CPI_0, \quad (3)$$

which means that the percentage of costs indexed to inflation has increased from 25% to 42%.

There is a different formula to index PB to the inflation but its impact on the concessionaire's revenue is very low.

Regarding availability risk, the regional government has transferred this risk through two ways: 1) If the plant interrupts its functioning, the concessionaire is penalized; 2) The public body in charge of supervising the operation of the plants controls every week the quality of the water that comes out of the plant. If it does not meet the standards set in the terms of reference, the concessionaire is penalized.

4 SOURCES OF DELAYS IN THE IMPLEMENTATION OF THE PROGRAM

The implementation of this PPP program has experienced significant delays. According to the information provided in the interviews conducted, there are three main reasons: the rush in implementing the program, inadequate transfer of some risks, and wrong allocation of some tasks.

4.1 The rush in implementing the program

The regional government wanted to build many plants in little time. On the one hand, they wanted to comply with the European Union regulation that made it compulsory by December 2005 to treat waste water in all municipalities that had over 2,000 equivalent inhabitants. On the other hand, there were political reasons—it was a way of getting votes. The rush in implementing the program had various negative consequences.

First, it represented too much work for the public officers. The public body of the regional government responsible for the implementation of this PPP program was Instituto Aragonés del Agua (IAA). However, many of the tasks of supervising the construction designs and the construction process were transferred from IAA to another public body called SODEMASA because the latter had more flexibility to hire personnel. In spite of hiring more people, the office in charge of implementing the program was overwhelmed and it was a source of delays.

Second, the terms of reference established unrealistic periods of time for some tasks, like getting the lands available (four months), getting all permits and authorizations, elaboration of the construction designs by the concessionaires (two months), and the supervision of each construction design by the regional government (one month).

4.2 Inadequate transfer of some risks

The risk transfer established in the concession contracts might be considered adequate in general. However, there were a few problematic issues regarding risk allocation in the contracts.

One problematic issue is that the concessionaires assume possible increases of some operating costs that are not under their control but have a great potential impact on their profits. For example, in waste water treatment plants, electricity cost has a great influence. However, the formula established in the terms of reference for the yearly update of the tariffs to be paid to the concessionaires does not reflect it. In addition, electricity cost has escalated in Spain in the last few years. The concessionaires have sued the regional government because they consider that the sharp increase of electricity costs is an unforeseeable risk and its consequences have to be assumed by the public sector. The court has not ruled yet.

Another problematic issue is that the concessionaires assume geotechnical risk. However, the concessionaires had no access to the land at the time of elaborating the construction design.

4.3 Wrong allocation of some tasks

The experience of the projects of the first phase shows that the regional government, not the municipalities, should have been in charge of getting the lands available. And the expropriation procedure should have been used from the beginning. Furthermore, according to some interviews conducted for this research, the role of the municipalities should have been taken over but the regional government. Most of the municipalities are small (under 5,000 inhabitants) and the people in charge of dealing with these projects lacked the preparation needed to deal with major infrastructure projects.

5 CONCLUSIONS

Overall, the implementation of the concession program of the regional government of Aragon might be considered successful. As of early 2014, the program has managed to build 102 waste water treatment plants and 14 more plants are under construction (out of 131 initially planned). And three more tenders will be launched in the near future to build and operate 15 more plants. The plants built are in operation and working normally. However, the implementation of this PPP program has been somehow problematic. Most of the projects have experienced significant delays, in many cases up to 40–50% of the period established from the contract signature until entering in operation. Furthermore, two contracts (out of 11 concession contracts) have been terminated by mutual agreement and another one has been renegotiated (but the plants were built and are in operation).

The analysis in this study shows that arguably the main reasons for the problems in the implementation of the program were the rush in implementing it, the inadequate transfer of some risks, and the wrong allocation of some tasks. Probably, all those were influenced by the lack of experience of the regional government—and everywhere else—in using the PPP formula to build and operate waste water treatment plants.

The analysis in this study also illustrates two features of this PPP program that arguably have strongly influenced its successful implementation. One of them is how demand risk can be mitigated in smart way. The mitigation of demand risk has had two positive consequences: 1) There was a lot of competition for all projects (although it decreased in the consecutive phases of the program); 2) Only one contract has been renegotiated, and the reason was not a financial problem of the concessionaire. Another relevant feature of the program was that the regional government carried out thorough studies to estimate the demand (flow) in each plant. The rigorous estimations of demand have had relevant positive consequences. Roughly in 95% of the cases, the concessionaires' revenues are at least 95% of the projected ones. Moreover, none of the projects has experienced financial problems once it has entered in operation.

ACKNOWLEDGEMENTS

The authors gratefully acknowledge the support of 'Cátedra Juan-Miguel Villar Mir in Business

Administration' of the Polytechnic University of Madrid, which provided funding for this research.

REFERENCES

Ali, M.H., Osman, H., Marzouk, M., Ibrahim, M. 2012, Valuation of Minimum Revenue Guarantees for PPP Wastewater Treatment Plants, Conference Proceeding Paper, Construction Research Congress: Construction Challenges in a Flat World, doi: http://dx.doi.org/10.1061/978078441 2329.161

Demirag, I., Khadaroo, I., Stapleton, P., Stevenson, C., 2012.The Diffusion of Risks in Public private Partnership Contracts, Accounting, Auditing and Accountability, 25, (8), 1317–1339.

Flyvbjerg, B., 2009. Survival of the unfittest: why the worst infrastructure gets built—and what we can do about it. Oxford Review of Economic Policy, 25(3), 344–367.

Grimsey, D., Lewis, M., 2002. Evaluating the risks of public private partnerships for infrastructure projects.International Journal of Project Management, 20, 107–118.

Kwak, Y. H.,Chih, Y., Ibbs, C.W., 2009. Towards a Comprehensive Understanding of Public Private Partnerships for Infrastructure Development. California Management Review, 51 (2), 51–78.

Marques, R., Berg, S., 2011. Risks, Contracts, and Private-Sector Participation in Infrastructure. Journal of Construction Engineering and Management, 137(11), 925–932.

Memon, M.A. (2002). Public-Private Partnerships for Urban Water Supply and Wastewater Treatment: An overview of the concept of PPP and its applications for urban water In 2nd Thematic, Conference Paper, Seminar: Kitakyushu Initiative Seminar on Public-Private Partnerships for Urban Water Supply and Wastewater Treatment

Ng, A., Loosemore, A. M., 2007. Risk allocation in the private provision of public infrastructure. International Journal of Project Management, 25, 66–76.

Shen, L., Platten, A., Deng, X.P., 2006. Role of public private partnerships to manage risks in public sector projects in Hong Kong. International Journal of Project Management, 24, 587–594.

Siemiatycki, M., Friedman, J., 2012.The Trade-Offs of Transferring Demand Risk on Urban Transit Public-Private Partnerships. Public Works Management Policy, 17(3), 283–302

Tang, L., Shen, Q., Cheng, E., 2010. A review of studies on Public-Private Partnership projects in the construction industry. International Journal of Project Management, 28, 683–694.

Whitehead, J.C. (2002). Incentive incompatibility and starting-point bias in iterative valuation questions, Land Economics, 78(2), 285–297.

Yescombe, E.R., 2011. *Public-private partnerships: principles of policy and finance.* Butterworth-Heinemann.

Zheng. S. and Tiong, L.K., 2010. First Public-Private-Partnership Application in Taiwan's Wastewater Treatment Sector: Case Study of the Nanzih BOT Wastewater Treatment Project, Journal of Construction Engineering and Management, 136, 913–922.

Civil Engineering and Urban Planning III – Mohammadian, Goulias, Cicek, Wang & Maraveas (Eds)
© 2014 Taylor & Francis Group, London, ISBN 978-1-138-00125-1

Welding process from the civil engineering point of view

M. Al Ali, M. Tomko & I. Demjan
Faculty of Civil Engineering, Technical University in Košice, Slovak Republic

ABSTRACT: It is generally known that the welding process results in creation of a stress, which is called welding or residual stress. It begins with the induction of a very high temperatures, formation of transient thermal stresses, uneven cooling and creation of plastic zones. The paper deals with the development of plastic zones, distribution of welding stresses and their influence on the resistance of welded steel elements during the welding process and after its finishing.

1 INTRODUCTION

The local influence of the weld has a global effect on the welded element. The welded element, during welding process, undergoes high temperature fluctuations which melts the material around the weld. The electrode induces high temperatures that cause a temporary reduction of Young's modulus and yield stress of the welded material. For detailed explanation, some results of local heat transfer are presented in the paper. Welding on the flange of steel element with profile IPE 200 made from steel S235 (yield stress 235 MPa, ultimate stress 360 MPa, Young's modulus 210 GPa and elastic strain is 0.0011) is used here to simulate the local heat transfer during the thermal cycle of welding.

For creation of the calculation model, electrodes with 1 and 2 mm thicknesses were used. The electrode caused local thermal effects of 2500°C in the flange's midpoint of the modeled IPE profile (the flange is 8.5 mm thick). At places which are more remote from the weld's location, the temperature effect is smaller and influenced areas have different material properties. This fact required the creation of a calculation model with a large number of finite elements, which takes into account the local variations of material properties. Regulations of the Standard (EN 1993-1-2, 2007) were used for modeling of temperature's influence.

2 TIME OF FIRE AND THE FIRE RESISTANCE OF STEEL STRUCTURES

Fire resistance of the steel structures is determined according to EN 1993-1-2, related standards and recommendations. Fire resistance is expressed by the time for which preservation of the stability and resistance of the structure is guaranteed. The intense heating of a steel structural member occur during the fire, as well as the decrease of its material strength characteristics, particularly decrease of the yield strength, ultimate strength and Young's modulus E. If the decrease of yield stress due to fire is high, which means that its value is less than the immediate working stress, so the structural member deform or break. The temperature at which deformation and/or break occurs, is the critical temperature. This temperature is achieved for the usual types of structural steels at approximately 550°C, but may also be different depending on the dimensions of a structural member. At this temperature the steel maintains only about 60% of the initial yield stress, compared with normal temperature.

Standard EN 1993-1-2 defines individual material characteristics ($X_{d,fi}$) for the determination of load carrying capacity of steel structures exposed to fire's effects:

$$X_{d,fi} = \frac{k_\theta \cdot X_k}{\gamma_{M,fi}} \qquad (1)$$

where X_k is the corresponding material characteristic at normal temperature, k_θ is the reduction factor applicable for this characteristic and material temperature at the fire and $\gamma_{M,fi}$ is the reliability factor for the given material characteristic and a fire situation.

The mentioned standard also contains reduction factors that characterize the decrease of corresponding material properties for the individual strength characteristics at enhanced temperatures, such as the yield strength, limit of proportionality and Young's modulus, in relation to temperatures and coefficient of thermal expansion, see Figures 1 and 2.

3 CALCULATION MODEL AND USED DIAGRAM

The process of transformation during which plastic deformations appear and develop is a plastic transformation, i.e. plastic deformations of the entity are

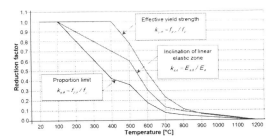

Figure 1. Reduction factors of strain-stress diagram at higher temperatures.

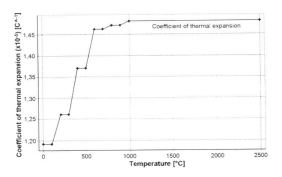

Figure 2. Considered coefficient of thermal expansion.

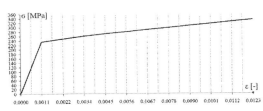

Figure 3. Bilinear stress-strain diagram used in the 3D model.

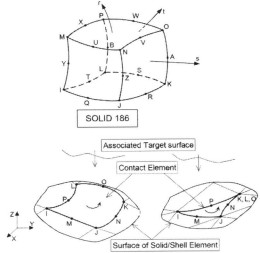

Figure 4. Used finite elements; SOLID 186 and CONTA 174.

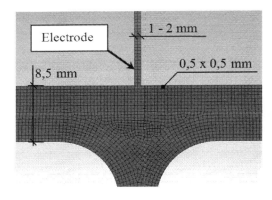

Figure 5. Part of the calculation model.

permanents. The criterion which defines the conditions for the entity point to get from elastic into plastic state is called plasticity condition and is formulated on the basis of experimental measurements. According to fluxion theory, the speed of deformation in the considered point of entity is given as the immediate existing stress and the speed of this stress. The resultant deformation needs to be solved using incremental method and the resulting state is the sum of individual increments (Kmet' S. et al. 2006). The real dependences between stress and deformation in the plastic transformation process are quite demanding, so in the technical applications of plasticity theory, the properties of materials are characterized by idealized stress-strain diagrams.

Non-linear relation between stress and transformation beyond the plasticity limit was considered as linear and stress-strain diagram with hardening characterized by bilinear approximation was used in the calculation model (steel S235, $E = 209821$ MPa, $E_{hard} = 8944$ MPa), as shown in Figure 3.

3D model of the mentioned IPE 200 steel profile was created using software ANSYS, taking into account the reduction factors and the coefficient of thermal expansion (ANSYS Inc., 2011; Al Ali M. et al., 2014). 3D finite elements *SOLID 186* of 0.5 mm were used to create the model, while 3D contact elements *CONTA 174* were used to transform the isolines of thermal influence into loading effects. Figures 4 and 5 illustrate the finite elements and a part of the calculation model.

4 SIMULATION OF TEMPERATURE EFFECT

Simulation of thermal loading was realized as following: first step of modeling was the introducing of local thermal effects at 2500°C by the electrode. Then the heat spread through the microstructure of the modeled profile's flange and isolines of the heat flow were obtained. This result was generated as a loading

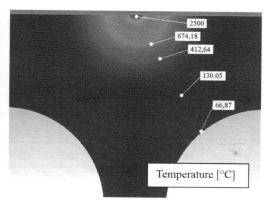

Figure 6. Part of the model; local heat transfer.

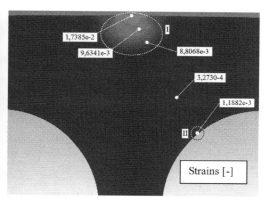

Figure 7. Part of the model; development of plastic zones.

5 LOCAL ISSUE WITH GLOBAL IMPLICATION

Because the welding process causes a temporary reduction of Young's modulus and yield stress of the welded material, it can be stated that the affected areas in the vicinity of heat source are weakened. The implication of that is the temporary local weakening of the steel cross-section, displacement of the original centroid and changes of cross-sectional characteristics, as shown in Figure 8.

When strengthening of existing members under load by welding a new components, additional

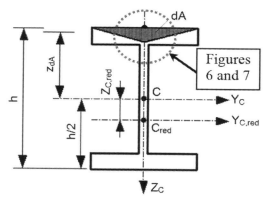

Figure 8. Section with weakened flange and shifted centroid.

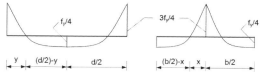

Figure 9. Size and distribution of welding stresses.

moments of internal forces are created from the displacement of the centroid, which increase the stresses by the length of the weakened segment. Although this condition is temporary, new stress state could cause immediate exceeding of the member's resistance in the relevant section.

The severity of analytical solutions relies on the complex identification of the actual stress pattern, which changes incessantly during the welding of new components to original cross-sections until finishing of the strengthening process and consecutive additional loading. This fact changes the local effect of the weld into a global stability problem which the civil engineer must take into account when designing the welding process.

After welding completion, the welding stresses take their final shape and size as a result of uneven heating and cooling processes and remaining plastic strains. First author of the paper modified an empirical formulae for determining the welding stresses in the web and the flanges of welded I-section in relative values of the welded material's yield stress f_y (Al Ali M., in press). Size and distribution of welding stresses are given by Figure 9.

Points at which the stress passes from compressive to tensile values is defined as follows:

$$x_0 = \frac{b}{2\sqrt{2}}, \quad y_0 = \frac{d}{2\sqrt{2}} \quad (2)$$

Modified formulae also assume that the beginning of y coordinates is placed at the edge of the web, while the beginning of x coordinates is at the centre of the flange. Taking into account the mentioned assumptions and distribution of welding stresses from Figure 9, the modified formulae took on the following form:

– *For the flanges:*

$$\sigma_{wf,m1} = \frac{3f_y}{4}\left[1 - 8\left(\frac{x}{b}\right)^2\right], \text{ for } 0 \leq x \leq \frac{b}{2\sqrt{2}} \quad (2)$$

$$\sigma_{wf,m2} = \frac{f_y}{4}\left[1 - 8\left(\frac{x}{b}\right)^2\right], \text{ for } \frac{b}{2\sqrt{2}} \leq x \leq \frac{b}{2} \quad (3)$$

– *For the web:*

$$\sigma_{ww,m1} = \frac{3f_y}{4}\left[1 - 8\left(\frac{y}{d}\right)^2\right], \text{ for } 0 \leq y \leq \frac{d}{2\sqrt{2}} \quad (4)$$

$$\sigma_{ww,m2} = \frac{f_y}{4}\left[1 - 8\left(\frac{y}{d}\right)^2\right], \text{ for } \frac{d}{2\sqrt{2}} \leq y \leq \frac{d}{2} \quad (5)$$

Determined welding tensile and compressive stresses are subsequently summed up along with existing stresses, caused by the loading of a structural member. This results in a change of the expected stress state of the member.

The above mentioned is especially significant in the case of welded cross-sections of classes 3 and 4, because the material premature enters into elastic-plastic zone. It is more manifested in the members subjected to bending.

6 CONCLUSIONS

Presented paper is a part of running theoretical, numerical and experimental research oriented to the influence of the welding process in relation to stability problems associated with the welding stresses. Preliminary results confirm and extend the base of existing knowledge.

Influence of the weld is significant during the welding process, but also after its completion. The consequences of welding stresses are more significant for steel cross-sections of classes 3 and 4 due to the premature entry into the elastic-plastic zone.

Presented partial results revealed that although the weld only affects a relatively small area, the influence exceeds its immediate point of action. Therefore, the weld must be regarded as a local issue with global consequences.

ACKNOWLEDGEMENTS

The paper is prepared within the research project VEGA 1/0582/13 "The elastic-plastic behavior of compressed thin-walled cold-formed steel elements and stress-strain analysis of welded steel beams", supported by the Scientific Grant Agency of The Ministry of Education of the Slovak Republic and the Slovak Academy of Sciences.

The paper was also funded by project No. 1/0788/12: "Theoretical and Experimental Analysis of Stability and Strength of Composite Members in Compression and Bending" of the grant agency VEGA of the Ministry of Education of the Slovak Republic

REFERENCES

Al Ali, M., Tomko, M. & Demjan I. 2014. Development of Plastic Zones During the Thermal Cycle of Welding. *Key Engineering Materials* 586(2014): 11–14.

Al Ali, M. 2014. The Welding Process as a Local Issue with Global Consequences. *Advanced Materials Research* – in press.

Ambriško, Ľ. & Pešek, L. 2009. Accuracy of strain measurement using ME 46 video-extensometric system. *Acta Metallurgica Slovaca*, 15/2(2009): 105–111.

ANSYS, Inc. 2011. *Release 11.0, Documentation for ANSYS*: http://www.kxcad.net.

EN 1993-1-2. 2007. *Design of steel structures, Part 1–2: Structural fire design*. CEN, Brussels.

Kmet, S., Tomko, M. & Brda, J. 2006. Probabilistic assessment of the reliability of systems based of suspension diffraction-solid elements. In: *Reliability of Construction*, Prague 2006.

Author index

Al Ali, M. 165, 541
Annunziata, A. 339
Annunziata, F. 339

Bai, L.M. 503
Bai, Y. 79
Bartošová, V. 203
Beke, P. 165
Bi, X.Y. 307

Călin, M.M. 493
Cao, Y. 87, 113
Cao, Z.M. 267
Carpintero, S. 535
Chan, C.-M. 473
Chen, D.K. 253
Chen, J.K. 53, 117
Chen, K.M. 425
Chen, K.Z. 401
Chen, S.G. 123, 239, 249
Chen, W.C. 407
Chen, X.H. 457
Chen, X.M. 33
Chen, Z.G. 109
Cheng, J.J. 483
Cheng, S.W. 443
Cheng, Y.B. 15, 335, 517
Chu, X.J. 451

Dai, S.Q. 177
Demjan, I. 541
Deng, X.H. 393
Diao, M.J. 257
Ding, K. 151
Ding, K.W. 245
Dong, H.G. 253
Dong, X.M. 383
Duan, J. 33
Dubecky, D. 165

Fang, H. 297
Frith, D. 207

Gao, D.S. 275
Gao, F. 151, 349
Gao, Y.F. 229
Ge, Y.J. 439
Gong, M.S. 159
Gu, H. 313
Guo, C. 267
Guo, Q. 279

Guo, X. 361
Guo, X.C. 307
Guo, Z. 9
Guo, Z.L. 335, 517

Han, N.N. 439
Han, Y. 213
He, F. 361
He, K. 53
He, Y. 101
Heo, G. 141
Hoo, L.-S. 473
Hou, J.M. 503
Hou, L.Q. 195
Hu, S.S. 499
Hu, X.W. 199, 421
Hu, Y.N. 499, 523
Huang, C. 129
Huang, J. 109
Huang, T. 155

Jeon, J. 141
Ji, J. 297
Jia, J.H. 487
Jiang, L. 151
Jiang, Y. 523
Jiang, Y.T. 401
Jiang, Z.W. 79
Jiao, L.X. 279
Jin, Y.H. 261

Kan, B. 69
Kang, H.Z. 447
Katunský, D. 203
Ke, S. 439
Kirik, E. 5
Kocurova, R. 165
Kuang, H.W. 93
Kvocak, V. 165

Labovský, M. 203
Lai, G.B. 47
Li, C.H. 279
Li, H. 407
Li, H.Y. 21
Li, J. 199
Li, L.X. 433
Li, M. 169, 173
Li, M.L. 417
Li, M.Q. 413
Li, P.F. 27, 457
Li, Q. 287, 439

Li, Q.S. 79
Li, W. 37
Li, X. 43
Li, X.K. 87
Li, Y.F. 479
Li, Y.G. 33
Li, Y.L. 53, 117
Li, Y.S. 239
Li, Y.Z. 87, 401
Li, Z. 325
Li, Z.Q. 325
Lian, W.B. 523
Liang, H. 177
Liang, S.S. 225
Liang, Y. 451
Liang, Y.D. 361
Lin, C.Y. 371
Lin, J. 297
Ling, X.J. 307
Liu, A.M. 283, 447
Liu, F. 467
Liu, H.B. 417
Liu, H.L. 291
Liu, Q.F. 159
Liu, T.Y. 447
Liu, Y. 1
Liu, Y.Y. 275
Liu, Z. 429
Lopušniak, M. 203
Luan, T.T. 225
Lv, F. 301

Ma, C.X. 283
Ma, H.Y. 93
Ma, L. 37
Ma, R.Y. 177
Ma, W.Y. 195
Ma, X.W. 195
Malyshev, A. 5
Man, D.W. 245
Mao, Y.H. 21
Meng, L.J. 195
Miao, R.L. 169, 173
Mo, P.C. 69
Morian, D. 207
Mu, X.W. 113

Ni, F. 287
Noda, T. 63

Ou, X.-P. 151

Pan, J.Y. 133
Pan, W. 93
Petersen, O.H. 535
Pun, K.-H. 473

Qi, D.C. 27
Qi, H. 33
Qi, Y.L. 377
Qian, R. 433
Qiao, G.H. 15, 335, 517
Qin, M.L. 407
Qiu, C. 257
Qiu, W. 155
Qu, T.J. 355

Ridout, A. 79
Ruan, F. 499

Schmitz, R.P. 183
Sebeşan, I. 387, 493
Seiyong, K. 531
Shafray, E. 531
Song, X.Y. 301
Spiroiu, M.A. 387
Su, M.H. 37
Sun, L.L. 313
Sun, X. 219

Tan, M.X. 137
Tang, Y.H. 87
Tian, Q.F. 229
Tomko, M. 541
Tong, L. 9
Tu, Y.F. 457

Wan, H.Y. 507
Wang, F.J. 467
Wang, G. 207
Wang, J. 199, 421
Wang, M.Y. 275

Wang, Q. 213, 297
Wang, W.Q. 117
Wang, X.M. 155
Wang, Y. 59, 425
Wang, Y.T. 97
Wang, Z. 365, 365
Wang, Z.D. 433
Wei, Z.T. 487
Wu, H.Y. 9
Wu, J. 365
Wu, Z.Y. 53, 117

Xia, L. 191
Xia, S.Y. 191
Xia, W.X. 507
Xiang, X.D. 279
Xiang, Y. 429
Xiang, Y.J. 101
Xie, Z.H. 225
Xu, B. 63
Xu, G.W. 365
Xu, J. 73, 393, 443
Xu, W.-L. 151
Xu, X. 291
Xu, Y.L. 169, 173

Ya, J. 407
Yamada, S. 63
Yamasaki, M. 83
Yang, B. 21
Yang, H. 105
Yang, H.Z. 355
Yang, J. 109
Yang, M. 307
Yang, Q.X. 267
Yang, S.B. 487
Yang, W.S. 361
Yang, Y.W. 133
Yao, Y. 73
Yao, Y.H. 479

Ye, S. 319
Yi, J. 261
Yin, D.Y. 159
Yin, Y.X. 335, 517
Yuan, L. 59, 83, 283
Yue, L. 147
Yue, S.B. 257

Zhang, G.Q. 377
Zhang, G.Y. 413
Zhang, H. 219
Zhang, H.B. 443
Zhang, H.N. 169
Zhang, H.S. 123
Zhang, M. 173
Zhang, P. 105
Zhang, P.W. 1
Zhang, Q.K. 413
Zhang, Q.Y. 377
Zhang, S.Y. 463
Zhang, T.J. 113
Zhang, X. 133
Zhang, Y. 219, 301, 329
Zhang, Y.P. 443
Zhang, Z.C. 275
Zhao, B.H. 47
Zhao, G. 355
Zhao, J. 261, 407
Zhao, S.Z. 229
Zhao, W.S. 513
Zhao, Y.B. 487
Zhen, Y.Y. 53
Zheng, M.M. 235
Zhong, Y. 483
Zhou, K. 421
Zhou, Y. 361
Zhou, Z.J. 117
Zhou, Z.L. 239, 249
Zong, F. 229
Zuo, Z.Y. 235